THE SEA

Ideas and Observations on Progress in the Study of the Seas

Volume 1	HILL/PHYSICAL OCEANOGRAPHY
Volume 2	HILL/COMPOSITION OF SEA WATER
Volume 3	HILL/THE EARTH BENEATH THE SEA
Volume 4	MAXWELL/NEW CONCEPTS OF SEA FLOOR EVOLUTION (IN TWO PARTS)
Volume 5	GOLDBERG/MARINE CHEMISTRY
Volume 6	GOLDBERG/MARINE MODELING
Volume 7	EMILIANI/OCEANIC LITHOSPHERE
Volume 8	ROWE/DEEP-SEA BIOLOGY
Volume 9	Le MÉHAUTÉ/OCEAN ENGINEERING SCIENCE
Volume 10	BRINK AND ROBINSON/THE GLOBAL COASTAL OCEAN: PROCESSES AND METHODS
Volume 11	ROBINSON AND BRINK/THE GLOBAL COASTAL OCEAN: REGIONAL STUDIES AND SYNTHESES
Volume 12	ROBINSON, McCARTHY, AND ROTHSCHILD/BIOLOGICAL-PHYSICAL INTERACTIONS IN THE SEA
Volume 13	ROBINSON AND BRINK/THE GLOBAL COASTAL OCEAN: MULTISCALE INTERDISCIPLINARY PROCESSES
Volume 14	ROBINSON AND BRINK/THE GLOBAL COASTAL OCEAN: INTERDISCIPLINARY REGIONAL STUDIES AND SYNTHESES (IN TWO PARTS)
Volume 15	BERNARD AND ROBINSON/TSUNAMIS
Volume 16	FOGARTY AND McCARTHY/MARINE ECOSYSTEM-BASED MANAGEMENT

THE SEA

Ideas and Observations on Progress in the Study of the Seas

Marine Ecosystem-Based Management

Edited by

MICHAEL J. FOGARTY and **JAMES J. McCARTHY**
National Oceanic and Atmospheric Administration
Harvard University

THE SEA

Ideas and Observations on Progress in the Study of the Seas

Volume 16

Harvard University Press
Cambridge, Massachusetts
London, England
2014

Printed in the United States of America

Library of Congress Cataloging-in-Publication Data

Marine ecosystem-based management / edited by Michael J. Fogarty and James J. McCarthy.
pages cm. -- (The sea : ideas and observations on progress in the study of the seas ; volume 16)
Includes bibliographical references and index.
ISBN 978-0-674-07270-1 (alk. paper)
1. Marine ecosystem management. 2. Marine resources--Management. 3. Ecosystem services.
I. Fogarty, Michael J. II. McCarthy, James J.
QH541.5.S3M273 2014
333.95'6--dc23 2013015772

CONTENTS

PREFACE

We wish to acknowledge the role of the late Allan R. Robinson in revitalizing this series of publications. Consistent with his forward looking view of ocean science, and with a new Editorial Board, THE SEA will continue to focus on topics where groups of ocean scientists can describe a new state of understanding of the sea and its uses, and draw attention to needs and opportunities for further advancement of knowledge in these areas.

All uses of the sea have intensified in recent decades as a disproportional fraction of human population growth has occurred in coastal regions. It is also evident that the role of science has become increasingly important in decision-making at all levels of governance relating to our uses of the sea. In multiple sectors initiatives are underway to assess the state of the ocean and marine life in order to guide us towards more sustainable management practices.

Driven by the lack of integration in efforts to sustain the broad array ecosystem services upon which many people depend and the need to assess tradeoffs among competing uses, the concept of ecosystem-based management has arisen. Management strategies now implemented by separate sectors—coastal zone and fisheries management, for example—could be replaced or supplemented by integrated management of human activities in defined ecological regions. This approach requires that important relationships among parts of the system and how they respond to environmental forcing factors be understood, including the ways in which humans both influence and are affected by changes in marine ecosystems. A synthesis of the research necessary to support this approach is timely.

To this end, Volume 16 of THE SEA examines progress in formulating and executing a Marine Ecosystem-Based Management approach, and identifies important opportunities for the ocean science community to contribute to its refinement and implementation world-wide.

We are indebted to the Editorial Advisory Panel that helped to shape these chapters and suggest and persuade authors that their contributions would be valued, and to the external reviewers whose comments improved both the consistency and the coordination among chapters. We are also in the debt of Ms. Helene Smith Ferranti who scrutinized every word and every comma, and raised questions that helped immeasurably to improve this volume. Throughout the process we have appreciated the encouragement and patience of Mr. Michael Fisher at Harvard University Press. We also thank Ms. Lauren K. Esdaile, also at HUP, for her valuable technical assistance and Ms. Cheryl Lincoln for overseeing the production of this volume.

We dedicate this volume to our late friend and mentor, Allan R. Robinson.

James J. McCarthy
Michael J. Fogarty
Kenneth H. Brink

CONTRIBUTORS

IRIT ALTMAN
Department of Biology
Boston University
5 Cummington Ave.
Boston, MA 02215

GABRIELLA BIANCHI
Food and Agriculture Organization of the United Nations
Via delle Terme di Caracalla
00100 Rome, Italy

ROEL BOUMANS
AFORDable Futures
4685 Greenbush Rd.
Charlotte, VT 05445

ANTHONY CHARLES
School of Business / School of the Environment
Saint Mary's University
923 Robie St.
Halifax, NS B3H3C3, Canada

KEVERN COCHRANE
Department of Ichthyology and Fisheries Science
Rhodes University
Grahamstown, 6140 South Africa

JON M. ERLANDSON
Museum of Natural and Cultural History
University of Oregon
Eugene, OR 97403

WARRICK FLETCHER
Department of Fisheries,
Government of Western Australia,
Research Division, P.O. Box 20,
North Beach, Western Australia 6920, Australia

DAVID FLUHARTY
School of Marine Affairs
University of Washington
3707 Brooklyn Ave NE
Seattle, WA 98105

MICHAEL J. FOGARTY
National Oceanic and Atmospheric Administration
Northeast Fisheries Science Center
166 Water St.
Woods Hole, MA 02543

ELIZABETH A. FULTON
Commonwealth Scientific and Industrial Research Organization
Marine and Atmospheric Research
GPO Box 1538
Hobart, TAS 7001, Australia

SARAH GAICHAS
National Oceanic and Atmospheric Administration
Northeast Fisheries Science Center
166 Water St.
Woods Hole, MA 02543

SUCHI GOPAL
Department of Geography and Environment
Boston University
675 Commonwealth Ave.
Boston, MA 02215

ANNE D. GUERRY
The Natural Capital Project
c/o Stanford Woods Institute
473 Via Ortega, MC: 4205
Stanford, CA 94305

CHRISTINA C. HICKS
James Cook University
Townsville, QLD 4811, Australia

ALISTAIR J. HOBDAY
Commonwealth Scientific and
Industrial Research Organization
Marine and Atmospheric Research
GPO Box 1538
Hobart TAS 7001, Australia

RICHARD C. HOFFMANN
Department of History
York University
Toronto, ON M3J 1P3, Canada

EDWARD HOUDE
University of Maryland Center for
Environmental Science
Chesapeake Biological Laboratory
P.O. Box 38
Solomons, MD 20688

LES KAUFMAN
Department of Biology
Boston University
5 Cummington Ave.
Boston, MA 02215

ANDREW KENNY
Centre for Environment, Fisheries &
Aquaculture Science
Pakefield Rd.
Lowestoft, Suffolk NR330HT
United Kingdom

SUE KIDD
Department of Civic Design
School of Environmental Sciences
University of Liverpool
Liverpool L69 3BX
United Kingdom

SUAM KIM
Department of Marine Biology
Pukyong National University
599-1 Daeyeon 3-dong, Nam-gu
Busan 608-737, Korea

MARIANO KOEN-ALONSO
Northwest Atlantic Fisheries Centre/
Centre des pêches de l'Atlantique
Nord-Ouest
Fisheries and Oceans Canada/
Pêches et Océans Canada
80 East White Hills Rd.
PO Box 5667
St. John's, NL A1C 5X1 Canada

J. ANTHONY KOSLOW
Scripps Institution of Oceanography
9500 Gilman Dr.
University of California–San Diego
La Jolla, CA 92093-0218

JACOB P. KRITZER
Environmental Defense Fund
18 Tremont St. Suite 850
Boston, MA 02108

PHILLIP S. LEVIN
National Oceanic and Atmospheric
Administration
Northwest Fisheries Science Center
2725 Montlake Blvd.
Seattle, WA 98112

JASON S. LINK
National Oceanic and Atmospheric
Administration
Northeast Fisheries Science Center
160 Water St.
Woods Hole, MA 02543

HEIKE K. LOTZE
Department of Biology
Dalhousie University
1355 Oxford St.
Halifax, NS B3H 4R2, Canada

ROBIN MAHON
Centre for Resource Management
and Environmental Studies
(CERMES)
University of the West Indies
Cave Hill Campus
St. Michael, Barbados
West Indies

BRUCE D. MAPSTONE
Commonwealth Scientific and Industrial Research Organization
Marine and Atmospheric Research
Hobart, TAS 7000, Australia

JAMES J. MCCARTHY
Museum of Comparative Zoology
Harvard University
26 Oxford St.
Cambridge, MA 02138

OLE ARVE MISUND
Institute of Marine Research
P.O. Box 1870 Nordnes
5817 Bergen, Norway

FABIAN PINA-AMARGÓS
Center for Coastal Ecosystems Research
Cayo Coco, CP 69400 Cuba

MARK L. PLUMMER
National Oceanic and Atmospheric Administration
Northwest Fisheries Science Center
2725 Montlake Blvd.
Seattle, WA 98112

JEFFREY J. POLOVINA
National Oceanic and Atmospheric Administration
Pacific Islands Fishery Science Center
2570 Dole St.
Honolulu, HI 96814

CHRISTIAN REISS
National Oceanic and Atmospheric Administration
Southwest Fisheries Science Center
3333 North Torrey Pines Ct.
La Jolla, CA 92037

JAKE RICE
Ecosystem Sciences Branch
Department of Fisheries and Oceans
200 Kent St.
Ottawa, ON K1A 0E6, Canada

JOE ROMAN
Gund Institute for Ecological Economics
University of Vermont
Burlington, VT 05405

VINCENT S. SABA
National Oceanic and Atmospheric Administration
Northeast Fisheries Science Center
Geophysical Fluid Dynamics Laboratory
201 Forrestal Rd.
Princeton, NJ 08540

PETER F. SALE
United Nations University
Institute for Water, Environment and Health
Hamilton, ON L8P 0A1, Canada

JAMEAL F. SAMHOURI
National Oceanic and Atmospheric Administration
Northwest Fisheries Science Center
2725 Montlake Blvd.
Seattle, WA 98112

ANTHONY D. M. SMITH
Commonwealth Scientific and Industrial Research Organization
Marine and Atmospheric Research
GPO Box 1538
Hobart TAS 7001, Australia

HEATHER TALLIS
Natural Capital Project
Stanford University
371 Serra Mall
Stanford, CA 94305

CHANG-IK ZHANG
Department of Marine Biology
Pukyong National University
Nam-gu, Busan 608-737, Korea

Shijie Zhou
Commonwealth Scientific and
Industrial Research Organization
Marine and Atmospheric Research
PO Box 2583
Brisbane, QLD 4001, Australia

EDITORIAL ADVISORY PANEL

EXTERNAL REVIEWERS

Jeremy Collie
University of Rhode Island
Narragansett, RI

Paul Dayton
Scripps Institute of Oceanography
San Diego, CA

Geret DiPiper
National Oceanic and Atmospheric Administration
Woods Hole, MA

Timothy Essington
University of Washington
Seattle, WA

Troy Hartley
University of Maryland
College Park, MD

Porter Hoagland
Woods Hole Oceanographic Institution
Woods Hole, MA

Robert Hofman
Marine Mammal Commission (retired)
Bethesda, MD

Isaac Kaplan
National Oceanic and Atmospheric Administration
Seattle, WA

Christopher Kelble
National Oceanic and Atmospheric Administration
Miami, FL

Christian Möllmann
University of Hamburg
Hamburg, Germany

Mark Monaco
National Oceanic and Atmospheric Administration
Silver Spring, MD

Andrew Rosenberg
Union of Concerned Scientists
Cambridge, MA

Patricia Pinto da Silva
National Oceanic and Atmospheric Administration
Woods Hole, MA

Sevrine Sailley
Woods Hole Oceanographic Institution
Woods Hole, MA

Michael Sissenwine
Woods Hole Oceanographic Institution
Woods Hole, MA

Timothy D. Smith
World Whaling History
Redding, CA

Gunnar Steffanson
University of Iceland
Reykjavík, Iceland

Howard Townsend
National Oceanic and Atmospheric Administration
Annapolis, MD

Yvonne Walther
Institute of Marine Research
Karlskrona, Sweden

Chapter 1. An Overview of Marine Ecosystem-Based Management

MICHAEL J. FOGARTY

National Oceanic and Atmospheric Administration

JAMES J. MCCARTHY

Harvard University

Contents

1. Introduction

Marine Ecosystem-Based Management (mEBM) fundamentally concerns the sustainable delivery of ecosystem services from the sea. Marine and coastal ecosystems have long provided human societies with high quality food resources; avenues for transportation and commerce; opportunities for recreation; sources of inspiration and the aesthetic appreciation of nature; and many other benefits. Healthy marine and coastal habitats afford protection from storms and can diminish potential climate change impacts. They also serve as buffers to pollution and other stressors. Emerging prospects for renewable energy from wind, tides, and ocean currents are presently under active development in many parts of the world and must be viewed in concert with other existing uses of the ocean. When different ocean use sectors invoke pre-emptive use of space or place other competing demands on the system, tradeoffs among them must be directly confronted.

A burgeoning human population is already heavily concentrated in coastal areas of the world ocean. Currently about 40% of the population resides within 100 km of the coast. As a result, provision of many valuable ecosystem services is increasingly threatened. To meet these challenges, management of human activities in coastal and marine ecosystems is undergoing a transformation that employs more fully integrated strategies. Longstanding calls for the adoption of holistic approaches to ocean management are now being translated into action on a global basis.

Although many definitions of mEBM have been proffered (see Text Box 1), most hold certain elements in common. mEBM is place-based and recognizes

The Sea, Volume 16, edited by Michael J. Fogarty and James J. McCarthy
ISBN 978-0-674-07270-1

1. Concepts and Definitions of Ecosystem-Based Management: A Sampler

COMPASS (Scientific Consensus Statement on Marine Ecosystem-Based Management)

Ecosystem-based management is an integrated approach to management that considers the entire ecosystem, including humans. The goal of ecosystem-based management is to maintain an ecosystem in a healthy, productive and resilient condition so that it can provide the services humans want and need. Ecosystem-based management differs from current approaches that usually focus on a single species, sector, activity or concern; it considers the cumulative impacts of different sectors. Specifically, ecosystem-based management:

- Emphasizes the protection of ecosystem structure, functioning, and key processes;
- Is place-based in focusing on a specific ecosystem and the range of activities affecting it;
- Explicitly accounts for the interconnectedness within systems, recognizing the importance of interactions between many target species or key services and other non-target species;
- Acknowledges interconnectedness among systems, such as between air, land and sea; and
- Integrates ecological, social, economic, and institutional perspectives, recognizing their strong interdependences.

McLeod, K. L., J. Lubchenco, S. R. Palumbi, and A. A. Rosenberg (2005).

US Commission on Ocean Policy

"U.S. ocean and coastal resources should be managed to reflect the relationships among all ecosystem components, including human and nonhuman species and the environments in which they live. Applying this principle will require defining relevant geographic management areas based on ecosystem, rather than political, boundaries."

USCOP(2004)

NOAA Ecosystem Goal Team

An Ecosystem Approach to Management (EAM) is one that provides a comprehensive framework for living resource decision making. In contrast to individual species or single issue management, EAM considers a wider range of relevant ecological, environmental, and human factors bearing on societal choices regarding resource use.

NOAA defines EAM as a geographically specified, adaptive approach that takes account of ecosystem knowledge and uncertainties, considers multiple external influences, and strives to balance diverse societal objectives. Implementation will need to be incremental and collaborative. NOAA recognizes that transition to and implementation of an ecosystem approach to management needs to be incremental and collaborative.

http:// http://ecosystems.noaa.gov/why_an_eam.htm

humans as an integral part of the ecosystem. It accounts for the dynamic interplay among the different parts of the system and with the environment. Further, mEBM specifically addresses the cumulative impact of anthropogenic stressors and highlights the critical importance of preserving the basic structure and integrity of the ecosystem as a whole. Finally, an mEBM strategy stresses the need to address potentially competing goals and objectives. Through mEBM, we seek to develop integrated management strategies for defined ecological regions rather than for individual ocean use sectors such as fisheries, aquaculture, shipping, or tourism and recreation. By recognizing humans as part of a coupled Social-Ecological System (SES), we see that we both affect other parts of the ecosystem by our activities and are affected by natural changes in these systems. Effective stakeholder engagement in identifying problem areas and overall goals and objectives is essential to the success of any mEBM initiative. Finally, while marine ecosystems exhibit varying levels of resilience to different threats and impacts, it is clear that through our actions, we have the capacity to fundamentally alter ecosystem structure and function. These alterations can, in turn, adversely affect the sustainable delivery of the ecosystem services so vital to human well-being.

The current framework for mEBM can be traced to seminal developments over the last several decades in Integrated Coastal Zone Management (ICZM), progress in identifying and implementing Ecosystem Approaches to Fisheries Management (EAFM), and the evolution and adoption of spatial management tools though Marine Spatial Planning (MSP). This triad of inter-related strategies (along with a host of other antecedents) is now being forged into a more integrated whole under the banner of mEBM.

In this volume, we provide an overview of the current state of play in implementing mEBM on the world stage. Implementation of mEBM is a clearly a work in progress whose development and evolution is still being charted (see also McLeod and Leslie 2009). We attempt to identify the key structural elements of mEBM as it now stands and call attention to some of the many challenges remaining. In the end, mEBM cannot be viewed as a monolithic enterprise. In practice, it will necessarily be tailored to the needs and objectives in different parts of the globe and will assume different forms in different places. Although the basic tenets described above provide a common underlying foundation for mEBM, local and regional characteristics will place a strong stamp on its ultimate form in these locations. Existing scientific infrastructure and the nature of governance and regulatory mechanisms will further shape the identity of mEBM as it is implemented in different areas. Both are strongly related to levels of affluence, since affluence is a strong predictor of the scale and intensity of threats and impacts in coastal and marine ecosystems in relation to coastal development and industrialization (MA 2005).

Framed in this way, mEBM is an eminently utilitarian construct. Because humans are defined as part of the ecosystem, sustainable human use of natural resources is viewed as part of an organic whole. All living organisms require and use resources in their environment; humans are, of course, no different. Humans do differ, however, in our ability to gauge the impacts of our actions and to understand their implications now and for the future. With this capability comes the responsibility to avoid the prospect of irreversible harm to the environment. We acknowledge the dynamic tension between ecocentric and anthropocentric worldviews that is an essential part of the dialogue on how best to think about and manage human activities in the oceans. Ultimately, mEBM involves an attempt to strike a balance that respects the intrinsic value of these ecosystems and their components, the human need for ecosystem services, and our responsibility to ensure sustainability to meet the needs of future generations.

2. Ecosystem Services, Threats and Impacts

Human health and security is intimately linked to the manifold services provided by nature. In the following we adopt the taxonomy of ecosystem services described in the Millennium Ecosystem Assessment Report (MA 2005) to provide an overall context and to identify prevalent threats to the sustainable supply of these services. Basic categories include Provisioning, Regulating, Supporting, and Cultural and Amenity Services (see Text Box 2). In this volume, we focus on these services and threats to the ongoing delivery of these services as a primary organizing theme. These threats translate into impacts on marine ecosystems and the human communities dependent on them. The mix of different services derived from different ecosystems, present and projected patterns of human utilization of each, and con-

2. Marine Ecosystem Services

- Provisioning Services
    - Commercial Fisheries
    - Recreational Fisheries
    - Aquaculture
    - Natural Products
    - Energy Resources
    - Bio-prospecting
- Supporting Services
    - Habitat Provision
    - Primary Production
    - Nutrient Cycling
- Regulating Services
    - Climate Regulation
    - Carbon Sequestration
    - Filtration/Detoxification
    - Nutrient Regulation
    - Water Supply/Hydrology
    - Shoreline Stabilization/Protection
    - Waste Disposal
- Cultural and Amenity Services
    - Tourism
    - Recreation
    - Historical/Archeological Sites

sideration of the anthropogenic and natural stressors important in these systems sets the stage for defining the key elements required for management strategies tailored to meet regional needs.

The Millennium Ecosystem Assessment Report (MA 2005) reveals that approximately 60% (15 out of 24) of global terrestrial and aquatic ecosystem services are degraded, including several such as capture fisheries that are predominately supplied by coastal and marine waters (see also UNEP 2006). Halpern et al. (2012) provided a subsequent global assessment of the sustainable flow of marine ecosystem services mapped to ten principal public goals. This approach is designed to be flexible and adaptable to accommodate new or improved data sources over time. In this analysis, scores for each coastal country were computed for each goal; a composite overall score was also provided. The global composite score over all coastal countries in this initial assessment was 60 out of a (nominal) maximum score of 100 (in practice because of interactions among goals, attaining a score of 100 is likely impossible). The exact score is perhaps less important than the general message that considerable room for improvement in management is evident at present.

Degradation of our ability to maintain the flow of ecosystem services at sustainable levels can be linked to a set of direct threats (Text Box 3). Identification of the global reach and intensity of these threats provides the first steps in specifying remedial action through management. Halpern et al. (2008) identified 17 stressors affecting marine ecosystems. In addition to the overarching issue of human population size, these can be classified into 6 major categories including (1) nutrient enrichment and pollution in (a) coastal and (b) oceanic systems; (2) destructive fishing practices in (a) commercial and (b) artisanal fisheries; (3) benthic structures

3. Threats to Marine Ecosystem Services

- Overfishing resulting in declines in abundance, alteration of food web structure, and reduced resilience to other stressors
- Eutrophication and increased incidence of anoxic and hypoxic events
- Loss in habitat due to coastal development, ocean mining, and fishing activity
- Declines in biodiversity due to anthropogenic stressors
- Increases in invasive species due to deliberate and incidental introductions
- Contamination due to runoff from land, industrial waste, hydrocarbon and other chemical spills
- Marine Litter and incidental ingestion by marine organisms
- Ocean noise disrupting animal communication
- Global change affecting temperature, salinity, stratification, currents, pH

such as oil rigs; (4) invasive species; (5) shipping; and (6) anthropogenic global change affecting (a) temperature, (b) pH, and (c) UV radiation. Although clear differences in the regional manifestation of the cumulative impacts of these drivers is evident (Figure 1a), virtually no part of the world ocean remains unaffected by some form of anthropogenic impact.

Our ability to understand the full ramifications of climate change, and in particular our ability to forecast future impacts on coastal and marine systems is still very much an evolving art. Nonetheless, if current projections hold and CO_2 emissions are not brought under control, we can anticipate the prospect of major alterations in coastal and marine ecosystems related to a number of factors including increasing temperature (Figure 1b), ocean acidification (Figure 1c) and sea level rise (Figure 1d). Temperature affects virtually every aspect of the biology and ecology of marine and coastal organisms from their physiology to distribution. It also exerts major effects on basic oceanographic processes. Increasing ocean acidification reflects the ongoing role of the ocean in buffering increasing CO_2 emissions, and points to constraints and limits to its buffering capacity. Decreasing pH levels hold the potential to directly or indirectly affect all living organisms in the sea with particularly dramatic direct effects on systems as diverse as coral reef and shellfish beds.

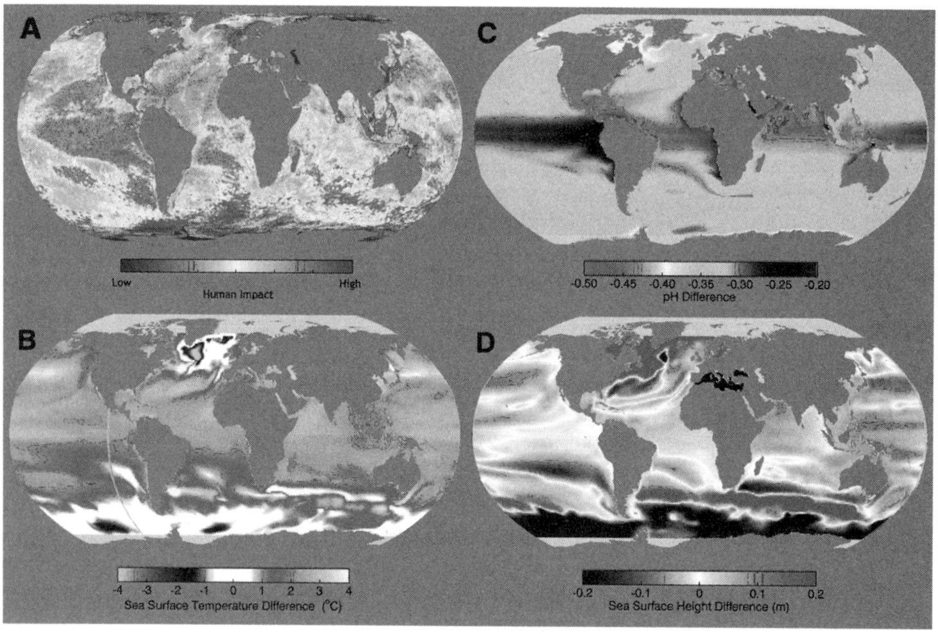

Fig. 1.1 – (See color insert) Maps of the global distribution of (a) cumulative anthropogenic impacts (data provided by Ben Halpern, UCSB and used with permission), (b) projected change in sea surface temperature by 2100 (V. Saba, NOAA), (c) projected change in pH by 2100, and (d) local difference relative to average global sea level rise by 2100". Data for mapping derived from the GFDL ESM2 Global Coupled Climate–Carbon Earth System Model (V. Saba, NOAA personal communication; for model details and caveats, see Dunne et al. 2012; Dunne et al. in press). Maps prepared by S. Lucey and Kim Hyde, NOAA. Note: no estimates for Arctic regions (grey) because of ice cover. Reference period for SST, pH, and SLR 1996–2005.

3. The Architecture of Marine Ecosystem-Based Management

Here we seek to provide a roadmap to the major elements of mEBM and its implementation that, in turn, provides a guide to the chapters to follow. We describe inter-relationships among the components of the mEBM process and attempt to connect elements that are sometimes conceived of (or at least treated) as separate processes altogether. One of many possible pathways to mEBM is depicted in Figure 1.2. The elements comprise specification of overarching aspirational goals, identification of relevant spatial management units, selection of strategic objectives, and assembling an integrated management plan incorporating consideration of a spectrum of ocean use sectors. Aspirational goals encompass the fundamental requirement to protect ecosystem structure and function to ensure the sustainable flow of services. Because ecosystems are typically defined in space, we require designation of spatial boundaries for management with the recognition that these units are open and interconnected. Strategic objectives translate a broad vision of overarching goals into specific statements providing a blueprint for action. Embedded in this concept is the need not only to clearly articulate objectives but to devise strategies to assess the state of the system as a whole and to specify reference points to guide management actions to meet these objectives. Some form of Integrated Ecosystem Assessment [IEA—sometimes designated simply as Integrated Assessment (IA)] is necessary to meet this requirement. With the development of an integrated management plan comes the need to incorporate sub-objectives for different ocean use sectors, to deal directly with the tradeoffs that will inevitably emerge, and to select the appropriate management tools to meet these needs. The latter will include spatial management strategies through some form of marine spatial planning and zoning as a major component. It will also require tactics to control and limit the effect of anthropogenic stressors to the extent possible and/or to modulate the extraction of resources.

We can visualize the mEBM process as comprising an inter-related set of elements starting with scoping and stakeholder engagement and leading to the specification of objectives and choice of management units; the development of IEAs to determine targets and limits for management action and evaluation of the current state of the system supported by ecosystem research and monitoring; the selection of tactical management tools to meet strategic objectives and culminating in the formulation of management advice (Figure 1.3). The IEA process itself can be further decomposed into components related to selection of informative ecosystem indicators to monitor the state of the system, implementation of some form of operating model that allows creation of a virtual world in which ecosystem, assessment and management processes can be simulated and the performance characteristics of different management strategies can be evaluated. It is then necessary to determine the risks associated with alternative management actions. In this construct, IEAs and MSP are viewed as inter-related parts of the overall mEBM process and not separate entities (or substitutes for the whole mEBM process). In the past, spatial strategies such as the designation of Marine Protected Areas (MPAs) have been discussed as if they were synonymous with mEBM rather than as a critical element of the toolkit to implement the broader dimensions of mEBM.

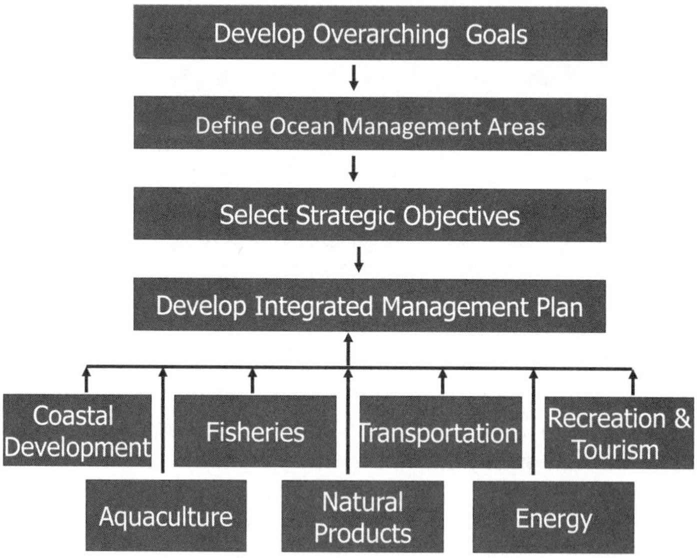

Fig. 1.2 – (See color insert) Roadmap to marine ecosystem-based management culminating in development of an integrated management plan incorporating sectoral consideration from a selection of ocean use communities.

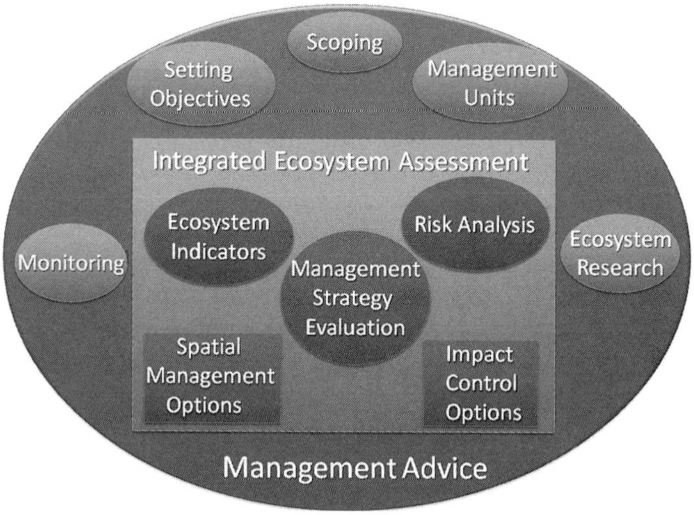

Fig. 1.3 – (See color insert) Elements of marine ecosystem-based management engagement with stakeholders (scoping and setting objectives); selection of spatial management units, ecosystem research and monitoring, integrated ecosystem assessment (incorporating selection of indicators, management strategy evaluation, risk analysis and choice of tactical management tools [including spatial management strategies and input or output controls affecting ecosystem impacts]); and development of scientific advice for management.

4. The Architecture of This Volume

In this volume we explore the philosophical framework and scientific underpinnings of mEBM as it is now unfolding and as we make the transition from concept to reality. We have arranged the volume into two main sections. The first addresses the overarching themes, tools, and considerations underlying mEBM. It is intended to put flesh on the bare bones of the preceding discussion to document the broader dimensions and requirements for successful implementation of mEBM. These chapters encapsulate issues as diverse as historical lessons for successful management, the human dimensions in mEBM, governance frameworks, modeling strategies, assessment and evaluation methods, marine spatial planning, and tools for identifying and resolving tradeoffs. The second section spans the globe from the tropics to the poles to provide an assessment of the status of implementation and the ongoing challenges of making mEBM a reality in selected systems.

4.1 Part I: Principles and Concepts

We begin with an homage to Santayana's apt and ever-relevant aphorism: 'Those who cannot remember the past are condemned to repeat it'. Lotze et al. (Chapter 2) set the stage by describing lessons from historical ecology, documenting along the way just how often we have continued to repeat past errors. They provide important proscriptions on how to break the cycle of historical error to make mEBM effective.

The western intellectual tradition has typically imposed strong barriers to interchange between the social and natural sciences (and indeed to sub-disciplines within each). The recognition that successful mEBM requires breaking down these barriers and forging new alliances serves as a major catalyst for change. Charles (Chapter 3) outlines the required steps and indicates what still needs to be done to develop a new transdisciplinary science to support mEBM for Social-Ecological Systems.

Implementation of mEBM cannot proceed without the appropriate regulatory and governance frameworks in place to enable it. Cochrane et al. (Chapter 4) describe the status of governance structures at national to international scales. Rapid adoption of the fundamental principles of mEBM is evident now on both scales. In an inversion of the usual order of things, regulatory and governance frameworks in some cases have surpassed the related necessary transformation of scientific structures to support mEBM. Often, scientific institutions now charged with (or at least encouraged to) support more holistic approaches to management remain fixed in traditional sectoral orientations that complicate the transition to mEBM.

The scientific support structures for mEBM entail not only the collection of relevant social and ecological data and research into socio-ecological processes but synthetic tools to connect the pieces, make forecasts, and evaluate alternative management options. Models of varying degrees of complexity serve as the integrative tools to meet this need. Fulton and Link (Chapter 5) describe the state of the art in ecosystem modeling to support mEBM. Ecosystem models have now been developed and applied in virtually all parts of the world ocean, providing a valuable jumping off point for further evaluation. Ecosystem models such as EcoPath with Ecosim, ATLANTIS and others have been applied in this context and can serve as key components of IEAs for Management Strategy Evaluation and risk analyses (see below).

Plummer and Levin (Chapter 6) describe key elements of the IEA process with a focus on social-ecological considerations. Many of the issues described in the preceding chapters are brought together under the IEA framework. The human dimension, reflected in effective stakeholder involvement to identify key issues and specifying goals and objectives, is critical in initiating the overall EBM process. The particular governance and regulatory frameworks in place in a given region shape the character of options for mEBM that must be evaluated through the IEA process. Models occupy the core of the analytical engine that is an IEA. We can view the IEA process as an iterative cycle stepping through these elements (Figure 1.4) although, depending on circumstances, some parts of this cycle may be altered or may indeed proceed concurrently rather than sequentially.

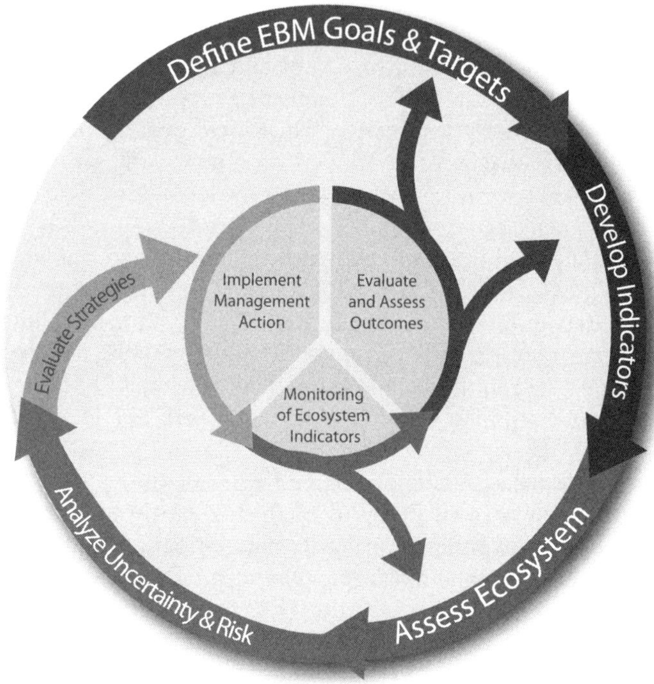

Fig. 1.4 – (See color insert) Developmental cycle for an integrated ecosystem assessment. (Courtesy of the NOAA IEA Program Office)

As emphasized above, mEBM is inherently place-based and options for spatial management flow naturally from this perspective. We can view the overall mEBM approach as comprising a nested hierarchical structure in space. It will inevitably and necessarily play out on multiple spatial scales from the local (e.g. an estuary or embayment) to the level of Large Marine Ecosystems (e.g. Sherman and Duda 1999) or even ocean basins for some applications. Evaluating options for spatial management is prominent in the IEA process. Designated ecosystem boundaries occupy the highest level of organization with various options for specifying marine protected areas, ocean zoning and other strategies falling within the basic unit. Rice et al. (Chapter 7) trace the development of these spatial strategies and their uses in meeting mEBM goals, including conflict resolution concerning the preemptive use of space.

We conclude this section with two complementary chapters describing decision support tools for evaluating the status of ecosystem services and for identifying and resolving tradeoffs. Guerry and Tallis (Chapter 8) describe the now widely applied Integrated Valuation of Environmental Services and Tradeoffs (InVEST) framework. The approach starts with a strong spatial basis to characterize the distribution and magnitude of ecosystem services and options for adjudicating competing uses of coastal and marine resources. Altman et al. (Chapter 9) apply an ecosystem accounting approach with both spatial and temporal dimensions within the Multiscale Integrated Model of Ecosystem Services (MIMES) format in an approach again designed to confront the issue of tradeoffs.

4.2 Part II: Case Studies

In the second part of this volume we turn to a sequence of case studies across major ecotypes including polar and subarctic, temperate, tropical, open ocean, and coastal and estuarine systems. The contrasts among these ecotypes are sharply defined by basic considerations such as human population density (Figure 1.5a) and levels of industrialization; water temperature (Figure 1.5b) and its attendant effects on ecosystem structure and function; ecosystem productivity as indicated by inputs at the base of the food web through primary production (Figure 1.5c); and levels of biodiversity in marine communities and ecosystems (Figure 1.5d). Many other factors of course serve to demarcate these ecotype boundaries. The polar regions are sparsely populated and many, but not all, anthropogenic impacts reflect far field forcing from other parts of the globe. In contrast, the temperate and tropical regions are often densely populated along the coasts but with generally much greater industrialization in the former. The coastal and estuarine areas of the world ocean bear the brunt of a diverse set of anthropogenic stressors as a result. The adjacent continental shelves are not immune from these impacts, however, particularly, but not exclusively, in the temperate zone. Human habitation in the major ocean basins away from continental land masses is of course primarily confined to individual islands and archipelagos. The overall distribution of human populations and the degree of industrial activities of various types sets the stage for the form and magnitude of anthropogenic disturbance in each of these systems—a principal determinant of the challenges to be met by mEBM in these areas.

Designation of traditional ecotypes such as polar, temperate, and tropical systems is typically defined by latitude (with demarcation points usually set at at 66.5° to 90° for the polar regions, 23.5° to 66.5° for the temperate zones in both hemispheres, and from 23.5°S to 23.5°N for the tropics). These cutpoints also naturally reflect distinct temperature regimes (Figure 1.5b) with all that this implies for ecosystem properties. Primary production is higher in coastal waters and in upswelling areas (Figure 1.5c) and marine biodiversity is highest by far in tropical locations (Tittensor et al. 2010; Figure 1.5d). Coupled with considerations related to human population density and industrialization as described above, these differences point to the overall utility of the ecotypes we have selected for characterizing the different needs and challenges for mEBM on a global scale.

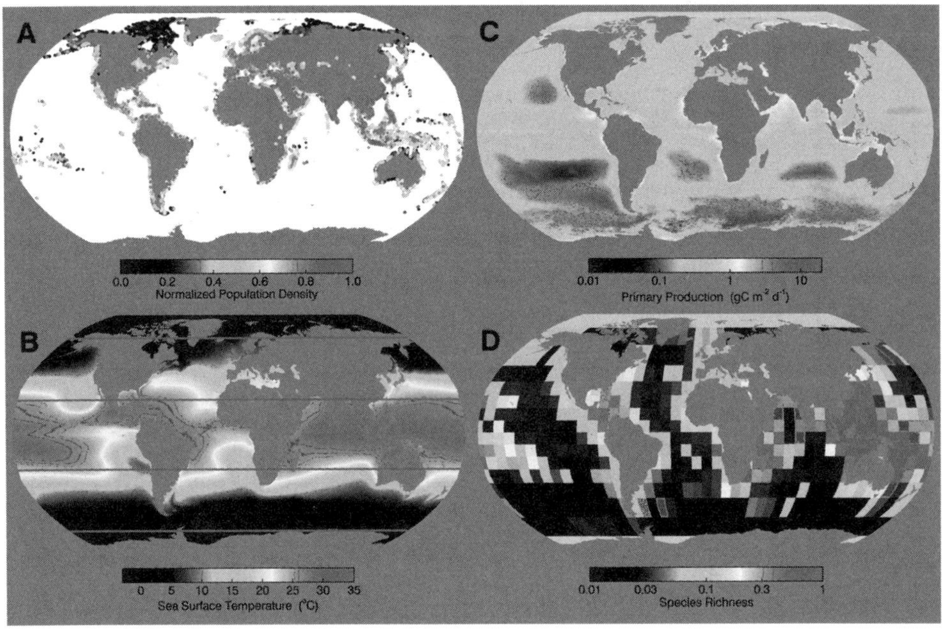

Fig. 1.5 – (See color insert) Maps of the global distribution of (a) human population density (data provided by Ben Halpern UCSB and used with permission), (b) sea surface temperature, (c) primary production, and (d) marine biodiversity (data provided by Derek Tittensor and Boris Worm, Dalhousie University and used with permission). (Maps prepared by S. Lucey and Kim Hyde, NOAA)

Within each chapter devoted to a particular ecotype, we provide a limited number of case studies (usually 3–5) to afford opportunities for comparison within ecotypes and reflecting contrasts related to differences in population density and industrialization, historical developments in resource use, regulatory and governance structures, cultural differences and other factors within each. Our approach therefore differs from perhaps more traditional approaches in which a single area (e.g. the Great Barrier Reef, Chesapeake Bay, etc.) provides the focal point for detailed treatment. We note that the selection of a restricted number of case studies (perhaps more accurately characterized as mini-case studies within a chapter) attempts to strike a balance between generalities representative of an entire ecotype vs. a more detailed, single-location emphasis. In essence, the 'experimental design' of this section involves comparisons within and between ecotypes (with admittedly non-random selection in the former).

In each of the chapters in Section II, we first provide an overview of the physical setting (physiography, oceanography etc.) of each system. In effect, start by describing the stage on which the social-ecological play unfolds. We highlight critical dimensions of human connections to the coasts and seas in each. The important ecological services provided and the past, present and emerging threats to the delivery of these services are also described. Finally we attempt an overview of the state of implementation of mEBM in each mini-case study.

Our evaluation of current activity levels of different ocean use sectors (and forecasts of future demand) provide important insights into potential spatial conflicts and tradeoffs that will require resolution under mEBM. Because these competing interests can reveal potential fault lines in the development of integrated management plans, we highlight the importance of major provisioning services in these systems and explore the implications for development of effective mEBM strategies.

Embedded within each of these broad ecotype boundaries are nested regional ecosystems that will provide the foci for actual implementation of mEBM. Although it is perhaps tempting to think that we can identify some platonic ideal of what characterizes an ecosystem, almost invariably ecologists have defined ecosystem boundaries to meet some specified objective or set of pragmatic considerations. One potential starting point is the global framework afforded by the Large Marine Ecosystem (LME) concept. An LME is defined as a large ocean area (typically on the order of 200,000 km^2) characterized by bathymetry, hydrography, productivity, and trophic relationships. A total of 66 LMEs have now been defined throughout the world ocean (Figure 1.6a). Significantly, the Global Environmental Fund has invested substantially in enhancing scientific capacity in the developing world to implement the LME approach, providing an important avenue for implementation of mEBM. The LME framework specifies a strategy for monitoring

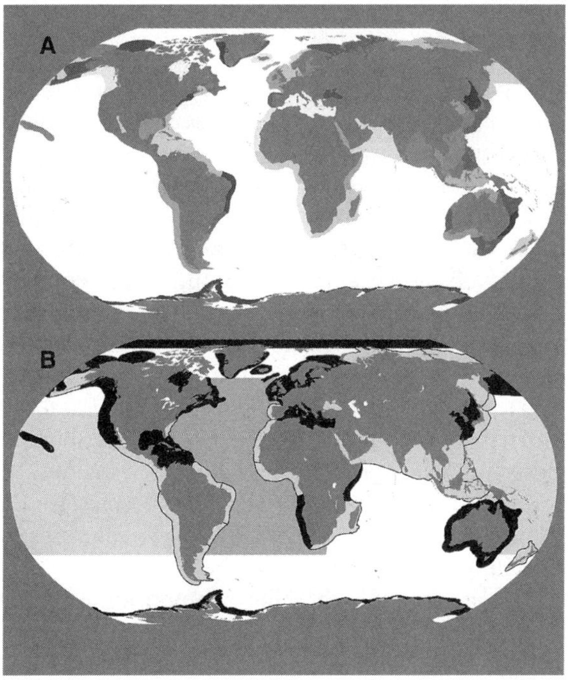

Fig. 1.6 – (See color insert) Location of (a) designated large marine ecosystems and (b) LMEs covered in part or in whole in this volume. Open ocean areas in the Atlantic and Pacific Oceans covered here also shown. (Maps prepared by Kim Hyde, NOAA)

and assessing changes in five basic modules (1) productivity, (2) fish and fisheries, (3) pollution and ecosystem health, (4) socioeconomics and (5) governance structures. The modules are then evaluated with respect to recovery, sustainability and management of marine resources and their habitats. It might prove desirable in some cases to specify further subregions within LMEs as operational management units, again building on the concept of nested hierarchical structures.

A representation of the case studies covered in whole or in part in both Parts I and II of this volume is provided in Figure 1.6b. We also show the principal open ocean areas in the Atlantic and Pacific Oceans represented in our case studies. The deep ocean basins outside the continental shelves of the globe are not represented within the LME construct. In a number of instances, parts or all of several LMEs are treated in several chapters in the volume, each offering different perspectives on different dimensions of the overall mEBM process.

We begin our overview of ecotypes with polar and subarctic systems (Gaichas et al. Chapter 10). The polar regions are, of course, the least densely populated region of the earth's surface but perhaps the most vulnerable to climate change. The polar regions are governed by international treaties and protocols and therefore involve very different regulatory processes than many other systems considered in this volume. The establishment of ecosystem principles in management of Antarctic marine resources was one of the earliest examples of mEBM on the world stage. The subarctic region has long played a key role in food provisioning through fisheries, natural products (and in some cases, food) through whaling and now is increasingly exploited for hydrocarbon resources.

Samhouri et al. (Chapter 11) provide contrasts in the development of mEBM in two North American continental shelf systems that differ markedly in physical characteristics (and key issues in the supply of ecosystem services) with a unique 'marine ranch' strategy off the Korean Peninsula. The focus on northern hemisphere mini-case studies reflects, in part, the much smaller land mass in the temperate zone of the southern hemisphere.

Coral reefs are among the largest biogenic ecosystems on earth and face some of the greatest threats due to human influences, from coastal development to climate change. Kritzer et al. (Chapter 12) accordingly center their treatment of tropical systems principally on coral reefs. The important connections to adjacent nearshore habitats (including mangrove systems etc.) are not, however, neglected.

Our world is of course dominated by the seven oceans, accounting for approximately 70% of surface of the globe. Of this total, by far the largest portion is open ocean, in deep water, and outside the jurisdiction of individual countries. Polovina et al. (Chapter 13) describe the unique challenges that accompany management of these systems in both the pelagic and demersal/benthic realms. The former principally involve consideration of services related to fishering and shipping while the latter encompasses seamount and ocean ridge ecosystems where vulnerabilities to disturbance are especially acute.

Houde et al. (Chapter 14) tackle the highly dimensional threats and impacts to be found in coastal and estuarine ecosystems in areas immediately adjacent to centers of human habitation. The heavily industrialized regions of the northern hemisphere naturally present the greatest challenges due to the diversity of human uses and the broad spectrum of accompanying stressors.

We close the volume with a brief overview of lessons learned and the challenges that remain.

In summary, in this volume we attempt to:

- Describe the fundamental structural elements of successful Marine Ecosystem-Based Management from setting objectives, through analysis and evaluation, to implementation and governance.
- Provide an evaluation of both the cumulative impacts of utilization of ocean services by humans and the benefits of devising sustainable use practices.
- Identify the tradeoffs inherent in management of ocean use sectors that often involve competing goals and objectives.
- Unite topics such as Coastal and Marine Spatial Planning and Integrated Ecosystem Assessment, often treated as separate but inter-related topics, under the overarching banner of Marine EBM.
- Show the unique challenges in devising and implementing Marine EBM in very different ecosystem types including polar, boreal, temperate, and tropical systems characterized by different levels of human influence, scientific understanding, biodiversity patterns, and environmental and climate forcing.

Acknowledgments

We express our sincere gratitude to the members of our Editorial Advisory Board, Keith Brander, Phil Levin, Andy Rosenberg, Tony Smith, and Boris Worm for their sage counsel in giving shape to this volume. We are deeply indebted to Kim Hyde, Sean Lucey, Vince Saba, Laurel Smith and Robert Gamble for their contribution to analysis, mapping and the graphic arts. Their work greatly assisted our portrayal of global processes relevant to mEBM.

References

Dunne, J. P., J. G. John, A. J. Adcroft, S. M. Griffies, R. W. Hallberg, E. Shevliakova, R. J. Stouffer, W. Cooke, K. A. Dunne, M. J. Harrison, J. P. Krasting, S. L. Malyshev, P. C. D. Milly, P. J. Phillipps, L. T. Sentman, B. L. Samuels, M. J. Spelman, M. Winton, A. T. Wittenberg, and N. Zadeh. 2012. GFDL's ESM2 Global Coupled Climate–Carbon Earth System Models. Part I: Physical Formulation and Baseline Simulation Characteristics. *J. Clim.* 25:6646–6665.

Dunne, J. P., J. G. John, E. Shevliakova, R. J. Stouffer, J. P. Krasting S. L. Malyshev, P. C. D. Milly, L. T. Sentman, A. J. Adcroft, W. Cooke, K. A. Dunne S. M. Griffies, R. W. Hallberg, M. J. Harrison, H. Levy, A. T. Wittenberg, P. J. Phillipps, and N. Zadeh. 2012. GFDL's ESM2 Global Coupled Climate–Carbon Earth System Models. Part II: Carbon system formulation and baseline simulation characteristics. *J. Clim, in press.*

Halpern, B. S., S. Walbridge, K. A. Selkoe, C. V. Kappel, F. Micheli, C. D'Agrosa, J. F. Bruno, K. S. Casey, C. Ebert, H. E. Fox, R. Fijita, D. Heinnemann, H. S. Lenihan, E. M. P. Madin, M. T. Perry, E. R. Selig, M. Spalding, R. Steneck, and R. Watson. A global map of human impact on marine ecosystems. *Science* 319:948–952.

Halpern, B. S., C. Longo, D. Hardy, K. L. McLeod, J. F. Samhouri, S. K. Katona, K. Kleisner, S. E. Lester, J. O'Leary, M. Ranelletti, A. A. Rosenberg, C. Scarborough, E. R. Selig, B. D. Best, D. R. Brumbaugh, F. S. Chapin III, L. B. Crowder, K. L. Daly, S. C. Doney, C. Elfes, M. J. Fogarty, S. D. Gaines, K. Jacobsen, L. B. Karrer, H. M. Leslie, E. Neeley, D. Pauly, S. Polasky, Bud Ris, K. St. Martin, G. S. Stone, U. R. Sumaila, and D. Zeller. 2012. An index to assess the health and benefits of the global ocean. *Nature* 488:615–622.

McLeod, K. and H. Leslie (Eds.). 2009. Ecosystem-Based Management for the Oceans. Island Press, Washington, D.C. 368 pp.

McLeod, K. J. Lubchenco, S. Palumbi, and A. Rosenberg. 2005. Consensus statement on marine ecosystem-based management. Signed by 221 academic scientists and policy experts with relevant expertise. http://www.compasssonline.org/pdf_files/EBM_Consensus_Statement_v12.pdf

MA (Millennium Ecosystem Assessment). 2005. Ecosystems and human well-being: synthesis. Island Press, Washington, D.C. 368 pp.

Sherman, K. and A. M. Duda. 1999. An ecosystem approach to global assessment and management of coastal waters. *Mar. Ecol. Prog. Ser.* 190:271–287.

Tittensor, C. Mora, W. Jetz, H. K. Lotze, D. Ricard, and W. V. Berghe. 2010. Global patterns and predictors of marine biodiversity across taxa. *Nature* 466:1098–1101.

UNEP (United Nations Environmental Program). 2006. Marine and coastal ecosystems and human well-being: A synthesis report based on the findings of the Millennium Ecosystem Assessment. Nairobi. UNEP.

USCOP (U.S. Commission on Ocean Policy) 2004. An ocean blueprint for the 21st century. Final Report. Washington, D.C. USCOP. ISBM 0–9759462 0 X.

Chapter 2. Lessons from Historical Ecology and Management

HEIKE K. LOTZE

Dalhousie University

RICHARD C. HOFFMANN

York University

JON M. ERLANDSON

Museum of Natural & Cultural History, University of Oregon

Contents

1. Introduction

Ecosystem-based management is a relatively new concept, yet throughout history and around the world people have responded to declining resource abundance and changing environments by adapting their actions and behaviors. This is what drove the evolution of technology, culture and management over millennia. Today, we do not need to invent a completely new idea of how to manage ocean ecosystems; we can build upon our successes and failures of the past, combine promising management strategies and identify what has been missing or needs improvement. In this chapter we provide an overview of what can be learned from the past to better understand and advance marine ecosystem-based management.

Such a historical perspective can be taken because of recent advances in the growing field of Marine Historical Ecology. Here, palaeontologists, archaeologists, historians, fishery scientists and marine ecologists have come to work together to unravel the history of change in marine resources over past centuries and millennia (Jackson et al., 2001; Pandolfi et al., 2003; Lotze et al., 2006, 2011a, b; Rick and Erlandson, 2008; Kittinger et al., 2011). They have combined records from their various disciplines, ranging from sediment core data, animal remains in archaeological middens, written accounts by early explorers, whaling and fishing log books, old maps, paintings and photographs to fisheries statistics, research surveys

The Sea, Volume 16, edited by Michael J. Fogarty and James J. McCarthy
ISBN 978-0-674-07270-1

and genetic data, to piece together the timeline and magnitude of change (Lotze and Worm, 2009; Lotze et al., 2011a). They have also tried to identify natural baselines and trace shifts in resource abundance and our perspectives of them over time (Pauly, 1995; Dayton et al., 1998; Jackson et al., 2011). This approach has spurred more in-depth studies on changes in marine resource abundance in different periods of the past and different regions of the world, and of different target species (Hoffmann, 2005; Rosenberg et al., 2005; Sáenz-Arroyo et al., 2005; Lotze, 2005, 2010; Hardt, 2009; McClenahan et al., 2006; McClenahan and Cooper, 2008; Rick and Erlandson, 2008). Overall, this growing body of research increasingly demonstrates that the depletion of marine resources started much earlier and was of much greater magnitude than previously thought (Lotze and Worm, 2009; Erlandson and Rick, 2010; Jackson et al., 2011). But how have people in the past dealt with changes in the natural resources they depended upon for survival?

Until recently, most resource users, managers, fishery scientists and marine ecologists would have agreed that today's depletion of marine resources and degradation of ocean ecosystems is the consequence of the industrialization of fisheries since the 1950s. It is clearly the most efficient and far-reaching way of exploiting marine resources ever seen in human history. Yet there have been many changes in how humans have exploited marine resources over past centuries and millennia, such as changes in gear, target species, the area fished and ways of local enhancement of production such as aquaculture (Hoffmann, 2005; Lotze, 2007; Rick and Erlandson, 2008). These changes in exploitation strategies reflect changes in resource availability as well as people's attempts and adaptations to meet a continuing demand for food and other products. In many cases, people were concerned about the declining abundance and health of natural resources and implemented measures to protect them or reverse negative trends, such as privatization and fisheries regulations (Lotze et al., 2011c). Some of these were successful, others failed.

In this chapter, we first review what historical studies can tell us about the responses of people to observed changes in resource availability in different periods of the past and how this has changed the general patterns of exploitation over time. Next, we will highlight what has and what hasn't worked in terms of managing resource abundance, securing the survival of species and maintaining functioning ecosystems. For millennia, humans have affected marine ecosystems through direct exploitation, probably the earliest and most pervasive impact. A growing number of studies point to early evidence of other human-induced changes, however, including habitat alteration, water pollution, sedimentation and invasive species. Because mitigating all these changes is an essential part of ecosystem-based management, we will also review how other human impacts in the past have changed resource conditions and been dealt with. Moreover, natural environmental fluctuations have always affected resource abundance. Thus, historical investigations may provide insight into the effects of climate change. Finally, today and in the past, ecosystems do not show a single-species response to the effects of fishing or other human influences, yet few studies have so far aimed at investigating changes in past biodiversity, food-web structure and ecosystem functioning. We therefore provide an overview of the ecosystem consequences of past human-induced changes. In our last section, we synthesize important lessons from the past that can inform and advance marine ecosystem-based management today and for the future.

2. Marine Resource Use and Management Through Time

2.1 Archaeological records

The earliest known evidence for intensive use of marine resources dates back to ~164,000 years ago by anatomically modern humans in South Africa and ~150,000 years ago by Neanderthals in southern Spain, and reveals shellfish gathering in coastal waters (Marean et al., 2007; Erlandson, 2010; Cortés-Sánchez et al., 2011). Since those early times, archaeological studies from around the world point to abundant and diverse uses of marine species, including marine mammals, birds, reptiles, fish and shellfish, by humans living near the sea (Rick and Erlandson, 2008). Coastal peoples used a variety of fishing gears, including nets, traps, hooks, weirs and spears as well as sophisticated harpoons, characteristic flatfish hooks and multipoint fishing spears (Pinnegar and Engelhard, 2008). In many cases, these people deposited their kitchen wastes, including the bones, shells, scales, hairs and teeth of marine animals, in garbage heaps or shell middens. If well preserved, they reveal components of human diets over different time periods. Interestingly, an increasing number of such studies have documented clear shifts in the relative abundance of different species, their size and age as well as changing gears and cultural practices, often indicating signs of local resource depletion that occurred much earlier than previously thought.

For example, on San Miguel Island, California's native people used >150 species over the past 12,000 years. Throughout the record, intertidal shellfish and nearshore fish appear to have been of highest importance, followed by marine mammals (Erlandson et al., 2009). A local depletion of sea otters appears to have led to the increase of shellfish populations as much as 7500 years ago (Erlandson et al., 2005). Subsequent reductions in the mean size of primary shellfish resources such as red abalone, California mussel, and limpets appear to reflect intensified predation pressure on intertidal resources by growing human populations (Erlandson et al., 2011). As a result, fishing increased in importance over time. About 1500 years ago, fisheries expanded into deeper waters targeting larger offshore species such as swordfish, tunas, sharks and barracudas with more advanced hooks (Figure 2.1) (Rick et al., 2008). These changes occurred at a time of marked human population growth and expanded regional trade networks, when it appears that traditional local resource supplies could not meet the increasing demand. Similar trends toward the increasing importance of marine fish over time, together with a spatial expansion of fishing from inshore to offshore areas, have been found in archaeological remains from southern Taiwan (Kuang-Ti, 2001), the Wadden Sea (Lotze, 2007), Great Britain (Barrett et al., 2004) and the North Atlantic (Perdikaris and McGovern, 2008).

In San Francisco Bay, another example points to local depression of preferred target species. From ~2600 to 700 years ago, the relative abundance of white sturgeon decreased together with its average size and age, suggesting local overexploitation (Broughton, 2002). During the same period, remains of geese decreased while smaller bird species increased indicating a gradual shift in target species.

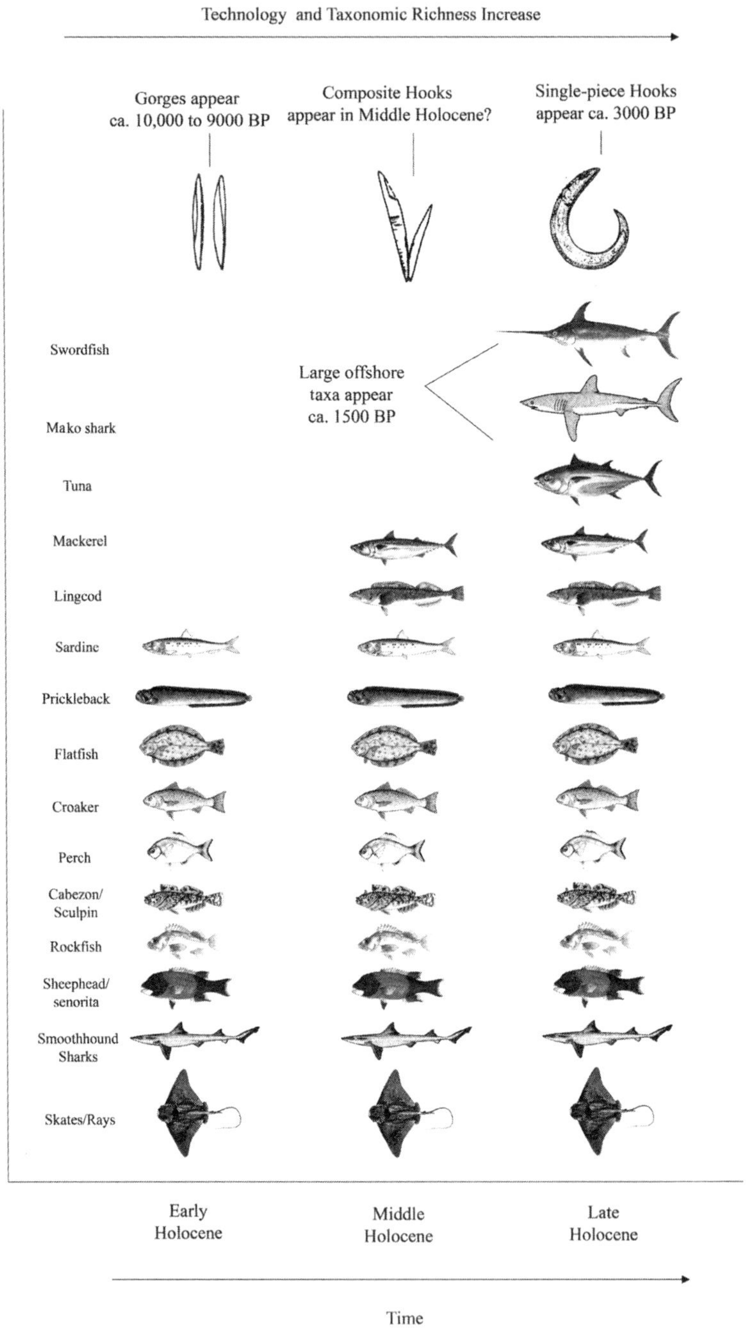

Fig. 2.1 – Changes in fishing technology and taxonomy from the early to the late Holocene on San Miguel Island, California. (From Rick et al., 2008)

Changes in adult-to-juvenile ratios in animal remains suggested the collapse of local breeding colonies of cormorants and sea otters (Broughton, 2002). People also travelled farther for hunting over time. This example illustrates human-induced changes in natural resources on a local scale, as well as the subsequent adaptations people made to deal with those changes in order to meet continuing demand.

More and more studies describe similar signs of resource depletion in archaeological records from around the world. Probably the earliest records stem from the Mediterranean and South Africa, where hunter-gatherers influenced the size and structure of nearshore shellfish populations beginning at least ~23,000 and 13,000 years ago (Stiner, 2001; Steele and Klein, 2008). In Puerto Marqués, Mexico, archaeologists revealed a sharp decline and later loss of sea turtle remains in shell middens between ~5500 and 2300 years ago and a shift from high to low trophic level fish (Kennett et al., 2008). Strong shifts in species composition, especially a decline in apex predators (cod, swordfish, sea mink) and increase in mesopredators (flatfish, sculpin), also occurred in the Gulf of Maine from ~4350 to 400 years ago (Bourque et al., 2008). In New Zealand, after the arrival of Maoris ~800 years ago, hunting of fur seals expanded around the North and South Islands and resulted in a severe range contraction of seals to the southern tip of the South Island by the late 18th century (Smith, 2005). Of course, there are not only human-induced changes in the archaeological records. In the Aleutian Islands, Causey et al. (2005) demonstrated both long-term climate-driven fluctuations in seabird populations and local overhunting of accessible breeding colonies (see also *Section 5.4 Climate change*).

Finally, some native people have aimed at increasing local resource supply by cultivating nearshore shellfish or fish (Rick and Erlandson, 2009). Along the Northwest Coast of North America, in British Columbia and Alaska before European contact, people created and cultivated "clam gardens". They constructed rock walls and terraces in the low intertidal, thereby expanding and maintaining the area in which clams can live and be harvested (Williams, 2006). In ancient Hawai'i, coastal people also farmed fish in brackish water ponds and enclosed large sections of shallow reef to create salt water ponds used for mariculture, resources that were generally controlled by elite members of society (Kirch, 1985). Thus, already in the archaeological record we find some common responses of humans to local resource depletion, including shifts in target species, expansion and control of hunting and fishing grounds, gear adaptations and cultivation of local resources to enhance supply.

2.2 *Roman times*

Many ancient civilizations have relied on resources from the ocean. One civilization with a particularly well-documented maritime influence was the Romans. The Roman Empire expanded around the Mediterranean Sea after 300 BC, reaching its greatest extent by ~100 AD. Increases in human population density, growing urbanization and the development of a market economy greatly enhanced the demand for food, luxury and fashion products, many of them from the sea. This spurred the first commercial and industrial-scale exploitation of marine resources. Fishing became a significant economic activity with more than 120 species of economic importance, including sharks and rays, anchovies and sardines, and a variety of flatfish and groundfish (Radcliffe, 1921; Hughes, 1994; Lotze et al., 2011b). Whale meat and a large variety of birds were sold in markets (Hughes, 1994), and industrial-scale exploitation occurred for purple dye snails (Koutsoubas et al., 2007), red coral for jewelry (Bresc, 2000) and bathing sponges (Pronzato and Manconi, 2008).

The Romans developed sea-going vessels for hook-and-line and net fishing as well as tuna traps and watchtowers for spotting tunas along the shores (Hughes, 1994). Apparently the only fishing gear they did not know or use was the trawl (Bekker-Nielsen, 2010). Technological advances posed few limits on fishing, yet the determining factor was preservation and storage (Bekker-Nielsen, 2005). To overcome this, some species were used to produce salt-fish ("tarichos"). Others were fermented into fish sauce ("garum"), especially mackerel and anchovies. Fish were also dried, pickled or kept alive in a pond, creel, and basket or a "well-box" (Trakadas, 2006; Bekker-Nielsen, 2010). These preservation techniques expanded the geographic range of fishing and marketing. Also, a cultural shift occurred whereby eating sea fish, formerly associated with lower-class fishermen, became fashionable among Roman elites and increased the demand for fish and fish sauce at all levels of society (Bekker-Nielsen, 2010).

By the first century AD, Italy had outfished its coastal waters and expanded to Sicily and Corsica (Radcliffe, 1921; Hughes, 1994; Trakadas, 2006). Garum processing sites expanded all around the Mediterranean and Black Sea and adjacent Atlantic coast (Trakadas, 2006; Bekker-Nielsen, 2010) and wild oysters were imported from the North Sea (Gunter, 1897). Other signs of coastal resource depletion included complaints about depleted fisheries and increasing prices on Roman markets (Hughes, 1994). In response, certain fishing techniques were prohibited (e.g. fishing with torch light at night), local fish and shellfish stocks were enhanced with transplants (e.g. parrot fish was fished in the Aegean and released in the Tyrrhenian Sea), and some fish were raised commercially (Hughes, 1994; Trakadas, 2006; Bullock, 2008). Romans cultured oysters, and fishponds became fashionable among the elite in Rome. Their holiday residences along the bay of Naples became the center of ostriculture and pisciculture (Bekker-Nielsen, 2010). This introduced private control and limited access to some fish resources that had previously been freely accessible in the ocean.

In summary, signs of local resource depletion led Romans to respond by spatially expanding exploitation, importing resources from afar, and stocking and cultivating resources to enhance local production. There are also first signs of restrictive management in prohibiting certain fishing gears, introducing private control and limiting access.

2.3 Medieval Europe

The use of marine resources during Roman times likely affected marine ecosystems at local and regional (e.g. Mediterranean) scales, but such impacts reached another level during medieval times (ca. 500–1500 AD). First, a growing human population throughout mainland Europe strongly increased the demand for food, a demand shaped by human numbers as well as nutritional and cultural needs (Hoffmann, 2005). Christian religion forbade eating "meat" (terrestrial animals) on ~130 days of the year, but allowed the substitution of "fish" (aquatic animals). Eating fresh meat or fish was also considered healthy and promoted over salted fish. Moreover, high social rank was displayed by eating large, fresh, expensive fish such as pike, salmon and sturgeon (Hoffmann, 2005). These cultural values strongly enhanced the demand for fresh fish. In addition, the rising human population also increased other pressures on aquatic ecosystems, including water pollution, sedimentation, damming and habitat alteration (see *Section 5;* Hoffmann, 2005). As a result, signs of depletion of preferred species, such as sturgeon and salmon, were evident by the 13th century (Hoffmann, 2001). Sturgeon remains dropped from 70% of fish consumed in the 7–9th centuries at 17 sites along the southern Baltic Sea to 10% in the 12–13th centuries (Benecke, 1986), and similar declines were noticed throughout Europe (Hoffmann, 2005). In response to dwindling numbers, landowners and authorities privatized some fisheries and on others implemented public regulations concerning quotas, gear, seasonal or other forms of restrictions (Hoffmann, 2002). Sturgeon was even declared a "royal fish" and reserved for the elites (Hoffmann, 2005). Along the coasts of 14th century France, people became concerned about the destruction of fish stocks, decline in average size, and the effect of poor-quality and high-priced fish on food supply; thus royal edicts imposed fisheries closures and regulated gear in an effort to stabilize valuable fish stocks (Hoffmann 1996; Roberts, 2007).

Another factor in medieval times was the formation and spread of more complex economic and political institutions influencing exploitation patterns. Early medieval fisheries were predominantly for subsistence, but small-scale commercial fisheries spread across Europe from the 10th century, joining a wider commercial revolution with continually expanding markets and rising prices (Hoffmann, 2005). This commercialization was not restricted to fisheries and included other resources, such as whales, birds and invertebrates that were exploited for food and other products needed in expanding markets for fuel, fashion and luxury items (Lotze, 2005; Lotze et al., 2011b).

Third, the development of fisheries spread from inland and coastal Europe to marine environments in pursuit of new resource frontiers, possibly in response to the depletion of formerly preferred freshwater and migrating fish (Hoffmann, 2005). In Great Britain, a profound switch in fish consumption from predominantly freshwater to marine fish occurred in the 11th century (Barrett et al., 2004). Around Europe, many former local subsistence fisheries for species including herring, cod, hake, pike, sardine and bluefin tuna became subject to marine exploitation (Figure 2.2) (Hoffmann, 2005). The intensification of exploitation on resource frontiers drove the commercialization of fisheries and their spatial expansion to more distant waters. Moreover, innovations in fishing methods, preservation, transport and marketing enlarged supply to and demand by distant consumers (Hoffmann, 2005).

A further response to dwindling freshwater fish was the introduction of carp ponds and aquaculture (Hoffmann, 2000; Hoffmann and Winiwarter, 2010). Before the 7th century, carp was limited in Europe to the Black Sea watershed, but live transport and storage of wild-caught fish in ponds or other water bodies spread the occurrence of carp across Europe by 1600 AD (Figure 2.3) (Hoffmann, 2005). Some entrepreneurs developed large-scale artificial aquaculture using innovative techniques to rear carp from eggs and larvae for subsistence and market sale.

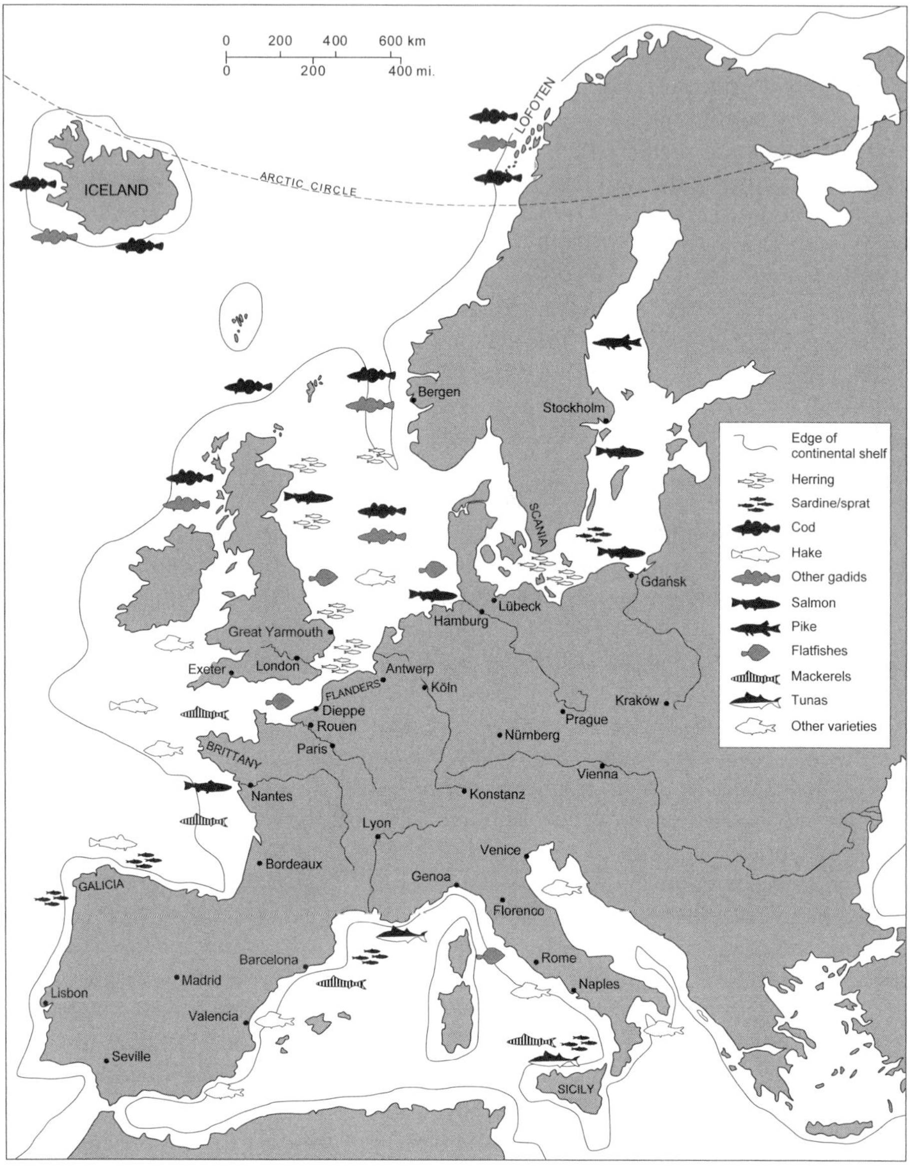

Fig. 2.2 – The development of major fisheries on marine frontiers around Europe in medieval times. (From Hoffmann, 2005, with kind permission from Springer Science & Business Media)

Around Italy, people also introduced valli culture to trap migrating fish on their way out of rivers or coastal lagoons until they reached market size (Hofrichter, 2002).

In summary, in medieval Europe, the depletion of preferred freshwater and migratory fish led to a shift in target species to marine resources, a spatial expansion from inland and coastal to marine fisheries, an intensification and commercialization of resource exploitation, and the rise of aquaculture to enhance local

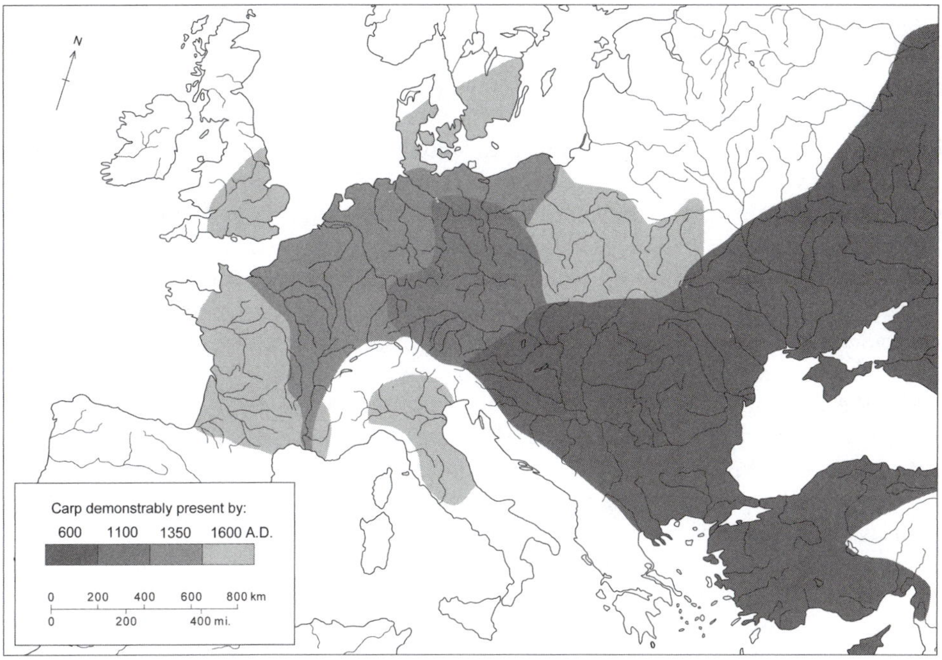

Fig. 2.3 – The spread of carp from southeastern to northwestern Europe during medieval times. (From Hoffmann, 2005, with kind permission from Springer Science & Business Media)

production. Although these responses to resource depletion were similar to those in previous periods, they reached a new magnitude and spatial scale all across Europe. In many cases, privatization, restrictions and regulations on resource use were imposed, with some successes and failures, yet such protective measures were easily forgotten at distant, seemingly infinite, resource frontiers (Hoffmann, 2005; Lotze, 2005).

2.4 *Colonial expansion*

By the end of the medieval period, fishing and whaling had encompassed all European waters and people started to look elsewhere for new frontiers. Beginning in the late 15th and early 16th centuries, early explorers and naturalists provided vivid descriptions of the incredible abundance of marine resources in the New World, something they hadn't seen in their European homeland during their lifetimes (Jackson et al., 2001; Lotze and Milewski, 2004; McClenahan et al., 2006; Hardt, 2009; Lotze, 2010). Whaling expanded from the coasts of mainland Europe to Spitzbergen, Greenland and Newfoundland in the 16th century (Bryant, 1995; Lotze, 2007), and fisheries for cod and other groundfish from the coastal to the open North Sea and further to Iceland and North America (Hoffmann, 2005; Roberts, 2007). Thus, the colonial period took marine resource exploitation to the next spatial level and European fishing practices expanded to America, Africa, Australia, Asia, and many remote island groups. At first, fishing and whaling camps were established over the summer, especially in eastern North America. All processed yields, including dried fish, oil and baleen, and products traded with Native Americans, especially furs, were transported back to the European mainland where they fed growing markets (Lotze and Milewski, 2004; Roberts, 2007; Lotze, 2010). Eventually, permanent settlements were established in the colonies, building their own markets and expanding agriculture, fishing and whaling along temperate and tropical coasts (Pandolfi et al., 2003; Lotze et al., 2006). Essentially, the resource use and market practices that had developed in Europe over the previous 500–1000 years were imposed on the New World in a matter of decades, but rarely with the measures of communal, private, and state control which medieval Europeans had developed at home (Hoffmann, 2000). In many areas, this colonial expansion coincided with industrialization in the 18th to early 20th centuries (see *Section 2.5*); thus, the imposed changes were even more accelerated.

In the colonies, where until then Native people had used marine resources mainly for subsistence and some trade, this meant an abrupt change in fishing technology, the kinds of marine species targeted and quantities taken. The rapid increase in European settlers and their demands for a wide variety of products as well as land clearing for agriculture and the introduction of different industries completely changed the magnitude, spatial scale and range of human influences on marine ecosystems. Thus, following European colonization, very similar trajectories of change emerged across coral reef ecosystems in the Atlantic, Red Sea and Australia (Pandolfi et al., 2003), estuaries and coastal seas in North America (Figure 2.4) and Australia (Lotze et al., 2006; Lotze, 2010), South Africa (Griffiths et al., 2004), Caribbean (McClenachan et al., 2006; McClenachan and Cooper, 2008; Hardt, 2009) and Hawaiian islands (Kittinger et al., 2011), with rapid declines particularly in large and highly valued species, including whales, pinnipeds, large waterfowl, seabirds, sea turtles, diadromous and groundfish. Large-scale land clearing increased sediment and nutrient loads and municipal and industrial developments discharged wastewater and chemical pollutants with adverse effects on water quality and habitat-building species such as seagrass (see *Section 5;* Cooper and Brush, 1993; McCulloch et al., 2003; Lotze et al., 2006). Other habitat-building species were directly harvested (e.g. oysters, mangroves) or mined for lime and shells (oysters). Drainage and peat mining intentionally destroyed wetlands. Finally, Europeans introduced a suite of species from their homeland or from the mainland to islands, some intentionally (cats, dogs, foxes, sheep, cattle, etc.) others unintentionally (rats, raccoons, pests, etc.), often with severe consequences for native flora and fauna (Lotze and Milewski, 2004; Lotze et al., 2006). These multiple new human influences strongly affected marine and island ecosystem structure and functioning in the New World, including Native resource use and culture.

In essence, the colonial expansion represented a huge new resource frontier with apparently unlimited abundance and diversity of resources, almost no restrictions on harvesting or other activities, and little consideration by governments that were mostly far away and primarily interested in further expansion, accumulation and fortification rather than restriction or precaution (Richards, 2003). Over time, however, the colonies established their own social and economic networks, culture and identity, and when some valuable resources became scarce some management plans and efforts were implemented. For example, only three years after settlement in Piscataqua in 1623, the Plimoth Plantation gave itself the legal authority to regulate fisheries to secure local fish supply (Anonymous, 1887; Rosenberg et al.,

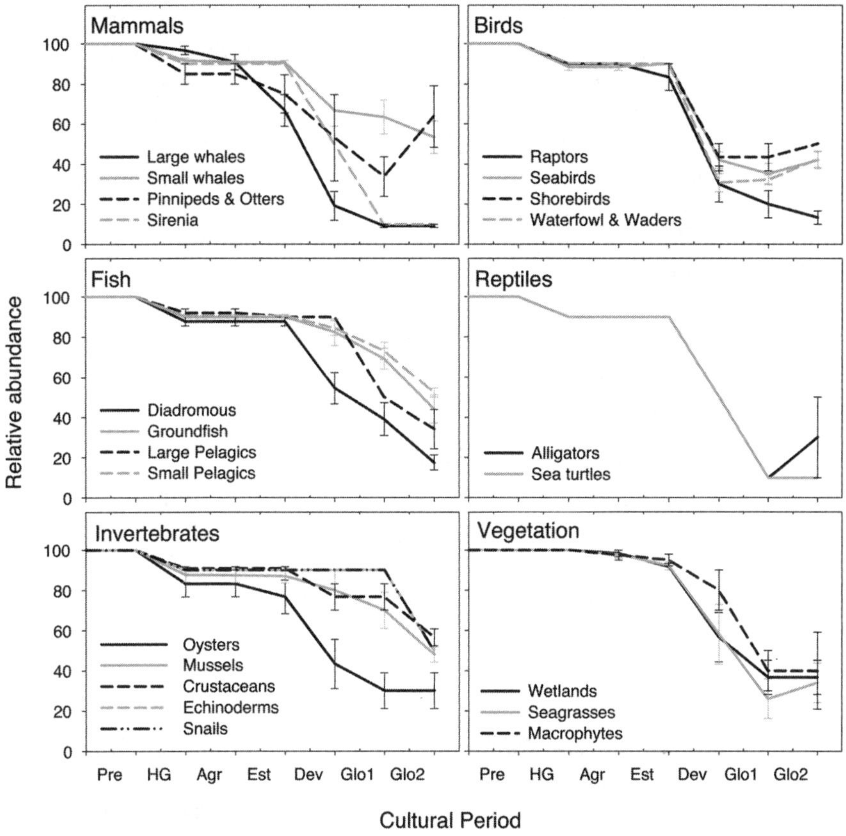

Fig. 2.4 – Trends in relative resource abundance (mean ± SE) across 6 estuaries in the United States, including Massachusetts Bay, Delaware Bay, Chesapeake Bay, Pamlico Sound, Galveston Bay and San Francisco Bay over different cultural periods from prehuman times until the year 2000 (redrawn from Lotze, 2010). Cultural periods refer to: Pre = Prehuman, HG = Hunter gatherer, Agr = Agriculture, Est = Colonial establishment, Dev = Colonial development, Glo1 = Early global (1900–1950), Glo2 = Late global (1950–2000). (From Lotze, 2010)

2011). They controlled the location of fish flakes, dams and weirs and licensed privileges to catch and preserve certain fish (Leavenworth, 2008). Other regulations imposed by the Pilgrims and Puritans settling New England included limits on the days fishing, fishing seasons, protection of spawning grounds, gear restrictions and even a landings moratorium, as in response to a shortage of cod near Boston in the 1650s (Anonymous, 1887; Rosenberg et al., 2011). Many of these early, local and adaptive management practices were later overrun by the rising industrialization that favored large-scale exploitation and undermined small-scale producers (Rosenberg et al., 2011).

2.5 *Industrial age*

In the 18th to early 20th centuries, countries around the world experienced the increasing commercialization and industrialization of resource use as well as the rise of luxury and fashion markets. Basically, every animal and plant species—or parts thereof—that could be marketed in one way or another was harvested, and the harvesting was performed on increasingly massive scales (Figure 2.5). In addition to being a source of food, whales were harvested for their oil and baleen; pinnipeds and otters for their fur and blubber; walrus for their ivory; birds for their feathers, eggs, and sometimes oil; reptiles for their skins, shells, and eggs; fish for their roe and as fertilizer; oysters for lime and shells for road construction; other invertebrates for jewelry and buttons; seagrasses for insulation and mattresses; seaweeds for fertilizer and alginates; and the list goes on (Lotze and Milewski, 2004; Lotze, 2010). Although harvested for similar purposes throughout history, the scale of harvesting increased in its magnitude and spatial extent and the mindset changed. Since everything could be turned into money, and more money was the goal of the market economy, whether for the harvester, trader, merchant or consumer, resource species were less and less valued for their products but more as items of monetary value. This drove a relentless exploitation on massive scales (Lotze and Milewski, 2004; Roberts, 2007; Lotze, 2010), sometimes destroying entire breeding colonies of pinnipeds or birds, sometimes pursuing the last individual of a now extinct species, such as the great auk, seemingly without any consideration of the ecological limits to renewal. The sea was still seen as inexhaustible.

Fig. 2.5 – Industrial-scale exploitation of whales, fish and invertebrates: (A) Whaling ships and oil casks at Central Wharf in New Bedford, Massachusetts in 1870 (photo by Stephen F. Adams, ca. 1870); (B) Sturgeon in Hamburg Harbour, Germany (drawing by Hans Petersen, ca. 1885); (C) Oyster shells in Hampton, Virginia (postcard, ca. 1913); and (D) Abalone shells in Santa Barbara, California (postcard, ca. 1920).

What allowed this mass-harvesting were the technological advances in gear (from spears to rifles to machine guns, from hook-and-line to long lines, gill nets, purse seines and trawls), vessels (from row to sail to motor to factory ships), preservation (from salting and drying to canning and refrigeration), transportation (railroads, trucks and airplanes), remote sensing and communication (sonar, radio, telephone) and mass-marketing and trade in food, fashion and luxury products (Lotze and Milewski, 2004; Roberts, 2007; Lotze, 2010). All of these advances were celebrated as progress and made it easier and faster to kill and sell giving rise to increasing numbers of commercial market hunters, whalers and fishers that supplied the growing appetite of merchants and markets. At the same time, increasing urbanization disconnected consumers from the natural origin of the products bought, making it easier to not know or not think about the mass-killing and ecological limits to renewal. When some fashionable ladies in New England became aware of the killing of thousands of birds to supply their tail feathers to the millinery trade, they started a movement to raise awareness and change fashion ideals to protect the birds from such unnecessary slaughter (Price, 1999; Forbes and Jermier, 2002).

Most marine resources could not withstand such relentless exploitation and populations of many species plummeted before the 20th century, sometimes to very low numbers or to local, regional or even global extinction (Dayton et al., 1998; Jackson et al., 2001; Pandolfi et al., 2003; Lotze and Worm, 2009). Global extinctions included Steller's sea cow in 1768, the great auk in 1844, Labrador duck in 1875, sea mink in 1894 and Atlantic gray whale sometime in the 1800s. Because the same or a similar species could be found elsewhere, however, exploitation continued to expand spatially. For example, whaling of right whales moved from the North to the South Atlantic in the late 18th century, then into the Pacific in the 19th century, depleting whale stocks within decades (Josephson et al., 2008). Fishing grounds in the Gulf of Maine moved from inshore to increasingly offshore on continental shelves in the 19th and early 20th century (Steneck, 1997; Rosenberg et al., 2005) and oyster fisheries moved down the coast of eastern North America and up the coast of western North America and eastern Australia as one estuary after the other was depleted (Kirby, 2004).

How did people respond to this massive exploitation and increasing depletion of local resources? By the end of the 19th and early 20th century, management agencies and governments began to step in to halt declines and prevent further extinctions, sometimes with drastic protection measures. For example, the near extinction of the great blue heron was prevented in the USA by the Federal Lacey Act in 1900, which prohibited the trade of highly valued feathers, and the Migratory Bird Treaty Act between the USA and Great Britain in 1918 protected a range of migratory birds from hunting, egg collection and nest destruction (Dorsey, 1998; Lotze, 2010). In 1911, the International Fur Seal Treaty protected severely decimated fur seals and sea otters in the Pacific from further hunting and the League of Nations in the 1930s banned commercial whaling of the remaining right, bowhead and gray whales (Dorsey, 1998; Lotze, 2010; Lotze et al., 2011c). These individual laws and treaties helped several species to survive and set the stage for the rise of a broader conservation movement in the 20th century. For fish, however, management bodies in the late 19th and early 20th centuries still promoted fisheries expansion and increased effort and efficiency despite growing concerns by fishermen about stock depletions (Lotze and Milewski, 2004).

2.6 Globalization and conservation

In the 20th and 21st centuries, especially after World War II, two major movements occurred. Marine exploitation and their markets, especially fisheries and whaling, further expanded to become global enterprises that rapidly covered this final new frontier. Concurrently, increasing reports of species depletions and extinctions, especially among marine mammals, birds and reptiles, led to the rise of conservation and awareness that natural resources, even those of the global ocean, are finite. Today, we are still in the midst of both movements, trying to further enhance marine resource supply for a growing human population while simultaneously aiming for sustainable exploitation and protection of marine biodiversity and ecosystems, as well as their services to human well-being (McNeill, 2000).

The expansion of fisheries and whaling in the 20th century occurred in three ways: spatially, taxonomically and technologically, reaching global limits in all of them. After the depletion of coastal whales in the 19th century, whaling expanded from coastal waters into the high seas and southern ocean, sequentially depleting larger whales, such as blue, fin and sperm whales, and newly targeting smaller species, such as sei and minke whales (Christensen, 2006), until commercial whaling was banned by the International Whaling Commission's Moratorium in 1986. Marine fisheries also expanded from coastal waters and continental shelves off the North Atlantic and Pacific into the high seas and the southern hemisphere (Swartz et al., 2010) and increasingly deeper waters (Morato et al., 2006). By the mid-1990s, one third of the world's ocean and two thirds of continental shelves were exploited, leaving only unproductive waters of the high seas and relatively inaccessible waters in the Arctic and Antarctic as the last remaining refuges. The growth in marine fisheries catches since 1950 was only possible by the expansion to new fishing grounds, which has reached its global limits (Swartz et al., 2010).

As in the history of commercial whaling, fishing also expanded taxonomically in pursuit of new species that could be harvested and marketed. In some regions, such as in the North Atlantic, this expansion occurred down the food web from high to low trophic level species (Pauly et al., 1998), while in other areas it expanded through the food web (Essington et al., 2006). Industrial fishing fleets moved into the open ocean in pursuit of tuna, billfish and other large pelagics (Myers and Worm, 2003; Worm et al., 2005), deep-sea fisheries targeted benthic fish on continental slopes and seamounts, such as orange roughy and roundnose grenadier (Clark et al., 2000; Devine et al., 2006), and invertebrate fisheries expanded to catch squids, cuttlefish, shrimps, crabs, sea urchins and sea cucumbers around the world (Berkes et al., 2006; Anderson et al., 2011a, b). Global markets, along with cheap energy and transportation, made it possible to catch something in one region, process it in another, and sell it in yet another region. Thus, over the past 50 years, many fish and invertebrate species have been depleted or collapsed (Worm et al., 2006, 2009; Anderson et al., 2011a, b; Branch et al., 2010), global catches have peaked and further expansion seems unlikely (Pauly et al., 2002; FAO, 2010).

This spatial and taxonomical expansion required a technological expansion and a massive capital investment. To catch more fish, global fishing effort increased dramatically since 1970 (Anticamara et al., 2011). To make fishing in the high and deep seas possible, gear types and depth ranges increased, advances in echo-location and spotter planes enabled fishers to find more fish, and onboard processing, refrigeration and marketing allowed fishing vessels to stay away from ports for extended periods of time. Finally, to catch an increasing variety of species, existing gears were modified or new gears developed. Some fishing gears, such as purse seines, were able to scoop up entire schools of herring, mackerel, tuna, or other pelagic fish, rapidly depleting overall stocks. Today, our technology is so powerful that we could theoretically fish the last fish—or restrain ourselves from doing so—both posing limits to further expansion.

The spatial, taxonomic and technological expansion has also occurred in marine aquaculture, which grew from mussels, oysters and seaweeds to shrimp and salmon and increasingly to other finfish such as tuna, sturgeon, cod and halibut (FAO, 2010). Aquaculture is spreading around the world and hugely increasing its production (FAO, 2010). As in massive terrestrial meat-growing operations, however, aquaculture is reaching limits in terms of disease outbreaks, environmental pollution with wastes and chemicals, and the necessary feed for producing high trophic level fish (Naylor et al., 1998, 2000; Deutsch et al., 2007).

While the expansion of industrial fisheries and aquaculture has been understood quite well, there is growing concern about the magnitude and impact of small-scale, subsistence and recreational fisheries, which often receive much less monitoring and management. We now know that even small-scale subsistence fishing can strongly alter marine ecosystem structure (Sandin et al., 2008) and that about 12 million fishers participate in small-scale fisheries worldwide with an estimated annual catch of 21 million metric tonnes of fish (Chuenpagdee et al., 2006). Often, such small-scale catches are not included in official fisheries statistics, leading to an underestimate of global fisheries landings. For example, shore-based subsistence catches in American Samoa have been estimated 16.6-fold higher than officially reported commercial landings by boats, and have experienced a 79% decrease since 1950 (Zeller et al., 2006). In other regions, recreational catches pose a similar problem. In Hawaii, for example, recreational catches have been estimated to be of similar magnitude as commercial catches, essentially doubling overall fisheries landings (Zeller et al., 2008). In the mainland USA, recreational catches amount to 4–10% of total fisheries catch, but they can reach 23% for species of conservation concern and 36–64% in certain regions, such as the Gulf of Mexico and Southern Atlantic coast (Coleman et al., 2004; Samhouri et al., this volume). To properly monitor and manage all fisheries these small-scale and recreational fisheries need to be accounted for.

The second big movement in the 20th and 21st centuries has been the rise of conservation. At first, this considered mainly terrestrial species and ecosystems, and high-profile marine species that were known to be heavily depleted, such as many marine mammals, birds and reptiles (Lotze et al., 2011c). Over time, marine conservation has also concerned fish and some invertebrates, as well as overall marine ecosystems. Efforts in the early 20th century and before were largely an attempt to protect individual species or groups of species from extinction or further depletion (Lotze et al., 2011c). This included laws and treaties banning exploitation and trade. Over time, management and conservation strategies were expanded to other species (e.g. USA Endangered Species Act in 1973) or entire taxa (e.g. USA Marine Mammal Protection Act in 1972) and refined to not only prohibit exploitation but also to protect habitats and ban pollutants or other negative stressors (see Cochrane et al., this volume). Such measures allowed many species to increase in numbers, albeit rarely to historical levels of abundance (Lotze et al., 2011c).

For other species, such as most fish and invertebrates, the interest in commercial exploitation has continued and management and conservation strategies have been increasingly directed towards setting harvest levels that would allow the stocks to remain productive or rebuild to former levels (Worm et al., 2009). In wealthy countries, this included fisheries-independent research surveys and stock assessments to estimate population abundance and set harvest levels, first in a single-species mode but increasingly in a multi-species management framework that takes predator-prey and competitor interactions into account. The amount of fishing has then been regulated by input controls, such as capacity, effort, gear, seasonal or area controls, or output controls, such as fishing quotas, size or sex restrictions (King, 2007). In countries that could not afford costly assessments, other strategies to control fishing have been used, such as community-based management, gear restrictions or closed areas (Defeo and Castilla, 2005; McClanahan et al., 2008). When implemented and properly enforced, several combinations of these strategies have proven successful in rebuilding fish populations in both wealthy and poor countries (Stefansson and Rosenberg, 2005; Worm et al., 2009). Marine protected areas have also enabled fish and ecosystem recovery (Worm et al., 2006; Rice et al., this volume), but enforcement is again key as no-entry areas have been much more successful than no-take areas (Robbins et al., 2006). However, there are continuing problems with illegal, unreported and unregulated fisheries (IUU) and discards (Pauly et al., 2002; Agnew et al., 2009), displacement of fishing fleets from wealthy to poor countries, such as European, Asian and Russian fleets fishing in African waters (Alder and Sumaila, 2004), lack of enforcement (Rosenberg et al., 2006) and management in the high and deep seas (Cullis-Suzuki and Pauly, 2010), many small-scale subsistence and recreational fisheries (see above; Coleman et al., 2004; Zeller et al., 2006, 2008), and many emerging fisheries such as invertebrates (Anderson et al., 2008, 2011a, b). Clearly, the blue movement of ocean conservation lags behind the green movement on land.

3. Common Patterns in Resource Depletion and Management

Looking back on the long history of human use and response to the depletion of marine resources, several common strategies emerge. In the face of depletion humans have usually:

- switched to a different species that was generally less valued (e.g. less nutritious, less tasty, culturally disfavored, smaller or more difficult or costly to catch);

- moved to a new unexploited or less exploited area (e.g. further offshore, deeper, different country or continent);
- increased effort and efficiency to increase catches (e.g. technological advances and capital investments in gears and vessels, preservation, transportation, communication);
- enhanced production on local scales through human-aided stocking, mari- or aquaculture (e.g. clam gardens, oyster culture, salmon and other fish farming).

These strategies can be witnessed in almost every period of the past. Essentially humans were always searching for the next new frontier, where one could tap into formerly unexploited resources without any restrictions or regulations in place and thus repeating the tragedy of the commons. Over time, the magnitude and scale of these taxonomic, spatial, technological, and aquaculture expansions, and thus the human footprint on the ocean, have accelerated until we are now reaching or even exceeding their global limits. In ancient times only nearshore ecosystems, such as local bays, reefs, or estuaries were affected, but the spatial scale of such impacts expanded around the Mediterranean in Roman times, around Europe in medieval times, to the New World in colonial times and globally to the continental shelves and some remote islands over the past 100–250 years, the high seas over the last 50 years and the deep sea over the last 20 years, leaving almost no place for further expansion (Pauly et al., 2003; Lotze and Worm, 2009; Swartz et al., 2010). This global expansion has been accompanied by a temporal acceleration of change over time. Whereas historical changes in coastal waters played out over centuries or millennia (Jackson et al., 2001; Rick and Erlandson, 2008), similar changes happened increasingly faster on continental shelves, in the open ocean and deep sea (Lotze and Worm, 2009). Today, many new fisheries cannot withstand these fast developments and show rapid boom-and-bust patterns, as in deep-water fish and some invertebrates (Koslow et al., 2000; Anderson et al., 2011b).

Yet, there have also been proactive responses to resource depletion in the past on local, regional, and sometimes global scales aimed at protecting marine resources, whether for selfish reasons to benefit an individual, community or nation or for the common good (Cochrane et al., this volume). These have included:

- governance in terms of legal or management authority (e.g. privatization or ownership of a species or area by a landowner, community or nation and international laws or treaties);
- restriction of who and how many can fish where, when and how (e.g. input controls such as limits on fishing licenses, seasons, areas, capacity, effort or gear);
- regulation of what and how much can be fished (e.g. output controls such as limits on target species, quotas, size and sex);
- management of other human impacts such as habitat protection and pollution controls (see *Section 5*).

In such cases what has been essential is that first someone needs to establish the legal or management authority that can implement restrictions and regulations on harvests or other human impacts to prevent a tragedy of the commons. Secondly, the authority must act to create and implement such restrictions and regulations and finally, it must enforce them.

4. What Has and Hasn't Worked in the Past

Throughout history, humans have responded to resource depletion not only by further spatial, taxonomic or technological expansion but also by implementing restrictions and regulations to halt further declines or secure continued supply. In this section, we summarize which strategies have and haven't worked in the past and what factors have influenced the use and management of marine resources, such as market drivers and cultural values.

4.1 Closed seasons

For most of history, periods of bad weather in cold, rainy or stormy seasons have prevented people from going out to sea in small boats to fish or hunt. Essentially, these were natural closed seasons that occurred regularly (e.g. winter in temperate latitudes) or irregularly (e.g. storms) and released many species from exploitation pressure for weeks or months at a time. As technology increased, however, and boats became bigger, faster and heavier, harvesters became increasingly independent of weather conditions, especially since the invention of steam power and combustion engines. Today's factory fishing vessels can be at sea year round, so there are no natural closed seasons anymore. Several unintended closed seasons in the 20th century revealed their power for replenishment, such as low exploitation rates during World War I and II in Europe and after the extension of the 200-mile limit in Canada and the United States in the 1970s. In each case groundfish stocks strongly increased in abundance until exploitation intensified once more after the Wars and in the 1980s by expanded national fishing fleets (Caddy and Agnew, 2004; Lotze and Milewski, 2004; Worm et al., 2007). Yet temporal or seasonal restrictions have also been used proactively as a way to manage fisheries. For example, in late 12th century Scotland, the king's council decreed a closed season on salmon and required nets to be lifted from Saturday night to Monday dawn (Hoffmann, 2000, 2009). Royal edicts in 14th century France imposed fishery closures along the coast to stabilize fish stocks (Roberts, 2007; Rosenberg et al., 2011) and the Pilgrims and Puritans in 17th century New England imposed limits on the days fishing and fishing seasons as well as a landings moratorium (Anonymous, 1887; Rosenberg et al., 2011). Today, many fisheries have clearly defined fishing seasons (King, 2007), often protecting their breeding season, and fishing moratoria have been used as emergency responses to fish declines, such as for several herring stocks in the Northeast Atlantic, haddock on Georges Bank and northern cod in Atlantic Canada, although the latter with so far limited success (Caddy and Agnew, 2004; Hutchings and Reynolds, 2004; Worm et al., 2007).

4.2 Closed areas

Similar to closed seasons, large areas of the ocean could not be easily reached or exploited with small boats and simple gears for most of history. These natural closed areas included distant, remote and offshore areas that were too far, too rough, too deep or too cold to harvest. Essentially, most of the open ocean and deep sea were natural protected areas until the expansion of industrial whaling and fishing in the 20th century. Yet closed areas have also been used proactively to protect critical habitats, such as spawning, breeding or nursery grounds or as measures to halt resource declines. For example, fisheries regulations imposed in 17th century New England included the protection of spawning grounds (Anonymous, 1887; Rosenberg et al., 2011). The 1918 Migratory Bird Treaty Act between the USA and Great Britain protected many bird species at their breeding grounds and prohibited nest destruction and egg collection (Lotze et al., 2011c). In the 20th century, many breeding beaches for sea turtles, breeding colonies for pinnipeds, and nursery and feeding areas for whales also were protected, and aided in the recovery of depleted populations (Lotze et al., 2011c). Closed areas have also been implemented for fish to protect critical nursery habitat and spawning grounds and are part of some fishery management strategies, such as the 'plaice box' in the North Sea (Piet and Rijnsdorp, 1998; Caddy and Agnew, 2004; Stefansson and Rosenberg, 2005). Increasingly, marine protected areas are used as more general marine conservation areas protecting not only target species but entire communities, ecosystems and biodiversity (Micheli et al., 2004; Worm et al., 2006; McClanahan et al., 2007). Here, no-go areas have been more successful than no-take areas (Robbins et al., 2006) pointing to the problems of enforcement, illegal fishing and general human disturbance. Yet closed areas have also been imposed as an emergency measure in fisheries management, such as the closure of Georges Bank to protect haddock stocks from further declines (Murawski et al., 2000).

4.3 Capacity and effort restrictions

Compared to the 20th and 21st centuries, people throughout most of history were few and had only small boats and simple fishing gear. Yet even small human populations can have negative impacts on local resource abundance albeit generally on local scales (see Section 2). Thus, in the face of resource depletion or to secure local supply, authorities in the past have aimed at limiting access and controlling harvest effort. Such input controls included certain privileges or licenses and restrictions on the number and size of boats or gear employed. For example, the Romans introduced private control and limited access to some fish resources that had been freely accessible before (Bekker-Nielsen, 2010). Many medieval examples involve privatization, such as the north Frisian oyster beds that were privatized by the Danish King in 1587 (Lotze, 2005), but also licensing of fishers, such as by French monarchs from 1289 on (Duplés-Agier, 1852; Jeulin, 1964). Fishers' guilds under municipal authority in upper Rhineland, along the Danube, and on Italian lakes and rivers also limited membership, and thus access, as well as the size or number of gear each licensed master could use (Mone, 1853; Scialoja, 1910; Petrin, 1979). Regional authorities set rules for fishers along French and Provencal coasts of the Mediterranean, and Sicilian rulers licensed onshore tuna traps (Aubenas, 1953; Bresc 1986). In 17th century New England, the Plimoth Plantation determined the location of fish flakes, dams and weirs and licensed privileges to catch and preserve certain fish (Leavenworth, 2008). In the 20th century, effort controls have become important parts of many hunting or fishery management plans, such as limits on the number of licenses, boats, traps or other gears (King, 2007). In recent decades, reductions in capacity and effort have been essential elements in reducing overall exploitation rate and rebuilding depleted fish stocks (Worm et al., 2009). Unfortunately, limiting or eliminating former privileges of open or unlimited access is rarely popular. In some cases, governments have bought back fishing licenses to reduce overall capacity, but concerns about job losses often spur governments to grant further access or even pay subsidies to fishers thereby stimulating overcapacity (Pauly et al., 2002; Caddy and Seijo, 2005).

4.4 Gear restrictions

In the deep past, fishing or hunting gears were simple and small and generally restricted by what people could hold, handle, or tow. However, humans have always been inventive in developing new and more efficient gears, including a large variety of hooks, lines, nets and traps, and power of fishing vessels (see Section 2). Over centuries and millennia, gear efficiency has strongly increased. At the same time technology has become less selective between target and non-target species and more destructive regarding habitat destruction and bycatch mortality. Yet throughout history, people voiced concerns about the negative effect of certain fishing gears. For example, in Roman times, some fishing techniques were prohibited, such as fishing with torch light at night (Bullock, 2008). In medieval Europe, certain fishing gears were regulated in estuarine and coastal waters, including mesh size, net design and the dimensions of seines and trawls, yet these regulations almost never extended to the open sea (Benecke, 1880; Aubenas, 1953; Bresc, 1986; Hoffmann, 2002). Fishing gears were also regulated by royal edicts along the coasts of 14th century France (Rosenberg et al., 2011), by 17th century Pilgrims and Puritans in New England (Anonymous, 1887; Rosenberg et al., 2011), and a 1811 law banned the use of dredges in the oyster fishery in Virginia (McCay, 1998). Over time, fishing technology became increasingly powerful and disruptive. In the 20th century, restrictions on gear type, design and the size of nets, hooks or mesh have become an essential part of fishery management both for limiting the number and size of target catch and reducing bycatch and habitat destruction (King, 2007; Worm et al., 2009). Some gears have been banned altogether: for example, the ban of beach seine nets in Kenya in combination with closed areas resulted in strong increases in fish abundance (McClanahan et al., 2008); in the Southern California Bight, the ban of gill nets in 1994 resulted in a slow recovery of target white sea bass as well as non-target soupfin and leopard sharks (Pondella and Allen, 2008). In 2006, the USA banned bottom trawling off most of its Pacific coast in an effort to protect severely depleted rockfish, but trawling and dredging are still allowed in most parts of the world with severe impacts on habitats and benthic diversity (Thrush and Dayton, 2002; Kaiser et al., 2006) and in many fisheries high bycatch rates continue to threaten populations of non-target species such as sea turtles and sharks (Lewison et al., 2004; Dulvy et al., 2008). With increasing pressure to protect endangered species and reduce collateral damage more selective and less destructive gears or ways of fishing are being developed, such as the use of turtle excluder devices (Brewer et al., 2006), but this is still a slow process.

4.5 Catch regulations

For most of history, humans fished and hunted for subsistence and the amount and size of catch was naturally limited by what one could handle, eat, share, store, transport or trade. The advent of a market economy encouraged catching more than needed for local use, whereas progress in preservation and storage (fermenting, drying, salting, smoking, canning, refrigerating), transport (ship, rail, trucks, air) and trade (local to global) increasingly removed natural catch limits. Thus, authorities have developed restrictions on the amount of catch that can be taken (quotas) and regulations on the size and sex of animals taken in an effort to prevent the depletion of large breeding individuals. For example, in 13th century France, authorities imposed species-specific size restrictions on catches (Duplés-Agier, 1852). A Scottish law from 1318 forbade fishing salmon parr or smolts, but also the fry of other species by means of size, seasonal and gear limits (RPS, 2007). In the 16th century, hunting restrictions were introduced for many waterfowl such as cranes, egrets and shellducks, which were heavily hunted (Lotze, 2005). In 1873, the decrease in the size of lobsters in the Bay of Fundy spurred the first regulations including size limits, prohibiting the take of soft-shelled lobsters and egg-bearing females, as well as closed seasons (Found, 1912). Imposing size or sex restrictions only requires an understanding of biology, while setting proper catch limits requires an estimate of stock abundance and the proportion that can be caught without depleting the stock. This spurred the development of fishery science in the 20th century with the aim to properly assess stock abundance, predict maximum sustainable yield (MSY) and set catch quotas. Today, many valuable fisheries in developed nations are managed by quotas (King, 2007; Anderson et al., 2008; Worm et al., 2009). Unfortunately, actual catches often far exceed proposed quotas because management agencies only partly follow scientific advice. In addition, high amounts of illegal, unreported and unregulated (IUU) fishing as well as discards complicate assessments of stock abundance and biomass removals (Pauly et al., 2002; Agnew et al., 2009; MacKenzie et al., 2009). As a result, many fish stocks have been heavily overfished and are below their management target (biomass at MSY) and require rebuilding (Rosenberg et al., 2006; Worm et al., 2009). Although enforcement of management targets can insure better management outcomes (see Section 4.7; Rosenberg et al., 2006), there are also efforts to revise the concept of MSY and use more precautionary estimates of maximum optimum yield (MOY) or maximum economic yield (MEY) and to account for many species in a multi-species maximum sustainable yield (MMSY) (Hilborn et al., 2006; King, 2007; Worm et al., 2009).

4.6 Species and legal protection

Unfortunately, many fisheries and commercial harvests of marine mammals and birds in the past have remained largely unregulated, or existing regulations have been ignored or overruled, leading to severe declines of many exploited populations around the world (Lotze and Worm, 2009). In many cases, agencies only stepped in after populations had reached very low levels in an effort to prevent further decline and extinction, yet occasionally that was too late. One early response to sharply declining sturgeon throughout medieval Europe was to declare it a "royal fish" that only the elites were allowed to eat (Hoffmann, 2005). Swans were protected in 16th century Netherlands because they were known to be easily extirpated (Lotze, 2005). The severe decline of many valuable species in the 19th and 20th centuries spurred more wide-ranging legal protections. For example, the USA Lacey Act in 1900 prevented the extinction of the great blue heron by prohibiting the trade of highly valued feathers (Davis et al. 1997; Lotze, 2010). In 1911, the International Fur Seal Treaty Act protected sea otters and fur seals from hunting in the North Pacific and the League of Nations in the 1930s banned the commercial whaling of severely decimated right, bowhead and gray whales (Davis et al. 1997; Lotze et al., 2011c). Over time, legal protection was expanded to entire species groups, such as the USA Marine Mammal Protection Act in 1972 and the International Whaling Commission's moratorium on all commercial whaling in 1986. Today, many countries have some form of endangered species legislation and the International Union for the Conservation of Nature (IUCN) aims to assess the global conservation status of all species (Lotze et al., 2011c). This raises awareness and pressure but has no positive legal power.

4.7 *Governance and enforcement*

To make any management possible, there first needs to be an authority in place that can implement and enforce restrictions and regulations. In the past, such governance came from landowners and public authorities (e.g. kings; Hoffmann, 2000, 2009), communities (e.g. Plimoth Plantation; Rosenberg et al., 2011) or corporate groups (e.g. fisher's guilds; Hoffmann, 1996) and over time entire national or territorial states (e.g. establishment of exclusive economic zones). Today, many nations have several different legal and management branches responsible for distinct species, areas or human activities, often with overlapping, ineffective mandates (Mora et al., 2009). Yet sometimes, community-based management has proven more successful in managing local resources (González et al., 2006; McClanahan et al., 2008). Such legal authority is still largely lacking on the high seas. Although the 1982 United Nations Convention on the Law of the Sea (UNCLOS) defines the rights and responsibilities of nations in their use of the world's oceans, the UN has no direct operational role in the implementation of the convention or the enforcement of management strategies. Nevertheless, some international treaties have worked in the past to protect whales, fur seals and migratory birds, whereas international management of other species, such as bluefin tuna and pelagic sharks, is so far less successful (Dulvy et al., 2008; MacKenzie et al., 2009). Global governance is also needed to prevent the displacement of fishing effort from rich to poor nations, such as from Europe, Japan and Russia to West Africa under distant-water fishing access agreements (Alder and Sumaila, 2004).

Even if a management authority is in place and can implement regulations, enforcement is needed to be successful. For example, under the USA Magnuson-Stevens Act 74 fish stocks have required recovery, yet from 1996 to 2004 only half showed some increase, three reached their recovery target, and 34 stocks still experienced overfishing (Rosenberg et al., 2006). Similarly, 35% of 166 assessed fish stocks experienced higher exploitation rates than recommended by management (Worm et al., 2009) and a large proportion of fishery catches are illegal, underreported or unreported (Agnew et al., 2009). Clearly, enforcement of management actions, such as the reduction and control of exploitation, is essential for success (Lotze et al., 2011c).

4.8 Economic changes and market restrictions

Another important regulating force in fisheries and resource exploitation are changes in market demand and prices. While high demand and high prices spur further exploitation, low demand and prices have the opposite effect. Probably the most important factor in saving the great whales was the introduction of cheap mineral oil that rapidly replaced expensive whale oil (Schneider and Pearce, 2004). Yet sometimes the depletion of a highly valued resource may increase the market price thereby providing an incentive for intensified exploitation rather than conservation. Until they were protected in 1918, prices for canvasback ducks in USA restaurants increased from $20–40 in the 1860s to $80–100 in the 1910s as ducks became scarcer in the wild (Jones, 2008). Today, bluefin tuna is threatened by the same problem. In general, fisheries have been largely profit-driven in terms of costs and revenues, initially targeting shallow-water species with high abundance, large catches, high prices and big body size, and over time expanding to less desirable species (Sethi et al., 2010). Dedicated catch shares or territorial fishing rights granted to individual fishers or fishing communities have often provided economic incentives to reduce fishing effort and exploitation rates to maintain viable fish stocks (Costello et al., 2008). One problem undermining the self-regulation of markets is government subsidies; many fisheries would not be profitable if fuel or boats were not intensively subsidized (Sumaila et al., 2007). Yet new markets can also open up, such as the Asian market importing sea urchin and sea cucumbers from countries around the world thereby providing new opportunities—and problems—for exploitation (Berkes et al., 2006; Anderson et al., 2011b).

4.9 Awareness, ethics and choice

Finally, public awareness, animal ethics and consumer choice can strongly influence the demand for certain natural resources, their exploitation and protection. In Roman times, medieval Europe and the industrial era the rise of new cultural ideas about fashion and luxury products as well as healthy, status-enhancing and religiously appropriate seafood choices promoted the commercial exploitation of a wide variety of species (see *Section 2*). In contrast, in the 20th century, there has been a dramatic cultural shift in our value and perception of marine mammals, birds and sea turtles, such that the commercial exploitation of these species is now prohibited in many nations. This was caused by shifts in fashion ideals (e.g. not wearing bird feathers on hats or whale bone corsets anymore) and animal ethics (e.g. recognizing the intelligence of whales and dolphins) which raised public support for wildlife and ocean conservation and political willingness to implement legal actions and management plans (Lotze et al., 2011c). More recently, public education on fisheries and ocean issues has strongly enhanced awareness on overfishing and destructive fishing practices and promoted consumer choice, thereby increasing pressure on fishers and the seafood industry to provide more sustainable products. Organizations in many countries now provide seafood choice cards to guide consumers and fisheries can apply to be certified for sustainable practices by the Marine Stewardship Council (MSC). Such fisheries certification can serve as incentive to promote better management practices (Worm et al., 2009).

5. History of Managing Other Human Impacts

Of course, marine resource use and exploitation has not been the only impact people have had on marine ecosystems, although it was probably the earliest and most pervasive (Jackson et al., 2001). People have also caused other forms of disturbance, transforming coastal habitats, creating erosion and physical, chemical, light and noise pollution, introducing non-native species, and contributing to a changing climate. Many of these impacts contributed to exploitation-driven species declines, creating additive or synergistic effects, and in some cases posed a major or primary threat (Lotze et al., 2006). In this section, we describe past signs of these other human impacts and management responses.

5.1 Habitat alteration and loss

Even the earliest people altered the habitat they lived in, simply by clearing space for settlement, depositing kitchen wastes, and burning the landscape. Yet with low population densities, their footprint remained on a local scale. There are several examples, however, where hunter-gatherers around North America have deposited vast quantities of shells and other wastes, creating large shell mounds or islands (Lightfoot and Luby, 2002; Voorhies, 2004; Thompson and Pluckhahn, 2010), while others constructed and cultivated clam gardens in intertidal areas (Williams, 2006). Evidence of large-scale habitat alteration comes from Italy, where Etruscans (5th century BC), Romans and later medieval Europeans drained large stretches of wetlands along the Po River and Delta and Venice lagoon (Marchetti, 2002; Lotze et al., 2011b). About 25,000 km^2 of wetlands existed in Italy during Roman times, of which 7000 km^2 remained in the 1900s and only 1000 km^2 today (Airoldi and Beck, 2007). Around the world, wetlands were actively reclaimed or filled with overall losses ranging from 95% in San Francisco Bay to 80% in the Boston area, 50% in Pamlico Sound (Figure 2.4) (Lotze, 2010) and 50–80% in many European countries (Airoldi and Beck, 2007). Only in the 20th century did our perception of wetlands change from wastelands to important habitats, leading to the rise of wetland protection and restoration. Seagrass beds and other submerged aquatic vegetation (SAV) were reduced by eutrophication, direct exploitation or destruction and disease (Figure 2.4) (Lotze et al., 2006; 2011b). Oyster reefs were directly and destructively harvested in many parts of the world, and other habitat builders such as corals, sponges, and hydrozoans were also directly harvested or destroyed with the expansion of trawling (Figure 2.4) (Lotze, 2005, 2010; Lotze et al., 2011b). Most efforts to protect or restore these habitats only began in the 20th century (Airoldi and Beck, 2007), although opposition to some habitat destruction, such as from bottom trawling, was recorded in 14th century eastern England (a Parliamentary petition of 1377) and southern France (Bourin-Derruau et al., 2001; PROME, 2005). Today, seafloor trawling increasingly affects offshore and deep sea areas, so far with few controls.

The damming of rivers reduced the access of migratory fish to their spawning grounds since at least medieval times, yet there is evidence that the problem of migration barriers was recognized and controlled early on in Italy, Scotland, England, and Germany (Benecke, 1880; Stolz, 1936; Hoffmann, 1996, 2000). River damming caused severe declines of migratory fish around the world, whereas the provision of fish passages in the 19th and 20th centuries helped many species (Lotze and Milewski, 2004; Lotze 2010). For birds, human settlements and activities on coastal islands led to the abandonment of many breeding colonies (Lotze and Milewski, 2004; Lotze, 2010) while the direct use of beaches reduced nesting habitat for sea turtles (McClenachan et al., 2006). Active protection of breeding, feeding and nursery habitats and human abandonment of some islands contributed to the recovery of many species (Lotze et al., 2011c). Overall, habitat loss was the second most important cause for the decline of marine species in coastal areas around the world, and habitat protection the second most important reason for recovery, often in combination with reduced exploitation (Lotze et al., 2006).

5.2 *Sedimentation and pollution*

Studies of stratified sediment cores from estuaries and coral reefs reveal that water quality has rapidly deteriorated in the course of human settlement and land transformation due to increased sedimentation, nutrient loading, eutrophication and other pollution (Figure 2.6A) (McCulloch et al., 2003; Lotze et al., 2006). In Italy, sediment erosion increased during Roman and Medieval times in the Po Delta and Venice lagoon (Marchetti, 2002, Lotze et al., 2011b). Across Europe, increased coastal erosion and sedimentation associated with land clearing and agricultural expansion has been documented in the 11–14th centuries (Hoffmann, 1996; Bork et al., 1998; Bourin-Derruau et al., 2001) and also affected some Pacific Island coral reefs after Polynesian settlement and land clearing (Kirch, 2005). In the New World, sedimentation strongly increased after European settlement and land clearing, while nutrient loading and eutrophication strongly increased in the 20th century (Lotze, 2010). In most parts of the world, pollution controls and wastewater treatment have only been introduced in the second half of the 20th century, sometimes producing rapid improvement. For example, 10 years after a ten-fold reduction in nitrogen loads in Tampa Bay, Florida, seagrass beds started to recover (Cloern, 2001). Pollution controls in Thames River estuary in the 1960s enhanced oxygen levels, enabling the return of >110 fish species that had disappeared due to bad water quality (Cloern, 2001). While sediment loading, waste water and industrial discharges are increasingly being controlled by management agencies, other forms of pollution, including many chemicals, dumping, noise, light and litter are often low on the agenda. An important factor in the recovery of many birds, however, was the ban of DDT during the 1970s (Lotze, 2010).

5.3 *Invasive species*

Another human activity that has impacted marine species and ecosystems since early times has been the active introduction of non-native but familiar species, either for company and security (dogs), production (domestic plants and animals)

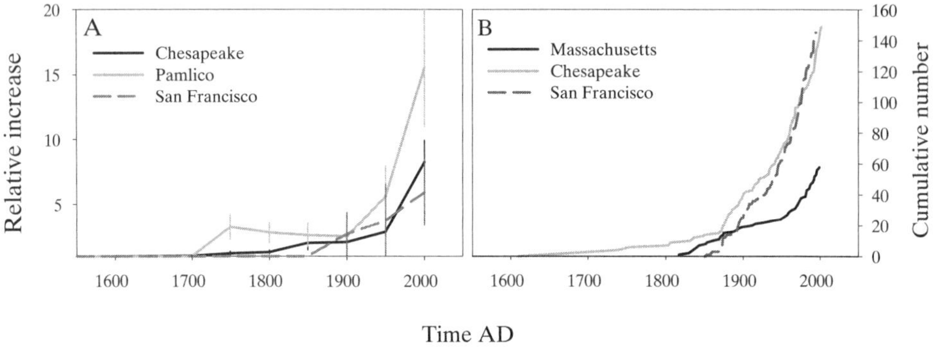

Fig. 2.6 – (A) Relative increase in water quality degradation, including sediment and nutrient loading and eutrophication signs, based on sediment core and hydrographic data in three U.S. estuaries (mean ± SE of n = 3–8 indicators) and (B) Cumulative increase in invasive species in three U.S. estuaries. (Redrawn from Lotze, 2010)

or other purposes (fur bearers), and the unintended introduction of exotic or nuisance species, such as rats, pests and many aquatic species (Figure 2.6B) (Lotze and Milewski, 2004; Lotze et al., 2006). The introduction of predatory animals onto islands had especially devastating consequences for many seabird colonies. Today, many efforts prevent such introductions (e.g. ballast water and border controls) and active removal of already introduced species is taking place. For example, the eradication of rats, foxes and raccoons helped to restore island seabird colonies in the Outer Bay of Fundy and New Zealand (Lotze and Milewski, 2004; Jones, 2010).

5.4 Climate change

Natural changes in climate and ocean productivity are known to have strong effects on the abundance and distribution of marine species over time scales of centuries to millennia (Baumgartner et al., 1992; Finney et al., 2002). In some cases it has been possible to distinguish between such natural versus human-driven changes in past marine populations. For example, archaeological records revealed that seabird remains on the Aleutian Islands showed long-term climate-driven fluctuations superimposed on local overhunting of accessible breeding colonies (Causey et al., 2005). In Newfoundland, changes in cod biomass between 1500 and 2000 were likely driven by climate-related declines in 1800–1880, large declines due to overfishing in the late 1960s and collapses in the late 1980s caused by both climate and overfishing (Rose, 2004). Over historical time scales, however, changes in climate have not been established as a main cause for severe depletions or extinctions in estuaries or coastal seas (Lotze et al., 2006). This may change with future warming: since ocean temperature is the main factor explaining marine diversity patterns (Worm et al., 2005; Tittensor et al., 2010) strong shifts in marine species are expected in a warming ocean (Cheung et al., 2009). Interestingly, in the North Sea, many fish species have shown climate-related fluctuations over the past

9000 years which mirror recently observed shifts related to climate warming (Enghoff et al., 2007).

6. Ecosystem Consequences

Humans have been part of marine food webs and ecosystems for tens of thousands of years through our effects as predators and consumers. An increasing number of studies demonstrate that even low levels of human exploitation, such as in the archaeological past or in remote areas today, can have strong impacts on marine populations by altering their size and age structure, abundance and distribution (Rick and Erlandson, 2008; Sandin et al., 2008; Lotze and Worm, 2009; Erlandson and Rick, 2010). The depletion of target species further affects the abundance of their prey, predators or competitors, thereby altering community structure (Simenstad et al., 1978; Heithaus et al., 2008; Ferretti et al., 2010). For example, the 19th century fur trade strongly decimated sea otters, a keystone predator in Aleutian kelp forests. Released from otter predation, sea urchins increased in abundance and reduced the kelp forests, their preferred food, causing large-scale kelp deforestation and urchin barrens (Steneck et al., 2002). Similarly, a proliferation of large red abalones on San Miguel Island between 7500–3000 years ago suggests that ancestors of the Chumash Indians may have kept sea otters, the abalones' main predator, in check by hunting (Erlandson et al., 2005; 2009). Such prey release effects and trophic cascades have been shown in several marine ecosystems and may have fundamentally altered the structure and functioning of marine communities over time (Heithaus et al., 2008; Ferretti et al., 2010).

Since all species are embedded in networks of species interactions and food webs, changing or losing some parts will affect others and alter emerging system properties. This can be analyzed with food-web or ecosystem modeling approaches. For example, in the Adriatic Sea, the historical expansion of human exploitation and the local or functional extinction of some species changed overall food-web structure and functioning (Lotze et al., 2011b). In particular, exploitation resulted in the depletion of higher trophic levels and simplification of overall food-web structure, and today's Adriatic food web is more degraded than marine food webs outside the Mediterranean (Coll et al., 2008). Interestingly, simulations of species losses indicate that the Adriatic ecosystem today may be less robust to further species extinctions than past ecosystems, thus they may have reduced resilience (Lotze et al., 2011b). Using a community matrix modeling approach, Britten (2011) found that the depletion of top and meso-predators and concomitant increase in resource species in the Ligurian Sea reduced community stability, indicated by decreased resistance and resilience and increased reactivity over time. Thus, human exploitation of marine ecosystems can severely alter their structure, functioning and resilience.

Ecosystem modeling can also help us understand the consequences of different levels of exploitation (Fulton and Link, this volume). For example, a size-based community model and more complex ecosystem models revealed similar effects of exploitation on species composition, size structure, biomass, and other ecosystem properties (Worm et al., 2009). Increasing exploitation rate resulted in an increase in total fish catch toward a multi-species maximum sustainable yield (MMSY) and a decrease thereafter. At the same time, total fish biomass and average body size steadily decreased and the proportion of collapsed species increased as exploitation rate increased. The models also showed that moderate reductions in exploitation rate below that at MMSY would rebuild total biomass, increase average body size, and reduce species collapses with little loss in long-term yield.

Over time, humans have adopted other roles in marine ecosystems and now directly and indirectly interact with a wide range of species not only through consumption but also through competition for food and space, habitat alteration, water pollution, introduction of exotic species, and climate change. Hence, we perform top-down and bottom-up control and have many side-in effects (Lotze and Milewski, 2004) that alter the structure and functioning of marine ecosystems and the services they provide for human well-being. Declines in coastal vegetation and seafloor habitats, such as wetlands, seagrass beds, kelp forests, oyster reefs and sponge beds, have caused substantial losses of nursery habitats, filtration capacity, and coastal protection (Worm et al., 2006; Lotze, 2010) and contributed to an increase in negative effects on human well-being such as harmful algal blooms, beach and shellfish closures, fish kills, and coastal flooding.

Over the past decade, our understanding of the ecosystem consequences of human-driven marine exploitation has dramatically improved, especially concerning the exploitation of high trophic level species such as marine predators. We are also increasing our understanding of the ecosystem effects of fishing lower trophic level forage fish, which are important prey for many fish, birds and mammals and form an essential part of marine food webs (Smith et al., 2011). Over the past 50 years, however, we have also increasingly exploited marine invertebrates, which perform a much wider range of ecosystem roles, including habitat providers, grazers, filter feeders, detritivores and scavengers (Anderson et al., 2011a). The ecosystem consequences of this expanding invertebrate exploitation are largely unknown.

7. Lessons Learned

What can we learn from the history of marine resource use and management for today's development and implementation of ecosystem-based management? Most of all, we can gain perspective on how humans have interacted with the ocean, how that has changed over time, where we are today, and what our best options are based on what we know from the past and what we see in our future. Therefore, the lessons learned include, that we:

- *Recognize that all human activities have a footprint on ocean ecosystems—we are active participants within them.* Even small levels of exploitation or other human activities affect populations and ecosystems on local scales, but so do protection and restoration efforts.
- *Recognize the main underlying drivers of resource depletion.* These include:
  - Human population growth and increasing demand for marine resources;
  - Cultural values that have shaped seafood and fashion choices and frequency of use in the past and can be influenced today;

  - Market economies that can create monetary value for everything and need checks and balances, especially in terms of paying full environmental costs without subsidies;
  - Technological progress that has created overly powerful vessels and gears largely outpacing the development of effective regulations and management.
- *Recognize the massive magnitude and spatial scale of today's human impacts compared to all other times in history.* For most of history, there have been large natural closed areas, closed seasons and restrictions on what and how much we could fish or process. Today, our technology allows us to efficiently fish everything everywhere. We therefore need:
  - Governance and enforcement on local, regional, national and international scales;
  - Restrictions and regulations on the spatial extent of harvesting (zoning, closed areas), technology used (selective and non-destructive gears) and species targeted (trophic levels, functional groups, size, sex);
  - Combinations of existing management tools for most effective management.
- *Recognize the cumulative impact of multiple human activities in today's ocean.* These include exploitation, habitat alteration, pollution, invasive species, and climate change that have increased over time and expanded from the coastal to the global ocean. To reduce their cumulative effects, more efforts should be directed towards:
  - Cumulative human impact management that requires coordination between management agencies dealing with individual human activities.
- *Protect biodiversity to keep all the parts.* Despite serious population depletions, we have prevented the extinction of most marine mammals, birds and reptiles in the 20th century by strongly restricting exploitation and protecting critical habitats (Lotze et al., 2011c). For fish we have started to move in that direction (Worm et al., 2009) but this is not true for most invertebrates (Anderson et al., 2011a, b). Besides the inherent value of each species, we don't know which species or genes will be most needed in a changing ocean, so we should keep all parts as insurance in the face of uncertainty.
- *Protect the structure and functioning of ecosystems.* All marine species and human well-being depend on the continued functioning of ocean ecosystems in terms of productivity, food and habitat provision, biogeochemical cycles and climate regulation, among others. Depleted or degraded marine ecosystems are less stable and resilient and more variable in the face of perturbations. Thus, restoring and protecting the structure and functioning of marine ecosystems should become a management priority to ensure their resilience and stability and the continued provision of ecosystem services.

Many of these lessons have been identified as key elements of ecosystem-based management (Pikitch et al., 2004; MacLeod and Leslie, 2009) while others go beyond the immediate marine resources or ecosystems of concern and include the underlying human drivers of resource depletion and ocean degradation. Without addressing the ultimate causes (e.g., human demands), managing proximate drivers will only go so far. We have become such a dominant force in marine ecosystems that we need to move beyond our traditional response system to secure our own future food supply and well-being. A key component to developing smarter approaches to managing and restoring ocean resources and ecosystems is recognizing that humans have a deep history of interacting and interfering in marine systems and that studying the past can provide important lessons for better managing our future.

Acknowledgments

We are grateful to the members of several working groups at the National Center for Ecological Analysis and Synthesis (NCEAS) and the Census of Marine Life's History (HMAP) and Future (FMAP) of Marine Animal Populations programs for inspiring discussions and inter-disciplinary work. Many thanks also to Karen Alexander and Mike Fogarty for comments on the manuscript. Financial support is acknowledged by the Natural Sciences and Engineering Research Council (NSERC) of Canada, the Lenfest Ocean Program and Dalhousie University to HKL; York University and the Leverhulme Trust to RCH, and NCEAS, the National Science Foundation, and the University of Oregon to JME.

References

Agnew, D. J., Pearce, J., Pramod, G., Peatman, T., Watson, R., Beddington, J. R. and Pitcher, T. J. 2009. Estimating the worldwide extent of illegal fishing. *PLoS ONE,* 4, e4570.

Airoldi, L. and Beck, M. W. 2007. Loss, status, and trends for coastal marine habitats of Europe. *Oceanogr. Mar. Biol. Ann. Rev.,* 45, 345–405.

Alder, J. and Sumaila, U. R. 2004. Western Africa: a fish basket of Europe past and present. *J. Environ. Dev.,* 13, 156–178.

Anderson, S. C., Lotze, H. K. and Shackell, N. L. 2008. Evaluating the knowledge base for expanding low-trophic-level fisheries in Atlantic Canada. *Can. J. Fish. Aquat. Sci.,* 65, 2553–2571.

Anderson, S. C., Mills Flemming, J., Watson, R. and Lotze, H. K. 2011a. Rapid global expansion of invertebrate fisheries: trends, drivers, and ecosystem effects. *PLoS ONE,* 6, e14735.

Anderson, S. C., Flemming, J. M., Watson, R. and Lotze, H. K. 2011b. Serial exploitation of global sea cucumber fisheries. *Fish Fish.,* 12, 317–339.

Anonymous, 1887. *Laws relating to inland fisheries in Massachusetts 1623–1886.* Wright and Potter Printing Company, State Printers, Boston.

Anticamara, J. A., Watson, R., Gelchu, A. and Pauly, D. 2011. Global fishing effort (1950–2010): Trends, gaps, and implications. *Fish. Res.,* 107, 131–136.

Aubenas, R. 1953. *Le droit de pêche de l'abbaye de Lérins.* Equip des Historiens Cannois, fascic. no. 2, Cannes.

Barrett, J. H., Locker, A. M. and Roberts, C. M. 2004. The origins of intensive marine fishing in medieval Europe: the English evidence. *Proc. Royal Soc. London B,* 271, 2417–2421.

Baumgartner, T. R., Soutar, A. and Ferreira-Bartrina, V. 1992. Reconstruction of the history of pacific sardine and northern anchovy populations over the past two millennia from sediments of the Santa Barbara basin, California. *CalCOFl Rep.,* 33, 24–40.

Bekker-Nielsen, T. 2005. The technology and productivity of ancient sea fishing. In *Ancient fishing and fish processing in the Black Sea region* (ed. Bekker-Nielsen, T.). Aarhus, 83–96.

Bekker-Nielsen T. 2010. Roman fishing technology in context. In *When humanities meet ecology: Historic changes in Mediterranean and Black Sea marine biodiversity and ecosystems since the Roman period until nowadays* (eds. Gertwagen, R., Fortibuoni, T., Giovanardi, O., Libralato, S., Solidoro, C. and Raicevich, S.). ISPRA Serie Atti, Rome, 25–33.

Benecke, B. 1880. Beiträge zur Geschichte der Fischerei in Ost- und Westpreussen. Altpreussisch Monatsschrift, 17, 300–322 and 385–416.

Benecke, N. 1986. Some remarks on sturgeon fishing in the southern Baltic region in medieval times. In *Fish and archaeology* (eds. Brinkhuizen, D. C., Clason, A. T.). BAR International, 9–17.

Berkes, F., Hughes, T. P., Steneck, R. S., Wilson, J. A., Bellwood, D. R., Crona, B., Folke, C., Gunderson, L. H., Leslie, H. M., Norberg, J., Nystrom, M., Olsson, P., Osterblom, H., Scheffer, M. and Worm, B. 2006. Globalization, roving bandits, and marine resources. *Science,* 311, 1557–1558.

Bork, H. R., Bork, H., Dalchow, C., Faust, B., Piorr, H. P. and Schatz, T. 1998. *Landschaftsentwicklung in Mitteleuropa: Wirkung des Menschen auf Landschaften.* Klett Perthes, Gotha & Stuttgart.

Bourin-Derruau, M., Le Blévec, D., Raynaud, C. and Schneider, L. 2001. Le littoral languedocien au Moyen Âge. In *Castrum 7: Zones côtières littorales dans le monde méditerranéen au Moyen Âge: defense, peuplement, mise en valeur* (ed. Martin, J-M.). *Collection de L'École Française de Rome,* 105/7, 345–423.

Bourque, B. J., Johnson, B. J. and Steneck, R. S. 2008. Possible prehistoric fishing effects on coastal marine food webs in the Gulf of Maine. In *Human impacts on ancient marine ecosystems: a global perspective* (eds. Rick, T. C. and Erlandson, J. M.). University of California Press, 165–185.

Branch, T. A., Watson, R., Fulton, E. A., Jennings, S., McGilliard, C. R., Pablico, G. T., Ricard, D. and Tracey, S. R. 2010. The trophic fingerprint of marine fisheries. *Nature,* 468, 431–435.

Bresc, H. 1986. *Un monde méditerranéen. Économie et société en Sicilie 1300–1450.* École Française, Rome.

Bresc, H. 2000. Pêche et commerce du corail en Méditerranée de l'Antiquitè au Moyen Age. In *Corallo di ieri, corallo di oggi. Scienze e materiali del patrimonio culturale 5* (eds. Morel, J. P., Rondi-Costanzo, C. and Ugolini, D.). Edipuglia, Bari, 41–53.

Brewer, D., Heales, D., Milton, D., Dell, Q., Fry, G., Venables, B. and Jones, P. 2006. The impact of turtle excluder devices and bycatch reduction devices on diverse tropical marine communities in Australia's northern prawn trawl fishery. *Fish. Res.,* 81, 176–188.

Britten, G. L. 2011. Modelling trophic dynamics and stability in a predator-depleted marine fish community using an autoregressive time-series model. Thesis, Dalhousie University, Halifax.

Broughton, J. M. 2002. Prey spatial structure and behavior affect archaeological tests of optimal foraging models: examples from the Emeryville Shellmound vertebrate fauna. *World Arch.,* 34, 60–83.

Bryant, P. J. 1995. Dating remains of gray whales from the eastern North Atlantic. *J. Mam.,* 76, 857–861.

Bullock, A. 2008. Lo *Scarus* degli antichi: la storia dello *Sparisoma cretense* nel secolo I. In *Il mare come era* (eds. Gertwagen, R., Raicevich, S., Fortibuoni, T. and Giovanardi O.). Proceedings II HMAP Mediterranean and Black Sea project, Chioggia, Italy, 94–105.

Caddy, J. F. and Agnew, D. J. 2004. An overview of recent global experience with recovery plans for depleted marine resources and suggested guidelines for recovery planning. *Rev. Fish Biol. Fish.,* 14, 43–112.

Caddy, J. F. and Seijo, J. C. 2005. This is more difficult than we thought! The responsibility of scientists, managers and stakeholders to mitigate the unsustainability of marine fisheries. *Phil. Trans.,* 360, 59–75.

Causey, D., Corbett, D. G., Lefevre, C., West, D. L., Savinetsky, A. B., Kiseleva, N. K. and Khassanova, B. F. 2005. The palaeoenvironment of humans and marine birds of the Aleutian Islands: three millennia of change. *Fish. Oceanogr.,* 14, 259–276.

Cheung, W. W. L., Lam, V. W. Y., Sarmiento, J. L., Kearney, K., Watson, R. and Pauly, D. 2009. Projecting global marine biodiversity impacts under climate change scenarios. *Fish Fish.,* 10, 235–251.

Christensen, L. B. 2006. Marine mammal populations: reconstructing historical abundances at the global scale. *UBC Fish. Cent. Res. Rep.,* 14.

Chuenpagdee, R., Liguori, L., Palomares, M. L. D. and Pauly, D. 2006. Bottom-up, global estimates of small-scale marine fisheries catches. *Fisheries Centre Research Reports,* 14(8).

Clark, M., Anderson, O., Francis, R. and Tracey, D. 2000. The effects of commercial exploitation on orange roughy (*Hoplostethus atlanticus*) from the continental slope of the Chatham Rise, New Zealand, from 1979 to 1997. *Fish. Res.,* 45, 217–238.

Cloern, J. E. 2001. Our evolving conceptual model of the coastal eutrophication problem. *Mar. Ecol. Prog. Ser.,* 210, 223–253.

Coleman, F. C., Figueira, W. F., Ueland, J. S. and Crowder, L. B. 2004. The impact of United States recreational fisheries on marine fish populations. *Science,* 305, 1958–1960.

Coll, M., Lotze, H. K. and Romanuk, T. N. 2008. Structural degradation in Mediterranean Sea food webs: Testing ecological hypotheses using stochastic and mass-balance modelling. *Ecosys.,* 11, 939–960.

Cooper, S. R. and Brush, G. S. 1993. A 2,500-year history of anoxia and eutrophication in Chesapeake Bay. *Estuaries,* 16, 617–626.

Cooper, S. R., McGlothlin, S. K., Madritch, M. and Jones, D. L. 2004. Paleoecological evidence of human impacts on the Neuse and Pamlico Estuaries of North Carolina, USA. *Estuaries,* 27, 617–633.

Cortés-Sánchez, M., Morales-Muniz, A., Simon-Vallejo, M. D., Lozano-Francisco, M. C., Vera-Pelaez, J. L., Finlayson, C., Rodriguez-Vidal, J., Delgado-Huertas, A., Jimenez-Espejo, F. J., Martinez-Ruiz, F., Martinez-Aguirre, M. A., Pascual-Granged, A. J., Bergada-Zapata, M. M., Gibaja-Bao, J. F., Riquelme-Cantal, J. A., Lopez-Saez, J. A., Rodrigo-Gamiz, M., Sakai, S., Sugisaki, S., Finlayson, G., Fa, D. A. and Bicho, N. F. 2011. Earliest known use of marine resources by Neanderthals. *PLoS ONE,* 6, e24026.

Costello, C., Gaines, S. D. and Lynham, J. 2008. Can catch shares prevent fisheries collapse? *Science,* 321, 1678–1681.

Cullis-Suzuki, S. and Pauly, D. 2010. Failing the high seas: A global evaluation of regional fisheries management organizations. *Mar. Pol.,* 34, 1036–1042.

Davis, L. E., Gallman, R. E. and Gleiter K. 1997. *In pursuit of Leviathan: Technology, institutions, productivity, and profits in American Whaling, 1816–1906.* University of Chicago Press.

Dayton, P. K., Tegner, M. J., Edwards, P. B. and Raiser, K. L. 1998. Sliding baselines, ghosts and reduced expectations in kelp forest communities. *Ecol. Appl.,* 8, 309–322.

Defeo, O. and Castilla, J. C. 2005. More than one bag for the world fishery crisis and keys for co-management successes in selected artisanal Latin American shellfisheries. *Rev. Fish Biol. Fish.,* 15, 265–283.

Deutsch, L., Graeslund, S., Folke, C., Troell, M., Huitric, M., Kautsky, N. and Lebel, L. 2007. Feeding aquaculture growth through globalization: Exploitation of marine ecosystems for fishmeal. *Glob. Environ. Change,* 17: 238–249.

Devine, J. A., Baker, K. D. and Haedrich, R. L. 2006. Deep-sea fishes qualify as endangered. *Nature,* 439, 29.

Dorsey, K. 1998. *The dawn of conservation diplomacy: U.S.-Canadian wildlife protection treaties in the progressive era.* University of Washington Press.

Dulvy, N. K., Baum, J. K., Clarke, S., Compagno, L. J. V., Cortés, E., Domingo, A., Fordham, S., Fowler, S., Francis, M. P., Gibson, C., Martínez, J., Musick, J. A., Soldo, A., Stevens, J. D. and Valenti, S. 2008. You can swim but you can't hide: the global status and conservation of oceanic pelagic sharks and rays. *Aqu. Cons. Mar. Freshw. Ecosys.,* 18, 459–482.

Duplés-Agier, H. 1852. Ordonnances inédites de Philippe-le-Bel et de Philippe-le-Long sur la police de la péche fluviale. *Bibliotheque de l'École des Chartes,* 14, 43–55.

Enghoff, I. B., MacKenzie, B. R. and Nielsen, E. E. 2007. The Danish fish fauna during the warm Atlantic period (ca. 7000–3900 BC): Forerunner of future changes? *Fish. Res.,* 87, 167–180.

Erlandson, J. M. 2010. Food for thought: the role of coastlines and aquatic resources in human evolution. In *Environmental influences on human brain evolution* (ed. Cunnane, S. and Stewart, K.). Wiley & Sons, 125–136.

Erlandson, J. M. and Rick, T. C. 2008. Archaeology, marine ecology, and human impacts on marine environments. In *Human impacts on ancient marine ecosystems: a global perspective* (eds. Rick, T. C. and Erlandson, J. M.). University of California Press, 1–19.

Erlandson, J. M. and Rick, T. C. 2010. Archaeology meets marine ecology: the antiquity of maritime cultures and human impacts on marine fisheries and ecosystems. *Ann. Rev. Mar. Sci.,* 2, 165–185.

Erlandson, J. M., Rick, T. C., Estes, J. A., Graham, M. H., Braje, T. J. and Vellanoweth, R. L. 2005. Sea otters, shellfish, and humans: A 10,000 year record from San Miguel Island, California. In *Proceedings of the Sixth California Islands Symposium* (eds. Garcelon, D. and Schwemm, C.). National Park Service Technical Publication CHIS-05-01. Institute for Wildlife Studies, Arcata, California, 9–21.

Erlandson, J. M., Rick, T. C. and Braje, T. J. 2009. Fishing up the food web? 12,000 years of maritime subsistence and adaptive adjustments on California's Channel Islands. *Pacific Sci.,* 63(4), 711–724.

Erlandson, J. M., Braje, T. J., Rick, T. C., Jew, N. P., Kennett, D. J., Dwyer, N. and Watts, J. 2011. 10,000 years of human predation and size changes in the owl limpet (*Lottia gigantea*) on San Miguel Island, California. *J. Archaeol. Sci.,* 38, 1127–1134.

Essington, T. E., Beaudreau, A. H. and Wiedenmann, J. 2006. Fishing through marine food webs. *Proc. Nat. Acad. Sci.,* 103, 3171–3175.

FAO 2010. The state of world fisheries and aquaculture 2009. Tech. Rep., Food and Agriculture Organization of the United Nations, Rome.

Ferretti, F., Worm, B., Britten, G. L., Heithaus, M. R. and Lotze, H. K. 2010. Patterns and ecosystem consequences of shark declines in the ocean. *Ecol. Lett.,* 13, 1055–1071.

Finney, B. P., Gregory-Eaves, I., Douglas, M. S. V. and Smol, J. P. 2002. Fisheries productivity in the northeastern Pacific Ocean over the past 2,200 years. *Nature,* 416, 729–733.

Forbes, L. and Jermier, J. 2002. The institutionalization of bird protection: Mabel Osgood Wright and the early Audubon movement. *Organ. Environ.,* 15, 458–465.

Found, W. A. 1912. The lobster fishery of Canada. In *Sea-fisheries of Eastern Canada* (ed. Commission of Conservation Canada). The Mortimer Co., Ottawa, 50–77.

González, J., Stotz, W., Garrido, J., Orensanz, J. M. L., Parma, A. M., Tapia, C. and Zuleta, A. 2006. The Chilean turf system: How is it performing in the case of the loco fishery? *Bull. Mar. Sci.,* 78, 499–527.

Griffiths, C. L., van Sittert, L., Best, P. B., Brown, A. C., Clark, B. M., Cook, P. A., Crawford, R. J. M., David, J. H. M., Davies, B. R., Griffiths, M. H., Hutchings, K., Jerardino, A., Kruger, N., Lamberth, S., Leslie, R., Melville-Smith, R., Tarr, R. and van der Lingen, C. D. 2004. Impacts of human activities on marine animal life in the Benguela: a historical overview. *Oceanogr. Mar. Biol. Ann. Rev.,* 42, 303–392.

Gunther, R. T. 1897. The oyster culture of ancient Romans. *J. Mar. Biol. Assoc. UK,* 4, 360–365.

Halpern, B. S., Walbridge, S., Selkoe, K. A., Kappel, C. V., Micheli, F., D'Agrosa, C., Bruno, J. F., Casey, K. S., Ebert, C., Fox, H. E., Fujita, R., Heinemann, D., Lenihan, H. S., Madin, E. M. P., Perry, M. T., Selig, E. R., Spalding, M., Steneck, R. and Watson, R. 2008. A global map of human impact on marine ecosystems. *Science,* 319, 948–952.

Hardt, M. J. 2009. Lessons from the past: the collapse of Jamaican coral reefs. *Fish Fish.,* 10, 143–158.

Heithaus, M. R., Frid, A., Wirsing, A. J. and Worm, B. 2008. Predicting ecological consequences of marine top predator declines. *Trends Ecol. Evol.,* 23, 202–210.

Hilborn, R., Annala, J. and Holland, D. S. 2006. The cost of overfishing and management strategies for new fisheries on slow-growing fish: orange roughy (*Hoplostethus atlanticus*) in New Zealand. *Can. J. Fish. Aquat. Sci.,* 63, 2149–2153.

Hoffmann, R. C. 1996. Economic development and aquatic ecosystems in Medieval Europe. *Am. Hist. Rev.,* 101, 631–669.

Hoffmann, R. C. 2000. Medieval fishing. In *Working with water in medieval Europe: technology and resource use* (ed. Squatriti, P.). Brill, Leiden, 331–393.

Hoffmann, R. C. 2001. Frontier foods for late medieval consumers: culture, economy, ecology. *Enviro & Hist,* 7, 131–167.

Hoffmann, R. C. 2002. Carp, cod, connections—New fisheries in the medieval European economy and environment. In *Animals in human histories—the mirror of nature and culture* (ed. Henninger-Voss, M. J.). University of Rochester Press, 3–55.

Hoffmann, R. C. 2005. A brief history of aquatic resource use in medieval Europe. *Helgol. Mar. Res.,* 59, 22–30.

Hoffmann, R. C. 2009. Strekfusz: A fish dish links Jagiellonian Kraków to distant waters. In *Central and Eastern Europe in the Middle Ages: A cultural history* (eds. Górecki, P. and van Deusen, N.). Tauris Academic Studies, Tauris, London, 116–124.

Hoffmann, R. C. and Winiwarter, V. 2010. Making land and water meet: the cycling of nutrients between fields and ponds in pre-modern Europe. *Agri. Hist.,* 84(3), 352–380.

Hofrichter, R. 2002. *Das Mittelmeer: Fauna, Flora, Oekologie.* Teil I—Allgemeiner Teil. Spektrum Akademischer Verlag, Heidelberg.

Hughes, J. D. 1994. *Pan's Travail. Environmental problems of the ancient Greeks and Romans.* Hopkins University Press, Baltimore.

Hutchings, J. A. and Reynolds, J. D. 2004. Marine fish population collapses: consequences for recovery and extinction risk. *BioScience,* 54, 297–309.

Jackson, J. B. C., Alexander, K. E. and Sala, E. (eds.) 2011. *Shifting baselines: the past and the future of ocean fisheries.* Island Press.

Jackson, J. B. C., Kirby, M. X., Berger, W. H., Bjorndal, K. A., Botsford, L. W., Bourque, B. J., Bradbury, R. H., Cooke, R., Erlandson, J., Estes, J. A., Hughes, T. P., Kidwell, S., Lange, C. B., Lenihan, H. S., Pandolfi, J. M., Peterson, C. H., Steneck, R. S., Tegner, M. J. and Warner, R. R. 2001. Historical overfishing and the recent collapse of coastal ecosystems. *Science,* 293, 629–638.

Jeulin, P. 1964. L'elaboration par la monarchie de l'ancienne reglementation de la peche fluviale. *Revue historique du Droit francais et étranger, ser.4,* 42, 728–730.

Jones, G. A. 2008. 'Quite the choicest protein dish': The costs of consuming seafood in American restaurants, 1850–2006. In *Oceans past: management insights from the history of marine animal populations* (eds. Starkey, D. J., Holm, P. and Barnard, M.). Earthscan, 47–76.

Jones, H. P. 2010. Prognosis for ecosystem recovery following rodent eradication and seabird restoration in an island archipelago. *Ecol. Appl.,* 20, 1204–1216.

Josephson, E., Smith, T. D. and Reeves, R. R. 2008. Depletion within a decade: The American 19th-century North Pacific right whale fishery. In *Oceans past: management insights from the history of marine animal populations* (eds. Starkey, D. J., Holm, P. and Barnard, M.). Earthscan, 133–147.

Kaiser, M. J., Clarke, K. R., Hinz, H., Austen, M. C. V., Somerfield, P. J. and Karakassis, I. 2006. Global analysis of response and recovery of benthic biota to fishing. *Mar. Ecol. Prog. Ser.,* 311, 1–14.

Kennett, D. J., Voorhies, B., Wake, T. A. and Martinez, N. 2008. Long-term effects of human predation on marine ecosystems in Guerrero, Mexico. In *Human impacts on ancient marine ecosystems: a global perspective* (eds. Rick, T. C. and Erlandson, J. M.). University of California Press, 103–124.

King, M. 2007. *Fisheries biology, assessment and management.* Blackwell.

Kirby, M. X. 2004. Fishing down the coast: historical expansion and collapse of oyster fisheries along continental margins. *Proc. Nat. Acad. Sci.,* 101, 13096–13099.

Kirch, P. V. 1985. *Feathered gods and fishhooks: an introduction to Hawaiian archaeology and prehistory.* University of Hawaii Press, Honolulu.

Kirch, P. V. 2005. Archaeology and global change: the Holocene record. *Ann. Rev. Environ. Res.,* 30, 409–440.

Kittinger, J. N., Pandolfi, J. M., Blodgett, J. H., Hunt, T. L., Jiang, H., Maly, K., McClenachan, L. E., Schultz, J. K. and Wilcox, B. A. 2011. Historical reconstruction reveals recovery in Hawaiian coral reefs. *PLoS ONE,* 6, e25460.

Koslow, J. A., Boehlert, G. W., Gordon, J. D. M., Haedrich, R. L., Lorance, P. and Parin, N. 2000. Continental slope and deep-sea fisheries: implications for a fragile ecosystem. *ICES J. Mar. Sci.,* 57, 548–557.

Koutsoubas, D., Galinou-Mitsoudi, S., Katsanevakis, S., Leontarakis, P., Metaxatos, A. and Zenetos, A. 2007. Bivalve and gastropod molluscs of commercial interest for human consumption in the Hellenic Seas. In *State of the Hellenic Fisheries* (eds. Papaconstantinou, C., Zenetos, A., Tserpes, G. and Vassilopoulou, V.). HCMR Publications, Athens, 70–84.

Kuang-Ti, L. 2001. Prehistoric marine fishing adaptation in Southern Taiwan. *J. East Asian Arch.,* 3, 47–74.

Leavenworth, W. B. 2008. The changing landscape of maritime resources in seventeenth-century New England. *Intern. J. Marit. Hist.,* 20, 33–62.

Lewison, R. L., Freeman, S. A. and Crowder, L. B. 2004. Quantifying the effects of fisheries on threatened species: the impact of pelagic longlines on loggerhead and leatherback sea turtles. *Ecol. Lett.,* 7, 221–231.

Lightfoot, K. G. and Luby, E. M. 2002. Late Holocene in the San Francisco Bay area: temporal trends in the use and abandonments of shell mounds in the East Bay. In *Catalysts to complexity: Late Holocene societies of the California Coast* (eds. Erlandson, J. M. and Jones, T. L.). Cotsen Institute of Archaeology, UCLA, 263–281.

Lotze, H. K. 2005. Radical changes in the Wadden Sea fauna and flora over the last 2000 years. *Helgol. Mar. Res.,* 59, 71–83.

Lotze, H. K. 2007. Rise and fall of fishing and marine resource use in the Wadden Sea, southern North Sea. *Fish. Res.,* 87, 208–218.

Lotze, H. K. 2010. Historical reconstruction of human-induced changes in U.S. estuaries. *Oceanogr. Mar. Biol. Ann. Rev.,* 48, 267–338.

Lotze, H. K. and Milewski, I. 2004. Two centuries of multiple human impacts and successive changes in a North Atlantic food web. *Ecol. Appl.,* 14, 1428–1447.

Lotze, H. K. and Worm, B. 2009. Historical baselines for large marine animals. *Trends Ecol. Evol.,* 24, 254–262.

Lotze, H. K., Lenihan, H. S., Bourque, B. J., Bradbury, R. H., Cooke, R. G., Kay, M. C., Kidwell, S. M., Kirby, M. X., Peterson, C. H. and Jackson, J. B. C. 2006. Depletion, degradation, and recovery potential of estuaries and coastal seas. *Science,* 312, 1806–1809.

Lotze, H. K., Erlandson, J. M., Newman, M. J., Norris, R. D., Roy, K., Smith, T. D. and Whitcraft, C. 2011a. Uncovering the ocean's past. In *Shifting baselines: the past and the future of ocean fisheries* (eds. Jackson, J. B. C., Alexander, K. E. and Sala, E.). Island Press, 137–161.

Lotze, H. K., Coll, M. and Dunne, J. 2011b. Historical changes in marine resources, food-web structure and ecosystem functioning in the Adriatic Sea, Mediterranean. *Ecosys.,* 14, 198–222.

Lotze, H. K., Coll, M., Magera, A. M., Ward-Paige, C. and Airoldi, L. 2011c. Recovery of marine animal populations and ecosystems. *Trends Ecol. Evol.,* 26, 595–605.

MacKenzie, B. R., Mosegaard, H. and Rosenberg, A. A. 2009. Impending collapse of bluefin tuna in the northeast Atlantic and Mediterranean. *Cons. Lett.,* 2, 25–34.

MacLeod, K. and Leslie, H. 2009. *Ecosystem-based management for the oceans.* Island Press.

Marchetti, M. 2002. Environmental changes in the central Po Plain (northern Italy) due to fluvial modifications and anthropogenic activities. *Geomorphology,* 44, 361–373.

Marean, C. W., Bar-Matthews, M., Bernatchez, J., Fisher, E., Goldberg, P., Herries, A. I. R., Jacobs, Z., Jerardino, A., Karkanas, P., Minichillo, T., Nilssen, P. J., Thompson, E., Watts, I. and Williams, H. M. 2007. Early human use of marine resources and pigment in South Africa during the Middle Pleistocene. *Nature,* 449, 905–908.

McCay, B. J. 1998. *Oyster wars and the public trust.* University of Arizona Press, Tucson.

McClanahan, T. R., Graham, N. A. J., Calnan, J. M. and MacNeil, M. A. 2007. Toward pristine biomass: reef fish recovery in coral reef marine protected areas in Kenya. *Ecol. Appl.,* 17, 1055–1067.

McClanahan, T. R., Hicks, C. C. and Darling, E. S. 2008. Malthusian overfishing and efforts to overcome it on Kenyan coral reefs. *Ecol. Appl.,* 18, 1516–1529.

McClenachan, L. and Cooper, A. B. 2008. Extinction rate, historical population structure and ecological role of the Caribbean monk seal. *Proc. Royal Soc. London B,* 275, 1351–1358.

McClenachan, L., Jackson, J. B. C. and Newman, M. J. 2006. Conservation implications of historic sea turtle nesting beach loss. *Front. Ecol. Environ.,* 4, 290–296.

McCulloch, M., Fallon, S., Wyndham, T., Hendy, E., Lough, J. and Barnes, D. 2003. Coral record of increased sediment flux to the inner Great Barrier Reef since European settlement. *Nature,* 421, 727–730.

McNeill, J. R. 2000. *Something new under the sun: An environmental history of the twentieth-century world.* Norton, New York.

Micheli, F., Halpern, B. S., Botsford, L. W. and Warner, R. R. 2004. Trajectories and correlates of community change in no-take marine reserves. *Ecol. Appl.,* 14, 1709–1723.

Mone, F. J. 1853. Ueber die Flussfischerei und den Vogelfang vom 14. bis 16. Jahrhundert. *Zeitschrift für die Geschichte des Oberrheins,* 4, 67–97.

Mora, C., Myers, R. A., Coll, M., Libralato, S., Pitcher, T. J., Sumaila, R. U., Zeller, D., Watson, R., Gaston, K. J. and Worm, B. 2009. Management effectiveness of the world's marine fisheries. *PLoS Biol.,* 7, e1000131.

Morato, T., Watson, R., Pitcher, T. J. and Pauly, D. 2006. Fishing down the deep. *Fish Fish.,* 7, 23–33.

Murawski, S. A., Brown, R., Lai, H.-L., Rago, P. J. and Hendrickson, L. 2000. Large-scale closed areas as a fisheries management tool in temperate marine systems: the Georges Bank experience. *Bull. Mar. Sci.,* 66, 775–798.

Myers, R. A. and Worm, B. 2003. Rapid worldwide depletion of predatory fish communities. *Nature,* 423, 280–283.

Naylor, R. L., Goldburg, R. J., Mooney, H., Beveridge, M., Clay, J., Folke, C., Kautsky, N., Lubchenco, J., Primavera, J. and Williams, M. 1998. Nature's subsidies to shrimp and salmon farming. *Science,* 282, 883–84.

Naylor, R. L., Goldburg, R. J., Primavera, J., Kautsky, N., Beveridge, M., Clay, J., Folke, C., Lubchenco, J., Mooney, H. and Troell, M. 2000. Effects of aquaculture on world fish supplies. *Nature,* 405, 1017–1024.

Pandolfi, J. M., Bradbury, R. H., Sala, E., Hughes, T. P., Bjorndal, K. A., Cooke, R. G., McArdle, D., McClenachan, L., Newman, M. J. H., Paredes, G., Warner, R. R. and Jackson, J. B. C. 2003. Global trajectories of the long-term decline of coral reef ecosystems. *Science,* 301, 955–958.

Pauly, D. 1995. Anecdotes and the shifting baseline syndrome of fisheries. *Trends Ecol. Evol.,* 10, 430.

Pauly, D., Christensen, V., Dalsgaard, J., Froese, R. and Torres Jr., F. 1998. Fishing down marine food webs. *Science,* 279, 860–863.

Pauly, D., Christensen, V., Guenette, S., Pitcher, T. J., Sumaila, U. R., Walters, C. J., Watson, R. and Zeller, D. 2002. Towards sustainability in world fisheries. *Nature,* 418, 689–695.

Pauly, D., Alder, J., Bennett, E., Christensen, V., Tyedmers, P. and Watson, R. 2003. The future for fisheries. *Science,* 302, 1359–1361.

Petrin, S. 1979. Das Archiv der Tullner Fischerzeche. *Mitteilungen aus dem Niederösterreichischen Landesarchiv,* 3, 29–34.

Perdikaris, S. and McGovern, T. H. 2008. Codfish and kings, seals and subsistence: Norse marine resource use in the North Atlantic. In *Human impacts on ancient marine ecosystems: a global perspective* (eds. Rick, T. C. and Erlandson, J. M.). University of California Press, 187–214.

Piet, G. J. and Rijnsdorp, A. D. 1998. Changes in the demersal fish assemblage in the south-eastern North Sea following the establishment of a protected area ('plaice box'). *ICES J. Mar. Sci.,* 55, 420–429.

Pikitch, E. K., Santora, C., Babcock, E. A., Bakun, A., Bonfil, R., Conover, D. O., Dayton, P., Doukakis, P., Fluharty, D., Heneman, B., Houde, E. D., Link, J., Livingston, P. A., Mangel, M., McAllister, M. K., Pope, J. and Sainsbury, K. J. 2004. Ecosystem-based fishery management. *Science,* 305, 346–347.

Pinnegar, J. K. and Engelhard, G. H. 2008. The 'shifting baseline' phenomenon: a global perspective. *Rev. Fish Biol. Fish.,* 18, 1–16.

Pondella, D. J. and Allen, L. G. 2008. The decline and recovery of four predatory fishes from the Southern California Bight. *Mar. Biol.,* 154, 307–313.

Price, J. 1999. *Flight maps: Adventures with nature in modern America.* Basic Books.

PROME. 2005. *The Parliament Rolls of Medieval England, 1275–1504* (eds. Given-Wilson, C., Brand, P., Curry, A., et al.). Boydell & Brewer, Woodbridge.

Pronzato, R. and Manconi, R. 2008. Mediterranean commercial sponges: over 5000 years of natural history and cultural heritage. *Mar. Ecol.,* 29, 146–166.

Radcliffe, W. 1921. *Fishing from the earliest times.* Ares, Chicago.

Richards, J. F. 2003. *The unending frontier: an environmental history of the early modern world.* University of California Press.

Rick, T. C. and Erlandson, J. M. 2008. Human impacts on ancient marine ecosystems: a global perspective. University of California Press.

Rick, T. C. and Erlandson, J. M. 2009. Coastal exploitation. *Science,* 325, 952–953.

Rick, T. C., Erlandson, J. M., Braje, T. J., Estes, J. A., Graham, M. H. and Vellanoweth, R. L. 2008. Historical ecology and human impacts on coastal ecosystems of the Santa Barbara Channel region, California. In *Human impacts on ancient marine ecosystems: a global perspective* (eds. Rick, T. C. and Erlandson, J. M.). University of California Press, 77–101.

Robbins, W. D., Hisano, M., Connolly, S. R. and Choat, J. H. 2006. Ongoing collapse of coral-reef shark populations. *Curr. Biol.,* 16, 2314–2319.

Roberts, C. M. 2007. *An unnatural history of the sea.* Island Press.

Rose, G. A. 2004. Reconciling overfishing and climate change with stock dynamics of Atlantic cod (*Gadus morhua*) over 500 years. *Can. J. Fish. Aquat. Sci.,* 61, 1553–1557.

Rosenberg, A. A., Bolster, W. J., Alexander, K. E., Leavenworth, W. B., Cooper, A. B. and MacKenzie, B. R. 2005. The history of ocean resources: modeling cod biomass using historical records. *Front. Ecol. Environ.,* 3, 84–90.

Rosenberg, A. A., Swasey, J. H. and Bowman, M. 2006. Rebuilding US fisheries: progress and problems. *Front. Ecol. Environ.,* 4, 303–308.

Rosenberg, A. A., Alexander, K. E. and Cournane, J. M. 2011. Management in the Gulf of Maine. In *Shifting baselines: the past and the future of ocean fisheries* (eds. Jackson, J. B. C., Alexander, K. E. and Sala, E.). Island Press, 177–191.

RPS 2007. *The Records of the Parliaments of Scotland to 1707* (eds. Brown, K. S. et al.). St. Andrews, 2007–2011, online edition http://www.rps.ac.uk/

Sáenz–Arroyo, A., Roberts, C. M., Torre, J., Cariño-Olvera, M. and Enríquez-Andrade, R. R. 2005. Rapidly shifting environmental baselines among fishers of the Gulf of California. *Proc. Royal Soc. London B,* 272, 1957–1962.

Sandin, S. A., Smith, J. E., DeMartini, E. E., Dinsdale, E. A., Donner, S. D., Friedlander, A. M., Konotchick, T., Malay, M., Maragos, J. E., Obura, D., Pantos, O., Paulay, G., Richie, M., Rohwer, F., Schroeder, R. E., Walsh, S., Jackson, J. B. C., Knowlton, N. and Sala, E. 2008. Baselines and degradation of coral reefs in the Northern Line Islands. *PLoS ONE,* 3, e1548.

Schneider, V. and Pearce, D. 2004. What saved the whales? An economic analysis of 20th century whaling. *Biodiv. Cons.,* 13, 543–562.

Scialoja, A. 1910. Statuta et ordinamento artis piscium civitatis Perusii (1296–1376). *Bollettino della Regia Deputazione di Storia patria per l'Umbria,*1910, 813–873.

Sethi, S. A., Branch, T. A. and Watson, R. 2010. Fishery development patterns are driven by profit but not trophic level. *Proc. Nat. Acad. Sci.,* 107, 12163–12167.

Simenstad, C. A., Estes, J. A. and Kenyon, K. 1978. Aleuts, sea otters, and alternate stable state communities. *Science,* 200, 403–411.

Smith, A. D. M., Brown, C. J., Bulman, C. M., Fulton, E. A., Johnson, P., Kaplan, I. C., Lozano-Montes, H., Mackinson, S., Marzloff, M., Shannon, L. J., Shin, Y.-J. and Tam, J. 2011. Impacts of fishing low-trophic level species on marine ecosystems. *Science,* 333, 1147–1150.

Smith, I. 2005. Retreat and Resilience: fur seals and human settlement in New Zealand. In *The exploitation and cultural importance of sea mammals* (ed. Monks, G.). Oxbow Books, 6–18.

Steele, T. E. and Klein, R. G. 2008. Intertidal shellfish use during the Middle and Later Stone Age of South Africa. *Archaeofauna,* 17, 63–76.

Stefansson, G. and Rosenberg, A. A. 2005. Combining control measures for more effective management of fisheries under uncertainty: quotas, effort limitation and protected areas. *Phil. Trans.,* 360, 133–146.

Steneck, R. S. 1997. Fisheries-induced biological changes to the structure and function of the Gulf of Maine ecosystem. *RARGOM Report* 91–1, 151–165.

Steneck, R. S., Graham, M. H., Bourque, B. J., Corbett, D., Erlandson, J. M., Estes, J. A. and Tegner, M. J. 2002. Kelp forest ecosystems: biodiversity, stability, resilience and future. *Environ. Cons.,* 29, 436–459.

Stiner, M. C. 2001. Thirty years on the "Broad Spectrum Revolution" and paleolithic demography. *Proc. Natl. Acad. Sci.,* 98, 6993–6996.

Stolz, O. 1936. *Geschichtskunde der Gewaesser Tirols.* Schlern-Schriften 32, Universitaets-Verlag Wagner, Innsbruck.

Sumaila, U. R., Khan, A., Watson, R., Munro, G., Zeller, D., Baron, N. and Pauly, D. 2007. The World Trade Organization and global fisheries sustainability. *Fish. Res.,* 88, 1–4.

Swartz, W., Sala, E., Tracey, S., Watson, R. and Pauly, D. 2010. The spatial expansion and ecological footprint of fisheries (1950 to Present). *PLoS ONE,* 5, e15143.

Thompson, V. D. and Pluckhahn, T. J. 2010. History, complex hunter-gatherers, and the mounds and monuments of Crystal River, Florida, USA: a geophysical perspective. *J. Isl. Coast. Arch.,* 5, 33–51.

Thrush, S. F. and Dayton, P. K. 2002. Disturbance to marine benthic habitats by trawling and dredging: Implications for Marine Biodiversity. *Ann. Rev. Ecol. Syst.,* 33, 449–473.

Tittensor, D. P., Mora, C., Jetz, W., Lotze, H. K., Ricard, D., Berghe, E. V. and Worm, B. 2010. Global patterns and predictors of marine biodiversity across taxa. *Nature,* 466, 1098–1101.

Trakadas, L. A. 2006. 'Exahusted by fishermen's nets': Roman sea fisheries and their management. *J. Medit. Stud.,* 16, 259–272.

Voorhies, B. 2004. *Coastal collectors in the Holocene: the Chantuto People of Southwest Mexico.* University Press of Florida, Gainesville.

Williams, J. 2006. *Clam gardens: aboriginal mariculture on Canada's West Coast.* New Star Books, Vancouver.

Worm, B., Sandow, M., Oschlies, A., Lotze, H. K. and Myers, R. A. 2005. Global patterns of predator diversity in the open oceans. *Science,* 309, 1365–1369.

Worm, B., Barbier, E. B., Beaumont, N., Duffy, J. E., Folke, C., Halpern, B. S., Jackson, J. B. C., Lotze, H. K., Micheli, F., Palumbi, S. R., Sala, E., Selkoe, K. A., Stachowicz, J. J. and Watson, R. 2006. Impacts of biodiversity loss on ocean ecosystem services. *Science,* 314, 787–790.

Worm, B., Barbier, E. B., Beaumont, N., Duffy, J. E., Folke, C., Halpern, B. S., Jackson, J. B. C., Lotze, H. K., Micheli, F., Palumbi, S. R., Sala, E., Selkoe, K. A., Stachowicz, J. J. and Watson, R. 2007. Biodiversity loss in the ocean: how bad is it? *Science,* 316, 1282–1285.

Worm, B., Hilborn, R., Baum, J. K., Branch, T. A., Collie, J. S., Costello, C., Fogarty, M. J., Fulton, E. A., Hutchings, J. A., Jennings, S., Jensen, O. P., Lotze, H. K., Mace, P. M., McClanahan, T. R., Minto, C., Palumbi, S. R., Parma, A. M., Ricard, D., Rosenberg, A. A., Watson, R. and Zeller, D. 2009. Rebuilding global fisheries. *Science,* 325, 578–585.

Zeller, D., Booth, S., Craig, P. and Pauly, D. 2006. Reconstruction of coral reef fisheries catches in American Samoa, 1950–2002. *Coral Reefs,* 25, 144–152.

Zeller, D., Darcy, M., Booth, S., Loweb, M. K. and S. Martell. 2008. What about recreational catch? Potential impact on stock assessment for Hawaii's bottomfish fisheries. *Fisheries Research,* 91, 88–97.

Chapter 3. Human Dimensions in Marine Ecosystem-Based Management

ANTHONY CHARLES

Saint Mary's University

Contents

1. Introduction

Ecosystem-based management (EBM) is widely accepted as a mechanism for more effectively managing human uses of natural resources and environmental services. The World Summit on Sustainable Development's (WSSD) Plan of Implementation "encourages the generalised adoption of ecosystem approaches" which is considered a key element of the WSSD's "decision of restoring marine ecosystems by 2015" (Nauen, 2006).

The logic of EBM is based on recognition of the need for a systems approach, for 'systems thinking', in the science and governance of natural resources. This is intended to rectify the failings of conventional natural resource management, which tended to focus narrowly on specific resources and resource users, while neglecting their interactions with broader natural and human systems. Such was the case, for example, with forest management in the first half of the twentieth century, when the focus was strictly on trees, specifically how many could or should be cut down at any given time. This has since broadened to managing multiple uses of the whole forest, reflecting the idea of an ecosystem approach.

However, while EBM as a framework is well-established in many terrestrial environments, it is still relatively new in marine settings (cf., Fanning et al., 2011; McLeod and Leslie, 2009; UNEP, 2011). Its rationale seems well accepted, though, reflected in international agreements and driven by an underlying recognition that

The Sea, Volume 16, edited by Michael J. Fogarty and James J. McCarthy
ISBN 978-0-674-07270-1

in the major marine sector of fisheries, the lack of systems thinking in conventional management (focused on single stocks and fleets) was among the causes of past fishery collapses (e.g., Charles, 2001). Accordingly, governments around the world, as well as multilateral and nongovernmental bodies of various forms, are now engaged in discussing and debating how EBM should be applied in specific economic sectors, ecosystem types and geographical regions of the marine environment. This is accompanied by a rapidly growing academic literature on the subject (e.g., Leslie and McLeod, 2007; Sanchirico et al., 2008).

A key element in the debate over implementation of marine EBM concerns the extent to which 'human dimensions' should be incorporated. Since EBM is explicitly an approach to managing human uses, in this case of the sea, it seems sensible that it should take into account the range of such social, cultural, economic, political, legal and institutional dimensions—to deal with aspects such as the values driving human behaviour, the human responses to regulations, and the structure of governance institutions. This is the logical conclusion of a shift to 'systems thinking': incorporating the human system as well as the natural one, rather than constraining EBM solely to consideration of ecosystems and the ecological interactions within them. Indeed, this broader perspective has been the trend for many years with EBM in terrestrial environments, and is featured prominently in the Millennium Ecosystem Assessment (2005a, b).

Within marine systems, there has been increasing attention over the past decade to the human dimensions of ecosystem-based management (e.g., De Young et al., 2008), perhaps reflecting a recognition that EBM in marine settings, as elsewhere, is as much about people as ecosystems. This perspective is highlighted in the FAO's Technical Guidelines on the Human Dimensions of the Ecosystem Approach to Fisheries (FAO, 2009). Indeed, EBM brings some key strengths as a vehicle to deal with human concerns (such as wealth generation, poverty alleviation and food security), while also moving marine systems toward ecosystem health. However, despite this, the extent of incorporation of human dimensions remains a subject of considerable debate in marine EBM, and examples can be found reflecting very different responses, as will be indicated below.

Section 2 of this chapter addresses the fundamental question "What are the human dimensions of marine EBM?" and reviews the various social, cultural, economic, political, legal and institutional aspects. Section 3 then explores how EBM has evolved in marine environments, specifically the matter raised above of how human dimensions are perceived and incorporated within marine EBM. Section 4 assesses the role of human dimensions in the practical implementation of EBM, specifically in *inducing, influencing and supporting* EBM. This highlights the relevant 'entry points' for developing and implementing EBM, and links to the related governance initiatives of integrated management and marine spatial planning. All of the factors *inducing, influencing and supporting* EBM arise at multiple scales, since ecosystems themselves exist at a range of scales, from the very local to the global. Accordingly, section 5 addresses the varying roles of human dimensions in marine EBM across scales, with emphasis on local-level marine environments. Recognizing and incorporating human dimensions in marine EBM has implications in terms of knowledge needs and monitoring processes—these aspects are examined in Section 6. Finally, the chapter ends with conclusions in section 7.

2. A Typology of Human Dimensions

In this section, I briefly review how social, cultural, economic, political, legal and institutional dimensions appear in marine systems, and how these may be relevant to EBM. For further details on the range of human dimensions relevant to EBM, see De Young et al. (2008), together with extended coverage of particular aspects in Charles and De Young (2009) and De Young and Charles (2009). As will be seen in subsequent sections, the analysis here informs the relevance of human dimensions in *inducing, influencing and supporting* EBM and its implementation.

Indeed, McConney and Salas (2011), in exploring the role of human factors in marine EBM, note that "It is unlikely that scientific and technical approaches to marine EBM will succeed on their own . . . if [these] considerations are ignored". McConney and Salas also note that consideration of these human dimensions "should also apply to conventional management", i.e., their role in EBM is important but no different from what their role should be in other forms of resource management. They list a series of human-oriented themes—stakeholders, institutions, community, power and politics, social structure, culture, participation, adaptive capacity, livelihoods and poverty, knowledge and conflict—most of which will be discussed at some point in this section.

Social. Social considerations, which involve the interaction of people and groups in society, have been addressed over many decades in social analyses to better understand fisheries and marine systems, drawing particularly on sociology and anthropology (e.g. McCay, 1981). Understanding social factors and social structure enables us to assess some of the key values and drivers underlying the need for EBM, and contributes to the prediction of the outcomes from policy interventions. For example, the 'social capital' in a given community measures the strength of social contacts, social networks and cooperation among the people involved, and thus will influence the extent to which they are likely to come together in support of an EBM initiative. At the same time, the crucial issue of power and its distribution within the system will affect who wins or loses in that initiative, thereby affecting livelihoods, poverty levels and the like.

Cultural. While some would think that EBM was invented only in the past few decades, in fact there are many examples worldwide of longstanding holistic management of humans' resource use through 'systems thinking' that takes ecosystem impacts and ecosystem health into account. This type of EBM particularly involves cultural factors—for example, local values and norms may be put in place to restrict use of resources and the environment for cultural reasons, and often with an ecological basis as well. Such cultural factors will affect the types of management measures seen as appropriate to the given context (e.g., Acheson, 1975); accordingly these are crucial to understand for effective implementation of EBM or any other governance approach. The knowledge base of local communities and resource users (e.g., as 'traditional ecological knowledge') is also a major cultural element, and can provide important input to EBM processes.

Economic. To the extent that economic considerations drive the behaviour of individual ocean users, and of governmental management agencies, the outcomes of economic processes and their distribution between groups in society, as well as over time and space, may well be key determinants to what is or is not feasible within EBM. At the same time, economic incentives that influence human behaviour (e.g., through taxes, fines, access rights and pricing policies) are considered among the major instruments for efficiently operating a management system. Accordingly, economic dimensions are especially important in considering the implementation of marine EBM (Holland et al., 2010; Schuhmann et al., 2011).

Political. Marine resource users, coastal communities and other stakeholders all interact within a complex system that is meant to be guided through mechanisms such as EBM. Power strongly influences how these interactions take place, often through the vehicle of politics—"the set of social relations that people use to gain and exercise power" (McConney and Salas, 2011). It is especially important to understand and take into account political processes (e.g., with political science methods and policy analysis) not only because implementation of new initiatives such as EBM often must go through the political process, but also because (a) inherent power disparities and inequities may affect the ability to achieve EBM objectives, and (b) "political changes can modify the direction of choices, actions and hence the implementation of EBM" (McConney and Salas, 2011).

Legal and Institutional. The concept of an 'institutional framework' refers to both the set of rules governing human use (in this case, of the ocean and marine resources), and the specific organizational arrangements involved in the formulation and implementation of laws, policies, strategies and programmes (De Young et al., 2008). Such frameworks range from global fora of the United Nations through to local-level community fora for discussion among decision-makers, as well as infrastructure such as cultural facilities and schools. Within modern thinking on 'good governance', institutional dimensions often focus on stakeholders and their participation in decision-making processes, as well as the nature of conflict and its resolution. Complementary to this are legal elements, which include laws and regulations adopted by governments, while the court system may be seen as both legal and institutional. Legal and institutional incentives (ranging from enforcement of formal laws through to participatory management systems and governance arrangements) can be important in motivating and influencing human behaviour, including support from stakeholders for EBM initiatives.

This section has presented only the briefest of outlines for each of the social, cultural, economic, political, legal and institutional dimensions relevant to marine systems and EBM. Every one of these could have (or have had) a full book written about them. For now, the key point to emphasize is that each of these elements can be important to understand, albeit to varying extents depending on the context, as human dimensions *inducing, influencing and supporting* EBM and its implementation. This is elaborated in section 4 below. In reality, however, these social, cultural, economic, political, legal and institutional elements do not tend to arise individually, but rather as a set of human dimensions that jointly impact on marine systems and affect marine EBM approaches. Accordingly, attention shifts for the remainder of the chapter from examining each of the constituent dimensions separately, as was done here, to discussions of the human dimensions in a collective manner.

3. Perspectives on Human Dimensions in Marine EBM

As the application of EBM in marine settings has evolved, a range of contrasting views about its scope and content has emerged (e.g. Berkes, 2011). This range is illustrated here by four particular perspectives, two from within the fishery sector specifically and two from ocean science and governance. For each of these pairs, one is relatively narrow in scope, while the other is broader, particularly in terms of the incorporation of human dimensions—these differences reflect a key area of contention.

(1) Fisheries responded to failures of the single-species approach through a version of EBM that added consideration of multi-species interactions of "target" resources with other species (as in predator-prey relationships) as well as ecosystem and fish habitat considerations. As Smith et al. (2007) noted, this reflected "a gradual evolution in fisheries management from a primary focus on sustainability of target species and resources to a much wider focus on ecosystems, and the impacts of fisheries on them." This is exemplified by the Commission for the Conservation of Antarctic Marine Living Resources, with its convention noting such principles as "maintenance of the ecological relationships between harvested, dependent and related populations of Antarctic marine living resources" (Commission for the Conservation of Antarctic Marine Living Resources, 1982; see Gaichas et al, this volume). Indeed, this is an example of how objectives can broaden over time, from managing human uses of resources solely for sustained yields of individual resources, to adding goals of ecosystem health.

However, overall, there has been a strong minimalist tendency within this framework, seeking the least possible disruption to conventional fishery science and management. In some cases, this has involved as little as supplementing a conventional single-species fish stock assessment with the addition of a single paragraph on ecosystem interactions (such as habitat impacts and predator-prey effects). Certainly EBM in fisheries of many developed countries is being viewed by governments as restricted to incorporating: (a) fishing effects on the ecosystem, (b) ecosystem influences on fisheries, and (c) ecosystem impacts on fisheries arising from other sectors. This approach avoids a full consideration of the human dimensions now seen as relevant to EBM, such as the behaviour of resource users, governance issues and use conflicts. People might be seen solely as among the 'predators' in the ecosystem, or considered solely in terms of the need for acceptance of ecosystem thinking among the various stakeholders, i.e., determining what is needed to induce people to implement EBM.

(2) The second fishery-focused perspective on ecosystem-based management is broader in nature, including consideration not only of multi-species, ecosystem and fish habitat impacts in the ocean but also the interaction of humans with ecosystems—arising in terms of social, cultural and economic values, and the human goals being met through marine uses. This perspective is reflected in the Ecosystem Approach to Fisheries (EAF)—see FAO (2003), Garcia et al. (2003) and Cochrane et al. (this volume). Indeed, FAO (2003) defines the EAF as an approach that *"strives to balance diverse societal objectives, taking account of the knowledge and uncertainties of biotic, abiotic and human components of ecosystems and their interactions and applying an integrated approach to fisheries within ecologically meaningful boundaries"*. As De Young et al. (2008) note, this perspective on an ecosystem approach focuses on "aspects within the ability of fisheries management bodies to implement", but it is nevertheless relatively comprehensive, incorporating natural, human and management aspects of the 'fishery

system' and beyond, and recognizing "the fisheries sector's responsibility in collaborating in a broader multi-sectoral [EBM]".

This is therefore a comprehensive view of ecosystem-based management that sees humans (societies, technologies, economies, needs and values) as integral to the system. Essentially this extends the focus on ecosystems to comprehensive 'systems' that incorporate human and ecological goals, as well as human and ecological components (Charles, 2001; Garcia and Charles, 2007). This reflects modern resource governance in addressing the 'bigger picture' by combining ecosystem-centred and people-centred thinking.

(3) A third conception of marine EBM tends to originate among those focused on ocean science, management and conservation. A key example is a major scientific consensus statement produced in the United States which defines marine ecosystem-based management as "An integrated approach to management that considers the entire ecosystem, including humans" (McLeod et al., 2005). This reflects the point made some time ago by Schramm and Hubert (1996) that EBM "recognizes that humans—including their societies, technologies, economies, needs and values—are part of the ecosystem". The consensus statement (McLeod et al., 2005) further states that: "The goal of ecosystem-based management is to maintain an ecosystem in a healthy, productive and resilient condition so that it can provide the goods and services humans want and need" and lists a number of attributes typical of ecosystem-based management (McLeod et al., 2005):

- "Emphasizes the protection of ecosystem structure, functioning, and key processes;
- Is place-based in focusing on a specific ecosystem and the human activities affecting it;
- Explicitly accounts for the interconnectedness within systems, recognizing the importance of interactions between many target species or key services and other non-target species;
- Acknowledges interconnectedness among systems, such as between air, land and sea; and
- Integrates ecological, social, economic, and institutional perspectives, recognizing their strong interdependences".

While relatively broad in scope, this perspective on EBM nevertheless focuses largely on ecological aspects, with the human dimensions restricted largely to the last point, of integrating "ecological, social, economic, and institutional perspectives".

(4) The fourth model sees EBM as tied closely to related governance frameworks, notably Integrated Management and Marine Spatial Planning (see Cochrane et al. this volume; Rice et al. this volume). Integrated management of oceans and coasts provides a unified arrangement for decision making in a multi-sectoral context (i.e. considering all marine uses, such as fisheries, shipping, and mining) while the closely related marine spatial planning is defined as "a public process of analyzing and allocating the spatial and temporal distribution of human activities in marine areas to achieve ecological, economic, and social objectives that are usually specified through a political process" (Douvere, 2008; Ehler and Douvere, 2009).

While ecosystem-based management originates in an ecological (and ecosystem health) perspective that emphasizes the need to manage resource use within a broader context of the ecosystem, integrated management and marine spatial planning come from an administrative and governance perspective, focused on processes and institutions able to deal with allocation challenges and conflict resolution. However, these approaches have many commonalities, sharing a focus on specific well-defined spatial areas of the ocean, on an 'integrated' and holistic approach to the management of human uses in the given area, and on paying attention to the ecosystem impacts of those human uses. The latter is obviously inherent in EBM but is typically incorporated in integrated management initiatives as well. Furthermore, while one might contrast integrated management as an organizational or institutional approach, with EBM as a 'systems' approach, there is nevertheless a high degree of complementarity and constructive overlap, and indeed signs of convergence between these, particularly given that within this model of EBM, human dimensions are explicitly incorporated.

In summary, with governments today essentially obliged to adopt an ecosystem approach of some form, there remains a tension between those advocating or pursuing each of the above perspectives. Fundamentally, all of them are rooted in 'systems thinking', albeit to varying extents (Garcia and Charles, 2007, 2008). The choice lies in the extent to which EBM remains restricted to ecosystem considerations (with minimal human dimensions) or adopts a holistic and integrated interpretation. The latter is more in keeping with modern use of analyses based on the concept of 'social-ecological systems' (e.g., Berkes et al., 2003; Berkes and Folke, 1998; Pollnac et al., 2010). However, regardless of the perspective adopted for marine EBM in specific situations, the reality is that certain human dimensions are bound to be relevant in EBM implementation. Subsequent sections of the chapter illustrate this by exploring the role of human dimensions in effectively developing and implementing marine EBM, across various spatial and organizational scales.

4. Inducing, Influencing and Supporting Marine EBM

The previous section described differing perspectives on EBM, through fishery-specific and broader marine lenses—some downplay human dimensions while others fully incorporate them. Given that, particularly for those who come from an ecological and/or natural science background, life may be simpler if we were to minimize human dimensions in EBM, it is helpful to focus on the 'value-added' resulting from incorporation of human dimensions in marine EBM. More specifically, what human dimensions need to be incorporated in marine EBM to improve and ensure its effectiveness, and how exactly should those crucial human dimensions be properly incorporated in practice?

To address these questions, this section provides a conceptual framework for assessing the role of human dimensions in *inducing, influencing and supporting* EBM (Figure 3.1). This role can be viewed as being comprised of four key *'entry points'* (De Young et al., 2008)—pathways by which EBM, its implementation and the corresponding human dimensions enter logically into the marine social-ecological system. These entry points are as follows:

1. ***Drivers*** underlying the rationale for EBM, and thus 'inducing' its adoption, are often in the form of human dimensions, notably social, cultural, economic, political, legal and/or institutional objectives and factors;
2. ***Tradeoffs*** to society and to individuals arising in applying EBM (and *influencing* the form it takes) relate strongly to human dimensions—including the costs and benefits inherent in each of the social, cultural, economic, political, legal and institutional components;
3. People-focused ***tools (or instruments)*** are crucial in *supporting* the implementation of EBM—these may be of social, cultural, economic, political, legal and/or institutional forms, and will involve relevant institutions and incentives;
4. The ***enabling environment*** can include supports but also constraints on the successful implementation of EBM, arising as human dimensions of various forms (social, cultural, economic, political, legal, institutional)—this environment plays a role in all three of *inducing, influencing and supporting* EBM.

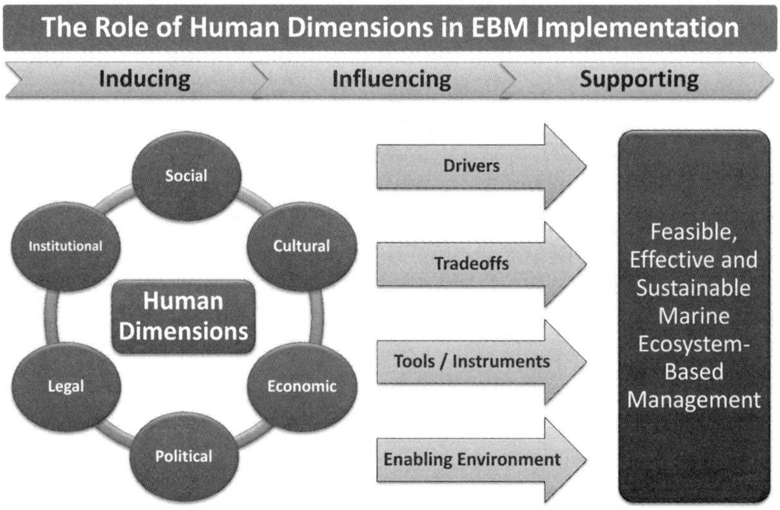

Fig. 3.1 – (See color insert) The human dimensions relevant to marine ecosystem-based management, and their role in inducing, influencing and supporting EBM implementation, acting through a set of *entry points:* drivers, tradeoffs (including costs and benefits of EBM), tools / instruments (as suitable for EBM), and the enabling environment (supporting and constraining factors within the corresponding marine system).

Drivers. Societal concerns over food security, rural well-being and livelihoods, biodiversity conservation and ecosystem health, among others, provide driving forces in the move to EBM. An illustration of this lies in the growing recognition that the value of the ocean to humans goes well beyond what is seen in the marketplace. Humans value ocean ecosystems and the services they provide in many different ways, so an intricate connection exists between those ecosystem services and human well-being (see Guerry and Tallis, this volume). In the valuation of marine ecosystems, certainly there is a conventional set of ***use values,*** including the standard net economic benefits of ocean use (e.g., income and employment from sectors such as fisheries, shipping and tourism), food provision and food security benefits, and the value of ocean uses in providing livelihoods and social interaction among its users. In addition, however, there are ***non-use and existence values,*** which may include (a) cultural benefits of marine ecosystems, such as art, ceremonies, and maintenance of traditional ways of life for fishermen, (b) aesthetic and existence benefits, such as the enjoyment of a seascape or the value of knowing that blue whales are living in the ocean, and (c) the option value in terms of maintaining, through conservation, possible but as yet unknown future benefits (such as potential medicinal products).

Another form of driver *inducing* EBM arises in the political system as specific legislative or policy thrusts. For example, Smith et al. (2007, p. 633) note that marine EBM developed in Australia as a result of policy initiatives such as "a national, government-wide approach to ecologically sustainable development, released in 1991" and "Australia's Oceans Policy, which adopts an explicit ecosystem-based approach to management, with explicit requirements for regional ocean planning for all uses and users of the marine environment. . . ."

Tradeoffs. Any human action has accompanying costs and benefits; the implementation of EBM is no exception. An assessment of these costs and benefits, perhaps along the lines of an extended economic cost-benefit analysis, can be important in practice (see Altman et al. this volume). Potential benefits often cited typically arise in the long term, including improvements in many of the drivers noted above (such as the ecosystem services provided by marine environments, including ecosystem health, biodiversity conservation and long term food security) as well as resource-focused benefits such as more sustainable resources and a reduction in the risk of fishery collapse. On the other hand, costs may be immediate—such as those for implementation of EBM (e.g., higher management costs) and those due to a loss of employment as resource use is curtailed to permit ecosystem restoration. Indeed employment loss may be not only a short term cost, but one that is permanent, if the sustainable flow of ecosystem services is unable to support as many people as previously. Thus tradeoffs can occur between the various marine uses, between different attributes of the social-ecological system, and between the present and the future.

Furthermore, potential benefits and costs may occur over a wide range of spatial / geographical or administrative scales (local, national, international). How the tradeoffs manifest themselves, and in fact whether or not each benefit or cost occurs in practice, will depend on the specifics of the marine system and the manner of EBM implementation. A key aspect in this is the matter of who receives the benefits and who incurs the costs of that implementation, i.e. the distributional impacts, which can arise between economic sectors, between geographical areas, and indeed between different social and cultural groups. In terms of human dimensions, issues of power, equality and inequality are important in assessing tradeoffs and resolving conflicts.

Tools/Instruments. A basic insight of natural resource governance highlights the reality that it is not the natural resources themselves but rather the people who utilize the resource who are to be managed. This is applied as well to spatial areas—notably in the sense that oceans are not managed, the people using them are. Thus if the word 'management' in EBM is meaningful, it implies that human dimensions are crucial to incorporate since humans are being managed. In other words, whether in an EBM context or otherwise, the tools for management and governance—referred to here as instruments—are inherently people-centered. Such instruments may be seen as applying to three levels of governance:

- High-level strategic governance approaches (policy, legal and institutional instruments, including global instruments such as international conventions, the Millennium Development Goals, and the precautionary approach)—e.g., the application of principles of good governance through participatory processes for reaching consensus, whether on human values to be highlighted or on cooperative mechanisms for decision making;
- Mid-level 'tactical' instruments that take as given the strategic governance directions, and provide practical tactics for moving the marine system in those directions—this will include specific categories of management tools, such as spatially-structured (e.g., zoning), market-based (e.g., labelling), rights-based (e.g., community access rights) and/or regulatory (e.g., licensing schemes);
- Operational management tools that enable the detailed implementation in practice of the accepted tactical instruments, governed by the chosen strategic instruments—e.g., designing and implementing a set of marine protected areas, or setting total allowable catches for species in the marine ecosystem targeted by fisheries.

Fundamentally, in implementing EBM, one must understand how the instruments are to be put in place, and how people will support, counteract or otherwise respond to them. For example, the operational task of choosing the location for and setting the rules within a marine protected area will be affected by the broader choice of tactical instruments, such as access rights arrangements or marine spatial planning schemes, while in turn those tactical choices, and how they are made, will be influenced by strategic governance, e.g., decisions to adopt participatory planning or co-management systems. Note that at each of these levels, implementation of the corresponding tools would be difficult if not impossible without an understanding and inclusion of suitable human dimensions.

A key requirement here is to understand how human behavior is or could be influenced by the range of possible incentives in place (whether purposely or inadvertently) to affect that behavior. The right incentives—whether social, cultural, economic (financial), political, legal or institutional—can *induce* a shift in behavior in desired directions, e.g., to support ecosystem rebuilding and other EBM principles. On the other hand, 'perverse' incentives push behavior in undesirable directions—for example, incentives to develop industrial activity, while well-meaning, may lead to loss of ecosystem well-being, leading perversely to economic decline. Developing the right institutions can also move incentives in a desired manner—see, e.g., Makino et al. (2009) on implementing EBM through the creation of co-management structures.

Enabling Environment. Human dimensions play *supporting or constraining roles* in implementing EBM, in many different ways. Here we differentiate between the major driving forces for EBM, discussed above, and human factors that may indirectly support EBM (as part of the enabling environment), or perhaps more importantly, may constrain its adoption. For example, the existence of strong organizations of ocean users (in fisheries, aquaculture, tourism, shipping, etc.) may be important in providing the leadership needed for acceptance of the change involved in shifting to EBM (Gutiérrez et al., 2011), although one could also envision such organizations *constraining* EBM in some contexts. Similarly, cultural and religious considerations may have either supportive or constraining effects—for example, local taboos (such as those that restrict resource use, with a cultural or religious basis) may provide a strong foundation for broader EBM initiatives. Some human factors may be relevant to EBM particularly in constraining its implementation—notably poverty and food insecurity make it very difficult to implement 'short-term pain for long-term gain' as may be desired for ecosystem rebuilding. At an institutional level, the inertia of government structures (e.g., departmental or ministry 'stove-pipes) has often been seen to constrain a comprehensive adoption of EBM, especially in marine settings where multiple parts of government often share jurisdiction.

While this indicates that human dimensions may well play supporting or constraining roles in the implementation of EBM, for marine environments or elsewhere, it is important to note that the precise role of such factors will in turn depend on the context—which itself has social, cultural, economic, political, legal and institutional dimensions. In other words, it is important to examine human dimensions both in terms of how they interact directly with EBM, and also in terms of the nature and context of the human environment within which EBM is being implemented. The latter could relate, for example, to major overarching elements in society, such as moves toward good governance, decentralization and devolution of governance, poverty alleviation and economic globalization. Notably, the political context for marine EBM will differ across nations, so that the role of human dimensions and their relative importance will be quite different in say, Cambodia (where community-based management has been a governance trend) and Canada (where the national government directs marine initiatives such as EBM).

5. Human Dimensions of Marine EBM across Scales

Just as ecosystems exist at many different scales, so too is EBM applicable at a variety of scales. The need to carefully assess the appropriate scales at which interventions should occur, referred to as 'scale-matching', has been identified by Costanza et al. (1998) as one of the six key principles needed for sustainable governance of the oceans.

Typically, a large-scale focus has predominated in 'official' (generally governmental) marine EBM initiatives—operating at large scales from regions of a country to national, multinational and global scales. Likely the most frequent of these initiatives are national in scale, taking place in a large proportion of the world's countries, with undoubtedly a high level of diversity across approaches (Cochrane et al. this column). Despite these national efforts, perhaps the best publicized scale overall is that of the world's Large Marine Ecosystems (Sherman and Hempel, 2008) in which a wide range of EBM initiatives are being undertaken. Many of these incorporate human dimensions, to various degrees; this is often in relation to issues of governance, participatory processes and stakeholder involvement (e.g., Fanning et al., 2011; Mahon et al., 2010; Volovik, 2010). In developing countries,

EBM in Large Marine Ecosystems may also seek to address societal concerns such as poverty and food security, where relevant.

At the other end of the spectrum, human dimensions are even more prominent in EBM initiatives at the local level, such as an individual bay or stretch of coastline. First, small-scale marine systems (e.g., small-scale fisheries) often have long histories of de facto ecosystem approaches, rooted in particular visions of nature, traditional ecological knowledge (TEK—e.g., Kliskey et al., 2009) and 'folk management'. In such situations, a grass-roots 'systems' approach may well form part of the cultural fabric in place-based coastal communities, integrating naturally across the uses of an area, and placing the health of ecosystems high on the agenda. Second, while there has been very limited attention paid in official circles to implementing the 'scientific' version of EBM, greater support for local-level and community EBM efforts is in keeping with the 'principle of participatory governance' (UNEP, 2009). Furthermore, local EBM can improve management success, for reasons that tend to revolve particularly around the human dimensions involved. Specifically, local EBM (a) can promote recognition and utilization of all sources of knowledge, notably traditional ecological knowledge (TEK) and local knowledge; (b) can build on community collaboration, as well as enthusiasm for local ecosystems and interest in their stewardship; (c) can draw on the strengths of existing human institutions in communities; and (d) can effectively utilize participatory research to meet knowledge needs. All of these possible benefits of small-scale marine EBM are inextricably linked to an understanding and appreciation of corresponding human dimensions.

For example, a successful case of local EBM implementation took place in the Atlantic Canadian coastal community of Eastport, Newfoundland. Following the collapse of the Atlantic cod fishery, local fishers sought to improve and sustain the remaining lobster fishery. They partnered with the government and universities to learn more about their lobster populations, using that knowledge to shut down their own fishing activity within certain key ocean habitats, in order to improve the reproductive and growth processes. This appreciation of the interactions between 'target resources' and the broader ecosystem, which is at the heart of EBM, led over the course of several years to stronger lobster stocks and greater livelihood security. At that point, the fishers saw advantages in extending their self-imposed fishery closure more comprehensively. This led to the creation by the Canadian government of an official marine protected area, which is strongly supported by the local community and now safeguards the ecosystem from not only fishing but also other potential economic activities (Charles and Wilson, 2009). The success in this EBM situation can be traced to several factors: an initial sense of crisis and risk (following the cod collapse), a willingness to learn and innovate on the part of the fishers, strong local leadership and community values, and supportive individuals in academia and government.

An appreciation of human dimensions, including values, goals and aspirations, suggests that in many locations, there needs to be a balance in implementation of marine EBM between large-scale and small-scale initiatives. There are now some interesting moves in this direction, such as the merging of community-based management and EBM in the Pacific islands (Secretariat of the Pacific Community, 2010). Achieving this cross-scale EBM seems to require overcoming certain challenges (Charles et al., 2010) including: (a) the 'scale mis-match' noted above (large-scale versus local ecosystem initiatives), (b) a potential values mis-match (requiring a 'meeting of the minds' between government and community), and (c) equity issues (relating to who wins and who loses, and achieving a fair mix of benefits and costs). This may be achieved through mutual learning, with higher-level governments demonstrating respect and support for local-level ecosystem-based efforts, providing suitable 'policy space', allowing for incorporation of pre-existing ecosystem approaches, and facilitating 'scaling up' and 'scaling down' to find the best mix of EBM across multiple scales.

6. Knowledge Needs for Marine EBM

A core element of EBM lies in assessing the state of marine social-ecological systems, and providing ongoing monitoring of those systems. Recognizing that this includes both natural and human systems, there will be certain knowledge needs corresponding to the relevant human dimensions in any given marine EBM scheme. This is sometimes seen as an onerous challenge, notably in marine management bodies that lack social science expertise, but in reality there are many mutually reinforcing pathways through which this can be achieved.

One of these pathways involves baseline studies of the human (social, cultural, economic, political, legal and institutional) dimensions in specific ocean areas. This includes various experiences globally to produce socio-economic or socio-cultural overviews and assessments in advance of new ocean management initiatives (Charles et al., 2007) as well as social impact assessments (Pollnac et al., 2006). Such studies often complement biophysical and ecological studies that are done in parallel, and involve assessing the human dimensions that contribute to management decision making, on scales from large (government) to small (community, NGO). Applications of these assessment approaches include large ocean management areas (LOMAs) in Canada, OSPAR in Europe, Large Marine Ecosystems (LMEs) worldwide, and on a smaller scale, coral reefs (e.g., Bunce et al. 2000; see also Cochrane et al. this volume; Houde et al. this volume; Kritzer et al. this volume). Interestingly, it seems that for larger-scale systems, the human dimensions assessed have been mostly economic, while cultural assessments have been done mostly at a local scale; furthermore, including governance aspects in the assessment appears crucial to success at all scales (Charles et al., 2007).

A second pathway for incorporating human dimensions is based on understanding and assessing the social and economic values of ecosystem services (De Young et al., 2008). Such ecosystem valuation involves examining use values, non-use values, existence values and option values in an ecosystem, as discussed earlier. This is often approached from a largely economic perspective (Holland et al., 2010) but has also been examined in terms of socio-cultural values (e.g., Sherrouse et al., 2011). Mapping approaches and use of GIS can be powerful tools for overlaying the spatial distribution of human uses and values, on the one hand, and the set of spatially-identified marine ecosystem services and valued ecosystem components. These approaches relate closely to the use of marine spatial planning (MSP), which is becoming popular as an integrating tool linked to marine EBM (e.g., Douvere, 2008; Ehler and Douvere, 2009).

A third avenue for incorporating human dimensions into monitoring and assessment is that of indicator frameworks. Integrated indicator frameworks link together biophysical, ecological, and environmental indicators (e.g., measures of ocean currents, fish biomass, trophic diversity and marine contaminants) with social, economic and institutional indicators (e.g., measures of net revenues, diversity of employment and participatory governance) [See Plummer and Levin (this volume)]. By integrating indicators relating to 'human dimensions' fully within such an indicator framework, this ensures that the human side of the social-ecological system receives equal attention to the ecological side. For example, Charles et al. (2009) compiled such a framework relating to fisheries and the marine environment for Nova Scotia, Canada. Note that to properly understand how humans envision and interact with marine ecosystems, it is helpful to include not only objective indicators (such as levels of employment and of participation in decision-making) but also subjective indicators where possible, to capture human values and perceptions of the marine environment, which may also affect behaviour of marine users (e.g., see Smith and Clay, 2010).

Overall, there is no doubt that some knowledge of human dimensions is needed in order to effectively utilize these aspects within EBM. Unfortunately, within marine systems especially, there has been too little official support for this knowledge creation to date, including a pervasive lack of personnel to do so within many governmental agencies. There will need to be some efforts to rectify this long-standing imbalance relative to provision of natural science knowledge. However, the above examples indicate an important point, namely that the required knowledge may be suitably obtained and incorporated using processes (assessment, ecosystem services studies, indicator frameworks) that operate in parallel with those used in generating biophysical and ecological knowledge. This may provide an efficient approach to creating effective marine EBM arrangements.

7. Conclusions

This chapter has sought to indicate (1) how the human dimensions of marine (and other) systems, comprising social, cultural, economic, political, legal and institutional components, are relevant to marine ecosystem-based management, (2) how an understanding and incorporation of these human dimensions can be utilized to more effectively and successfully implement EBM, and (3) how the role of human dimensions in EBM will depend on the context, and particularly the scale of the EBM initiative. The following conclusions may be drawn from the analysis presented in this chapter, providing guidance on effectively incorporating human dimensions into marine ecosystem-based management, and thereby improving its success and effectiveness:

Human Dimensions: Social, Cultural, Economic, Political, Legal and Institutional. The various human dimensions each play a role in marine systems, and accordingly each has its own potential relevance both in achieving an acceptable understanding of the marine system, and in facilitating the implementation of EBM approaches. Some examples of possibly relevant considerations include the role of power and of politics in shaping EBM schemes, the occurrence of culturally appropriate local forms of EBM, the various economic, legal and other incentives that can act to reinforce EBM goals, and the institutional frameworks in which principles of good governance support stakeholder participation in decision-making processes.

Inducing, Influencing and Supporting EBM. It is important to understand how human dimensions are involved in *inducing, influencing and supporting* marine EBM, notably in terms of drivers, tradeoffs (costs and benefits), implementation instruments, and enabling environments (supporting and constraining factors). These entry points enable human considerations to arise naturally in discussions of marine ecosystems and EBM. For example, recognition of the importance of ecosystem services can be a driver of EBM, and thus an entry point to implementation. Frameworks of integrated ocean management and marine spatial planning can contribute to the enabling environment as an entry point to support EBM implementation. Suitable incentive structures and institutions can provide the instruments needed to effectively implement EBM, and help to ensure its financial sustainability.

Seamless Integration of Human Dimensions. Given that EBM is often still seen (erroneously) as being about ecosystems more than about people, it is crucial that human dimensions are not (and are seen not to be) after-the-fact 'add-ons'. Comprehensive 'seamless' approaches are useful—for example, in assessment and monitoring these may include integrated indicator frameworks and assessment systems that have human as well as ecological aspects built fully into the process from the start. Seamless integration also helps in the common challenge in EBM of addressing underlying conflicts between ecological and human objectives, and between ecological and human boundaries.

Context Matters for EBM. While in almost all situations, marine EBM will require or at least benefit from having human dimensions incorporated, the precise extent to which this should occur in implementing marine EBM will depend on the ecosystem and the 'people context'. For example, if we compare Antarctic marine resource systems with those involving local aboriginal communities or small-scale fisheries in Canada or Norway, the extent and form of human dimensions needed to assess the marine system and to implement EBM will be quite different between these two scenarios (see also Gaichas et al. this volume). Furthermore, there will typically be a need to coordinate EBM with the existing policy, legal and institutional context. Thus any EBM initiative would do well to first, recognize the need to incorporate human dimensions into the process, and second, to assess what exactly is needed on a context-specific basis.

Multiple Scales of EBM. Just as ecosystems and human systems exist at multiple scales, so too can EBM be envisioned at these different scales. Indeed, in some cases, local and/or traditional forms of EBM may already exist, created over time by local communities—these should be recognized and supported. In other cases, it will be necessary to learn how to implement EBM at suitable scales, and how to 'scale up' and 'scale down' EBM approaches to fit appropriate scales. In particular, local-level EBM may be desirable, notably in coastal settings with relatively dense human population and economic activity; in these cases, principles of good governance indicate the desirability of providing suitable 'policy space' to enable these initiatives, participatory approaches for its implementation, and cross-scale linkages to connect small-scale EBM to other levels.

In closing, this chapter has highlighted the need to pay attention to human dimensions in marine ecosystem-based management, albeit at a scale appropriate to the context and to an extent justifiable by the situation at hand. Thus, in reflecting on the four differing perspectives on marine EBM presented earlier, the analysis of this chapter shows that there is no single 'correct' perspective on human dimensions, but rather that the extent and form by which such considerations should be incorporated will depend both on the objective reality of the ecosystem and the human system, and on the governance context, including elements such as the available financial and human capacity. Meeting the challenge of incorporating human dimensions into EBM in a suitable manner is admittedly not a trivial task and not without costs, but done judiciously, this can be expected to improve the performance of marine ecosystem-based management, and certainly to make the difference, in some cases, between failure and success.

Acknowledgments

I am grateful to two anonymous referees for very helpful comments, and to Michael Fogarty for his support in the production of this contribution. Financial support is acknowledged from the Social Sciences and Humanities Research Council of Canada and the Natural Sciences and Engineering Research Council of Canada (operating grant 6745).

References

Acheson, J. M. 1975. The lobster fiefs: economic and ecological effects of territoriality in the Maine lobster industry. Human Ecology, **3,** 183–207.

Berkes, F. 2011. Implementing ecosystem-based management: evolution or revolution? *Fish and Fisheries,* doi: 10.1111/j.1467–2979.2011.00452.x

Berkes, F., J. Colding and C. Folke. 2003. *Navigating Social-Ecological Systems: Building Resilience for Complexity and Change.* Cambridge University Press, Cambridge, UK.

Berkes, F. and C. Folke (editors). 1998. *Linking Social and Ecological Systems: Management Practices and Social Mechanisms for Building Resilience.* Cambridge University Press, New York.

Bunce, L., P. Townsley, R. Pomeroy and R. Pollnac. 2000. Socioeconomic manual for coral reef management. Global Coral Reef Monitoring Network. Australian Institute of Marine Science, Townsville Q 4810. Australia

Charles, A. 2001. *Sustainable Fishery Systems.* Blackwell Science, Oxford.

Charles, A. and C. De Young. 2009. Benefits and costs of implementing the ecosystem approach to fisheries. In: *The Ecosystem Approach to Fisheries* (G. Bianchi and H. R. Skjoldal, editors). CAB International.

Charles, A. and L. Wilson. 2009. Human dimensions of marine protected areas. *ICES Journal of Marine Science,* **66,** 6–15.

Charles, A., M. Wiber, K. Bigney, D. Curtis, L. Wilson, R. Angus, J. Kearney, M. Landry, M. Recchia, H. Saulnier and C. White. 2010. Integrated management: A coastal community perspective. *Horizons,* **10,** 26–34.

Charles, A., C. Burbidge, H. Boyd and A. Lavers. 2009. *Fisheries and the Marine Environment in Nova Scotia: Searching for Sustainability and Resilience.* GPI Atlantic, Nova Scotia, Canada [Available online at www.gpiatlantic.org/pdf/fisheries/fisheries_2008.pdf].

Charles, A., M. Winterbottom, L. Wilson and S. Mills. (2007). *Social, Economic and Cultural Overview and Assessment for Ocean Management.* Report prepared for Department of Fisheries and Oceans, Government of Canada.

Commission for the Conservation of Antarctic Marine Living Resources 1982. *Text of the Convention on the Conservation of Antarctic Marine Living Resources.* Available on-line at: <http://www.ccamlr.org/pu/e/e_pubs/bd/pt1.pdf>.

Costanza, R., F. Andrade, P. Antunes, M. van den Belt, D. Boersma, D. F. Boesch, F. Catarino, S. Hanna, K. Limburg, B. Low, M. Molitor, J. G. Pereira, S. Rayner, R. Santos, J. Wilson and M. Young. 1998. Principles for sustainable governance of the oceans. *Science,* **281,** 198–199.

De Young, C. and A. Charles. 2009. Creating incentives for the ecosystem approach to fisheries management: A portfolio of approaches. In: *The Ecosystem Approach to Fisheries* (G. Bianchi and H. R. Skjoldal, editors). CAB International.

De Young, C., A. Charles and A. Hjort. 2008. *Human Dimensions of the Ecosystem Approach to Fisheries: An Overview of Context, Concepts, Tools and Methods.* Fisheries Tech. Paper 489. FAO, Rome.

Douvere, F. 2008. The Importance of Marine Spatial Planning in Advancing Ecosystem-Based Sea Use Management. *Marine Policy,* **32,** 762–771.

Ehler, C. and F. Douvere. 2009. *Marine Spatial Planning: A step-by-step approach toward ecosystem-based management.* Intergovernmental Oceanographic Commission and Man and the Biosphere Programme. IOC Manual and Guides No. 53, ICAM Dossier No. 6. Paris: UNESCO.

Fanning, L., R. Mahon and P. McConney. 2011. *Towards Marine Ecosystem-Based Management in the Wider Caribbean.* Amsterdam University Press. Amsterdam, Netherlands.

FAO 2003. *The Ecosystem Approach to Fisheries.* FAO Technical Guidelines for Responsible Fisheries No. 4 (Suppl. 2). Food and Agriculture Organization, Rome, Italy, 112 p.

FAO 2009. *Fisheries Management. 2. The Ecosystem Approach to Fisheries. 2.2 Human Dimensions of the Ecosystem Approach to Fisheries.* FAO Technical Guidelines for Responsible Fisheries, No. 4, Supplement 2, Add. 2. Food and Agriculture Organisation of the United Nations, Rome.

Garcia, S. M. and A. Charles. 2007. Fishery systems and linkages: From clockwork to soft watches". *ICES Journal of Marine Science,* **64,** 580–587.

Garcia, S. M. and A. Charles. 2008. Fishery systems and linkages: Implications for science and governance. *Ocean and Coastal Management,* **51,** 505–527.

Garcia, S. M., A. Zerbi, C. Aliaume, T. Do Chi and G. Lasserre. 2003. *The Ecosystem Approach to Fisheries—Issues, Terminology, Principles, Institutional Foundations, Implementation and Outlook.* FAO Fisheries Technical Paper. No. 443. Rome, FAO. 71pp.

Gutiérrez, N. L., R. Hilborn and O. Defeo. 2011. Leadership, social capital and incentives promote successful fisheries. *Nature* **470:**386–389.

Holland, D., J. N. Sanchirico, R. Johnston and D. Joglekar. 2010. *Economic Analysis for Ecosystem-Based Management: Applications to Marine and Coastal Environments.* Resources for the Future Press. Washington, D.C.

Kliskey, A., L. Alessa and B. Barr. 2009. Integrating Local and Traditional Ecological Knowledge, pp. 145–161. In: K. McLeod and H. Leslie, eds. *Ecosystem-Based Management for the Oceans.* Washington, DC: Island Press. 368p.

Leslie, H. M. and K. L. McLeod. 2007. Confronting the challenges of implementing marine ecosystem-based management. *Frontiers in Ecology and the Environment* **5:**540–548.

Mahon, R., L. Fanning, P. McConney and R. Pollnac. 2010. Governance characteristics of Large Marine Ecosystems, *Marine Policy,* **34,** 919–927.

Makino, M., H. Matsuda and Y. Sakurai. 2009. Expanding fisheries co-management to ecosystem-based management: A case in the Shiretoko World Natural Heritage Area, Japan. *Marine Policy,* **33,** 207–214.

McCay, B. J. 1981. Optimal foragers or political actors? Ecological analyses of a New Jersey fishery. *American Ethnologist,* **8,** 356–382.

McConney, P. and S. Salas. 2011. Why incorporate social considerations into marine EBM? pp. 99–109. In: L. Fanning, R. Mahon and P. McConney (eds). *Towards Marine Ecosystem-Based Management in the Wider Caribbean.* Amsterdam University Press. Amsterdam, Netherlands.

McLeod, K. and H. Leslie (editors) 2009. *Ecosystem-Based Management for the Oceans.* Washington, DC: Island Press. 368p.

McLeod, K. L., J. Lubchenco, S. R. Palumbi and A. A. Rosenberg. 2005. *Scientific consensus statement on marine ecosystem-based management.* Communication Partnership for Science and the Sea (COMPASS).

Millennium Ecosystem Assessment. 2005a. *Ecosystems and Human Well-being: A Framework for Assessment.* URL:<http://www.millenniumassessment.org/en/Framework.aspx>.

Millennium Ecosystem Assessment. 2005b. *Ecosystems and Human Well-being: Synthesis.* URL:<http://www.millenniumassessment.org/documents/document.356.aspx.pdf>.

Nauen, C. E. 2006. Implementing the WSSD decision of restoring marine ecosystems by 2015—scientific information support in the public domain. *Marine Policy,* 30, 455–461.

Pollnac, R. B., S. Abbott-Jamieson, C. Smith, M. L. Miller, P. M. Clay and B. Oles. 2006. Toward a model for fisheries social impact assessment. *Marine Fisheries Review,* **68,** 1–18.

Pollnac, R., P. Christie, J. E. Cinner, T. Dalton, T. Daw, G. E. Forrester, N. A. J. Graham and T. R. McClanahan. 2010. Marine reserves as linked social–ecological systems. *Proc. Natl. Acad. Sci.* **107,** 18251–18255.

Sanchirico, J. E., M. D. Smith and D. W. Lipton. 2008. An empirical approach to ecosystem-based fishery management. *Ecological Economics* **64,** 586–596.

Schramm, H. L. and W. A. Hubert. 1996. Ecosystem management: Implications for fisheries management. *Fisheries,* **21,** 6–11.

Schuhmann, P. W., J. C. Seijo and J. Casey. 2011. Economic considerations for marine EBM in the Caribbean. pp. 111–121. In: L. Fanning, R. Mahon and P. McConney (eds). *Towards Marine Ecosystem-Based Management in the Wider Caribbean.* Amsterdam University Press. Amsterdam, Netherlands.

Secretariat of the Pacific Community 2010. *A Community-Based Ecosystem Approach to Fisheries Management: Guidelines for Pacific Island Countries.* Secretariat of the Pacific Community, Noumea Cedex, New Caledonia.

Sherman, K. and G. Hempel (Editors) 2008. *The UNEP Large Marine Ecosystem Report: A Perspective on Changing Conditions in LMEs of the World's Regional Seas.* UNEP Regional Seas Report and Studies No. 182. United Nations Environment Programme. Nairobi, Kenya.

Sherrouse, B. C., J. M. Clement and D. J. Semmens. 2011. A GIS application for assessing, mapping, and quantifying the social values of ecosystem services. *Applied Geography* 31:748–760. [See also: Sherrouse, B. C. 2012. *Social Values for Ecosystem Services.* Ecosystem-Based Management Tools Network. http://ebmtoolsdatabase.org/tool/social-values-ecosystem-services-solves.]

Smith, A. D. M., E. J. Fulton, A. J. Hobday, D. C. Smith and P. Shoulder. 2007. Scientific tools to support the practical implementation of ecosystem-based fisheries management. *Journal of Marine Science,* **64,** 633–639.

Smith, C. and P. M. Clay. 2010. Measuring Subjective and Objective Well-being: Analyses from five commercial fisheries. *Human Organization,* **69,** 158–168.

UNEP (United Nations Environment Programme) 2009. Rio Declaration on Environment and Development, 1992 reaffirming the Declaration of the UN Conference on the Human Environment, adopted at Stockholm, 1972. Available on-line at: <www.unep.org>.

UNEP (United Nations Environment Programme) 2011. *Taking Steps toward Marine and Coastal Ecosystem-Based Management—An Introductory Guide.* (Authors: T. Agardy, J. Davis, K. Sherwood, O. Vestergaard) UNEP Regional Seas Reports and Studies No. 189, 67p.

Volovik, Y. S. 2010. *Agulhas and Somali Current Large Marine Ecosystems Project (ASCLME). Mid-Term Evaluation Report.* ATLAS ID 47255 PIMS-2205.

Chapter 4. Regulatory and Governance Frameworks

KEVERN COCHRANE

Rhodes University, South Africa

GABRIELLA BIANCHI

Food and Agriculture Organization of the United Nations

WARRICK FLETCHER

Department of Fisheries, Western Australia

DAVID FLUHARTY

University of Washington

ROBIN MAHON

University of the West Indies, Barbados

OLE ARVE MISUND

Institute of Marine Research, Norway

Contents

1. Introduction

From the human perspective, natural resources have always been seen as utilities to satisfy human needs and natural resources management aims at maximizing the benefits that we can derive from them. The recognition that greater benefits could be obtained in the long run through judicious use of aquatic resources developed in some locations long before the technological era started. Traditional tenure systems such as those that still exist in some rivers in Europe and in Pacific Island

The Sea, Volume 16, edited by Michael J. Fogarty and James J. McCarthy
ISBN 978-0-674-07270-1

States are good examples of these age-old systems (Johannes, 1978; Lotze et al., this volume). These systems were often based upon a broad knowledge of the interrelationships among resources and their environment. In one form or another, awareness of the need for "ecosystem-based management" has existed for a long time.

These traditional systems developed where there was a close contact between the user and the natural system being exploited and the primary resources were generally confined to the local system. In the case of fisheries, the development of technology and the transformation of fishing into an organized and sometimes large-scale industry operating in larger oceanic areas weakened this intimate contact. The enormous strides in technology, a shift in incentives from subsistence to profit, and the difficulty for the users in the wider more open systems to define limits, gain the necessary ecological knowledge and monitor use patterns have been important factors underlying widespread resource overexploitation (e.g. Ostrom, 2009; Caddy and Cochrane, 2001).

From the end of the 19th Century, the number of fishing units and their efficiency grew rapidly as did an understanding of the limits to fish production and the need to constrain fishing mortality to sustainable levels. This improved understanding did not always translate into new, sustainable practices, however. In marine fisheries in particular, the drive for sustainable use focused on the development of single-species, population dynamics models to advise managers how they could maximize the long term yield from fish stocks (e.g. Caddy, 1999). Modern, science-based fisheries management made some progress under this single-species approach (e.g. Mace, 2004; Worm et al., 2009), but widespread and sometimes high profile failures in fisheries resulted in the major resource collapses of the second half of the 20th Century (e.g. California sardine, Canadian cod, Atlanto-Scandian herring and Peruvian anchoveta),[1] and demonstrated the limitations of fisheries management as commonly practised at that time (see e.g. Pitcher et al., 2009).

One of the several limitations recognised in fisheries management in the second half of last century was a common reliance on single-species approaches which considered only the impacts of fisheries on the abundance and productivity of harvested stocks and did not take into account broader ecosystem, including environmental, influences and interactions. As a result, and, in line with the adoption of the principles of sustainable development, an ecosystem approach to fisheries management (EAF) was proposed. EAF is an approach which recognises the underlying ecological and human societal realities of fisheries management and living natural resource use in general, and also aims to incorporate best practices in resource management in general (Garcia and Cochrane, 2005). Similar lessons were being learned in other sectors too, for example within the Convention on Biological Diversity,[2] leading to the development of Ecosystem-Based Management (EBM) which considers, in an integrated manner, the ecological, social and economic objectives associated with all activities undertaken or influencing a specific region.

At the ecological end of this continuum, recognition of the need for EBM has

[1] http://www.irgc.org/IMG/pdf/Fisheries_Depletion_full_case_study_web.pdf
[2] See e.g. http://www.cbd.int/ecosystem/background.shtml Accessed 27 June 2012

been prompted by increasing knowledge and awareness of the cumulative impacts of all human activities on the ecosystem. These can often have implications for the structure and the productivity of the ecosystem thereby impacting on a range of goods and services beyond the specific sector having the impact. From a societal perspective, there is increasing recognition of a wide range of often conflicting societal interests that need to be taken into account in use of marine resources, habitats and ecosystems, as well as the need for these stakeholder groups to be an integral part of the management system (Jentoft and McCay, 1995; Berkes, 2009; Charles, this volume).

EBM is a multi-sectoral approach to management. It does not replace sectoral management but integrates and completes it. An ecosystem approach at the sectoral level is a necessary but not sufficient condition for EBM. The sectoral components of EBM—for example, an ecosystem approach to fisheries, to coastal zone development, or to offshore mining, oil and gas extraction—will obviously vary from ecosystem to ecosystem according to the human activities taking place or being considered there. This chapter examines the role of regulatory and governance frameworks in implementation of EBM. Utilising the experiences of five case studies, it considers the existing frameworks and how they have helped or hindered progress. It then considers how these frameworks may need to be adjusted in order to assist with the broader adoption of EBM. The chapter gives particular attention to the fisheries sector, where ecosystem approaches have become a high priority within the last decade. While some authors, for example Christie et al. 2007, argue that there are differences between ecosystem-based fish-eries management (EBFM) and ecosystem approaches to fisheries (EAF), the authors of this chapter consider the overall goals and approach to be essentially the same and the two terms are treated as synonyms.

2. The Global Legal Framework for EBM

As the theory and practice of sustainable use of oceans have developed, the international legal framework has also been evolving. The UN Conference on the Human Environment in 1972[3] laid the basis for the concept of ecologically sustainable development while the 1987 World Commission on Environment and Development[4] presented the concept of sustainable development as one "which meets the needs of the present without compromising the ability of future generations to meet their own needs".

The current legal framework related to EBM consists of both binding law and 'soft law'. Soft law refers to the non-binding guidelines, standards, codes of practice and similar agreements which have been an important feature of international policy. While weaker than enforceable instruments, soft law is considered to play an important role in facilitating and encouraging implementation of EBM. Countries are often more likely to come to agreement on complex and challenging

[3] http://www.unep.org/Documents.Multilingual/Default.asp?documentid=97. Accessed on 13 November 2011.

[4] http://www.un.org/documents/ga/res/42/ares42–187.htm. Accessed on 13 November 2011.

issues through soft law agreements than if they were negotiating a binding agreement. The FAO Code of Conduct for Responsible Fisheries gives an example of this (FAO, 1995). Another benefit of soft law is that even its development can facilitate and encourage a common understanding and increase the priority for implementing positive changes amongst participants. The countries involved generally participate in good faith which is often continued in their efforts to implement the agreement. Soft law can lead to the subsequent development of 'hard law' at national, regional and global levels (FAO, 2006).

The central binding instrument relevant for EBM within marine environments is the United Nations Convention on the Law of the Sea of 10 December 1982[5] (the LOS Convention; Burke, 1994). The LOS Convention has been referred to as the 'constitution for the oceans' (Kuemlangan, 2009) and provides the legal basis for any other legal instruments addressing the use of the oceans and seas. Issues addressed within the LOS Convention include: Part V Exclusive Economic Zone (EEZ) including Article 61 Conservation of the living resources and Article 62 Utilization of the living resources; Part VII High Seas including Conservation and management of the living resources of the high seas (Section 2); and Part XII Protection and Preservation of the Marine Environment.

While agreement on the text of the LOS Convention predated global recognition of the full importance of EBM, the Convention still incorporates many important aspects. For example, in relation to fisheries, it recognises the importance of reconciling conservation targets and human well-being (e.g. Articles 61.3 and 61.4), the need to take into account interdependence of stocks, effects of fishing on species associated with or dependent on harvested species, and the importance of cooperation and communication between States fishing for particular species through competent international organizations (Articles 61.3–61.5).

Another binding instrument related to EBM is the 1992 Convention on Biological Diversity (CBD) which entered into force in 1993. It is built around the three pillars of conservation of biodiversity, the sustainable use of its components and the fair and equitable sharing of benefits from the use of biodiversity. The primary principle (Article 3, CBD 1992)[6] of the CBD is that:

"States have, in accordance with the Charter of the United Nations and the principles of international law, the sovereign right to exploit their own resources pursuant to their own environmental policies, and the responsibility to ensure that activities within their jurisdiction or control do not cause damage to the environment of other States or of areas beyond the limits of national jurisdiction."

Article 8 of the Convention sets out the obligations of Contracting Parties "as far as possible and as appropriate". These obligations include, for example, where necessary, establishment of a system of protected areas; the need for regulation or management of biological resources important for the conservation of biological diversity both inside and outside protected areas; promoting the protection of ecosystems, natural habitats and maintaining viable species populations in natural

[5] Oceans and Law of the Sea. United Nations
http://www.un.org/Depts/los/convention_agreements/convention_overview_convention.htm. Accessed on 12 June 2012.
[6] http://www.cbd.int/convention/text/. Accessed on 17 November 2011.

surroundings; and restoration of degraded ecosystems. Article 11 deals with economic and social incentives for the conservation and sustainable use of biological diversity. In 1995, decision II/10 of the 2nd Conference of the Parties to CBD[7] addressed the conservation and sustainable use of marine and coastal biological diversity and "Encourages the use of integrated marine and coastal area management as the most suitable framework for addressing human impacts on marine and coastal biological diversity and for promoting conservation and sustainable use of this biodiversity."

In terms of soft law instruments relevant for EBM, Chapter 17 of Agenda 21 of the 'Rio Conference'(UNCED, 1992) is commonly acknowledged as launching modern interest in ecosystem approaches to oceans and coastal areas (e.g. Cochrane et al., 2004). This includes the reminder that the marine environment "forms an integrated whole that is an essential component of the global life-support system and a positive asset that presents opportunities for sustainable development." Another example of soft law, the FAO Code of Conduct for Responsible Fisheries (the Code), was developed shortly after the 1992 Rio Conference and while the Code does not explicitly refer to the ecosystem approach or EBM, it includes their major features and requirements (FAO, 1995). The World Summit on Sustainable Development (WSSD) Plan of Implementation, was adopted 10 years after Agenda 21 and 7 years after the Code. It includes several targets that are directly related to EBM. In particular, the Plan includes agreement by signatory countries to "Encourage the application by 2010 of the ecosystem approach".[8]

Also contributing to EBM, the International Maritime Organization (IMO) allows for the designation of a Particularly Sensitive Sea Area (PSSA) to control potential damage from international shipping by identifying and implementing suitable protective measures. In order to be designated as a PSSA, an area must meet specified ecological, socio-economic or scientific criteria, it must be vulnerable to damage by international shipping, and the IMO must be able to adopt measures to protect the area from the threat posed by international shipping. The International Convention for the Prevention of Pollution from Ships (MARPOL) also allows for the designation of "special areas" in the ocean which are defined as "a sea area where for recognized technical reasons in relation to its oceanographical and ecological conditions and to the particular character of its traffic, the adoption of special mandatory methods for the prevention of sea pollution by oil, noxious liquid substances, or garbage, as applicable, is required."[9]

Other examples of binding and non-binding instruments relevant to EBM are shown in Table 4.1.

In addition to global instruments there are a large number of regional treaties, protocols, action plans and other agreements and institutions that are relevant to EBM (see, e.g., FAO, 2006; Kuemlangan, 2009; Meltzer, 2009). These include the UNEP Regional Seas Commissions and a number of Regional Fisheries Manage-

[7] Convention on Biological Diversity. http://www.cbd.int/marine/pow.shtml. Accessed on 21 September 2011.

[8] Paragraph 29d; http://www.johannesburgsummit.org/html/documents/summit_docs/2309_planfinal.htm

[9] NOAA Office of the General Council. http://www.gc.noaa.gov/gcil_mpa-sad.html. Accessed on 12 June 2012.

TABLE 4.1.
Selected binding and non-binding agreements relevant to EBM (after Kuemlangan, 2009).

BINDING INSTRUMENTS

i) *General and Sectors other than Fisheries*
   (d) Convention on Biological Diversity
   (e) United Nations Framework Convention on Climate Change
   (f) Agreement relating to the implementation of Part XI of the United Nations Convention on the Law of the Sea of 10 December 1982
   (g) Convention on the Conservation of Migratory Species of Wild Animals
   (h) Convention concerning the Protection of the World Cultural and Natural Heritage
   (i) Convention on the Prevention of Marine Pollution by Dumping Wastes and Other Matter, 1972
   (j) International Convention for the Prevention of Pollution from Ships, 1973, as modified by the Protocol of 1978 relating thereto
   (k) International Convention for the Control and Management of Ships' Ballast Water and Sediments

ii) *Fisheries specific*
   (a) United Nations Convention on the Law of the Sea of 10 December 1982
   (b) Agreement to Promote Compliance with International Conservation and Management Measures by Fishing Vessels on the High Seas
   (c) Agreement for the Implementation of the Provisions of the United Nations Convention on the Law of the Sea of 10 December 1982 relating to the Conservation and Management of Straddling Fish Stocks and Highly Migratory Fish Stocks

NON-BINDING INSTRUMENTS

i) *General*
   (d) United Nations Conference on Environmental Development 1992:
      • Rio Declaration on Environment and Development, 1992
      • Chapter 17 of Agenda 21: Protection of the Oceans, All Kinds of Seas, Including Enclosed and Semi-Enclosed Seas, and Coastal Areas and the Protection, Rational Use and Development of Their Living Resources
   (e) United Nations Millennium Declaration, UNGA Resolution 55/2 (A/55/L.2)
   (f) World Summit on Sustainable Development (WSSD) Declaration (Johannesburg Declaration)
      • WSSD Plan of Implementation (Johannesburg Plan of Implementation)
   (g) Global Conference on the Sustainable Development of Small Island Developing States
   (h) Programme of Action for the Sustainable Development of Small Islands
   (i) United Nations Conference on the Human Environment.

ii) *Fisheries specific*
   (a) 1995 Code of Conduct for Responsible Fisheries, including the four international plans of action (IPOAs) associated with it i.e.
      • IPOA for Reducing Incidental Catch of Seabirds in Longline Fisheries
      • IPOA for the Management of Fishing Capacity
      • IPOA for the Conservation and Management of Sharks
      • IPOA to Prevent, Deter and Eliminate Illegal, Unreported and Unregulated Fishing
   (b) The 2001 Reykjavik Declaration on Responsible Fisheries in the Marine Ecosystem.[1]
   (c) The 2005 Rome Declaration on Illegal, Unreported and Unregulated Fishing

1. fao.org/fi/DOCUMENT/reykjavik/y2198t00_dec.pdf

ment Organizations. Two examples of regional bodies with an explicit EBM mandate are the Commission for the Conservation of Antarctic Marine Living Resources (CCAMLR), which focuses on EAF or EBFM (see Gaichas et al., this volume), and the very new Benguela Current Commission which is described in Section 6.4 of this paper and is multi-sectoral in design.

3. Institutional Implications of Scaling-up from Sectoral Management to EBM

3.1 Institutional structure

In addition to the various hard and soft law global requirements outlined above, the regulatory and governance frameworks at national and international levels should take into account the social, economic, cultural, ecological, political and legal characteristics of the countries concerned. The details and specific emphases will therefore vary considerably among countries so it is impossible to provide meaningful blueprints of appropriate frameworks but there are some general patterns in what is likely to be required (Schechter et al., 2008).

States are the basic institution, conferring legitimacy on the other institutions involved in ocean management and use (Feral, 2009). These include the public institutions which act on behalf of the State to serve the interests of the public as a whole, and non-public institutions which can serve the economic or social interests of specific groups. Institutions can also be categorised into those related to the use itself, supporting activities and enterprises, representative groups, institutions for negotiation and co-management, etc. (Feral, 2009; see also Table 4.2).

The institutional structure used to implement EBM must reflect the overall purpose of taking an ecosystem approach. Both the CBD and the FAO descriptions highlight the need to use an integrated approach. In most countries, however, human activities are categorised, governed and regulated within designated sectors such as fisheries, tourism, shipping, coastal zone development, mining, and oil exploration and extraction (FAO, 2006). Having underlying sectoral approaches is appropriate and necessary as sectoral management must remain a core element of implementing EBM, but these sectoral approaches alone are not sufficient (Fletcher, 2006; McLeod and Leslie, 2009). EBM is needed to create the linkages, fill the gaps and reconcile the conflicting objectives that arise from a series of independent sectoral approaches to management.

The institutional design and implementation of EBM is best underpinned by having supporting and empowering laws and regulations, although some of the case studies in this chapter demonstrate that progress can still be made through soft laws. In order to enable EBM, the laws and regulations should at least ensure that there are no legal obstacles in the mandates of public bodies to enable cross-sectoral decision-making and that there are laws to prevent unrestricted development of sectoral activities operating in the oceans (FAO, 2006). At the same time some multi-sectoral framework or system is required to facilitate and enforce cross-sectoral collaboration and planning to enable implementation of EBM. In combination, these legal and institutional frameworks need to facilitate development and agreement of the full set of mutually compatible management objectives for the region under consideration and the complimentary management measures and actions to achieve those objectives (Rosenberg et al., 2009).

3.2 Institutional challenges for EBM

The transition from independent sectoral approaches to EBM requires changes in the institutional structure and processes in national and international governance

TABLE 4.2.
Functional typology of EBM institutions
(after Feral, 2009)

Function
1. Resources harvesting
a. economic functions
b. technical function of harvesting
2. Servicing enterprises
e.g. boat building, gear manufacturing, processing, trading
3. Representation of interest groups
professional and non-professional
4. Negotiation and co-management
5. Organization, control and professional discipline
a. regulating sectoral activities—setting the rules
b. regulating access to the resources
c. control (enforcement)
d. jurisdiction—taking action against infringements
e. arbitration—resolution of conflicts and disputes

regimes for oceans. The present institutional system in any particular region is a result of its history and that of the sectors within it, as well as the prevailing administrative customs. Change may therefore be difficult and studies on radical changes in aquatic-based, social-ecological systems have demonstrated the importance of using a 'window of opportunity' for change (Gelcich et al., 2010). Institutions, like most human cultures, are often resistant to change and there are clear parallels with the adaptive cycles of resilience theory. Resilience theory, in the context of ecosystems, considers the dynamics of stability, resilience and change in the structure and function of social-ecological systems. Holling (2001) describes adaptive cycles as alternating between longer periods of slow accumulation of ecological, economic, social, or cultural capital with more rapid periods of breakdown and reorganization. Major changes may therefore result from a collapse of the existing policies but the existing capital to fuel the change, including existing social structures, local and conventional scientific knowledge, and available skills are all important in the process (Gelcich et al., 2010).

Primary challenges in managing the transition from sector-based to ecosystem-based management are: i) to bring about the transformation without the need of a major crisis and ii) to generate reorganization through managed rearrangement of the existing capital rather than through uncontrolled breakdown. These could be facilitated by application of the three-phase model described by Gelcich et al. (2010) as facilitating important reforms in the governance of Chilean coastal resources: preparation, transition, and building resilience of the new regime. Existing social mechanisms are crucial elements for successful transformation and the Chilean example highlights the importance of making use of existing social networks amongst the stakeholders for dissemination of new knowledge.

In a transition to EBM a major related challenge is dealing with the increased complexity that results from incorporating the additional interactions between different scales (spatial and temporal) and diverse stakeholders that were often ignored under sectorally divided management. In cases of high complexity, progress is often significantly impacted when there is a high level of turnover in the representatives from participating institutions or stakeholders. This problem is even more magnified when multiple sectors are involved. This often results in revisiting previously discussed issues and even agreed resolutions which can lead to a high level of frustration.

Within fisheries, explicit and implicit conflicts in objectives are one of the major challenges to sustainable use (Sumaila, 2010) and these will be escalated when multiple sectors are being considered. Scaling up to EBM will almost always require recognising and reconciling additional layers of conflicts, between different sectors, that may have been ignored or managed only by crisis management in the past. This will be a major challenge for EBM. By way of example, Nielsen et al. (1997) described a process of 19 meetings of various planning committees, taking up more than 2,500 person-hours in one year in the development of the Lake Winnebago Comprehensive Management Plan. This was followed by a 'more expansive public participation process' the human costs of which are not quantified. Nevertheless, the Plan was finalized and adopted in three years and was reported by the authors as being successfully implemented because of "the deep ownership held by the public in the Winnebago plan as 'their plan'". National and regional authorities need to be prepared to invest time and resources in consultation and joint planning with all the legitimate stakeholders in order to achieve the benefits of EBM.

3.3 The critical role of user rights

Open access to any resource is likely to lead to undesirable conservation and economic outcomes and it is widely recognised that some form of regulation of access to natural resources is essential for their responsible and sustainable use (Charles, 2009; World Bank, 2009). Different forms of use rights have been applied in many fisheries for hundreds of years (Charles, 2009; Lotze et al., this volume) but inadequate or inappropriate systems of user rights remain a primary factor contributing to unsustainable fishing (Cochrane and Doulman, 2005).

The role and properties of user rights for governance of EBM can be summarised as (from Kearney et al., 2007):

1. the legal environment needs to specify the powers and responsibilities of the state and stakeholder groups respectively;
2. the legitimate and relevant stakeholder institutions need to be given legal recognition;
3. responsibilities and rights must be clearly linked to specific geographical areas and resources;
4. secure rights need to be granted based on explicit and legitimate criteria, for example, dependence on the resource or historical access.

Charles (2009) argues that appropriate systems of use rights are essential for sustainable and responsible fisheries but cautions that use rights can be highly controversial, in particular because they involve allowing some individuals access to a resource and excluding others. The choices made in implementation of use rights, such as the nature of the rights and their duration, who will and will not receive rights and whether or not they are transferable will have a significant influence on the economic and social characteristics and performance of the fishery and resource sustainability. As with any fisheries management decisions, these influences are likely to have knock-on effects for the ecosystem as a whole and other users. Charles (2009) therefore advises that introducing or revising a system of use rights must be approached with care and that the system implemented needs to be appropriate to the social-ecological characteristics of the fishery.

4. Progress in Implementation

While there is limited reliable information on progress at the global level, there is little doubt that implementation is lagging considerably behind global policies (e.g. Murawski, 2007; Rosenberg et al., 2009). Within the fisheries sector, Pitcher and his colleagues (2009) evaluated progress in implementation of ecosystem-based management in fisheries of 33 countries which together accounted for 90% of the global annual fisheries catch. They did not identify any countries that had made good progress and only four countries were scored as being adequate. Over half of the 33 countries were scored as failing. The top four countries, in order of their scores on Overall EBFM Performance, were the USA, Norway, New Zealand and Canada. The only developing country in the top ten was South Africa but others such as Malaysia, Peru, Chile and Mexico were ranked within the top fifteen countries and higher than some developed countries. In the next section of this chapter, progress in implementation in five case studies is examined in more detail.

Looking at the fisheries sector alone, there has been some progress in rebuilding of fish stocks (Worm et al., 2009) and application of an ecosystem approach but this has been limited. Many countries and regional fisheries bodies are implementing aspects of an ecosystem approach but the authors do not know of a single maritime state that would claim to have fully implemented EBFM in its oceans. There is still a long way to go to achieve the World Summit on Sustainable Development target for implementation of an ecosystem approach. Similarly, while there are some encouraging examples of well-managed and healthy fish stocks and fisheries, it is extremely unlikely that the WSSD target of "Maintaining or restoring stocks to levels that can produce the maximum sustainable yield where possible and not later than 2015" will be fully realized. Similarly, the Millennium Development Report (UN DESA, 2010) indicates that "The world has missed the 2010 target for biodiversity conservation, with potentially grave consequences."

This slow progress is forcing a shift in targets. For example, in its tenth meeting, the Conference of the Parties (CoP10) of the Convention on Biological Diversity (CBD, 2011) concluded that its own 2010 biodiversity target had not been met in full. In response, CoP10 adopted "Decision X/2—Strategic Plan for Biodiversity 2011–2020", including the Annex "Strategic plan for biodiversity 2011–2020 and the Aichi Biodiversity Targets" which includes new targets for, for example, keeping the impacts of use of natural resources within safe ecological limits (Target 4), sustainable use of all fish, invertebrate stocks and aquatic plants in accordance with an ecosystem approach (Target 6), and restoration and safeguarding of ecosystems that provide essential services (Target 14).

At the time of writing this paper, the United Nations Conference on Sustainable

Development (UNCSD) commonly referred to as Rio+20 had just been completed.[10] The Conference marked the 20th anniversary of the 1992 United Nations Conference on Environment and Development (UNCED) and the 10th anniversary of the 2002 WSSD. The meeting agreed that in the twenty years since the 1992 Rio Conference progress towards sustainable development, and poverty eradication, had been 'uneven.'[11] Contrary to the hopes and expectations of many, the political agreement did not include major new ideas or initiatives, but it did reaffirm the commitment of the participating countries to the Rio Principles and related action plans. It is to be hoped that greater progress will be made after Rio+20 than has been made up until now.

5. EBM in Practice: Five Case Studies

5.1 Australia

5.1a Introduction

Following the adoption by the Australian Government of the World Commission on Environment and Development view of sustainable development (WCED, 1987), in 1992 the leaders of all Federal, State and Territory governments endorsed the National Strategy for Ecologically Sustainable Development (ESD; Commonwealth of Australia, 1992). Successive levels of governments in Australia have remained committed to the sustainable use of resources with a series of ESD-based initiatives having been underway. These include the adoption of ESD principles into the management of individual fisheries (e.g. Fletcher et al., 2002), the development of a national system of marine protected areas (e.g. ANZECC, 1998), Commonwealth's Ocean Policy (NOO, 2003) and state level marine area planning (e.g. DEH S. A., 2004) and, more recently, bioregional-level fisheries planning (Fletcher et al., 2010). These initiatives were, however, not coordinated which often resulted in duplication of activities and sometimes even competing processes in a region. The clear challenge was therefore how to effectively manage the collective 'use' of marine, coastal and estuarine resources in order to achieve the most appropriate ESD-based outcomes for the entire community. The urgency has increased given the growing call by conservation groups for 'non-use areas' to cover significant portions of marine waters combined with an increasing loss of community support for the activity of fishing.

For the past decade, the Marine and Coastal Committee (MACC) had the responsibility to coordinate national policy development among all Australian governments and agencies with jurisdiction of marine and coastal areas. The MACC reported to the Natural Resource Management Ministerial Council (NRMMC) and included the heads of all federal, state and territory agencies from marine based portfolios, including fisheries, environment and transport. In 2008, a report to MACC noted the progressive incorporation of ESD-related approaches into marine policy in Australia with a number of such initiatives underway nationally.

[10] http://www.uncsd2012.org/rio20/. Accessed on 13 November 2011.

[11] The Future We Want", text of the agreement reached at the Rio+20 Conference. http://www.uncsd2012.org/content/documents/727The%20Future%20We%20Want%2019%20June%201230pm.pdf. Accessed on 27 June 2012.

The report concluded that to provide a more coordinated response to issues facing marine biodiversity, a multi-sectoral, Ecosystem-Based Management (EBM) approach should be adopted by Australia. This was considered a necessary contribution to achieving ESD in the marine environment because it would integrate management of all uses into a single, comprehensive and coordinated system. For further details, see Houde et al. (this volume).

With agreement on the main principles of EBM by the NRMMC in 2009, the MACC was instructed to develop a National EBM framework suitable for use within the Australian context. A MACC working group with representatives from different jurisdictions, portfolios and expertise in sectoral-level ecosystem approaches was established to generate the framework.

5.1b How was EBM defined and how does it fit with sectoral and ESD frameworks?

The key principles for EBM within the marine environment in Australia were that it must consider ecological, social, economic and cultural objectives using an adaptive approach to decision making that must recognise any uncertainty. The overall goal was to ensure that all marine industries are economically sustainable, meet societal needs, and that the individual or cumulative impacts of all ocean and relevant land-based activities do not generate unacceptable threats to marine ecosystem integrity or intergenerational equity (MACC, 2010).

It was recognised that there needed to be a common basis for EBM assessment and planning processes used by all institutions and jurisdictions. It was agreed, therefore, that the geographical basis for undertaking an EBM approach would be the meso-scale ecosystem units for the Australian continent developed through the Integrated Marine and Coastal Regionalisation of Australia process (CoA, 2006). This divided the marine environment of Australia into a series of 60 bioregions each representing a relatively distinct, high level ecosystem.

In addition to having a consistent and pragmatic unit for planning, the EBM process needed to integrate and add value to the sector-level based processes already underway. Given the previously outlined initiatives, many sectors already had environmental policies and procedures that operated in conjunction with economic and social policies consistent with ESD principles (Fletcher, 2008). For example, before they can export product, all commercial fisheries in Australia have to be assessed against a comprehensive set of guidelines consistent with the ecosystem approach (CoA, 2007). While each of these sectoral level approaches and frameworks are already contributing to achieving the Government's overall goal of ESD (Fletcher, 2006) it was recognised that there needed to be an EBM approach to integrate and link the significant work that had already occurred for the ecosystem-based management of the fisheries sector with that being proposed for the other sectors and users of the marine environment to improve consistency of policy outcomes, and minimize conflicts and wasteful actions. Each of the various management frameworks should form an integrated hierarchy within an overall ESD context, with each level providing the building blocks for the next level (Fletcher, 2006). EBFM and EBM are, therefore, just some of the strategies needed to enable the full implementation of ESD principles and objectives across all levels of government.

5.1c The National Ecosystem-Based Management approach to marine environments—"the eight steps approach"

The steps outlined below define the key elements of the draft National EBM framework which was developed by the MACC working group for use in developing EBM based plans (MACC, 2010). It is largely based on the EBFM framework (Fletcher et al., 2010) and involves the use of a step-wise, risk-based approach to identify the relevant issues and determine the appropriate priority and focus for any management response or actions for all fishing activities operating within a region. Limited knowledge of the ecosystems or any other component does not preclude initiating the planning process or taking early actions where there are clear measurable benefits. Once implementation of the EBM plan begins and a monitoring, evaluation and reporting program is in place, any additional ecosystem or other knowledge that is required to improve outcomes or additional actions can be identified. This was considered to be a pragmatic and practical method for implementing EBM. The eight steps are:

i. Establish a group with overall responsibility for implementing EBM.
ii. Define the scope, including the boundaries of the ecosystem, and establish the overall ecological, social and economic values.
iii. Agree on all relevant objectives for the ecosystem and each asset based on the values.
iv. Generate individual risk values and consolidate to asset level.
v. Prioritise assets across the ecosystem.
vi. Determine actions to meet the objectives of the governing body and establish a monitoring evaluation and reporting framework for the ecosystem and assets.
vii. Develop and implement an action plan.
viii. Regularly review outcomes, make necessary changes and communicate.

Communication with stakeholders is essential throughout the entire process in order to maintain engagement and ownership.

5.1d What progress has been made?

Significant progress has been made in implementing an ecosystem approach at the fishery sector level, including some recent progress in adopting the coordinated EBFM-based management for all activities of this sector at a bioregional level (Fletcher, 2008; Fletcher et al., 2010). In the other marine sectors, the regional marine planning initiative currently being undertaken by the Commonwealth (e.g. CoA, 2011) has largely been restricted to the establishment of series of marine protected areas. In recognition of the political need to minimise social and economic impacts within the final design, this planning included some level of consultation with fishing and other marine based sectors about potential impacts on their activities from the imposition of a series of proposed protected areas. The level of recognition and influence among sectors was, however, highly patchy, with areas important to the oil and gas sectors largely excluded from having any restrictions imposed whereas concerns about material impacts on fishing activities were often not heeded. Moreover, these planning activities were not part of an overall process to develop an integrated EBM plan for these regions that included clear, measurable, outcome-based objectives such as would be generated using the EBM framework outlined above.

The final development phase for the national EBM framework was to have the MACC endorse the completion and evaluation of a series of pilot studies to examine how effective EBM could be and what were the likely benefits and costs for government, stakeholders and the environmental assets involved. These case studies were to explicitly examine links between and among the sectors and ecosystem elements involved, including how an activity undertaken by one sector affected other sectors to accommodate these links when setting objectives and for decision-making. Unfortunately, in early 2011 a review of all national committees by the Council of Australian Governments resulted in both the MACC and the NRMMC being disbanded. A replacement national forum has yet to be established, which has resulted in cessation of nearly all national level EBM initiatives (at least in the short term).

There is still a high level of ecosystem approach activities underway within the fishery sector but only a few formal bi-lateral activities continue on EBM. Given that the fisheries sector generally has the most to lose from poor coordination of the management and access arrangements in the marine environment, the most likely path to the practical adoption of EBM in the near future will come from greater adoption of the multi-fishery, regional ecosystem-based fisheries management (EBFM) frameworks which are now operational in Western Australia (Fletcher et al. 2012). These could naturally evolve to EBM through time as they incorporate other sectoral issues but this will only happen where these are deemed to provide clear and direct benefits to all the sectors involved. It has been recognised that not all institutions and stakeholder groups may be as enthusiastic about the benefits from adopting a comprehensive framework with clear, measurable outcome-based objectives as the basis of integrated decision making. Such an approach could also be seen by some as relinquishing their level of control on planning activities and outcomes and therefore raising accountability and transparency.

5.2 Norway

5.2a Introduction

Norway's coastal and oceanic waters are biologically rich and among the most productive in the world. The oil and gas industry, developed exclusively in marine areas, is the country's largest industry and amounts to 27 per cent of the GDP. Fisheries, including fish farming and fish processing, is the second most important sector in the country as regards its contribution to GDP. The state of Norway's marine waters is generally good, but there is growing pressure from human activities such as aquaculture, extensive fishing, oil and gas production, shipping and tourism.

The legal regime for Norway's marine regions is based on the LOS Convention (1982). Other international treaties and instruments relevant to the protection of the marine environment and to which Norway has committed are the 1995 UN Fish Stocks Agreement, Convention on Biological Diversity, the 1972 London Dumping Convention (regulating discharge of waste from vessels), the 1973

MARPOL Convention (stipulating standards for vessels engaged in shipping) and the International Maritime Organization (IMO) agreements on reduction of negative impacts of marine transport on the environment. "Soft law" arrangements, such as Chapter 17 of Agenda 21 and the Plan of Implementation of the 2002 WSSD, have also gained importance (Hoel, 2009). Norway is also part of numerous regional and bilateral collaborations and treaties for the sustainable management of its marine waters.

Norway held the chairmanship of the North Sea Ministerial Conference in the period 1995 to 2002 during which an agreement was reached at the regional level for the application of the ecosystem approach. Through White Paper no. 12 (2001–2002) "A Clean and Rich Sea", the Norwegian Parliament recognized the need for an integrated and ecosystem-based management of the Norwegian marine regions (Barents Sea, Norwegian Sea and North Sea-Skagerrak) and the decision was taken that integrated management plans should be developed for these regions. So far two integrated management plans have been developed and approved by Parliament, one for the Barents Sea and one for the Norwegian Sea. Because of the more advanced stage of the Barents Sea management plan, this case study will focus on the developments in this region (see also Gaichas et al., this volume).

5.2b Evolution of the existing EBM governance regime in the Barents Sea

There is long tradition in Norway in sector-based ocean governance with well developed regulatory frameworks. Fisheries authorities are responsible for the management of fishery resources. For the petroleum sector, regulation of ongoing activities is with relevant ministries and directorates, and any decision to open new areas for petroleum activity can only be made by the Storting (Parliament).

The management plan for the Barents Sea is the first integrated plan developed, mainly because of the urgency to consider the expansion of oil and gas activities in the area (Winsnes and Skjoldal, 2008). It should be noted that the plan only covers the Norwegian part of the Barents Sea ecosystem as the remaining area is within the Russian EEZ or jurisdiction.

The Barents Sea is a shallow ocean (average depth of 230 meters), covering the area from the Norwegian Sea in the west to the coast of Novaja Semlja in the east, and from the coast of Norway and Russia and up to about 80° N. It holds important resources, including fisheries and large reservoirs of gas and oil. It also represents a critical habitat for the life cycle of many marine species.[12] The influx of Atlantic water flowing north-eastwards makes this area ice-free and accessible most of the year, distinguishing it from other Arctic regions. The challenge is to ensure that existing fisheries activities together with increasing maritime transport and new petroleum activities can sustainably co-exist.

This account is largely based on the documentation related to the formulation of the first Management Plan for the Barents Sea (2006), in particular the document "Integrated Management of the Marine Environment of the Barents Sea and the Sea Areas off the Lofoten Islands", Recommendation of 31 March 2006 by the Ministry of the Environment, approved in the Council of State the same day

[12] http://www.npolar.no/en/the-arctic/the-barents-sea/

(White paper from the Stoltenberg II Government),[13] and the first integrated management plan developed for Norwegian marine areas.[14]

The Government appointed a steering group, chaired by the Ministry of Environment and with representatives from relevant ministries, to lead and coordinate the planning work. So far responsibility for the marine environment had been split between several ministries. The Norwegian Polar Institute (*www.npolar.no*) which is a directorate under the Ministry of Environment was assigned the project leadership leading up to the Management Plan. The work included three main phases (Olsen et al., 2007).

1. Preparation of status reports on the environment and the resources, identification of valuable ecological areas, socio-economic aspects, economic activities, geographic scope of the plan and overall aims. This included a transparent stakeholder consultation process.
2. Four extensive government-funded Environmental Impact Assessments (EIAs) were carried out with a time frame extending to 2020, covering the impact of fisheries, shipping, hydrocarbon extraction, and external pressures (e.g. pollution) on the environment, resources, and local communities.
3. Further analyses of impacts were undertaken, particularly on the combined impacts of all human activities, which proved to be particularly challenging, and on identification of existing or potential conflicts between activities, definition of high level goals common to all sectors and identification of knowledge gaps. A set of operational indicators of ecosystem state was developed based on high-level management goals and to allow ecosystem monitoring.

Area-based management/spatial planning approaches are considered key tools for the implementation of the plan. Area-based management is used to resolve conflicts and protect identified ecologically valuable or vulnerable areas. The plan represents a synergy of previously separate management regimes: management of fisheries, shipping, and the hydrocarbon industry.

5.2c Current regulatory and governance frameworks

Based on the recommendations contained in the 2006 Management Plan, more knowledge should be gathered on this marine ecosystem as a basis for revision of the plan. As a result, a large seismic data collection campaign was carried out from 2007–2009 on the narrow shelf area off the Lofoten Islands under the leadership of the Norwegian Petroleum Directorate (www.npd.no), and a large seabed mapping program (the MAREANO program) (www.mareano.no) was established. These areas were targeted because of their importance for oil exploration and exploitation and fisheries.

A revised Management Plan for the Lofoten-Barents Sea area was presented as

[13] http://www.regjeringen.no/Upload/MD/Vedlegg/Svalbard%20og%20polaromraadene/Forvaltningsplan%20Barentshavet/PDF0080506_engelsk-TS.pdf

[14] http://www.regjeringen.no/en/dep/md/Selected-topics/hav—og-vannforvaltning/integrated-management-of-the-barents-sea.html?id=87148

a white paper to the Parliament in March 2011. Special attention is placed on resolving conflicts between petroleum activity and protecting fisheries interests, of which seismic activity related to oil exploration and exploitation has been the most controversial. Representatives from the Norwegian Petroleum Directorate and the Directorate of Fisheries worked together to revise laws and regulations covering seismic activities in order to reduce conflicts. These make extensive use of sector-specific area-based measures. A central concept of the plan is that it is based on science and takes a precautionary approach.

Implementation of the plan is ensured through the establishment of a system of multi-sectoral coordinating groups headed by a steering group in turn coordinated by the Ministry of Environment.

- The Management Forum group, led by the Norwegian Polar Institute.
- The Forum on Environmental Risk Management, led by the Norwegian Coastal Administration Coast Guard (www.kystverket.no).
- The Advisory Group on monitoring the Barents Sea, led by the Institute of Marine Research (www.imr.no).

While the plan contains specific regulations related to environmental targets relevant to all the sectors involved, the responsibility for management within each sector still rests with respective, sector-based management institutions. The 2006 Management Plan Chapter 11.2.3 states:

"The measures proposed in this management plan will call for closer cooperation between agencies, but no changes will be made in the formal organizational structure. The measures will also call for closer general coordination between research and management".

This reflects the essence of the governance implications of the new system, i.e. that sector-based management will continue being the main pillar of EBM, but take account of goals and overall decisions reflected in the management plan at the multi-sectoral level. Overall responsibility for implementation of the plan resides with the Ministry of the Environment. Final decisions on allocation of user rights across sectors, such as whether new oil exploitation activities should be allowed in a given area or not, are taken by the Government and the *Storting*.

5.2d Progress in policy implementation and major challenges

Collaboration and consensus building based on developing a common knowledge base have been instrumental for the positive results so far achieved, as well as inclusivity and transparency. Important challenges, however, still need to be addressed for its successful implementation and are briefly discussed below.

Conflicting interests

Of the various activities taking place in the Barents Sea, probably the oil and gas versus fisheries are those prone to generate the main conflicts if not effectively managed. Available information indicates that the area is rich in oil while for more than 1000 years it has been an important fishing ground. Consistent with the precautionary approach, the Norwegian government decided to gather more facts relevant to a possible future impact assessment of petroleum activity for this area.

However, this is being challenged by the oil and gas interests while conservation-oriented and fisheries interests argue that the area should be regulated as a marine protected area not to be opened for oil and gas exploration.

Transboundary collaboration for EBM of the Barents Sea

There has long been a tradition of good collaboration between Norway and Russia on fisheries and the marine environment (Haug et al., 2009). Fish stocks have been co-managed by a Joint Norwegian-Russian Fisheries Commission since 1976 (Jakobsen and Ozhigin, 2011). Recently, Russia and Norway have agreed on the borders between the two countries' EEZs, which have been an issue for more than 30 years, and this can only enhance ongoing collaboration. Nevertheless, key challenges to joint application of integrated management in the Barents Sea may include harmonization of environmental objectives, legislation and decision making at different geographic, sectoral and administrative scales (Denisov and Mikhalylichenko, 2009).

Assessing the performance of implementation of the management plan

The Management Plan and its implementation are science-based and the importance of knowledge for decision making is highlighted and considered key to successful implementation. Given the wide range of issues to be dealt with, and the need to provide knowledge that is as accurate as possible to optimize maximum sustainable use, applying EBM will imply higher research costs. Winsnes and Skjoldal (2008) also warn of the challenges of ensuring that scientific advice is transparent and independent. The present arrangement, with multi-agency groups responsible for producing management recommendations, may blur the interface between science and management potentially resulting in weakened accountability and less transparent decision-making.

However, during the six years since the first Lofoten-Barents Sea Management plan was adopted by the Storting, it has functioned quite well scientifically, administratively and politically. The scientific basis of the plan is recognized as generally solid, and the administrative regulations have been followed by the authorities. Maybe the most critical issues have been the political aspects, but the plan was developed by a liberal government coalition, and ratified by a social democratic lead coalition with a liberal and a left wing partner. The ability to withstand quite different political interests is possibly the best proof of a reasonably well functioning plan.

5.3 United States of America

5.3a Introduction

United States coastal and marine environments are extremely diverse. They range from the tropical ecosystems of the Western Pacific, Caribbean Sea and the Gulf of Mexico, include the temperate ecosystems of the West Coast including the Gulf of Alaska and the Eastern Seaboard and extend into the Arctic ecosystems of the Bering Sea, Aleutian Islands and the Arctic Ocean. Fishing is the most extensive and intensive human use of marine waters and substrate although there are regional and local differences depending on such things as the amount of oil and gas exploration and production taking place, the patterns of marine transportation, location of fishing, and other activities.

Implementing any comprehensive EBM plan is further complicated by the fact that the Submerged Lands Act, in general, gives individual states jurisdiction over fishery resources within three nautical miles of a coastal baseline (nine miles in the case of Texas and the Florida Gulf Coast), while the federal government has control over the area extending from the three nautical mile limit to the outward boundary of the Exclusive Economic Zone (200 nautical miles). There are significant interactions between the diverse environments and across federal, state and international borders (NOAA SAB, 2006).

At the federal level, the United States has designated eight regional ecosystems that roughly correspond to large scale ecosystems. The regional diversity requires a more area-specific and flexible approach. Finding the appropriate scale for application of EBM requires a coordinated integration of estuarine and coastal ecosystems most relevant to coastal communities with broader ecosystems that encompass fishing, oil and gas exploration, coastal development, marine conservation, marine transportation, and other related sectors that cover a wide geographical area. EBM in the United States is being explored in different ecosystems, multiple scales and under widely divergent circumstances (see also Houde et al.; Samhouri et al., this volume). It is a work in progress that emphasizes incremental empirical progress toward a theoretical goal of EBM (*in sensu* MacMahon, 1998).

5.3b History of the existing EBM governance regime

EBM has been a focus of terrestrial resource management for a considerable time in the United States (Boyce and Haney, 1996; Yaffe et al., 1996) for the coastal zone and for the Great Lakes which are considered in US ocean policy. Spencer Baird, the first Commissioner of the US Commission on Fish and Fisheries (1871), stated that our understanding of fish "... would not be complete without a thorough knowledge of their associates in the sea, especially of such as prey on them or constitute their food" (Hobart, 1995). However, ecosystem principles were not seriously considered in fisheries management until the mid-1990s when Congress initiated an investigation into the extent of application of ecosystem principles in fisheries management. This led to separate reports by the National Research Council (NRC) and the National Marine Fisheries Service (NMFS) assessing the status of ecosystem-based fishery management (EBFM) in the US. At the same time practitioners and academics were discussing the same theme in earnest (University of Alaska Sea Grant, 1999). Without a specific mandate for using EBFM it was not surprising that both reports found that implementation was mixed. Some regions like the North Pacific incorporated many of the elements of EBM (Witherell et al., 2000) while others were at earlier stages or just getting started. In order to encourage the development of EBM, NMFS (1999) recommended that fishery management regions develop Fisheries Ecosystem Plans (FEP) (NMFS, 1999). A pilot FEP was initiated for Chesapeake Bay to provide a case example.

Over time, substantial progress towards EBM was made in the waters adjacent to Alaska, for example, through a series of management measures. They included a cap on total removals from the ecosystem, a ban on forage fish harvests, conservative total allowable catch (TAC) rates, assessment of ecosystem considerations when setting TACs, counting of bycatch levels against TACs, designation of trawl closure areas, and a significant level of industry-funded observer coverage for catch accounting. These measures were taken largely independent of each other but the cumulative effect was to take ecosystem limits and dynamics into account. Thus, EBM adoption progressed as an evolutionary approach as opposed to a revolutionary approach (Pikitch et al., 2004; Sissenwine and Murawski, 2004; Fluharty, 2005).

As a result of broader concerns about management, two influential ocean commissions were formed: the Pew Oceans Commission (2003) and the US Commission on Ocean Policy (2004). Both advocated science-based management through EBM. In response, the National Oceanic and Atmospheric Administration (NOAA) convened a small group of experts to study and make recommendations on how it could implement EBM. This group recommended that NOAA develop Integrated Ecosystem Assessments (IEA) for each of its designated regions (NOAA SAB, 2006). These IEAs would be the scientific foundation for coastal and ocean management and a cornerstone to developing EBM (Levin et al., 2009). Another important development for EBM occurred in 2004 when President Bush created at Cabinet-level, the Commission on Ocean Policy by Executive Order 13366 to coordinate the activity of all agencies with ocean management responsibilities so as to implement EBM. However, it does not mandate any specific steps or actions to achieve this goal.

The action at state and regional levels was generating discussion about regional approaches to ocean management that included EBM. The West Coast Governors' Agreement on Ocean Health (2006), the Gulf of Mexico Alliance (2006), and the Gulf of Maine Council (2007), which includes Canada, are examples (Hershman and Russell, 2006; Carollo et al., 2010). Because both NOAA's approach and those recommended by the Ocean Commission and the Commission on Ocean Policy emphasize voluntary, regional implementation of EBM, these initiatives can be seen as strategic on the part of the states and regions.

Simultaneously, there was considerable interest in the concept of coastal and marine spatial planning (CMSP) as a way to implement EBM (Crowder et al., 2009; ELI, 2009). For many advocates of marine protected areas (MPAs) as fisheries and ocean management tools, CMSP is seen as a preferred approach to obtaining consideration of MPAs in the broader management context (NRC, 2001). The CMSP approach found favour with President Obama, who after a national review, repealed his predecessor's executive order for the Commission on Ocean Policy and instead established the National Ocean Council to coordinate the efforts of federal agencies to develop coastal and marine spatial plans for each of the designated ocean regions by 2015 (Executive Order 11547 www.whitehouse.gov /administration/eop/oceans).

The proliferation of all these management concepts—EBM, MPA, IEA, and CMSP- in fairly rapid succession—is potentially confusing. In the view of the authors of this paper, the elements of this alphabet soup all lead to a common sense understanding that they are similar in intent and the real issue is about choosing to act on the advice. Each management approach in its own way can lead towards implementation of EBM. However, EBM remains the overarching framework (NOAA SAB, 2011; NOAA, 2011) while MPAs, CSMP and IEAs, for example, can contribute to implementing that framework.

5.3c Current regulatory and governance frameworks

Despite participating in and having significant influence on the outcome of the Law of the Sea Convention, the CBD and the WSSD Declaration and Implementation Plan, the United States is not a signatory to these agreements. Arguably, the United States is developing EBM approaches similar to those called for in these international instruments. While there have been calls for ratification of these agreements, the sentiment in Congress is not in favour of doing so. Congress also seems averse to legislation which may better integrate the various elements of US ocean policy. Therefore, US ocean policy at the federal level remains fragmented and sector-based with at least 147 separate law entities with ocean management responsibilities. Eleven of the fifteen cabinet level departments have influence over ocean and coastal policies and are members of the National Ocean Council.

Despite this patchwork of laws and regulatory agencies, progress is being made in achieving EBM in the US for three reasons. First, most of the body of laws applies to specific activities or sectors. Even if there is not a clear mandate to apply EBM in a certain context, an agency may choose to do so if that produces the desired outcomes. EBM can be applied within a single sector, for example as EBFM, but the ideal approach would apply a comprehensive and integrated EBM across all sectors (Arkema et al., 2006). Thus, existing law does not prohibit the practice of EBM even if it does not expressly require it. Further, the combination of laws protecting the environment when taken together, amounts to a significant body of law that can be effective at supporting EBM (Parenteau et al., 2009; ELI, 2009a) The National Environmental Policy Act (1969), for example, requires that all federal actions be supported by an Environmental Impact Statement, which means that the individual and cumulative environmental impacts of a federal action be assessed and considered in the decision to allow a particular action.

Fisheries law under the Magnuson-Stevens Act, for example, does not require EBM but it does require that ecosystem considerations should be part of each Fishery Management Plan. More importantly, the combined effect of prohibitions on overfishing, requirements to rebuild overfished stocks of fish, and obligations to protect essential fish habitat cumulatively amount to a set of regulations that promote EBM (NMFS, 2009). Fishery Management Plans constitute major federal actions and therefore environmental impact statements must be prepared under the National Environmental Policy Act (NEPA) (42 USC 4321). This means that consideration must be given to both the separate and cumulative impacts of a plan on the ecosystem and the agency must mitigate these impacts to the extent feasible. Meeting these NEPA requirements may not demonstrate that EBM has been achieved but in many cases it provides a good start. Second, the application of these laws has a cumulative effect, even though coordination may be lacking. NMFS is responsible for implementation of the Marine Mammal Protection Act, the Migratory Bird Treaty Act, and the Endangered Species Act. These three laws cumulatively result in NMFS utilizing EBM principles, *de facto*.

Taking a similar approach to the Coastal Zone Management Act, the Outer Continental Shelf Lands Act, Clean Air and Clean Water Acts, National Park Act, Marine Sanctuaries Act, National Wildlife Refuge Act and National Monuments (Antiquities Act), all contain provisions that are permissive with respect to the use of EBM. When taken together this body of laws offers significant opportunities to move in the direction of EBM. There are some significant overlaps in management measures and conflicts among the laws but the consensus seems to be that the US is not prepared to develop a comprehensive set of ocean legislation to resolve these issues (Parenteau et al., 2009). In this impasse, the action by President Obama to issue an Executive Order requiring his administration to implement coastal and marine spatial planning is a bold and positive step. It remains to be seen how the President's Executive Order is implemented to know how far it can be seen as EBM. At present, the CSMP approach is being met with opposition in Congress and among ocean and coastal user communities.

5.3d Progress in policy implementation and major challenges

Progress in developing EBM has taken place on many fronts at the regional, state and local levels, especially with respect to fisheries and marine protected area management. As noted above, the Alaska region has continued to apply an ecosystem approach through its conservative management. Various steps have also been taken in most other regions. Progress is being made, for example, in Chesapeake Bay (Chesapeake Bay Fisheries Ecosystem Advisory Panel, 2006) and the Puget Sound region of Washington State (Ruckelshaus et al., 2009). For further detail, see Houde et al. (this volume).

Within the context of MPAs, the National Marine Sanctuaries (Channel Islands, Florida Keys) have applied EBM in their management plans. President Bush used the Antiquities Act to designate large areas in the US Pacific as Marine National Monuments with very high levels of protection. While the management direction is set at preservation and restoration of these areas, an EBM approach can inform these actions.

Coastal and Marine Spatial Planning is commencing but faces daunting challenges from a hostile Congress, a distressed economy and efforts to reduce and restrict government actions. At the same time, strong proponents are pushing for full implementation of a comprehensive, integrated, EBM science-based planning effort to be completed by 2015. Experience from other countries shows these expectations to be optimistic (NOAA SAB, 2011). Promising results are seen at the state level, for example, in Massachusetts and Oregon with their spatial planning for territorial sea waters. However, none of these efforts are comprehensive and fully integrated EBM. California is close to completing its marine reserve network, which can be seen as a down payment of part of CMSP and EBM (Sivas and Caldwell, 2008, Osmond et al., 2010).

5.4 The Benguela Current region

5.4a Introduction

The Benguela Current region, as described here, is that area falling under the jurisdiction of the Benguela Current Commission (BCC) and is defined as the area of the Benguela Current Large Marine Ecosystem to the extent that it falls within the internal waters, territorial seas or exclusive economic zones of the Contracting Parties: Angola, Namibia and South Africa. It therefore covers the southwest Atlantic coast of Africa from central Angola through Namibia to the south coast of South Africa. Legally it is described as "the relatively large marine ecosystem associated with the Benguela Current and characterised by distinct bathymetry, hydrography, productivity and trophically dependent populations that is bounded by the high water mark along the coasts of South Africa, Namibia and Angola" with a northern boundary at the line of latitude 5° South, a southern boundary at the southern limit of South Africa's EEZ, an eastern boundary of 27° East and a western boundary at the 0° meridian (BCC, 2007).

The region is characterized by one of the more productive upwelling regions in the world and supports productive fisheries as well as containing rich deposits of diamonds, other precious minerals and large oil and gas reserves (O'Toole, 2010). The utilization of these resources provides important social and economic benefits for the three coastal states, all of which are developing countries. The relative importance of the different sectors varies among the three countries. In Namibia, fisheries was reported as being economically the third-largest sector after agriculture and mining, with the industrial fishery generating up to 10% of the country's gross domestic product (GDP) in recent years (Sumaila et al., 2005). In South Africa, the fishing industry contributes less than 1% of the country's GDP but is locally important particularly along the west coast of the country. In Angola, the fisheries sector ranks third in importance after oil and diamond mining.

5.4b Evolution of the existing EBM governance regime

All three countries are still in transition following the upheavals of colonization and subsequent independence after prolonged liberation struggles in each of them. One of the consequences of the history of the region is a marked gradient in capacity from south to north (UNDP, 2001). In relation to EBM, the primary problems that have been identified (UNDP, 2001) include: habitat loss and pollution arising from inadequately controlled tourism development and the ongoing expansion of rural areas, degradation of coastal areas adjacent to urban centers in the southern part of the region, over-exploitation of many of the fishery resources, and pollution and habitat degradation from mineral and petrogenic energy exploration and production (offshore and coastal).

All three countries have modern and appropriate policies and legislation that encourage responsible use of marine and coastal resources. In all of them, though, the responsibility for governance of marine areas is spread amongst different ministries with insufficient coordination and cooperation between them. For example, in Namibia the Ministries of Environment and Tourism, Fisheries and Marine Resources, Works and Transport, Mines and Energy, and of Agriculture, Water and Forestry have responsibilities that are relevant to EBM of the oceans. The Namibian authorities have recognized that there are gaps in the current legislation that make integration across these departments difficult and impede full implementation of EBM. Similar situations are likely to exist in South Africa and Angola too. In addressing the problem, Namibia produced a Green Paper for the Coastal Policy of Namibia in 2009 which is currently under consideration (Namibia Ministry of Environment and Tourism, 2009). It provides rigorous analysis of the current governance and institutional arrangements and puts forward four options as a starting point for further discussion. Those four options are: a Namibian Coastal Coordinating Commission; Assignment to an Existing Planning, Budget or Co-coordinating Office; Designation of an Existing Line Ministry to Act as Lead Ministry; and Creation of Strategic Alliances with a National Lead Agency.

Rather surprisingly, more visible progress in implementation of EBM is probably taking place at the regional level. This was facilitated by long-standing cooperative research programmes among the three countries, in particular the Benguela Environment Fisheries Interaction and Training Programme (BENEFIT), which started in 1997, and the Benguela Current Large Marine Ecosystem programme (BCLME), which started in 2002. Both of these programmes contributed to and have been largely absorbed into the Benguela Current Commission (BCC).

5.4c Current regulatory and governance frameworks

The Benguela Current Commission was established on 31 January 2007 when an Interim Agreement was signed as the first step towards the development of "a binding legal instrument that will establish a comprehensive framework to facilitate the implementation by the Contracting Parties of an ecosystem approach to the conservation and development of the Benguela Current Large Marine Ecosystem". At the time of writing this paper, that instrument, the 'Benguela Current Convention', was being finalized and it was anticipated that it would be ratified by the three countries at a ceremony in March 2013.

Under the Interim Agreement, the major objective of the BCC was to constitute a "formal institutional structure for co-operation between the Contracting Parties that would facilitate the understanding, protection, conservation and sustainable use of the Benguela Current Large Marine Ecosystem" (BCC, 2007). The Commission is headed by a Ministerial Conference which includes ministers from the three countries representing the range of relevant sectors, including petroleum, fisheries, water, marine transport, environment, mines and energy. A Management Board serves to interpret and apply the policy decisions of the Ministerial Conference and includes three Committees (still to be implemented): a minerals and petroleum committee; a marine living resources committee; and an ecosystem health committee. There is also a permanent Secretariat located in Swakopmund, Namibia, and an Ecosystem Advisory Committee to "support decision-making by the Management Board, the Ministerial Conference and the Contracting Parties by providing them with the best available scientific, management, legal and other information, and expert advice" (BCC, 2007).

During the interim phase, the Commission had an advisory function only and had the mandate to consider and make recommendations on matters such as: the determination of optimum levels of harvesting in respect of shared or straddling stocks (still to be implemented); conservation of the biological diversity of the BCLME; implementation of integrated coastal management and of the ecosystem approach ("in accordance with international law and non-binding international undertakings made by the Contracting Parties") (BCC, 2007); prevention of the introduction of harmful and invasive alien species; matters related to the environmental impacts of marine prospecting, mining and dredging and from the exploration and development of oil and gas fields; guidelines on water quality standards within the BCLME; and others. Once signed and ratified, the Benguela Current Convention should play an important role in regulating management of human activities impacting on the Benguela Current ecosystem, helping to ensure that the three countries cooperate in responsible industrial development of the system in sectors such as fishing, marine mining and oil and gas extraction. The establishment of the BCC and the agreement on the Benguela Current Convention therefore demonstrate a serious commitment by the three countries to effective ecosystem-based management of the oceans within their EEZs and provide an institutional framework to facilitate this.

5.4d Progress in policy implementation and major challenges

There have been no detailed studies of progress in implementation of EBM as a whole at the national level in the three countries. However, a recent BCLME and FAO project did consider progress and constraints in implementation of governance and management for an ecosystem approach to fisheries in the Benguela Current region (Cochrane et al., 2009). The study found that while the three countries all tended to focus on conventional approaches to fisheries with a largely centralized management system, they had made some progress in addressing wider ecosystem issues. One key conclusion was that all countries needed to modify their resource management structures to make them suitable for an ecosystem approach to fisheries management (EAF), including means to ensure direct involvement of stakeholders in the decision-making process. There were particular problems with insufficient scientific and management capacity for EAF in all countries, although the study concluded that existing capacity should be sufficient to make some progress. It was recognized that implementing EAF would require a broadening in attitudes of researchers, managers, and industry. In relation to EBM, there was considered to be a need to improve communications with the oil industry in Angola and the marine diamond mining industry in Namibia, noting that these sectors tended to be given precedence over fisheries at the political level. Authorities in South Africa consider that most of the elements and structures are in place in that country to address EAF issues, including an EAF Steering Committee at top management level and an active EAF Scientific Working Group, regular advice on ecosystem issues from fishing sector scientific working groups and regular updates of EAF reviews (J. Augustyn, Department of Agriculture, Forestry and Fisheries, pers. Comm.).

At the regional level, the BCC has only been in place for 3 years and it is difficult to evaluate progress at this stage. There has undoubtedly been good cooperation in research and the BCC oversees a strong regional research programme, which includes, for example, regional fish stock surveys, a study on spatial biodiversity assessment and spatial management including MPAs, assessment of the impacts of harmful algal blooms on the inshore environment, joint research cruises, and stock assessments for shared or transboundary fisheries stocks (BCC, 2011). The Commission has also strengthened regional and inter-sectoral discussion at the highest political levels through Ministerial Conferences in 2007, 2010 and 2011. In addition to continuing these scientific and consultative activities and processes, the coastal States plan to go further and realign and harmonise national policies and laws in order to allow for regional integrated transboundary management of the large marine ecosystem. If they are successful with this goal, it could be an excellent example of EBM in practice.

5.5 The Wider Caribbean Region

5.5a Introduction to the case study area

The Wider Caribbean Region (WCR) extends from the mouth of the Amazon River, Brazil, in the south, through the insular Caribbean, Central America, the

Gulf of Mexico and north along the east coast of North America to Cape Hatteras. The WCR is a recognized geopolitical entity defined by the Cartagena Convention and is covered by the activities and programs of several UN and regional intergovernmental agencies (Fanning et al., 2009). There are four Large Marine Ecosystems (LMEs) within the WCR: The Caribbean Sea LME, the Gulf of Mexico LME, the North Brazil Shelf LME and the Southeast US Continental Shelf LME (Sherman et al., 1986). The WCR includes 29 countries and 15 territories that are dependencies of France, United Kingdom, United States and The Netherlands. The EEZs of the WCR cover the entire region. Consequently, there is a high incidence of transboundary resource management issues, even at relatively small spatial scales.

The marine environment of the WCR is highly variable both spatially and temporally. The North Coast of South America is dominated by the effluence of the Amazon, Orinoco and several other large rivers into the North Brazil Current (Longhurst, 1998). Most Caribbean islands are more influenced by the nutrient-poor North Equatorial Current. Islands with appreciable shelf area exhibit significant coral reef development. Coral reefs at shallow depths are extensive on the continental shelf from Isla Margarita west to Mexico. Seagrass and mangroves are also common coastal habitats (Burke and Maidens, 2004).

The fisheries of the Caribbean region exploit a diversity of resources. Most coastal resources are considered to be overexploited and there is increasing evidence that pelagic predator biomass has been depleted (Mahon and McConney, 2004). The majority of fisheries are artisanal, or small-scale. Notable exceptions are the shrimp and groundfish fisheries of the Brazil-Guianas and US continental shelves and the tuna fisheries of Venezuela and the USA.

Marine-based tourism is a major contributor to the economy in many Caribbean countries. Healthy marine ecosystems are critical for tourism; however, the tourism industry has generally not seen marine ecosystem management as its business. Many local, national and regional assessments have concluded that the Caribbean Sea has been severely impacted by human uses: overexploitation of most coastal and offshore living marine resources; destruction of coastal habitats by tourism; industrial and urban development; and degradation of the marine environment by pollution from land and ship-based sources (Mahon and Escobar, 2009). Coral reef ecosystems in particular have been noted as experiencing significant impacts from a variety of sources and are especially important for both fisheries and tourism. The cumulative and often synergistic effects of the human activities impacting marine ecosystems in the Wider Caribbean have led to an increased recognition of the need for an ecosystem approach to management at the regional level.

5.5b Current policies

No comprehensive definition of, or implementation strategy for, marine EBM has been adopted for the WCR, or can be said to be in general usage. Many local, national and regional attempts to promote activities could be considered as EBM (Fanning et al., 2009). Aspects of EBM are increasingly reflected in regional agreements and, of course, in the key international multilateral environmental agreements that many countries have signed.

5.5c Evolution of the existing EBM governance regime

The need for marine EBM at the regional level gained prominence through the Caribbean Large Marine Ecosystem and Adjacent Areas (CLME) Project (Fanning et al., 2009). The need to elaborate on what EBM means in the variety of resource contexts and geographical scales in the WCR was recognized early in project development and led to a regional symposium on 'Marine Ecosystem-Based Management in the Caribbean' with the aim of promoting a regional level discussion on marine EBM (Fanning et al., 2011). The principles and strategies that emerged as important to participants are closely aligned with the FAO Ecosystem Approach to Fisheries (EAF) (Bianchi and Cochrane, 2011). At its outset in January 2010, the CLME Project also reoriented its work towards the EAF and identified the three major marine ecosystems of the WCR as a working context: open sea pelagic ecosystem, coral reef ecosystem and continental shelf ecosystem. Recognizing that EBM or any other kind of management cannot be pursued effectively unless governance institutions are in place and operational, the CLME project emphasizes governance (Mahon et al., 2009).

5.5d Current regulatory and governance frameworks

The region is characterized by a diversity of national and regional governance institutional arrangements. There are about 30 subregional, regional and international organisations with a mandate for various aspects of living marine resource governance in the WCR (fisheries, biodiversity, coastal zone management, land-based sources of pollution, etc.). These operate at a diversity of geographical scales (Chakalall et al., 2007). This situation mirrors the global level proliferation and fragmentation of environmental governance that is of concern to many (Young, 2008).

Despite this complex and diverse arena, institutional arrangements for the management of transboundary living marine resources in the WCR have been emerging in practice from the ongoing efforts of various organisations, such as the Organization of Fishing and Aquaculture in Central America, the Caribbean Regional Fisheries Mechanism and the Caribbean Fishery Management Council (Chakalall et al., 2007). There is reluctance in the region to establish a single regional ocean governance organization. This reflects: (1) the fact that the WCR has few major fish stocks attracting large commercial fleets, revenues from which can be expected to support a regional organisation (2) pre-emption of jurisdiction by existing organisations, and (3) the geopolitical and ecosystem multi-scale nature of ocean governance issues in the WCR. The fact that the tourism industry has not yet tuned in to the need to invest in marine ecosystem-based management is a major impediment as it is the primary beneficiary.

The emerging institutional arrangements in the Caribbean are potentially more suited to the large diversity of resources that are already mostly exploited by indigenous fleets, so the issues relate primarily to conservation, optimization and intra-regional equity (Chakalall et al., 2007; McConney et al., 2007). The situation in which there is a suite of institutions that are networked is similar to that proposed for the Arctic by Young (2010). This approach has been endorsed by the countries of the region at two meetings of the Western Central Atlantic Fisheries Commission (WECAFC) (FAO, 1999; 2001) in the Caribbean Large Marine Ecosystem Project (Fanning et al., 2011) and in the Caribbean Sea Commission of the Association of Caribbean States (ACS/CERMES-UWI, 2010).

The LME Governance Framework was developed for the WCR with a view to capturing the key features of the emerging governance complex in such a way that they could be assessed and strengthened (Fanning et al., 2007). The framework (Figure 4.1c) draws on key common governance concepts from the major schools of natural resource and environmental governance (Mahon et al., 2011). Based on linked policy cycles at multiple scales (local, national, regional and global) it emphasizes governance architecture, processes and interactions. However, the full complexity of the multilevel intersecting governance arrangements cannot be fully depicted graphically. The framework is normative in that it argues that the policy cycles that are needed for governance must be complete and functional and must be linked vertically and laterally if the governance arrangements that address specific issues are to be effective. The suite of arrangements needed to address specific issues and the interactions among these arrangements make up the overall ocean governance architecture or regime complex in the WCR. The long-term goal is a suite of functional arrangements with appropriate policy cycles and interactions within and between arrangements. The LME governance framework seeks to depict and communicate this in a way that can be used by stakeholders to understand the governance arrangements and to get involved in improving them.

The governance architecture and the performance of the arrangements that comprise the framework are the focus of assessment with a view to developing interventions that improve the structure and performance of the system and thus governance effectiveness. Assessments may target different areas of the framework: either entire governance arrangements as in the case of the pilot projects and case studies of the CLME Project, which is designed around the LME Governance Framework (Fanning et al., 2009); or types of interaction, for example the interface between the national and regional levels of governance (Mahon et al., 2010). An initial obvious outcome of the regional governance architecture was the recognition that the system was lacking an overarching ocean policy arena. The Caribbean Sea Commission which was established in 2006 but has yet to become fully operational was noted as having the potential to play that role. This potential role was endorsed and refined at a consultation held by the Association of Caribbean States in 2010 to engage the key regional organisations that would make up the perceived WCR ocean governance regime (ACS/CERMES-UWI, 2010).

5.5e Progress in policy implementation and major challenges

Progress in building an ocean governance regime that is effective and encompasses agreed principles and practices is slow. It must ensure that all actors, both national and regional, are of a similar mind. Initiatives aimed at strengthening the regime must ensure that their inputs are equitably distributed and that all concerns are considered. The WCR is the most geopolitically complex region in the world. The countries range from among the largest (e.g. Brazil, USA) to among the smallest (e.g. Barbados, St. Kitts and Nevis), and from the most developed (e.g. USA, France) to the least developed (e.g. Haiti, Guyana). Consequently, there is an extremely wide range in their capacities for governance. There are also considerable differences in perspectives on ocean governance among the countries of the WCR. The establishment of good ocean governance, and through it EBM, in the

WCR is a long-term process of building processes and interactions until partners are ready to engage in binding arrangements.

6. Discussion

The overview in Section 3 of this chapter outlined some of the more important and well-known international instruments relevant to EBM and demonstrated that recognition of the need for an ecosystem-based approach to ocean management is well entrenched in international legislation and agreements. The requirement that these principles be applied to multi-sectoral management through use of EBM is also well recognised (as, for example, the CBD; Table 4.3).

Notwithstanding the good understanding of the concept of an ecosystem approach and agreement on its importance at an international level, the case studies presented here and a comparable study by Rosenberg at al. (2009) demonstrate that progress in its implementation even at the sectoral level has in general been mixed. Greatest progress in all the case studies has been within the fisheries sector. The reasons why fisheries have moved more rapidly in implementing an ecosystem approach have not been rigorously analyzed here but there are clearly strong driving forces within the sector. In particular, with many fishery resources under pressure, the sector has a lot to lose if it does not move rapidly to ensure sustainability and, at the same time, there has been strong public pressure for improvements in fisheries impacts on resources and ecosystems in many countries and at an international level. While progress in implementation of EBFM is occurring, important obstacles remain and even among the best examples, none can yet be considered to have reached the stage of full implementation of EBFM let alone EBM.

From the perspective of regulatory and governance frameworks, this review demonstrates several key factors that are hindering implementation of EBM and need to be taken into account in promoting its more rapid and widespread implementation. These are discussed below in no particular order of priority.

6.1 Institutional structures

The optimal institutional structure for EBM will vary from case to case and it would be misleading to promote a one-size-fits-all system. Nevertheless, the experiences in the case study areas together with general understanding (e.g. FAO, 2003) provide broadly consistent and intuitive concepts in terms of the processes involved and likely suitable institutional structures (Figure 4.1). The processes and structures needed for planning, decision-making and implementation at a sectoral level are outlined in Figure 4.1a. A potential mechanism to integrate the various institutions to scale-up to a multi-sectoral approach within a single ecosystem-based management system is shown in Figure 4.1b. The structure and processes of Figure 4.1a will need to be followed for each sector shown in Figure 4.1b, as well as in the process of integration and reconciliation of objectives and management measures across sectors. The range of functional types of institutions listed in Table 4.2 will be present at both levels.

In most cases, there will rarely be a single, homogeneous management unit for EBM. The more typical case, resulting from political characteristics and historical fragmented development, will be multiple, nested and overlapping units as de-

TABLE 4.3.
Convention on Biological Diversity Principles of the Ecosystem Approach (CBD, 2002)

Principle 1:	The objectives of management of land, water and living resources are a matter of societal choice.
Principle 2:	Management should be decentralized to the lowest appropriate level.
Principle 3:	Ecosystem managers should consider the effects (actual or potential) of their activities on adjacent and other ecosystems.
Principle 4:	Recognizing potential gains from management, there is usually a need to understand and manage the ecosystem in an economic context. Any such ecosystem management programme should: a) reduce those market distortions that adversely affect biological diversity; b) align incentives to promote biodiversity conservation and sustainable use; c) internalize costs and benefits in the given ecosystem to the extent feasible.
Principle 5:	Conservation of ecosystem structure and functioning, in order to maintain ecosystem services, should be a priority target of the ecosystem approach.
Principle 6:	Ecosystems must be managed within the limits of their functioning.
Principle 7:	The ecosystem approach should be undertaken at the appropriate spatial and temporal scales.
Principle 8:	Recognizing the varying temporal scales and lag-effects that characterize ecosystem processes, objectives for ecosystem management should be set for the long term.
Principle 9:	Management must recognize that change is inevitable.
Principle 10:	The ecosystem approach should seek the appropriate balance between, and integration of, conservation and use of biological diversity.
Principle 11:	The ecosystem approach should consider all forms of relevant information, including scientific and indigenous and local knowledge, innovations and practices.
Principle 12:	The ecosystem approach should involve all relevant sectors of society and scientific disciplines.

scribed in the USA and WCR for example. In these cases additional management layers are required to coordinate and integrate across those units, as shown in Figure 4.1c. The institutions of Figure 4.1b, including the nested processes of Figure 4.1a, are also required in these cases for the integration of the different geopolitical management units. While this scaling-up and integrating process is conceptually simple, it has high transaction costs, both human and financial, which escalate at every scaling-up point, as do the potential conflicts and uncertainties. These are discussed further below.

6.2 Role of international and national law

International law is a product of countries coming together to agree on a common way forward. It should therefore simultaneously reflect current understanding of best practice and promote application of that understanding. The case studies demonstrate the important role that both hard and soft international law has played in promoting implementation of EBM. Sectoral management and integrated EBM in Norway are underpinned by hard and soft international law. In Australia the initial impetus for EBM was provided by the 1987 World Commission on Environment and Development. In the USA the failure of successive governments to sign relevant international laws including UN LOS and the WSSD Plan of Implementation is considered to be a disadvantage and that country is considered not to be addressing the provisions of these instruments to the same

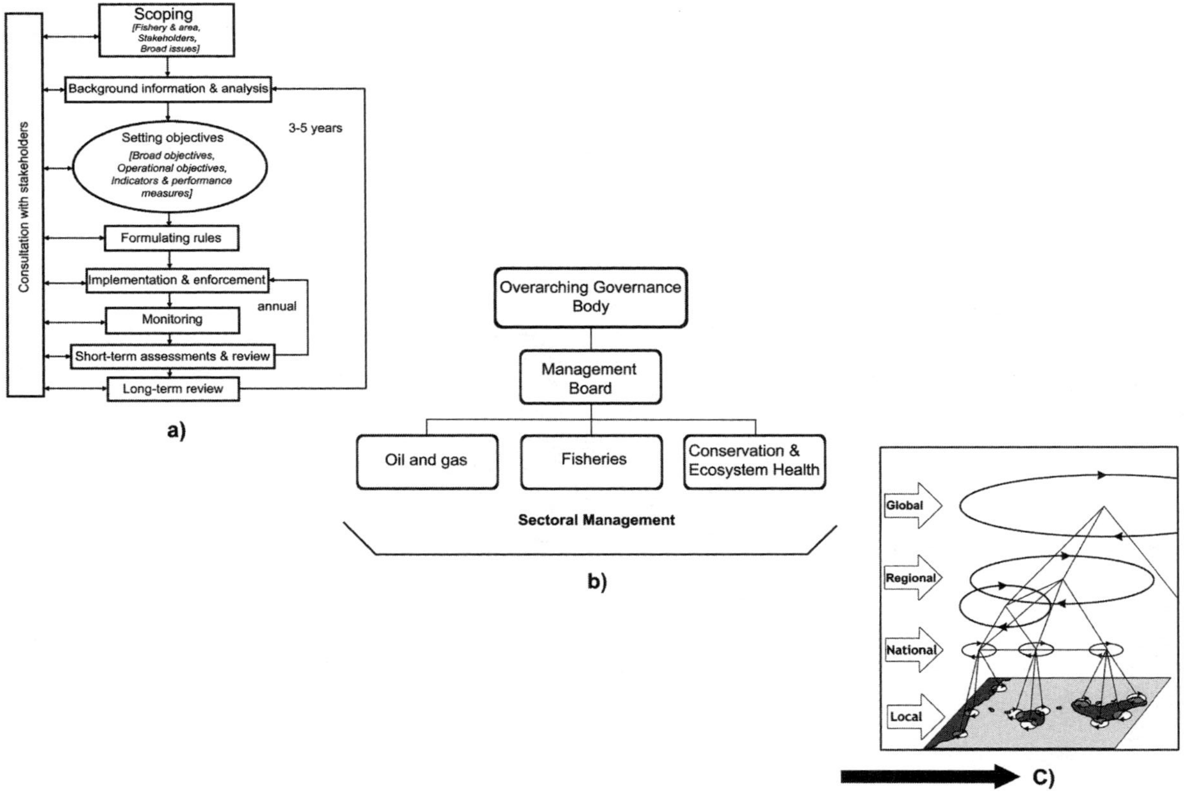

Fig. 4.1 – A generalised process and institutional structure for ecosystem-based management. The figure shows the processes involved in scaling up (indicated by the direction of the arrow) from (a) sectoral management; to (b) a hypothetical case of inter-sectoral management; which (c) will typically need to be repeated at multiple geo-political scales. The processes shown in figure (a) will need to be repeated in all sectors and at each integrating step in scaling-up from one to multiple sectors and from smaller areas to larger areas within a nested geo-political area. (Figure (a) from FAO (2003), Figure (c) from Mahon et al. (2010))

extent as would have been likely if it was a signatory. In the Benguela and Wider Caribbean regions international law is not explicitly recognised as driving EBM but the timing and nature of the processes there too demonstrate a consistency with and therefore likely response to current international law and thinking.

At the same time, the case studies reveal that practical implementation can proceed in advance of the development of hard laws. The USA and the WCR case studies, for example, report progress in EBM that goes beyond the existing hard laws on EBM. The USA case study describes the cumulative effects of individual laws leading to a result that goes beyond their original legal intentions. Progress in the WCR, while slow, is similarly moving forward ahead of the current hard laws in the region. In contrast, in Norway the legal framework and practice appear to be progressing largely in parallel. These examples suggest that as the political complexity and heterogeneity of the system to be governed increases it becomes increasingly difficult to achieve hard law solutions. Where there is a need and incentive to find solutions, soft law or simply pragmatic arrangements may be the only way forward or at least an initial way forward. One key prerequisite for such ahead-of-the law solutions, as demonstrated in the USA, must be that the existing laws do not in some way prohibit EBM.

While important progress is being made under a combination of hard law, soft law and pragmatism, it is questionable how far this could be taken under circumstances involving conflict. Rosenberg et al. (2009) concluded that progress in EBM is dependent on the willingness of the participants to drive its implementation. As long as it is in everyone's interest, and there are no hard law obstacles, pragmatic solutions can be found. When conflicts arise and hard solutions are required, whatever their limitations, the existing hard laws may still provide the best means of obtaining good solutions by avoiding unjust use of influence and power. Rosenberg et al. (2009) proposed that the lack of legislation addressing the need for cross-sectoral action was an important obstacle to implementation of EBM.

6.3 Contribution and constraints of the sectoral approach

The recognition that the economic context is important within the context of EBM (e.g., Table 4.3) requires recognition of the different sectors making use of and impacting on the oceans and of the different institutions set up to manage each of those sectors. The case studies clearly demonstrate the key role of sectoral implementation of EBM in overall progress. This recognises that EBM is being approached as an evolutionary development, building on what is already there, rather than a revolutionary sweeping away of the old and replacing it with something completely new (e.g. Fletcher, 2006).

While sectoral approaches are necessary, ultimately they are insufficient on their own (see also Rosenberg et al., 2009). There is always a likelihood that even if each sector is being adequately managed with an ecosystem approach, the sum of the independent approaches will fail to take into account and address all of the significant interactions and conflicting objectives between the different sectors. A primary role of and challenge for EBM is to create linkages across the sectors that will allow for the evaluation and reconciliation of the objectives of the full set of stakeholders, and of the methods and measures that are being used at the sectoral level to achieve the objectives.

At the same time as making progress at the sectoral level, sectoral approaches can create obstacles to multi-sectoral EBM. For example, the sectoral approach, without coordination, has led to the wide-spread fragmentation that predated promotion of EBM as best-practice and has generated inward-looking attitudes hindering progress towards EBM in all the case studies. The reluctance of some sectors to embrace EBM may be influenced by self-interest and the different objectives of each of these groups (e.g. Murawski, 2007). Their attitude is also likely to be influenced by how they perceive the likelihood of favourable outcomes for themselves from maintaining a non-integrated planning process compared to embracing integrated EBM. Ultimately, free choice will come down to the nature and direction of the incentives for each group (see below).

Overall, it is clear that efforts to implement EBM must continue at both the multi-sectoral and sectoral levels. Nevertheless, an important general lesson from the nature and extent of progress to date described in the case studies is that, at the current stage of implementation, investments in improving ecosystem approaches within sectors are likely to be more effective than investing in efforts at the multi-sectoral level. This observation should not, however, detract from the need to implement EBM as the multi-sectoral approach it is intended to be, especially in cases where important inter-sectoral conflicts already exist and require urgent solutions.

6.4 The challenge of creating a robust governance structure

The task of creating a robust governance structure that brings together and coordinates the sectors, different agencies and multiple levels of government is a major challenge, as demonstrated in the Australian, USA and WCR examples. In the USA and WCR, very little progress has been made in establishment of an overarching structure with real power and there is significant hostility and indifference to attempts to do so from within governments and other stakeholders. In the WCR, obstacles include inertia or turf protection from the existing organizations and the complexity of the geo-political and ecosystem structure of the region. In Australia, the slow but reasonable progress made through the NRMMC and MACC appears to have been lost in the recent policy changes, while in the USA changes in leadership at the highest level had substantial influences on the approaches taken. These examples demonstrate the political challenges to creating an overarching institution with such broad and far-reaching influence. The success within the BCLME to gain support for regional EBM at the highest political levels is significant although it is still to be fully tested. Clearly, success in establishing and maintaining an integrative governance policy and structure will depend on the sustained strength of political will and the existence of robust incentives for the different stakeholders.

6.5 Incentives and disincentives for the stakeholders

While at the conceptual level the long-term, common benefits of EBM are widely recognized and supported in terms of the collective good, serious challenges in actual implementation occur at the level of the individual stakeholders and stakeholder groups. Where resources, including geographical space, are insufficient to satisfy all stakeholders, actions to reduce the negative impacts and ensure overall sustainability will result in winners and losers. Those threatened with losses will rarely give way without a struggle (Hilborn, 2007). There are many examples in natural resource management of failures to overcome strong, short-term interests in order to sustain wider, long-term benefits (e.g. Cochrane, 2000).

A sobering example of the limits to consensus approaches is the on-going effort to ensure sustainable use of a valuable abalone (*Haliotis midae*) fishery in South Africa threatened by serious illegal fishing. These efforts have included attempts to implement a range of best-practices in fisheries management but without success, leading to a controversial closure of the fishery in 2008. The authors of a recent study point to some weaknesses in the management response but add that "Notwithstanding these shortcomings, it remains questionable whether any formal process to institute good governance (and by implication limit fishing effort) would have succeeded in the face of the incentives created by the high price of abalone and fluid socio-political setting in which access to the resource was highly contested" (Raemaekers et al., 2011).

Information from the case studies also demonstrates the influence of vested interest groups in, for example, the simmering conflicts between the oil and gas industry on the one hand and the fishing industry on the other in the Barents Sea and the rising support for conservation groups in Australia to the detriment of fishing. As stated in principle 4 of the CBD principles for an ecosystem approach (Table 4.3), any EBM approach needs to align incentives to achieving the agreed EBM goals or progress. As social and political complexity increases, the likelihood of win-win scenarios dependent only on the resources within the system must diminish, and compensation or opportunities outside the system will be required to avoid strong disincentives for EBM for some stakeholders.

6.6 *Ecological, political and social complexity*

The case studies cover a range of political and ecological diversity from Norway, which is the probably structurally the simplest of the examples, to the WCR as the politically most complex and the USA which is probably the ecologically most complex. The challenges of increasing complexity are clear. In relation to social and political complexity they include the need to reconcile diverse political views and perspectives, coping with inequalities in power and wealth of different stakeholders (where stakeholders can range from national governments to sub-sectoral associations), differences in the capacities of stakeholders to negotiate and to manage, and different priorities. Ecological complexity arises through diversity between and within ecosystems, unclear boundaries and overlapping boundaries and the issue of appropriate scaling in nested ecosystems.

Coping with complexity in natural resource management is a widely discussed challenge (e.g., Healey and Hennessey, 1998; Cochrane, 1999, 2000; Fletcher, 2008; Kooiman et al., 2005) and cannot be reviewed here. A primary approach to coping with complexity must be to prioritise goals and problems in a systematic way and ensure that the available resources, including political will, are directed towards the agreed priorities (Fletcher et al., 2010). Approaches that are oriented towards enabling capacity for resilience to the unpredictability that is characteristic of complex systems and which have been discussed for other resource types such as forestry and water are only recently finding their way into the discourse on ocean management (Kooiman et al., 2005; Mahon et al., 2009; Garcia et al., 2008; Jentoft and Chuenpagdee, 2009; Cochrane et al., 2011). These remain to be tried and integrated with other approaches.

6.7 The role of scientific advice

The important role of scientific advice clearly emerges from several of the case studies. The two Ocean Commissions in the USA recommended IEAs as the scientific foundation and a cornerstone for developing EBM. The management plan for the Barents Sea and its implementation are science-based and the important role of knowledge is emphasised in the plan. Knowledge generation and awareness raising from the BENEFIT and BCLME scientific programmes played a very important role in developing and building political consensus for the creation of the Benguela Current Commission (Cochrane et al. 2009). The 'eight steps approach' in Australia appears to be less heavily dependent on science than the other case studies and can be applied within a risk-based framework with whatever knowledge is available. While progress in implementation of EBM is clearly linked to good scientific knowledge, the principles of using the best available information and the precautionary approach must be applied in decision-making to avoid the absence of knowledge from unnecessarily hindering progress in EBM.

It also needs to be recognised that EBM will always encounter substantial uncertainty about conditions and trends across all dimensions. This requires an adaptive approach to management allowing for review and modification of management measures and, where necessary, objectives in response to new information (Rosenberg et al. 2009).

Science can also be accused of hindering progress in EBM, for example, in the confusion caused by the many diverse and overlapping concepts and acronyms that have flooded the market in recent decades (Fletcher, 2006). The problem is highlighted in the case studies from Australia and the USA but also impacts progress in many regions around the world. In recent decades, politicians, managers and stakeholders have been bombarded with integrated coastal zone management; EAF, EBM and their many subtle variants; MPAs; CMSP; and, by the time this is published, they will have learned from Rio+20 that they also need to take on board green economies.[15] EBM needs to be clearly identified as consistent and timeless and the core approach for improving the coordination of management among agencies and sectors using a pragmatic risk-based approach, not as some highly technical, science-based initiative (Fletcher 2008).

6.8 Spatial management

Australia, Norway and the USA have emphasised and made use of spatial management in implementation of EBM. Many fisheries in Australia already have a level of spatially-based management restrictions; the federal government has implemented a national system of MPAs and most states are implementing marine

[15] http://www.uncsd2012.org/rio20/. Accessed on 13 November 2011.

area planning. In the USA, there are eight regional councils the jurisdiction of which corresponds approximately to large scale ecosystems, while planning and action also take place at a variety of other scales according to the needs and characteristics in each case, especially added habitat protection. Norway too uses area-based management or spatial planning as a key management measure for EBM, for example in helping to resolve conflicts and to protect designated areas. Marine spatial planning is not synonymous with EBM but is clearly an important tool in planning and implementation of an ecosystem approach (Crowder et al., 2009; Rice et al., this volume).

6.9 Crises and tipping points

Finally, the role of tipping points that lead to increased urgency for EBM implementation is an interesting consideration which may provide important catalysts in countries and regions where implementation has been slow. The five case studies demonstrate proactive implementation of EBM and crises do not seem to have been a primary driving force, although serious concerns arising from widespread negative human impacts in the Caribbean have prompted consideration of the need for a regional ecosystem approach in the WCR. These studies include some of the most advanced examples of EBM in practice but progress has nevertheless been slow. In this context, the description by Gelcich et al. (2010) of how a collapse of existing policies can provide a trigger or opportunity for positive change may yet be relevant. It remains to be seen whether large-scale crises will be required before the governments and stakeholders of the world will give the needed priority to implementation of EBM and an adequate response is stimulated and sustained.

7. Conclusions

At the global level there is a good policy basis to promote and guide EBM that has now been adopted, whether formally or informally, by nearly all countries. This framework is being implemented and tested in a number of places and at a range of scales with a variety of tools. There is progress at the sectoral level and there are some success stories at the multi-sectoral level. Even though progress has been slow and taken place largely within sectors, particularly the fisheries sector, these examples provide lessons to facilitate further implementation.

The slow progress does not stem from a lack of awareness of the importance of EBM or its principles, or understanding of the institutional framework required to implement it. The obstacles lie in the implementation itself and in large part stem from two inter-related problems: i) the substantial practical difficulties, including time and cost, of integrating across diverse stakeholders and interest groups; and ii) the vested interests and different levels of commitment amongst the different groups of stakeholders, including politicians and those within existing institutions. Some see themselves as being better served by the status quo than by the transition to a more sustainable integrated approach in which there will often be direct winners and losers at least in the short-term. Thus there remains a common lack of shared objectives at all political levels. This increases the influence of parochial opinion and unclear public perceptions in driving public policy to the detriment of optimal marine management outcomes.

These problems are not surprising but are no less challenging as a result. A key solution lies in greater attention to working with the stakeholders at all levels to improve awareness of the need for and benefits of EBM for present and future generations. This awareness alone will not result in change though unless the disincentives for cooperation and integration can be reduced and the incentives increased. As pointed out by Jentoft (2007) that will require that the governing system and the governed social system are compatible and responsive to each other. That in turn will require focused planning and action to agree on and implement suitable governance and regulatory frameworks, incorporating positive social, regulatory and market incentives. This will not happen on its own, neither will it be driven by the users and markets alone. Considerably strengthened attention, resources and action by governments, working together with other stakeholders, will be required to bring about meaningful change.

Acknowledgments

The contributions of the following people to the case studies are gratefully acknowledged: Erik Olsen for the Norway case study; Eric Young, Cherie Wagner, Kiry Nelson and Beena Apgar-Kurtz for that on the USA; Johann Augustyn and Hashali Hamukuaya for the Benguela Current region; and Patricia von Baumgarten in Australia. Andy Rosenberg and an anonymous reviewer provided helpful comments on an earlier version of this manuscript.

References

ACS/CERMES-UWI. 2010. Report of the Expert Consultation on the Operationalisation of the Caribbean Sea Commission: building a science-policy interface for ocean governance in the Wider Caribbean. University of the West Indies, Cave Hill Campus, Barbados, 7–9 July 2010. *CERMES Technical Report* No. 33 (English), 90 p.

ANZECC. 1998. Guidelines for Establishing the National Representative System of Marine Protected Areas. Australian and New Zealand Environment and Conservation Council, Task Force on Marine Protected Areas. Canberra Australia.

Arkema, K., Abramson, S. & Dewsbury, B. 2006. Marine ecosystem-based management: from characterization to implementation. *Frontiers in Ecology and the Environment,* 4(10), 525–532.

BCC. 2007. Interim Agreement. Benguela Current Commission, Windhoek. 15 pp. Accessed at http://www.benguelacc.org/admin/interim_agreement/BCC%20INTERIM%20AGREEMENT%20E.pdfhttp://www.benguelacc.org/admin/interim_agreement/BCC%20INTERIM%20AGREEMENT%20E.pdf, 3 September 2009.

BCC. 2011. Benguela News. Issue, July 2011. Benguela Current Commission, Windhoek. 44 pp. Accessed at http://www.benguelacc.org/admin/publications/BCC%20Newsletter%20lores%20final.pdf, 3 September 2009.

Berkes, F. 2009. Social aspects of fisheries management. In: *A Fishery Manager's Guidebook.* Cochrane, K. L. & Garcia, S. M. (eds), pp. 52–74. FAO, Rome and Blackwell Publishing.

Bianchi, G. & Cochrane, K. L. 2011. Meeting the challenge of applying an ecosystem approach to fisheries management: some experiences and considerations based on FAO's work. In: *Proceedings of Marine Ecosystem Based Management in the Caribbean: An essential component of Principled*

Ocean Governance. University of the West Indies, Cave Hill Campus, Barbados, 10–12 December 2008.

Boyce, M. & Haney, A. 1996. *Ecosystem Management: Applications for Sustainable Forest and Wildlife Resources.* Yale University Press, New Haven. 361 pp.

Burke, W. T. 1994. *The New International Law of Fisheries: UNCLOS 1982 and Beyond.* Clarendon Press, Oxford. pp. 382.

Burke, L. & Maidens, J. 2004. *Reefs at Risk in the Caribbean.* Washington DC: World Resources Institute. 80 p.

Caddy, J. F. 1999. Fisheries management in the twenty-first century: will new paradigms apply? *Reviews in Fish Biology and Fisheries,* 9, 1–43.

Caddy, J. F. & Cochrane, K. L. 2001. A review of fisheries management past and present and some future perspectives for the third millennium. *Ocean & Coastal Management,* 44, 653–682.

Carollo, C., D. Reed, J. Ogden & D. Palandro. 2010. The importance of data discovery in advancing ecosystem-based management. *Marine Policy.* 33(4), 651–653.

CBD. 2002. Decision V/6 Ecosystem approach. Fifth Ordinary Meeting of the Conference of the Parties to the Convention on Biological Diversity, 15–26 May 2000—Nairobi, Kenya. http://www.cbd.int/decisions/cop/?m=cop-05. Accessed on 22 September 2011.

CBD. 2011. Decision X/2 Strategic Plan for Biodiversity 2011–2020. Tenth meeting of the Conference of the Parties to the Convention on Biological Diversity, 18–29 October 2010—Nagoya, Japan. http://www.cbd.int/decision/cop/?id=12268. Accessed on 22 September 2011.

Chakalall, B., Mahon, R., McConney, P., Nurse, L. & Oderson, D. 2007. Governance of fisheries and other living marine resources in the Wider Caribbean. *Fisheries Research,* 87, 92–99.

Charles, A. 2009. Rights-Based Fisheries Management: The Role of Use Rights in Managing Access and Harvesting. In: *A Fishery Manager's Guidebook.* Cochrane, K. L. & Garcia, S. M. (eds), pp. 253–282. FAO, Rome and Blackwell Publishing.

Chesapeake Bay Fisheries Ecosystem Advisory Panel. 2006. *Fisheries Ecosystem Planning for the Chesapeake Bay.* American Fisheries Society. Trends in Fisheries Science and Management 3. Bethesda, MD.

Christie, P., D. L. Fluharty, A. T. White, L. Eisma-Osoriod & W. Jatulan. 2007. Assessing the feasibility of ecosystem-based fisheries management in tropical contexts. *Marine Policy,* 31 (3), 239–250.

Cochrane, K. L. 1999. Complexity in fisheries and limitations in the increasing complexity of fisheries management. *ICES Journal of Marine Science,* 56, 917–926.

Cochrane, K. L. 2000. Reconciling sustainability, economic efficiency and equity in fisheries: the one that got away? *Fish and Fisheries,* 1, 3–21.

Cochrane, K. L. & Doulman, D. 2005. The rising tide of fisheries instruments and the struggle to keep afloat. *Philosophical Transactions of the Royal Society B.,* 360 (1453), 77–94.

Cochrane, K. L., Augustyn, C. J., Cockcroft, A. C., David, J. H. M., Griffiths, M. H., Groeneveld, J. C., Lipińnski, M. R., Smale, M. J., Smith, C. D. & Tarr, R. J. Q. 2004. An Ecosystem Approach to Fisheries in the Southern Benguela Context. *Afr. J. mar. Sci.,* 26, 9–35.

Cochrane, K. L., Augustyn, C. J., Fairweather, T., Japp, D., Kilongo, K., Iitembu, J., Moroff, N., Roux, J. P., Shannon, L., Van Zyl, B. & Vaz Velho, F. 2009. Benguela Current Large Marine Ecosystem—Governance and Management for an Ecosystem Approach to Fisheries in the Region. *Coastal Management,* 37, 235–254.

Cochrane, K. L., Andrew, N. L. & Parma, A. M. 2011. Primary fisheries management: a minimum requirement for provision of sustainable human benefits in small-scale fisheries. *Fish and Fisheries,* 12 (3), 275–288.

Commonwealth of Australia (CoA). 1992. *The national strategy for ecologically sustainable development,* AGPS, Canberra.

Commonwealth of Australia (CoA). 2006. *A guide to The Integrated Marine and Coastal Regionalisation of Australia—Version 4.0.* Department of Environment and Heritage, Canberra, Australia.

Commonwealth of Australia (CoA). 2007. *Guidelines for the ecologically sustainable management of fisheries—Edition 2.* http://www.environment.gov.au/coasts/fisheries/publications/pubs/guidelines.pdf.

Commonwealth of Australia (CoA). 2011. *Marine bioregional plan for the North-west Marine Region.* Department of Sustainability, Environment, Water, Population and Communities, Canberra, Australia.

Crowder, L, Osherenko, G., Young, O., Arimé, S., Norse, E., Baron, N., Day, J., Douvere, F., Ehler, C., Halpern, B., Langdon, S., McLeod, K., Odgen, J., Peach, R., Rosenberg, A. & Wilson, J. 2009. Resolving mismatches in US ocean governance. Policy Forum. *Science,* 314 Aug. 4.

Denisov, V. V. & Mikhaylichenko, G. 2009. Management of the Russian Arctic Seas, *In:* Hoel, A. H. (ed.), pp. 20–35. *Best practices in Ecosystem-based Oceans Management in the Arctic.* 129. Rapportserie, Norsk Polarinstitutt. 113 p.

Department of Environment and Heritage. 2004. *Living coast strategy for South Australia.* Adelaide, Government of South Australia, 80 pp.

Environmental Law Institute (ELI). 2009. *Expanding the Use of Ecosystem-Based Management in the Coastal Zone Management Act.* 17 pp. January.

Environmental Law Institute (ELI). 2009a. *Marine Spatial Planning in US Waters: An Assessment and Analysis of Existing Legal Mechanisms, Anticipated Barriers, and Future Opportunities.* Environmental Law Institute, Washington, DC. 104 pp. December.

Fanning, L., Mahon, R. & McConney, P. (eds). 2011. *Towards Marine Ecosystem-based Management in the Wider Caribbean.* Amsterdam University Press, Amsterdam, 426 p.

Fanning, L., Mahon, R. & McConney, P. 2009. Focusing on living marine resource governance: the Caribbean Large Marine Ecosystem and Adjacent Areas Project. *Coastal Management,* 37, 219–234.

FAO. 1995. *Code of Conduct for Responsible Fisheries.* FAO, Rome. 41 pp.

FAO. 1999. *Report of the ninth session of the Western Central Atlantic Fishery Commission and the sixth session of the Committee for the Development and Management of Fisheries in the Lesser Antilles.* St Lucia, Castries, 27–30 September 1999.

FAO. 2001. *Report of the tenth session of the Western Central Atlantic Fishery Commission and of the seventh session of the Committee for the Development and Management of Fisheries in the Lesser Antilles.* Bridgetown, Barbados, 24–27 October 2001.

FAO. 2003. *Fisheries Management. 2. The Ecosystem Approach to Fisheries. FAO Technical Guidelines for Responsible Fisheries. 4 (Suppl. 2).* FAO, Rome. 112 pp.

FAO. 2006. *Integrated coastal management law. Establishing and strengthening national legal frameworks for integrated coastal management.* FAO Legislative Study 93. FAO, Rome. 274 pp. http://www.fao.org/docrep/012/a0863e/a0863e00.pdf. Accessed on 22 September 2011.

Feral, F. 2009. The Fishery Management Institutions. In: *A Fishery Manager's Guidebook.* Cochrane, K. L. & Garcia, S. M. (eds), pp. 135–163. FAO, Rome and Blackwell Publishing.

Fletcher, W. J. 2006. Frameworks for managing marine resources in Australia through ecosystem approaches: Do they fit together and are they useful? *Bulletin of Marine Science,* 78, 691–704.

Fletcher, W. J. 2008. Implementing an ecosystem approach to fisheries management:lessons learned from applying a practical EAFM framework in Australia and the Pacific. In: *The Ecosystem Approach.* Bianchi, G. & Skoldal, H. R. (eds), pp. 112–124. FAO CABI, Rome.

Fletcher, W. J., Chesson, J., Fisher, M., Sainsbury, K. J., Hundloe, T., Smith, A. D. M. & Whitworth, B. 2002. *National ESD Reporting Framework for Australian Fisheries: The 'How To' Guide for Wild Capture Fisheries.* FRDC Project 2000/145, Canberra, Australia. 120 pp.

Fletcher, W. J., Shaw, J., Metcalf, S. J. & Gaughan, D. J. 2010. An Ecosystem Based Fisheries Management framework: the efficient, regional-level planning tool for management agencies. *Marine Policy,* 34, 1226–1238.

Fletcher, W. J., Gaughan, D. J., Metcalfe, S. J. & Shaw, J. 2012. Using a regional level, risk based framework to cost effectively implement Ecosystem Based Fisheries Management (EBFM). In: *Global Progress on Ecosystem-Based Fisheries Management.* Kruse, G. H., Browman, H. I., Cochrane, K. L., Evans, D., Jamieson, G. S., Livingston, P. A., Woodby, D. & Zhang, C. I. (eds.), pp. 129–146. Alaska Sea Grant College Program doi:10.4027/gpebfm.2012.07.

Fluharty, D. 2005. Evolving ecosystem approaches to management of fisheries in the USA. *Mar. Ecol. Prog. Ser.,* 300. 248–253.

Garcia, S. M., Allison, E. H., Andrew, N. J., Béné, C., Bianchi, G., de Graaf, G. J., Kalikoski, D., Mahon, R. & Orensanz, J. M. 2008. *Towards integrated assessment and advice in small-scale fisheries: principles and processes.* FAO Fisheries Technical Paper. No. 515. 84 p.

Garcia, S. M. & Cochrane, K. L. 2005. Ecosystem approach to fisheries: a review of implementation guidelines. *ICES Journal of Marine Science,* 62: 311–318.

Gelcich, S., Hughes, T. P., Olsson, P., Folke, C., Defeo, O., Fernández, M., Foaleb, S., Gunderson, L. H., Rodríguez-Sickerth, C., Schefferi, M., Steneck, R. S. & Castilla, J. C. 2010. Navigating transformations in governance of Chilean marine coastal resources. *Proceedings of the National Academy of Sciences,* 107, 16794–16799.

Gulf of Maine Council on the Marine Environment. 2007. *Gulf of Maine Council on the Marine Environment Action Plan 2007–2012.* gulfofmaine.org.

Haug, T., Røttingen, I., Gjøsæther, H. & Misund, O. A. 2009. Fifty years of Norwegian—Russian collaboration in marine research. *Marine Biology Research,* 5(1), 1–3(1).

Healey, M. C. & Hennessey, T. 1998. The paradox of fairness: the impact of escalating complexity on fisheries management. *Marine Policy,* 22, 109–118.

Hershman, M. & Russell, C. 2006. *Regional ocean governance in the United States: concept and reality.* Duke Environmental Law & Policy Forum. 16, 227.

Hilborn, R. 2007. Defining success in fisheries and conflicts in objectives. *Marine Policy,* 31, 153–158.

Hobart, W. 1995. *Baird's Legacy: The History and Accomplishments of NOAA's, National Marine Fisheries Service 1871–1996.* USDOC, NOAA, NMFS Technical Memorandum NMFS—F/SPO 18.

Hoel, A. H. (ed.). 2009. *Best practices in Ecosystem-based Oceans Management in the Arctic.* 129. Rapportserie, Norsk Polarinstitutt. 113 p.

Holling, C. S. 2001. Understanding the Complexity of Economic, Ecological, and Social Systems. *Ecosystems,* 4, 390–405.

Jakobsen, T. & Ozhigin, V. 2011. *The Barents Sea. Ecosystem, Resources, Management. Half a century of Russian—Norwegian cooperation.* Tapir, Trondheim. 820 pp.

Jentoft, S. & Chuenpagdee, R. 2009. Fisheries and coastal governance as a wicked problem. *Marine Policy,* 33, 553–560.

Jentoft, S. & McCay, B. J. 1995. User participation in fisheries management. Lessons drawn from international experience. *Marine Policy,* 19, 227–246.

Jentoft, S. 2007. Limits of governability: institutional implications for fisheries and coastal governance. *Marine Policy,* 31, 360–370.

Johannes, R. E. 1978. Traditional marine conservation methods in Oceania and their demise. *Annual Review of Ecology and Systematics,* 9, 349–64.

Kearney, J., Berkes, F., Pinkerton, E., Charles, A. & Wiber, M. 2007. The Role of Participatory Governance and Community-Based Management in Integrated Coastal and Ocean Management in Canada. *Coastal Management,* 35, 79–104.

Kuemlangan, B. 2009. Legal aspects. In: *A Fishery Manager's Guidebook.* Cochrane, K. L. & Garcia, S. M. (eds), pp. 105–134. FAO, Rome and Blackwell Publishing.

Kooiman, J., Bavinck, M., Jentoft, S. & Pullin, R. (eds). 2005. *Fish for life: Interactive governance for fisheries.* MARE Publication Series No. 3, University of Amsterdam Press.

Levin, P., Fogarty, M., Murawski, S. & Fluharty, D. 2009. Integrated Ecosystem Assessments: developing the scientific basis for ecosystem-based management of the ocean. *PLoS Biol.* 7(1): e1000014 .d0i:10.1371/journal/pbio/1000014.

Longhurst, A. 1998. *Ecological geography of the seas.* Academic Press. San Diego. 398 p.

MACC. 2010. *Ecosystem-based management approach for managing uses of the marine environment.* Biodiversity Working Group Report to the Marine & Coastal Committee. April 2010.

Mace, P. M. 2004. In defence of fisheries scientists, single species models and other scapegoats: confronting the real problems. *Marine Ecology Progress Series,* 274, 285–291.

MacMahon, J. 1998. Empirical and theoretical ecology as a basis for restoration: An ecological success story. Chap. 9. In: *Successes, Limitations, and Frontiers in Ecosystem Science,* Pace, M. & Groffman, P. (eds), pp. 220–247. Springer Verlag, New York.

Mahon, R. & Escobar, E. 2009. The Wider Caribbean Region. In: *An Assessment of Assessments, Findings of the Group of Experts. Start-up Phase of a Regular Process for Global Reporting and Assessment of the State of the Marine Environment including Socio-economic Aspects.* UNEP and IOC-UNESCO (eds), pp. 231–242. Progress Press Ltd., Malta.

Mahon, R. & McConney, P. 2004 (eds). *Management of large pelagic fisheries in CARICOM.* FAO Fisheries Technical Paper No. 464, 149 p.

Mahon, R., Fanning, L. & McConney, P. 2009. A governance perspective on the large marine ecosystem approach. *Marine Policy,* 33, 317–321.

Mahon, R., Fanning, L. & McConney, P. 2011. *Observations on governance in the Global Environment Facility (GEF) International Waters (IW) Programme. Version 3.* July 2010. CERMES Technical Report No. 45. 40 p.

Mahon, R., McConney, P. & Roy, R. 2008. Governing fisheries as complex adaptive systems. *Marine Policy,* 32, 104–12.

Mahon, R., McConney, P., Parsram, K., Simmons, B., Didier, M., Fanning, L., Goff, P. Haywood, B. & Shaw, T. 2010. *Ocean governance in the Wider Caribbean Region: Communication and coordination mechanisms by which states interact with regional organisations and projects.* CERMES Technical Report No. 40, 84 p.

McConney, P., Oxenford, H. A. & Haughton, M. 2007. Management in the Gulf and Caribbean: Mosaic or melting pot? *Gulf and Caribbean Research,* 19(2), 103–112.

McLeod, K. & Leslie, H. (eds). 2009. *Ecosystem-Based Management for the Oceans. Island Press.* Washington, DC. 368 pp.

Meltzer, E. 2009. *The Quest for Sustainable International Fisheries: Regional Efforts to Implement the 1995 United Nations Fish Stocks Agreement.* NRC Research Press, Ottawa. 448 pp.

Murawski, S. A. 2007. Ten myths concerning ecosystem approaches to marine resource management. *Marine Policy* 31, 681–690.

Namibia Ministry of Environment and Tourism. 2009. *The Green Paper for the Coastal Policy of Namibia. Ministry of Environment and Tourism,* Windhoek. 92 pp.

National Marine Fisheries Service. 1999. *Ecosystem-Based Fishery Management.* Report to Congress by the Ecosystem Principles Advisory Panel. National Oceanic and Atmospheric Administration, Department of Commerce. Silver Spring, MD. 66 pp. www.nmfs.noaa.gov/sfa/EPAPrpt.pdf.

National Marine Fisheries Service. 2009. *National Standards and Guidelines* (50 CFR 600.310 et seq.).

National Oceanic and Atmospheric Administration (NOAA). 2011. *Clarifying the relationships among ecosystem-based management, integrated ecosystem assessments, and coastal and marine spatial planning.* NOAA response to NOAA SAB letter of April 5, 2010. March 2, 2011. Washington, D.C. 14 pp. [available at www.sab.noaa.gov/Reports/Reports.html].

National Oceanic and Atmospheric Administration, Science Advisory Board (NOAA SAB). 2006. *Evolving an Ecosystem Approach to Science and Management Throughout NOAA and its Partners.* The External Review of NOAA's Ecosystem Research and Science Enterprise. Silver Spring, MD. 85 pp. [available at www.sab.noaa.gov/Reports/Reports.html].

National Oceanic and Atmospheric Administration, Science Advisory Board (NOAA SAB). 2011. *Strategic Advice on Designing and Implementing Coastal and Marine Spatial Management Plans.* NOAA. Silver Spring, MD 36 pp. [available at www.sab.noaa.gov/Reports/Reports.html].

National Research Council (NRC). 1999. *Sustaining Marine Fisheries.* National Academy Press. Washington, DC. 164 pp.

National Research Council (NRC). 2001. *Marine Protected Areas: Tools for Sustaining Ocean Ecosystems.* National Academy Press. Washington, DC. 268 pp.

Nielsen, L. A., Knuth, B. A., Ferreri, C. P., McMullin, S. L., Bruch, R., Glotfelty, C. E., Taylor, W. W. & Schenborn, D. A. 1997. The stakeholder satisfaction triangle: a model for successful management. In: *Developing and sustaining world fisheries resources: the state of science and management.* Hancock, D. A., Smith, D. C., Grant, A. & Beumer, J. P. (eds), pp. 183–189. Proceedings of the 2nd World Fisheries Congress, CSIRO Publishing, Melbourne.

NOO. 2003. *Oceans Policy: Principles and Processes.* National Oceans Office, Hobart, Tasmania, Australia. 20 pp. http://www.environment.gov.au/coasts/oceans-policy/publications/pubs/principles-processes.pdf

Olsen, E., Gjøsæter, H., Røttingen, I., Dommasnes, A., Fossum, P. & Sandberg, P. 2007. The Norwegian ecosystem-based management plan for the Barents Sea. *ICES J. Mar. Sci.,* 64 (4), 599–602.

O'Toole, M. J. 2010. Ocean Governance in the Benguela Large Marine Ecosystem—Establishing the Benguela Current Commission. Chapter 9. In: *Sustainable Development of the World's Large Marine Ecosystems During Climate Change.* Sherman, K. & Adams, S. (eds), pp. 139–150. IUCN, Gland, Switzerland.

Osmond, M, Airame, S., Caldwell, M. & Day, J. 2010. Lessons for marine conservation planning: A comparison of three marine protected area planning processes. *Ocean and Coastal Management,* 53, 41–49.

Ostrom, E. 2009. A general framework for analyzing sustainability of socio-ecological systems. *Science,* 325, 419–422.

Parenteau, P., Baur, D. & Schorr, J. 2009. Legal authorities for ecosystem-based management in US coastal and ocean areas. Chapter 19. In: *Ocean and Coastal Law and Policy.* Baur, D., Eichenberg, T. & Sutton, M. (eds). American Bar Association, Section of Environment, Energy, and Resources. New York. 708 pp.

Pikitch, E., Santora, C., Babcock, E., Bakun, A., Bonfil, R. & Conover, D. et al. 2004. Ecosystem-based fishery management. *Science,* 305, 346–7.

Pitcher, T. J., Kalikoski, D., Short, K., Varkey, D. & Pramoda, G. 2009. An evaluation of progress in implementing ecosystem-based management of fisheries in 33 countries. *Marine Policy,* 33, 223–232.

Raemaekers, S., Hauck, M., Bürgener, M., Mackenzie, A., Maharaj, G., Plagányi, É. E. & Britz, P. J. 2011. Review of the causes of the rise of the illegal South African abalone fishery and consequent closure of the rights-based fishery. *Ocean & Coastal Management,* 54, 433–445

Rosenberg, A. A., Mooney-Seus, M. L., Liessling, I., Mogensen, C. B., O'Boyle, R. & Peacey, J. 2009. Lessons from national-level implementation Across the World. Chapter 16. In: *Ecosystem-Based Management for the Oceans.* Karen McLeod & Heather Leslie (eds). 2009. Island Press, Washington, DC. 294–313.

Ruckelshaus, M., Essington, T. & Levin, P. 2009. Puget Sound, Washington, USA. Chapter 12. In: *Ecosystem-Based Management for the Oceans.* McLeod, K. & Lester, H. (eds). Island Press. Washington, DC. 368 pp.

Schechter, M., Leonard, N. & Taylor, W. (eds). 2008. *International Governance of Fisheries Ecosystems: Learning from the Past, Finding Solutions for the Future.* American Fisheries Society, Bethesda, MD. 458 pp.

Sherman, K. & L. M. Alexander. 1986. *Variability and Management of Large Marine Ecosystems.* AAAS Selected Symposium 99, Westview Press, Boulder.

Sissenwine, M. & Murawski, S. 2004. Moving beyond "intelligent tinkering": advancing an ecosystem approach to fisheries. *Marine Ecology Progress Series.* 274: 291–295.

Sivas, D. A. & Caldwell, M. 2008. A new vision for California ocean governance: comprehensive ecosystem-based marine zoning. *Stanford Environmental Law Journal.* 27, 209–270.

Sumaila, U. 2010. A cautionary note on individual transferable quotas. *Ecology and Society,* 15(3), 36–44.

Sumaila, U. R., Munro, G. & Keith, H. 2005. Benguela Current Commission (BCC) Economic Study. BCLME,Windhoek. 61 pp. Available at http://www.bclme.org/projects/docs/PCU-BCC-04-02-Final%20Report.doc (accessed 17 November 2011).

UN DESA. 2010. *The Millennium Development Goals Report.* United Nations Department of Economic and Social Affairs, New York. 80 pp. Accessed at http://www.un.org/millenniumgoals/pdf /MDG%20Report%202010%20En%20r15%20-low%20res%2020100615%20-.pdf on 22 September 2011.

UNCED. 1992. Protection of the oceans, all kinds of seas, including enclosed and semi-enclosed seas, and coastal areas and the protection, rational use and development of their living resources. Chapter 17. *Report of the United Nations Conference on Environment and Development* (Rio de Janeiro, 3–14 June 1992) http://www.un.org/depts/los/consultative_process/documents/A21-Ch17.htm accessed on 4 October 2011.

UNDP. 2001. Project Document RAF/00/G32/A/1G/31: *Integrated management of the Benguela Current large marine ecosystem.* United Nations Development Programme, New York. 125 pp.

University of Alaska Sea Grant Program. 1999. *Ecosystem approaches for fisheries management.* AK-SG-99-1, Fairbanks. 756 pp.

WCED. 1987. *Our Common Future.* World Commission on Environment and Development. Oxford University Press. 400 pp.

Winsnes, I. & Skjoldal, H. R. 2008. Management Plan for the Norwegian Part of the Barents Sea Ecosystem. In: *The Ecosystem Approach to Fisheries.* Bianchi, G & Skjoldal, H. R. (eds), pp. 228–246. FAO, Rome.

Witherell, D., Pautzke, C. & Fluharty, D. 2000. An ecosystem-based approach for Alaska groundfish, *ICES Journal of Marine Science,* 57, 771–777.

World Bank (WB). 2009. *The Sunken Billions. The Economic Justification for Fisheries Reform.* World Bank, Washington DC. 130 pp.

Worm, B., Hilborn, R. & Baum, J. and 18 other co-authors. 2009. Rebuilding global fisheries. *Science,* 325, 578–585.

Yaffe, S., Philips, L., Frentz, I., Hardy, P., Maleki, S. & Thorpe, B. 1996. *Ecosystem Management in the United States: An Assessment of Current Experience.* Island Press, Washington, DC. 352 pp.

Young, O. R. 2008. The architecture of global environmental governance: bringing science to bear on policy. *Global Environmental Politics,* 8, 13–42.

Young O. R. 2011. If an Arctic Ocean treaty is not the solution, what is the alternative? *Polar Record.* Cambridge University Press. 47, 327- 334. doi:10.1017/S0032247410000677.

Chapter 5. Modeling Approaches for Marine Ecosystem-Based Management

ELIZABETH A. FULTON

Commonwealth Scientific and Industrial Research Organization
Division of Marine and Atmospheric Research

JASON S. LINK

National Oceanic and Atmospheric Administration
Northeast Fisheries Science Center

Contents

1. Introduction

Models play a central role in Ecosystem-Based Management (EBM) as a vehicle for synthesis and integration of physical, ecological, social, and economic data to inform management decisions. One of the chief challenges of EBM is the inherent difficulty of elucidating ecosystem dynamics and establishing threshold levels and reference points for management. This has made using management performance measures particularly challenging. Models are a way to handle these inherent challenges. As such modeling is an essential tool in understanding the dynamics of social-ecological systems, defining targets and limits for management, and testing and predicting the outcome of alternative management actions to meet these challenges. Models can also play a valuable role in education and outreach to stakeholders.

Coping with uncertainty is a critical consideration in marine EBM. One form of management that is expressly intended for use when system state is uncertain is adaptive management (Walters, 1986). This form of management relies on feedback control mechanisms that use new information to adjust management settings or approaches. Successful EBM employs this evidence-based approach. Thus tools used in support of EBM must provide either information for setting targets or the ability to explore management constraints cognizant of uncertainty. While supplying such tools and defining forms of EBM that can deliver sounds like a daunting task, experience is showing that there are a wide range of useful EBM approaches

The Sea, Volume 16, edited by Michael J. Fogarty and James J. McCarthy
ISBN 978-0-674-07270-1

(including analytical methods, indicators, frameworks and governance considerations) and that such decision support tools do exist; we note that modeling approaches are chief among them.

In this chapter, we highlight several aspects of ecosystem modeling with particular relevance to the main themes of this volume. To provide a context for the remainder of the chapter, we begin with an illustration of the power of modeling to delineate choices for management and as a vehicle for communication and interaction with stakeholders. We then provide an overview of some basic tenets of effective modeling and their value in addressing management issues. One of the most important roles of modeling in this context is to permit evaluation of alternative management strategies. Management strategy evaluation (MSE) is a key element in the development of an important tool for EBM, Integrated Ecosystem Assessments (see Plummer and Levin, this volume). Effective use of MSE and models in support of EBM requires careful choice of a model that is "fit for purpose". Space does not permit a complete description of the full range of models relevant to marine EBM (see Supplemental Materials). Accordingly, we've selected example applications of the most widely used model types to highlight how they can address full multisectoral marine EBM. Throughout, we provide several case studies based on our own experience to illustrate key points and concepts.

2. Approaches to Facilitate Decision Support for Marine EBM

Before launching into a description of methods for modeling complex marine or coastal social-ecological systems, it is useful to discuss how models can be used to inform management and facilitate decisions. There are several approaches that can be used to give a system perspective to managers. Qualitative models (Dambacher et al., 2003; Dambacher et al., 2009), risk analysis (e.g., Canter, 2001; Hobday et al., 2007; Hobday et al., 2011), formal decision support systems (DSS), and multi-model inference are all approaches that have been used for vetting ecosystem model outputs and prioritizing model needs. Some of these approaches help to prioritize what warrants modeling and some approaches help to winnow down information from model outputs that would be most useful for making decisions. In terms of EBM, experience has shown that multi-model inference that is performed in a structured, iterative and hierarchical way can incorporate many of the most attractive features among the plethora of approaches to lead to the most robust and informative end product.

Different kinds of models can take on complementary roles in EBM, each with their own specific role, but building toward a larger common understanding. The application of multiple model types in an analysis of management options offers important advantages. If multiple models based on differing premises or methods provide similar responses, the confidence in the robustness of the general result is heightened. This approach is analogous to ensemble modeling commonly used to address uncertainty in atmospheric and climate sciences (Krishnamurti et al., 2000; Gneiting and Raftery, 2005; Tebaldi and Knutti, 2007).

As an example, we note an effective multi-model method that has been used in support of multiple use management in Ningaloo Reef region of northwest Australia (Fulton, 2011). This multi-model approach uses five different model types to inform policy-making in the region and de-coupling stakeholder engagement from the development of a full-system model aimed at answering core management questions. This separation allows for continuous engagement and considerable flexibility in addressing specific questions as they arise during project development while simultaneously providing a basis for teaching stakeholders about the problems that arise when trying to manage complex nonlinear systems (something few people have an intuitive feel for; Sorrentino and Roney, 2000; Boschetti et al., 2010).

To fulfill the educational as well as decision support roles for the models, five broad classes of models were employed in the Ningaloo case:

(i) Conceptual models, which summarize system understanding and focus on the main drivers of a system and highlight what needs to be incorporated in the other models. These are typically represented diagrammatically and are often drawn up in conjunction with stakeholders and topic experts.

(ii) Toy models,[1] which are used to present a highly simplified version of the problem that runs rapidly, can be modified quickly, and can be used in training sessions with stakeholders (i.e., hands-on) so that they can understand how models are constructed and how complex systems respond (e.g., stocks and flows causing lags or feedbacks leading to counterintuitive system responses). This understanding then helps stakeholders start to construct questions for the more complicated models and to identify intervention points and the implications of system interconnectedness. These toy models are also incredibly useful for verifying the soundness of assumptions about system behavior adopted from the conceptual models and for showing modelers how people make decisions in certain circumstances, allowing for a better representation of the human components in the more complex models.

(iii) Single-system (i.e., ocean-use sector) models, which provide a fairly detailed representation of a single component of the system (e.g., fishing or tourism). These models can be used to provide preliminary results about a single activity in the system, which subsequently feed into the development of the final full-system model, or can be used in their own right in support of sector-specific management issues (e.g., fisheries regulations).

(iv) Partial-system (or "shuttle") models, which are based on the "minimum realistic model" concept, incorporating the minimum number of processes that capture all the crucial components for a basic understanding of the overall problem. These models provide a route of communication between stakeholders and modelers, allow further training to build upon experience gained with the toy models (increasing trust, understanding and democratization of knowledge). They also provide further understanding on key mechanics that can be shuttled into the full system model.

[1] For those readers who feel uncomfortable with the word toy, this nomenclature is not intended to trivialize these kind of simple models, rather to highlight that humans learn most effectively when actively engaged (i.e., playing) with a tool.

(v) Full-system models, which synthesize all the information collected on the system, include some representation of all the major parts of the system (physical, ecological, economic and social) and address all the major stakeholders concerns (whose definition has been greatly eased by the use of 'simpler' models). A highly simplified version of the model structure illustrating the potential mix of agent types is given in Figure 5.1.

Other versions of multi-model inference differ slightly (e.g., they may not employ such distinct model types, but may be a comparison of multiple partial system models). Irrespective of their form, multi-model approaches have many advan-

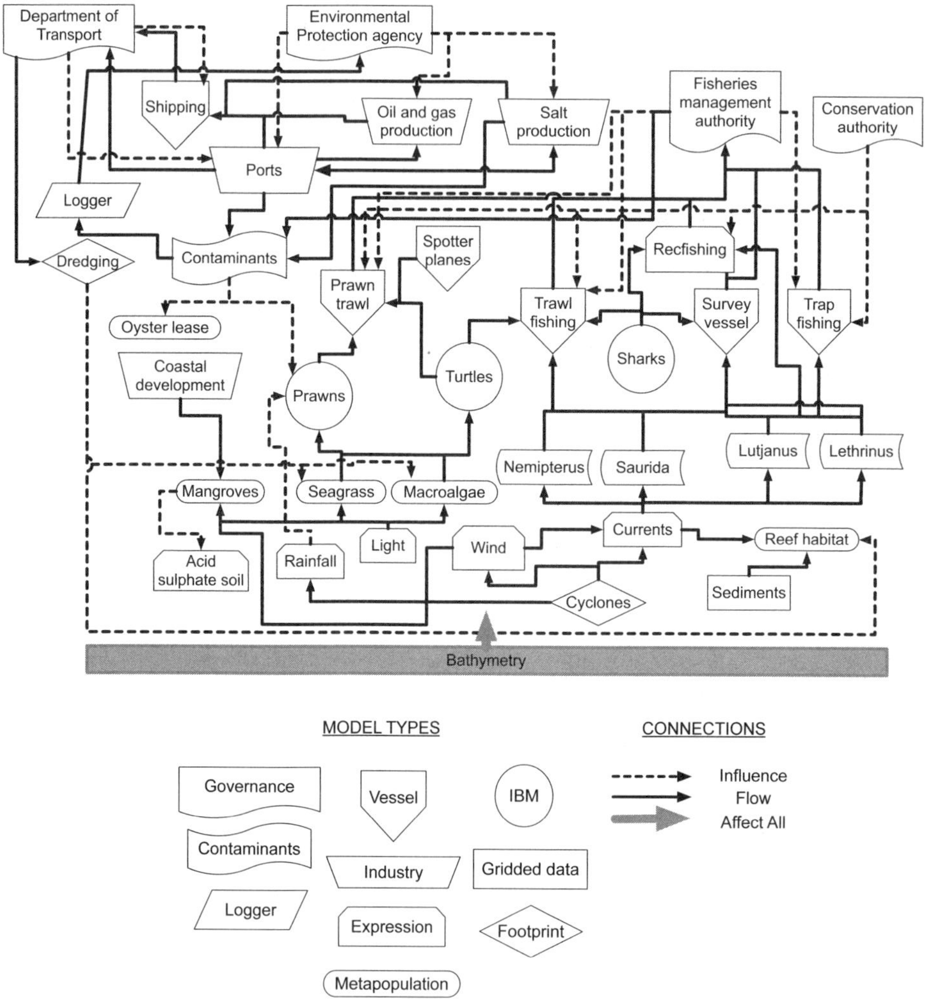

Fig. 5.1 – An example of a simple InVitro model structure showing how different kinds of agents (representing different formulation types from analytical models through to true individual-based models that are driven by decision trees) can be linked together (from Fulton, 2010, with permission from Elsevier). Model structures are often far more complex than this if being applied to EM questions at a region scale.

tages: such an approach can increase the robustness of decision making—by supplying supporting tools, providing information around the robustness of understanding or management methods and by increasing stakeholder experience with models and complex systems so they can better use the information provided. In situations where the system is used by many sectors, the multi-model approach (particularly if used in an MSE framework) provides for a greater understanding of how adaptive management works, how objectives can be in conflict (or require trade-offs) and what the implications of alternative futures for a system may be.

3. The Value of Models

Whether used in such a complementary multi-model approach or not, models form a solid foundation for EBM in a number of ways. Statistical models can be used for the evaluation of longstanding fisheries and oceanographic monitoring surveys and the development of systemic indicators. Process-oriented models can be used to elucidate ecological and environmental relationships pertinent to the marine resources or ecosystems of interest. Such models can be used to make comparisons amongst ecosystems or to determine unique and general marine ecosystem properties. Finally, models can be used to explore the range of EBM options using "in silico" virtual studies—e.g., management strategy evaluations that explore the relative importance of ecosystem processes and the robustness of various management strategies. Together these different uses of models collate and integrate a broad range of data types and sources; evaluate the relative importance of several concurrent processes; and allow for the formulation and testing of hypotheses concerning ocean system structure, functioning, and efficacy of management actions.

Within the marine realm the use of models has been most widespread in support of fisheries. However, over the last decade modeling in support of multiple use management has begun. Modeling studies spanning ecosystem services, such as CO_2 storage and sequestration or nutrient cycling, and the cumulative activities and effects of industries such as fisheries, conservation, energy, shipping, tourism are becoming a feature of integrated coastal zone management in some locations (e.g., tropical Australia).

While many of the examples of model use in marine systems are fisheries-focused, many of the concepts map naturally to EBM. Perhaps the earliest and most widespread reason for the development of marine models was the estimation of fishery production potential, particularly when attempting to set meaningful Biological Reference Points (BRPs; e.g., such as maximum sustainable yield). Evaluation of production potential of a given region is critical whether considering single stock or ecosystem-based fisheries management. Originally these calculations were done on a species-by-species basis, but they are increasingly being performed at an aggregate or ecosystem level (Lucey et al., 2012). This shift in focus is due to acknowledgment of the shared dependence of the ecosystem on energy delivered by the lowest trophic levels (Pauly and Christensen, 1995; Pauly et al., 1998; Pauly et al., 2002); the differential productivity of fish stocks, which makes it difficult to simultaneously attain single stocks objectives in multispecies fisheries (May, 1975) let alone a multiple use context; and the complex nature of taking management decisions when there are biological or technological interactions

(Garrod, 1973; May, 1975; Pope, 1979; Mayo et al., 1992). Extending the concept in this way not only builds on existing familiarity with the concept of reference points, but also benefits from the fact that aggregated indices of this form tend to be more stable than those at the population level (Link et al., 2012a).

The next most common use of modeling in the context of marine management is to establish ecosystem-level perturbation (i.e., overfishing) thresholds and criteria for evaluating the status of the system or fish communities from a broader perspective. The rationale and value of this use of models has been copiously noted previously (Murawski, 2000; Fulton et al., 2005; Jennings, 2005; Link, 2005; Link, 2010; Samhouri et al., 2010; Shin et al., 2010a), but briefly this use provides a way to detect major changes that can lead to irreversible regime shifts in ecosystems and food webs, it coordinates across a broader range of ecosystem components, it establishes trigger points for control rules and allows for factors associated with other ocean-use sectors to be directly considered. The natural companion to the use of models to identify system thresholds is to use models to assess current system state (to judge against the thresholds). This use of models for providing support for tactical management advice, with respect to ecosystem terms of reference, is becoming more common, expanding beyond simply providing context to become integral data streams in stock assessments (e.g., by providing estimates of predation mortality, or providing for the inclusion of environmental drivers of growth and stock-recruit relationships) to developing BRPs for aggregated or system-level resource considerations.

Another widespread use of models has been in the context of cumulative impacts and multiple use management for EBM. Cross-jurisdictional and cross-sectoral mandates, such as the Australian Environmental Protection and Biodiversity Conservation Act or the US Presidential Executive Order of July 2010, require consideration of the trade-offs across multiple ocean use sectors (see Cochrane et al.; Rice et al., this volume). Globally, there are few examples of true multisectoral agencies. The few examples that do exist tend to be multi-national in nature (e.g., HELCOM, which has charge of the Baltic—http://www.helcom.fi/; and OSPAR which has oversight for the management of the NE Atlantic—http://www.ospar.org/) (see Cochrane et al., this volume; Houde et al., this volume; Gaichas et al., this volume). At national and regional levels integrated assessments that pull together the influences of multiple industries are beginning to coalesce (McDonald et al., 2008; Levin et al., 2009; McLeod and Leslie, 2009; Plummer and Levin, this volume); many of these use models to evaluate the viability of various management, trade-off or development options (often using the Management Strategy Evaluation approach, detailed below). Use of models in this role is likely to grow in the future, particularly in response to societal needs to understand the consequences of pressures from growing populations and large-scale global climate change.

4. Management Strategy Evaluation

One of the most effective means of using models in support of EBM is Management Strategy Evaluation (MSE), a simulation technique based on modeling each part of the adaptive management cycle—the biophysical world, resource users, as well as the management decision process (see also Plummer et al. this volume).

Thus MSE frameworks involved creating a representation of the resource(s), but also of the human components of the system, how they operate and the feedbacks between the two—effectively a virtual world. Critically the human operators in this virtual world have imperfect knowledge of the state of the biophysical components. This means these model worlds are perfect test beds for exploring the implications, for both the biophysical realm and associated resources and stakeholders, of alternative management strategies (defined by combinations of monitoring data, assessment procedures, and decision rules).

MSE was originally developed in the late 1980s to support consideration of the implications of alternative management strategies for whale stocks (IWC, 1992). This approach subsequently expanded from applications for whales to fish stocks and other natural resources (Kirkwood, 1997; de la Mare, 1998; Punt and Smith, 1999; Butterworth and Rademeyer, 2005). The method is now widely accepted as a best practice approach for single stock (Butterworth, 2007) and EBM questions (FAO, 2007) and is being increasingly adopted for cross-sector multiple use questions and integrated coastal zone management (McDonald et al., 2008; Fulton et al., 2011a). A regional MSE was a key part of the Ningaloo multi-model example outlined above. MSE is also used in the context of sustainability or EBM analyses at many jurisdictional levels, from international bodies [e.g., the International Whaling Commission (IWC), Commission for the Conservation of Antarctic Marine Living Resources (CCAMLR), Marine Stewardship Council (MSC)] to national fisheries departments (e.g., in South Africa, Australia and New Zealand) and even state, regional or local level governments (e.g., state regulatory bodies in Australia).

The approach has many strengths:

(i) Instead of trying to find an optimal solution, MSE evaluates alternative hypotheses (typically in the form of management strategies) using multiple candidate models to represent the uncertainty about the function of, and relationships between, different steps of the adaptive management cycle (resource dynamics, human use, monitoring, assessment, management decision making, social and economic filtering of management regulations shaping new human activities).

(ii) The MSE approach is consultative, with the potential for all interested parties to have input into the candidate models and management scenarios. This participatory facet also means it works best when used iteratively. Ultimately, the approach demands clear objectives to judge the performance of the alternative strategies, forcing participants to be clear about their objectives and to specify relevant performance indicators. In practice such clarity takes time to emerge, with people first having to come to terms with the concept of a virtual world and how to use it.

(iii) The outputs of MSE highlight unanticipated consequences of management actions and system responses and trade-offs between objectives held by different stakeholders. Explicitly allowing for an analysis of outputs against a range of competing or conflicting objectives means that the method is not crippled by existing difficulties in determining management objectives that take into account diverse societal goals or ecosystem metrics (Marasco et al., 2007).

Used to its fullest capacity MSE allows a common interpretation of available information; provides a basis for the development of skills and attitudes needed when facing complex problems; and offers an avenue for communication, negotiation and collaboration. Here the model forms a common talking point, helping to diffuse conflict between groups with different worldviews who may interpret and draw very different conclusions from the same information (Leviston and Walker, 2010). Thus the approach effectively acts as a 'flight simulator,' giving interested people insight into system dynamics. This is an important management aid, as without it managers may act on theoretically optimal solutions that are beyond what is actually feasibly achieved given the system's current state. As a result well-intentioned management actions can have counter intuitive and counterproductive outcomes. The history of fisheries management is replete with such examples (Smith and Link, 2005; Fulton et al., 2010) and EBM runs the same risk. Useful reviews of the MSE approach can be found in Butterworth and Punt (1999), Smith et al., (1999), Sainsbury et al., (2000), Plagányi et al., (2007) and Rademeyer et al., (2007). The ultimate point of MSE is to test management strategies in a virtual world before they are actually implemented.

5. Modeling Approaches

A wide range of model types can be used in support of EBM, from physico-chemical environmental models (e.g., Chen et al., 2001; Franks and Chen, 2001; Townsend et al., 2006; Hu et al., 2008) to ecologically focused multispecies (Horbowy, 1989; Little et al., 2008) or food web based models (e.g., Field et al., 2006; Mackinson and Daskolov, 2007) through to socio-economic models and approaches (e.g., Holland and Sutinen, 1999; Edwards et al., 2004) and whole ecosystem models (which incorporate both biophysical and social-economic components). Here we provide an introduction to some of these kinds of models. We will start with a brief description of types of models that consider only part of the system. However, the majority of this chapter will focus on examples of the application of system models for EBM. Our intention is to provide an entry point to the broader literature. However, this list is far from exhaustive and readers interested in sourcing further information on models used in support of EBM should seek out reviews such as Hollowed et al. (2000), Plagányi (2007), Townsend et al., (2008) and FAO (2008). Further, to assist readers trying to decide which approach is most useful in their situation, we have also provided a summary in an appendix to this chapter that discusses the major model types, their individual focus, strengths, weaknesses, data needs and uses. The examples in the rest of this chapter represent only a minor subset of the range of possibilities.

5.1 Conceptual and qualitative models

Qualitative models remain an informative means of gaining insights into system functioning (Levins, 1966) and EBM (Dambacher et al., 2003; Dambacher et al., 2009). This form of model has no specific data requirements and can be drawn together based on surveys, industry time series, expert information or any other available data or understanding (although having access to some validating data to confirm interactions is helpful). This means that this approach can easily be applied at system-level scales and is perfect for EBM applications. The method can also be deployed rapidly—a modeling exercise can take as little as a few hours. The rapidity of qualitative models also means that a large number of alternative model structures and configurations can be readily explored, or key system features filtered out for further consideration (e.g., as done by Dambacher et al., 2011; Hosack and Dambacher, 2010). This combination of speed of application and flexibility means stakeholder advice and input can be incorporated easily, which aids engagement and ensures their requirements are relatively easily addressed. In turn this improves uptake, and can be very useful for identifying key system components, processes and potential feedbacks. These approaches can be used alone, which gives the method a systematic and analytical basis, or as a form of triage for further modeling approaches, informing subsequent choices regarding model structure.

While qualitative models can be used conceptually, loop analysis provides the potential to gain additional insights from these models. Loop analysis uses the conventions of signed digraphs (from network theory) to assign positive and negative signs to each connection, or interaction, in the qualitative model. Combined with the algebraic methods described by Dambacher et al. (2003), the digraphs provide significant insight into potential system-level evolution and responses to press perturbations. The approach does have some constraints, particularly in that the complicated nature of large networks means that the results of any loop analysis are ambiguous. Moreover qualitative models cannot anticipate alternative system structures (e.g., before and after a regime shift), instead those need to be defined ahead of time and analyses run on both versions.

EBM Modeling Example 1—Qualitative EBM. An illustrative example of the EBM use of qualitative modeling is Dambacher et al. (2007). This modeling study synthesized findings from social and biological research programs on Lihir Island, Papua New Guinea, in order to understand the interaction of the human and ecological systems; particularly, the complex interactions between gold mining activity, marine resources, and the island communities. Dambacher et al. (2007) then used the qualitative models developed (combined into a single system in Figure 5.2) to consider potential future impacts of mining and artisanal fishing on the island's natural and social systems. The models indicated that expansion of mining operations would expand the workforce and the island's cash economy (allowing for more artisanal fishing), as well as lead to increased sewage inputs, sediment flow, feasting and associated land clearing. These activities affect primary production (via nutrients from sewage), suppress corals (via smothering by sediment from land erosion and mine wastes) and lead to a restructuring of reef fish communities (due to fishing pressure). The only aspect of the system the model suggested would not change was the intensity of subsistence farming, as the cash income is used to buy "luxury foods" (e.g., meat and sugar) rather than replace the woman's cultural role as "gardeners" producing starchy staples. Just as with quantitative models (discussed further below), qualitative analyses of social-ecological system models can then be used to plan future multiple use development, management and research.

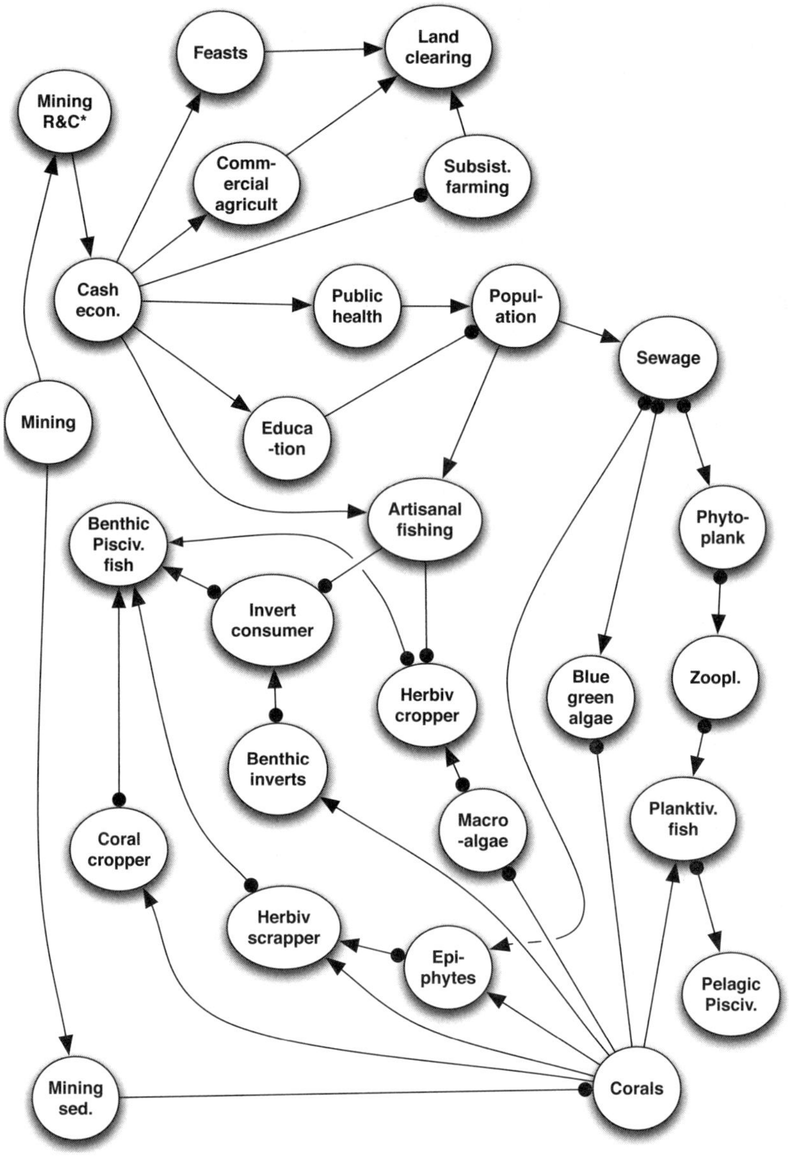

Fig. 5.2 – Signed digraph model of the shallow-water reef-edge fish community and socioeconomic system of Lihir Island, Papua New Guinea (drawn by combining Figure 2a and 2b in Dambacher et al., 2007). Note that R&C stands for royalties and compensation paid by the local gold mining industry. The self limiting negative feedbacks of each component on itself (representing density dependent effects) have been omitted to avoid clutter in the diagraph. Only the mining components (sediments and R&C do not have such a self limiting feedback).

5.2 *Single system (sector) models*

Not all models used to inform management give equal attention to all parts of the system. Whether because of a disciplinary focus or partial information availability, models focused on a sub-set of the system have been used to provide information. These may be focused on physical elements of a system—like nutrient state, oxygen levels, temperature, salinity or pollution levels—or specific target species, economic considerations or social connections. In each case the other parts of the system are represented as simple drivers or closure terms (e.g., fishing mortality, or time series of nutrient inputs). While these models do not include as many feedbacks as full system models they can still be informative, especially if the sub-system represented is strongly correlated with the system components of particular interest or if the number of activities being considered is constrained—one doesn't need to worry about modeling things that are not actually present in the system.

Biophysical models of the physical environment (either of the climate or hydrodynamic flows) have been used in EBM to explore physical state-production relationships. While the correlation of environmental drivers and fisheries production has been of interest to fisheries science for some time (Holden and Raitt, 1974; Beamish and Bouillon, 1993; Beamish et al., 2004; Keyl and Wolff, 2008; Hollowed et al., 2009), it has become a particular focus once the potential magnitude of climate change over the next few decades to centuries was realized. Over the last decade the output of general circulation models has been used to provide time series of physical drivers (e.g., temperature, salinity or oxygen fields) for ecological models, which in turn have been used estimate distributions of species or catch (e.g., Sarmiento et al., 2004; Cheung et al., 2008; Cheung et al., 2009; Cheung et al., 2010; Barange et al., 2011; Blanchard et al., 2011).

The dynamic coupling of physical models and full ecosystem representations is under development, but not currently in regular use. Yet coupling with biogeochemical plankton models does occur fairly regularly (Cox et al., 2000; Friedlingstein et at., 2001; Dufresne et al., 2002; Stock et al., 2011), particularly for the assessment of water quality or the exploration of alternative flow or catchment management regimes, outfall levels (e.g., from sewage plants) or coastal developments (e.g., aquaculture, channel modification or effluent release). In some cases these models have been used to produce potential prey fields to feed single species fish population models and so provide predictions of potential biomass, or fisheries production (e.g., Kishi et al., 2007; Oguz et al., 2008; Fennel, 2009). The greatest potential drawbacks of these models are (i) their tendency to have significant computational costs and (ii) no explicit representation of feedbacks from the highest trophic levels (or social and economic considerations).

At the other extreme, bioeconomic models concentrate on economic drivers, but often with very little explicit biology included beyond a "resource pool" (e.g., Kennedy, 2003). Although there are a few more complex instances that include industry decision-making and management combined with multispecies stock models (Sparre, 2008). Moreover, few (to date) consider non-economic social drivers of human behavior. A review of these kinds of models is beyond the scope of this chapter, but interested readers are encouraged to read thorough reviews like those of Prellezo et al. (2009). Consideration of intrinsic, market, and non-market value of ecosystem services is essential in evaluating alternative management choices and in tradeoff resolution. These topics are accorded separate treatment in Guerry and Tallis (this volume) and Altman et al. (this volume) and will not be further considered here.

Not all models that focus on the human aspects of marine ecosystems focus on the economic aspects. Because of the centrality of the coupled social-ecological construct for marine EBM, we provide a more detailed description of nascent efforts to model social dynamics. Social processes are now also being more fully integrated into full system models (see below). While it is more common for social information to be presented descriptively, there are a few different ways in which social information is captured in models. Social process models are not as common as biophysical or economic models. They are typically agent-based models, which are better suited to capturing the information networks, drivers and characteristics that are most relevant to social systems (Weisbuch and Duchateau-Nguyen, 1998; Wilson et al., 2007). Other forms of social models focus on networks (social, communication or quota trading) or the assessment of sociocultural variables relevant to the management questions—e.g., the effect of a management action on social equity, individual health and the population's well-being (Pollnac et al., 2006; Lasseter, 2008). While these kinds of models are capable of handling processes not easily captured in any other way—such as communication networks—these models can be difficult to validate (Janssen and Ostrom, 2006) and reliable data can be hard to access (Hall-Arber et al., 2009). The specific data requirements are dependent on the processes being models, but will need to cover the characteristics of the agents in the model, the process parameters and exchange rates. Formal frameworks for describing the parameterization, and fostering repeatability, have also been defined (Smajgl et al., 2011).

One of the most difficult aspects of modeling or assessment of social aspects of EBM is the selection of appropriate variables. Social Impact Assessments (SIA) are an attempt at creating an assessment method that could reflect social implications of fisheries in a way that is similar to biophysical and economic assessments. These assessments begin with a scoping process to determine the sociocultural variables relevant to the management questions. The social entities captured in the assessments can include individuals, firms, families, and communities, or all of these (in which case they must be treated as distinct units of analysis). Like all modeling approaches, the objective is not to include every element in the system, but to identify critical components of the system. In an SIA these are populations that have a significant stake in the management action, or related issues that can affect the population's well-being (Pollnac et al., 2006). After the scoping step the relevant variables are operationalized and the relationships between them specified; either within single levels of analysis (e.g., the community) or across levels of analysis (e.g., individual, family, or community). The SIA data demands are concerned with operationalizing the variables of interest—specifying the external forces, management regulations and structures, all the attributes of the fishing (or other industry) activities, measures of satisfaction associated with the activities, community attributes (e.g., social stratification, solidarity, occupational structure), individual attributes (e.g., health, personality), potential social problems and well being. While this approach has the advantage of directly addressing the social aspects of the system missed by most other EBM approaches, it can be quite difficult to establish defensible indices of the specific variables (Hall-Arber et al., 2009).

One particularly useful approach, network analysis, has been adapted from other areas of social science. It has less rigorous data demands, but remains highly informative—as social or financial interactions between resource users influence decision-making and system dynamics. These models have been used to consider social networks in fishing communities or cooperatives (Lasseter, 2008), information networks (Maiolo et al., 1992), or changes in quota trading patterns through time (van Putten et al., 2011). These studies have provided understanding about how fishermen make decisions and adapt to changing availability of resources and alternative fisheries policies. The strength of the approach is its conceptual simplicity—it is simply the connections between actors in the system, which means it is easily extendible to other resource users in an EBM context. Although in actuality reliable data on this can be quite difficult to access.

5.3 Partial system models: The simpler end of the spectrum

At the simpler end of complex models, Minimum Realistic Models (MRMs) (also known as Models of Intermediate Complexity for Ecosystem assessments (MICE)) only include those ecosystem components most likely to be important for capturing the main effects of the management question under consideration. In addition, the formulation focus is to reduce complexity by employing simpler representations, but without sacrificing so much detail that fits to the available data degrade.

In general, MRMs attempt to explain the underlying ecological processes for a population, or group of populations and their anthropogenic interactions. These models focus on essential subwebs, such as target species and their predators and prey, relevant ecological processes, environmental drivers or habitat and dominant industries (e.g., fishing, tourism or land use). Decision-making processes of those involved in the industries, and their management, are given as much attention as ecological processes. There are many different forms of MRM depending on their disciplinary backgrounds. For example, some have been developed from ecological models (e.g., the riparian restoration-salmon population models of Watanabe et al., 2005) while others come from "stocks and flows" approaches (e.g., the tourism-community model of Dutra et al., 2011). A good review of the MRM concept, and model development decisions required when creating such a model for EBM purposes, is given in Plagányi et al. (2012). A summary of the key considerations are given in Table 5.1, as similar decisions and logic must be used when pulling together more complicated full system models (discussed further in following sections).

While no model can be exhaustingly inclusive, MRMs are intentionally constrained in an attempt to provide information in support of EBM while still incorporating the best features of existing single-species models of relative simplicity, such as the ability to apply standard statistical methods for parameter estimation. This is what makes them attractive as a tactical (day-to-day or annual) management tool as they are easily incorporated into existing management arrangements (e.g., their outputs are amenable to calculating BRPs). Moreover they are amenable to the use of standard data fitting methods, which are already familiar to fisheries and coastal zone scientists and managers.

TABLE 5.1.
Key considerations when defining the contents of system models, with examples discussed in the main text.

System Component	Logic
General complexity[1]	
Process detail	Only include key processes or effects-relationships. Where an uncertain process is key, assess the sensitivity of model outputs to the use of alternative formulations.
Temporal resolution	If rapid processes or short lived groups, use finer time steps.
Spatial resolution	Include explicitly if a critical resource (i.e., strong spatial structure in processes).
Biophysical	
Environmental drivers	Include where data streams at the correct resolution are available and where there is clear evidence that the physical state (e.g., temperature) is a significant influence on other system processes of interest (e.g., ecological processes).
Trophic composition	Include key commercial and conservation species and their main sources of prey and predation. For each modelled group aim to capture >90% of their mortality within the model.
Population structure	If there is data available and key ecological processes are governed by age/size then use an age/size structure model. Otherwise use aggregate numbers or biomass.
Animal movement & behaviour	Represent using time invariant distributions or dynamically if driven by potentially changing processes.
Species interactions	Inclusion of predation, competition and habitat association should all be considered carefully. Similarly whether they will be one-way or two-way connections (i.e., feedback or not, note though that feedback is often important). Where information for parameterisation is limited, use effects rather than process models (e.g., effects of consumption on survival, growth and reproduction rather than modelling consumption itself).
Recruitment	Use of standard stock-recruitment relationships is an option, but empirical relationships with other system components are also possible (or a combination of the two where the state of one model component, for example a prey field, modifies the stock-recruitment relationship). Spatial transport of recruits can be a key consideration in some circumstances.
Anthropogenic	
Sectors	Include key sectors of interest. If not a full multiple use case then it is still important to include the effects of other human activities at the location (even if only represented as invariant patterns or forced time series).
Human behaviour	As with animal behaviour this can be represented using simple invariant rules or rates (e.g., fishing mortality) or highly detailed and process based if dynamic responses are anticipated or of interest.

1. More complex representations should only be used where needed for accuracy in capturing key mechanics and where other dimensions of the model allow for a realistic representation. For example, there is no point putting in a highly complex animal movement model if there is only a very simple spatial distribution with annual time steps and no habitat, environmental or prey representations.

Minimum realistic models are also natural candidates for use in MSE and multimodel inference in support of EBM. They can be flexibly tailored to specific applications in collaboration with stakeholders, they capture the key system components (biophysical, industry, management) and they can formally deal with uncertainty—whether parametric, structural or in terms of uncertainty of the potential effectiveness of the implementation of a management action (implementation uncertainty; Link et al., 2012b). These features, in combination with the fact they build on methods already familiar with stakeholders, make MRMs a robust method that is likely to be accepted by most stakeholders. Thus there is significant potential to rapidly advance EBM using these approaches.

MRMs also have a second role in MSE and EBM. There is the potential to use them as an assessment model. That is, instead of making them the system model at the core of the MSE, they can become the "simple" assessment model used in the management process, with a more complicated full system model providing the system representation. In this way the potential benefits (or pitfalls) of using MRMs for tactical decision-making can be formally assessed. This opens the way to using them as a means of accounting for ecosystem considerations in real world management processes.

In their simplest form MRMs are simply extended single species assessment models, where predation, consumptive demands, or the environment is added to a single species assessment model (Plagányi, 2007; Townsend et al., 2008; Link et al., 2012c). These models are not only useful for providing estimates of stock biomass, but also for delineating sources of other mortality (e.g., predation) and serving as "reality checks"—either for estimates of allowable catch (e.g., checking recommended target species quotas will not adversely impact predator or prey species) or for estimates of biomass, making sure they are consistent with what is feasible given the estimated ecosystem state (e.g., Link and Garrison, 2002; Tyrrell et al., 2007; Overholtz et al., 2008; NEFSC, 2010). Technical interactions between human uses of the system can also be considered (Pelletier and Mahévas, 2005).

The critical issue to balance is selecting appropriate complexity (see e.g., Fulton et al., 2003; Pinnegar et al., 2005; Quince et al., 2005). While keeping the number of species (or species groups) considered low reduces the complexity of the food web and the parameterization, it can introduce significant artifacts if the original web was highly diffuse (Yodzis, 1998; Yodzis, 2000). This is why more and more MRM approaches include multiple species. These can take many forms: whether they are effectively multi-species forms of age-structured population models linked by feeding relationships (e.g., GADGET, Stefansson and Palsson, 1998; MSVPA, ICES, 1991) or other multispecies extensions of simple single species approaches (e.g., yield per recruit analyses or surplus production models [Murawski, 1984; Gamble and Link, 2009]); or take more novel approaches by structuring fish communities and ecosystems via relationships based on size (e.g., Jennings et al., 2002. Hall et al., 2006; Blanchard et al., 2009) or trophic levels (e.g., Gascuel et al., 2008; Gascuel and Pauly, 2009).

Some MRMs are true system models that consider the ecological system (potentially including physical drivers), how it is impacted by human activities (potentially including explicit models of human behavior), and how the modeled processes are impacted by management, but also how they feed back to effect management in return. These kinds of models are intermediate in complexity between single-species assessments and whole-of-ecosystem models and have been expressly developed in an attempt to provide a tactical EBFM (and ultimately EBM) capability (Plagányi et al., 2012)—such as providing reference points based on ecosystem considerations. Consequently, they are designed based on the problem being addressed in tight combination with data available to parameterize the model, with parameter estimation based on standard optimization methods (e.g., Fournier et al., 2011) and parameter uncertainty quantified using methods such as Markov Chain Monte Carlo (MCMC) sampling, inversion of the Hessian matrix, or likelihood profiles (Plagányi et al., 2012).

EBM Modeling Example 2—MRM: EBM on a Budget: There are many good examples of partial system models (or MRMs) applied in fisheries or other single sector applications around the world, making the selection of a single example quite difficult. While early EBM examples are beginning to take shape, a fisheries example is perhaps best to showcase the flexibility and utility of the approach. One region that has seen extensive use of the partial system model approach is the Barents Sea. Through the 1980s and 1990s there were many multispecies models developed for the region (e.g. Bulgakova et al., 1995; Bogstad et al., 1997; Tjelmeland and Bogstad, 1998) and since the late 1990s there has been a priority put on the development of MRMs, which are more tractable in a management context. Tools such as GADGET (Globally applicable Area-Disaggregated General Ecosystem Toolbox; Stefansson and Palsson, 1998) have put a premium on including multispecies interactions while also explicitly including statistical handling of uncertainty (Taylor et al., 2007). While any of the MRM approaches could easily be extended in an EBM context, the Barents Sea GADGET-FLR implementation by Howell and Bogstad (2010) is a very good illustrative example of how selective inclusion of parts of the system that directly influence stock dynamics can be particularly informative about future trends, uncertainty and management implications.

Howell and Bogstad (2010) use an MSE approach to consider the performance of alternative harvest control rules under a range of enforcement regimes, environmental conditions and uncertainties about the magnitude of system processes (such as the effect of temperature on cod recruitment). Building a partial system model involves making pragmatic decisions about how much of the system to include and the links between the selected components. In the case of the Barents Sea a monthly time step and four spatial areas were represented. The interaction of cod, herring, capelin and minke whales has been identified as critical for ecological understanding of stock and fisheries dynamics in the region, as was the feedback nature of the fisheries and their management (Figure 5.3). Howell and Bogstad (2010) also chose to use assessment models available in the "Fisheries Libraries in R" (FLR) package so that standard ICES stock assessments could be represented along with potential mismatches between system structure or state and the assumptions underlying the assessment, as well as any management process lags. The final model created by Howell and Bogstad (2010) is a good example of a well made partial system model as its structure includes key system components, but process detail is kept to fundamentals with potential system connections that are poorly understood represented by simple formulations that can be easily varied or modified, allowing for maximum understanding. For instance, instead of trying to define some form of process-driven environmental influence on stock-recruitment relationships, observed deviations for a Beverton-Holt curve were drawn from historical time series and applied during projections so that realistic autocorrelation within and between species is captured.

Howell and Bogstad (2010) tried a range of scenarios, from ones considering changed harvest control rules, to scenarios considering varying levels of herring recruitment, through to scenarios that vary possible environmental influences on the system. The projections indicated that the currently agreed harvest control rules were sufficient to maintain viable herring and cod stocks, though little could

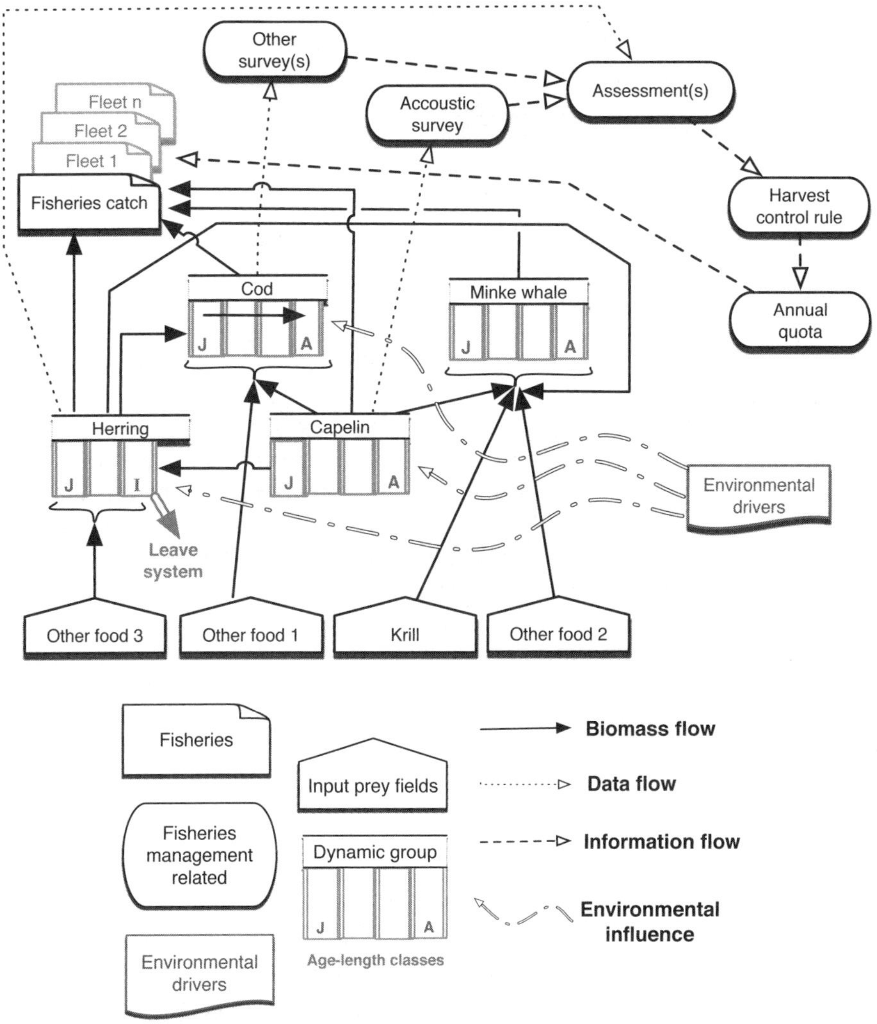

Fig. 5.3 – Overview of the GADGET-FLR integrated model (redrawn from Figure 1 in Howell and Bogstad, 2010; note that the direction of arrows have been reversed from Howell and Bogstad to be consistent with the convention used in the other figures in this chapter). The ecological processes not shown are individual growth, maturation, reproduction, prey selection (and fisheries selectivity) and movement. The biophysical and fishing components are contained in GADGET and the fisheries management components in FLR.

be done at current herring stock levels to reduce the vulnerability of capelin to stock collapse (with herring consumption of capelin larvae a key mechanism in that outcome). Varying the degree of compliance, or target fishing pressure levels, affected the stability of catches and long term averages, but did not see the cod stocks collapse.

The representation of the anthropogenic components in this example model was relatively simple, and in its current form other human sectors can only be represented via the impact they have on system function (e.g. climate change impacts on the environment or industrial run-off impacting productivity or recruitment). While using drivers in this way would be a straightforward EBM addition to the simple MSE style of this particular implementation, more explicit multi-sector considerations could be addressed by applying the MRM principles to selectively extend the model to explicitly incorporate any EBM relevant activities in the area (e.g., shipping, oil and gas extraction, etc).

5.4 *Partial system models: A fuller system perspective*

In contrast to the partial models discussed above that are quite selective in terms of the system components included, other system models more fully represent some aspects of the system and how they are connected without quite taking the final step to "full system models". For instance, they may represent food webs (attempting to be broadly inclusive of all trophic levels and major functional group types), but not always the habitat connections, environmental forcing or finely resolved human activities. These have most often been used to consider fisheries, but can be extended to cover the many other uses of marine and coastal systems. Qualitative versions exist, but in terms of quantitative models these models are generally categorized as moderately or highly complex. It is this level of complexity that distinguishes them from MRMs, though in reality it is an artificial split as there is really a continuum of partial system models (on to full system models) of varying degrees of complexity.

As these models typically address many of the interactions at the core of EBM, they may also be known as network models—sharing some of the history and concepts of the network analyses discussed above for social aspects of EBM. Ecopath with Ecosim (EwE) has been applied in over 200 systems (Christensen pers. comm.) and is one of the most widely used model-based EBM tools. EwE is composed of a mass balance model (Ecopath; Polovina, 1984; Pauly et al., 2000; Christensen and Walters, 2004; Christensen et al., 2005) from which temporal (Ecosim) and spatial (Ecospace) dynamic simulations can be developed (Walters et al., 1997). The simple, but versatile way in which EwE encompasses the representation of marine food webs does mean it runs the risk of misuse (particularly if there is limited data, though this is true of all the models discussed in this chapter). Nevertheless, the ease of use of EwE has also been a major strength. It has been used to describe and explore ecosystem structure and functioning, to provide ecosystem as a basis for comparative studies (spatial and temporal), and to explore the impacts of environmental drivers and fishing on ecosystems (Christensen and Pauly, 1992; Christensen et al., 2005; see also Polovina et al., this volume). The software is under continuous development and has grown to include economic market chains, management strategy modules, and parameter sensitivity analyses. It is now beginning to be coupled to Nutrient-Phytoplankton-Zooplankton-Detritus (NPZD) models, hydrodynamic models, and even General Circulation Models (GCMs) for forecasting climate effects—taking it closer to the realms of a "full system model".

The EwE software outputs a wide range of system indices (especially for the static Ecopath model), but the main outputs usually considered are biomass and catch through time (and space) per group and fishery. The input data requirements scale with the size of the web considered, with estimates of biomass, production and consumption rates, catch and diets required for each group specified in the model. Additionally, if environmental or facilitative relationships are to be included in Ecosim the form of these must be specified and spatial distributions and habitat preferences are required for Ecospace.

EBM Modeling Example 3—Trophic model: EwE example.

A key fisheries management question is whether it is possible to optimize yield simultaneously for all stocks. This issue has repeatedly brought ecosystem considerations to the fore in the NW Atlantic region. In an ecosystem context one key question is, "can we manage all species that we have at their single-species biomass at maximum sustainable yield (B_{MSY}) levels at the same time?" This question is driven by the result of fisheries management decisions (1994–2005) for a range of 19 groundfish stocks, which has resulted in the resurgence of some, but not all, depleted fish populations (NEFSC, 2008). This differential recovery raised a topic of some concern among stakeholders as to whether the ecosystem can support biomass at estimated B_{MSY} levels simultaneously for all the groundfish stocks, and more broadly, the entire fish community in the region.

The total system biomass and the total biomass under fishery management targets were computed for the U.S. Northeast large marine ecosystem. Information on the single species biological reference points for groundfish, other demersal species, and small pelagic components of the U.S. Northeast Large Marine Ecosystem (LME) were summarized and compared to historical studies, recent energy budgets for the region and to similar estimates from comparable worldwide temperate, marine systems. The technical basis for estimating biological reference points for groups of species has been well established, both classically (Garrod, 1973; May, 1975; Pope, 1975; Fukuda, 1976; Pope, 1979) and in more recent modeling and empirical studies (Mayo et al., 1992; Pauly and Christensen, 1995; Pauly et al., 1998; 2002).

Although a wide range of modeling approaches was applied to address questions concerning simultaneous MSY, here we focus on results from EwE models (Link et al., 2008). Results for pelagic and demersal biomass were compared to those for nine temperate and boreal systems, including Canadian, European and west coast U.S. systems (NEFSC, 2008).

Overall, these and other aggregate production model results (NEFSC, 2008) suggest that the estimated MSY level for all species is lower than the sum of individual species MSY estimates and overall fishing mortality should be lower. Therefore, all species are unlikely to simultaneously be at B_{MSY} if interactions among species are important, a conclusion supported by several lines of evidence and multiple approaches. One key piece of evidence was the mass-balance constraint elucidated by EwE models, highlighting that there are indeed tradeoffs among the fish community resultant from differential management practices and inherent ecologies.

5.5 *Full system (end-to-end) models*

Cross-scale and cross-sector interactions are more explicit in full system, or end-to-end, models, which attempt to include all major relevant processes in the system (physical, ecological and increasingly, social-economic features). These highly complex models have been developed specifically as simulation test beds for EBM and integrated multi-sector management. As such they have attempted to include all relevant critical processes, components and scales. Abiotic processes incorporated can include: catchments, riverine and atmospheric inputs; currents and other major water body features like eddies, upwelling, downwelling, turbulence, stratification and resuspension; wind; irradiance; precipitation and water column properties such as temperature, salinity, acidity and resulting zones such as the mixed layer and photic zone. Ecologically these types of models span all trophic levels and processes such as: nutrients and biogeochemical cycling, including the microbial loop; different types of detritus and their breakdown pathways; benthos and benthic dynamics (both in and on the sediments), bioturbation, bioirrigation and particulate entrainment; pelagic and benthic primary production; and the major processes associated with invertebrate and vertebrate consumers (pelagic and demersal) like movement, reproduction, ontogeny, feeding, waste production. These models also include habitat-based interactions, not just trophic connections. The representation of the different ecological groups can be of many different types, from size-based to biomass pools, or full age or staged based models. In some cases even separate genetic stocks are represented per species or functional group. The aspect of end-to-end models to receive the most recent attention is human activities, which can now span competing and cumulative impacts of: terrestrial run-off; coastal development, ports and shipping; fishing (recreational, artisanal and commercial); tourism and other recreational activities; oil and gas exploration and extraction (including pipelines and other infrastructure needs); and less obvious uses such as defense (e.g., bombing ranges). These activities may only be represented via impact functions and external drivers, or they may be represented by detailed dynamic sector models.

The complicated nature of these models is intimidating and highly uncertain. Moreover, they are time consuming and costly to use and have extensive data requirements spanning many disciplines (an example of the kind of data required for Atlantis, for instance, is given in Table 5.2). Nevertheless these models have proven to be useful for highlighting critical feedbacks and non-linear responses that may not be discernible any other way. While the general trends, patterns and relative distributions they produce are informative, the absolute values are far less reliable. This makes them inappropriate for use as tools for setting tactical management measures, like fisheries quotas or nutrient load caps, which is the role of more single-sector focused models like fisheries stock assessment or biogeochemical models. Instead end-to-end models are best used to consider system-level 'what-if' management or impact scenarios of a more strategic nature (Fulton et al., 2010).

Over the last 15 years marine system models have developed from a largely ecological focus, with a simple treatment of fisheries, to become truly end-to-end models by incorporating the suite of components and processes listed above. Many of these take a hybrid approach, representing each part of the system in a different

TABLE 5.2.

Key information needs of a typical Atlantis model. Physical properties include things like currents, light, temperature, salinity etc. Fisheries data are only needed if progressing the model beyond ecological only stages. Socioeconomic data are only required if using socioeconomically driven effort models. Critical ecological data is marked with an *.

	Physical properties	Nutrients	Primary producers	Benthic invert.	Plankton	Nektonic invert.	Fish	Shark and rays	Mammals and birds	Humans
Biophysical										
Time series of states or flows	X									
Period of activity				X	X	X	X	X	X	X
Habitat dependency			X	X	X	X	X	X	X	X
Distributions and abundances*		X	X	X	X	X	X	X	X	X[1]
Swimming speeds							X	X	X	
Migration times and routes				X[2]		X[2]	X	X	X	
Diets[1]*				X	X	X	X	X	X	
Consumption (clearance) and growth rates*			X	X	X	X	X	X	X	
Length-weight relationships*							X	X	X	
Size at age*							X	X	X	
Non-predation mortality rates			X	X	X	X	X	X	X	
Maximum age*							X	X	X	
Age at maturity							X	X	X	
Spawning area and season				X[2]		X[2]	X	X	X	
Larval or gestation period						X	X	X	X	
Location of recruits							X			
Fecundity				X[2]		X[2]	X	X	X	
Size of recruits (young of the year)					X		X	X	X	
Fisheries related										
Catchability (and/or selectivity)				X		X	X	X	X	
Targeting										X
Discarding										X
Catch and effort (time series and distributions)										X
Management rules										X

1. Human population per port; for defining home ports for fisheries as well as the level of recreational fishing pressure in an area
2. Using age structured biomass

way, to best capture the desired resolution for that system component. Extensive examples could be drawn from the implementation of models incorporating relationships between climate variability, recruitment and biomass fluctuations (Lehodey, 2005; Maury, 2010), the Object-oriented Simulator of Marine ecOSystems Exploitation (OSMOSE) model (which is an agent-based model that has been applied to the dynamics of exploited fish populations; Shin and Cury, 2004), or instances where EwE has been coupled with climate or catchment models, or the latest generation of coupled models like those being developed by the Bering Sea Integrated Ecosystem Research Program (http://bsierp.nprb.org/) in the Bering Sea (linking climate scenarios, a Regional Ocean Modeling System (ROMS) hydrodynamic model, a nutrient and plankton biogeochemical model, a trophically coupled fish community model and finally an economic driven fisheries model; all resolved to a 10km geographic grid). Space precludes an exhaustive treatment of examples; thus we have chosen to give only two illustrative examples of these large-scale models—Atlantis, for system models focused on fisheries that still include an EBM context, and InVitro, for a true EBM modeling perspective.

EBM Modeling Example 4—Atlantis: Transitioning to EBM. Atlantis is an end-to-end model that integrates physical, chemical, ecological and fisheries dynamics in a 3D spatially explicit domain (Fulton et al., 2011b). In addition to ecological interactions and habitat-related processes and dependencies, it contains environmental components, such as a simulated ocean (with complex current fields, temperature, salinity, alkalinity and the potential to include estuarine or catchment components, riparian or edging vegetation, ice or terrestrial rookeries), a simulated monitoring and assessment process, a simulated set of ocean-uses (primarily fishing and aggregate dredging, but with the capacity to represent time series of pollutant release, shoreline habitat degradation), and a simulated dynamic adaptive management process. The dynamics represented in the model range from solar radiation to hydrodynamics, nutrient processes, growth (with age structure), feeding, settling, sinking, migration, fishery captures (including the differentiation of target and bycatch), effort allocation, market valuation, regulation, quota trading, investment/disinvestment, port activity levels, friendship networks and feedback among the various components of the model as appropriate.

Atlantis is built around the MSE concept. Thus it can incorporate multiple forms of myriad processes, can emphasize those considerations and processes most appropriate for a given system, and can be used to evaluate management decisions. Atlantis can be used to consider the indirect effects of the removal of species or functional groups, or to explore all major forms of fisheries management tools currently in use—such as area closures (spatial or seasonal), effort and technical controls, catch limits, and quotas (tradable or aggregate). Ecosystem scenarios can be explored separately or in conjunction with any of the fisheries scenarios to provide information on the robustness of management actions in the context of broad development, climate change, ocean acidification, eutrophication, hypoxia or similar phenomena.

Atlantis has been used to consider the effects of river flow on estuarine populations (Condie et al., 2012) and the cumulative effects of future change (Fulton, 2011), but has most often been used for exploring alternative fisheries management schemes (Fulton et al., 2011b), general fisheries questions (Worm et al., 2009; Fulton et al., 2012) or robust ecological indicators (Fulton et al., 2005; Branch et al., 2010; Samhouri et al., this volume), often as part of multi-model analyses (e.g., in combination with Ecopath with Ecosim models) when testing indicators. The most comprehensive Atlantis application to date is Atlantis-SE (Fulton et al., 2007), which was the basis of an MSE used to inform Australia's Fisheries Management Authority about strategic restructuring options for the South East Scalefish and Shark Fishery (SESSF)—a complex of multispecies, multigear fisheries that must meet an extended list of environmental, conservation, economic and social objectives. Atlantis-SE covers an area of 3.7 million km^2 and includes shifting climate-related environmental drivers, quite complex ecology (including a 61 member size-resolved microbial and food web from primary producers to top predators, with 37 age-structured groups, some resolved not only to the species level but to the level of genetic stocks) and the explicit representation of recreational and commercial fishing fleets; with the commercial fleets resolved to sub-fleets (defined by homeports, crew and vessel sizes) and driven by social and economic drivers that determine investment, disinvestment, quota trading, information updating, and effort allocation. All of this was also set against a background of river flows and the influences of other sectors, such as pollutant outfalls, coastal habitat modification due to development and spatial management zones imposed by the Department of Environment.

The set of potential management strategies explored with Atlantis-SE were defined by stakeholders in the fishery (including scientists working in the region, industry members, economists, resource managers and environmental NGO representatives). Each of these strategies involved an integrated package of management measures, although most focused on a particular type of fisheries management. Of the many strategies considered, four are particularly illustrative. One of these strategies, which acted as a reference baseline, represented the management measures in place in 2003 (limited entry and vessel restrictions, some largely ineffective gear restrictions, generally unconstraining quotas, and largely ineffective spatial management); another strategy focused on enhanced quota management. A third strategy was focused on spatial management while a fourth strategy represented a balanced combination of quotas, spatial management and gear controls. When considered across the different classes of objectives (summarized in Figure 5.4), none of these strategies outperformed all of the others across all objectives. Each strategy had its own strengths and weaknesses. The status quo strategy performed poorly economically and environmentally, though there was still significant investment in the fisheries (e.g. new vessels, gear, etc,) under this scenario as the management system was understood and the fishers were comfortable with access levels. The enhanced quota system had the greatest management efficiency, while the spatial management strategy the greatest level of stock recovery (though the fishery was actually economically unviable). The integrated strategy did not out-perform the other strategies across all objectives, but it had the fewest shortcomings, consistently ranking in the top two strategies trialed across all objectives and avoiding the pitfalls that undermined the performance of the other strategies. As such it was considered to be the scenario with the most desirable outcomes overall.

Importantly, no single management lever (i.e., tool) emerged as a solution by itself. Spatial management was insufficient as the dominant lever because it proves difficult to strike the balance between protecting sufficiently large conservation

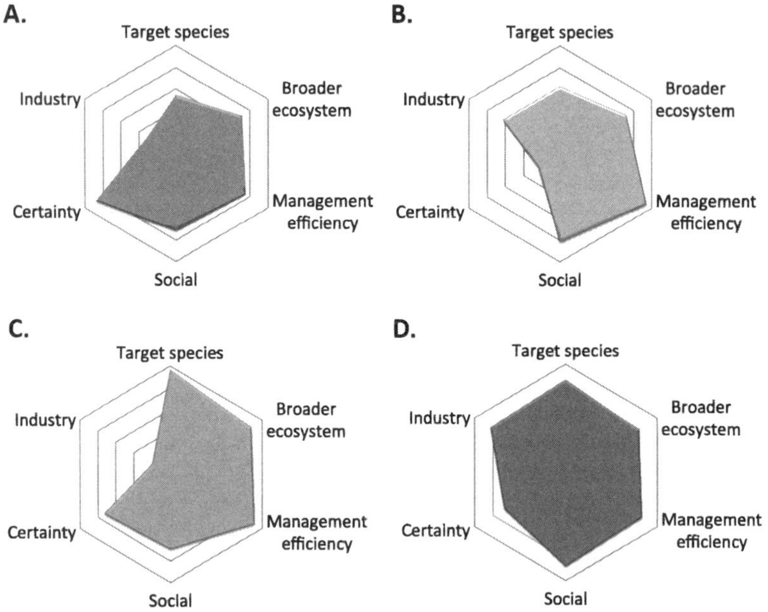

Fig. 5.4 – Example radar (or "amoeba") plots of four of the alternative management strategies assessed in the SESSF MSE based on Atlantis-SE: (a) is the status quo as of 2003, (b) is enhanced quota based management, (c) is focused on spatial management and (d) employs a balance of management levers. The objectives were grouped into rough categories, with a different category assigned to each axis of the radar plot; higher values are closer to the outer edge of the plot.

areas and the provision of sufficiently large fishing grounds to sustain an economically viable industry (Fulton et al., 2007). Similarly, quota management was insufficient as the intent of the ITQ-based management system was undermined by incentives and human behaviors—such as high grading, misreporting of quota species catches and the targeting of species not in the quota management system. These fisher behaviors—along with other issues around quota implementation such as lags in the quota setting cycle, non-constraining quotas, unaccounted sources of mortality (e.g., discards), partial spatial or life history coverage and stock influences not addressed by quotas (e.g., habitat)—led to unsustainable biomass levels when enhanced quota management was the basis of the strategy. In that case less than half the target groups had biomasses at or above 40% of unfished levels.

Notably the inclusion of explicit human behavior was important for capturing unintended consequences. Atlantis-based analyses often highlighted the potential for management strategies not to play out as intended. For example, small operators often found it difficult to remain profitable in the long term, despite lower cost structures, due to social constraints degrading adaptive capacity. These kinds of responses resulted in differential impacts of management among fleets (Fulton et al., 2007) and demonstrated that the addition of management levers can create new management issues—particularly pertaining to compliance, enforcement and the interaction of responses to management in one sector affecting the system in such a way that it impacted on objectives for other sectors. For example, management actions imposed to conserve offshore species increased pressure on inshore target species. Inclusion of human behavior and expanding the scope of the models beyond fisheries has also been important for ongoing analyses of cumulative impacts of global change and fisheries on ecosystem (e.g., Griffith et al., 2011; Griffith et al. in press) and barriers to adaptation (Fulton et al., 2012).

The output of the Atlantis-SE MSE for the SESSF and a comparative qualitative modeling based exercise (Smith et al., 2004) contributed information to a strategic management assessment that led to major changes in the breadth and scope of management tools being adopted in the fishery. A library of Atlantis runs is also available on a day-to-day basis to provide insight on new strategic management issues as they arise, at www.csiro.au/seaview. This degree of uptake has been possible for two reasons. First, Australia's Environment Protection and Biodiversity Act requires that all export and federally managed fisheries demonstrate that they are ecologically sustainable, which requires explicit consideration of the ecological effects of fishing. Second, Australia has a well-developed, participatory fishery management system (Smith et al., 1999; Smith et al., 2001) that has made extensive use of MSE and models. Thus the demand for EBM and ecological indicators, in combination with stakeholder familiarity with MSE, has led to a level of comfort that models can reflect their own 'on the water' experience.

EBM Modeling Example 5—InVitro: True EBM. Although Atlantis does address fisheries in an EBM context, InVitro is a true EBM model. InVitro (Gray et al., 2006) uses a modular agent-based modeling approach to allow for the simultaneous representation of many aspects of coastal and marine systems. These include environmental drivers, ecological ecosystem components, fine scale human decision models of local communities and industry operators (across multiple industries), regional economics, management, monitoring and assessment. This modularity not only provides flexibility, by providing a choice of alternative representations for each part of the system, but also simulates the many different processes using a mix of analytical equation-based formulations (particularly for system-scale, physics or lower trophic level processes) and algorithmic, individual-based behavioral models (for rare species, top-level predators or human activities occurring at the finest scales). Each sub-model runs at the appropriate time and spatial scale.

The exact form of the model is tailored to the system of interest, but can become quite complex when addressing sustainable multiple use management questions. For example, a version of InVitro that has been implemented in northwestern Australia (focusing on Ningaloo Reef and Exmouth Gulf; Little et al., 2006; Fulton et al., 2011a), has:

- an oceanographic and climate sub-model, including rainfall, temperature, salinity, geomorphology, nutrient and contaminant layers;
- 60 ecological components that make up the food-web and habitats, spanning primary producers, benthic habitats, benthic invertebrate communities, pelagic forage fish, the main target fin-fish and crustacean species, top predators and species of special interest like turtles and whale sharks;
- 1:1 representations of each of the resident communities and main industries active in the region, including commercial and recreational fisheries, tourism, oil and gas extraction, salt production, coastal development, urban services,

pastoral leases, horticulture, ports, shipping and road transport, regional economics, catchment use, recreation, conservation and infrastructure like roads and pipelines.

Providing details on all facets of this model would require a book in its own right (interested readers are referred to Gray et al., 2006; Fulton et al., 2011a; Gray et al. in press). Suffice it to say that as much attention has been given to the anthropogenic components as to the biophysical aspects of the system; the simplified version of the model structure in Figure 5.1 illustrates the potential mix of agent types in the model. Both biophysical and anthropogenic components are represented using a combination of mechanistic process models, analytical decision models, response functions, specified rules, historical data and scenarios. The model is also stochastic, so as to capture natural ambiguity in human and animal behavior, and the influence of missing or incorrect information and unpredictable and catastrophic events. While this array of agent types is fairly comprehensive, the computational costs can be significant (taking more than a week to simulate a thirty year period).

The InVitro model of Ningaloo was just one part of a multi-model inference MSE that not only introduced stakeholders to models, but used models as tools to provide stakeholders with an appreciation for complex systems, elucidate understanding of the structure and function of the Ningaloo-Exmouth region, explore system behavior, project alternative futures, and provide a common basis to foster collaboration and communication. As noted above, the full suite of models used in this multi-model exercise included qualitative and conceptual models, simple stocks and flows representations, single sector models (e.g., ELFsim, Little et al., 2011; and a tourism destination model, Jones et al., 2011), Ecopath with Ecosim and the full system InVitro model. While a lot was learned from the full suite of models here we will concentrate on the outputs of the InVitro model.

Extensive and iterative interactions with groups interested in the Ningaloo-Exmouth region produced close to 100 combinations of management strategies and contextual scenarios that could describe alternative futures for the region. Between them, these spanned many management methods (e.g., spatial management, fishing regulations, increased emphasis on education and enforcement) and addressed key development questions for the region, from the existing Ningaloo Regional Coastal Strategy, to hypothetical developments of various sizes, to significant infrastructure investments and residential developments driven by the growth of the oil and gas industry in the region.

Multivariate cluster analysis and principal components analysis performed on the simulation results showed that there were eight major classes of outcome or dynamic systems (Fulton et al., 2011a). While there was variation in specific details within each of these classes there were some common signifying features per class, summarized in Table 5.3. The performance of each class of outcome against stakeholder-defined system objectives is provided in Figure 5.5. The dynamics of particular system components were fairly consistent across simulations. For instance, the proportion of the population in older age groups consistently increases relative to today, while large and potentially vulnerable megafauna like whale sharks and turtles decrease (to differing extents depending on the management strategy in place).

TABLE 5.3.
A summary of the major classes of system behaviour observed in the InVitro-Ningaloo MSE

Class	Base case	Reduced growth	Changed management, no developments	Ecolodges and reduced growth	Introduction of modified bag limits	Large developments (resorts or resource sector)	Large developments and modified bag limits	Changed climate
Description	• Little if any population growth. • Stagnation of the economy of the region, rising unemployment and an aging population. • The visitor mix slides more toward recreational fishing, seeing a further decline in stocks and the state of the natural system. • There is some reduction in demand for services and infrastructure.	• In gross terms similar in trend to the base case, but more extreme. • There is a contraction in the state of the system away from nearly all objectives (environmental, social or economic).	• The changes to spatial management zones or user education typically mitigate some of the environmental impacts (e.g., on habitats), but have little overall effect on the system.	• Development of Ecolodges (even with reduced growth in the rest of the economy), sees a significant per capita increase in spending. • Also stronger environmental outcomes – due to the different visitor profile attracted. • Still does not prevent strong competition for local housing. • There is not enough development to lead to widespread land release, but there is sufficient tourism to see a strong labour market for that sector (meaning in-migration of workers and competition for existing dwellings).	• Introduction of modified bag limits (in line with the wilderness fishing concept) provides significant protection for fish stocks, which increase relative to 2006–2010 levels • Stocks rebuild to a point where the probability of catching trophy fish per trip significantly increases. • Recreational fishers attracted to the region help maintain the local economy (though they do compete with residents for the local houses).	• Major expansion of the resource sector, tourism or other industry leading to a major increase in the economy. • Related increases in road transport, resident population and demand on services and infrastructure. • All of this comes at the cost of the environment.	• As for large developments but with less of an impact on fish stocks (to the point they remain at the levels of 2006–2010 or even increase).	• Agriculture contracts with some other aspects of the local economy (e.g., the peak tourism season contracts through the cooler months and there is a drop off in tourism segments unwilling to pay for air conditioned accommodation). • Slower growth in the population. • Decline in available services. • Ecological impacts are strong, with the habitats becoming more vulnerable (due to the combined effects of storms and acidification) and turtle nesting beaches heavily impacted by sea level rise.

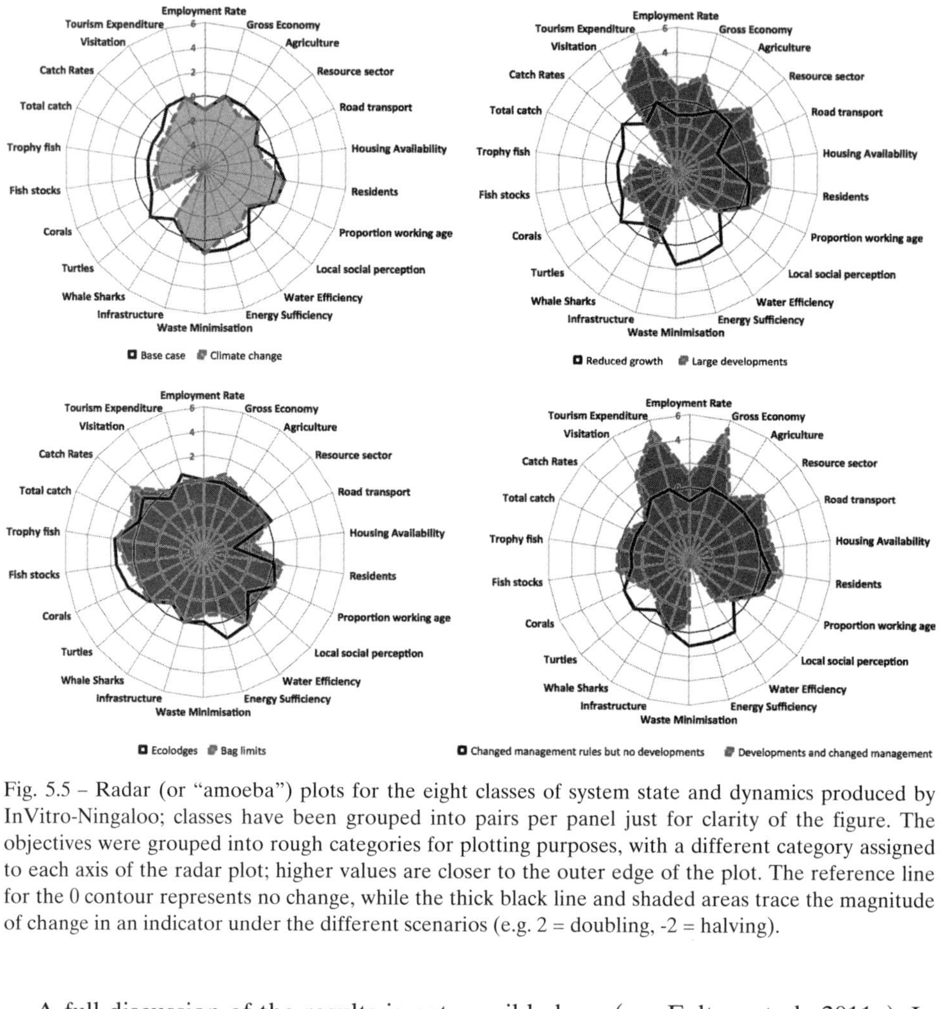

Fig. 5.5 – Radar (or "amoeba") plots for the eight classes of system state and dynamics produced by InVitro-Ningaloo; classes have been grouped into pairs per panel just for clarity of the figure. The objectives were grouped into rough categories for plotting purposes, with a different category assigned to each axis of the radar plot; higher values are closer to the outer edge of the plot. The reference line for the 0 contour represents no change, while the thick black line and shaded areas trace the magnitude of change in an indicator under the different scenarios (e.g. 2 = doubling, -2 = halving).

A full discussion of the results is not possible here (see Fulton et al., 2011a). In summary, the simulations highlighted the complex relationships between development and environmental status in the region. The need for system-level management and decision-making across sectors is especially evident given the interactions possible between different sectors. For example, fish stocks in the region are already under sufficient fishing pressure (especially from recreational fishers) that any further increases in pressure lead to significant declines in biomass. Consequently, additional recreational pressure applied by a growing population, attracted to the region by significant developments or industrial expansion, may be sufficient to cause a collapse in some key target species (e.g., the spangled emperor, *Lethrinus nebulosus*). However, the resident human population is vulnerable to issues associated with a steadily rising median age and high labor force turnover and without growth the region faces a significant risk of economic and population decline and contraction of services. Sustainable futures that better match outcomes to stakeholder desired objectives, available resources and services and can reduce impacts are possible—e.g., via modification of recreational fisheries regulations and targeted developments that minimize increases in waste production, demand for utilities and regional housing and labor constraints. Among the management strategies explored, fishing regulations, housing plans, infrastructure development and marketing and catering for specific tourist types are likely to provide the largest impact. Consequently, informed decision-making around these system features can lead to a much better match of outcomes and EBM objectives. Achieving these futures is made more difficult, however, by the effects of ocean acidification and climate change-driven shifts in sea-level and weather patterns.

The picture obtained from the MSE of the Ningaloo-Exmouth region is one of a typical "complex" system, that is, a system which is controlled by large-scale external pressures as much as by local processes. Thus the regional future is affected by global drivers, like climate change and external industrial development (e.g., in the neighboring Pilbara region), but also by local points of intervention (e.g., the availability of housing), sanctuary zone boundaries, infrastructure (e.g., boat ramps or utilities) and road access. While stakeholders familiar with regional issues had already identified some of these, the full extent of the potential interactions could only be assessed via extensive exploration of the alternative futures. Simpler models included in the suite of models applied to the system (see Jones et al., 2011; Little et al., 2011) highlighted some aspects of the potential futures, but the interconnection between the industrial developments and larger marine state (even though not physically co-located) was only evident once the system information was integrated in InVitro. As with Atlantis, the use of multi-model inference and a participatory MSE-based approach to the use of models in support of EBM has seen greater uptake of the final results. Information from the suites of models, but particularly the single sector, shuttle and full system models, is being used in management decision processes in the region (e.g., reviews of proposed developments and regional strategic plans).

5.6 Other spatial approaches

The modeling tools outlined so far have grown in complexity, culminating with the full system models discussed above. However, we would be remiss not to draw attention to two other simpler, more empirical methods before drawing to a close. Marine EBM is inherently place-based and spatial management is an important tool for its implementation. This is why models such as Atlantis and InVitro were designed to incorporate spatial structure and to permit evaluation of spatial management strategies. However, fully dynamic system models are not the only means of considering spatial information. There are other simpler and complementary approaches (see below and Rice et al., this volume).

Approaches based on geographical information systems and other constructs can also be used to evaluate spatial management strategies. Management of marine areas, as well as coastal land use management, requires identification of the distribution of key ecological features, ecosystem processes, existing human uses and potential development sites. Geographic information systems (GIS) can provide relevant spatial information to decision makers and other stakeholders so that they can make informed decisions. GIS-based packages are by far one of the more widespread EBM tools. In some instances static GIS layers are overlaid, or statistically analyzed using more sophisticated techniques, to provide a combined map. In other cases static layers are coupled with dynamic spatial impact or development models, which provide time varying maps dependent on proposed system changes. These integrated GIS packages may be either simply linked (i.e., a human user exports, imports data between components) or they may be fully coupled, such as the ALCES mapper, which projects outcomes of the ALCES model[2] (Carlson et al., 2010; Stevens et al., 2010) spatially so that spatial dynamics can be incorporated and changes in the system can be visualized through time.

The strengths, weaknesses and data requirements of GIS-based applications are specific to the details of the models included. Initially they were largely static or statistical, but coupled forms are becoming more dynamic and include more processes explicitly (at the cost of increased data requirements and complexity). The majority of data (barring any process-related parameterization in impacts or dynamic model components) will need to be spatial and the list can become quite extensive. For example, if evaluating potential developments, input would be required from local stakeholders, scientific experts, resource managers, and political officials regarding current land use policy, development trends, conservation elements, conservation goals, pollutant sources, run-off, catchments, habitat distributions, species vulnerability and species viability. In addition, all GIS approaches have a dependence on GIS software and there is an increasing number of reliable open source alternatives (e.g., quantum GIS (QGIS), http://www.qgis.org) now available.

5.7 Other means of doing risk assessment

Risk assessment sits at the heart of EBM analyses. Risk assessment effectively allows for the evaluation of risk that a stock or ecosystem will be perturbed beyond what is sustainable. This approach also has the ability to compare across stocks, regions, gears and fisheries, and is increasingly being used in a fisheries context. In effect what risk assessment does is identify those attributes of the ecosystem that are deemed worth evaluating, identify salient features or measures of those attributes that will indicate responses to potential perturbation, identify those factors that can perturb the ecosystem attributes of interest and then rank, weight, or assign probability to the risk of those attributes responding to the driving or perturbing factors such that they end up in a negative or otherwise undesired state (Suter, 1993; Canter, 2001; Peterman, 2004; Fletcher, 2005; Smith et al., 2007).

All of the models discussed above can be used in that context, but there is a wide range of targeted risk assessment (RA) approaches of varying degrees of quantitative complexity. Risk assessment is a rich field in its own right (e.g. Suter, 1993; Canter 2001; Hobday et al., 2007; Hobday et al., 2011) with methods ranging from categorical ranking (e.g. Milton, 2001; Stobutzki et al., 2001a; Stobutzki et al., 2001b; Fletcher, 2005; Braccini et al., 2006; Griffiths et al., 2006; Zhou and Griffiths, 2008; Patrick et al., 2009) to full Bayesian modeling (e.g. Ellison, 1996;

[2] ALCES (named for the taxonomic nomenclature for moose) uses the Stella modeling software to implement a land use and impacts MRM, originally primarily terrestrial and freshwater, it is being expanded into coastal and marine EBM.

Punt and Hilborn, 1997; Peterman and Anderson, 1999; Peterman, 2004). Semi-quantitative methods have been used for over two decades in fisheries to address the fishery impacts on bycatch and targeted species (Francis, 1992; Lane and Stephenson, 1998; Stobutzki et al., 2001a; Stobutzki et al., 2001b), extinction risk (Musick, 1999; Roberts and Hawkins, 1999; Cheung et al., 2005; Mace et al., 2008), and ecosystem viability (Jennings et al., 1999; Fletcher et al., 2005; Astles et al., 2006). One of the main reasons semi-quantitative approaches are used is because they permit the inclusion of a wider range of factors (which may not all have rigorously quantitative information). They can also complement both stock and ecosystem assessments.

EBM Modeling Example 7—Productivity-Susceptibility Analysis (PSA). Patrick et al., (2009; 2010) used the semi-quantitative PSA risk assessment method to assist evaluation of the vulnerability of stocks to overfishing, especially data-poor stocks. This analysis focused on easily obtainable and readily rankable information, with overall vulnerability as a composite measurement of a stock's productivity and its susceptibility to a fishery. Productivity refers to the capacity of the stock to recover if depleted and to produce optimal levels of yield (e.g. MSY). Susceptibility is the potential for the stock to be impacted by the fishery. In general, vulnerability is an important factor to consider when organizing stock complexes; developing buffers between biological reference points like target and limit fishing mortalities; and determining which stocks should be managed explicitly under a fishery management plan. Using the hierarchical methods of Hobday et al. (2011), it is possible to take this concept further and evaluate not only target and bycatch species, but also habitats and fish communities and even how the PSA results may change under alternative management systems.

The main data inputs for PSA are categorical ranking scores of a set of productivity and susceptibility measures, most loosely related to life-history characteristics and the stock's relationship to the fishery. Regional experts typically assign the scores. This simplicity is one of the main benefits of the approach, as it provides semi-quantitative evaluations of risk without requiring detailed and sophisticated information. This makes it particularly useful in a triage exercise, where there is a question of which stocks to focus on either because time or resources are short. Perhaps the greatest drawback of the approach is that it can lead to a false sense of security in data-poor situations. It can also minimize the use of some extant and routinely collected data types.

6. Discussion

The previous sections have attempted to provide a brief overview for a few of the different kinds of models currently being used in support of EBM for marine and coastal ecosystems. We would like to stress that this discussion has not been exhaustive and that there is no exact formula nor a specific "model" that can be applied universally. We have focused here on MSE, but there are other approaches—e.g., multi-model inference and risk assessments—which can be used stand-alone or in a complementary fashion (e.g., a multi-model approach that includes MSE and risk assessment models, as was done the for Ningaloo example). Regardless of the approach selected, these model-based decision frameworks are procedural in nature, with the exact combination of approaches and specific details varying among particular applications and regions. Nevertheless we can glean a few general lessons on model use that need to be understood when employing models as decision support tools for EBM.

The first of these lessons is to try multiple models, remembering to select ones that are best matched for the intended use of the tool (it is not usually advisable to default to the most complicated model). In particular, we strongly recommend carefully deciding whether tactical tools (that give specific assessments or set the exact levels of specific regulations) are required or whether a more strategic model is warranted (that provides global, or regional, contextual understanding or provides theoretical considerations for setting guidelines). EBM requires both tactical and strategic tools at different times depending on the advice being provided. For instance, tactical questions refer to the setting of biological reference points or determining specific quotas, while strategic advice pertains to what kind of management strategies are feasible while satisfying management objectives. Models like the MRMs better address issues such as specific environmental drivers, trophic interactions, or technical interactions. Full system models are better placed to address strategic trade-offs among multiple system components, not just solely biota or fisheries. Note that care in the selection of models has a dual role, not only in guaranteeing the most appropriate tool is used, but also because many stakeholders are not familiar with models and so inappropriate use can severely hamper further efforts to use models in support of EBM (potentially with ultimately detrimental management outcomes as decisions will still be made).

The second major modeling lesson is that it is important to be forthright about uncertainty, but not let it lead to paralysis that prevents a move to EBM until everything is fully understood. Uncertainty is pervasive in ecosystems and will remain so no matter how much monitoring or modeling is done. However, by using standard uncertainty handling methods (e.g., Markov Chain Monte Carlo simulations to characterize confidence intervals; Link et al., 2010) and multi-model approaches, managers can be given a useful sense of uncertainty that helps (rather than derails or misguides) risk-based adaptive management decisions. This is particularly important given that structural uncertainty (which pertains to uncertainty about what is in the system and how it is connected) is far more important than parametric uncertainty in EBM; alternative, potentially equally plausible, system structures can lead to significantly different ranking of management options. The only way to address this is to either run all plausible model structures or to try and simplify the set by bringing in additional information or running preliminary analyses to see if it really is a problem. Loop analysis (qualitative models; Dambacher et al., 2003) is a rapid screening technique that can show how many system structures do need to be maintained through the entire analysis, identifying system structures that appear to be fundamentally different that do not actually lead to substantially different outcomes due to their internal feedback structure (Fulton, 2010). While maintaining multiple model structures can be a significant exercise, both in terms of build and maintenance, software for building system models (e.g., Stella, Vensim or Repast) and publicly available toolboxes (e.g., FLR, http://flr-project.org/, or the NMFS toolbox, http://nft.nefsc.noaa.gov/) are simplifying the issue. Moreover, if different stakeholder groups have proposed different system structures, maintaining all of them can provide for greater uptake, discussion and understanding (Fulton et al., 2011a).

The third lesson is that the modeling tools are ultimately useless without venues, audiences, and demands for the model outputs. Stakeholder engagement is critical for ongoing tool use in EBM. There need to be open and effective channels of communication and established pathways or forums where information derived from EBM tools can be fed into the management process—e.g., uptake of EBM tool outputs has been facilitated in Australia by enabling legislation (the Environment Protection and Biodiversity Act, which requires industries to demonstrate they are ecologically sustainable), and the existence of a well-developed, participatory fishery management system (Smith et al., 1999; Smith et al., 2001). The need for regular communication between all interested EBM parties is critical for successful EBM outcomes. If one group is neglected then there are adverse outcomes, as epitomized by many historical fisheries failures (Fulton et al., 2010).

As modeling in support of EBM matures, one thing that may assist uptake is an appropriate means for reviewing these tools. Reviews of the use of single sector, single stock assessment models are now standard practice in many countries (e.g., routine reviews held in USA, EU, South Africa), but such reviews are less common for the other classes of models. This lack of a review process may be hampering uptake of EBM-relevant model output in management systems that see review as a critical step in data uptake. For example, it is relatively easy to see the value of MRMs as immediately deployable EBM tools. They have been used to consider some strategic questions. There have been demonstrations of how they could also be used in tactical fisheries management. There is a good understanding of underlying mechanisms and sufficient data which are really no less certain than data already used in the assessment and management process (Link et al., 2011). It has also been demonstrated that increases in uncertainty that may result from the inclusion of additional data is largely outweighed by decreases in other forms of uncertainty (e.g., around the inclusion of critical processes; Link et al., 2010). Yet, despite the growing number of applications, MRMs have only rarely been utilized in tactical fisheries or ecosystem-based management (Plagányi et al., 2012). Thus it is likely that a lack of broader familiarity with the approach and the lack of a transparent review process are preventing MRMs being taken up formally as commonly used tactical EBM tools within existing management bodies. To facilitate the uptake of MRMs (or other models) as "ancillary information" or "contextual assessments" along with standard assessments, it would be advantageous if venues for evaluating all forms of EBM models were available, analogous to those available for stock assessment or biogeochemical nutrient-load models. Outputs from the more complicated model types (e.g., full system models) would need to be evaluated by a different set of expertise that is familiar with the nuances of a broader array of considerations that go beyond single sectors.

The interactions between sectors evident in many EBM models suggest that advice drawn from models may not always fit easily into existing sector-focused management arrangements. There may also need to be a restructuring of existing management bodies to more easily allow for multi-species, cross-sectoral and even cross jurisdictional advice. How such policy choices evolve is beyond the scope of this chapter, but clearly warrants attention for models to be effectively used for EBM.

References

Arkema, K. K., Abramson, S. C. and Dewsbury, B. M., 2006. Marine ecosystem-based management: From characterization to implementation. *Frontiers in Ecology and the Environment,* **10,** 525–532

Astles, K. L., Holloway, M. G., Steffe, A., Green, M., Ganassin, C. and Gibbs, P. J., 2006. An ecological method for qualitative risk assessment and its use in the management of fisheries in New South Wales, Australia. *Fisheries Research,* **82,** 290–303

Barange, M., Allen, I., Allison, E., Badjeck, M.-C., Blanchard, J., Drakeford, B., Dulvy, N. K., Harle, J., Holmes, R., Holt, J., Jennings, S., Lowe, J., Merino, G., Mullon, C., Pilling, G., Rodwell, L., Tompkins, E. and Werner, F., 2011. Predicting the Impacts and Socio-Economic Consequences of Climate Change on Global Marine Ecosystems and Fisheries: The QUEST_Fish Framework. In *World Fisheries: A Social-Ecological Analysis,* First Edition, R. E. Ommer, R. I. Perry, K. Cochrane and P. Cury (eds). Blackwell Publishing Ltd., Oxford

Beamish R. J. and Bouillon D. R., 1993. Pacific salmon production trends in relation to climate. *Canadian Journal of Fisheries and Aquatic Sciences,* **50,** 1002–1016

Beamish R. J., Schnute J. T., Cass A. J., Neville C. M. and Sweeting R. M., 2004. The influence of climate on the stock and recruitment of pink and sockeye salmon from the Fraser River, British Columbia, Canada. *Transactions of the American Fisheries Society,* **133,** 1396–1412

Blanchard, J. L., Jennings, S., Law, R., Castle, M. D., McCloghrie P., Rochet, M. J. and Benoît, E., 2009. How does abundance scale with body size in coupled size-structured food webs? *Journal of Animal Ecology,* **78,** 270–280

Blanchard, J. L., Law, R., Castle, M. D. and Jennings, S., 2011. Coupled energy pathways and the resilience of size-structured food webs. *Theoretical Ecology,* **4,** 289–300

Bogstad, B. and Tjelmeland, S. (Eds)., 1992. Interrelations Between Fish Populations in the Barents Sea. *Proceedings of the Fifth PINRO-IMR Symposium, Murmansk, 12–16 August 1991.* Institute of Marine Research, Bergen, Norway. 238 pp.

Bogstad, B., Hiis Hauge, K. and Ulltang, Ø., 1997. MULTSPEC—A Multi-Species Model for Fish and Marine Mammals in the Barents Sea. *Journal of Northwest Atlantic Fishery Science,* **22,** 317–341

Boschetti, F., de La Tour, A., Fulton, E. A. and Little, L. R., 2010. Interactive modeling for natural resource management. *Environmental Modeling and Software,* **25,** 1075–1085

Botsford, L. W., Castilla, J. C. and Peterson, C. H., 1997. The management of fisheries and marine ecosystems. *Science,* **277,** 509–515

Braccini, J. M., Gillanders, B. M. and Walker, T. I., 2006. Hierarchical approach to the assessment of fishing effects on non-target Chondrichthyans: a case study of Squalus megalops in south-eastern Australia. *Canadian Journal of Fisheries Aquatic Sciences,* **63,** 2456–2466

Branch, T. A., Watson, R., Fulton, E. A., Jennings, S., McGilliard, C. R., Pablico, G. T., Ricard, D. and Tracey. S. R., 2010. The trophic fingerprint of marine fisheries. *Nature,* **468,** 431–435

Browman, H. I. and Stergiou, K. I., 2004. Perpectives on ecosystem-based approaches to management of marine resources. *Marine Ecology Progress Series,* **274,** 269–303

Bulgakova, T. I., Vasilyev, D. A., Korzhev, V. A. and Tretyak, V. L., 1995. The results of multispecies analysis for the Barents Sea fishery community (cod, capelin, shrimp and herring). ICES Document CM 1995/D:14. 24 pp.

Butterworth, D. S., 2007. Why a management procedure approach? Some positives and negatives. *ICES Journal of Marine Science,* **64,** 613–617

Butterworth, D. S. and Punt, A. E., 1999. Experiences in the evaluation and implementation of management procedures. *ICES Journal of Marine Science,* **56,** 985–998

Butterworth, D., and Rademeyer, R., 2005. Sustainable management initiatives for the Southern African hake fisheries over recent years. *Bulletin of Marine Science,* **76,** 287–320

Canter, L. W., 2001. *Environmental impact assessment.* McGraw-Hill Inc., New York

Carlson, M., Antoniuk, T., Farr, D., Francis, S., Manuel, K., Nishi, J., Stelfox, B., Sutherland, M., Yarmoloy, C., Aumann, C. and Pan, D., 2010. Informing Regional Planning in Alberta's Oilsands Region with a Land-use Simulation Model. *In International Environmental Modeling and Software Society (iEMSs) 2010, International Congress on Environmental Modeling and Software Modeling for Enviornment's Sake, Fifth Biennial Meeting, Ottawa, Canada* D. A. Swayne, W. Yang, A. A. Voinov, A. Rizzoli and T. Filatova (Eds.). http://www.iemss.org/iemss2010/index.php?n=Main.Proceedings

Chen, C., Beardsley, R. C. and Franks, P. J. S., 2001. A 3-D prognostic model study of the ecosystem over Georges Bank and adjacent coastal regions. Part I: physical model. *Deep Sea Research,* **48,** 419–456

Cheung, W. W. L., Pitcher, T. J. and Pauly, D., 2005. A fuzzy logic expert system to estimate intrinsic extinction vulnerabilities of marine fishes to fishing. *Biological Conservation,* **124,** 97–111

Cheung, W. L. L., Lam, V. W. Y. and Pauly, D., 2008. Modeling present and climate-shifted distribution of marine fishes and invertebrates. Fisheries Centre Research Reports 16(3). 72pp.

Cheung, W. W. L., Close, C., Lam, V. W. Y., Watson, R. and Pauly, D., 2008. Application of macroecological theory to predict effects of climate change on global fisheries potential. *Marine Ecology Progress Series,* **365,** 187–197

Cheung, W. W. L., Lam, V. W. Y., Sarmiento, J. L., Kearney, K., Watson, R. and Pauly, D., 2009. Projecting global marine biodiversity impacts under climate change scenarios. *Fish and Fisheries,* **10,** 235–251

Cheung, W. W. L., Lam, V. W. Y., Sarmiento, J. L., Kearney, K., Watson, R., Zeller, D. and Pauly, D., 2010. Large-scale redistribution of maximum fisheries catch potential in the global ocean under climate. *Global Change Biology,* **16,** 24–35

Christensen, V. and Pauly, D., 1992. ECOPATH II: a software for balancing steady-state models and calculating network characteristics. *Ecol. Model.,* **61,** 169–185

Christensen, V. and Walters, C., 2004. Ecopath with Ecosim: methods, capabilities and limitations. *Ecological Modeling,* **172,** 109–139

Christensen, V., Walters, C. J. and Pauly, D., 2005. *Ecopath with Ecosim: a User's guide.* Fisheries Centre of University of British Columbia, Vancouver, Canada. 154 pp.

Condie, S. A., Hayes, D., Fulton, E. A. and Savina, M., 2012. Modeling ecological change in a subtropical estuary over half a century: Impacts of land-use, urbanization, freshwater extraction and climate change. *Marine Ecology Progress Series,* **457,** 43–66

Cox, P. M., Betts, R. A., Jones, C. D., Spall, S. A. and Totterdell, I. J., 2000. Acceleration of global warming due to carbon cycle feedbacks in a 3D coupled model. *Nature,* **408,** 184–187

Dambacher, J. M., Li, H. W. and Rossignol, P. A., 2003. Qualitative predictions in model ecosystems. *Ecological Modeling,* **161,** 79–93

Dambacher, J. M., Brewer, D. T., Dennis, D. M., Macintyre, M. and Foale, S., 2007. Qualitative modeling of gold mine impacts on Lihir Island's socioeconomic system and reef-edge fish community. *Environmental Science & Technology,* **41,** 555–562

Dambacher, J. M., Gaughan, D. J., Rochet, M.-J., Rossignol, P. A. and Trenkel, V. M., 2009. Qualitative modeling and indicators of exploited ecosystems. *Fish and Fisheries,* **10,** 305–322

Dambacher, J. M., Hosack, G. R. and Rochester, W., 2011. *Ecological Indicators for the exclusive economic zone of Australia's East Marine Region.* Report for the Australian Government Department of Sustainability, Environment, Water, Population and Communities, Canberra

de la Mare, W. K., 1998. Tidier fisheries management requires a new MOP (management oriented paradigm). *Reviews in Fish Biology and Fisheries,* **8,** 349–56

Dufresne, J. L., Friedlingstein, P., Berthelot, M., Bopp, L., Ciais, P., Fairhead, L., Treut H. L. and Monfray, P., 2002. On the magnitude of positive feedback between future climate change and the carbon cycle. *Geophysical Research Letters,* **29,** 1405, doi:10.1029/2001GL013777

Dutra, L. X. C., Haworth, R. J. and Taboada, M. B., 2011. An integrated approach to tourism planning in a developing nation: A case study form Beloi (Timor-Leste). In *Stories of Practice: Tourism policy and planning,* D. Dredge and J. Jenkins (eds). Ashgate Publishing: Farnham Surrey. Pg: 269–294

Edwards, S. F., Link, J. S. and Rountree, B. P., 2004. Portfolio management of wild fish stocks. *Ecological Economics,* **49,** 317–329

Ellison, A. M., 1996. An introduction to Bayesian inference for ecological research and environmental decision-making. *Ecological Applications,* **6,** 1036–1046

FAO (Food and Agriculture Organization of the United Nations), 2007. *Best practices in ecosystem modeling: Modeling ecosystem interactions for informing an ecosystem approach to fisheries.* FAO Technical Guidelines for Responsible Fisheries: Fisheries Management—The Ecosystem Approach to Fisheries. 44pp.

FAO (Food and Agriculture Organization of the United Nations), 2008. Fisheries Management. 2. The ecosystem approach to fisheries. 2.1 Best practices in ecosystem modeling for informing an ecosystem approach to fisheries. *FAO Technical Guidelines for Responsible Fisheries* **4** (Suppl. 2, Add. 1), 79 pp.

Fennel, W. 2009. A nutrient to fish model for the example of the Baltic Sea. *Journal of Marine Systems,* **81,** 184–195

Field, J. C., Francis, R. C. and Aydin, K., 2006. Top-down modeling and bottom-up dynamics: Linking a fisheries-based ecosystem model with climate hypotheses in the Northern California Current. *Progress In Oceanography,* **68,** 238–270

Fletcher, W. J., 2005. The application of qualitative risk assessment methodology to prioritize issues for fisheries management. *ICES Journal of Marine Science,* **62,** 1576–1587

Fletcher, W. J., Chesson, J., Sainsbury, K. J., Hundloe, T. J. and Fisher, M., 2005. A flexible and practical framework for reporting on ecologically sustainable development for wild capture fisheries. *Fisheries Research,* **71,** 175–183

Fournier, D. A., Skaug, H. J., Ancheta, J. Ianelli, J., Magnusson, A., Maunder, M. N. Nielsen, A. and Silbert, J., 2011. AD Model Builder: using automatic differentiation for statistical inference of highly parameterized complex nonlinear models. *Optimization Methods and Software,* **27,** 233–249

Francis, R. I.C. C., 1992. Use of risk analysis to assess fishery management strategies: a case study using orange roughy (*Hoplostethus atlanticus*) on the Chatham Rise, New Zealand. *Canadian Journal of Fisheries and Aquatic Sciences,* **49,** 922–930

Franks, P. J. S. and Chen, C., 2001. A 3-D prognostic numerical model study of the Georges Bank ecosystem. Part II: biological-physical model. *Deep-Sea Research II,* **48,** 457–482

Friedlingstein, P., Bopp, L., Ciais, P., Dufresne, J.-L., Fairhead, L., LeTreut, H., Monfray P. and Orr, J., 2001. Positive feedback between future climate change and the carbon cycle. *Geophysical Research Letters,* **28,** 1543–1546

Fukuda, Y., 1976. *A note on yield allocation in multi-species fisheries.* Far Seas Fish. Res. Lab., Shimizu, Japan

Fulton, E. A., 2010. Approaches to end to end ecosystem models. *Journal of Marine Systems,* **81,** 171–183

Fulton, E. A., 2011. Interesting times: winners and losers and system shifts under climate change around Australia. *ICES Journal Marine Science,* **68,** 1329–1342

Fulton, E. A., Smith, A. D. M. and Johnson, C. R., 2003. Effect of complexity on marine ecosystem models. *Marine Ecology Progress Series,* **253,** 1–16

Fulton, E. A., Smith, A. D. M. and Punt, A. E., 2005. Which ecological indicators can robustly detect effects of fishing? *ICES Journal of Marine Science,* **62,** 540–551

Fulton, E. A., Smith, A. D. M. and Smith, D. C., 2007. *Alternative Management Strategies for Southeast Australian Commonwealth Fisheries: Stage 2: Quantitative Management Strategy Evaluation.* Australian Fisheries Management Authority Report. Canberra. 378pp.

Fulton, E. A., Smith A. D. M., Smith D. C. and van Putten, I. E., 2010. Human behaviour—the neglected source of uncertainty in fisheries management. *Fish and Fisheries,* **12,** 2–17

Fulton, E. A., Gray, R., Sporcic, M., Scott, R., Little, R., Hepburn, M., Gorton, B., Hatfield, B., Fuller, M., Jones, T., De la Mare, W., Boschetti, F., Chapman, K., Dzidic, P., Syme, G., Dambacher, J. and McDonald, D., 2011a. *Ningaloo Collaboration Cluster: Adaptive Futures for Ningaloo.* Ningaloo Collaboration Cluster Final Report No. 5.3

Fulton, E. A., Link, J. S., Kaplan., I., Savina-Rolland, M., Johnson., P., Ainsworth, C., Horne, P., Gorton, R., Gamble, R. J., Smith A. D. M. and Smith, D. C., 2011b. Lessons in modeling and management of marine ecosystems: The Atlantis experience. *Fish and Fisheries,* **12,** 171–188

Fulton, E. A., Johnson P. and Gorton R., 2012. *SE Fisheries & Climate Simulations: Barriers to Adaptation. 2010/023—Quantitative testing of fisheries management arrangements under climate change using Atlantis:* Milestone Report Appendix 3. CSIRO, Australia

Gamble, R. J. and Link, J. S., 2009. Analyzing the Tradeoffs Among Ecological and Fishing Effects on an Example Fish Community: a Multispecies (Fisheries) Production Model. *Ecological Modeling,* **220,** 2570–2582

Garcia, S. M., Zerbi, A., Aliaume, C., Do Chi, T. and Lasserre, G., 2003. *The ecosystem approach to fisheries: issues, terminology, principles, institutional foundations, implementation and outlook.* FAO Technical Paper 443. FAO, Rome. 71 pp.

Garrod, D. J., 1973. *Memorandum on the mixed fishery problem in Subarea 5 and Statistical Area 6.* ICNAF Research Document. 73/6

Gascuel, D., Morissette, L., Palomares, D. and Christensen, V., 2008. Trophic flow kinetics in marine ecosystems: toward a theoretical approach to ecosystem functioning. *Ecological Modeling,* **217,** 33–47

Gascuel, D. and Pauly, D., 2009. EcoTroph: modeling marine ecosystem functioning and impact of fishing. *Ecological Modeling,* **220,** 2885–2898

Gneiting, T. and Raftery, A. E., 2005. Weather forecasting with ensemble methods. *Science,* **310,** 248–249

Gray, R., Fulton, E. A., Little, L. R., and Scott, R., 2006. *Operating Model Specification Within an Agent Based Framework.* North West Shelf Joint Environmental Management Study Technical Report, vol. 16. CSIRO, Hobart, Tasmania. 127pp.

Gray, R., Fulton, E. A. and Little, L. R., in press. Human–ecosystem interaction in large ensemble-model systems. In *Empirical Agent-Based Modeling,* A. Smajgl and O. Barreteau (eds). Springer

Griffiths, S. P., Heales D. S., Milton D. A. and Stobutzki I. C., 2006. Validating ecological risk assessments for fisheries: assessing the impacts of turtle excluder devices on elasmobranch bycatch populations in an Australian trawl fishery. *Marine and Freshwater Research* **57,** 395–401

Griffith, G. P., Fulton, E. A. and Richardson, A. J., 2011. Effects of fishing and acidification-related benthic mortality on the southeast Australian marine ecosystem. *Global Change Biology,* **17,** 3058–3074

Griffith, G. P., Fulton, E. A., Gorton, R. and Richardson, A., in press. Predicting multiple human effects on the marine ecosystem. *Conservation Biology*

Hall, S. J., Collie, J. S., Duplisea, D. E., Jennings, S., Bravington, M. and Link, J. S., 2006. A length-based multispecies model for evaluating community responses to fishing. *Canadian Journal of Fisheries and Aquatic Sciences,* **63,** 1344–1359

Hall-Arber, M., Pomeroy, C. and Conway, F., 2009. Figuring out the human dimensions of fisheries: illuminating models. *Marine and Coastal Fisheries: Dynamics, Management, and Ecosystem Science,* **1,** 300–314

Hobday, A. J., Smith, A. D. M., Webb, H., Daley, R., Wayte, S., Bulman, C., Dowdney, J., Williams, A., Sporcic, M., Dambacher, J., Fuller M. and Walker, T., 2007. *Ecological risk assessment for the effects of fishing: methodology.* CSIRO Marine and Atmospheric Research.

Hobday, A. J., Smith, A. D. M., Stobutzki, I. C., Bulman, C., Daley, R., Dambacher, J. M., Deng, R. A., Dowdney, J., Fuller, M., Furlani, D., Griffiths, S. P., Johnson, D., Kenyon, R., Knuckey, I. A., Ling, S. D., Pitcher, R., Sainsbury, K. J., Sporcic, M., Smith, T., Turnbull, C., Walker, T. I., Wayte, S. E.,

Webb, H., Williams, A., Wise, B. S. and Zhou, S., 2011, Ecological Risk Assessment for the effects of fishing. *Fisheries Research,* **108,** 372–384

Holden, M. J. and Raitt, D. F. S. (ed), 1974. *Manual of Fisheries Science: Part 2—Methods of Resource Investigation and their Application.* Food and Agriculture Organization of the United Nations: Rome

Holland, D. S. and Sutinen, J. G., 1999. An empirical model of fleet dynamics in New England trawl fisheries. *Canadian Journal of Fisheries and Aquatic Sciences,* **56,** 253–264

Hollowed, A. B., Bax, N., Beamish, R., Collie, J., Fogarty, M., Livingston, P., Pope, J. and Rice, J. C., 2000. Are multispecies models an improvement on single-species models for measuring fishing impacts on marine ecosystems? *ICES Journal of Marine, Science* **57,** 707–719

Hollowed, A. B., Bond, N. A., Wilderbuer, T. K., Stockhausen, W. T., A'Mar, Z. T., Beamish, R. J., Overland, J. E. and Schirripa, M. J., 2009. A framework for modeling fish and shellfish responses to future climate change. *ICES Journal of Marine Science,* **66,** 1584–1594

Horbowy, J., 1989. A multispecies model of fish stocks in the Baltic Sea. *Dana,* **7,** 23–43

Hosack G. R. and Dambacher, J. M., 2010. *Ecological Indicators for the Exclusive Economic Zone of Australia's South East Marine Region.* Report for the Australian Government Department of Sustainability, Environment, Water, Population and Communities, Canberra.

Howell, D. and Bogstad, B., 2010. A combined Gadget/FLR model for management strategy evaluations of the Barents Sea fisheries. *ICES Journal of Marine Science,* **67,** 1998–2004

Hu, S., Townsend, D. W., Chen, C., Cowles, G., Beardsley, R. C., Ji, R. and Houghton, R. W., 2008. Tidal pumping and nutrient fluxes on Georges Bank: A process-oriented modeling study. *Journal of Marine Systems,* **74,** 528–544

ICES (International Council for the Exploration of the Sea), 1991. Multispecies models relevant to management of living resources. *ICES Marine Science Symposium,* **193,** 358 pp.

IWC (International Whaling Commission), 1992. Report of the Scientific Committee, Annex D. Report of the Sub-Committee on Management Procedures. *Reports of the International Whaling Commission,* **42,** 87–136

Janssen, M. A. and Ostrom, E., 2006. Empirically based, agent-based models. *Ecology and Society,* **11,** 37

Jennings, S., 2005. Indicators to support an ecosystem approach to fisheries. *Fish and Fisheries,* **6,** 212–232

Jennings, S., Reynolds, J. D. and Polunin, N. V. C., 1999. Predicting the vulnerability of tropical reef fishes to exploitation with phylogenies and life histories. *Conservation Biology,* **13,** 1466–1475

Jennings, S., Pinnegar, J. K., Polunin, N. V. C. and Warr, K. J., 2002. Linking size-based and trophic analyses of benthic community structure. *Marine Ecology Progress Series,* **226,** 77–85

Jones, T., Glasson, T., Wood, D. and Fulton E. A., 2011. Regional Planning and Resilient Futures: Destination Modeling and Tourism Development—The Case of the Ningaloo Coastal Region in Western Australia. *Planning Practice & Research,* **26,** 393–415

Kennedy, J. O. S., 2003. Scope for Efficient Multinational Exploitation of North-East Atlantic Mackerel. *Marine Resource Economics,* **18,** 55–80

Keyl, F. and Wolff, M., 2008. Environmental variability and fisheries: what can models do? *Reviews in Fish Biology and Fisheries,* **18,** 273–299

Kirkwood, G. P., 1997. The Revised Management Procedure of the International Whaling Commission. In *Global trends: fisheries management,* pp. 41–99. Ed. by E. K. Pikitch, D. D. Huppert, and M. P. Sissenwine. American Fisheries Society Symposium, 20, Bethesda, Maryland

Kishi, M. J., Kashiwai, M., Ware, D. M., Megrey, B. A., Eslinger, D. L., Werner, F. E., Aita, M. N., Azumaya, T., Fujii, M., Hashimoto, S., Huang, D., Iizumi, H., Ishida, Y., Kang, S., Kantakov, G. A., Kim, H.-C., Komatsu, K., Navrotsky, V. V., Smith, S. L., Tadokoro, K., Tsuda, Yamamura, O., Yamanaka, Y., Yokouchi, K., Yoshie, N., Zhang, J., Zuenko, Y. I. and Zvalinsky, V. I., 2007. NEMURO—a lower trophic level model for the North Pacific marine ecosystem. *Ecological Modeling,* **202,** 12–25

Krishnamurti, T. N., Kishtawal, C. M., Zhang, Z., LaRow, T., Bachiochi, D., Williford, E., Gadgil, S. and Surendran, S., 2000. Multimodel Ensemble Forecasts for Weather and Seasonal Climate. *Journal of Climate,* **13,** 4196–4216

Lane, D. E. and Stephenson, R. L., 1998. A framework for risk analysis in fisheries decision-making. *ICES Journal of Marine Science,* **55,** 1–13

Larkin, P. A., 1996. Concepts and issues in marine ecosystem management. *Reviews Fish Biology and Fisheries* **6,** 139–164

Lasseter, A., 2008. Predicting Adaptability with Social Network Analysis in a Small-scale Lobster Fishery. *Proceedings of the 61st Gulf and Caribbean Fisheries Institute,* November 10–14, 2008, Gosier, Guadeloupe, French West Indies. GCFI: **6,** 50–56

Lehodey, P., 2005. Reference Manual for the Spatial Ecosystem and Populations Dynamics Model—SEAPODYM. WCPFC-SC1, ME IP-1

Leslie, H., Rosenberg, A. A. and Eagle, J., 2008. Is a new mandate needed for marine ecosystem-based management? *Frontiers in Ecology and the Environment,* **6,** 43–48

Levin P. S., Fogarty, M. J., Murawski, S. A. and Fluharty, D., 2009. Integrated Ecosystem Assessments: Developing the Scientific Basis for Ecosystem-Based Management of the Ocean. *PLoS Biol* **7,** e1000014. doi:10.1371/journal.pbio.1000014

Levins, R., 1966. The strategy of model building in population biology. *American Scientist,* **54,** 421–431

Leviston, Z. and Walker, I. A., 2010. *Baseline Survey of Australian attitudes to climate change.* Climate Adaptation Flagship. CSIRO. Perth

Link, J. S., 2002. What Does Ecosystem-Based Fisheries Management Mean? *Fisheries,* **27,** 18–21

Link, J. S., 2005. Translation of Ecosystem Indicators into Decision Criteria. *ICES Journal of Marine Science,* **62,** 569–576

Link, J. S., 2010a. *Ecosystem-based fisheries management: confronting trade-offs.* Cambridge University Press. Cambridge, UK. 224 pp.

Link, J. S., 2010b. Adding Rigor to Ecological Network Models by Evaluating a Set of Pre-balance Diagnostics: A Plea for PREBAL. *Ecological Modeling,* **221,** 1582–1593

Link, J. S. and Garrison, L. P., 2002. Changes in piscivory associated with fishing induced changes to the finfish community on Georges Bank. *Fisheries Research,* **55,** 71–86

Link, J., Overholtz, W., O'Reilly, J., Green, J., Dow, D., Palka, D., Legault, C., Vitaliano, J., Guida, V., Fogarty, M., Brodziak, J., Methratta, E., Stockhausen, W., Col, L., Waring, G. and Griswold, C., 2008. An Overview of EMAX: The Northeast U.S. Continental Shelf Ecological Network. *Journal of Marine Systems,* **74,** 453–474

Link, J. S., Gamble, R. J. and Fogarty, M. J., 2011. *An Overview of the NEFSC's Ecosystem Modeling Enterprise for the Northeast US Shelf Large Marine Ecosystem: Towards Ecosystem-based Fisheries Management.* NEFSC Center Res. Doc. CRD 11-23. Woods Hole, MA

Link, J. S., Ihde, T. F., Townsend, H., Osgood, K., Schirripa, M., Kobayashi, D., Gaichas, S., Field, J., Levin, P., Aydin, K., Watters, G. and Harvey, C. (Eds), 2010. *Report of the 2nd National Ecosystem Modeling Workshop (NEMoW II): Bridging The Credibility Gap—Dealing With Uncertainty In Ecosystem Models.* NOAA Tech. Memo. NMFS-F/SPO-102, 72p.

Link, J. S., Gaichas, S., Miller, T. J., Essington, T., Bundy, A., Boldt, J., Drinkwater, K. F. and Moksness, E., 2012a. Synthesizing Lessons Learned from Comparing Fisheries Production in 13 Northern Hemisphere Ecosystems: Emergent Fundamental Features. *Marine Ecology Progress Series,* **459,** 293–302

Link, J. S., Ihde, T. F., Harvey, C. J., Gaichas, S. K., Field, J. C., Brodziak, J. K. T., Townsend, H. M. and Peterman, R. M., 2012b. Dealing with Uncertainty in Ecosystem Models: The Paradox of Use for Living Marine Resource Management. *Progress in Oceanography,* **102,** 102–114

Link, J. S., Lucey, S. M. and Melgey, J. H., 2012c. Examining cannibalism in relation to recruitment of silver hake *Merluccius bilinearis* in the U.S. northwest Atlantic. *Fisheries Research,* **114,** 31– 41

Little, L. R., Fulton, E. A., Gray, R., Hayes, D., Lyne, V., Scott, R., Sainsbury, K. and McDonald, A. D., 2006. *Multiple Use Management Strategy Evaluation for the North West Shelf: Results and Discussion.* North West Shelf Joint Environmental Management Study Technical Report—Vol 18, CSIRO, Hobart, Tasmania

Little, L. R., Begg, G. A., Goldman, B., Ellis, N., Mapstone, B. D., Punt, A. E., Jones, A., Sutton S. and Williams, A., 2008. *Modeling multi-species targeting of fishing effort in the Queensland coral reef fin fish fishery.* James Cook University Fishing & Fisheries Research Centre, Townsville. 291 pp.

Little, L. R., Thébaud O., Boschetti, F., McDonald, A. D., Marriott, R., Wise, B. and Lenanton, R., 2011. *An Evaluation of Management Strategies for Line Fishing in the Ningaloo Marine Park.* Final Report for Ningaloo Reef Project 3.2.3 Biodiversity Assessment, Ecosystem Impacts of Human Usage and Management Strategy Evaluation. CSIRO. 106 pp.

Lucey, S. M., Cook, A. M., Boldt, J. L., Link, J. S., Essington, T. E. and Miller, T. J., 2012. Comparative analyses of surplus production dynamics of functional feeding groups across 12 northern hemisphere marine ecosystems. *MEPS,* **459,** 219–229

Mace, G. M., Collar, N. J., Gaston, K. J., Hilton-Taylor, C., Akcakaya, H. R., Leader-Williams, N., Milner-Gulland, E. J. and Stuart, S. N., 2008. Quantification of extinction risk: IUCN's system for classifying threatened species. *Conservation Biology,* **22,** 1424–1442

Mackinson, S. and Daskalov, G., 2007. *An ecosystem model of the North Sea to support an ecosystem approach to fisheries management: description and parameterisation.* Science Series Tech Report, Cefas Lowestoft. **142,** 196 pp.

Maiolo, J. R., Johnson, J. and Griffith, D., 1992. Applications of social science theory to fisheries management: three examples. *Society and Natural Resources,* **5,** 391–407

Marasco, R. J., Goodman, D., Grimes, C. B. Lawson, P. A., Punt, A. E. and Quinn, T. J. II., 2007. Ecosystem-based fisheries management: some practical suggestions. *Canadian Journal of Fisheries and Aquatic Sciences,* **64,** 928–939

Maury, O., 2010. An overview of APECOSM, a spatialized mass balanced "Apex Predators ECOSystem Model" to study physiologically structured tuna population dynamics in their ecosystem. *Progress in Oceanography,* **84,** 113–117

May, A. W., 1975. *Report of standing committee on research and statistics.* ICNAF Seventh Special Commission Meeting—September 1975

Mayo, R. K., Fogarty, M. J. and Serchuk, F. M., 1992. Aggregate fish biomass and yield on Georges Bank, 1960–87. *JNAFS,* **14,** 59–78

McDonald, A.D, Little, L. R., Gray, R., Fulton, E., Sainsbury, K. J. and Lyne, V. D., 2008. An agent-based modeling approach to evaluation of multiple-use management strategies for coastal marine ecosystems. *Mathematics and Computers in Simulation* **78,** 401–411

McLeod, K. and Leslie, H., 2009. *Ecosystem-based management for the oceans. Island Press.* Washington D.C. 392 pp.

Milton, D. A., 2001. Assessing the susceptibility to fishing of populations of rare trawl bycatch: Sea snakes caught by Australia's northern prawn fishery. *Biological Conservation,* **101,** 281–290

Murawski, S. A., 1984. Mixed-species yield per recruitment analyses accounting for technological interactions. *Canadian Journal of Fisheries and Aquatic Science,* **41,** 897–916

Murawski, S. A., 2000. Definitions of overfishing from an ecosystem perspective. *ICES Journal Marine Science,* **57,** 649–528

Musick, J. A., 1999. Criteria to define extinction risk in marine fishes. *Fisheries,* **24,** 6–14

NEFSC (Northeast Fisheries Science Center, NOAA) 2008. Assessment of 19 Northeast Groundfish Stocks through 2007. Report of the 3rd Groundfish Assessment Review Meeting (GARM III), Northeast Fisheries Science Center, Woods Hole, Massachusetts, August 4–8, (2008). Northeast Fisheries Science Center Reference Document 08–15

NEFSC (Northeast Fisheries Science Center, NOAA) 2010. 50th Northeast Regional Stock Assessment Workshop (50th SAW) Assessment Report. Northeast Fisheries Science Center Reference Document 10–17, part A, C

NMFS (National Marine Fisheries Service) 1999. *Ecosystem-based fishery management.* A report to Congress by the Ecosystems Principles Advisory Panel. U.S. Department of Commerce, Silver Spring, MD

Oguz, T., Salihoglu, B. and Fach, B., 2008. A coupled plankton-anchovy population dynamics model assessing nonlinear controls of anchovy and gelatinous biomass in the Black Sea. *Mar. Ecol. Progr.* Ser. 369, p. 229

Overholtz, W. J., Jacobson, L. D. and Link, J. S., 2008. Developing an ecosystem approach for assessment advice and biological reference points for the Gulf of Maine-Georges Bank herring complex. *North American Journal of Fisheries Management,* **28,** 247–257

Patrick, W. S., Spencer, P., Ormseth, O., Cope, J., Field, J., Kobayashi, D., Gedamke, T., Cortés, E., Bigelow, K., Overholtz, W., Link, J. and Lawson, P., 2009. *Use of productivity and susceptibly indices to determine the vulnerability of a stock: with example applications to six U.S. fisheries.* Final Report of the National Marine Fisheries Service Vulnerability Evaluation Work Group. Silver Spring, MD. 117 pp.

Patrick, W. S., Spencer, P., Link, J., Ormseth, O., Cope, J., Field, J., Kobayashi, D., Gedamke, T., Cortés, E., Bigelow, K., Overholtz, W. and Lawson, P., 2010. Assessing the vulnerability of U.S. fisheries to becoming overfished or undergoing overfishing through the use of productivity and susceptibility indices. *Fisheries Bulletin,* **108,** 305–322

Pauly, D. and Christensen, V., 1995. Primary production required to sustain global fisheries. *Nature,* **374,** 255–257

Pauly, D., Christensen, V., Dalsgaard, J. R. and Froese, F. T. Jr., 1998. Fishing down marine food webs. *Science,* **279,** 860–863

Pauly, D., Christensen, V. and Walters, C., 2000. Ecopath, Ecosim and Ecospace as tools for evaluating ecosystem impact of fisheries. *ICES Journal of Marine Science,* **57,** 697–706

Pauly, D., Christensen, V., Guénette, S., 2002. Towards Sustainability in World Fisheries. *Nature,* **418,** 689–695

Pelletier, D. and Mahévas, S., 2005. Spatially explicit fisheries simulation models for policy evaluation. *Fish and Fisheries,* **6,** 307–349

Peterman, R. M., 2004. Possible solutions to some challenges facing fisheries scientists and managers. *ICES Journal of Marine Science,* **61,** 1331–1343

Peterman, R. M. and Anderson, J. L., 1999. Decision analysis: a method for taking uncertainties into account in risk-based decision making. *Human and Ecological Risk Assessment,* **5,** 231–244

Pikitch, E. K., Santora, C., Babcock, E. A., Bakun, A., Bonfil, R., Conover, D. O., Dayton, P., Doukakis, P., Fluharty, D., Heneman, B., Houde, E. D., Link, J., Livingston, P., Mangel, M., McAllister, M., Pope, J. and Sainsbury, K. J., 2004. Ecosystem-based fishery management. *Science,* **305,** 346–347

Pinnegar, J. K., Blanchard, J. L., Mackinson, S., Scott, R. D. and Duplisea, D. E., 2005. Aggregation and removal of weak-links in food-web models: system stability and recovery from disturbance. *Ecological Modeling,* **184,** 229–248

Pitcher T. J., Kalikoski, D., Short, K. Varkey, D. and Pramoda, G., 2009. An evaluation of progress in implementing ecosystem-based management of fisheries in 33 countries. *Marine Policy,* **33,** 223–232

Plagányi, É. E., 2007. *Models for an Ecosystem Approach to Fisheries.* FAO Fisheries Technical Paper. No. 477. Rome, Italy: FAO, pp. 108

Plagányi, É.E., Punt, A. E., Hillary, R., Morello, E. B., Thebaud, O., Hutton, T., Pillans, R., Thorson, J. T., Fulton, E.A, Smith, A. D. M., Smith, F., Bayliss, P., Haywood, M., Lyne, V. and Rothlisberg, P. C., 2012. Multispecies fisheries management and conservation: tactical applications using models of intermediate complexity. *Fish and Fisheries.* doi: 10.1111/j.1467–2979.2012.00488.x

Pollnac, R. B. Abbott-Jamieson, S., Smith, C., Miller, M. L., Clay, P. M. and Oles, B., 2006. Toward a Model for Fisheries Social Impact Assessment. *Marine Fisheries Review,* **68,** 1–18

Polovina, J. J., 1984. Model of a coral reef ecosystem I: the Ecopath model and its application to French Frigate Shoals. *Coral Reefs,* **3,** 1–10

Pope, J. G., 1975. *The application of mixed fisheries theory to the cod and redfish stocks of subarea 2 and division 3K.* ICNAF Research Document. 75/IX/126

Pope, J., 1979. *Stock assessment in multistock fisheries, with special reference to the trawl in the Gulf of Thailand.* South China Sea Fisheries Development Programme. SCS/DEV/79/19 FAO. Manila, Philippines. 106 pp.

Prellezo, R., Accadia, P., Andersen J. L., Little, A., Nielsen R., Andersen, B. S., Röckmann C., Powell J. and Buisman, E., 2009. *Survey of existing bioeconomic models: Final report.* Sukarrieta: AZTI-Tecnalia. 283 pp.

Punt, A. E. and Hilborn, R., 1997. Fisheries stock assessment and decision analysis: the Bayesian approach. *Reviews in Fish Biology and Fisheries,* **7,** 35–63

Punt, A. E. and Smith, A. D. M., 1999. Harvest strategy evaluation for the eastern gemfish (Rexea solandri). *ICES Journal of Marine Science,* **56,** 860–875

Punt, A. E., Deng, R. A., Dichmont, C. M. Kompas, T., Venables, W. N., Zhou, S., Pascoe, S., Hutton, T., Kenyon, R., van der Velde, T. and Kienzle, M., 2010. Integrating size-structured assessment and bioeconomic management advice in Australia's northern prawn fishery. *ICES Journal of Marine Science,* **67,** 1785–1801

Quince, C., Higgs, P. G. and McKanec, A. J., 2005. Topological structure and interaction strengths in model food webs. *Ecological Modeling,* **187,** 389–412

Rademeyer, R. A. Plagányi, E. E. and Butterworth, D. S., 2007. Tips and tricks in designing management procedures. *ICES Journal of Marine Science,* **64,** 618–625

Roberts, C., 2007. *The Unnatural History of the Sea.* Island Press. Washington, D. C. 465 pp.

Roberts, C. M. and Hawkins, J. P., 1999. Extinction risk in the sea. *Trends in Ecology and Evolution,* **14,** 241–248

Sainsbury, K. J., Punt, A. E. and Smith, A. D. M., 2000. Design of operational management strategies for achieving fishery ecosystem objectives. *ICES Journal of Marine Science,* **57,** 731–741

Samhouri, J. F., Levin, P. S. and Ainsworth, C. H., 2010. Identifying thresholds for ecosystem-based management. *Public Library of Science (PLOS) ONE* 2010; 5, e8907

Sarmiento, J. L., Slater, R., Barber, R., Bopp, L., Doney, S. C., Hirst, A. C., Kleypas, J., Matear, R., Mikolajewicz, U., Monfray, P., Soldatov, V., Spall, S. A. and Stouffer, R., 2004. Response of ocean ecosystems to climate warming. *Global Biogeochemical Cycles,* **18,** GB3003, doi:10.1029/2003GB002134

Shin, Y.-J. and Cury, P., 2004. Using an individual-based model of fish assemblages to study the response of size spectra to changes in fishing. *Canadian Journal of Fisheries and Aquatic Sciences,* **61,** 414–431

Shin, Y.-J., Bundy, A., Shannon, L., Simier, M., Coll, M., Fulton, E. A., Link, J. S., Jouffre, D., Ojaveer, H., Mackinson, S., Heymans J. J. and Raid, T., 2010a. Can simple be useful and reliable? Using ecological indicators for representing and comparing the states of marine ecosystems. *ICES Journal of Marine Science,* **67,** 717–731

Smajgl, A., Brown, D. G., Valbuena, D. and Huigen, M. G. A., 2011. Empirical characterisation of agent behaviours in socio-ecological systems. *Environmental Modeling and Software,* **26,** 837–844

Smith, T. D. and Link, J. S., 2005. Autopsy your dead . . . and living: a proposal for fisheries science, fisheries management and fisheries. *Fish and Fisheries,* **6,** 73–87

Smith, A. D. M., Sainsbury, K. J. and Stevens, R. A., 1999. Implementing effective fisheries management systems—management strategy evaluation and the Australian partnership approach. *ICES Journal of Marine Science,* **56,** 967–979

Smith, D. C., Smith, A. D. M. and Punt, A. E., 2001. Approach and process for stock assessment in the South East Fishery: a perspective. *Marine and Freshwater Research,* **52,** 671–681

Smith, A. D. M., Sachse, M., Smith, D. C., Prince, J. D., Knuckey, I. A., Baelde, P., Walker, T. J., and Talman, S., 2004. Alternative management strategies for the Southern and Eastern Scalefish and Shark Fishery—qualitative assessment report. Australian Fisheries Management Authority, Canberra

Smith, A. D. M., Fulton, E. A., Hobday, A. J., Smith, D. C. and Shoulder, P., 2007. Scientific tools to support the practical implementation of ecosystem-based fisheries management. *ICES Journal of Marine Science,* **64,** 633–639

Sorrentino, R. and Roney, C., 2000. *Uncertain Mind: Individual Differences in Facing the Unknown Philadelphia.* Psychology Press

Sparre, P. J., 2008. *User's Manual for the EXCEL Application "TEMAS" or "Evaluation Frame".* DTU-Aqua Report 190–08: 182 pp.

Stefansson, G. and Palsson, O. K., 1998. Points of view: A framework for multispecies modeling of Arcto-boreal systems. *Reviews in Fish Biology and Fisheries,* **8,** 101–104

Stevens, C. E., Council, T. and Sullivan, M. G., 2010. Influences of human stressors on fish-based metrics for assessing river condition in central Alberta. *Water Quality Research Journal of Canada,* **45,** 35–46

Stobutzki, I. C., Miller, M. J., Jones, P. and Salini, J. P., 2001a. Bycatch diversity and variation in a tropical Australian penaeid fishery: the implications for monitoring. *Fisheries Research,* **53,** 283–301

Stobutzki, I., Miller, M. and Brewer, D., 2001b. Sustainability of fishery bycatch: a process for assessing highly diverse and numerous bycatch. *Environmental Conservation,* **28,** 167–181

Stock, C. A., Alexander, M. A., Bond, N. A., Brander, K., Cheung, W. W. L., Curchitser, E. N., Delworth, T. L., Dunne, J. P., Griffies, S. M., Haltuch, M. A., Hare, J. A., Hollowed, A. B., Lehodey, P., Levin, S. A., Link, J. S., Rose, K. A., Rykaczewski, R. R., Sarmiento, J. L., Stouffer, R. J., Schwing, F. B., Vecchi, G. A. and Werner, F. E., 2011. On the use of IPCC-class models to assess the impact of climate on living marine resources. *Progress in Oceanography,* **88,** 1–27

Suter, G. W., 1993. *Ecological risk assessment.* Chelsea (MI), USA: Lewis.

Taylor, L., Begley, J., Kupca, V. and Stefansson, G., 2007. A simple implementation of the statistical modeling framework Gadget for cod in Icelandic waters. *African Journal of Marine Science,* **29,** 223–245

Tebaldi, C. and Knutti, R., 2007. The use of the multi-model ensemble in probabilistic climate projections. *Philosophical Transactions of the Royal Society A: Mathematical, Physical and Engineering Sciences,* **365,** 2053

Tjelmeland, S. and Bogstad, B., 1998. MULTSPEC—a review of a Multi-species modelling project for the Barents Sea. *Fish. Res.,* **37,** 127–142

Townsend, D. W., Thomas, A. C., Mayer, L. M. and Thomas, M. A., 2006. Oceanography of the Northwest Atlantic Continental Shelf (1,W). p 119–168. In *The Sea: The Global Coastal Ocean: Interdisciplinary Regional Studies and Syntheses* in A. R. Robinson and K. H. Brink (Eds.). vol 14. Harvard Univ. Press.

Townsend, H. M., Link, J. S., Osgood, K. E., Gedamke, T., Watters, G. M., Polovina, J. J., Levin, P. S., Cyr, N. and Aydin, K. Y., (eds), 2008. *Report of the NEMoW (National Ecosystem Modeling Workshop).* NOAA Technical Memorandum NMFS-F/SPO-87. 93 pp.

Tyrrell, M. C., Link J. S., Moustahfid, H. and Smith, B. E., 2007. The dynamic role of pollock (*Pollachius virens*) as a predator in the Northeast US Atlantic ecosystem: a multi-decadal perspective. *Journal Northwest Atlantic Fisheries Science,* **38,** 53–65

van Putten, I., Hamon, K. G. and Gardner, C., 2011. Network analysis of a rock lobster quota lease market. *Fisheries Research,* **107,** 122–130

Walters, C., 1986. *Adaptive Management of Renewable Resources.* New York, Macmillan.

Walters, C., Christensen, V. and Pauly, D., 1997. Structuring dynamic models of exploited ecosystems from trophic mass-balance assessments. *Reviews in Fish Biology and Fisheries,* **7,** 139–172

Watanabe, M., Adams, R. M., Wu, J. J., Boltec, J. P., Cox, M. M., Johnson, S. I., Lisse, W. J., Boggess, W. G. and Ebersole, J. L., 2005. Toward efficient riparian restoration: integrating economic, physical, and biological models. *Journal of Environmental Management,* **75,** 93–104

Weisbuch G. and Duchateau-Nguyen, G., 1998. Societies, cultures and fisheries from a modeling perspective. *Journal of Artificial Societies and Social Simulation* **1(2)**

Wilson, J., Yan, L. and Wilson, C., 2007. The precursors of governance in the Maine lobster fishery. *PNAS,* **104,** 15212–15217

Worm, B., Hilborn, R., Baum, J., Branch, T., Collie, J., Costello, C., Fogarty, M., Fulton, E. A., Hutchings, J., Jennings, S., Jensen, O., Lotze, H., Mace, P., McClanahan, T., Minto, C., Palumbi, S., Parma, A., Ricard, D., Rosenberg, A., Watson, R. and Zeller, D., 2009. Rebuilding global fisheries. *Science,* **325,** 578–585

Yodzis, P., 1998. Local trophodynamics and the interaction of marine mammals and fisheries in the Benguela ecosystem. *Journal of Animal Ecology,* **67,** 635–658

Yodzis, P., 2000. Diffuse effects in food webs. *Ecology,* **81,** 261–266

Zhou, S. and Griffiths, S. P., 2008. Sustainability Assessment for Fishing Effects (SAFE): A new quantitative ecological risk assessment method and its application to elasmobranch bycatch in an Australian trawl fishery. *Fisheries Research,* **91,** 56–68

Appendix

SUPPLEMENTAL MATERIALS.

Summary of major model types, their focus, strengths, weaknesses, data needs and example references. Note that Management Strategy Evaluation (MSE) is in theory possible for most models, but only those types where MSE has been used are indicated here.

Model Type	Focus	Strengths	Weaknesses	Data Needs	Uses
Climate and General Circulation Model (GCM)	Physical circulation of atmosphere & oceans in long term. Can be linked to NPZD.	Reflect large scale drivers.	Represent fine scale processes through parameterisation. Poor in coastal regions. High computational cost.	Physical initial conditions & process parameterisations.	Identify state-production relationships. Generate drivers for smaller scale models.
Biophysical					
Hydrodynamic	Advection, diffusion, mixing, tracer transport. Some include sediment processes. Can be linked to NPZD.	Fine scale—at scales relevant to ecology and industry activities. Faster computers making them basis of "physics to fish to fisheries" analyses.	High computational cost.	Physical initial conditions, process parameterisations, boundary conditions.	Consider physical drivers of marine systems. Identify state-production relationships. Generate drivers for smaller scale models.
Nutrient-Phytoplankton-Zooplankton-Detritus (NPZD)	Nutrient cycling; plankton & benthic invertebrate production dynamics.	Fine scale. Useful for real time forecasting to long-term productivity analyses.	High computational cost. If one-way link to higher trophic levels then lack feedback.	Initial conditions, process parameters (physical; plus growth & mortality), exogenous forcing time series (e.g. catchment flows).	Water quality. Implications of coastal/catchment activities & management at regional scale. Provide prey fields for single species (or fisheries production) distribution estimates, or prey sub-model for fish (& fisheries) models.
Minimum Realistic					
Extended Single species Assessment Model (ESAM)	Single species and its immediate major drivers (e.g. environment, time varying predation mortality).	Improved biological realism while easily incorporated into existing management arrangements.		Single species data (stock sizes, time series of abundance and removals, population parameters). Time series of the driver (predation mortality, environmental index) at appropriate scales.	Reality check on single species assessments (e.g. whether stock production feasible). Competition between fisheries & fish predators. Implications of habitat modification/ climate change for environmentally driven stocks.

Model Type	Focus	Strengths	Weaknesses	Data Needs	Uses
Multispecies	Trophic or habitat dependencies.	Capture critical system components; potentially end-to-end, but small enough can do statistical analysis and characterise uncertainty. Can often be easily incorporated into existing management arrangements.	Correctly determining complexity can be problematic.	Population initial conditions (biomass & abundance) & rate parameters, size-at-age, landings, connections between components.	Consideration of population or system dynamics (particularly trophic or habitat related); MSE
Size-based	Size spectra of communities, community dynamics.	Captures community dynamics (especially for fish) with minimal assumptions, parameter & computational demands.	Can miss broader drivers and obfuscate dynamics of key taxa (e.g. benthos may need alternative representation).	Biomass distribution (preferably through time), growth, consumption and reproduction relationships & basal production.	Response of marine (often fish) community to perturbation (e.g. fisheries, environment).
Length-based	Community dynamics.	Retains species identity while dealing with size structure of populations & communities.	Not fit to data per se and can take significant time to implement.	Initial biomasses, biological rate parameters, fishing rates.	Response of fish community to perturbation (e.g. fisheries, environment).
Trophic	Trophic spectra of communities, community dynamics.	Same logic as size based models, but uses trophic levels instead.	Misses species-specific details (some times important). Assumes constant growth rates even as system structure changes.	Initial biomasses and rate parameters.	Response of community to perturbation (e.g. fisheries, environment).
Energy & Networks					
Energy Transfer	Production potential (either system or fisheries level).	Accommodates shifting patterns in productivity. Easy to explain.	Highly aggregated, can obscure nonlinear processes.	Biomasses, diet composition, vital rates, fisheries removals.	Production potential under alternative development or harvesting policies.
Topological webs	Mapping of food web connections.	Simple to construct & analyse.	Static & requires diet data (which can be difficult to access in sufficient quantities).	Diet data.	Elucidating ecosystem structure, function, stability & vulnerability.

Qualitative (& Loop Analysis)	Can span as much of the system as desired.	Inclusive & easy to implement, explain. Loop analysis allows for perturbation analysis.	Loop analysis can become ambiguous for larger networks. Cannot anticipate alternative system structures.	No specific requirements.	Descriptive, help identify the implications of alternative model structures. Investigate system perturbations.
Eco-networks (including EwE)	Mainly trophic connections, but growing in end-to-end capability.	Conceptually simple, versatile, accessible & adaptable.	Some simplifying assumptions that are not always appropriate. Ease of misuse high[1].	Estimates of biomass, mortality, production, consumption, diet & catch.	System understanding. Effects of fishing on trophic dynamics and ecosystem properties.
Bioenergetics	Individual physiology & population dynamics.	Can be used in (or to constrain) ecosystem models.	Largely based on laboratory information.	Detailed physiological parameters.	Food required to support population (or community) activities.
Bioeconomics					
Portfolio analysis	Value of species in multispecies fisheries.	Explicit consideration of economics.	Ecological details simplified. Little consideration of social drivers.	Biomass, landings and value per species & costs of fishing.	Elucidate trade offs among cost and profit in a multi-species context. Evaluate ecosystem goods and services.
Effort allocation	Effort allocation & fleet dynamics.	Can consider some non-economic drivers.	Can have strong simplifying assumptions on ecology & decision drivers. Often assumes perfect knowledge.	Detailed cost structures, historical fishing patterns, management regulations, market data; biological stock size & production.	Effects of fishing; MSE.
Management sim.	Dependent on question, but cover to some degree socioecological system.	Can be tailored to specific systems & questions. Can include social aspects.	Dependent on specific applications, but often simplify ecological details.	Fleet details (gear, discarding rates, historical effort patterns, vessel & crew details), costs, prices, management regulations; biological rate parameters & biomass estimates.	Evaluate (economic) implications of management regimes (including via MSE).
Social Models					
Social processes	Social drivers of behaviour.	Consider social processes.	Difficult to validate.	Data on characteristics of individuals in the system and their behaviour.	Represent socially driven behaviour & information sharing.

Model Type	Focus	Strengths	Weaknesses	Data Needs	Uses
Social impact Assessment	Social impacts of industries & management.	Directly addresses social aspects of marine systems.	Difficult to establish defensible metrics for all the relevant variables.	External forces, management details, industry attributes, community attributes, individual attributes, social problems & well being.	Consider social implications of industry activities & management interventions.
Social network analysis	Social drivers of behaviour.	Conceptually simple.	Data can be hard to access.	Network structure and potential connection strength.	Understand decision making & information sharing.
Spatial Planning	Distributions of ecological features & human uses.	Provide relevant spatial information in easily interpretable form. In many cases can be rapidly deployed.	Implementation specific depending on methods/sub-models used.	Spatial distributions of all features of interest. If dynamic model included then all initial conditions, drivers & parameters for the model components should also be included.	Inform spatial planning & management decisions—evaluate implications or highlight points of overlap or tension points.
Full system—End to end (E2E) socio-econ-ecological models of an entire system					
Atlantis	Whole of system dynamics (primarily climate and fisheries relevant system dynamics).	Multiple representations of sub-models available. Highlights trade-offs and non-linearities. Covers all components of the adaptive management cycle as intended for use in MSE. Can include some social drivers of human behaviour.	Highly complex & computationally expensive. Validation typically rudimentary. Hard to work with & can be difficult to interpret.	Depends on specific implementation, but usually considerable & includes environmental drivers, biomass distribution per group (and age), rate parameters, diet, habitats, fleet details, landings, cost & prices.	Ecosystem state & function (especially under changing environmental drivers). MSE

Hybrids (& Agent Based Models)	Whole of system dynamics.	Use best means of representing each part of the system.	Can be highly complex & computationally expensive. Validation can be problematic for some ABMs.	Depends on specific implementation, but typically includes biomass distribution per group (and age), rate parameters, diet, habitats, and also potentially environmental drivers and details of human activities (potentially across multiple human industries).	Ecosystem state & function (especially under disturbance or changing drivers); MSE
BEST-BSERIP (FEAST)	Climate and fisheries relevant system dynamics.	State-of-the-art consideration of each system component.	Can be highly complex & computationally expensive. Validation can be problematic.	Environmental drivers, biomass distribution per group (and age), rate parameters, diet, habitats, fleet details, landings, cost & prices.	Implications of climate change and fisheries management.
Biodiversity					
Bioclimate envelope & species distributions	Distribution of models & how may change with changed environment.	Prediction of large-scale potential ecological changes.	Some lack relevant or non-linear ecological processes. Biased towards to well sampled/easily characterised taxa.	Extensive data requirements on species distributions & environmental covariates (more sophisticated ones also require biological rate parameters).	Predict changes in geographic distributions due to changed biophysical conditions.
Trait-based	Biodiversity shifts under changing environmental conditions (spatially or temporally).	Simple conceptual framework applicable across many scales and taxa.	No standardised means of determining environment-trait relationships.	Extensive data requirements on species distributions & environmental covariates.	Map or predict species presence-absence. Predict changes in geographic distributions due to changed biophysical conditions.
Macroecology	Biodiversity shifts under changing environmental conditions (spatially or temporally).	Conceptually simple, good theoretical grounding, strong empirical support.	Can miss some important ecological processes.	Quantitatively defined food web data.	Understanding food web structural constraints.

Model Type	Focus	Strengths	Weaknesses	Data Needs	Uses
Community-level	Biodiversity shifts under changing environmental conditions (spatially or temporally).	Simpler than mechanistic models.	Complex statistical methods & optimisations at fine scales over large areas makes this computationally expensive. Sufficient data usually only available for few taxa or locations.	Extensive survey & environmental data sets.	Estimation of species richness. Prioritisation of survey sites. Identification of effective spatial management locations. Test macroecology. Input data for metacommunity models.
Statistical					
Multivariate ordinations	Dependent on question.	Can deal with composite data sets. Can make interpretation easier.	Can require large data sets to be powerful. Can be difficult to feed into management unless a role for contextual information.	Depends on specific implementation, but can include surveys, landings, diet, oceanography, environmental, social & economic data.	Evaluate status & trends.
Multivariate time series	Dependent on question.	Relatively simple structure, but increased short-term predictive capability.	Doesn't always capture non-linearities. Requires long time series of data.	Depends on specific implementation.	Evaluate status & trends. Consideration of implications of fishing or climate scenarios.
Emulators	Dependent on question, but started with hydrodynamics and production.	Run rapidly, rapid sensitivity & uncertainty analysis of larger system.	Can take many runs of the original model to create emulator (so partial uptake might only be feasible approach for E2E).	Original model to emulate (and feasible parameter ranges for that model).	Dependent on original model. Provide parameters for other models.
Risk Assessments	Risk to the system or its components due to different pressures (e.g. environmental change or human activities).	Include wide range of factors, complement both stock & ecosystem assessments. Useful for prioritisation of attention.	May underestimate uncertainty in data-poor situations. Does not always see maximal use of routinely collected data types.	Categorical ranking of productivity & susceptibility measures (e.g. life-history characteristics & stock's relationship to the fishery).	Risk assessment for target species & other ecosystem components. Estimate extinction risk or ecosystem viability. Identify useful indicators of the key system properties & response to pressures.

1. Misuse is a criticism of all models, but the laudable effort that the EwE developers have put into making the model available and easy to use does make this risk of misuse higher. This does not mean the model should not be used, just that (as with any model) model construction must be done with care and with detailed parameter pedigrees and documentation.

Chapter 6. Integrated Ecosystem Assessments and Ecosystem-Based Management: A Social-Ecological Perspective

MARK L. PLUMMER AND PHILLIP S. LEVIN

National Oceanic and Atmospheric Administration
Northwest Fisheries Science Center

Contents

1. Introduction

In 1999, a preeminent group of ecologists and fisheries scientists produced a blue-ribbon report urging the US Congress to adopt the principles of ecosystem-based management (EBM) (e.g., Grumbine 1994) in U.S. marine waters (NOAA 1999). This report was well-grounded in decades of thinking about ecosystem approaches to resource management (e.g., Shelford 1933; Wallis 1958; Watt 1971; Sloan 2002). At the same time, the treatise was visionary and an important departure from the status quo of single-species and single-sector management. Nevertheless, the report seemed to us a lot like a fairy tale—an engaging story that ultimately requires a sequence of far-fetched events.

In this case, it might be that fairy tales can come true. After years of hand wringing over what ecosystem-based management (EBM) means, whether it is possible, and if it is even needed, we have arrived at a time when comprehensive, large-scale EBM is emerging (JSOST 2007; Lubchenco and Sutley 2010), manifested in efforts to recover depleted marine species and the ecosystems that sustain them (Pew Oceans Commission 2003; U.S. Commission on Ocean Policy 2004). To solidify the transition of EBM from principles to practice, we must develop approaches that provide an efficient, transparent means of summarizing the status of ecosystem components, screening and prioritizing potential risks, and evaluating alternative management strategies against a backdrop of environmental (e.g., climatic, oceanographic, seasonal) variability. Because there are often tradeoffs for the delivery of the varied ecosystem services society wants and needs, and between those services and the other valued aspects of human economies and cultures, EBM approaches must also provide a means of evaluating such tradeoffs. EBM requires that we expand our concept of what services we obtain from marine ecosystems beyond provisions, such as food and biochemicals, to include cultural

The Sea, Volume 16, edited by Michael J. Fogarty and James J. McCarthy
ISBN 978-0-674-07270-1

services, such as recreational opportunities, and regulating services, such as coastal protection (Daily 1997; MEA 2005; Tallis and Kareiva 2005).

Even though resource managers and policy makers have accepted marine EBM, practical advice on how to implement the tenets of EBM is scarce. This is not to suggest that we do not have a thorough scientific understanding of many of the social and biogeophysical components of marine ecosystems—we do. Rather, a major impediment to the implementation of EBM is the lack of a widely accepted process for using science to inform decision making about multiple interacting management objectives in an EBM context. Here, we describe such an approach for organizing science in order to inform decisions in natural resource management. It uses the Integrated Ecosystem Assessment (IEA) framework (Levin et al. 2009) as a foundation, aiming to guide the process of synthesizing and analyzing relevant scientific information needed for an ecosystem approach to management.

IEAs are a process for organizing diverse information to inform EBM decisions. A number of IEA frameworks exist (Levin et al. 2009; ICES 2010), including:

1) The European Union's Marine Strategy Framework Directive (MSFD), which aims to maintain or achieve "good environmental status" of European marine waters by 2020, while allowing for the sustainable use of marine goods and services. In support of this goal, the European Environment Agency produced an assessment of the state of the European marine and coastal environment (EEA 2010). This includes a summary of the status of European marine ecosystems, protected species and habitats, and the impacts of climate change, pollution, fisheries, aquaculture, shipping and energy development on marine ecosystems;

2) The Northeast Atlantic IEA, in which Robinson and colleagues (2009) developed a qualitative assessment of the status of broad ecosystem components as well as the pressures affecting them. In the North Sea, an IEA was developed by the regional ecosystem study group (ICES 2006). The assessment, based on the compilation and analyses of a comprehensive integrated data set, has provided some valuable insights into the significance of the relationships between different human pressures (e.g., nutrient inputs and fisheries) and state changes (e.g., plankton, fish, and seabirds) at different spatial scales and the time scales over which changes take place; and

3) The framework set out in Canada's Oceans Act, which requires the development of Integrated Management Plans for each of five large ocean management areas (DFO 2002). This Act envisions extensive consultation where stakeholders, agency representatives, and academics would develop a matrix of social/economic outcomes that could be achieved while ensuring productive ecosystems. These consultations are informed by "Ecosystem Assessment and Overview Reports" that describe the basic structure and function of the biophysical system (e.g. Lucas et al. 2007). Conservation objectives are then identified based on these reports. These objectives are designed to guide decision makers to ensure that the health of the ecosystem in the large ocean marine area is not compromised and to guide management approaches for conservation and human activities.

These approaches differ in the degree to which pressures are linked to ecosystem states, the degree of integration across human and natural dimensions, and the regional consistency that the frameworks promote. They all share a motivation to describe the status of the ecosystem relative to some desired state, however. Below, we focus on NOAA's IEA framework, while acknowledging that IEAs are an emerging tool that is being approached differently by different countries and agencies.

NOAA defines an IEA as a formal synthesis and quantitative analysis of information on relevant natural and socioeconomic factors relative to specified ecosystem management goals (Levin et al. 2009). It involves and informs citizens, industry representatives, scientists, resource managers, and policy makers through formal processes to contribute to attaining the goals of EBM. Below, we first outline the steps that constitute an IEA, as proposed by NOAA. We then note that IEAs, like other elements of EBM, must confront some complexities involved in incorporating human dimensions—a crucial component of any EBM enterprise. Finally, we argue that if IEAs and EBM as a whole are to be successful, they must evolve to fully incorporate human systems (see also Charles, this volume).

2. IEAs in Principle

IEAs fundamentally provide answers to three questions: 1) What is a healthy ecosystem? 2) What are the risks faced by an ecosystem? 3) What management actions should be taken to restore or maintain ecosystem health? Answering these questions requires the development of indicators and reference levels to measure progress toward ecosystem objectives; approaches that quantify the risk confronting the ecosystem from a variety of human activities and natural perturbations; and a toolbox of conceptual and quantitative models to forecast the potential outcome of different management options or environmental trends, which are evaluated against EBM goals (Figure 6.1). Below we describe in more detail these three IEA activities.

2.1 Indicator development

A critical step in the IEA process is to select indicators that capture the key ecosystem states and processes that underlie healthy ecosystems. Indicators are quantitative measures that serve as proxies for characterizing key attributes of biogeochemical and human systems (Heinz Center 2008). Indicators serve as measures of the many ecosystem services about which policy makers and stakeholders are concerned (Link 2005), and are one of the primary contact points between policy and science.

Hundreds of indicators have been proposed for use in EBM (Kershner et al. 2011). These range from physical indices to abundance of single species (Goericke et al. 2007), to community-level metrics such as the size structure of the community, biomass ratios, and indices of diversity. There is clearly no shortage of potential indicators of ecosystem status, but the real work is to select wisely from among long lists of potential indicators (Rice and Rochet 2005).

Rice and Rochet (2005) outline a valuable framework for building a portfolio of informative indicators for EBM. They argue that indicators should be directly observable; based on well-defined theory while also being understandable to the general public; cost effective to measure; supported by historical time series; sensitive and responsive to changes in ecosystem state (and management efforts); and

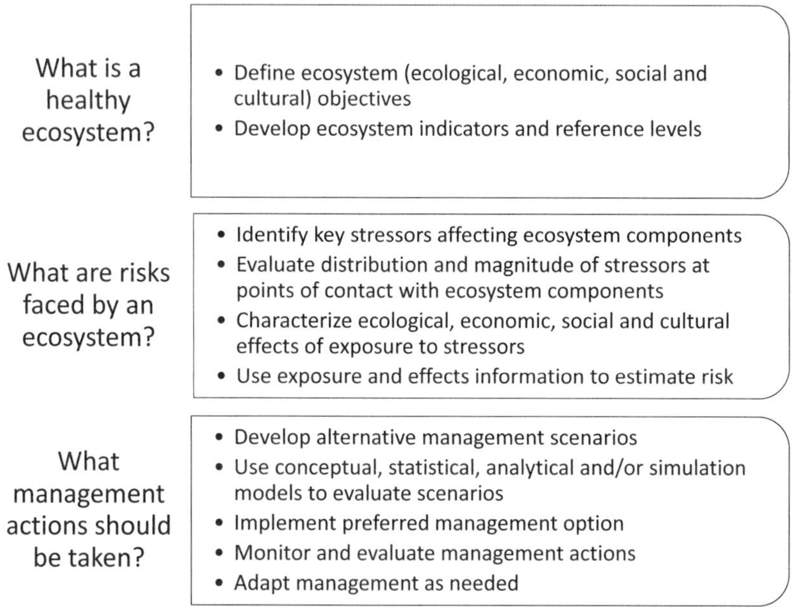

Fig. 6.1 – NOAA's Integrated Ecosystem Assessments rely on an iterative process that uses a step-wise process to answer three key questions: what is a healthy ecosystem, what risks do ecosystems face, and what management actions should be taken. An IEA begins with a scoping process to identify key management objectives and constraints, identifies appropriate indicators and reference levels, and determines the risk that indicators face. The potential of different management strategies to alter ecosystem status is evaluated, and then management actions are implemented and their effectiveness monitored. The cycle is repeated in an adaptive manner. (Adapted from Levin et al. 2009)

responsive to properties they are intended to measure. Levin et al. (2011) adapted this framework for use in Puget Sound. They solicited and organized expert judgment from the scientific community concerning potential ecosystem indicators for the region. Using a framework comparable to that proposed by Rice and Rochet (2005), a team of scientists representing different agencies and areas of expertise then worked through proposed indicators, determining how well they met criteria related to public awareness, cost effectiveness, theoretical foundation, measurability, and availability of historical data (Kershner et al. 2011). A qualitative screening process such as this can be an appropriate start to an indicator screening process, but cannot rigorously evaluate key indicator traits such as sensitivity, responsiveness, or specificity.

Building on the work of Rice and Rochet (2005), the IndiSeas project (http://www.indiseas.org) developed a set of indicators that are ecologically significant, measurable, sensitive to fishing pressure, and understandable by the general public. The indicators address issues relevant to biodiversity, ecosystem stability, structure and function, and resource potential. This project summarized a set of eight indicators for 30 countries spanning Africa, Europe and North America.

Rigorous evaluation of indicator performance requires empirical analyses or simulation modeling. Empirical evaluation of indicator sensitivity, responsiveness, or specificity requires extensive data as well as knowledge about the statistical properties of the indicators and processes structuring the ecosystem. For instance, Trenkel and Rochet (2003) examined the performance of a series of population and community-level indicators in the Celtic Sea fish assemblage. They used data from a groundfish survey to test hypotheses related to pre-selected reference points. The performance of indicators was evaluated using estimates of precision (e.g. coefficient of variation) and statistical power.

Computer simulation modeling provides another approach for evaluating indicators. For example, Samhouri et al. (2009), used Ecopath with Ecosim (EwE) food web models to determine the degree to which potential indicators reflected changes in ecosystem attributes. Fulton and colleagues (2004) conducted a similar analysis using the end-to-end ecosystem model, Atlantis (see also Fulton and Link, this volume). In this approach, an operating model (EwE or Atlantis in these cases) is used to simulate the dynamics of the system over time, producing a time series of pseudo-data. These pseudo-data are processed using standard techniques to generate time series of indicators. Indicators are then evaluated by their ability to detect or predict changes in "true" values of key ecosystem attributes, which are known from the simulation model.

2.2 *Reference levels*

Establishing a set of specific indicator values that reflect progress towards specific management objectives is critical for successful EBM. Such reference levels provide context for evaluating performance and progress. Reference levels can include both ecosystem state variables of interest (e.g., population size, productivity, etc.) as well as metrics of ecosystem pressures (e.g., fishing rate, nutrient or contaminant input, etc.). These levels can be drawn from the underlying properties of the natural and human systems or they can be designated as part of the process of setting management goals.

2.2a Types of reference levels

For an IEA, reference levels serve a variety of purposes. The most obvious is the role they play in the establishment of targets for restoring and protecting the ecosystem (Sainsbury et al. 2000; Bottrill et al. 2008). A reference target is the level of an ecosystem attribute (or by proxy an ecosystem indicator) that signals the desired state of the ecosystem. A reference limit, on the other hand, is the attribute level that marks an ecosystem state to be avoided (Samhouri et al. 2011). Because some ecosystem indicators respond slowly to management action or natural drivers, there may be a need for benchmarks, or intermediate indicator values that demonstrate progress or regression toward those levels. For example, for a case where an ecosystem goal is to recover an endangered species, a benchmark might be achieving an annual population growth rate greater than one.

Baseline values of indicators are another type of reference level. These are indicator levels derived from time periods or locations that are chosen to reflect a "baseline condition" of the ecosystem. In fisheries management the status of populations is often expressed as a ratio of current biomass to the biomass expected in an unfished system. It is similarly possible to express the status of an ecosystem indicator relative to that which might exist in a system free from human pressures (Samhouri et al. 2011). Using a human-free system as an appropriate baseline for coupled natural-human systems is problematic, however (Denevan 2011). Thus, while baseline reference levels can provide important context for IEAs, their use in management applications should be carefully considered.

2.2b Setting reference levels

Establishing a reference level is clearly a policy matter, but an IEA can play an important role in supporting this task. The key is distinguishing between *descriptive* and *normative* reference levels. A descriptive reference level is an indicator value that provides important information about the natural or human systems that are the subject of an EBM effort, and can be established in an objective way. In contrast, if a reference level is set in the context of social, economic, or cultural values, it is then a normative reference level. A target reference level, for example, is inherently normative as it is derived from an assessment of the "desired" state of the ecosystem, which is fundamentally grounded in the values held by the humans who are an integral part of the system. A normative reference level can be derived from and even equal to a descriptive reference level, but that choice lies outside the scope of an IEA, at least in terms of the scientific underpinnings of an IEA.

Nonlinearities in the relationship between pressures and ecosystem states are an important source of descriptive reference levels for an IEA (Samhouri et al. 2010). For instance, single non-indigenous species may have little or no impact on community structure. Above some threshold number of non-indigenous species, however, changes in attributes of community structure can be rapid (Simberloff and von Holle 1999). Such nonlinear functional relationships are commonly used to inform single-sector management. For example, the concept of maximum sustainable yield is based on the expectation that fisheries productivity peaks at intermediate levels of population size.

While a descriptive reference level may also be chosen as a target or limit, there are many instances where the two are likely to be distinct. For example, the maximum sustainable yield of a fishery is a key descriptive reference level for fishery management, but it is not necessarily (and in fact rarely is) the "optimum" yield, which is the (normative) target level under the Magnuson-Stevens Fishery Conservation and Management Act (50 CFR Ch. VI, §600.310(a)). Similarly, an ecological threshold that, if exceeded, produces a "significant" change in the natural system does not automatically establish a normative reference limit (Perrings and Pearce 1994). The social and economic values of that change (or of avoiding that change) also play a role in determining whether such a change should be avoided, and so setting a descriptive threshold as a normative limit is fundamentally a policy choice.

2.3 Risk assessment

Risk assessment is an analytical approach for describing risk—the likelihood that a subject will experience adverse consequences due to exposure to particular hazards (Burgman 2005; Suter 2007). In the context of an IEA, a risk assessment quantifies the degree to which human activities or natural processes impact the delivery of ecosystem goods and services (Levin et al. 2009). Risk assessment must include activities that occur on land (e.g., coastal development, agriculture, etc.) and on the ocean itself (e.g., shipping, naval exercises, fishing, energy extraction) (Halpern et al. 2009). Thus, an ecosystem risk assessment ideally requires an understanding of the distribution and intensity of land- and sea-based activities, as well as the impacts these activities will have on ecosystem components.

While there are a number of approaches to risk assessment, most forms of risk assessment can be applied within the ecological risk assessment framework used in Australia and described by Hobday and colleagues (2011). Briefly, this is a hierarchical approach that moves from qualitative but comprehensive analyses to a less comprehensive, semi-quantitative analysis, to a focused, fully quantitative analysis. The initial analysis may be a qualitative assessment of the impact of each human activity on focal ecosystem components of the IEA. Those pressures receiving a high impact score move onto more detailed, semi-quantitative analyses.

The most commonly used framework for semi-quantitative ecological risk assessment is the US EPA framework (EPA 1998). This framework consists of 3 main steps: problem formulation, analysis, and risk characterization. In the EPA model, problem formulation assembles and summarizes information, operationally defines assessment end points, and develops a plan for analysis. In large part, the problem formulation step in the EPA framework overlaps with the previous steps in the IEA framework. The analysis step of the EPA framework evaluates data relating to the nature, distribution, and magnitude of stressors at points of contact with ecosystem components (or their indicators), and then characterizes the ecological effects of the exposure. This includes results from measurements or observations indicating the responses of ecological endpoints to variation in exposure to stressors. The final step of the EPA process uses the exposure and effects information to estimate risks and analyze underlying uncertainty.

Samhouri and Levin (2012) applied the analysis phase of the EPA process for a variety of ecosystem indicators in Puget Sound. They defined risk (R_i) to indicator *i* as the Euclidean distance from the origin in a space defined by exposure (E) and sensitivity (S) of the species to particular human activities:

$$R_i = \sqrt{(E-1)^2 + (S-1)^2} \tag{1}$$

Using this definition, the risk to an indicator increases with distance from the origin and each axis receives equal weight in estimating risk.

The exposure was estimated as a function of spatial and temporal overlap of activities, also incorporating management factors that could aggravate or mitigate the exposure. For instance, fish species targeted by fisheries may be more likely to be affected by stressors because of the potential for cumulative impacts. Sensitivity was estimated as a function of the degree to which life history attributes or behavior affected an indicator's ability to resist or recover from exposure to a human activities.

The third, fully quantitative step in the Australian risk analysis framework (Hobday et al. 2011) uses quantitative methods such as those used in stock assessments and population viability analyses. A number of modeling approaches lend themselves to such analyses (e.g., Morris and Doak 2002).

2.4 *Management strategy evaluation*

A critical step in an IEA is the development and evaluation of management scenarios representing suites of proposed strategies designed to achieve specific ecosystem objectives. EBM strategies are complex, driven by interactions between the natural system, human behavior and decision-making, and uncertainty throughout the coupled natural-human system. Consequently, evaluating management strategies must consider variability and uncertainty in ecosystem properties.

Following the publication of the Millennium Ecosystem Assessment (MEA 2005), systematic scenario analysis is increasingly being used as an approach to evaluate management options. Scenario analysis generates multiple alternative descriptions of potential outcomes, including processes of change, thresholds, and uncertainties (Alcamo 2008). Scenarios explore alternative perspectives about underlying system processes and can illuminate key issues by using a consistent set of assumptions about the system state to broaden perspectives (Raskin 2005; Refsgaard et al. 2007). They generate alternative, internally consistent, logical descriptions of the future. Scenarios can be qualitative, in which "storylines" are developed, or quantitative, in which the outcomes of numerical models are explored (Refsgaard et al. 2007). Scenarios typically include assessments of the ecosystem state variables and driving forces, descriptions of critical uncertainties, and approaches for resolving them (Swart et al. 2004). One unique attribute of scenarios is that they acknowledge the interdependencies of system components. The advantages of qualitative scenarios include more flexibility to incorporate multiple stakeholder perspectives and greater capacity for creative thinking. Quantitative scenarios can provide geographical and numerical specificity to the concepts provided by qualitative scenarios (Alcamo et al. 2005).

The United Kingdom National Ecosystem Assessment, for example, developed a range of scenarios based on six storylines employing very different policy priorities (Watson 2012). Scenario assessments compared management strategies in terms of agricultural food production, net change in greenhouse gases, open access recreation, urban green space amenity, and biodiversity. The UK National Ecosystem Assessment demonstrates the ability of methods to assess the relative economic and conservation implications of different storylines in a manner that is useful in ecosystem-scale decision-making.

In addition to scenario analysis, IEAs can benefit from formal Management Strategy Evaluation (MSE). MSE uses simulation models to compare alternative strategies in a virtual world. The foundation of the MSE approach is a model (or models) that depicts 'true' ecosystem dynamics. The aim of this operating model is to capture key physical, chemical, ecological, and socio-economic processes that generate system dynamics. Data are sampled from the operating model to simulate research surveys, which results in simulated measurements of key ecosystem indicators such as biomass of species, diversity, mean trophic level, and so forth. Research surveys are simulated, incorporating typical biases and error inherent in a monitoring program. These pseudo-data are then passed to a management model (e.g., risk analysis and assessment models). Management models estimate the predicted status of individual species or indicators and the ecosystem as a whole. Ideally, management models also include reference points that are linked to simulated management actions (e.g., fishing levels). Human response to this simulated decision is modeled, incorporating uncertainty in compliance and behavioral responses to regulations. Human actions are then linked back to the operating model, thus potentially influencing the simulated ecosystem state. Repeating these steps creates a simulation of the full management cycle. For more on Management Strategy Evaluation, see Fulton and Link (this volume).

By evaluating a range of management scenarios using multiple performance indicators (and potentially multiple operating models), a formal MSE can be used to test the utility of modifying indicators and management targets, assessments, monitoring plans, management strategies, and decision rules. Critically, the usual objective of formal MSE is not optimality. Rather, MSE is typically used to screen out poorly performing management strategies and to identify approaches that are robust to various types of uncertainty.

3. Incorporating Human Dimensions into IEAs

Despite the clear goal to include economic, social and cultural objectives in IEAs, the focus to date of IEAs in practice has been on the ecological components of coupled natural-human systems (e.g. Levin and Schwing 2011; Puget Sound Science Update 2011). This reflects the development of the concept of EBM itself. In its early incarnations, EBM focused on "ecological systems that may cross administrative and political boundaries, incorporating a 'system' perspective sensitive to issues of scale, and managing for ecological integrity" (Endter-Wada et al. 1998). This initial definition was an ecologically centered view with human systems incorporated simply as political boundaries or more complexly as impacts on the system to be controlled or reduced.

Although the purely ecocentric view still exists, there has been increasing recognition of the need to integrate humans and our social systems more completely into the EBM framework. How this is done influences the conduct of an IEA that supports the management effort. Here, we briefly discuss two issues. First, selecting the goals of an EBM must necessarily address humans, their economic and social systems, and more broadly and importantly, their well-being. An IEA must similarly grapple with these issues because they can influence the selection and interpretation of indicators used to assess both the ecological and human systems. Second, one must confront some unique features of potential indicators that assess human systems and their well-being, more commonly known as social indicators. Social indicators can span a wide variety of attributes that reflect the multiple dimensions of human well-being, even down to the level of an individual. While this creates the opportunity for selecting a rich set of human well-being (HWB) indicators, it also presents challenges in how to interpret changes in those indicators.

3.1 What do we humans want?

A typical initial step in developing an IEA is aligning its scope with goals of the associated EBM effort, which supports the development of indicators and other elements of the IEA (Figure 6.1, Levin et al. 2009). Recognizing that the selection of the EBM's goals is inherently a policy process, we nevertheless consider the implications of that process for the development of the associated IEA framework.

The initial phase of an EBM often produces a set of goals that have both biophysical and human dimensions. For the purpose of conducting the accompanying IEA, understanding the nature of EBM goals is important for selecting indicators and evaluating management strategies. In particular, if strategies produce conflicts among goals, the questions arise of whether and how an IEA can address tradeoffs among goals (more than just describing them) and contribute to their resolution.

The consideration of HWB in the context of an EBM sometimes takes the form of creating a "human well-being" goal, which then stands alongside other, ecologically-oriented goals (e.g., Puget Sound Partnership 2009). This approach is viewed as one way of resolving a long standing tension between adopting a wholly ecocentric or wholly anthropocentric viewpoint in EBM (Endter-Wada et al. 1998). Setting these two types of goals apart, however, effectively characterizes the ecological goals as having something in the nature of intrinsic value. Although hard to define precisely, intrinsic value means that something is an end in itself—it "counts" independent of any human doing the counting (Rolston 1989).

In the context of an EBM, this would mean that goals that are focused on biophysical elements such as species, habitat, water quality, and water quantity (somehow) have value independent of the human well-being they support and generate. It could be that a particular biophysical goal supports aspects of other natural goals (e.g., the value of nearshore habitat may be derived from its support for certain species). This supporting value may then ultimately find expression in human-oriented values such as commercial and recreational fishing. But such connections need not exist if these goals have intrinsic value. In this (admittedly extreme) view, ecological goals are, in principle, an end in themselves, obviating the need to draw any link to human well-being. Indeed, one can legitimately pursue such goals even if human well-being is not improved by doing so or even diminished.

A problem with this construction is the difficulty it creates when intrinsically valued goals conflict with one another or, in this case, with HWB (Justus et al. 2009). Little guidance is given about which goal should take precedence, and so the resolution of conflicts is hard to assess in a consistent, reasoned way. While it is certainly possible to establish quantitative indicators for these goals, granting them the nature of intrinsic value undercuts the ability to consider whether changes in indicators across various management strategies are "worth it"—a conclusion that intrinsic value effectively forecloses.

An alternative view of EBM goals is to consider them explicitly in terms of their connections, direct or indirect, to human values and more generally to human well-being (see Guerry and Tallis, this volume; Altman et al., this volume). The goals then have the nature of instrumental value. As noted by Justus et al. (2009), something has instrumental value to the extent that it is "considered valuable by valuers"—that is, in the context of an IEA, it is something that humans value about the environment. Instrumental does not mean material or based solely on monetary values, but includes values that are independent of consumption or the use of a resource, for example, and can even involve actions that are to the material detriment of the valuer.

Viewing EBM goals as instrumental creates an opportunity for an IEA to evaluate strategies with indicators that have a common ultimate policy metric, because each goal and its associated indicators are viewed as "instruments" in achieving some higher, over-arching goal. One such approach (but not the only one) is to view HWB as an over-arching goal. Ecological goals can be viewed as instrumental in supporting HWB, which then becomes in principle a common metric by which to assess management actions.

Using this approach, it is straightforward to consider different pathways that connect ecological goals to HWB. For example, consider an EBM effort that sets a goal of improving the viability of certain species populations. Such a goal has direct connections to HWB but through possibly multiple types of values. A species may be harvested commercially, and so have a market-based economic value; it may be harvested recreationally, and so have non-market economic and cultural values; it may be valued simply for its existence as an iconic species; or it may be a forage species for others that have any of the preceding values and so be valued for its support of these species and their values. In all of these cases, strategies that affect the goal of improving the species' viability can be assessed in an IEA using indicators that ultimately can be related to HWB, and so tradeoffs among the different values identified above or even between this goal and other goals that have a similar ultimate metric can be evaluated within the IEA.

Even if an EBM effort exclusively adopts ecological goals as its guiding principles, assessing the goals in terms of human well-being is still a likely and desirable part of that effort. The guiding principle of the MSFD, for example, is the achievement or maintenance of a "good environmental status" (GES) of European seas by 2020. While this concept could, in principle, be defined and indicators chosen without reference to human well-being, the MSFD requires EU member states to account for the social and economic aspects of their strategies for achieving or maintaining GES (COWI 2010). It even requires a cost-benefit analysis of the measures they intend to take, a requirement that necessarily calls for linking GES to human welfare, a touchstone of cost-benefit analysis (Zerbe 2001).

Whether an EBM explicitly embraces HWB as an ultimate metric or adopts myriad goals, only some of which are directly measured in HWB terms, an EBM effort certainly must ponder the links between HWB and ecological goals. An IEA can support this by documenting and, where sufficient data exist, even modeling these links. In this way, MSE and other evaluation strategies can provide a full set of effects and tradeoffs.

3.2 Selecting social indicators for human systems

An important difference between humans and other species, at least for the purposes of an EBM, is that the well-being of humans depends on more than just biological attributes such as population abundance and growth. While those simple metrics, along with other more demographic-based measures, can be addressed in a straightforward fashion, HWB is much broader. It encompasses material wealth; relationships with family and friends; emotional and physical health; work and recreation; how one feels about one's community; and personal safety. A second challenge in incorporating HWB into EBM, then, is accounting for this broad and complicated nature in the selection of social indicators.

An important feature of a social indicator is one that stems from another distinction between humans and other species: We talk back. This gives rise to the possibility of having both objective and subjective indicators (Diener and Suh

1997; Costanza 2007; Cummins 2000). Objective indicators are those that can, in principle, be measured and verified in the "public domain," as expressed by Cummins (2000). Objective indicator data can be gathered by observation or other forms of impersonal measurement, or by surveys that seek objective information from individual respondents. The key feature of an objective indicator is the perspective: in principle, they measure attributes of HWB that are publicly visible and have a uniform interpretation across individuals.

Objective social indicators help us understand how specific individuals, communities, or economic sectors and systems utilize resources or interact with the environment, but they do not measure psychological satisfaction, happiness, and life fulfillment; how people feel about their place; or how their subjective experience is influenced by the health of the environment. Measures of these aspects of HWB are called subjective indicators. These are private attributes of HWB in the sense of not being amenable to public, independent observation and verification. By necessity, subjective social indicators are gathered by talking (or more generally, communicating) with people directly, through survey research and other instruments that ascertain the subjective reality in which people live.

The social science literature that has built up around the subjective measurement of well-being is generally known as life satisfaction or "happiness" studies. The types of measures used to assess HWB in this way fall into two categories: (1) measures that reflect an individual's self-reported well-being in a global or holistic sense; and (2) measures that reflect an individual's self-reported well-being in the moment (Frey and Stutzer 2002; Vitarelli 2010).

Data on self-reported individual well-being can be used to explore its determinants, an approach that has been taken for income (Hsieh 2003; Solberg et al. 2002; Vera-Toscano et al. 2006; Warr 1999; and many others), housing and homeownership (Diaz-Serrano 2006), and education (Hayo and Wolfgang, 2003), to name just a few possible determinants. This approach has also been used to examine the relations between environmental conditions and subjective well-being (Welsch and Kühling 2009; Ferreira and Moro 2010), including climate (Becchetti et al. 2007; Frijters and van Praag 1998; Rehdanz and Maddison 2005), air pollution (Welsch 2002; Welsch 2006; Di Tella and MacCulloch 2007; Luechinger 2009), flood hazards (Luechinger and Raschky 2009), and water pollution (Israel and Levinson 2003).

While research on measuring subjective HWB directly and exploring its determinants is growing, the literature has not yet expanded to cover a broader set of goals more relevant to EBM. Nevertheless, these studies and this method provide an interesting perspective on how links between ecological conditions and HWB can be assessed. If changes in these conditions have progressed to the point of having serious impacts on human systems, viewing the impacts through the lens of direct, subjective measurement of HWB would seem a fruitful avenue.

A second important feature of a social indicator is one that is drawn above for indicators in general, the difference between descriptive and normative. If a social indicator has an unambiguous relationship to HWB, the indicator can then be used as a normative indicator, or one that can be directly tied to a social policy goal and used to gauge progress toward that goal (Land 1983; Sharpe 1999). If an indicator does not have a clear policy relation, it can still be used as a descriptive indicator (Land 1983).

An example of how this distinction complicates matters comes from the possibility of using activities that take place in a natural setting (e.g., beach visitation, fishing, etc.) as indicators of the HWB derived from those places. Measures of those activities can obviously be used as descriptive indicators but may not be good candidates for normative indicators. Between 1991 and 2006, for example, U.S. participation in freshwater and saltwater fishing, and in wildlife viewing away from home decreased (U.S. Fish and Wildlife Service 2009 and 2010). Can one infer that this change signals a decrease in HWB? Unfortunately, no. Outdoor activities compete with other activities for an individual's leisure time. A decrease in opportunities for outdoor recreation or in environmental quality could result in a decline in participation. Such a case would also signal a decline in HWB (other things equal). But the same change in participation could be the result of an expansion in the availability or quality of other leisure activities, which would correspond to an increase in HWB. Thus, indicators based on human behavior that is linked to environmental opportunities and qualities can faithfully reflect changes in these factors but can also change for non-ecological reasons, which complicates their interpretation.

An alternate approach can resolve some of these problems. The level of human activities is the outcome of two factors: the supply of opportunities for that activity and the demand for engaging in that activity (Walls et al. 2009). Demand depends on myriad factors that lie outside the scope of an IEA, such as income, tastes, the availability of substitute activities, and so on. This is, in part, what makes the level of the activity itself a relatively less useful normative indicator. Supply may be a useful indicator, however, because an IEA is more likely to consider strategies that will affect its level (or its quality), and variation in this level, other things constant, is a good predictor of variation in the value humans can derive from that activity.

3.3 EBM goals and IEA social indicators

Assessing the goals of EBM in terms of their connections to HWB and considering how these connections can be reflected in both ecological and social indicators provides some guidance for IEAs, then, in several ways. First, where sufficient evidence exists to indicate the strength of a connection such evidence can highlight potential indicators associated with relatively strong connections. Second, the evidence can at least give some insights into the overall effect on HWB in cases where proposed management actions have multiple effects and potential tradeoffs. The evidence might indicate where such tradeoffs are likely to be "one-sided," in the sense of one value or connection being significantly stronger than any other; or it might indicate where such tradeoffs might be "closer," in that they involve multiple connections with some value but which move in opposite directions in response to a proposed action. And finally, collecting and cataloging evidence of this sort can highlight the (unfortunately many) areas where evidence is sparse, particularly for the connections among biophysical conditions, human behavior and values, and overall human well-being.

This is not to suggest that every ecological indicator should be converted into monetary terms, and that the IEA effort should be reduced to a simple benefit-cost calculus. Explicitly connecting EBM goals to HWB terms, acknowledging where evidence exists or is absent, and being cognizant of the complexities of choosing appropriate social indicators is a good start. In these ways, an IEA can provide science-based insights into how different strategies are reflected in trade-offs and provide some data—not determinative but at least supportive—of how the goals in play could be compared in terms of a common metric.

4. Conclusions

There is little question that ocean and coastal ecosystems are beset with a range of pressures that, left unattended, could significantly impact the delivery of ecosystem services. IEAs can provide this support but to do so effectively requires that they continue to evolve to encompass the full system that EBM addresses. The challenge is more than just adding indicators that cover humans. As we have discussed above, the nature of HWB creates both opportunities and challenges for the development of EBM and an IEA.

Because humans talk back, the set of social indicators available for an IEA is richer in some ways than those for other species. But that richness also makes the interpretation of social indicators more challenging, and so IEAs must embrace or at least endure the sometimes messy world of human well-being. Doing so will better inform the performance of an EBM and avoid simplistic platitudes about what is and is not "good" for humans. The real issue of moving EBM forward (or perhaps the dragon in our fairy tale of EBM) may not simply be threats arising from human use of the sea, but may also include our failure to adequately consider the complexities of human systems. While we do not yet know how this fairy tale will end, we contend that by fully and rigorously considering the coupled human-natural system, the story is more likely to have a happy ending.

References

Alcamo, J. 2008. *Environmental futures, Volume 2: the practice of environmental scenario analysis.* Elsevier, Amsterdam. 212p.

Alcamo, J., D. van Vuuren, C. Ringler, W. Cramer, T. Masui, J. Alder, and K. Schulze. 2005. Changes in nature's balance sheet: model-based estimates of future worldwide ecosystem services. *Ecology and Society* 10(2): 19. [online] URL: http://www.ecologyandsociety.org/vol10/iss2/art19/.

Becchetti, L., S. Castriota, and D. A. Londono Bedoya. 2007. Climate, happiness and the Kyoto protocol: someone does not like it hot. Departmental Working Paper No. 247, Tor Vergata University, 2007.

Bottrill, M. C., L. N. Joseph, J. Carwardine, M. Bode, C. Cook, E. T. Game, H. Grantham, S. Kark, S. Linke, E. Donald-Madden, R. L. Pressey, S. Walker, K. A. Wilson, and H. P. Possingham. 2008. Is conservation triage just smart decision making? *Trends in Ecology & Evolution* 23:649–654.

Burgman, M. 2005. *Risks and decisions for conservation and environmental management.* Cambridge University Press, Cambridge, UK. 504p.

Canadian Department of Fisheries and Oceans (DFO). 2002. *Canada's ocean strategy: policy and operational framework for integrated management of estuarine, coastal and marine environments in Canada.* Ottawa, Ontario. 43p.

Costanza, R., B. Fisher, S. Ali, C. Beer, L. Bond, R. Boumans, N. Danigelis, J. Dickinson, C. Elliott, J. Farley, D. E. Gayer, L. M. Glenn, T. Hudspeth, D. Mahoney, L. McCahill, B. McIntosh, B. Reed, S. A. T. Rizvi, D. M. Rizzo, T. Simpatico, and R. Snapp. 2007. Quality of life: an approach integrating opportunities, human needs, and subjective well-being. *Ecological Economics* 1:267–276.

COWI. 2010. Scoping study on the requirements for economic assessment in the Marine Strategy Framework Directive. Final Report, European Commission DG Environment, June 2010. 81p.

Cummins, R. 2000. Objective and subjective quality of life: an interactive model. *Social Indicators Research* 52: 55–72.

Daily, G. C. 1997. *Nature's services: societal dependence on natural ecosystems.* Island Press, Washington, D.C. 412p.

Denevan, W. M. 2011. The "pristine myth" revisited. *Geographical Review* 101: 576–591.

Di Tella, R., and R. J. MacCulloch. 2007. Gross national happiness as an answer to the Easterlin paradox? *Journal of Development Economics* 16: 22–42.

Diaz-Serrano, L. 2006. Housing satisfaction, homeownership and housing mobility: A panel data analysis for twelve EU countries. IZA Discussion Papers 2318, 2006. Institute for the Study of Labor (IZA).

Diener, E., and E. Suh. 1997. Measuring quality of life: economic, social, and subjective indicators. *Social Indicators Research* 40:189–216.

Endter-Wada, J., D. Blahna, R. Krannich, and M. Brunson. 1998. A framework for understanding social science contributions to ecosystem management. *Ecological Applications* 8: 891–904.

European Environment Agency (EEA). 2010. *The European environment: state and outlook, 2010.* Copenhagen, Denmark.

Ferreira, S., and M. Moro. 2010. On the use of subjective well-being data for environmental valuation. *Environmental Resources and Economics* 46:249–273.

Frey, B. S., and A. Stutzer. 2002. What can economists learn from happiness research? *Journal of Economic Literature* 40: 402–435.

Frijters, P., and B. van Praag. 1998. The effects of climate on welfare and well-being in Russia. *Climate Change* 39: 61–81.

Fulton, E. A., A. D. M. Smith, and C. R. Johnson. 2004. Effects of spatial resolution on the performance and interpretation of marine ecosystem models. *Ecological Modelling* 176: 27–42.

Goericke, R., E. Venrick, T. Koslow, W. J. Sydeman, F. B. Schwing, S. J. Bograd, W. T. Peterson, R. L. Emmett, J. R. L. Lara, and G. G. Castro. 2007. The state of the California Current, 2006–2007: regional and local processes dominate. *California Cooperative Oceanic Fisheries Investigations Reports* 48:33–66.

Grumbine, R. E. 1994. What is ecosystem management? *Conservation Biology* 8:27–38.

Halpern, B. S., C. M. Ebert, C. V. Kappel, E. M. P. Madin, F. Micheli, M. Perry, K. A. Selkoe, and S. Walbridge. 2009. Global priority areas for incorporating land-sea connections in marine conservation. *Conservation Letters* 2:189–196.

Hayo, B., and S. Wolfgang. 2003. Subjective economic well-being in Eastern Europe. *Journal of Economic Psychology* 24: 329–348.

Heinz Center. 2008. The state of the nation's ecosystems 2008: measuring the land, waters, and living resources of the United States. The H. John Heinz III Center for Science, Economics, and the Environment, Washington, DC. 44p.

Hobday, A. J., A. D. M. Smith, I. C. Stobutzki, C. Bulman, R. Daley, J. M. Dambacher, R. A. Deng, J. Dowdney, M. Fuller, D. Furlani, S. P. Griffiths, D. Johnson, R. Kenyon, I. A. Knuckey, S. D. Ling, R. Pitcher, K. J. Sainsbury, M. Sporcic, T. Smith, C. Turnbull, T. I. Walker, S. E. Wayte, H. Webb, A. Williams, B. S. Wise, and S. Zhou. 2011. Ecological risk assessment for the effects of fishing. *Fisheries Research* 108: 372–384.

Hsieh, C. M. 2003. Income, age and financial satisfaction. *International Journal of Aging & Human Development* 56: 89–112.

ICES. 2006. *Report of the Regional Ecosystem Study Group of the North Sea (REGNS), 15–19 May 2006.* ICES Headquarters, Copenhagen. ICES CM 2006/RMC:06. 111p.

ICES. 2010. *Report of the Working Group on Ecosystem Effects of Fishing Activities (WGECO), 7–14 April 2010.* Copenhagen, Denmark. ICES CM 2010/ACOM:23. 22p.

Israel, D., and A. Levinson. 2003. Examining the relationship between household satisfaction and pollution. Mimeo, Indiana State University, Terre Haute, Indiana.

Justus, J., M. Colyvan, H. Regan, and L. Maguire. 2009. Buying into conservation: intrinsic versus instrumental value. *Trends in Ecology and Evolution* 24:187–191.

Kershner, J., J. F. Samhouri, C. A. James, and P. S. Levin. 2011. Selecting indicator portfolios for marine species and food webs: a Puget Sound case study. *PLoS One* 6: e25248. doi:10.1371/journal.pone.0025248.

Land, K. C. 1983. Social indicators. *Annual Review of Sociology* 9: 1–26.

Levin, P. S., M. J. Fogarty, S. A. Murawski, and D. Fluharty. 2009. Integrated ecosystem assessments: developing the scientific basis for ecosystem-based management of the ocean. *PLoS Biology* 7:e14 doi:10.1371/journal.pbio.1000014.

Levin, P. S., A. James, J. Kershner, S. O'Neill, T. Francis, J. Samhouri, C. Harvey, M. Brett, and D. Schindler. 2011. What is our desired future and how do we measure progress along the way? In: *Puget Sound Science Update,* April 2011 version. Accessed from http://www.psp.wa.gov/. Puget Sound Partnership. Tacoma, Washington, 12 September 2012.

Levin, P. S., and F. Schwing. 2011. *Technical background for an IEA of the California Current: ecosystem health, salmon, groundfish and green sturgeon.* U.S. Dept. of Commerce, NOAA Tech Memo., NMFS-NWFSC-109. 330p.

Levin, P. S., M. J. Fogarty, G. C. Matlock, and M. Ernst. 2008. *Integrated ecosystem assessments.* U.S. Dept. of Commerce, NOAA Tech. Memo., NMFS-NWFSC-92. 20p.

Link, J. S. 2005. Translating ecosystem indicators into decision criteria. *ICES Journal of Marine Science: Journal du Conseil* 62:569–576.

Lubchenco, J., and N. Sutley. 2010. Proposed US Policy for ocean, coast, and Great Lakes stewardship. *Science* 328:1485.

Lucas, B. G., S. Verrin, and R. Brown, eds. 2007. *Ecosystem overview: Pacific North Coast Integrated Management Area (PNCIMA).* Canadian Technical Report of Fisheries and Aquatic Sciences 2667, Fisheries and Oceans Canada, Sidney, British Columbia. 120p.

Luechinger, S. 2009. Valuing air quality using the life satisfaction approach. *Economic Journal* 119: 482–515.

Luechinger, S., and P. Raschky. 2009. Valuing flood disasters using the life satisfaction approach. *Journal of Public Economics* 93: 620–633.

Millenium Ecosystem Assessment (MEA). 2005. Linking ecosystem services and human well-being, In: *Ecosystem and Human Well-being: Multiscale Assessment.* Ed. D. Capistrano, C. K. Samper, M. J. Lee, and C. Raudsepp-Hearne, 45–60. Island Press, Washington, D.C.

Morris, W. F., and D. F. Doak. 2002. *Quantitative conservation biology.* Sinauer Associates, Sunderland, Massachusetts. 480p.

Perrings, C., and D. Pearce. 1994. Threshold effects and incentives for the conservation of biodiversity. *Environmental and Resource Economics* 4:13–28.

Pew Oceans Commission. 2003. *America's living oceans: charting a course for sea change.* A report to the Nation. Pew Oceans Commission, Arlington, Virginia. 166p.

Puget Sound Partnership. 2009. Puget Sound Action Agenda. Olympia, WA. 213p.

Puget Sound Science Update, April 2011 version. Accessed from http://www.psp.wa.gov/. Puget Sound Partnership. Tacoma, Washington, 12 September 2012.

Raskin, P. D. 2005. Global scenarios: background review for the Millennium Ecosystem Assessment. *Ecosystems* 8:133–142.

Refsgaard, J. C., J. P. van der Sluijs, A. L. Hojberg, and P. A. Vanrolleghem. 2007. Uncertainty in the environmental modelling process—a framework and guidance. *Environmental Modelling and Software* 22:1543–1556.

Rehdanz, K., and D. Maddison. 2005. Climate and happiness. *Ecological Economics* 52: 111–125.

Rice, J. C., and M. J. Rochet. 2005. A framework for selecting a suite of indicators for fisheries management. *ICES Journal of Marine Science* 62:516–527.

Robinson, L. A., C. Karman, S. Rogers, and C. L. J. Frid. 2009. *Methodology for assessing the status of species and habitats at the OSPAR Region scale for the OSPAR Quality Status Report 2010.* Contract No: C-08-0007-0085 for the Joint Nature Conservation Committee. University of Liverpool, Liverpool and Centre for the Environment, Fisheries and Aquaculture Science, Lowestoft. 33pp.

Rolston, H. 1989. *Philosophy gone wild: environmental ethics.* Prometheus Books, Buffalo, New York. 269p.

Sainsbury, K. J., A. E. Punt, and A. D. M. Smith. 2000. Design of operational management strategies for achieving fishery ecosystem objectives. *ICES Journal of Marine Science* 57:731–741.

Samhouri, J. F., P. S. Levin, and C. J. Harvey. 2009. Quantitative evaluation of marine ecosystem indicator performance using food web models. *Ecosystems* 12:1283–1298.

Samhouri, J. F., and P. S. Levin. 2012. Linking land-and sea-based activities to risk in coastal ecosystems. *Biological Conservation* 145:118–129.

Samhouri, J. F., P. S. Levin, and C. H. Ainsworth. 2010. Identifying thresholds for ecosystem-based management. *PLoS ONE* 5: e8907. doi:10.1371/journal.pone.0008907.

Samhouri, J. F., P. S. Levin, C. A. James, J. Kershner, and G. Williams. 2011. Using existing scientific capacity to set targets for ecosystem-based management: A Puget Sound case study. *Marine Policy* 35:508–518.

Sharpe, A. 1999. A survey of indicators of economic and social well-being. Background paper prepared for Canadian Policy Research Networks. Ottawa, Ontario.

Shelford, V. E. 1933. The preservation of natural biotic communities. *Ecology* 14:240–245.

Simberloff, D., and B. von Holle. 1999. Positive interactions of nonindigenous species: invasional meltdown? *Biological invasions* 1:21–32.

Sloan, N. A. 2002. History and application of the wilderness concept in marine conservation. *Conservation Biology* 16:294–305.

Solberg, E. C., E. Diener, D. Wirtz, R. E. Lucas, and S. Oishi. 2002. Wanting, having, and satisfaction: examining the role of desire discrepancies in satisfaction with income. *Journal of Personality and Social Psychology* 83:725–734.

Suter, G. W. 2007. *Ecological risk assessment.* CRC Press, Boca Raton, FL. 560p.

Swart, R. J., P. D. Raskin, and J. Robinson. 2004. The problem of the future: sustainability science and scenario analysis. *Global Environmental Change* 12:137–146.

Tallis, H., and P. Kareiva. 2005. Ecosystem services. *Current Biology* 15:746–748.

Trenkel, V. M., and M. J. Rochet. 2003. Performance of indicators derived from abundance estimates for detecting the impact of fishing on a fish community. *Canadian Journal of Fisheries and Aquatic Sciences* 60:67–85.

U. S. Commission on Ocean Policy. 2004. *An ocean blueprint for the 21st century.* Final report to the Present and Congress. Washsington, D.C. 676p.

U.S. Environmental Protection Agency (EPA). 1998. Guidelines for ecological risk assessment. *Federal Register* 63:26846–26924.

U.S. Fish and Wildlife Service. 2009. *Wildlife Watching Trends 1991–2006: Addendum to the 2006 National Survey of Fishing, Hunting, and Wildlife-Associated Recreation.* Report 2006–3. Washington, D.C. 96p.

U.S. Fish and Wildlife Service. 2010. *Trends in Fishing and Hunting 1991–2006: A Focus on Fishing and Hunting by Species Addendum to the 2006 National Survey of Fishing, Hunting, and Wildlife-Associated Recreation.* Report 2006–8. Washington, D.C. 72p.

U.S. National Ocean and Atmospheric Administration (NOAA). 1999. *Ecosystem-based fishery management.* A report to Congress by the Ecosystem Principles Advisory Panel. National Marine Fisheries Service, Silver Springs, Maryland.

U.S. National Science and Technology Council, Joint Subcommittee on Ocean Science and Technology (JSOST). 2007. Charting the course for ocean science in the United States for the next decade. Washington, D.C. 100p.

Vera-Toscano, E., V. Ateca-Amestoy, and R. Serrano-del-Rosal. 2006. Building financial satisfaction. *Social Indicators Research* 77:211–243.

Vitarelli, A. 2010. Happiness metrics in federal rulemaking. *Yale Journal on Regulation* 27:115–158.

Wallis, O. L. 1958. Research and interpretation of marine areas of the U.S. National Park Service. National Park Service, Washington, D.C.

Walls, M., S. Darley, and J. Siikamäki. 2009. *The state of the great outdoors: America's parks, public lands, and recreation resources.* Resources for the Future, Washington, D.C. 100p.

Warr, P. 1999. Well-being and the workplace. In *Well-being: the foundations of hedonic psychology.* Eds. D. Kahneman, E. Diener and N. Schwarz. 392–412. Russel Sage Foundation, New York.

Watson, R. T. 2012. The science–policy interface: the role of scientific assessments—UK National Ecosystem Assessment. *Proceedings of the Royal Society A* in press.

Watt, K. E. F. 1971. *Ecology and resource management: a quantitative approach.* McGraw-Hill, New York. 450p.

Welsch, H. 2002. Preferences over prosperity and pollution: environmental valuation based on happiness surveys. *Kyklos* 55:473–494.

Welsch, H. 2006. Environment and happiness: valuation of air pollution using life satisfaction data. *Ecological Economics* 58:801–813.

Welsch, H., and J. Kühling. 2009. Using happiness data for environmental valuation: issues and applications. *Journal of Economic Surveys* 23:385–406.

Zerbe, Jr., R. O. 2001. *Economic Efficiency in Law and Economics.* Edward Elgar, Cheltenham, U.K. 336p.

Chapter 7. Marine Spatial Planning

JAKE RICE

Department of Fisheries and Oceans Canada

SUE KIDD

University of Liverpool

ANTHONY D. M. SMITH

Commonwealth Scientific and Industrial Research Organization
Marine and Atmospheric Research

Contents

1. Definition

Definitions of Marine Spatial Planning (or Coastal and Marine Spatial Planning—hereafter MSP) abound. They differ in detail reflecting differences in the terminology embedded in both the legal and institutional frameworks of the agencies accountable for planning and in the culture of the scientific and governance experts actually driving the planning processes. A few illustrative definitions are presented in Box 1.

2. Key Characteristics of MSP

By definition, marine spatial planning is inherently spatial, planning where (and sometimes when) activities in the sea can occur. The planning is intended to deal with competing multiple uses of the marine environment, implicitly assuming that at least some activities are inherently incompatible to pursue in the same place at the same time. To this extent it differs from within-sector regulatory approaches to marine management that tend to deal with questions of how much and in what

The Sea, Volume 16, edited by Michael J. Fogarty and James J. McCarthy
ISBN 978-0-674-07270-1

Box 1: Examples of the Varying Definition of Marine Spatial Planning

UNESCO: "Marine spatial planning is a *public process* of analyzing and allocating the spatial and temporal distribution of human activities in marine areas to achieve ecological, economic and social objectives that have been specified through a political process." http://www.unesco-ioc-marinesp.be/

United States – Interagency Ocean Policy Task Force: "CMSP is a comprehensive, adaptive, integrated, ecosystem-based, and transparent spatial planning process, based on sound science, for analyzing current and anticipated uses of ocean, coastal, and Great Lakes areas." http://www.whitehouse.gov/sites/default/files/microsites/091209-Interim-CMSP-Framework-Task-Force.pdf

United Kingdom – DEFRA: "A strategic plan for regulating, managing and protecting the marine environment that addresses the multiple, cumulative and potentially conflicting uses of the sea." http://jncc.defra.gov.uk/page-2848

South Australia: "The Marine Planning Framework for South Australia is a new large-scale, ecosystem-based zoning policy for management of development and use in the marine environment." (Day et al., 2008)

Canada uses the UNESCO definition and observes that "MSP is successful if there is a shared vision, if information gathered supports taking action, and if a central plan sets specific directions." http://www.mar.dfo-mpo.gc.ca/e0010285

Notwithstanding differences in detail, some emergent common properties are apparent. MSP is a process as well as the product the process produces. It is intended to be participatory and transparent, forward-looking, and integrative across multiple uses of the oceans, and it tends to focus strategically (on *what* outcomes should be pursued) rather than tactically (on *how* the outcomes should be achieved).

ways a single activity can be pursued, such as regulating overall levels of effort or catch in a fishery and the gears that can be used, or specifying rights and conditions of access for marine tourism operations. Within-sector management can also have spatial provisions, intended to achieve any combination of ecological outcomes (regulating impacts of an activity on ecosystem features) and socio-economic outcomes (regulating how benefits from ocean uses are distributed across society). Two features make marine spatial plans a step beyond within-sector management plans. First, they will typically encompass or at least comprehend arrangements for multiple individual sectors. Second, MSP usually follows rather than precedes development of individual sectoral arrangements for at least previously established activities in the sea. Therein lies one of its greatest challenges, as it may seek to change arrangements with which established sectors are comfortable, in order to accommodate new or expanding sectors.

Another key aspect of MSP is that it is a process intended to lead to a product (the "Plan"). Both the process and the product are important. The journey involved in development of the plan has many benefits other than the plan itself. For example, commonly a comprehensive inventory of both environmental assets and uses of the marine environment is absent prior to formally undertaking the MSP process. Often the MSP process has stimulated national and regional initiatives to identify bioregional ecological structure and better define the nature, location and scale of human uses of the sea. Such initiatives increase understanding of the entire area and human interactions within it and thereby improve the capability of spatial and sectoral planning to meet the challenges of ecosystem-based management.

The benefits of having an integrated Plan available depend on: the quality of the information available to the process; the breadth of industry sectors and societal interests included in the process; their collective buy-in to the provisions captured in the Plan; the specificity of those provisions; and the degree to which the provisions of the Plan successfully resolve the conflicts among different interests. Some of the conflicts may be among different ocean users competing for space within which they need to operate, whereas some are between sectors of society with different values regarding economic pursuits and protection of marine ecosystems. Resolving such conflicts requires both knowing how the various ocean industries provide and distribute benefits, and treading a careful line between resource use and long term ecosystem health. A well informed process that brings together all the competing ocean uses and representatives of major social values and treats them fairly should result in a Plan with greater potential for real benefits to be delivered. In addition, if the provisions of the Plan clearly guide implementation towards explicit, inter-compatible objectives, reflecting agreements on trade-offs among conflicts rather than on only high-level concepts, then the Plan has greater potential to produce the outcomes expected by all the participants in the process.

While a completed Plan is clearly a key benchmark in MSP, the process itself should be cyclic. Implementation should include monitoring and performance assessment leading to periodic review of both the appropriateness of the objectives of the Plan and the success of the plan at moving towards its objectives. Appropriate adjustment of strategies and tactics should follow each review. Framing MSP in an adaptive management context is both crucial in its own right and also makes it congruent with best practices in individual sectoral management, though the time frames for review and adjustment may differ. Full cycles from planning, through implementation, to monitoring, evaluation, and adaptive change are likely to occur on time scales of a decade or so, although often the ambition is for a shorter plan-making and review cycle. Despite the relatively recent adoption of MSP in most jurisdictions, in a few cases MSP is already in the second or third phases of the adaptive cycle (e.g., management planning for the Great Barrier Reef in Australia; see Kritzer et al., this volume).

MSP inevitably involves consideration of trade-offs, even if they are not confronted explicitly. There are two different types of tradeoffs that have to be addressed in relation to outcomes of MSP:

1. The ecological-economic-social consequences of choices; and
2. How the costs and benefits of the balance between protection and use, and among alternative uses, are distributed among sectors and communities.

The fact that activities are conducted in ocean space ensures that such trade-offs will exist and should be considered in developing plans, and where sectoral management includes the spatial allocation of user rights these trade-offs will become central to planning. See Guerry et al. (this volume) and Altman et al. (this volume) for examples. Obtaining universal buy-in to spatial allocation of use rights is often challenging even within sectoral management, and potential for conflicts and trade-offs multiplies quickly when the number of competing sectors increases and new uses may have to be accommodated. The planning process itself can be considered as a formal mechanism for identifying, considering, and helping to resolve such trade-offs and a number of tools exist to facilitate potential conflict identification, evaluation, and resolution. The resolution itself will be the outcome of a social and political process, but this can be informed constructively by scientific approaches to conflict identification and trade-off evaluation.

The longer-term adaptive nature of the planning process can result in shifts in emphasis over time across both elements of the trade-off space, so revisiting the trade-off considerations must be part of the cyclic process. First attempts at MSP in a region tend to be highly contentious as they can result in significant shifts in resource allocation and changes in patterns of use rights both within and among sectors. The difficulty of MSP increases with the number of sectors being considered, but so does the potential for greater benefits from resolving conflicts and making informed trade-offs among the full range of ocean users.

The process of MSP needs institutional leadership and commitment, as well as engagement and commitment from the breadth of interests and sectors involved. Institutional leadership can be problematic since, as already noted, MSP often arrives late on the scene. Strong institutional interests (agencies and jurisdictions) may already be firmly entrenched and unwilling to cede power or influence. This can be a problem even where there is clear legislative backing for the planning process, and can be even more difficult without such backing. Often a period of "education" of the decision-makers in existing agencies and jurisdictions is needed, so they understand why there is a need to change existing patterns of practice. In addition, institutional leadership should include political buy-in, not just bureaucrats keeping busy. If any key parties to the planning process believe they can get a "better deal" for their constituency by going around the MSP process and directly to the decision-makers, then commitment of other parties to the process is also likely to weaken. This would run counter to the broad engagement that is an asset in building buy-in to the outcomes. However, that same broad engagement can become a liability when the goals of some participants interfere with their ability to acknowledge the legitimacy of goals of other participants. Progress from the planning process is most secure if the broad engagement can be maintained through the implementation and monitoring steps as well.

While MSP is ideally informed by scientific analysis and mobilization of existing data, in practice such information is often limited. In any case, participants in the process often prefer not to be "overly constrained by data" (one is reminded of Lord Keynes's observation that governments don't like to be too well informed—it makes decision making so much harder), and experience suggests that significant progress can be made in delineating trade-offs with far less than comprehensive data sets. Even if data do not support quantification of the most likely outcomes of the resultant Plan, they may be sufficient to lay out the general considerations in the trade-offs, help identify strategies likely to lead to various combinations of social, economic, and ecological outcomes, and scale the *relative* costs and benefits likely to be experienced by various parties involved.

While MSP may succeed with inadequate underpinning data, it is unlikely to succeed without sufficient clarity on objectives. The process needs agreement on explicit ("operational" or "concrete") objectives to ensure that outcomes are understood in the same way by all participants. (An agreed "objective" to "conserve biodiversity" may be considered a fine concept, but can have the industry participants and the environmental non-government organization (ENGO) participants leaving the room with very different ideas of what the "agreed" outcome will look like). This gives the trade-off analyses (formal or informal) two roles:

- One is exploratory—when all participants have put their general goals on the table, trade-off analyses can explore the compatibility of different suites of goals, and identify where pursuit of some goals is likely to require concessions on others. This can often be done with relatively qualitative approaches.
- The other is implementational—when final candidate suites of goals are identified, trade-off analyses can inform how likely various suites of concrete objectives are to be mutually attainable. Iterative processes of adding or subtracting concrete objectives to the mix, and adjusting concrete objectives to reflect concessions in the sizes of the benefits to be obtained or impacts to be allowed, should converge on broadly acceptable balances across the two trade-off axes discussed above. These analyses may be quantitative (such as management strategy evaluation).

The confluence of all these characteristics makes MSP both complex and challenging. The challenges and their (partial) resolution are dealt with further in section 5 below.

3. How Does the *Process* Work?

Experts almost always present MSP as a stepwise process. This is legitimate in that to some extent there is a logical order to many of the things that are part of MSP. However, in practice many elements are likely to be considered alongside each other and may develop in an iterative manner. It is therefore not surprising that many of the most successful efforts have been flexible rather than formulaic, with individual cases often deviating significantly from any prescriptive pathway. Perhaps a key point to emphasise, though, is that it can be helpful for stakeholders to have an overview of the various steps in the process in mind, which can act as a reference framework as the process progresses.

3.1 The establishment phase

There are many alternative flow diagrams of the MSP process, generally variations on a few shared themes. Perhaps the most commonly cited and certainly one of the most comprehensive is that presented by UNESCO (Ehler and Douvere, 2009). See Figure 7.1. This sets out ten steps in a cyclic process. These include several that relate to the initial establishment phase of MSP (Steps 1–4) as well as those that are proposed as regular features of MSP cycles. The establishment phase is particularly critical in providing a solid foundation for MSP activity. Figure 7.1 high-

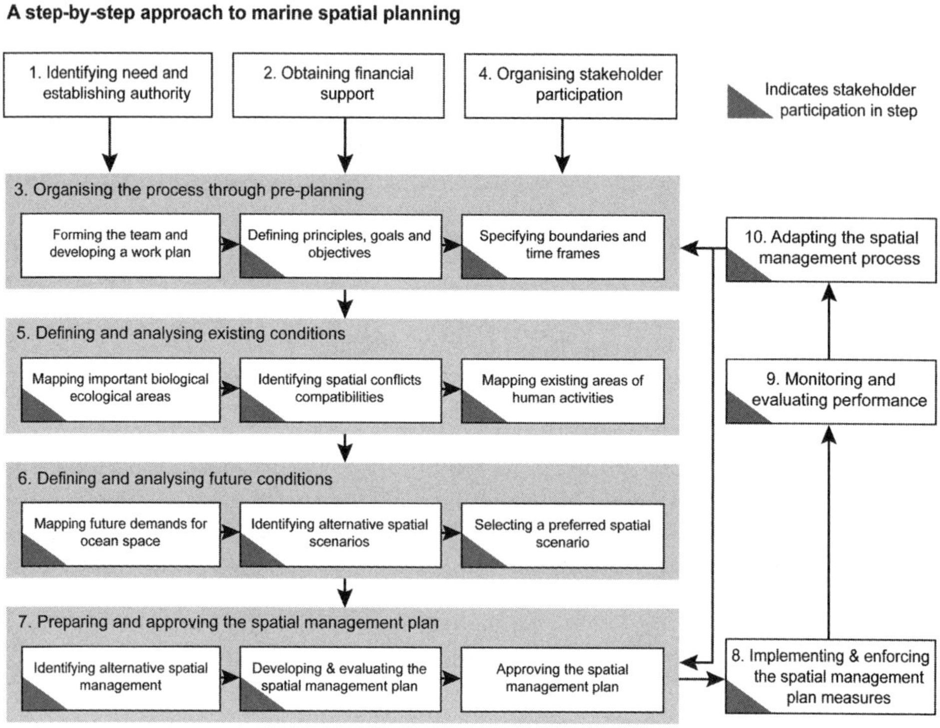

Fig. 7.1 – A step-by-step approach to marine spatial planning

lights that gaining agreement on the need for MSP, establishing the authority to act, obtaining appropriate financial support, forming a team with an adequate mix of skills and expertise, and defining a work plan are not surprisingly key elements here. In addition, organising stakeholder participation underpins potentially all subsequent steps in the MSP process and needs to be considered at the outset. Much has been written about the importance of stakeholder engagement in MSP and reflections on MSP experiences emphasise its significance (e.g. Pomeroy and Douvere, 2008; Flannery and O Cinneide, 2008). Reasons for this include: the delivery of better understanding of marine ecosystems and of human activities that relate to the sea; promoting democratic debate about MSP principles, goals and objectives; identifying, predicting and resolving conflicts; imaginative generation of options; improving the prospects of compliance for plan implementation; and involvement in plan monitoring. A recurring message is that the quality of stakeholder engagement is likely to greatly influence the quality of decision-making, the levels of stakeholder buy-in to the final plan output, and the relative success of subsequent implementation. As a result, careful stakeholder mapping and facilitation of stakeholder engagement, including the design of institutional structures supporting their involvement in MSP, are called for. Such considerations are inevitably highly context-specific. They are also likely to be time consuming and will inevitably require regular review and refinement to reflect shifting patterns of stakeholder activity and areas of interest, changing personalities, stakeholder fatigue, etc.

Specification of MSP boundaries is critical to many parts of the MSP process, including stakeholder identification, need for ecological, social, and economic information, and scope of trade-offs. In many instances the boundaries are determined by the legal jurisdictions of the participating governments or agencies, but these often bear little relationship to the scales of some sea uses or to ecosystem boundaries which may well be trans-national in scale. There is also the difficult issue of land-sea interaction and integration of MSP activity with terrestrial spatial planning and management, so comprehensive stakeholder engagement may necessitate a tiered and differentiated approach. Ultimately all this will influence definition of the principles, goals and objectives that will frame plan development.

3.2 Defining and analysing existing conditions

Step 5 in UNESCO's MSP process addresses the question *Where are we now?* It involves defining and analysing existing ecological conditions and patterns of human use and identifying key spatial conflicts and compatibilities. This is a major challenge for the MSP process as, in contrast with terrestrial contexts, information resources tend to be poorly consolidated and often incomplete. Ecological data can be extremely costly to obtain in the sea and are often patchy, dated, uncertain, and contested. Interestingly, some commentators indicate this situation can equally relate to the human aspects of sea use which is the main focus of MSP (St. Martin and Hall-Arber, 2008; Curtin and Prezello, 2010). In addition to data scarcity, data availability and compatibility are likely to be issues. This can arise from ill-developed data collection and sharing protocols across administrative boundaries or from commercial confidentiality of some data. In some instances where data on human use exist, they may not be disaggregated across the land-sea divide, or sufficiently geographically explicit for MSP purposes, with the absence of agreed sea-based units of analysis presenting further problems (Kidd and Shaw, forthcoming). There are, of course, notable exceptions to this picture. The Baltic Sea is a shining example in the European context of the benefits of years of collaborative information gathering and analysis.

More typical for MSP is the need for painstaking piecing together of information on ecological conditions and patterns of human use with, as indicated earlier, imaginative use being made of softer and more qualitative data sources. As a consequence MSP can be viewed as a positive force in promoting improvements in the quality, compatibility and availability of marine data offering the prospect of better quality MSP outputs over time. It is through this data collection and analysis process, in association with stakeholder discussion and reference to plan goals and objectives, that a clearer understanding of existing spatial conflicts and compatibilities can emerge.

3.3 Defining and analysing future conditions

Some MSP definitions draw attention to the forward looking aspect of spatial planning (e.g. Pomeroy and Douvere, 2008). This is the focus of Step 6 in UNESCO's process, addressing the question *Where do we want to be?* Key tasks here include projecting future trends associated with current ecological processes (e.g. climate change) and human activities; estimating the spatial and temporal demands of new marine uses; identifying possible alternative future scenarios for the area; and selecting a preferred scenario. The time scale of the plan dictates the need for projections of various types. Many existing initiatives plan for up to 10 years ahead but others such as the Strategic Plan for the Great Barrier Reef, which considers a 25 year time scale, cover much longer periods (Gilliand and Laffoley, 2008). Given the rapid acceleration in the level and range of sea use, together with the scale of predicted impacts of climate change, it seems sensible to consider more extensive time horizons. It is not uncommon, for example, for terrestrial flood defense plans to cover a 100 year horizon, reflecting the need for expensive infrastructure investment to be built for the long term and able to accommodate predicted long term changes in sea level, rainfall and storm events. The planning horizon of a plan is fundamental to its thematic scope and orientation and can significantly affect the framing of its goals and objectives as well as the scope of scenario building work. Drawing upon experience of terrestrial planning, shorter time horizons tend to promote less ambitious, more tactical plans focussed on relatively small adjustments to present circumstances, while longer time horizons facilitate more radical and creative thinking and encourage stakeholders to contemplate a more ambitious and phased approach to change. The benefits of the former, however, include a focus on immediate deliverability, while long term plans can provide a reason to put off hard decisions to a later date.

3.4 Preparing and approving the spatial management plan

Having identified a preferred MSP scenario, the question arises of *How do we get there?* This is the focus of Step 7 in UNESCO's MSP process, which entails identifying and evaluating alternative management measures that can help deliver the selected scenario, developing a zoning plan, and then approving a combined spatial management plan as a basis for guiding implementation activities. A key point to emphasise is that spatial and temporal management measures, which focus on determining how, where, and when human activities can occur, may sit alongside other forms of management action that influence human pressure on the sea in other ways. These include: input measures, for example specifying the number and size of fishing vessels; process measures, for example specifying particular waste treatment technology; and output measures, such as tonnage limits on sand and gravel extraction. Spatial and temporal management measures may include: designation of areas closed to fishing or other human activities; zoning of areas for specific uses, e.g., wind farms or military operations; or zoning of areas by objective, e.g., development areas, conservation areas, and multiple use areas. Although spatial zoning is perhaps the most obvious output of the MSP process, it is not without its controversy, with many arguing that a nuanced, multi-use approach should be favoured as much as possible. These arguments are reinforced by the very different approaches to zoning evident in marine plans (see, for example, Douvere, 2008). Moreover, consideration should be given to the multiple physical dimensions of marine space, as well as to the multiple uses. In Portugal, for example, the National Sea Strategy identifies the sea bottom, water column, surface and atmosphere as different dimensions requiring coherent and integrated planning (Calado et al., 2010), while planning and management activities related to the

Great Barrier Reef cover all areas from 915 meters above the sea surface to 1000 meters below the seabed. The control of airspace can relate, for example, to air traffic and potential conflicts with fixed structures such as wind farms, while growing interest in under-sea carbon storage illustrates the case for a below-sea surface perspective. In addition, the time dimension within a plan is particularly significant in the marine context. Both environmental sensitivity and human use may be highly time specific and depending on the factors involved may suggest that managers consider use of seasonal, cyclical, or even diurnal controls when implementing the Plan.

In UNESCO's MSP process, evaluation of the emerging plan proposals is put forward as a key aspect of Step 7. This should encompass an assessment of short and longer term economic, social and environmental impacts of the plan proposals and any alternatives, as well as matters related to implementation. Links back to plan goals and objectives are important here and it is often at this point that lack of clarity in their specification becomes apparent, as well as the significance of political considerations. Evaluation of this type is intended to refine the content of the plan and improve the quality of the output. The extent and nature of plan evaluation will be very much determined by local requirements and resource availability, but in many instances will include the need for a formal Strategic Environmental Assessment (SEA). If this is the case, then SEA good practice would suggest that evaluation should be treated in a similar way to stakeholder engagement, i.e., as a recurrent element throughout the plan-making process rather than an element which appears during the final stages of plan production. This is how SEA is expressed in flow diagrams of the UK's MSP process for example (Gilliland and Laffoley, 2008).

3.5 Implementation, monitoring and review

Adoption of a plan is by no means the end of the matter and the much harder task of putting it into action commences. Most models of the MSP process clearly define implementation, monitoring and review (Steps 8, 9 and 10 in UNESCO's diagram) as key parts of a circular process which ultimately leads to plan review. However, a frequent criticism of marine plans is that there is a gap between plan content and implementation (e.g. Agardy et al., 2011). This often arises from an unhelpful separation of plan-making and plan-implementation functions which can be reinforced by institutional and staffing structures as well as inadequate resource allocation. Lessons from terrestrial planning and the associated development of implementation theory suggest that plan delivery is enhanced where implementation, monitoring and review are considered at the outset of plan development (Kidd and Ellis, 2012), including approaches to stakeholder engagement, data gathering, the design of management measures, etc.

Implementation in many instances will need to extend beyond reactive licensing activities and encompass proactive programmes of activity across many fronts. Often perhaps regarded as the Cinderella aspect of MSP, these stages should be seen as the heart of the process. This is where the adaptive management dimensions of MSP really come into play. By linking implementation very closely to plan monitoring and review, MSP processes can facilitate sensitive experimentation in planning and management approaches. Imagination here can also help deliver accountability and act as a focus for ongoing stakeholder engagement. The Great Barrier Reef Marine Park Authority, for example, has used participatory monitoring and evaluation programmes to good effect (Day, 2008). All this should inform periodic formal reviews of the planning approach. A 5 year plan-making cycle is not uncommon.

4. What Does the *Product* Look Like?

The marine spatial planning process typically produces a number of different kinds of products. However, as the recent review of MSP experience from around the world undertaken by the United States National Oceanic and Atmospheric Administration reveals, there is no single standard for how the documentation of these products is organised (NOAA, 2011). At one end of a continuum, an MSP process can produce a single comprehensive document, possibly with a number of technical annexes, whereas at the other end of the continuum a single MSP process may produce a very large number of individually concise and theme-centred documents, possibly even with different variants of a single thematic document targeted at audiences with different degrees of technical expertise or different roles in governance. The plan may also reference individual sectoral arrangements and regulations.

As the documentation multiplies, what constitutes "the Plan" may not always be clear. Moreover, with an increasing trend towards web-based publication, the website for an authority overseeing the MSP process may include a diverse array of documents and links to other sites and documents, with the official status of various documents relative to "the Plan" also unclear. An additional complication arises as the MSP process matures and cycles begin to repeat, so that various parts of the documentation comprising "the Plan" may be updated or revised on varying schedules. Keeping this background of variation in documentation across MSP authorities in mind, however, there are several key products that usually can be found in the body of documentation arising from the MSP process, whether they are consolidated into a single Marine Spatial (or sometimes "Integrated Management") Plan or not. These include a summary of the legislative, policy, and jurisdictional context for the MSP; the background data and information on the ecosystem(s) and industries which the MSP process has considered; the goals and objectives of the MSP initiative; and some guidance on implementation of the contents of the Plan.

4.1 Legislative, policy and jurisdictional context

A fundamental property of MSP is that it considers where and how more than one use of the sea will occur (Douvere and Ehler, 2009). Hence, almost always any MSP Plan will have to be consistent with the regulations and policies applicable to more than one industry sector, and often social and cultural policies as well (Morrison and Snow-Cotter, 2008). Moreover, even when MSP is undertaken on relatively small scales there may be bylaws, regulations and policies of multiple levels of government—municipal, regional, states, provincial, and federal—that all have to be considered. If the MSP initiative crosses jurisdictional boundaries, the regulatory frameworks of all the parties have to be considered, and if the area to which MSP is applied extends beyond the boundaries of a single country, international agreements and commitments for biodiversity conservation and the international regulatory standards of all relevant industries enter the mix (e.g., the Wadden Sea Plan http://www.waddensea-secretariat.org/management/Plan.html.)

One of the typical products of the MSP process is a summary of the full jurisdictional and regulatory context for the activities to be carried out in the Region. If there is a single Plan, these are often merely listed in a section of the Plan, with the major goal(s) of each relevant policy or regulatory act presented along with the name of the policy or act. However, in some cases there is a more complete technical annex or separate document with substantial detail on the individual provisions of relevant policies and legislation. This practice is sometimes (but not always) followed when either there are several levels of government or adjacent jurisdictions involved in the process, or when several industries are affected by the outcome of the MSP process, and each industry has its own legislation and regulatory frameworks.

The initial purpose of assembling this part of the MSP information is primarily to ensure all participants in the process are aware of the legal constraints within which their discussions occur. Toward that end, there may have been some form of analysis of the compatibility and gaps among the various relevant policies and regulations as an early part of the planning process to ensure the process could proceed within a coherent context. However, it is very rare to find any record of the compatibility and gap analysis of policies and regulations in the MSP documentation. Nonetheless, including some summary of the jurisdictional and regulatory information in the MSP products serves more than just an archival function. It can have an important role in guiding implementation through documenting which agency(ies) is (are) competent and/or responsible for carrying out which parts of the Plan.

4.2 Background data and information on the ecosystem(s) and industries

This information is assembled to inform the dialogue central to the planning process (Katsanevakis et al., 2011; Nobre, 2011). Naturally, both the amount and nature of the information available will vary greatly among areas where MSP is undertaken. For example, in the Canadian Arctic quantitative monitoring data were very few and largely opportunistic in space and time, but extensive traditional knowledge of Elders engaged in subsistence livelihoods was collected systematically (http://www.beaufortseapartnership.ca/bsimp.html). On the other hand, for MSP in southwest Australia (http://www.environment.gov.au/coasts/mbp/publications/south-west/sw-region-profile.html) the Wadden Sea, and the Belgian coasts (http://www.unesco-ioc-marinesp.be/uploads/documentenbank/a36a8df232f6525808325812468196ee.pdf), quite extensive amounts of scientific survey work on ecosystem components were consolidated and made available. Likewise, the degree of quantitative information on the locations and intensities of historical uses was highly variable among MSP initiatives. The important component of this part of the MSP products, though, is that whatever amount and type of information is assembled, it is considered by the participants in the process to be the best available information, and to cover all major and reliable sources. That ensures participants in the MSP process that important knowledge known by at least some participants to exist is not left out of their discussions and (in many cases) formal or informal analyses of trade-offs.

Assembling the corpus of relevant data and information can be one of the most time-consuming and costly parts of the MSP process, and geo-referencing the information for analytical uses or even visual inspection often represents another large incremental cost (Ban and Klein, 2009; Levin et al., 2009; NOAA, 2011). Once done, it is important that the information and maps are archived and kept available for the cyclic and adaptive parts of the process. There is, however, huge variation among Plans in the details of how this is done, but variants of web links to data bases (with or without ongoing updating and custodianship) and electronic (and rarely paper) annexes to the documents constituting the MSP Plan are common approaches. Plans that stop at a fairly high "strategic" level tend to be particularly careful to ensure the information that has been assembled is kept available, because many of the focused discussions on trade-offs or options necessary to make the "strategic" components of the Plan operational still lie in the future.

4.3 Goals and objectives of the MSP initiative

Among the most important products of the MSP process are the goals and objectives on which agreement among participants is reached (Klein et al., 2010; Olsen et al., 2011). Given their central role in how the MSP Plan is implemented, we consider the nature of "objectives" more thoroughly than the other products of the MPS process. Notwithstanding their central role as a key product, however, there is still great diversity in what are considered "objectives" from an MSP process and how they are produced and communicated (Appendix A).

Some MSP processes have simply taken as their objectives the existing policies and legislation, consolidated the intended outcomes of each applicable one, and listed those outcomes as the objectives of the MSP Plan (e.g., South West Australia and Norway—"Strategic Goals", Appendix A). Consequently the "objectives" in the Plan were actually inputs to rather than outputs of the MSP process. However, in these cases the processes still had many issues to resolve since there are many pathways to achieve any particular objective or policy outcome. Among those pathways some may be readily compatible with other objectives from other policies and regulations, whereas other pathways may be directly or indirectly in conflict—perhaps achieving the objectives of prosperity and job creation in economic policies can only come at cost to the ecological objectives embedded in environmental policies. In addition, even if there are few direct conflicts to resolve among the objectives embedded in the relevant suite of policies and regulations, different pathways may distribute the costs and benefits in different ways among sectors of society and components of the ecosystems. Hence, in those cases where the objectives are taken directly from policies and legislation, the MSP process focuses on trade-offs inherent in the options to pursue in the mandated objectives, more than on the objectives themselves.

Many MSP processes, while taking note of relevant policies and legislation, rely on the dynamics of the MSP process itself to identify suites of objectives (e.g., Canada—Beaufort Sea and Rhode Island, Appendix A). The suites have to be consistent with the provisions of relevant legislation and policies, but are not as a rule taken directly from them. Of course, members of the suites of objectives have to be consistent with each other as well. Achieving such compatible sets of objectives may be the focus of formal or informal trade-off analyses, and require substantial time and dedication of participants to find broadly acceptable compromises.

When the MSP process is developing the objectives as part of its work, the process can stop at several levels of specificity. A number of MSP Plans are observed to stop at quite a high level, with all participants agreeing on general statements about management striving for high environmental quality, economic prosperity, and social well-being, but with the details to be worked out in subsequent steps in the cyclic process (e.g., Wadden Sea and Norway, Appendix A). In cases where the agreed objectives were at this high level, the resultant Plans usually were referred to as "strategic", and the result of consultation and debate rather than analyses of options. In turn, these high level objectives then set the parameters to be explored with trade-off analyses or negotiation of applied questions about which activities occur at what intensities and at which places.

Just because an MSP process includes the setting of objectives, it does not ensure the resultant objectives will be high-level. Some MSP initiatives took the objective-setting process through several more levels of specificity of outcomes resulting in much more explicit statements of the outcomes which management under the Plan was to achieve (e.g., US—Delaware and Canada—Beaufort Sea, Appendix A). In some cases formal trade-off analyses had been conducted as part of setting the more specific objectives, but often the more specific objectives were still the result of a consultation and negotiation process, although one much more likely to be using directly the data and information discussed above. Even in these cases, though, it was very rare for the objectives in an MSP Plan to have quantitative benchmarks for the outcomes associated with achievement of the objective. Hence, some and often a lot of the more analytical work on trade-offs still remained to be done, even when the objectives were relatively concrete.

There is no "right" level of specificity of objectives for an MSP. Rather the level of specificity depends on the degree of implementation guidance expected to be present in the MSP Plan. When the "Plan" is to be used as a conceptual framework, under which all discussions of trade-offs and implementation are to be conducted, then strategic high-level objectives may be sufficient for the Plan. However, when the Plan itself is intended to reflect the compromises and concessions that were arrived at by the MSP process, relative to the weight given to the various outcomes desired by different participants and to the preferred pathways to reach the outcomes, then the objectives in the Plan have to be specific enough to reflect those compromises unambiguously. But whether the MSP process invests the time and confronts the difficulties associated with setting operational objectives up front with specific objectives set as a key product, or only focuses on agreement on strategic objectives, sooner or later it will be necessary to undertake hard-nosed evaluations of compatibilities of very specific social, economic, and ecological outcomes; of necessary tradeoffs to resolve incompatibilities; and analyses of the distributions of costs and benefits on the alternative pathways available for achieving the outcomes (Garcia and Charles, 2008; Rice, 2011). The outcomes of those evaluations need to be captured somewhere in the documentation associated with the MSP initiative, whether they are part of the MSP Plan or not.

4.4 Guidance on implementation of the contents of the Plan

Just as there is great variability in the specificity of objectives included in the MSP Plans that have been produced, there is corresponding variability in the specificity of guidance provided for implementation of the contents of the plans. Some of the most operational Plans include specific management guidance on multiple industry sectors, laying out quite specific conditions under which they can operate. This is the case for the Belgian coastal and Norwegian plans, for example (http://www.unesco-ioc-marinesp.be/uploadsdocumentenbank/98f9ab04c6f054e434d506dbeba3f65e.pdf). However, other Plans with comparably specific objectives focus implementation guidance on marine places rather than industry sectors, with the focus on ocean zoning and allocation of space among competing uses, but implicitly assuming that sectoral management will ensure that operations within the allocated places will be consistent with the objectives of the Plan (e.g., in South West Australia and Canada—Beaufort Sea, Appendix A).

By contrast, for Plans which contain only more strategic objectives, rarely is there direct guidance on how particular sectors will be managed, or exactly which areas will be allocated to which uses. Exceptions are found to these generalisations, of course, particularly when special areas (e.g., areas meeting the Convention on Biological Diversity (CBD) criteria as ecologically and biologically significant) have been identified in the Plan, or particular industries are prominent in the planning process. Commonly though, these higher level Plans include guidance more on how the implementation process will be further developed to eventually cover the necessary consideration of specific trade-offs of particular competing uses (or non-use values) and mitigation of potential impacts on particular places, for example, the Baltic Sea Action Plan (http://www.helcom.fi/press_office/news_helcom/en_GB/BSAP_full/).

As with the inclusion of objectives in the MSP Plans, there is no "right" level of specificity of guidance for implementation of the plan. Rather, guidance on implementation and objectives must work in harmony, each adequate for the full realisation of the other. Detailed operational guidance on where and how uses are allowed and where conservation efforts are concentrated without clear objectives to justify the measures would make a Plan appear arbitrary (especially to those being regulated). Quite specific suites of objectives to deliver comparably specific outcomes, without guidance on which industries will have to change their activities in what ways, has not addressed the *how* in management yet, and therefore left much of the difficult work still to be done. Strategic Plans can be very useful, even if they do leave many challenging discussions and analyses of trade-offs yet to be undertaken. They can provide buy-in of all participants to the process by which those tasks will be undertaken, and increase the likelihood of compliance of all parties in the outcomes of the remaining parts of the MSP Process.

5. What Is the Relevance of MSP to EBM?

Marine spatial planning is essential to Ecosystem-Based Management (EBM) since, by definition, ecosystem-based management has a strong spatial dimension (Pikitch et al., 2004; Rice, 2011). Whether one subscribes to the view that EBM is managing human activities so they are sustainable at the ecosystem scale, or is managing sustainably the whole ecosystem with humans as a part of the ecosystem, the ecosystem is place-based and is the centre of management planning.

As reviewed in other chapters of this book, international commitments to apply EBM (or its variants) abound, in UN General Assembly Resolutions at the highest level, and the overarching policies of intergovernmental bodies relevant to the oceans, including the Food and Agriculture Organization (FAO), the Organization for Economic Cooperation and Development (OECD), the United Nations Educational, Scientific and Cultural Organization (UNESCO), the United Nations Environment Program (UNEP), the Convention on Biological Diversity (CBD), and many others (Cochrane et al., this volume; Gaichas et al., this volume; Samhouri et al., this volume; Polovina et al., this volume; Kritzer et al., this volume; Houde et al., this volume). These commitments have percolated down to the policies of Regional Fisheries Management and Regional Seas Organisations, to countries, and to governance below the national level. Throughout, the focus on ecosystems and therefore on space, is maintained. If EBM is going to meet all these commitments at these diverse levels, then it must ensure that, *inter alia:*

a) Cumulative and aggregate effects of all industry sectors operating in a given area are collectively sustainable (Crain et al., 2008)
b) Impacts of each industry sector on opportunities and outcomes of other sectors are taken into account in sectoral planning (Foden et al., 2011)
c) Actions of sectoral regulatory agencies should work in harmony and not antagonistically (Deboudt et al., 2008)
d) Decision-making within each sector should be inclusive of all legitimate (interpreted broadly) stakeholders, and be taken in the broad context of livelihoods of dependent communities, not the narrow context of just one dimension of sustainability (Gray and Hatcher, 2008; Jones at al., 2011).

All four elements can only be achieved if there is some forum for dialogue among industry sectors and among the agencies which regulate them, for scientific and technical experts to engage with the industries and their regulators, and for the diverse perspectives of civil society to participate in dialogue on trade-offs where different sectors of society may place different values on the various components in the trade-offs being considered (Ridgeway and Rice, 2010). MSP has emerged over the past decade as a preferred forum for dialogue addressing all those concerns (in fact, it can be argued that any forum supporting dialogue that covers all those dynamics is *de facto* an MSP forum). Although the MSP contribution to EBM may be greatest when the outcomes of MSP are binding on all parties, it is the credibility and legitimacy of the MSP processes and products that really allow EBM to progress. In fact, if MSP outcomes were made binding, many of the dynamics of building trust among participants with competing interests could be dramatically altered, and the transparency and inclusiveness of the processes would be placed under pressure from both institutional concerns about jurisdiction and legal competence, industry concerns about value of corporate intelligence, and even the relationships of stakeholder groups to their donors.

MSP is essential to EBM because both are fundamentally about seeking consensus and compromise among different interests, seeking outcomes which are sustainable socially, economically, ecologically and institutionally, and about evaluating options in a cyclic and adaptive framework, and because both are also fundamentally about how ocean and coastal space is used and/or protected. The *process* of MSP is a process ideally matched to the needs of EBM, and the *products* of MSP are the products on which EBM can proceed. However, even though they are ideally suited for each other, MSP still faces challenges, as does EBM.

6. Challenges and Benefits

This chapter has identified a number of challenges faced by MSP that have been dealt with successfully in a number of ways (Appendix B).

MSP does not start with a blank slate. Established sectors may be comfortable with a status quo which is unsustainable or cannot accommodate new or changing uses. At least the first attempts at MSP in a region usually result in shifts in resource allocation and changes in patterns of use rights both within and among sectors, which can be highly contentious. Strong institutional interests (agencies and jurisdictions) may already be firmly entrenched and unwilling to cede power or influence.

MSP addresses inherently difficult questions. It is fundamentally about resolving conflicts among sets of management objectives and pathways to achieve them, and the potential for conflicts and trade-offs multiplies quickly when the number of competing sectors and society perspectives increases. Participants in the processes will receive different distributions of benefits and bear different distributions of costs under alternative mixes of objectives and choices of pathways, and they may differ in their perceptions of what distributions of costs and benefits are equitable. Moreover, if the MSP initiative crosses jurisdictional boundaries, multiple regulatory frameworks may all constrain choices of objectives and pathways. Together, these factors may make finding broadly supported compromises difficult.

The knowledge to inform the MSP process is likely to be incomplete, uncertain, and possibly contested among participants. It can be time consuming and expensive to consolidate and geo-reference all the relevant data sources and traditional knowledge, and to develop the tools such as scenario exploration needed to inform the trade-off analyses (formal or informal). Even when the best information is available, data gaps and uncertainties often mean outcomes of various options cannot be quantified adequately to satisfy participants in the process. The knowledge challenges can be amplified when boundaries are determined by the legal jurisdictions of the participating governments or agencies, but have little relationship to the scales of some sea uses or to ecosystem boundaries.

The take-home message, though, is that none of these challenges are created by MSP. They are all created by society using the ocean in multiple ways, having a diversity of values and goals, and by governance processes having evolved according to political jurisdictions and focused on individual sectors. The value of MSP is that it offers a single, structured but flexible framework in which to find solutions to these challenges which society will support. If conducted carefully, patiently, and with respect for all participants, compromises can be found in which industry and society sectors may cooperate in achieving compatible and sustainable outcomes using coherent strategies and tools. It is exactly these sustainable outcomes—sustainable across ocean uses and for ecosystems as well as ocean users—that are the goal of ecosystem-based management. So the true challenges are to adequately resource the MSP processes, and follow the steps, which are well known and robustly validated.

References

Agardy, T., Notarbartolo di Sciara, G. and Christie, P. 2011. Mind the gap: addressing shortcomings of marine protected areas through large scale marine spatial planning. *Marine Policy* 35: 226–232

Ban, N. C. and Klein, C. J. 2009. Spatial socioeconomic data as a cost in systematic marine conservation planning. *Conservation Letters* 2: 206–215

Calado, H., Ng, K., Johnson, D., Sousa, L., Phillips, M. and Alves, F. 2010. Marine spatial planning: lessons from the Portuguese debate. *Marine Policy* 34: 1341–1349

Crain, C. M., Kroeker, K. and Halpern, B. S. 2008. Interactive and cumulative effects of multiple human stressors in marine systems. *Ecology Letters* 11: 1304–1315

Curtin, R. and Prezello, R. 2010. Understanding marine ecosystem based management: a literature review. *Marine Policy* 34: 821–830

Day, J. 2008. The need and practice of monitoring, evaluation and adapting planning and management—lessons from the Great Barrier Reef. *Marine Policy* 32: 823–831

Day, V., Paxinos, R., Emmett, J., Wright, A. and Goecker, M. 2008. The Marine Planning Framework for South Australia: A new ecosystem-based zoning policy for marine management. *Marine Policy* 32: 535–543

Deboudt, P., Dauvin J.-C. and Lozachmeur, O. 2008. Recent developments in coastal zone management in France: The transition towards integrated coastal zone management (1973–2007). *Ocean & Coastal Management* 51: 212–228

Douvere, F. 2008. The importance of marine spatial planning in advancing ecosystem-based sea use management. *Marine Policy* 32: 762–771

Douvere, F. and Ehler, C. N. 2009. New perspectives on sea use management: Initial findings from European experience with marine spatial planning. *Journal of Environmental Management* 90: 77–88

Ehler, C. and Douvere, F. 2009. Marine Spatial Planning: a step-by-step approach toward ecosystem-based management. Intergovernmental Oceanographic Commission and Man and the Biosphere Programme. IOC Manual and Guides No. 53, ICAM Dossier No. 6. UNESCO, Paris

Flannery, W. and O Cinneide, M. 2008. Marine spatial planning from the perspective of a small seaside community in Ireland. *Marine Policy* 32: 980–987

Foden, J., Rogers, S. I. and Jones, A. P. 2011. Human pressures on UK seabed habitats: a cumulative impact assessment. *Marine Ecology Progress Series* 428: 33–47

Garcia, S. M. and Charles, A. T. 2008. Fishery systems and linkages: implications for science and governance. *Ocean and Coastal Management* 51: 512–529

Gililand, P. M. and Laffoley, D. 2008. Key elements and steps in the process of developing ecosystem-based marine spatial planning. *Marine Policy* 32: 787–796

Gray, T. and Hatcher, J. 2008. A complicated relationship: stakeholder participation in the ecosystem-based approach to fisheries management. *Marine Policy* 32: 158–168

Jones, P. J. S., Qiu, W. and De Santo, E. M. 2011. Governing marine protected areas—Getting the balance right. Technical Report, United Nations Environment Programme, Nairobi

Katsanevakis, S., Stelzenmüller, V., South, A., Sorensen, T. K., Jones, P. J. S., Kerr, S., Badalamenti, F., Anagnostou, C., Breen, P., Chust, G., D'Anna, G., Duijn, M., Filatova, T., Fiorentino, F., Hulsman, H., Johnson, K., Karageorgis, A. P., Kroencke, I., Mirto, S., Pipitone, C., Portelli, S., Qui, W., Reiss, H., Sakellariou, D., Salomidi, M., van Hoof, L., Vassilopoulou, V., Fernandez, T. V., Voge, S., Weber, A., Zenetos, A. and ter Hofstede, R. 2011. Ecosystem-based marine spatial management: Review of concepts, policies, tools, and critical issues. *Ocean and Coastal Management* 54: 807–820

Kidd, S. and Shaw, D. (forthcoming) Reconceptualising territoriality and spatial planning: insights from the Sea. *Planning Theory and Practice*

Kidd, S. and Ellis, G. (2012) From the land to sea and back again? Using terrestrial planning to understand the process of marine spatial planning. *Journal of Environmental Policy and Planning* 14: 49–66

Klein, C., Steinback, C., Watts, M., Scholz, A. and Possingham, H. 2010. Spatial marine zoning for fisheries and conservation. *Frontiers in Ecology and the Environment* 8: 349–353

Levin, P. S., Fogarty, M. J., Murawski, S. A. and Fluharty, D. 2009. Integrated ecosystem assessments: developing the scientific basis for ecosystem-based management of the ocean. *PLoS Biology* 7: 23–28

Morrison, K. K. and Snow-Cotter, S. 2008. Toward more integrated ocean governance in Massachusetts: a progress report. *Coastal Management* 36: 412–430

NOAA. 2011. *Strategic Advice on Designing and Implementing Marine Spatial Management Plans.* Report of the Ecosystem Science and Management Working Group of the Science Advisory Board, NOAA, Silver Springs MD. 36 pp + Annex

Nobre A. M. 2011. Scientific approaches to address challenges in coastal management. *Marine Ecology Progress Series* 434: 279–289

Olsen, S., Olsen, E. and Schaefer, N. 2011. Governance baselines as a basis for adaptive marine spatial planning. *Journal of Coastal Conservation* 15: 313–322

Pikitch, E. K., Santora, C., Babcock, E. A., Bakun, A., Bonfil, R., Conover, D. O., Dayton, Doukakis, P., Fluharty, D., Heneman, B., Houde, E. D., Link, J., Livingston, P. A., Mangel, M., Pope, J. and Sansbury, K. J. 2004. Ecosystem-based fisheries management. *Science* 305: 346–347

Pomeroy, R. and Douvere, F. 2008. The engagement of stakeholders in the marine spatial planning process. *Marine Policy* 32: 816–822

Rice, J. C. 2011. Managing fisheries well: delivering the promise of an ecosystem approach. *Fish and Fisheries* 12: 209–231

Ridgeway, L. and Rice, J. C. 2010. International organisations and fisheries governance. Pp. 485–504 in Q. Grafton (ed.) *Handbook of Fisheries.* Oxford University

St. Martin, K. and Hall-Arber, M. 2008. The missing layer: geo-technologies, communities, and implications for marine spatial planning. *Marine Policy* 32: 779–786

Appendix A

Illustrative list of goals and objectives identified in Marine Spatial Planning initiatives of various national and regional jurisdictions. All entries are direct quotations from sources listed in the text, to illustrate the differing nature of what may be considered a goal, objective or target. Unless an entry is labelled as a "list" all entries are subsets of longer suites of goals or objectives.

Jurisdiction: Australia
Initiative: Southwest Regional Management Plan
Types of Objectives: Overarching goal, conceptual high level objectives, and strategic objectives

Examples of objectives:

The overall goal is "to help ensure that the marine environment of the region remains healthy and resilient".

The high level objectives include:

- Conserving biodiversity and maintaining ecosystem health
- Ensuring the recovery and protection of threatened species
- Improving the understanding of the region's biodiversity and ecosystems and the pressures they face

There are 7 strategic objectives including:

- Establish and manage a south-west Commonwealth marine reserve network as part of a National Representative System of marine Protected Areas
- Develop partnerships with relevant marine industries to increase understanding of the impacts of anthropogenic disturbance on the region's key ecological features and protected species
- Develop targeted collaborative programs to coordinate species recovery and environmental protection efforts across Australian Government and state agencies with responsibilities for the marine environment

Jurisdiction: Canada
Initiative: Beaufort Sea
Types of Objectives: Thematic goals and strategic objectives for each goal
Examples of objectives:

There are 6 high-level "thematic goals":

Governance—To achieve effective governance for the sustainable use of the Beaufort Sea

Economic—To foster sustainable economic opportunities and options for Canadians, northerners and coastal communities

Cultural—To maintain and increase peoples' sense of place and preserve cultural identity and spiritual connections as they relate to oceans and coastal areas

Social—To improve human capacity, health, quality of life and opportunities as they connect to oceans and coastal areas

Traditional and Local Knowledge—To promote the value, credibility and use of Traditional Knowledge (TK) and Local Knowledge (LK) to current and future generations

Ecosystem—To understand the Beaufort Sea ecosystem, to identify important areas and priority species and to maintain or enhance ecosystem integrity

There are 24 strategic objectives of the Integrated Ocean Management Plan (IOMP) to achieve the 6 goals. Examples include:

- Establish collaborative inter-governmental and inter-departmental structures and processes
- Ensure Aboriginal organizations have the capacity to be involved in the Integrated Ocean Management Plan (IOMP)
- Promote a vibrant local subsistence economy
- Promote the respect, value, and sharing of TK and LK

- Maintain ecosystem integrity within the Large Ocean Management Plan (LOMA)

Jurisdiction: Canada
Initiative: Scotian Shelf Integrated Management Plan
Types of Objectives: Conceptual goals and strategic objectives for each goal
Examples of objectives:

Goal 1: Collaborative governance and integrated management; with *9 Strategic Objectives,* such as:

- Collaborative structures and processes with adequate capacity, accessible to community members, are established
- Ocean users and regulators are compliant and accountable

Goal 2: Sustainable human use; with 7 *Strategic Objectives,* such as:

- Communities are sustainable
- Wealth is generated sustainably from renewable ocean resources

Goal 3: Healthy ecosystems; with 14 *Strategic Objectives,* such as:

- Diversity of benthic, demersal and pelagic community types is conserved
- Incidental mortality of all species is within acceptable levels

Jurisdiction: Norway
Initiative: Barents Sea—Lofoten Ecosystem Plan
Types of Objectives: Overarching goal, high-level objectives and sectoral strategic objectives
Examples of objectives:

Overall goal: The plan is intended to provide a framework for activities in the area and facilitate co-existence between the industries involved, such as the fisheries, maritime transport and petroleum industries, while at the same time taking environmental considerations into account.

There are 8 high-level objectives, such as:

- Management of the Barents Sea–Lofoten area will promote sustainable use of the area and its resources to the benefit of the region and the country in general.
- The management regime will ensure that activities in the area do not threaten the natural resource base and thus jeopardise opportunities for future value creation.
- The management regime will facilitate economically viable commercial activities and as far as possible promote value creation and employment in the region.

Strategic objectives for industry sectors are phrased as "The Government has set the following objective for. . . ."

- Preventing and combating pollution in the Barents Sea–Lofoten area: "Releases and inputs of pollutants to the Barents Sea–Lofoten area will not result in injury to health or damage the productivity of the natural environment and its capacity for self-renewal. Activities in the area will not result in higher levels of pollutants".

Strategic objectives for ecosystem status and their targets are phrased as "The Government has set the following objective for"

- Management of biodiversity in the Barents Sea—Lofoten area:
  "Management of the Barents Sea—Lofoten area will ensure that diversity at ecosystem, habitat, species and genetic levels, and the productivity of ecosystems, are maintained. Human activity in the area will not damage the structure, functioning, productivity or dynamics".

The Government has set the following targets for the management of biodiversity in particularly valuable and vulnerable areas and habitats:

- Activities in particularly valuable and vulnerable areas will be conducted in such a way that the ecological functioning and biodiversity of such areas are not threatened.
- Damage to marine habitats that are considered to be threatened or vulnerable will be avoided.
- In marine habitats that are particularly important for the structure, functioning, productivity and dynamics of ecosystems, activities will be conducted in such a way that all ecological functions are maintained.

Jurisdiction: Tri-Lateral Commission for the Wadden Sea
Initiative: Southwest Regional Management Plan Wadden Sea Regional Plan
Type of Objectives: Strategic
Examples of objectives: There are 13 strategic objectives and associated targets for industry sectors and social well-being. Illustrations of strategic objectives and associated targets include:

- Safety (3 targets at scales of)
  - Safety at sea
  - Safe living conditions and economic development through coastal protection
  - Protection of the landscape and nature against hazards
- Social (3 targets)
- Health (3 Targets)
- Regional Identity (3 Targets)

Jurisdiction: United Kingdom
Initiative: Irish Sea Regional Plan
Types of Objectives: Conceptual—for the planning process, not the ecosystem
Examples of objectives:

- Describe the environmental, social and economic characteristics and natural resources of the UK's coastal zone
- Describe the key laws, institutions and stakeholders that influence the management of the UK's coast at different scales

- Analyse how these laws, institutions and stakeholders interact and integrate with each other vertically, horizontally, spatially and temporally
- Identify the key gaps, overlaps and inconsistencies in these interactions, which act as obstacles to practical Integrated Coastal Zone Management in the UK

Jurisdiction: United States—Rhode Island
Initiative: Rhode Island Ocean Special Area Management Plan (SAMP)
Types of Objectives: Conceptual, starting process for strategic and operational:
Examples of objectives:

Conceptual: One of the objectives of the Ocean SAMP is to encourage marine-based economic development that considers the aspirations of local communities, and is consistent with and complementary to the state's overall economic development, social, and environmental needs and goals.
Strategic and Operational: Setting and adjusting objectives will be part of an adaptive management process which is still in the future

Jurisdiction: United States—Delaware
Initiative: Delaware Estuarine Program
Type of Objectives: Conceptual, strategic, guidance for development of tactical
Examples of objectives:

Conceptual: One of the objectives of the Delaware Estuary Program is to ensure that sufficient, appropriate, and safe waterfront open space and public access areas will be reserved and preserved for existing and future public use, and that adequate facilities will be provided to meet the recreational demands of the public without impairing the natural system

Strategic: There are 6 objectives, such as:

- To restore population levels of harvestable species of finfish and invertebrate species to levels that will support sustainable recreational and commercial fisheries
- To maintain or restore an assemblage of organisms and their habitat throughout the Delaware Estuary and tidal wetlands that contribute to the ecological diversity, stability, productivity, and aesthetic appeal of the region
- To preserve acreage and enhance quality of shoreline and littoral habitat to sustain a balanced natural system

Guidance on Tactical:

Estuary-wide objectives for species management plans would be identified as well as differences in those objectives. Examples of objectives could be:

- Managing acreage of land as a certain habitat type;
- Identifying and maintaining target population levels;
- Temporal habitat manipulations (controlling water levels, etc.)

Jurisdiction: United States—Florida
Initiative: Tampa Bay Estuarine Program
Types of Objectives: Strategic matched to tactical
Examples of objectives:

Strategic Management Objectives for Ecosystem Status

- Increase the number, diversity and health of the bay's fish and shellfish populations, and restore a self-sustaining bay scallop population
- Restore and protect wildlife habitats and food sources, and promote regional wildlife habitat planning
- Minimize physical impacts to bay wildlife and habitats

There are 6 Tactical Actions / Objectives, such as

- Increase on-water enforcement of environmental regulations
- Establish and enforce manatee protection zones
- Support bay scallop restoration

Tactical Actions matched to Strategic Objectives for factors including

- General monitoring objectives for the water quality, fisheries, benthic and habitat components of the program

Appendix B

Illustrative list of properties of six Marine Spatial Management processes

Name	Eastern Scotian Shelf Integrated Management Plan
Geographic location	Eastern Scotian Shelf, Canada
Area / coverage	325,000 km^2; coastal area (12 km from shoreline), excluded
Key enabling legal basis	Canada's Oceans Act
Lead Jurisdiction	Fisheries and Oceans Canada
Other jurisdictions that participated	Province of Nova Scotia, Federal Ministry of Natural Resources, Parks Canada, Federal Ministry of Environment
Major Industries included or excluded	All major offshore industries included in process, coastal development excluded
Engagement of ENGOs and civil society	Regular workshops and public meetings; Individuals from some industry organizations and ENGOS included in work teams; formal Stakeholder Advisory Council
How long the process took	Over 6 years with information consolidation and consultation running in parallel/iterative

Cost	Uncertain because of indirect costs, but in the few millions of $ Canadian
Outcome / Product	Scotian Shelf Integrated Management Plan. Implementation stalled because of inter-provincial boundary dispute.
Provision for review	Every 5 years

Name	Beaufort Sea Ocean Management Plan
Geographic location	Eastern Beaufort Sea, Canadian Arctic
Area / coverage	1,750,000 km^2
Key enabling legal basis	Canada's Oceans Act
Lead Jurisdiction	Fisheries and Oceans Canada
Other jurisdictions that participated	Federal Departments of Environment, Natural Resources, Indian and Northern Affairs, and Transportation; Yukon and Northwest Territorial governments, all tribal Councils and Management Boards recognized by Land Claims Agreements, Treaties, and enabling court decisions
Major Industries included or excluded	All major industries and subsistence uses included in process, coastal development excluded. Treaty Rights, Co-Management Land Claims Agreements, and court decisions on rights of Aboriginal Peoples to self-government in the North all supersede any integrated management (IM) decisions.
Engagement of ENGOs and civil society	Regional Coordination Committee, Beaufort Sea Partnership, and a number of working groups were created to lead and implement the management plan. Extensive consultation with Elders and Aboriginal Communities
How long the process took	4 years with information consolidation and consultation running in parallel / iterative
Cost	Uncertain because of indirect costs, but in the few millions of $ Canadian
Outcome / Product	Beaufort Sea Integrated Management Plan. Implementation announced 2010
Provision for review	First review in 3 years, thereafter every 5 years

Name	Great Barrier Reef Marine Park Management Plan
Geographic location	North eastern shelf, Australia
Area / coverage	344,000 km^2

Key enabling legal basis	Great Barrier Reef Marine Park Act 1975
Lead Jurisdiction	Great Barrier Reef Marine Park Authority (Department of Sustainability, Environment, Water, Population and Communities)
Other jurisdictions that participated	State of Queensland
Major Industries included or excluded	All major offshore industries included in the process, no mining allowed in the Marine Park
Engagement of ENGOs and civil society	Extensive engagement with a wide set of stakeholders; four reef advisory committees—catchment and coastal, ecosystem, indigenous, and tourism and recreation
How long the process took	Initial plan 1975 to 1979; rezoning 1998 to 2004
Cost	Annual management and monitoring costs of about A$30 million
Outcome / Product	25 year Strategic Plan for Great Barrier Reef World Heritage Area
Provision for review	Formally, every 25 years, with periodic review about every 10 years

Name	Marine bioregional plan for the South-west Marine Region
Geographic location	South Western Australia
Area / coverage	1.3 million km^2 excluding State waters and out to 200 nm
Key enabling legal basis	Environmental Protection and Biodiversity Conservation Act 1999
Lead Jurisdiction	Department of Sustainability, Environment, Water, Population and Communities (Federal)
Other jurisdictions that participated	States of Western Australia and South Australia
Major Industries included or excluded	All major offshore industries included in the process
Engagement of ENGOs and civil society	Public consultation on the final Plan closed in September 2012
How long the process took	5 years to date, starting with release of a bioregional profile
Cost	Not known

Outcome / Product	Bioregional plan including marine reserve network
Provision for review	10 years

Name	Belgium Part of North Sea Master Plan
Geographic location	North Sea, Belgium
Area / coverage	3600 km^2 (encompassing territorial sea, EEZ/fishery zone/continental shelf)
Key enabling legal basis	Continental Shelf Act 1969, EEZ Act 1999 and Marine Protection Act 1999, plus various Royal Decrees
Lead Jurisdiction	Belgium Federal Government North Sea Minister
Other jurisdictions that participated	Flemish Region, West Flanders Province, 10 coastal municipalities
Major Industries included or excluded	Phased inclusion of different industries and interests into the master plan
Engagement of ENGOs and civil society	Establishment of advisory body but wider engagement variable. No overall participation framework
How long the process took	Phase 1 2003/4, Phase 2 2005/6
Cost	Not known
Outcome / Product	Belgium Part of North Sea Master Plan, innovative exploration of future scenarios, but limitations in relation to stakeholder engagement and transnational integration in the wider North Sea
Provision for review	No formal review period is established as yet

Name	Spatial Plan for North Sea, 2010 part of National Water Plan
Geographic location	Dutch part of the North Sea (encompassing territorial sea and EEZ)
Area / coverage	57,000 km^2
Key enabling legal basis	Law on Spatial Development (revision 2008)
Lead Jurisdiction	Dutch Ministry of Infrastructure and Environment
Other jurisdictions that participated	Integration with other jurisdictions in the North Sea part of next phase of work.
Major Industries included or excluded	All major offshore industries included in process

Engagement of ENGOs and civil society	Intensive stakeholder participation at different scales, joint fact finding, open negotiations and investigation of alternatives, development of sub processes in areas with conflicting spatial claims
How long the process took	2 years from revision of Law on Spatial Development in 2008, but able to build on 2005 Integrated Management Plan for the North Sea 2015
Cost	Not known
Outcome / Product	Strategic framework for the coordination of sea uses. Next step is to developed more detailed plans for particular areas and better connections with wider North Sea picture
Provision for review	5 years

Chapter 8. Marine Ecosystem Services: A Framework and Practical Set of Tools for Ecosystem-Based Management

ANNE D. GUERRY AND HEATHER TALLIS

Natural Capital Project, Stanford University

Contents

1. Introduction

The oceans provide us with a rich and varied stream of benefits. Historically, the benefits we rely on most directly have been individually governed by separate entities and managed as if they flow from the oceans to people in independent streams. Fisheries are managed independently of oil and gas production, tourism is managed independently from coastal development, shipping is managed independently from wind and wave energy production, and so on. Furthermore, other, more foundational benefits provided from ocean processes—climate regulation, nutrient cycling, coastal protection—are not managed at all. However, growth in our biophysical, ecological and social understanding of the oceans and our connections to them has made clear the many interactions among these benefit streams and the compounding or synergistic effects of human decisions on each. Above all, we have learned how our ignorance of these interactions in management leads to unintended consequences, or externalities, for other sectors and the environment. This growing knowledge has led to the global scientific support of ecosystem-based management as a more sustainable approach to the management of oceans and coasts.

One of the most concrete ways that we can advance towards ecosystem-based management is by revealing the magnitude and nature of interactions among sectors in the decision-making process. All managers are implicitly making tradeoffs among their goals and the ability of other sectors to function, but these tradeoffs are seldom made based on a clear understanding of outcomes for or likely decisions made by those other sectors. Managers need a way to understand interacting goals and consequences of activities in a common language and a way to transparently assess tradeoffs among these goals. Ecosystem services provide such a platform.

The myriad direct and indirect benefits derived from natural systems are broadly characterized as "ecosystem services" (Daily et al. 1997; Millennium Eco-

The Sea, Volume 16, edited by Michael J. Fogarty and James J. McCarthy
ISBN 978-0-674-07270-1

system Assessment 2005). The ecosystem services concept applies equally to terrestrial, freshwater and marine systems anywhere on the spectrum from relatively untouched to intensively managed. Indeed, all ecosystems provide, to differing degrees, a set of ecosystem services. Marine ecosystems in particular provide a wealth of benefits to humans (Peterson and Lubchenco 1997; Costanza 2000; Agardy et al. 2005; United Nations Environment Program 2006; Patterson and Glavovic 2008; Wilson and Liu 2008; Guerry et al. 2011; Guerry et al. 2012a). To name a few, oceans and coasts provide seafood, protection from coastal hazards, opportunities for renewal and inspiration, feed for animals, materials for food additives, cosmetics and pharmaceuticals as well as livelihoods. Underpinning these benefits that flow directly to people, oceans are central to the functioning of the planet and thus provide innumerable "supporting services" (as defined by the Millennium Ecosystem Assessment (MA) 2005). For example, the oceans are responsible for approximately 40% of global net primary productivity (Schlesinger 1991; Melillo et al. 1993) and play a critical role in transforming, detoxifying, and sequestering wastes (Peterson and Lubchenco 1997 and refs therein). They are a primary driver of the atmosphere's temperature, moisture content, and stability (Colling 2001) and are central to global cycles of water and numerous major elements including carbon, nitrogen, and oxygen (Gleick 1996; Peterson and Lubchenco 1997).

Capturing all of these benefits in one concept, ecosystem services provide a framework within which to discuss unintended consequences of single-sector management such as negative impacts on other sectors and the overall suboptimal delivery of a wide range of benefits that people care about. Although there has been some movement toward incorporating costs and benefits associated with multiple ecosystem services into natural resource decision making, decisions that consider costs often reflect only a small portion of the full suite of values and focus on those that accrue to particular sectors, rather than to society as a whole. Cost-benefit analyses tend to focus on values that are easy to express in monetary terms, but many ecosystem services are not currently marketed. Therefore, services that provide non-market and non-use values are often ignored or assumed to have lesser value. This misrepresentation of the full social value set derived from an ecosystem can guide decision makers toward actions with negative repercussions for society. For example, mangrove areas are often cleared for shrimp aquaculture. The high market price of shrimp motivates this action, but the net value of shrimp production (market value minus subsidies and social costs) pales in comparison to the social and ecological benefits that intact mangroves provide, such as protection from storms, materials for construction, nursery habitat, etc. More complete accounting by Barbier and colleagues (e.g., Sathirathai and Barbier 2001) indicates that standing mangroves provide more benefits for society than shrimp farms do.

By definition, ecosystem services are things that people care about and need. They can bring diverse communities together with shared goals. For example, existence values (the values people place on simply knowing something exists, even without directly experiencing it) of a healthy ecosystem have been found to be among the 'most important' benefits provided by the Puget Sound system (Iceland et al. 2008).

An understanding of the values held and shared across diverse communities can help managers communicate in language that resonates with stakeholders and thus encourage participation in public processes and generate support for management activities.

In sum, an ecosystem services approach reveals, incorporates and communicates the connections among components of natural and social systems. These connections are often important not only for a single sector making management decisions, but also for the maintenance of ecosystem health and human well-being. Mapping, modeling and measuring ecosystem services can thus serve as a unifying approach for ecosystem-based management.

1.1 Conceptual foundations

Central to the concept of ecosystem services is the notion that people are part of, not separate from, natural systems (see Charles, this volume). Ecosystem services research has its foundations in research on complex adaptive systems (Levin 1998; Levin and Lubchenco 2008), coupled social-ecological systems (Ostrom 2009) and coupled human and natural systems (Liu et al. 2007a; Liu et al. 2007b). In all of these schools of thought, people are considered to be integral parts of natural systems. Much of the research focus in these themes is aimed at understanding the dynamics of coupled social-ecological systems and the effects of different management approaches on ecological (e.g., biomass of fished species, biodiversity) and social (e.g., efficiency, equity, accountability) outcomes through ecosystem services.

An explicit awareness of ecosystem services and the negative repercussions of their loss originated long ago. Plato recognized the role of the natural environment in providing social benefits and mourned its degradation. He wrote of hydrological ecosystem services of the deep past:

> What now remains of the formerly rich land is like the skeleton of a sick man with all the fat and soft earth having wasted away and only the bare framework remaining. . . .The soil was deep, it absorbed and kept the water. . . , and the water that soaked into the hills fed springs and running streams everywhere.

More modern concern with the delivery of benefits from ecosystems and their degradation can be traced to George Perkins Marsh who, in Man and Nature (1864), described a wide array of services and their ongoing losses. Many others followed in this vein, emphasizing human impact on the environment, the resultant social consequences of this decline (Osborn 1948; Vogt 1948; Leopold 1949; Ehrlich 1968; Holdren and Ehrlich 1974; Daily et al. 1997) and their economic expression (Krutilla 1967; Krutilla and Fisher 1975; Dasgupta and Heal 1979; Beattie and Taylor 1985; Clark 1990; Freeman III 1993).

The Millennium Ecosystem Assessment (MA) alerted global audiences to ecosystem services and provided the first comprehensive global assessment of the status and trends of all of the world's major ecosystem services. Called for by United Nations Secretary General Kofi Annan, it was conducted by over 1360 experts worldwide between 2001–2005. The MA emphasized the connections between human decisions and the natural environment that feed back to the human condition via changes in the flow of ecosystem services. The key finding of the MA was that two-thirds of the world's ecosystem services were declining (2005). This sobering statistic captured the attention of world leaders.

Since the MA, new methods and efforts to model, map, and value ecosystem services have emerged (e.g. UNEP and IOC-UNESCO 2009; Kareiva et al. 2011; UK National Ecosystem Assessment 2011). Much of the MA was focused on terrestrial systems, but recent scientific and political developments have catalyzed the expansion of the ecosystem services framework to marine systems. For example, the UK conducted a country-wide assessment of both terrestrial and marine ecosystem services and their values (Stokstad 2011; UK National Ecosystem Assessment 2011). In the marine realm, fisheries have received the most attention. Fisheries models are by far the most sophisticated marine ecosystem service models and have the longest history of use in managing marine systems. A growing focus on ecosystem-based fisheries management has begun to put the production of fish in the context of other important ecosystem services (e.g. supporting food webs for fished species (Constable 2004)). More generally, the MA and subsequent assessments have raised awareness of ecosystem services as well as the human dependence upon and threatened status of many of them.

1.2 Links to marine EBM

As discussed above, an ecosystem services framework provides important insights for the pursuit of ecosystem-based management. The consequences of changes in ecosystem structure and function flow to people through changes in ecosystem services. Ecosystem services and their values can thus be used as meaningful metrics for assessing alternative management interventions and their potential impacts on economic or social well-being (Charles, this volume).

In the marine realm, an ecosystem services framework provides the opportunity to enact ecosystem-based management and to overcome problems arising from traditional sector-by-sector management. An ecosystem services framework can help make connections by highlighting trade-offs among multiple objectives and allowing for decisions that aim to resolve or minimize those tradeoffs to be made transparently. Scientists and managers are recognizing links between diversity, productivity and resilience of marine systems and their response to management actions such as conserving, restoring and using marine ecosystem services (Liu et al. 2007a; Levin and Lubchenco 2008; Murawski et al. 2009; Chan and Ruckelshaus 2010). The ability to model and assess trade-offs among objectives of multiple management sectors (e.g., fisheries, energy extraction, recreation) is necessary for informing complex cases of ecosystem-based management that include a wide range of actors (Foley et al. 2010; Altman et al., this volume).

In marine systems, efforts to use the framework of ecosystem services to understand the connections between activities in one sector and their impacts on others on national and global scales are underway. Decision-makers at various levels of government from around the world have articulated the need for new approaches to ensure the continued delivery of benefits that people derive from oceans and coasts (Cochrane et al., this volume). Some examples include the U.S. Ocean Policy Task Force (2010), the European Commission (2010) and the recently approved UN Intergovernmental Science-Policy Platform on Biodiversity and Ecosystem Services (IPBES) (UN General Assembly (UNGA) 2010). Marine spatial planning (see Rice et al., this volume) is one approach to comprehensive management of marine and coastal systems that is in the toolbox of the ecosystem-based manager (Guerry et al. 2012a). Marine spatial planning is used to analyze current and potential future uses, identify areas most suitable for particular activities, and provide a process to better determine how oceans are used now and in the future (Douvere and Ehler 2009; Ehler and Douvere 2009). Ecosystem services provide a framework and a common language for marine spatial planning. The framework allows stakeholders and managers to articulate their goals transparently using common terms. In addition, it fosters a better understanding of how single decisions interact with those made by others.

Integrated Ecosystem Assessments (IEAs) (Levin et al. 2009; Plummer and Levin, this volume) are another example of current approaches to the management of marine ecosystems that draw upon similar foundations and can use ecosystem services in a modeling framework. An IEA uses ecosystem modeling and other quantitative analyses to inform a wide range of management decisions that affect natural, social, and economic conditions. Ecosystem services can provide a set of metrics for assessing these conditions (Tallis et al. 2012). Modeling changes in the delivery of ecosystem services to people can enable decision makers to compare alternative policies. IEAs have been successfully applied in fishery-based ecosystem management (Caddy 1999; Sainsbury et al. 2000; Smith et al. 2007). They are designed to be iterative, expanding in scope and sophistication as information develops over time (e.g. Dennison et al. 2007; Tallis et al. 2010; Samhouri et al. 2011).

In the remainder of this chapter, we describe key methodological issues and challenges, present several tools for incorporating ecosystem services in marine ecosystem-based management and highlight examples of the use of those tools in real decision contexts. We close with a discussion of emerging frontiers for advancing the science and practical advice for accounting for marine ecosystem services in the decisions of those responsible for the management of coastal and marine resources.

2. Methodological Issues and Challenges

2.1 The ecosystem services supply chain

All ecosystem services are generated by functions of a natural or managed system in a supply chain best considered in three discrete steps: supply, service, and value (Tallis et al. 2012). Supply is determined by ecosystem structure and function alone, service includes delivery to people, and value represents the contribution of the service to human well-being. For example, many different types of coastal habitats (e.g., coral reefs, mangroves, oyster beds, barrier islands) protect lives, infrastructure and property from storms by attenuating waves (Danielsen et al. 2005; Das and Vincent 2009; Barbier et al. 2011; Kritzer et al., this volume). A map of wave attenuation by these habitats represents the supply of potential protection from storms. However, people do not receive storm surge protection from all of these locations because some are far from human infrastructure and settlement. The distribution of human infrastructure and settlement, together with the location and amount of supply, are required to determine how much "service" is actually delivered at a given time. Finally, the service delivered to human communities contributes differently to human well-being depending on the social context. For example, flood protection may contribute more to well-being in places where people are not able to retreat from flood waters (e.g. elderly communities or areas with poor roads). The process of calculating change in well-being is discussed in the following section.

Metrics at each point in the ecosystem service supply chain can be useful in different ways during an ecosystem-based management process. Metrics of supply can shed light on the sustainability of natural resource use patterns and inform predictions about the delivery of services in future years. Previous research has focused most on supply metrics, so we generally have a better understanding of how this part of the ecosystem service delivery chain functions and how to measure its response. Service metrics reveal what society is actually receiving from ecosystems. Approaches for tracking service delivery are well established and commonly applied to provisioning services such as fisheries (landings) and aquaculture (production rates), but are not well developed or seldom tracked for many other important services (tourism visitation rates, avoided flood or erosion rates, etc). Value metrics reveal how important changes in ecosystem service provision are to human well-being. In many cases, these metrics are most useful because they resonate with people and mirror the currencies commonly used in cost-benefit analysis, national accounting and other common marine management decision-making processes. Value metrics are readily available only for marketed goods. Even for those tracked in markets, the market values often do not reflect well the full social value of the good. Overall, marine ecosystem services are severely under-represented in both research and monitoring at all stages of the supply chain.

2.2 *Production functions*

In economics, a production function represents the relationship between inputs (like labor (L)) and outputs (goods and services (Q)). The simplest economic production function accounts only for labor and specifies the shape of the relationship (e.g. linear, or if including diminishing returns, exponential). For example, if one worker can produce 10 widgets in one day, one can forecast the number of widgets produced by knowing only the number of workers and the time horizon in question. In this case, the production function would be $Q = 10L$ if linear, or might be $Q = 10L^{0.5}$ with significant diminishing returns. The same concept is used in ecological production functions.

Ecological production functions can be used to explore how changes in ecosystem structure and function lead to changes in the flows of services (National Research Council 2005). Ecological production functions are fundamentally process-based. They represent the relationship between inputs (instead of labor and raw materials, these might be the density of mangroves) and outputs (no longer widgets, but perhaps the degree of protection from storms). The production function approach has been used to model various ecosystem services on land (Ricketts et al. 2004; Kareiva et al. 2011) and in marine systems (Barbier et al. 2008; Holland et al. 2010; Guerry et al. 2011; Guerry et al. 2012a).

A production function approach allows estimation of ecosystem service delivery for services that are difficult to observe directly. For example, common water quality observations (e.g. the concentration of contaminants in a particular place) do not, in themselves, represent the contribution of natural capital to water quality regulation. Production function models that incorporate the presence of vegetation (in addition to topography, precipitation and other factors) into the production of clean water can reveal this marginal contribution.

If valuation is of interest, production function approaches can be combined with both market prices and non-market valuation methods to estimate value and to show how monetary value or other value currencies are likely to change under different conditions (see valuation discussion below). Using a production function approach to estimate changes in ecosystem service delivery is an important step toward appropriate valuation of ecosystem services because ecosystem service value is a combination of the supply of services and their contribution to well-being. When there is no market to do this integration for us, we need good estimates of both changes in the supply of the service and changes in well-being to realistically assess value. Appropriate valuation techniques can then be used to translate changes in supply to changes in well-being. Several of the options for this translation step are discussed below.

2.3 Values and valuation

Since the publication of the provocative and hotly contested estimate of the total economic value of the world's ecosystems (Costanza et al. 1997), the concept of "ecosystem services" has been unfairly simplified in the minds of many as nothing more than an attempt to privatize and monetize nature. Indeed, valuing ecosystem services in monetary terms is of interest in some contexts, but the framework of ecosystem services and the available approaches to valuation are significantly richer and more nuanced than often perceived.

The most common misconception about ecosystem service valuation is that total value and monetary value are the same thing. They are not. Why is this? What economists are trying to capture with value is the total impact on well-being of a change in a product or service, in this case, a change in an ecosystem service (National Research Council 2005). Markets should reveal this value for us because in a perfectly functioning market economy, people 'vote' with dollars and decide what to consume based on how much each good or service contributes to their well-being relative to its price. In such an ideal system, we could combine market prices and quantities purchased to reflect the well-being benefits people receive from ecosystem services. But in the current global economic system not all benefits are captured in markets. Many ecosystem services are public goods (e.g., climate regulation, coastal protection), making it difficult to provide these services through markets. Other ecosystem services are enjoyed primarily by the poor (such as subsistence fishing, charcoal harvest, etc). These are not well-represented in market systems precisely because the people enjoying them do not have much money with which to vote. Still other services, such as cultural benefits, provide intangible experiences that are not captured well in current valuation approaches.

Given the complexities and imperfections of our current economic system, we must use a diversity of approaches to capture the value of ecosystem services (see Liu et al. 2010 and Conte 2012 for recent reviews). Regardless of the approach, each measure should represent society's preference for how much service is delivered, and care needs to be taken to avoid either missing benefits or double-counting benefits through multiple metrics. Each of the available methods has strengths and weaknesses and needs to be carefully matched to the questions of interest, the data available, and other contextual information.

As described above, ecosystem services whose products are traded in markets and have explicit prices can be valued by observing what people pay for those goods and the quantities purchased (e.g. fish). For services that are not traded in markets, various non-market valuation techniques can be used to assign monetary value.

Three types of these methods are generally easier to infuse into decision-making processes because they rely in some way on existing markets or behaviors to reveal value. First, revealed preference methods include travel cost methods and hedonic pricing. The travel-cost method values site-based amenities based on the costs people incur to enjoy them (Brown and Mendelsohn 1983). Similarly, hedonic methods derive value estimates from what people actually pay for the service through purchases in related markets, such as housing (e.g. water clarity effects on property values in the Chesapeake Bay) (Leggett and Bockstael 2000). Second, production economics use the marginal contribution of a service to the production of a marketed good to extract the value of that service (e.g. the marginal contribution of fish nursery habitat to fish market value) (Boyd and Banzhaf 2007). Finally, the cost-based approach uses the cost of replacing a lost service (e.g., coastal protection by seawalls instead of coastal wetlands) to represent the value of the service. Avoided damages can also be used—with this approach, the estimated cost of the damage that would have occurred without the service (e.g. avoided flood damage) stands in for the value of the service.

Another major approach, stated preference analysis, is often more difficult to support in decision-making processes. The stated preference approach elicits information about values through surveys. For example, contingent valuation involves assessing people's willingness to pay or accept compensation for changes in services (e.g., willingness to pay for cleaner water). Similarly, conjoint analysis explores people's preferences by asking them to rank different scenarios. Often, values derived in this way are seen as too subjective and too dependent on the details of the valuation approach to be informative.

Once any of these approaches has been used to estimate the value of an ecosystem service, benefits transfer can be used to adapt an existing ecosystem valuation to new contexts that lack data. This approach, with careful matching of contexts (see Plummer 2009) can be used to estimate the value of a service in one place based on the calculated value of that service in another place. Also, it can be used to approximate the type, number, and level of services provided in a new location. The second, broader application of this method has significant limitations and should be used with caution. This approach is simple and can be used when primary data collection is not feasible. However, given the limited set and scale of existing valuation studies, this approach must often be based on the assumption that most (or all) hectares of a given habitat type are of equal value. Information about the rarity, spatial configuration, spatial context, size, and quality of habitats, and about the numbers, practices, and values of nearby people can be difficult, if not impossible, to include. Both of these limitations can lead to problematic estimates of value. Also, this approach does not address marginal changes. It tallies up the values of services, but does not provide information about how those values might change with different management actions.

Measuring benefits in monetary terms using any of the above approaches has limitations. Some arise from a lack of data required to apply useful approaches, others from approaches that are inappropriate for particular contexts, and still others from the simple idea that some things just cannot be measured in monetary terms. Various types of biophysical data needed to understand the delivery of ecosystem services in a particular place or social data on the benefits of particular services to specific groups (all residents, vulnerable people, visitors, or society as a whole) often do not exist, making it difficult to use otherwise well-grounded approaches or enticing to use approaches inappropriately (e.g. benefits transfer without careful matching of the contexts). Finally, nonmonetary indicators of ecosystem benefits are better suited for intangibles such as spiritual and cultural values that many stakeholders feel cannot be represented appropriately in monetary terms. Interviews, surveys, and other analyses by social scientists can help to articulate the beliefs and values of individuals and groups and the benefits they derive from ecosystems. When outcomes of different management decisions that affect nature's benefits are not measured in a common currency, dialogue and explicit discussion that integrates the various metrics to inform decision making is critical.

In sum, estimating the monetary value of ecosystem services is practical, appropriate, and useful in some marine management applications and difficult or impossible in others. However, the ability (or lack thereof) to value all services in monetary terms in no way diminishes the utility of the ecosystem service framework to environmental decision making and as a framework for conveying people's links to and dependence on functioning ecosystems.

3. Mainstreaming Ecosystem Services into Decisions

Conducting ecosystem-based management approaches in marine systems requires the capacity to design, implement, and iterate a science-policy process in a defensible manner. To do this using ecosystem service accounting, managers and scientists need decision support and analytical tools to help them organize information and quantify changes in ecosystem service values as described above. A number of tools have been developed that can fill this need (see a summary in Center for Ocean Solutions 2011), but the utility of many of these in real-world decision-contexts in marine environments has yet to be demonstrated.

The Natural Capital Project (NatCap) (www.naturalcapitalproject.org), is an international partnership developing tools that make incorporating ecosystem service information into decisions easy and replicable and demonstrating the utility of these tools in various decision contexts around the world. A critical component of NatCap's work is the development and application of InVEST (Integrated Valuation of Environmental Services and Tradeoffs), a family of tools designed to help incorporate nature's benefits into natural resource decision making. We focus on this tool and its application to a variety of decision contexts in marine systems for the remainder of this chapter.

InVEST helps decision makers visualize the impacts of potential policies by modeling and mapping the delivery, distribution and, in some instances, economic value of ecosystem services under alternative scenarios (for more information see: Kareiva et al. 2011; Guerry et al. 2012b). At its core, InVEST employs a production function approach (equations for all models are available in Tallis et al. 2011). Model outputs can help identify tradeoffs and compatibilities among environmental, economic, and social benefits. InVEST is designed to be used in real decision making processes and to be applied at a range of scales from local to global. To facilitate its use in real decision making contexts, a key design feature of InVEST is the relative simplicity of models, availability of data required and the free availability of models on the web. The first step of a practical InVEST application usually involves working with stakeholders to identify critical management decisions and to develop scenarios that represent alternative future management choices and climate (Fig. 8.1). These serve as inputs for the models to project (quantitatively and/or spatially) how the provision of services might change in response. The outputs of these models provide decision makers with information about costs, benefits, tradeoffs, and synergies of alternative natural resource management decisions.

InVEST is currently available as a set of tools for the Geographic Information System software ArcGIS (ESRI 2011). However, to minimize barriers to its use, particularly in developing countries, we are also developing a growing number of modules on non-proprietary software. NatCap and partners are using ecosystem

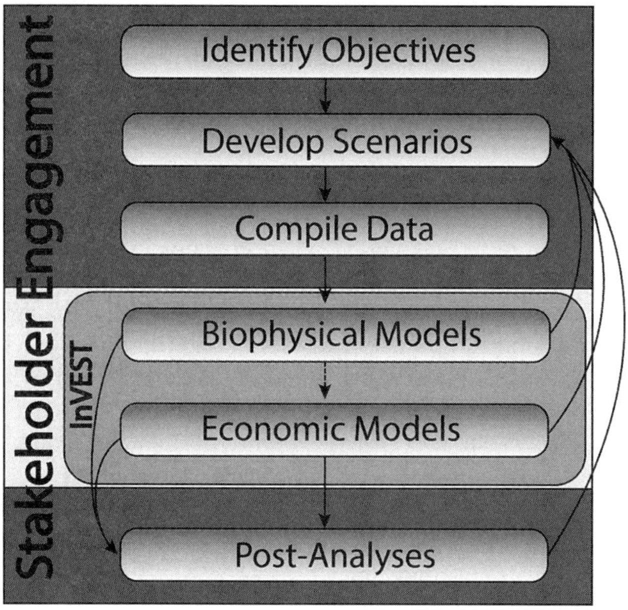

Fig. 8.1 – Using InVEST to inform natural resource decision making. To be most effective, InVEST modeling efforts need to be embedded in and in constant dialogue with stakeholders and decision makers. The identification of goals, development of alternative scenarios for consideration, and compiling of data are all important precursors to running the models. After running biophysical and economic models, outputs require some post-analysis to ensure the effective communication of results in metrics, maps, and graphs that resonate in the local context. Finally, results often spur the changes in goals, scenarios, and data; new iterations of models; and new results.

service approaches and the InVEST tool to inform natural resource decisions in diverse contexts around the world. Some examples include land-use planning in China, setting up payment for ecosystem service programs in the form of water funds in Latin America, and a number of marine applications. After a brief introduction to the marine models, we outline a few marine applications below and offer some lessons learned from employing the ecosystem services framework to inform management in marine systems.

3.1 Marine InVEST models

The current set of marine InVEST models (those released on-line or in advanced stages of development) includes: renewable energy (from waves and offshore wind), food from fisheries (numerous key commercial species) and aquaculture (finfish and shellfish), coastal protection, the provisioning of aesthetic views, recreation, and carbon storage and sequestration. Marine InVEST also includes two supporting service models that account for ecological linkages between the processes that generate changes in the ecosystem services listed above: water quality and habitat quality. Lessons from real applications feed directly into refinements of existing models and the development of new ones. Short descriptions of each model are available in Guerry et al (2012b), full documentation is available in Tallis et al. (2011), and additional information and the models themselves are available at www.naturalcapitalproject.org.

A critical component of marine InVEST is the ability to link service models to represent how a change in one service can impact delivery of other services. We achieve this through the water quality and habitat quality models (Fig. 8.2). These models reflect the importance of supporting services (MA 2005) as important endpoints in their own right for many constituents and also as links to the various final service models. For example, the aquaculture model can evaluate how operations of the aquaculture farming process (e.g. habitat alterations due to farm management) and climate change can affect the quantity and monetary value of farmed fish available for harvest. The model also produces outputs estimating the amount of waste generated or processed at the aquaculture facilities, as well as the propensity for pathogens (e.g. sea lice for salmon) to exist on the farm. Effects of pathogens and wastes or water filtration (by bivalves) can then be reflected in the water quality and habitat risk assessment models, and in turn those changes can be reflected in the fisheries and recreation models. It is important to note that we do not directly value changes in water quality or habitat quality. Changes in the values of these supporting or intermediate services are captured in changes in the values of the final services they support (e.g. fisheries, recreation). It is also important to note that this linking of service models is not seamless—one needs to take outputs from one model and use them as inputs to other models.

Marine InVEST is a deliberately simple suite of tools. Cognizant of the needs and timelines of real decisions, we strive to make the simplest models that capture the essence of the processes in question. This means that any one ecosystem service is likely to be better addressed by a locally built model or by a more detailed approach. For example, our coastal protection model can be used to explore the effects of nearshore vegetation (e.g. mangroves, coral reefs, marshes, seagrasses)

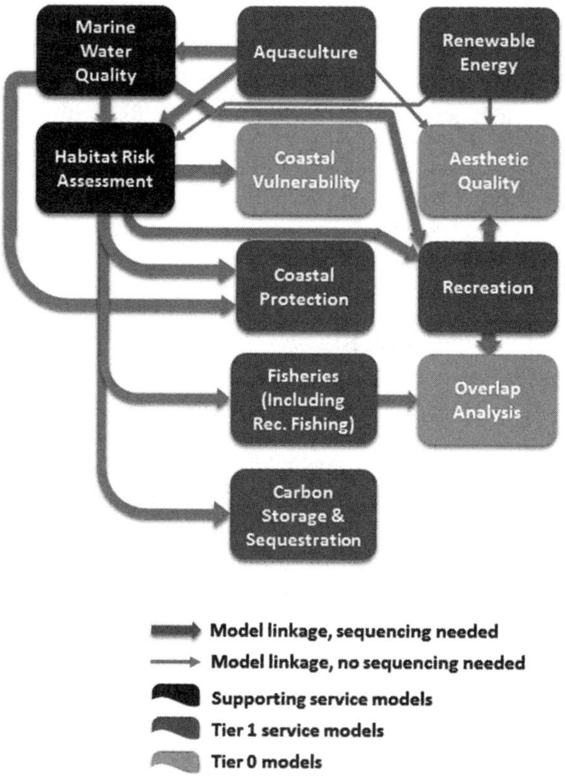

Fig. 8.2 – Marine InVEST models and their (optional) linkages. The supporting service models (shown in black) and the final ecosystem service models (shown in blue) can be directly affected by management and/or climate scenarios or they can be affected by the outputs of other models (possible interactions shown in red and grey arrows). To explore the effects of some scenarios (e.g. those that alter habitats), one can use outputs of one model (e.g. habitat risk assessment) as inputs to others (e.g. coastal protection and recreation). Models can be linked as shown, or used independently.

on erosion and flooding after a large storm and can compare that effect to that of a seawall or other structure. However, a detailed engineering study of the local area would be better equipped for determining design features of a seawall and choosing amongst particular detailed alternatives at a particular site. Similarly, current InVEST models do not model dynamic, complex interactions within the natural system or between the natural system and the human system with which it is coupled. While we use biophysical models to predict how changes in habitat lead to changes in the population of a key fished species, unless a foodweb model already exists for the region, InVEST cannot be used to explore how those changes might indirectly affect other species. Nor can it be used to explore how changes in the population of that fished species might lead to changes in fisheries, changes in demand, or changes in prices. This sort of detailed, bioeconomic modelling, with its roots in relatively simple, equilibrial models (e.g. Gordon 1954; Schaefer 1957), can dynamically explore how changes in harvest (timing, location, gear) affect the linkages between species and the impacts of harvest on other ecosystem services (e.g. Costello and Polasky 2008; Smith et al. 2008; Holland et al. 2010).

In recognition of the broad spectrum of models that can be used to inform decisions, InVEST has a tiered structure. Our simplest, or "Tier 0" models, map relative levels of services and can be used to highlight areas where particular services are in high demand. Our "Tier 1" models require more data than the "Tier 0" models, but still have relatively simple data requirements. These models give outputs in absolute terms and often provide the option for economic valuation. "Tier 2" models (many of which, for terrestrial and freshwater systems are described in Kareiva et al. (2011)), provide increasingly precise estimates of ecosystem services on increasingly resolved time scales. At NatCap, we do not intend to create new "Tier 3" models, but rather are working to develop the capability of InVEST to communicate with existing, complex models so that InVEST inputs (e.g. scenarios) can be fed in, and outputs from those complex models can be compared with other InVEST outputs. We expect users to mix and match tiers to create the best suite of models given past work, existing data, and the questions of interest.

We developed each of the InVEST models through an iterative process in the context of real policy processes. This co-development of models that have both academic and real-world policy application masters was critical to the generation of useful models. Also, early application to decision making demonstrates that while not yet easy, an ecosystem services framework can indeed inform ecosystem-based management in marine systems. In the remainder of this chapter, we briefly describe four of our engagements in decision contexts and highlight some lessons learned through our experience using ecosystem service approaches to inform decisions.

3.2 Marine Spatial Planning on the West Coast of Vancouver Island

The oceans continue to be viewed as a frontier, a place that has defied mapping of features and uses. Growing knowledge of the state of the oceans has emphasized the need for tools that allow comprehensive mapping and planning for the many human uses of the sea. Marine spatial plans attempt to do this kind of mapping. Current approaches are stymied by a lack of knowledge and tools that allow rapid, consistent assessment of how any given spatial plan will play out in terms of returns for major sectors and for society as a whole.

This challenge has been brought to the fore along the west coast of Vancouver Island, Canada. There, multiple, often competing interests are struggling to define the future character of the region. Existing extractive (timber, fishing, mining), industrial (wood products), and commercial uses (shipping, aquaculture); traditional First Nations subsistence and ceremonial uses; recreation and tourism; and emerging ocean uses such as the extraction of wave energy are critical pieces of the culture, society, and economy. The West Coast Aquatic Management Board (WCA) is working to create a marine spatial plan for the region. WCA is a public-private partnership with participation from four levels of government (Federal, Provincial, local, and First Nations), and diverse stakeholders. Ultimately, WCA's vision is to manage resources for the benefit of current and future generations of people and other species. More information is available at www.westcoastaquatic.ca.

Through the marine spatial planning process, WCA aims to:

1. Use a precautionary, ecosystem-based approach to protect, maintain, and restore marine and coastal resources.
2. Respect and protect First Nations' food, social and ceremonial requirements and treaty obligations.
3. Integrate expertise and knowledge from First Nations, local, scientific, and other sources; ensure broad participation in the planning process.
4. Foster initiatives that maintain or enhance opportunities for coastal communities to benefit from local resources, while achieving sustainable social, cultural, and economic benefits for the region.

In 2009, WCA began partnering with NatCap to use ecosystem service approaches and some InVEST models to help explore the effects of alternative spatial plans. Together, WCA and NatCap have created a number of spatially explicit scenarios—each representing alternative configurations and intensities of activities on the coast and in the ocean. Since marine spatial planning involves a diversity of decisions made by different industries and government agencies at a range of spatial scales, the NatCap/WCA team has needed to be flexible with the scale of analysis. In particular, we are developing scenarios at two distinct spatial scales (local and regional), each with a different mix of stakeholders and uses.

One of our first discoveries in this partnership was that the development of alternative scenarios for marine spatial planning is extremely difficult. Even gathering spatially explicit data about marine habitats and human uses of the marine environment was a significant challenge. At local scales, WCA and NatCap are collaborating with nine First Nations in Barkley and Clayoquot Sounds to develop a number of alternative management scenarios. The spatial scale of these scenarios matches the extent of each Nation's traditional territory. Thanks to considerable work with stakeholders on the part of WCA, these scenarios reflect the visions and values of each Nation and consist of zones for a range of human uses and activities. For example, in plans that are furthest along, the team identified zones to accommodate important income-generating activities (i.e. finfish farms) as well as zones for cultural and spiritual activities (i.e. 'culturally managed areas'). Each scenario represents an alternative arrangement of these zones in space (e.g., moving finfish farms from one inlet to another) or varying intensities or spatial extent of each activity (e.g., more or fewer finfish farms).

The InVEST habitat risk assessment (HRA) model has been a critical tool in the team's work on Vancouver Island. The HRA model provides a transparent and repeatable method for assessing the exposure of marine habitats to risk from human activities and the potential consequences of that exposure on habitats. When habitats have high exposure and significant consequences, the continued ability of that habitat to deliver a broad range of ecosystem services is likely impaired. Outputs include maps of risk "hotspots" where habitats' capacities to provide services are likely to be reduced. WCA is using the HRA model to compare alternative management scenarios and for use in marine spatial planning trade-off discussions.

In addition to the HRA, the team has used the aesthetic quality, aquaculture, recreation, and water quality models. A primary outcome of these analyses thus far is that we have been able to inform cross-sector dialogues by quantifying the effects of individual activities on the ability of other sectors and stakeholders to achieve their desired outcomes. The ecosystem services approach generally and

InVEST models in particular allowed us to explore alternative uses at various locations throughout the region. For example, in one instance, we were able to identify management options that increased some controversial uses (e.g. float homes (floating cabins with often untreated wastes) and shellfish harvest) while minimizing threats to benthic habitats (e.g. eelgrass) (Fig. 8.3). These results are helping to ease polarized planning discussions by encouraging those opposed to particular uses to consider that careful siting could reduce environmental impacts, making those uses more palatable to a wider range of stakeholders. While some of our results were intuitive to ecologists, our ability to generate quantitative outputs and to document connections between often separately considered activities has contributed to the dialogue and thus helped move the planning process forward (Guerry et al. 2012b).

Whereas the local-scale zoning maps primarily reflect the interests of the First Nations, a number of other sectors (e.g. commercial fisheries and aquaculture, shipping, tourism, Federal and Provincial government regulators) use the ocean at more regional scales (10–100km). To generate regional scale plans, the team is first 'stitching together' each of the small-scale traditional territory zoning maps to create regional-scale maps for each of Barkley and Clayoquot Sounds. Next, the team is poised to overlay the unified small-scale map with other ocean uses (e.g. shipping lanes, potential areas for wave energy conversion facilities) to create regional scale scenarios that represent the intersection of the interests of large-scale ocean users and the small-scale users in First Nations. The regional scenarios will be used to identify compatibilities and conflicts among multiple ocean uses. One example of early work at the regional scale is an analysis of the feasibility of potential wave energy conversion facilities and the ways in which economic returns from such facilities might best be balanced with minimal impacts on fishing and recreation (Kim et al. 2012).

The NatCap/WCA team is working to develop "soundwide metrics" that represent the baseline conditions (some measured directly, some modeled using InVEST). Example metrics include the number of user-days by tourists, the percent of the shoreline in highest categories of risk for coastal hazards, and the percent of community development areas that can see visual disamenities. These metrics will ultimately be used to compare the expected (and eventually realized) performance of alternative options for the marine spatial plan.

Ultimately WCA's Board of Directors, an entity with wide representation from stakeholder and sector groups, will review various scenarios. They will likely pay particular attention to conflicts and compatibilities with the activities of commercial sectors and Federal and Provincial government. WCA will then facilitate discussions among and within the sectors and the First Nations to try to resolve conflicts where possible, and come to a common understanding around the best marine spatial plan that balances different interests.

One critical lesson reinforced by our engagement in the marine spatial planning process on Vancouver Island is that model results and other scientific information can only be useful in the real world when carefully co-produced by scientists, decision makers, and the communities they represent in an iterative process. A lot of listening—and learning from things that fell flat—allowed the NatCap team to come to a better understanding of what approaches, models and metrics might best connect to the decisions on the table.

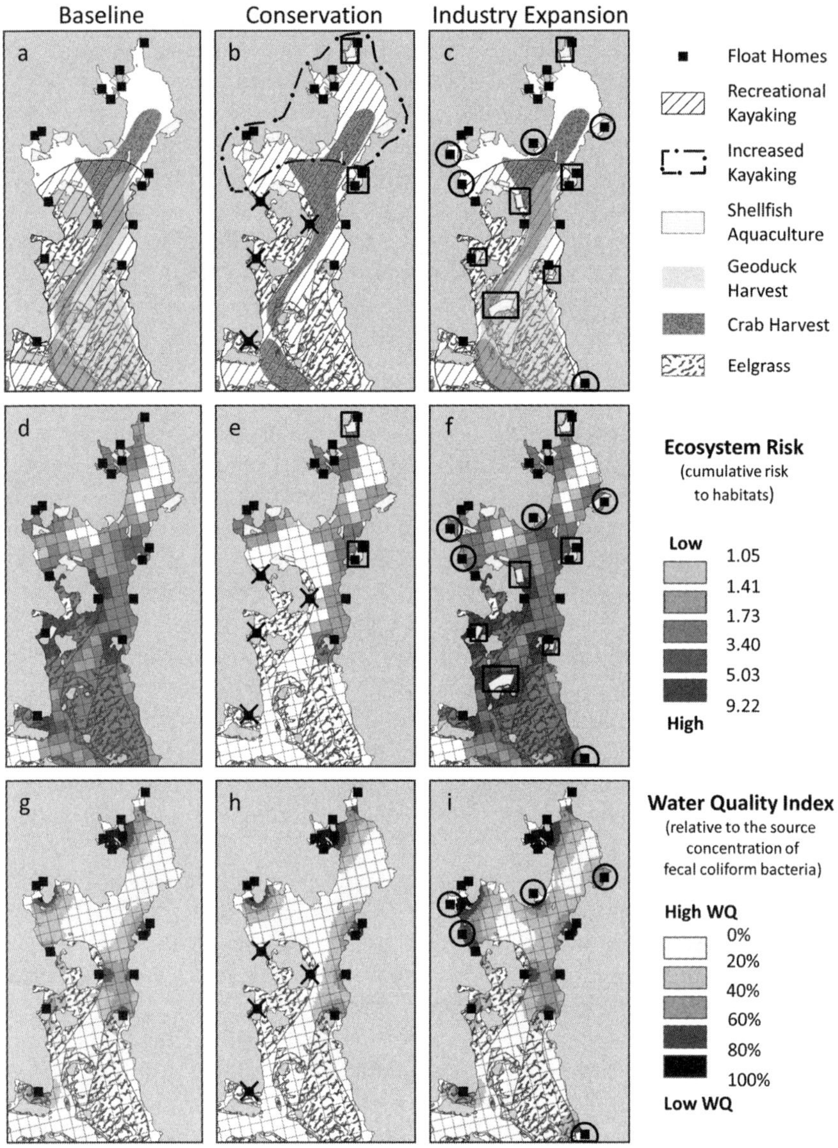

Fig. 8.3 – (See color insert) Three alternative management scenarios for Lemmens Inlet, BC identified by West Coast Aquatic (a–c) and some InVEST outputs (d–i). The three scenarios explored were: (a) Baseline (no changes to current uses or zones); (b) Conservation (zoning rules restrict float homes and aquaculture in areas near eelgrass beds—four float homes are removed from areas where they overlap with eelgrass (shown under black X's); two new oyster deepwater tenures are located outside of sensitive habitat zones (shown in black squares); kayaking routes are expanded into previously unused areas (shown in dashed line); and geoduck harvest is prohibited throughout the Inlet; (c) Industry expansion (five new float home leases are added, as shown in black circles; five new oyster tenures are added, as shown in black squares; and wild geoduck harvest is allowed). InVEST model outputs are shown in the remaining 6 panels. (d–f) Ecosystem risk, which is the cumulative risk of human activities (float homes, aquaculture, geoduck harvest) to nearshore habitats (eelgrass, soft, and hard bottom) under the three alternative management scenarios. (g–i) Concentration of fecal coliform bacteria relative to concentrations at float home sources under the three scenarios. (Reprinted from Guerry et al. 2012b, by permission of the publisher, Taylor & Francis Ltd., http://www.tandf.co.uk/journals)

On Vancouver Island, the general framework of ecosystem services and ecosystem service modeling, in particular, are helping to articulate connections between human activities that are often considered in isolation, to align diverse stakeholders around common goals, and to make implicit decisions explicit. Ecosystem service modeling results have informed early iterations of the local-scale marine spatial plans and will inform the creation of the final regional plans in 2012.

3.3 Coastal Zone Management in Belize

Belize is home to a dazzling array of marine life and coastal habitats, including the longest barrier reef in the western hemisphere. Belizeans depend on their ocean and coast as a critical source of food, fiber, inspiration, economic prosperity and cultural heritage. However, rapid development to support tourism, over-fishing, and population growth threaten local marine and coastal systems and the broad array of benefits they provide to Belizeans and their guests. To address this, the Belizean government established the Coastal Zone Management Authority and Institute (CZMAI). Its mandate is to develop an integrated management plan that guides sustainable economic growth while protecting the natural heritage that is the foundation for that growth. The ecosystem-based plan is to incorporate the interests of a broad coalition of stakeholders from Coastal Advisory Committees to government agencies and leaders in the private sector. One critical challenge that CZMAI faced was insufficient scientific capacity for assessing trade-offs among various marine uses and demonstrating potential win-wins where emphasis on the provisioning of one service yields benefits for another.

In 2010, CZMAI began partnering with NatCap and World Wildlife Fund (WWF) to support their effort to create a comprehensive, ecosystem-based national coastal zone management plan. The plan will recommend locations for coastal development and marine protected areas and lay out spatial and use guidelines for other types of activities.

In the initial phases of the partnership, the CZMAI/NatCap team collected and synthesized data from various sources to identify and map existing habitats, values, and uses of the coastal and marine region and to envision future ones. As was our experience on Vancouver Island, these were difficult but critical steps. After generating maps of current habitats and human uses, the team used historical trends, input from extensive regional meetings with stakeholders, and existing regional management and sustainable development plans to create three zoning options for the country's nine planning regions.

We then examined how these alternative scenarios would likely affect three critical benefits provided by the marine environment: (a) provision of commercial and artisanal fisheries for lobster, (b) flooding and erosion protection provided by mangroves, corals and seagrasses and (c) maintenance of tourism attractions such as snorkeling and diving. Using InVEST, the team mapped the delivery (e.g., the biomass of harvested fish, reduction in land flooded or eroded due to mangroves) and economic value (e.g., the net present value of harvest or avoided damages to property) of these services. In addition, the team used InVEST's habitat risk assessment model to quantify the potential loss of habitats based on the risk of human activities (e.g. coastal development) and sensitivities of those habitats (e.g. corals, mangroves, seagrasses) to the activities.

Through our analyses of different management scenarios, we have seen that combining biological and economic information can help to make clear the ways in which ecosystems support local livelihoods and culture. For example, in one region, model results revealed that a conservation-focused scenario led to more land protected from erosion and flooding and higher coastal protection value. In a middle-of-the-road scenario, loss of some habitat led to less land protected from coastal hazards, but habitat conservation targeted to the most critical habitats could minimze damages and improve the overall protective value of habitats. In addition, quantifying the value of coral, mangrove and seagrass habitat in supporting fisheries (Fig. 8.4) and recreation allowed stakeholders to explicitly see the contributions of these habitats to the provisioning of these critical benefits.

CZMAI staff presented the initial three zoning scenarios and quantitative model outputs to Coastal Advisory Committees and the public in a series of meetings in the nine planning regions throughout the country. These meetings served several key purposes including: learning to improve the communication of results

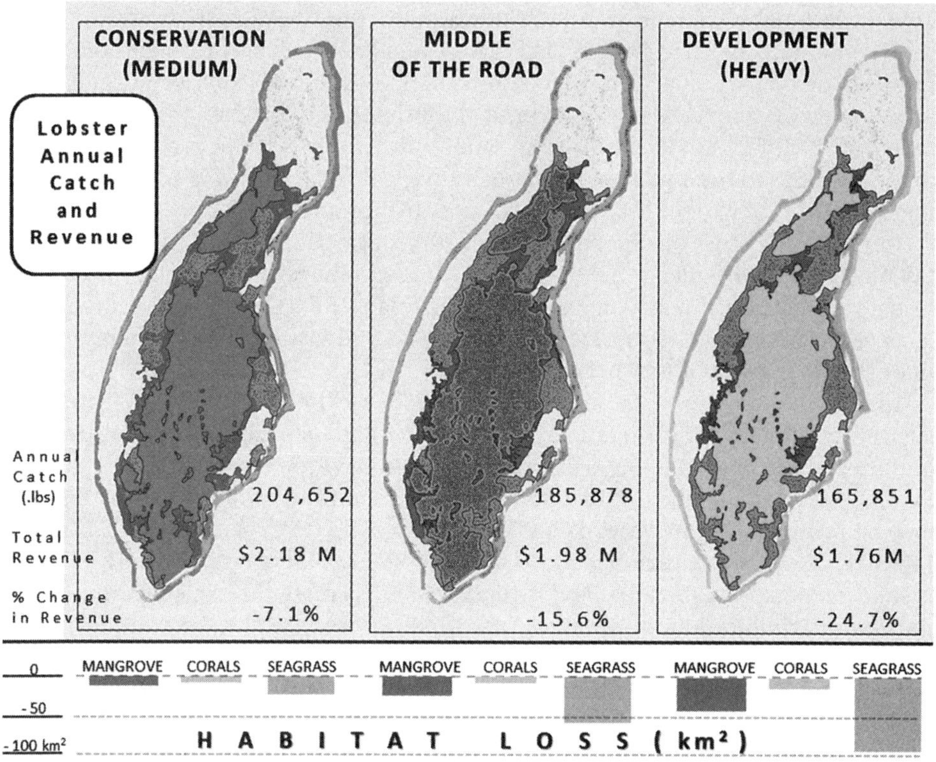

Fig. 8.4 – (See color insert) Outputs from the InVEST lobster fishery and habitat risk assessment model for Turneffe Atoll in Belize. The three panels show outputs from scenarios representing significant conservation efforts, a middle-of-the-road scenario, and a future with heavy coastal development. Darker pink colors indicate more harvest of lobster—with landings, total lobster revenue, and change from current revenue indicated. Colored bars along the bottom represent summarized losses, for each scenario, of each habitat from the habitat risk assessment model. These outputs from the habitat model served as inputs into the fishery model to produce differences in fishery yield.

to stakeholders, eliciting stakeholder feedback on how to iterate and refine the scenarios to better reflect local and regional realities, and, perhaps most importantly, using science to guide the discussion about what Belizeans want for their future.

The next steps with this work include developing a second iteration of the scenarios for the nine regions and combining the nine regions to create a national scale plan. At the time of writing, these are both works in progress. Over the next year, the team will review and improve the national coastal zone plan by incorporating feedback from stakeholders and from CZMAI's Advisory Council and Board and by iterating InVEST analyses. We will use this new knowledge to refine the recommendations for coastal zoning and permitting presented in the final plan. At the same time, we will help communicate the value and methods of the plan to diverse stakeholders. Ultimately, we will support CZMAI's presentation of the final plan to the Belizean Cabinet for approval.

3.4 Climate adaptation planning in California and Texas

In order to minimize the social and economic losses associated with climate change in coastal systems, local, regional and national governments are beginning to incorporate strategies for response and adaptation into their planning processes. The options include engineered solutions (e.g. seawalls), natural solutions (e.g. dune or marsh restoration and conservation) and planned retreat. Making informed decisions requires a systematic comparison of the costs and benefits of natural and engineered solutions. By strategically choosing alternatives that are cost-effective, plans can ensure the protection of people and property while also possibly protecting or restoring dwindling habitats and the full suite of benefits those habitats provide to people. Including climate adaptation considerations in existing planning and investment decisions is a new challenge, and the kind of information and ways in which it is best reported to be influential to decisions need to be tested.

We are testing the utility of these ideas and approaces in California and in Texas. In California, we have partnered with Stanford University's Center for Ocean Solutions and the Nature Conservancy's California state program to work with decision makers at three levels of governance in Monterey Bay. Climate adaptation planning is in its infancy in the region. When climate change is being considered in planning processes, it is often voluntary or the guidance is relatively broad and informal. Accordingly, there is a real opportunity to influence the ways in which decision makers approach this problem. In addition, adaptation decisions are currently dispersed among public and private entities, rather than concentrated in one or a few coastal planning processes focused singularly on adapting to climate change. Also, various state, regional and local bodies and initiatives are evaluating ways to include climate change in their planning processes. We are working at different scales (e.g., with county planners) to explore potential outcomes of management strategies aimed at addressing both their immediate mandates and also to illustrate how alternative strategies will affect system response to climate change.

Similar climate change and adaptation issues are arising in a very different social and ecological context in Galveston Bay, Texas. Despite the differences in setting and context, it is our hope that the same general approaches to including ecosystem services in climate adaptation decisions we are using in California can be successfully applied in Texas. There, we have partnered with the Nature Conservancy's Texas program to inform natural hazard mitigation and climate adaptation planning in the Galveston Bay region. Substantial habitat loss has occurred in the area as a result of coastal development, shoreline erosion, and land subsidence (e.g., the land surface has sunk as much as 10 feet since 1906 (Yoskowitz et al. 2009)). These shoreline changes make this area more vulnerable to sea-level rise, increased storm surge levels and inland inundation (Leatherman 1984). Indeed this region is among the most vulnerable in all of the US to projected sea-level rise (Weiss et al. 2011). In response to these predicted changes and the experience of several major hurricanes in recent decades, Galvestonians are identifying options to protect themselves from future risks. Some are considering a massive levee system aimed at protecting the ship channel and important industrial facilities in the region from future storms. However, green infrastructure options are also under consideration. One promising example is the proposed creation of the Lone Star Coastal National Recreation Area that would explicitly recognize not only the coastal protection benefits of natural areas, but also emphasize their critical role in providing opportunities for recreation and tourism and resulting economic benefits.

3.5 Terrestrial and marine linkages

Many scientific assessments of declining marine species and habitats along U.S. coasts point to upland and freshwater sources of imperilment (e.g. Rabalais and Turner 2001; Pew Oceans Commission 2003; US Commission on Ocean Policy 2004). And many have argued that the best hope for protecting marine resources for multiple uses is to consider larger-scale processes when designing management strategies (McLeod and Leslie 2009; Houde et al., this volume; Samhouri et al., this volume). Yet our scientific understanding of how land- and water-use practices affect outcomes for marine resources remains poor. Adding the effects of upland processes to the existing complexity of marine ecosystem models is a daunting prospect. However, an ecosystem service approach provides a clear and novel path forward—one that integrates ecological processes with socioeconomic behavior and values.

We are working to estimate and compare the influence of watershed practices on a suite of key marine and coastal ecosystem services, asking under what conditions including upland activities in management scenarios increases the likely effectiveness of marine management strategies. To this end, we have linked marine InVEST with its terrestrial and freshwater counterparts to create a set of watershed-marine models with ecosystem service outputs capable of evaluating management strategies on land and in the water. We are applying these models in three US locations: Puget Sound, Chesapeake Bay and Galveston Bay (see also Houde et al., this volume).

We are focusing on how changes in system function lead to changes in the provisioning of food from fisheries and aquaculture, recreation and coastal protection. Some watershed and marine management strategies under consideration include land- and water-use practices, fishery and aquaculture management, and protection and restoration of nearshore habitats such as eelgrass, oyster reefs, marshes and wetlands. We also use the linked InVEST models to evaluate the degree to which effects of climate change (e.g., changes in precipitation and sea surface temperature) exacerbate or mute changes in our focal ecosystem services and values. Initial results from Puget Sound's Hood Canal region show, in this area with limited scope for development in the foreseeable future, that plausible alternative future land-use and climate scenarios for the year 2030 have limited impacts on marine water quality and shellfish production (Toft et al. 2013).

Ecosystem-based management of marine systems is indeed a daunting task. Expanding the purview of the marine manager to include activities in coastal watersheds might seem virtually impossible to some. We are hopeful that one of the outcomes of this work exploring the role of terrestrial and marine linkages will be some rules of thumb that guide those responsible for the management of marine systems to those aspects of neighboring terrestrial systems that are most likely to affect the sustained delivery of services from marine systems.

4. Conclusions

Here, we have reviewed some of the services provided by marine environments, discussed potential audiences for information about how changes in marine ecosystems are likely to lead to changes in services, outlined approaches to modeling marine ecosystem services and described some examples of how these approaches can and are being used in real decision making. However, the road to adoption of ecosystem service accounting approaches in marine decisions is still long and winding. We highlight some of the key challenges below.

Modeling ecosystem services for natural resource decision-making requires strategic testing and validation. In order for models of ecosystem services to inform real decisions, they must simultaneously meet two, often competing objectives. First, they must be relatively simple to parameterize and use with existing or easy to obtain data. Second, they must be credible and transparent and include useful information about the uncertainty of model outcomes. With InVEST, we have worked hard to balance these two objectives, but significant work remains—particularly in the effective modeling and communication of uncertainty.

Modeling ecosystem services is no panacea. It cannot substitute for other ways of assessing marine and other ecosystems. Concerns about biodiversity of marine ecosystems, for example, may be poorly addressed by a pure ecosystem services approach. Some assert that protection and restoration of marine systems will improve both biodiversity and the ability of the oceans to provide ecosystem services; others point out that trade-offs can occur (Balvanera et al. 2006; Worm et al. 2006; Naeem et al. 2009; Palumbi et al. 2009; Klein et al. 2010; Brander 2010). Ecosystem modeling coupled with empirical observations of real systems continue to provide needed evidence for the conditions under which biodiversity and ecosystem service provision are positively related and where trade-offs are likely to occur.

Integrated multi-sector planning is hard and the perfect is the enemy of the good. Some realistically worry that coordinating multiple management bodies and stakeholder groups in one united negotiating and decision-making process is a practically impossible. However, experience is starting to show that even small steps towards integration at local scales or with single sectors can represent critical incremental progress. We highlighted the example on the west coast of Vancouver Island where local plans created by single entities (First Nations) are guiding the way towards a truly integrated, multi-sector regional plan. The same approach of starting small and simple helped the Great Barrier Reef Marine Protected Area process work up to a complex, multi-stakeholder result (e.g. Kenchington and Day 2011). Patience and learning will be required on all fronts for this challenging approach to become common practice. EBM is an iterative process and managers must not wait for the perfect, integrated multi-sector process to get started.

Monetary valuation is not the be-all and end-all of an ecosystem services approach. Although we are making significant progress in developing methods for revealing the monetary value of ecosystem services, there are many contexts where other metrics of value are more relevant and more elusive. Many political decisions, especially in the current economic climate, are made on the basis of job creation. Estimating the change in jobs related to a change in ecosystem services is still very challenging and seldom done. Other metrics, such as human health and equity, are often equally (if not more) important and are currently out of reach analytically. Incorporating a rich array of value currencies into ecosystem service accounting approaches—recognizing that those that are often most widely held are the non-market ones (e.g, cultural, spiritual and existence values)—has the potential to provide unique opportunities for using the EBM process to help diverse and often at-odds sectors find shared objectives.

Quantifying, mapping, and valuing marine ecosystem services have the potential to fundamentally change decision making in marine and coastal environments. Making explicit the connections between human activities in one sector and their effects on a broad range of other sectors leads people to think about whole ecosystems and to manage them accordingly. Ultimately, results from marine ecosystem service modeling can help society perceive the critical services oceans and coasts provide, appropriately value marine natural capital, and help human communities make better choices about the use of these life-sustaining environments.

Acknowledgments

We would like to thank all of our colleagues at the Natural Capital Project for helping to envision this work and to make it all happen. In particular, we are grateful to NatCap's founders (Gretchen Daily, Peter Kareiva, Taylor Ricketts, and Steve Polasky), NatCap's Managing Director (Mary Ruckelshaus) and the Marine Team (Greg Verutes, Jodie Toft, Katie Arkema, Greg Guannel, CK Kim, Spencer Wood, Joey Bernhardt, Jess Silver, Suzanne Langridge, Rob Griffin, and Joe Faries). Also, we would like to thank the Gordon and Betty Moore Foundation, The David and Lucile Packard Foundation, The Tides Foundation, The Summit Foundation, and NOAA/NSF's CAMEO program for funding the marine work described here.

References

Agardy, T., J. Alder, P. Dayton, S. Curran, A. Kitchingman, M. Wilson, A. Catenazzi, J. Restrepo, C. Birkeland, S. Blaber, S. Saifullah, G. Branch, D. Boersma, S. Nixon, P. Dugan, N. Davidson, and C. Vorosmarty. 2005. Coastal Systems. Assessment report chapter 19. *Millennium Ecosystem Assessment.*

Balvanera, P., A. Pfisterer, N. Buchmann, J. He, T. Nakashizuka, D. Raffaelli, and B. Schmid. 2006. Quantifying the evidence for biodiversity effects on ecosystem functioning and services. *Ecology Letters* 9:1146–1156.

Barbier, E., S. Hacker, C. Kennedy, E. Koch, A. Stier, and B. Silliman. 2011. The value of estuarine and coastal ecosystem services. *Ecological Monographs* 81:169–193.

Barbier, E., E. Koch, B. Siliman, S. Hacker, E. Wolanski, J. Primavera, E. Granke, S. Polasky, S. Aswani, L. Cramer, D. Stoms, C. Kennedy, D. Bael, C. Kappell, G. Perillo, and D. Reed. 2008. Coastal Ecosystem-Based Management with Nonlinear Ecological Functions and Values. *Science* 319:321–323.

Beattie, B. and C. R. Taylor. 1985. The Economics of Production. Wiley, New York.

Boyd, J. and S. Banzhaf. 2007. What are ecosystem services? The need for standardized environmental accounting units. *Ecological Economics* 63:616–626.

Brander, K. 2010. Reconciling biodiversity conservation and marine capture fisheries production. *Current Opinions in Environmental Sustainability* 2:416–421.

Brown, G. and R. Mendelsohn. 1983. The Hedonic Travel Cost Method. *Review of Economics and Statistics* 66:427–433.

Caddy, J. F. 1999. Fisheries management in the twenty-first century: Will new paradigms apply? *Reviews in Fish Biology and Fisheries* 9:1–43.

Center for Ocean Solutions. 2011. Decision-Guide: Selecting Decision-Suport Tools for Marine Spatial Planning. The Woods Institute for the Environment, Stanford University, Stanford, CA.

Chan, K. and M. Ruckelshaus. 2010. Characterizing changes in marine ecosystem services. *F1000 Biology Reports* 2:54–60.

Clark, C. 1990. *Mathematical Bioeconomics,* 2nd Edition. Wiley, New York.

Colling, A. 2001. *Ocean Circulation.* Butterworth Heineman and The Open University, Milton Keynes, England.

Constable, A. 2004. Managing fisheries effects on marine food webs in Antarctica: Trade-offs among harvest strategies, monitoring, and assessment in achieving conservation objectives. *Bulletin of Marine Science* 74:583–606.

Conte, M. 2012. Valuing Ecosystem Services.in S. Levin, editor. *Encyclopedia of Biodiversity.* Elsevier.

Costanza, R. 2000. The ecological, economic and social importance of the oceans.in C. R. C. Sheppard, editor. *Seas at the millenium: an environmental evaluation, vol. 3: global issues and processes.* Pergamon Press, New York.

Costanza, R., R. d'Arge, R. de Groot, S. Farber, M. Grasso, B. Hannon, K. Limburg, S. Naeem, R. V. O'Neill, J. Paruelo, R. G. Raskin, P. Sutton, and M. van den Belt. 1997. The value of the world's ecosystem services and natural capital. *Nature* 387:253–260.

Costello, C. and S. Polasky. 2008. Optimal harvesting of stochastic spatial resources. *Journal of Environmental Economics and Management* 56:1–18.

Daily, G. C., S. Alexander, P. R. Ehrlich, L. Goulder, L. Lubchenco, P. A. Matson, H. A. Mooney, S. Postel, S. H. Schneider, D. Tilman, and G. M. Woodwell. 1997. Ecosystem services: Benefits supplied to human societies by natural ecosystems. *Issues in Ecology* 2:1–16.

Danielsen, F., M. Sorensen, M. Olwig, V. Selvam, F. Parish, N. Burgess, T. Hiralshi, V. Karunagaran, M. Rasmussen, L. Hansen, A. Quarto, and N. Suryadiputra. 2005. The asian tsunami: a protective role for coastal vegetation. *Science* 370:643.

Das, S. and J. R. Vincent. 2009. Mangroves protected villages and reduced death toll during Indian super cyclone. *Proceedings of the National Academy of Sciences* 106:7357–7360.

Dasgupta, P. and G. Heal. 1979. *Economic Theory and Exhaustible Resources.* Cambridge Univ. Press, Cambridge, UK.

Dennison, W., T. Lookingbill, T. Carruthers, J. Hawkey, and S. Carter. 2007. An eye-opening approach to developing and communicating integrated environmental assessments. *Frontiers in Ecology and the Environment* 5:307–314.

Douvere, F. and C. Ehler. 2009. New perspectives on sea use management: Initial findings from European experience with marine spatial planning. *Journal of Environmental Management* 90:77–88.

Ehler, C. and F. Douvere. 2009. Marine spatial planning: a step-by-step approach toward ecosystem-based management. UNESCO, Paris.

Ehrlich, P. R. 1968. *The Population Bomb.* Sierra Club/Ballantine Books.

ESRI. 2011. ArcGIS Desktop: Release 10. Environmental Systems Research Institute, Redlands, CA.

European Commission. 2010. Maritime Spatial Planning in the EU: Achievements and Future Development. European Commission, Brussels, Belgium.

Foley, M., B. Halpern, F. Micheli, M. Armsby, M. Caldwell, C. Crain, E. Prahler, N. Rohr, D. Sivas, M. Beck, M. Carr, L. Crowder, J. E. Duffy, S. Hacker, K. McLeod, S. Palumbi, C. Peterson, H. Regan, M. Ruckelshaus, P. Sandifer, and R. S. Steneck. 2010. Guiding ecological principles for marine spatial planning. *Marine Policy* 34:955–966.

Freeman III, A. M. 1993. The Measurement of Environmental and Resource Values: Theory and Methods. Resources for the Future, Washington, DC.

Gleick, P. 1996. Water resources. Pages 817–823 in S. Schneider, editor. Encyclopedia of climate and weather. Oxford University Press, New York.

Gordon, H. S. 1954. The economic theory of a common-property resource: The fishery. The *Journal of Political Economy* 62:124–142.

Guerry, A., M. Plummer, M. Ruckelshaus, and C. Harvey. 2011. Ecosystem service assessments for marine conservation.in P. Kareiva, H. Tallis, T. Ricketts, G. Daily, and S. Polasky, editors. *Natural Capital: Theory and Practice of Mapping Ecosystem Services.* Oxford University Press.

Guerry, A., M. Plummer, M. Ruckelshaus, and D. S. Holland. 2012a. Modeling Marine Ecosystem Services. In S. Levin, editor. *Encyclopedia of Biodiversity.* Elsevier.

Guerry, A., M. Ruckelshaus, K. Arkema, J. Bernhardt, G. Guannel, C. K. Kim, M. Marsik, M. Papenfus, J. Toft, G. Verutes, S. Wood, M. Beck, F. Chan, G. Gelfenbaum, B. Gold, B. Halpern, W. Labiosa, S. Lester, P. Levin, M. McField, M. Pinsky, M. Plummer, S. Polasky, P. Ruggiero, D. Sutherland, H. Tallis, A. Day, and J. Spencer. 2012b. Modeling benefits from nature: using ecosystem services to inform coastal and marine spatial planning. *International Journal of Biodiversity Science, Ecosystem Services and Management* 8:107–121.

Holdren, J. P. and P. R. Ehrlich. 1974. Human population and the global environment. *American Scientist* 62:282–292.

Holland, D. S., J. Sanchirico, R. J. Johnston, and D. P. Joglekar. 2010. Economic Analysis for Ecosystem Based Management: Applications to Marine and Coastal Environments. Resources for the Future, Washington DC.

Iceland, C., C. Hanson, and C. Lewis. 2008. Identifying important ecosystem goods and services in Puget Sound. World Resources Institute.

Inter-agency Ocean Policy Task Force (IOPTF). 2010. Final Recommendations of the Interagency Ocean Policy Task Force., Council on Environmental Quality, Washington, DC.

Kareiva, P., H. Tallis, T. Ricketts, G. Daily, and S. Polasky, editors. 2011. *Natural Capital: Theory and Practice of Mapping Ecosystem Services.* Oxford University Press, Cambridge.

Kenchington, R. and J. Day. 2011. Zoning, a fundamental cornerstone of effective Marine Spatial Planning: lessons learnt from the Great Barrier Reef, Australia. *Journal of Coastal Conservation* 15:271–278.

Kim, C. K., J. E. Toft, M. Papenfus, G. Verutes, A. D. Guerry, M. H. Ruckelshaus, K. K. Arkema, G. Guannel, S. A. Wood, J. R. Bernhardt, H. Tallis, M. L. Plummer, B. S. Halpern, M. L. Pinsky, M. W. Beck, F. Chan, K. M. A. Chan, P. S. Levin, and S. Polasky. 2012. Catching the right wave: Evaluating wave energy resources and potential compatibility within existing marine and coastal uses. *PLoS ONE* 7(11).

Klein, C., N. Ban, B. Halpern, M. Beger, E. Game, H. Grantham, A. Green, T. Klein, S. Kinninmonth, E. Treml, K. Wilson, and H. Possingham. 2010. Prioritizing land and sea conservation investments to protect coral reefs. *PLoS One*:5: e12431.

Krutilla, J. V. 1967. Conservation reconsidered. The American Economic Review 57:777–786.

Krutilla, J. V. and A. C. Fisher. 1975. *The Economics of Natural Environments: Studies in the Valuation of Commodity and Amenity Resources.* Johns Hopkins University Press, Baltimore, MD.

Leatherman, S. R. 1984. Coastal Geomorphic Responses to Sea Level Rise: Galveston Bay, Texas. Page 238 in M. C. Barth, and Titus, J. G., editor. *Greenhouse Effect and Sea Level Rise: A Challenge for this Generation.* Van Nostrand Reinhold Company Inc., Washington, DC.

Leggett, C. and N. Bockstael. 2000. Evidence of the effects of water quality on residential land prices. *Journal of Environmental Economics and Management* 39:121–144.

Leopold, A. 1949. *A Sand County Almanac and Sketches Here and There.* Oxford University press.

Levin, P. S., M. J. Fogarty, S. A. Murawski, and D. Fluharty. 2009. Integrated ecosystem assessments: Developing the scientific basis for ecosystem-based management of the ocean. *PLoS Biology* 7:23–28.

Levin, S. 1998. Ecosystems and the biosphere as complex adaptive systems. Ecosystems 1.

Levin, S. and J. Lubchenco. 2008. Resilience, Robustness, and Marine Ecosystem-based Management. *BioScience* 58:1–6.

Liu, J., T. Dietz, S. R. Carpenter, M. Alberti, C. Folke, E. Moran, A. N. Pell, P. Deadman, T. Kratz, J. Lubchenco, E. Ostrom, Z. Ouyang, W. Provencher, C. Redman, S. H. Schneider, and W. W. Taylor. 2007a. Complexity of Coupled Human and Natural Systems. *Science* 317:1513–1516.

Liu, J., T. Dietz, S. R. Carpenter, C. Folke, M. Alberti, C. L. Redman, S. H. Schneider, E. Ostrom, A. N. Pell, J. Lubchenco, W. W. Taylor, Z. Ouyang, P. Deadman, T. Kratz, and W. Provencher. 2007b. Coupled Human and Natural Systems. *Ambio* 36:639–648.

Liu, S., R. Costanza, S. Farber, and A. Troy. 2010. Valuing ecosystem services; Theory, practice, and the need for a transdisciplinary synthesis. *Annals of the New York Academy of Sciences* 1185:54–78.

Marsh, G. P. 1864. *Man and Nature.* Scribner.

McLeod, K. and H. Leslie, editors. 2009. *Ecosystem-based Management for the Oceans.* Island Press, Washington.

Melillo, J. M., A. D. Mcguire, D. W. Kicklighter, B. Moore, C. J. Vorosmarty, and A. L. Schloss. 1993. Global Climate-Change and Terrestrial Net Primary Production. *Nature* 363:234–240.

Millennium Ecosystem Assessment. 2005. Ecosystems and human well-being: synthesis. Island Press.

Monterey Bay Sanctuary Foundation and the Association of Monterey Bay Area Governments. 2009. Southern Monterey Bay: Evaluation of erosion and mitigation alternatives.

Murawski, S. A., J. Steele, P. Taylor, M. J. Fogarty, M. P. Sissenwine, M. Ford, and C. Suchman. 2009. Why Compare Marine Ecosystems? *ICES Journal of Marine Science* 67:1–9.

Naeem, S., D. Bunker, and A. Hector. 2009. *Biodiversity, Ecosystem Functioning and Human Wellbeing: An Ecological and Economic Perspective.* Oxford University Press, New York, NY.

National Research Council. 2005. *Valuing Ecosystem Services: Toward Better Environmental Decision-Making.* National Academies Press, Washington, DC.

National Research Council. 2011. *Approaches for Ecosystem Services Valuation for the Gulf of Mexico After the Deepwater Horizon Oil Spill: Interim Report.* National Academies Press, Washington, DC.

Osborn, F. 1948. *Our Plundered Planet.* Little, Brown, and Co., Boston.

Ostrom, E. 2009. A general framework for analyzing sustainability of social-ecological systems. *Science* 325:419–422.

Palumbi, S., P. Sandifer, J. Allan, M. Beck, D. Fautin, M. Fogarty, B. Halpern, L. Incze, J.-A. Leong, E. Norse, J. S. JJ, and D. W. DH. 2009. Managing for ocean biodiversity to sustain marine ecosystem services. *Frontiers in Ecology and the Environment* 7:204–211.

Patterson, M. and B. Glavovic, editors. 2008. Ecological economics of the oceans and coasts. Edward Elgar, Cheltenham.

Peterson, C. and J. Lubchenco. 1997. Marine Ecosystem Services.in G. Daily, editor. *Nature's Services.* Island Press, Washington, DC.

Pew Oceans Commission. 2003. America's living oceans: charting a course for sea change. A report to the nation. Pew Oceans Commission, Arlington, VA.

Plummer, M. L. 2009. Assessing benefit transfer for the valuation of ecosystem services. *Frontiers in Ecology and the Environment* 7:38–45.

Rabalais, N. N. and R. E. Turner. 2001. Hypoxia in the northern Gulf of Mexico: description, causes and change. In N. N. Rabalais and R. T. Turner, editors. *Coastal Hypoxia: Consequences for Living Resources and Ecosystems.* American Geophysical Union, Washington, DC.

Ricketts, T., G. Daily, P. R. Ehrlich, and C. Michener. 2004. Economic value of tropical forest to coffee production. *Proceedings of the National Academy of Sciences of the United States of America* 101:12579–12582.

Sainsbury, K. J., A. E. Punt, and A. Smith. 2000. Design of operational management strategies for achieving fishery ecosystem objectives. *ICES J. of Marine Science* 57:731–741.

Samhouri, J., P. Levin, C. A. James, J. Kershner, and G. Williams. 2011. Using existing scientific capacity to set targets for ecosystem-based management: A Puget Sound case study. *Marine Policy* 35:508–518.

Sathirathai, S. and E. B. Barbier. 2001. Valuing mangrove conservation in southern Thailand. *Contemporary Economic Policy* 19:109–122.

Schaefer, M. B. 1957. Some considerations of population dynamics and economics in relation to the management of marine fishes. *Journal of the Fisheries Research Board of Canada* 14:669–681.

Schlesinger, W. 1991. Biogeochemistry: An analysis of global change. Academic Press.

Smith, A., E. Fulton, A. Hobday, D. Smith, and P. Shoulder. 2007. Scientific tools to support the practical implementation of ecosystem-based fisheries management. *ICES Journal of Marine Science* 64:633–639.

Smith, M. D., J. Zhang, and F. C. Coleman. 2008. Econometric modeling of fisheries with complex life histories: avoiding biological management failures. *Journal of Environmental Economics and Management* 55:265–280.

Stokstad, E. 2011. Appraising UK's ecosystems, Report envisions greener horizons. *Science* 332:1139.

Tallis, H., S. Lester, M. Ruckelshaus, M. Plummer, K. McLeod, A. Guerry, S. Andelman, M. Caldwell, M. Conte, S. Copps, R. Fujita, S. Gaines, G. Gelfenbaum, P. Kareiva, M. Papenfus, B. Silliman, L. Wainger, and C. White. 2012. New metrics for managing and sustaining the ocean's bounty. *Marine Policy* 36:303–306.

Tallis, H., P. S. Levin, M. Ruckelshaus, S. E. Lester, K. L. McLeod, D. L. Fluharty, and B. S. Halpern. 2010. The many faces of ecosystem-based management: Making the process work today in real places. *Marine Policy* 34:340–348.

Tallis, H., T. Ricketts, A. D. Guerry, E. Nelson, D. Ennaanay, S. Wolny, N. Olwero, K. Vigerstol, D. Pennington, G. Mendoza, J. Aukema, J. Foster, J. Forrest, D. Cameron, E. Lonsdorf, C. Kennedy, G. Verutes, C. K. Kim, G. Guannel, M. Papenfus, J. Toft, K. Arkema, M. Marsik, J. Bernhardt, S. Wood, and R. Sharp. 2011. InVEST 2.1 beta User's Guide. The Natural Capital Project, Stanford.

Toft, J. E., J. L. Burke, M. P. Carey, C. K. Kim, M. Marsik, D. A. Sutherland, K. K. Arkema, A. D. Guerry, P. S. Levin, T. J. Minello, M. Plummer, M. H. Ruckelshaus, and H. M. Townsend. From mountains to sound: modelling the sensitivity of Dungeness crab and Pacific oyster to land–sea interactions in Hood Canal, WA. *ICES Journal of Marine Science,* doi:10.1093/icesjms/fst072.

UK National Ecosystem Assessment. 2011. The UK National Ecosystem Assessment: Synthesis of the Key Findings. UNEP-WCMC, Cambridge.

UN General Assembly (UNGA). 2010. Sustainable development: Report of the Governing Council of the United Nations Environment Programme on its eleventh special session.

UNEP and IOC-UNESCO. 2009. An Assessment of Assessments, Findings of the Group of Experts. Start Up Phase of a Regular Process for Global Reporting and Assessment of the State of the Marine Environment including Socio-Economic Aspects.

United Nations Environment Program. 2006. Marine and coastal ecosystems and human wellbeing: A synthesis report based on the findings of the Millennium Ecosystem Assessment. United Nations Environment Program.

US Commission on Ocean Policy. 2004. An ocean blue-print for the 21st century. US Commission on Ocean Policy, Washington DC.

Vogt, W. 1948. *Road to Survival.* William Sloan Associates, New York.

Weiss, J., J. Overpeck, and B. Strauss. 2011. Implications of recent sea level rise science for low-elevation areas in coastal cities of the coterminous U.S.A. *Climatic Change* 105:635–645.

Wilson, M. A. and S. Liu. 2008. Evaluating the nonmarket value of ecosystem goods and services provided by coastal and nearshore marine system. In M. G. Patterson and B. C. Glavovic, editors. *The ecological economics of the oceans and coasts.* Edward Elgar, Cheltenham.

Worm, B., E. Barbier, N. Beaumont, J. Duffy, C. Folke, B. Halpern, J. Jackson, H. Lotze, F. Micheli, S. Palumbi, E. Sala, K. Selkoe, J. Stachowicz, and R. Watson. 2006. Impacts of biodiversity loss on ocean ecosystem services. *Science* 314:787–790.

Yoskowitz, D. W., J. Gibeaut, and A. McKenzie. 2009. The Socio-Economic Impact of Sea Level Rise in the Galveston Bay Region. Harte Research Institute of Gulf of Mexico Studies Texas A&M University-Corpus Christi.

Chapter 9. An Ecosystem Accounting Framework for Marine Ecosystem-Based Management

IRIT ALTMAN

Department of Biology,
Boston University

ROEL BOUMANS

AFORDablefutures LLC

JOE ROMAN

Gund Institute for Ecological Economics,
University of Vermont

SUCHI GOPAL

Department of Geography & Environment,
Boston University

LES KAUFMAN

Boston University Marine Program and
Conservation International

Contents

1. Introduction

At the heart of ecosystem-based management lies the notion that ecological patterns and processes are essential to human populations and that people, in turn, strongly influence the natural world. The idea is not a new one. George Perkins Marsh was likely the first to introduce the concept of linked systems in a formal and comprehensive way through his book *Man and Nature* (1864). This seminal account ushered in a new era in resource management and environmental thinking—Marsh's writings led the American Association for the Advancement of Science to petition the government for a national forestry commission in 1873

The Sea, Volume 16, edited by Michael J. Fogarty and James J. McCarthy
ISBN 978-0-674-07270-1

(Lowenthal 2000) and laid the foundation for more recent scientific work seeking to understand the effect of humans on Earth (Thomas 1956; Turner et al. 1990; Vitousek et al. 1997). While the philosophical foundations of linked social-ecological systems (SES, also known as coupled human and natural systems) have long been in place, the broad synthesis required for a formal, disciplined science on the topic has only recently emerged. Important advances in the field have been made (McPeak et al. 2006; Liu et al. 2007; Ostrom 2007); however, a deeper theoretical and operational understanding is urgently needed to address the large-scale, persistent, and varied threats to ecosystems and the services they provide.

In marine environments, the need for ecosystem-based approaches to management, and hence for a better understanding of the social-ecological interactions, is great. Here, challenges of EBM design and implementation are compounded by the prevalence of common-pool resources, the inherent complexity associated with ocean environments (Worm et al. 2006), and the historic and present day overuse and abuse of marine resources (Lotze et al. 2006; Halpern et al. 2008). This is a serious problem. Nearly half of the world's human population lives near the coast and depends directly upon the services that marine systems provide. The processes that generate these services are connected and the means by which they are accessed are often enmeshed in conflicts and tradeoffs. Only through a whole-system perspective can we gain a rational understanding of these issues and identify sustainable solutions (Ruckelshaus et al. 2008; Lubchenco and Sutley 2010). For example, traditional species and sector-based fisheries management have often not performed well. While this is due in part to a tendency to avoid following the advice of experienced watermen and scientists whenever it calls for restraint, part of the blame must also rest with the naïve operating assumptions that fish stocks operate in equilibrium and in isolation from ecosystem processes as well as each other (Caddy 1999; Sugihara et al. 2011).

Marine EBM faces three principal needs owing to deficits both in data and theory. First, we must work toward deeper understanding of the hierarchical complexity of the linked natural-human dynamics that characterize specific marine ecosystems. This understanding will directly inform the design and implementation of EBM in specific locations. Second, we must develop a generalized framework that supports the task above and at the same time allows for direct comparisons among individual systems that vary widely in their scale, character, and behavior. Only with such a framework in place can we identify the interplay between unique features of an ecosystem and their commonalities. Such a framework will support the emergence of broad theoretical understanding. Lastly, tools are needed that help translate the complexity of SES across disciplines and to a diversity of stakeholders in support of ecosystem-based decision-making.

The first two challenges described above can be addressed through the creation of a model that coalesces whole-system knowledge and helps yield a transportable theory of linked systems. We assume that SESs will prove to be lawful to some degree: that is, if we study and compare enough of them, broadly applicable principles will emerge to support decision-making toward consistently favorable outcomes. The third challenge, to communicate and grow understanding, requires the translation of model behavior to a larger stakeholder community in a way that reveals the resource trajectories associated with alternative decision pathways.

In this chapter, we introduce the concept of ecosystem accounting and outline how the approach can be used to advance EBM in association with the critical needs described above. Ecosystem accounting provides a structure for organizing and integrating information about how natural and human elements of an ecosystem interact. Using an accounting framework to detail interactions of a specific marine ecosystem, we can begin to characterize the ecosystem service flows, evaluate tradeoffs, and compare projected outcomes of alterative management strategies. An ecosystem accounting approach can also be made spatially explicit to inform marine spatial planning (for more on MSP, see Rice et al., this volume). We describe the use of the Multiscale Integrated Model of Ecosystem Services (MIMES) framework for doing ecosystem accounting and the Marine Integrated Decision Analysis System (MIDAS) for decision support, visualization, and communication with stakeholders. Along the way we discuss how these tools complement other ways of incorporating natural capital perspectives into marine EBM (for one example, see Guerry and Tallis, this volume). Finally, as a concrete example we describe the application of MIMES-MIDAS to a case study for Massachusetts Bay involving competing resource objectives. Our goal for this case study was to inform EBM planning in the region by evaluating tradeoffs among a set of ecosystem services that support a variety of commercial fishing sectors, offshore wind energy installations, and ecotourism (whale watching) in a spatially and temporally explicit manner.

2. Ecosystem Accounting

2.1 Background

Ecosystem (or environmental) accounting (EA) provides a general framework to understand the interplay of natural and human systems in terms of both natural and human currencies (species as well as specie). The approach hinges on incorporating measures of natural capital into a traditional economic accounting framework, the goal being to provide a more comprehensive evaluation of wealth and well-being and to help answer crucial policy questions regarding sustainability. The need for such an approach was identified during the first United Nations Conference on Environment and Development (UNCED) in 1992:

> A first step towards the integration of sustainability into economic management is the establishment of better measurement of the crucial role of the environment as a source of natural capital and as a sink for by-products generated during the production of man-made capital and other human activities. As sustainable development encompasses social, economic and environmental dimensions, it is also important that national accounting procedures are not restricted to measuring the production of goods and services that are conventionally remunerated (UNCED 1992; Chapter 8).

The approach was further developed by the United Nations System of Environmental-Economic Accounts (SEEA) as well as through the European Environmental Agency (EEA). Ecosystem accounting currently provides a robust conceptual framework for tracking changes in the quantity and quality of ecosystem services over time and guiding effective action on a range of environmental problems (SEEA 2003; EEA 2010). Whereas some assume that a standardized EA approach can be developed, others argue that regional idiosyncrasies necessitate an individualized, step-by-step path in accordance with the needs of a particular system (Maler et al. 2008, 2009). Part of this tension is because EA is still a relatively new practice and the theoretical basis that could support a generalizable framework is still being developed.

Ecosystem accounting expands on traditional approaches by recognizing a reinvestment cost for both human-built *and* natural capital. For the latter, reinvestment takes the form of maintenance and restoration of natural systems degraded through human activities. Recent financial crises in the U.S. and Europe provide dramatic and cautionary examples of how insufficient accounting practices can undermine the integrity of large-scale systems. One factor contributing to the recent crises was "hidden" debt that, by various means, was not adequately considered in accounting activities (Reinhart and Rogoff 2011) such as, for example, the aggregation of subprime lending debts into mortgage backed securities. Over time, the gap between real economic flows and those accounted for on the books widened, compromising the integrity of economic links and eventually leading to a broad-scale collapse of the markets. From an environmental standpoint, failing to account for the debts owed to natural systems, which become degraded as a result of human activity, means risking the collapse of production flows that humans depend on.

2.2 *Making ecosystem accounting operational*

In the work described here, we rely on the principles of EA and expand the approach to consider ecosystem service flows across both time and space. Such an expansion is necessary to realize fully the linkages and feedbacks that occur in actual social-ecological systems and to inform the policies that promote sustainability, including EBM. To better understand why this is true, let us think of humans' give-and-take with nature in terms of resource withdrawals, as from a bank, and environmental impacts as activities that deplete the account without providing any benefits. When considering a decision about removing funds, it may be simpler to think of this financial resource in terms of its static state, that is, how much is available now as a result of the past history of debts and credits. In reality, however, the state of both resources and impacts are transient features of an ongoing *process*. Thus, their accurate characterization is in terms of flows that occur over time and space. Extending this example to forms of natural capital, fish represent a resource or state, and provisioning and renewal of this resource represent an ecosystem service process or flow. If we are interested in having fish around for future human generations—in the form of a commercial service, supporting service, or otherwise—we don't just want to know how many fish are left to catch today, we also want to understand where, when, and how fish populations grow, regenerate, and respond to episodic disturbances or interventions, such as the institution of a new set of policies. Consequently, the accounting framework described here is both dynamic and spatially explicit, yielding predictions about how ecosystem service flows will respond to changing conditions resulting from management actions or other factors (e.g., climate change).

Naturally, the first step of the EA approach is to define the boundaries of a focal ecosystem. Because ecosystems are composed of nested units, boundary delineation for accounting is no easy task but must be based upon what is ecologically justifiable and workable (Rosenberg 2006). Whereas some aspects of delineating the boundary will always be a bit arbitrary, we can acknowledge that the area of greatest interest is not a closed system by identifying forms of exchange between it and adjacent areas. Within the bounded area, the ecosystem (including its human dimensions) is described in terms of three types of information: (1) that which gives an ecosystem its characteristic features and function (Lester et al. 2010); (2) processes germane to priority management issues (Gregr and Chan 2011); and (3) information that places the ecosystem in a broader, comparative context. These elements provide the foundation for the development of spatial management strategies.

Ecosystem accounting is initiated by gathering and organizing spatially explicit data that paint a multilayered portrait of an ecosystem, its geography and anatomy. By placing this information into a common GIS framework and using simple map overlays, we can begin to show how the distribution and intensities of various ecosystem elements intersect, including selected features of physical regimes, biological habitats, and human activities. Combining these snapshots with basic GIS techniques, it is a relatively simple task to determine areas of overlap and begin identifying ways of minimizing incompatible uses. This is an excellent starting point for ecosystem-based marine spatial planning, but it is important to keep in mind that GIS data by itself lacks information about the functional relationships that occur through time and space. A process-based understanding of these dynamics is critical to the development of EBM policies that actually protect the sustained flow of the full range of ecosystem services.

With initial spatial data in place, the next step is to define the production functions that generate potential ecosystem services in the system (this is well described in Guerry and Tallis, this volume). There is a rich literature on ecosystem services, their measurement in time and space, and their relationships in an ecological landscape (Tallis et al. 2008; Granek et al. 2010; Raudsepp-Hearne et al. 2010). Moreover, because production functions derive from basic principles that are general and transportable, we should be able to draw on understanding gained from similar systems. Both theory and on-the-ground information will guide the elucidation of production functions; however, we should also recognize that the science of ecosystem services is still in its infancy and there is much to be learned about how these flows arise and are maintained over time and space. Given this uncertain state of understanding, EA contributes an essential framework for formalizing assumptions and posing linked sets of hypotheses about ecosystem components and their interactions. The framework helps guide our thinking about the mechanistic relationships that underlie ecosystem service flows and identify the most promising working hypotheses that should become the focus of empirical research activities in the future.

Of course, there is a critical need to link our understanding of ecosystem service flows to policies that ensure their sustainable delivery (Daily et al. 2009). To this end we can use an EA model to explore the long-term outcomes of alternative management scenarios focusing on ecosystem service production flows. At the scenario development and analysis stage the fact that any one issue is attached to many others can become especially clear and this can cause tension as we seek to maintain efficiency and focus. Consequently there is a danger at this step in adopting too narrow a view. The ultimate goal should be to build a multifaceted understanding of the coupled system. This can be achieved with spare resources by making the exercise of adaptive management a cumulative one, with all information going into a common analytical framework, or "model", that builds in data and process richness over time while being continually vetted in all its elements.

The steps described above—defining the boundaries of an SES, pulling together and making relational a wealth of information to characterize a system, mapping production functions, defining alternative scenarios—all contribute to a synthesis of knowledge that can inform a common understanding. Ostrom et al. (1999) identified the need for "Users who depend on a resource . . . to share an image of how the resource system operates and how their actions affect each other and the resource" as a key element in participatory decision-making. An EA framework can serve as the nexus for this synthesis; however, the resulting model, no matter how complex or realistic, is by itself insufficient for coalescing a common vision (Dietz et al. 2003; Vervoort et al. 2010). Additional tools are necessary to support decision-making and help communicate the model's inner workings and insights. Tools must also demonstrate how resources and human activities are projected to change over time to bring questions of long-term sustainability to the forefront (Ostrom et al. 1999). Moreover, the design of visualization tools must resonate with a large and diverse stakeholder audience. This is especially true in marine systems, which are often large, complex, and associated with a very high diversity of user groups.

2.3 *Models to support ecosystem accounting*

EA is essentially an exercise in model construction, since it requires that we formally describe the functional relationships among ecosystem components. The type of model developed is highly dependent on the level of understanding available and the capacity in which the model will serve. Here, we can imagine a continuum of knowledge about how a particular system of interest works. At one end, we might have a very good on-the-ground understanding of a system that has been well studied over time. In this case, a model formulated within an accounting framework can be highly useful to understand the consequences of specific management options in an applied context. On the other end of the knowledge continuum, we rely more strongly on theoretical understanding to characterize relationships. In this case, the resulting model may have less applied utility but can advance theoretical predictions about the behavior of SESs. In most cases, ecosystem accounting will depend on some mixture of knowledge sources; this is highly appropriate and likely to produce the types of models that help to advance both applied and theoretical understanding.

Many modeling approaches can and have been used for ecosystem accounting. As its name suggests, the widely used InVEST framework (Guerry and Tallis, this volume) accounts for the cost of reinvesting in natural capital necessary to support the delivery of ecosystem goods and services. Other approaches include the Artificial Intelligence for Ecosystem Services (ARIES) that employs a Bayesian framework and relies on the data itself to inform functional relationships rather than defining the relationships *a priori* as with deterministic models. Cumulative Human Impact Analysis (Halpern et al. 2008) uses a weighting technique (based on expert opinion) to understand various human-derived impacts on supporting ecosystem services like habitats. Tradeoff Optimization (White et al. 2012) explores multiple ecosystem services and identifies optimal/suboptimal spatial configurations of human activities in a marine system. Another approach is the Multiscale Integrated Model of Ecosystem Services (MIMES), which will be the main focus for the remainder of this chapter. MIMES considers ecosystem service flows that result from natural and human behaviors in a time-evolving system. By focusing on the *dynamics* of system behavior, MIMES provides an opportunity to understand how ecosystem services change over the long-term. Exploring long time horizons is essential for understanding sustainability since systems can shift into phases that are not easily moved back to a more desirable state (Scheffer and Carpenter 2003; Walker and Meyers 2004; Kortsch et al. 2012).

Note that all the approaches mentioned above describe a landscape populated with spatially explicit ecosystem service production functions and most reveal tradeoffs that result from changes in land or ocean use. Features of each are also highly complementary in their ability to inform a particular class of problem or stage of EBM planning. For example, InVEST is very useful for understanding the consequences of alternative decisions when little information exists about a system (or when it is otherwise necessary to rely on more generalized functional relationships); however, in its current state of development the tool provides limited insight into the long-term nature of these tradeoffs. MIMES on the other hand can be more useful when a high level of knowledge exists for a system as it provides a framework for integrating many different types of data into a common and relational format. In addition, MIMES can provide insight into how interactions in coupled systems change over long time horizons. Both InVEST and MIMES, support critical goals of ecosystem accounting and the two approaches could be used in sequence (InVEST during early stages and MIMES at later stages of EBM planning) to understand ecosystem service tradeoffs and the consequences of different management alternatives. In the next section we focus in detail on the MIMES approach and highlight how this EA model is linked to the MIDAS user interface to promote learning and guide decision-support.

2.4 Multiscale Integrated Model of Ecosystem Services (MIMES)

The Multiscale Integrated Model of Ecosystem Services (MIMES) is a modeling framework designed to address the magnitude, dynamics, and spatial patterns of ecosystem service production and values. Theoretical principles, expert knowledge, and data-derived relationships are used to link diverse types of information about a system. The MIMES framework is comprehensive in its recognition that the following interactions play a crucial role in determining system behavior: ecosystem services originate from natural processes; these processes are encountered or intercepted by humans who receive their benefits; humans also respond to ecosystem service flows and their behavior in turn impacts the natural processes. Considering these interactions and feedbacks is critical to a full realization of system dynamics and to support ecosystem-based policy and management activities.

The architecture of MIMES consists of a set of linked domains that together encompass the full suite of dynamics underlying biophysical and climatological processes (Boumans et al. 2002). Each domain emphasizes a particular class of behaviors and considers a different type of capital including natural, human, built, and social forms (Mulder et al. 2007). The anthroposphere is composed of human interactions and the biosphere by interactions among non-human species. Additional domains include the atmosphere, hydrosphere, and lithosphere and are included as necessary depending on the scope and nature of each case study. Ecosystem services flow from the non-human domains to the anthroposphere and feedbacks within and across all the domains are considered. It is the dynamic behavior that results from these linkages that is of critical interest, revealing how system processes unfold over time and space and thus capturing the dimensions that characterize real SESs (Fig. 9.1). A variety of case studies focused at different scales and geographies have been developed using MIMES, including the Patuxent River Watershed model, which examines nutrient management and land-use changes (Costanza et al. 2002), the Puget Sound model, which examines salmon production flows and habitat, the coastal Louisiana model which examines coastal erosion, and a world model (Boumans et al. 2002). This body of work provides an important resource to modelers who can draw upon descriptions of processes from previous case studies and adapt them to new systems of interest. 'Modules' or sets of equations that describe a common process (e.g., watershed run-off) allow for the transfer of information across case studies (Voinov et al. 2004) and, in time, can be drawn upon to develop and test theoretical principles that advance our understanding of coupled systems.

The operating laws that control system dynamics in MIMES follow general theories of ecosystem function; however, the approach also emphasizes the inclusion of additional knowledge streams from a broad array of sources. Such diverse information (often linked indirectly to focal ecological processes) might at first be considered distracting to the primary questions of ecosystem service flows; however, its inclusion achieves two important objectives. First, integration of diverse knowledge seeds opportunities to discover unexpected outcomes since these links may capture important indirect influences and feedbacks. Here, we must remember that our understanding of SESs is still limited and answers to many of the most pressing questions regarding the generation, delivery, and effect of ecosystem services on human behavior still await discovery. Unexpected outcomes can also identify which empirical data gathering activities are needed to support/refute key hypotheses posed by the model. Second, building richness into MIMES ensures that the model and its outputs can be viewed through a diversity of lenses that resonate with different user groups. This is important because a common vision of the ecosystem is necessary to support collaborative decision-making towards sustainability.

2.5 *Decision support tools*

Ecosystem accounting provides an essential analytical framework from which to determine the outcome of alternative decisions, yet additional tools are needed to translate model outputs to a diverse stakeholder audience (including managers) and support the decision-making process. In this section we present Marine Inte-

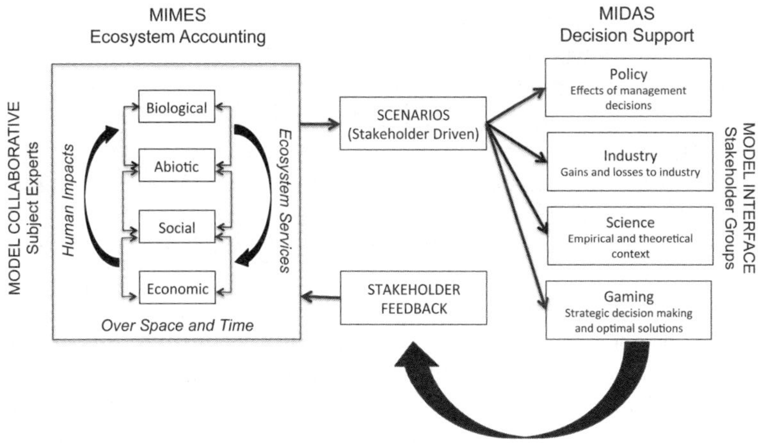

Fig. 9.1 – The MIMES-MIDAS approach supports the development and implementation of ecosystem-based management. The conceptual diagram shows the main features of MIMES for ecosystem accounting (left), MIDAS for decision support (right), and their integration. MIMES ecosystem accounting activities are led by subject experts and result in a model describing the ecosystem service flows and human impacts over space and time. The model is developed through collaborative, interdisciplinary work that describes both natural and human elements of the ecosystem and their interaction. Scenarios are developed with stakeholder input and designed to explore various conditions regarding management options (or scientific assumptions). Decision support is obtained through the MIDAS interface that simplifies the model output into digestible results. The interface design may vary according to the specific character and needs of various user groups. Feedback from users regarding preferences and values can be used to enrich the model in an iterative progression.

grated Decision Analysis (MIDAS), a visualization tool and interactive learning environment that supports collaborative decision-making. First developed to support marine spatial planning in Belize (Patel et al. 2011), the latest version of MIDAS was designed in parallel with MIMES to communicate the model's results and support understanding of ecosystem service tradeoffs related to alternative management decisions.

MIDAS is a user-friendly, web-based interface that incorporates features of open source GIS, participatory mapping, and social collaboration. Upon entering the MIDAS environment, a user selects his/her motivation for what the predominant uses of the ecosystem should look like. Motivations are tied to the individual's values, beliefs, and emotions and describe varied perspectives about the role of humans in the ecosystem. Motivations are fully described in MIDAS and include perspectives of commercial fishing, recreation, conservation, energy development and more (a neutral motivation is also available). Each motivation is accompanied by a set of relevant GIS layers that provide a backdrop for visualizing model outputs. While the selection of motivation does not influence the outputs from the MIMES model, motivations help cast visualizations from MIMES into a GIS field that resonates with the interests of individual users. In this way MIDAS provides a singular platform to engage a diversity of stakeholders and tailor their experience. At the same time, all users interact with outputs from a single MIMES model and thus all parties are forced to come to grips with the same basic understanding of ecosystem dynamics, tradeoffs, and system limits. After selecting a motivation, MIDAS directs users to step through a sequence of interactive, web-based windows that provide a general depiction of the ecosystem accounting model and its linkages and demonstrate how alternative management scenarios affect ecosystem service flows (Fig. 9.1).

Overall, MIDAS embraces a Web 2.0 perspective incorporating features of participatory information sharing, interoperability, user-centered design, and collaboration (O'Reilly 2005). Through MIDAS, users can debate alternative policy options, propose new management scenarios, and contribute to an ongoing dialogue about the future of coastal and marine systems. Paths through MIDAS can be shared among users with similar or differing motivations helping to identify competitive or cooperative outcomes and generally fostering an appreciation for the disparate perspectives related to human use of marine resources. Inevitably this will bring to the surface some tensions among stakeholders, but promoting engagement in a common forum is also necessary to pave the way towards better marine stewardship in the future.

In the next section we describe the MIMES-MIDAS application to a portion of the coastal and nearshore area of Massachusetts Bay. The example is meant to demonstrate how the approach works for a real world case study including how outputs of alternative management scenarios inform our understanding of natural and human behavior over time and space. We begin below with a brief overview of recent policy actions that make Massachusetts well suited for developing a dynamic SES model that helps to support decisions of marine EBM.

3. Case Study: Massachusetts Bay

3.1 Background and study area

In the US, The Commonwealth of Massachusetts has been at the forefront of coastal-marine EBM planning and implementation through the passage of the state's Oceans Act (2008). In 2009, a mere 18 months after the passage of this legislation, a first-cut integrative ocean management plan was completed that demonstrated a clear commitment to tackling the management of ocean resources from a comprehensive, ecosystem-based perspective (EOEEA 2009a, b). Over a very short time period the planning process in Massachusetts has brought together available data to map the distribution of the ecosystem's natural features and understand the spatial and temporal overlap of competing human uses. What remains is an urgent need to look beyond these static data overlays and come to grips with the more complex and counter-intuitive aspects of system dynamics. This means working to identify the interdependent pathways of ecosystem flows and human activities and their response to management interventions. Our work develops the ecosystem accounting framework to do this, while also helping to formulate a common ecosystem vision that supports participatory decision-making.

Following the steps outlined earlier in this chapter (see section 2.2), we first focused on delineating the boundaries of a focal study area in the coastal and nearshore ecosystem that encompasses important ecosystem service flows as well as a diverse concentration of human uses. Realizing that our objectives for this work—to develop a spatially explicit, dynamical model of this coupled system and construct a user interface—were ambitious, we sought to ground and limit this MIMES-MIDAS application by selecting a relatively small but significant portion of coastal and nearshore marine waters directly associated with Massachusetts Bay.

The selected study area is located strategically around the town of Gloucester, an iconic fishing community whose history, culture, and economy have been tied to marine resources for centuries. Anchoring the study area to Gloucester provides a clear geographic point of coupling between ecosystem service flows generated in the nearshore and a specific human coastal community. Overall, the study area encompasses 3,304 km^2 of coastal and nearshore waters that include the jurisdictional area of the MA Oceans Act (0.3–3 nm) and beyond (Fig. 9.2). A diverse set of human activities and regulated use areas are found within the study area, including commercial fishing activities, recreational activities, liquid natural gas (LNG) ports, shipping lanes and two of the State's provisional wind development sites. The Oceans Act provides regulatory authority within state waters; however effective EBM must consider dynamics occurring within ecological as well as political boundaries. For this reason we included both state and federal marine waters to encompass the focal areas of ecological flows. These are the offshore banks—sites of upwelling that support diverse communities of marine species and are also locations of high human activity overlap (e.g., Stellwagen Bank and Jeffreys Ledge, Fig. 9.2).

Essential habitat features for the study area were first characterized in MIMES using spatial data layers from a variety of sources, including Massachusetts Ocean Resource Information System. In addition, we derived a number of new data layers. For example, a map of sediment types (i.e., mud, sand, gravel, boulder) was created by applying ground-truthed cutpoints to sonar backscatter values obtained from multibeam surveys. We also included spatial information on bathymetry, slope, and a variety of other spatial data. These features are integral spatializing key ecosystem patterns and processes like species distribution and upwelling dynamics as well as human behavioral and economic processes that drive patterns of resource use.

At least as important as the aggregation of spatial data layers is the accumulation of knowledge on ecological, economic, and social processes in the study area. In contrast to many data-poor systems around the globe, the Massachusetts coastal and marine environment is well studied and many individuals have expert knowledge on various aspects of the system. To broaden and enrich the accounting framework we drew extensively on this pool of knowledge. Through formal discussions with regional science experts and other stakeholders, such as representatives from the Office of Coastal Zone Management, we outlined critical ecosystem interactions associated with priority management issues. We found it useful to describe in narrative form what is known about the key parameters at the heart of a particular management issue, including the relevant drivers and secondary relationships in both the natural and human-built environments. Narratives provide a non-mathematical point of entry for all participants, consolidating key insights of scholarship and discourse. This verbal script also provides the basis for the model's system of equations.

3.2 MIMES framework for Massachusetts Bay

In developing the Massachusetts Bay MIMES model we focused on interactions occurring in the marine hydrosphere and coastal anthroposphere. These spheres

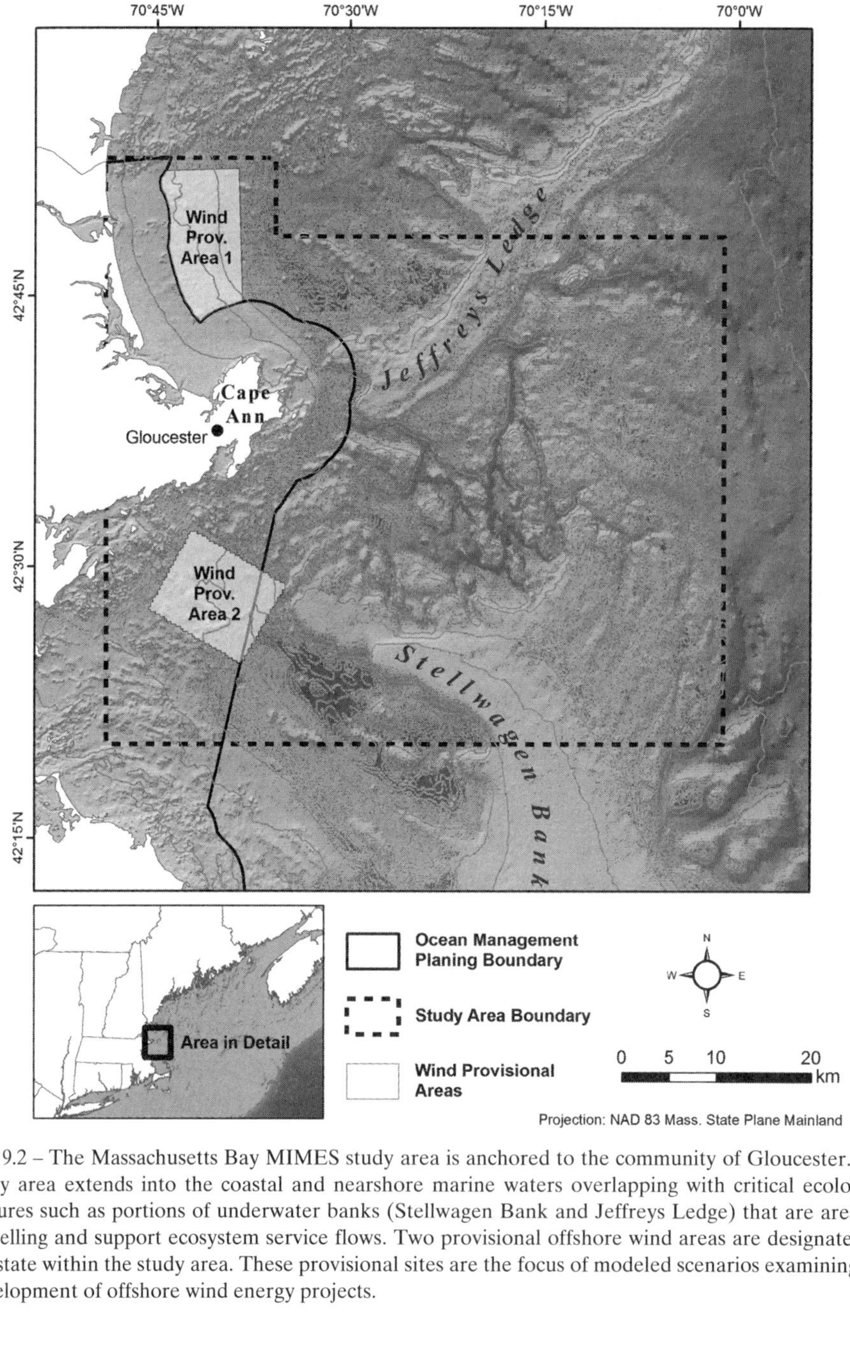

Fig. 9.2 – The Massachusetts Bay MIMES study area is anchored to the community of Gloucester. The study area extends into the coastal and nearshore marine waters overlapping with critical ecological features such as portions of underwater banks (Stellwagen Bank and Jeffreys Ledge) that are areas of upwelling and support ecosystem service flows. Two provisional offshore wind areas are designated by the state within the study area. These provisional sites are the focus of modeled scenarios examining the development of offshore wind energy projects.

are of course embedded within a larger global framework of MIMES; however for the current phase of work we limited the scope of consideration to these two realms. Operationally, we first artificially divided the natural (biosphere) and human (anthroposphere) subsystems to consider their individual attributes, then recombined them through equations that describe (1) how natural resources contribute to human economies and (2) how human activities affect the natural system. Model equations describing ecosystem processes are based on knowledge and insight gained through formal discussions with regional experts as well as information from the literature. These insights are captured in narrative documents that compile and organize our best understanding of ecosystem processes related to key management decisions in the Massachusetts marine environment. Within the study area, dynamics are modeled across a 1km^2 grid cell; however, the model also captures some processes occurring beyond the study area whose dynamics are not spatially explicit, allowing for the inclusion of multiscale processes.

A comprehensive overview of the Massachusetts Bay MIMES is beyond the scope of this chapter; we do, however, describe its important features and demonstrate the breadth of interactions included (for further details, including relevant technical documents, the interested reader may refer to http://www.seaplan.org/project/mimes/). The model is a work in progress designed to support our evolving understanding of the ecosystem-level consequences of alterative management decisions. Thus, while the results presented here provide a critical first cut to understanding the ecosystem interactions in this SES, future efforts are needed to vet, hone, and refine the model for increased realism and enhanced predictive capacity.

3.2a Natural subsystem

A time series of observed chlorophyll *a* concentration (remotely sensed from within the study area) is used to set the baseline for seasonal primary production dynamics in Massachusetts Bay MIMES. These baseline phytoplankton dynamics are then made more ecologically relevant by limiting productivity as a function of bottom slope, supplied as a habitat layer in the model. Overall, the equations simulate upwelling dynamics around Stellwagen Bank and Jeffreys Ledge within a reasonable and observed range of production for the area. Zooplankton production follows phytoplankton dynamics through a time-lagged function. The zooplankton dynamics are highly simplified and will be an area of focus for development in the future.

In addition to these lower trophic levels, the natural subsystem describes dynamics of 40 marine species or functional groups of higher trophic order selected because of their ecological, economic, and cultural importance: Atlantic cod, American lobster, winter flounder, sand lance, and humpback whales to name a few. Population dynamics of these species are modeled through a logistic production function, influenced by habitat specific features including depth, substratum, and benthic community maturity. The latter is determined by the extent to which the bottom has experienced recent disturbance, such as, for example, from mobile bottom-tending fishing gear, or is in an early recovery stage. Species' dynamics are also influenced by availability of appropriate food sources, which consist of phytoplankton, zooplankton, and the pool of higher trophic level species. For the last group, potential predator-prey relationships across the 40 modeled species are specified by parameters informed by the literature, theory, and expert knowledge.

In this MIMES application, species are considered either resident or migrant. Dynamics of residents (mostly demersal bottom dwelling fish, such as yellowtail flounder and goosefish) are confined to the study area while migrants (mostly pelagic fish and marine mammals, such as Atlantic herring and humpback whale) enter and leave the bounds as a function of parameters describing seasonal movement periods and food availability. When present in the study area, migratory species are influenced by the same factors as residents and are fully integrated into the study area food web (either as predators or prey). However, habitat parameters for migratory species are only lightly constrained as compared to residents since migratory species tend to be highly mobile with distributions governed more by food availability and migratory routes rather than by depth or substratum.

Initial biomass for each species within the study area is a data-driven parameter informed by a number of sources. For demersal fish species that are well represented using bottom trawl methods, we used data from two long-term fisheries independent surveys that monitor demersal fish populations in the study area—these are the bottom trawl surveys conducted by the National Marine Fisheries Service and the Massachusetts Department of Marine Resources. Population biomass of modeled species was averaged across a recent ten-year period on a per tow basis and scaled to the study area. For pelagic fishes and marine mammals that are not well captured in bottom trawl surveys, initial biomass conditions were informed by model output from the Northeast US ATLANTIS model across a recent ten-year period (Link et al. 2010). Only ATLANTIS outputs from a single polygon (polygon #12) that encompasses the MIMES study area were used and biomass from this polygon was scaled to the MIMES study area.

3.2b Human subsystem

Massachusetts Bay MIMES emulates the dynamics and inter-relationships of seven human-use sectors essential to the character and economy of coastal Massachusetts. These include five types of commercial fishing classified by gear type (bottom trawling, hand and long lining, midwater trawling, lobster fishing, and gillnet fishing), whale watching, and offshore-wind energy development. Human use dynamics are a function of potential economic costs and benefits, natural resource availability, interactions with other human uses, and regulatory decisions. Human uses operate at two scales in the model. Spatially explicit dynamics occur within the marine portion of the study area. Activities associated with human-use sectors also move across the landward boundary of the study system when conditions do not support their presence in the marine environment, such as, for example, if the target species for a given commercial fishing sector falls below a threshold abundance. This scale-dependent behavior of humans is similar to the dynamics associated with migrant species that move in and out of the study area.

Prior to building our model, we participated in an early phase of preparation for Massachusetts Oceans Act implementation called a "compatibilities analysis". This work produced a qualitative road map for understanding the likely compatibilities/incompatibilities associated with relevant human uses. Development of the MIMES model built upon this foundation and considered two types of incompatibilities: regulatory incompatibilities where management decisions limit or exclude one use in the presence of another; functional incompatibilities where the operating capacity of one use is limited in the presence of another. As an example of the latter, some types of mobile and stationary fishing gear are incompatible when operating at the same place and time. A parameter designating human use priority (that can be informed by stakeholder preferences, regulatory history, or economic value) determines which human use takes precedence when incompatibilities arise in MIMES.

3.2c Ecosystem services and economic valuation

Natural resources provide the basis for economic productivities and are the main driver of human-use dynamics in MIMES. For each human-use sector we specify the natural resource populations that contribute to its productivity. For commercial fishing sectors, this includes biomass of all species targeted by the particular gear in use (informed by NMFS 2011). For whale watching, the natural resource of interest is humpback whales showing up consistently and close to ports with whale watching vessels. For wind power, natural resources are a favorable wind field and the presence of soft sediments suitable for placement and construction of turbines.

Economic parameters characterizing operating capital (which determines the potential for activities that move from land into the marine system), depreciation rate, and investment are informed by a variety of sources. For fishing activities, we relied on NOAA Fishing Vessel Cost Survey data from Massachusetts over a recent 10-year time period to determine economic parameters on a gear-specific basis. Parameters associated with offshore wind development were informed by analyses from the European Wind Energy Association on this topic (Morthorst et al. 2009).

Economic valuation for commercial fishing sectors is based on the dollar value of fish landings, where market price is determined by Massachusetts-specific data on landed value (averaged over a recent 10-year period; data available through NOAA Fisheries Office of Science and Technology). The dollar value of the wind sector is determined by price generated per megawatt of wind energy as informed by published reports of costs from Massachusetts (available through the US Energy Information Administration). Whale-watching value is determined by a curvilinear relationship with humpback abundance developed by regional experts.

3.2d Ecosystem impacts

Ecosystem impacts characterize the effects of human activities on species populations and ecosystem processes. A total of four types of impacts are considered in this MIMES model: mortality on targeted fish populations, bottom habitat disturbance, species deterrence, and the artificial addition of hard substratum to the sea bottom. Impacts are assigned to relevant human activities and can be designated to occur during only a limited phase of an economic sector's dynamics. For example, the construction phase of offshore wind turbines will be associated with bottom disturbances but this impact is temporary and does not occur during the extended operation phase for this sector.

3.3 Tradeoffs involving offshore wind energy

The Massachusetts Bay MIMES model was developed to enhance our understanding of tradeoffs associated with focal management issues, taken within the context of overall ecosystem function (including its human dimensions). The assumptions underpinning these tradeoffs were consolidated through discussions with scientific experts and stakeholders and outlined in narrative documents during the initial planning phase. Our team has thus far developed two case studies to explore the ecosystem cost and benefits of different sets of management decisions. The first examines tradeoffs resulting from the development of offshore wind energy projects, presented below. The second case study examines tradeoffs linked to changes in fishing intensity of forage species, including expansion of the fishery to capture the northern sand lance (*Ammodytes dubius*), which is currently not subject to commercial fishing mortality. The same MIMES model serves as the engine to drive scenarios for both case studies, and it has the capacity and flexibility to investigate ecosystem-level responses related to additional management interventions, such as establishing no-take marine protected areas or marine zoning programs.

Offshore wind projects are an emerging issue in ocean management throughout the world (Gaudiosi 1996; Pelc and Fujita 2002). In the United States, energy plans at the federal and state level call for increased reliance on renewable energy sources, of which offshore wind energy has the potential to yield high returns (Bailey and Freedman 2008; Sheridan et al. 2012). Actualizing offshore wind energy, however, has faced both regulatory hurdles and aggressive challenges from other sectors, such as fisheries and real estate development. The Cape Wind project, proposed for southern Massachusetts, underscores the complexity and contentious nature of these issues; the project only recently won approval after ten years of protracted debates. The controversies surrounding Cape Wind make clear the need for a stronger participatory process to help stakeholders explore the costs and benefits associated with alternative-energy development plans in a meaningful and transparent way.

In recognition of the opportunities and regulatory challenges associated with offshore wind, the Commonwealth of Massachusetts designates a number of provisional sites for development of commercial-scale wind energy projects in its recent Ocean Plan. These areas are located within state and nearby federal waters and have passed an initial exclusionary screening process to avoid areas where conflicting uses are highest (based on GIS map overlays); provisional areas, however, may be associated with significant technical or other challenges that limit their practical development. In the MIMES study area, two provisional wind areas exist (Figure 9.2), but there are no current plans by the energy industry to develop these sites. In addition to commercial scale energy developments, the Massachusetts Ocean Plan supports smaller scale community wind projects. These are developments located close to shore and designed to meet the energy needs of a local community.

3.3a Scenario development

Scenarios for offshore wind were designed in close association with resource managers and other stakeholders and focus on two factors: (1) scale of wind energy projects and (2) potential restrictions of human uses in close proximity to turbines. The scale of wind energy developments was examined at four levels. At the smallest scale, wind development proceeds within a single community project located close to shore and in the northern portion of the study area. Second, we expand the development of wind fields to another community scale project in the south. Third, we expand wind development to include a commercial scale project within the northern provisional area. Lastly, we consider development of two commercial-scale projects in addition to the two community sites described for the other scenarios.

In addition to issues of scale, our MIMES-MIDAS wind scenarios examine potential incompatibilities that arise from regulatory decisions restricting human uses in close proximity to the wind energy projects. In Massachusetts, there is much concern from commercial fishing and other industries about the possibility of restricting marine activities close to turbines due to security and liability concerns. Also, depending on turbine placement, wind developments could impede maneuverability of some fishing boats and larger ocean vessels. We developed scenarios to examine the case where all human uses are fully restricted from proximity with wind turbines (within 1 km^2 buffer) and the case where human uses are allowed at full capacity. (We also examined the consequences of intermediate levels of restriction for some human uses as these scenarios were of strong interest to stakeholders; for brevity, we do not include results of these additional scenarios here.)

In MIMES, wind developments may only be constructed in areas legally designated for this activity, including the two provisional areas (designated for possible commercial development) and two smaller areas that could support community-scale wind projects located to north and south of Gloucester. Within these areas construction of wind turbines is further limited to areas of soft substrata (i.e., sand and mud), as unconsolidated bottoms are necessary to drive the monopoles deep into the ground. In MIMES the major ecosystem consequence of wind field development is to add hard substratum to areas that were formerly soft-bottom habitat. This impact results from both the presence of turbine structures as well as from rock armor that is emplaced as hard material around the foundation of monopoles to protect them from hydrodynamic forces. Increases in hard substrate can affect community composition because demersal species are often adapted to a specific sediment type. These changes can also affect the rest of the community through trophic and competitive interactions. Other impacts from wind development in MIMES are limited to the construction phase of wind projects that lasts for a single year at the beginning of the model run. These include deterrence of marine mammals from a 1km^2 buffer around the development site and bottom disturbance that results from construction at turbine locations.

Wind scenarios at four development scales and two levels of incompatibilities were run for eight years using daily time steps. We had the model output time series of net profits, species biomass, and species landings. Net profits associated with fishing sectors consider the dollar value of landings (based on species-specific market price) and incorporate costs of operation and maintenance. For whale watching, net profits are directly related to the abundance of humpback whales and boat maintenance costs.

3.3b Tradeoff results

Tradeoffs for human-use sectors are calculated as the change in net profits from the baseline scenario (i.e,. no wind energy projects) across the scales of wind energy development. Tradeoffs result from regulatory decisions to limit human-use access and as a function of the ecosystem impacts that result from wind field development. Ecosystem services to support human-use sectors are defined as target fish populations for fishing sectors and by the abundance of humpback whales for whale watching. Tradeoffs were assessed across two incompatibilities and four wind development scales. We focus on describing results in terms of the percent change in net profit from the baseline in order to compare sectors on equal footing.

As expected, tradeoffs for each human-use sector are highly dependent on regulatory decisions about whether to allow human activities inside wind areas. Under the scenarios where wind areas fully exclude all other uses, negative tradeoffs were observed for all fishing sectors based on average annual change in net profits compared to baseline conditions. At the largest and most profitable scale of wind, the average annual loss in net profits to fishing sectors excluded from wind areas was 3.7% per year (see results for fishing sectors in Fig 9.3 at 100% wind scale, which is about the same as the proportion of the study area that is developed for wind (i.e., 4.5%). These negative tradeoffs indicate that fishing sectors are largely unable to compensate for the loss in landings due to wind restrictions by fishing elsewhere in the study area. In other words, the natural populations that support fishing are being utilized at full capacity in the baseline scenario. In the model, the maximum fishing mortality rate is set at a fairly conservative value of 0.2. Under real world conditions restricted access to wind areas could result in increasing fishing pressure above this rate outside wind areas as fishers look to compensate for lost fishing grounds. This could introduce a variety of knock-on effects, both positive and negative in terms of fishing yields.

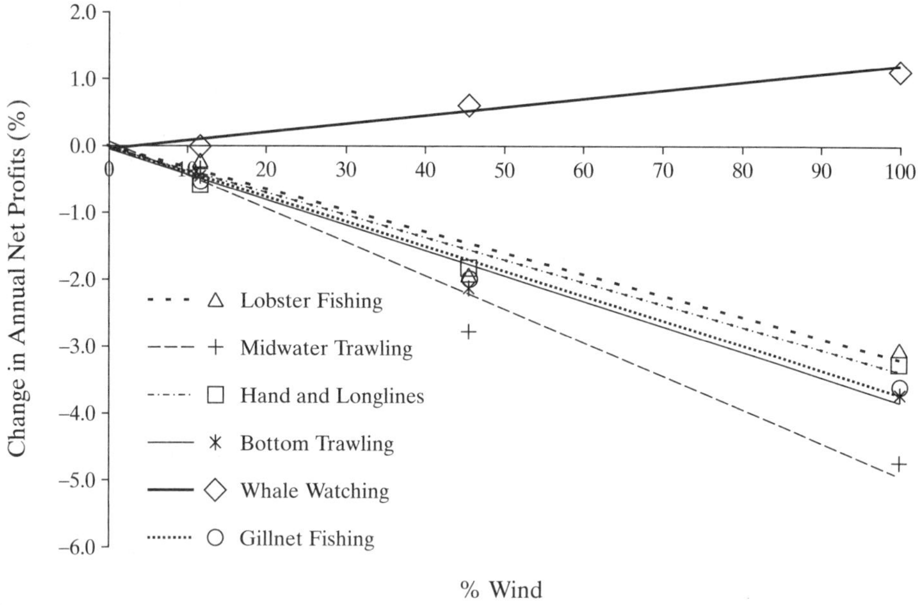

Fig. 9.3 – MIMES model results examining offshore wind developments in the Massachusetts Bay study area. Results demonstrate changes in profitability of different economic sectors for scenarios in which wind areas are designated as no-use zones and all human activities are restricted from operating in close proximity (within a 1 km^2 buffer) from wind turbines due to security or liability concerns. The average annual change in net profit (compared to baseline conditions of no wind) is shown for five types of commercial fishing and whale watching. The changes in net profits are shown across four scales of offshore wind field development where the smallest development represents a single community wind project (15 turbines, ~5% wind development) and the largest represents the development of a two commercial-scale and two community-scale projects (278 turbines total, 100% wind development). Note, these results may be compared to tradeoffs presented in Fig. 9.5 which demonstrate wind-energy developments across the same scales but for which human activities are allowed full access to wind areas

In contrast to fishing sectors, whale watching exhibits a positive tradeoff under the scenario where all uses are excluded from wind areas, exhibiting an average annual gain of 1% in net profits at the largest wind scale (Fig. 9.3). Whale watching in MIMES is directly related to humpback whales, whose seasonal presence, abundance, density, and behavior in the study area depend strongly on two focal prey species: the northern sand lance (*Ammodytes dubius*), which is not commercially targeted by fishing activities,[1] and Atlantic herring, a fishing target for midwater trawlers. At the largest wind scale for these scenarios, northern sand lance population exhibits a decrease in average annual biomass of 3% in wind areas as a result of changing community interactions and increased predation pressure. In contrast, prey populations of Atlantic herring, which are targeted by commercial fishing, increase 6% in wind areas mostly as a result of reduced fishing pressure. The increase in this second prey species is likely the strongest factor driving the small increases in whale abundance.

To better understand the distribution and abundance of humpback whales and their prey for the scenario described above (i.e., largest wind development scale where all human uses are restricted from wind areas) we examine maps of the average monthly change in biomass for humpback whales (Fig. 9.4a) and Atlantic herring (Fig. 9.4b). These spatial patterns indicate that some of the highest increases in biomass for both species are located within wind areas, suggesting that humpbacks are attracted to higher abundance of this prey species there.[2] Moreover, within wind areas the slope of the relationship between Atlantic herring and humpbacks is more strongly correlated than outside (Fig. 9.4c), suggesting that the predator-prey relationship is more tightly coupled there. We postulate that humpbacks receive strong benefits from increased concentration of food in the wind areas. These migratory species are then able to distribute gains in biomass across the study area to the benefit the whale watching industry. Once outside the wind areas, the distribution of humpbacks appears to be driven by additional factors than just Atlantic herring (hence the slight decrease in slope outside versus inside wind areas; Fig. 9.4c), including additional prey sources and habitat restrictions such as minimum depth.

Compared to the effect of regulatory decisions that restrict human uses close to wind developments, the ecosystem effects of offshore wind that result from the addition of hard substratum, deterrence of marine mammals, and bottom disturbance (the latter two occur only during the construction phase of wind development in MIMES) have only minimal effects on human uses when considered at the scale of the whole study area. This is indicated by tradeoffs where human activities are allowed unrestricted access to wind areas. Under these conditions, gillnet and lobster fishing exhibits slight gains (0.2% and 0.1%, respectively) in net annual profits (Fig. 9.5). These small increases result from complex changes in species population *and* human activity dynamics (discussed in more detail below). In con-

[1] The familiar "sand eel" of the bait trade in this area is usually the nearshore species *Ammodytes americanus,* raked from beaches.

[2] The positive change in whales is the average over an eight-year model run. Humpback whales in fact are deterred from wind areas during the initial construction period occurring during year one as this is one of the impacts associated with the turbine construction and placement phase. Yet, the initial loss of whales from wind areas is offset by gains realized over the course of model run.

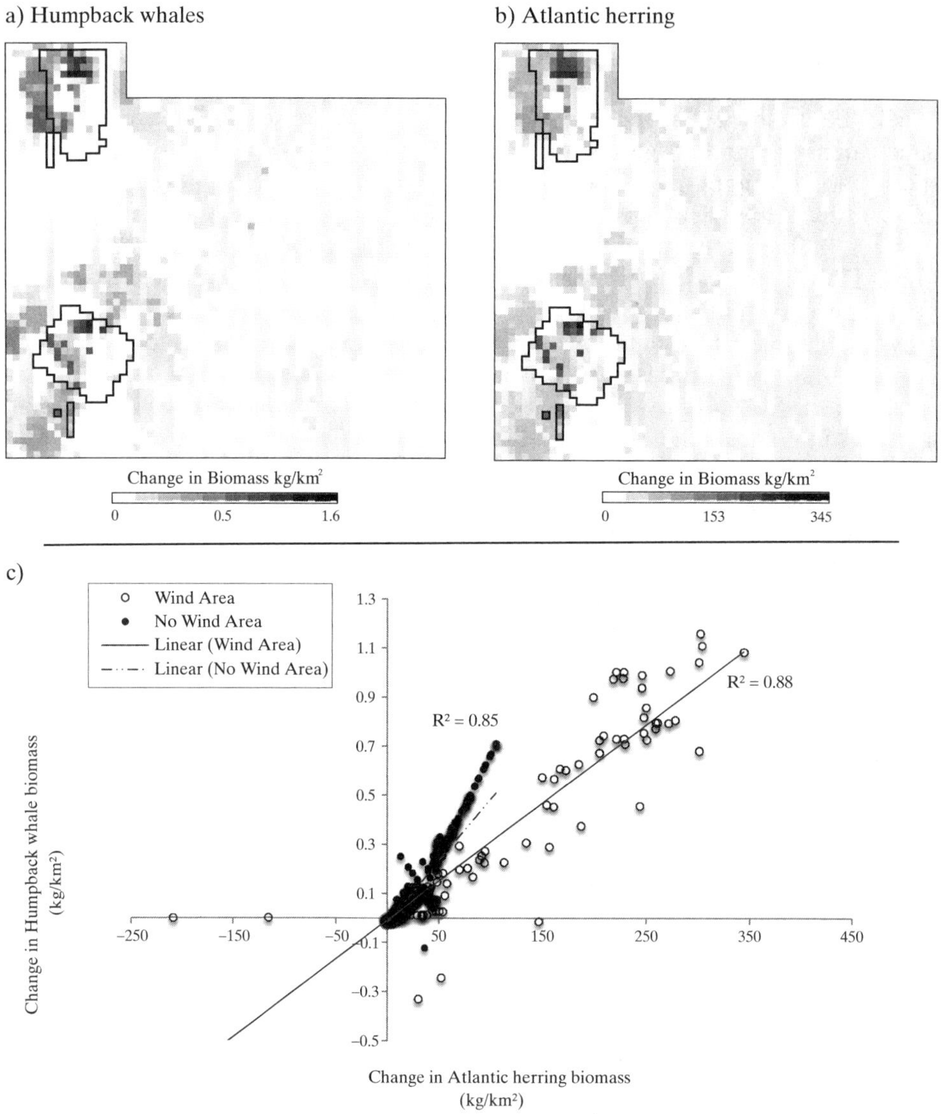

Fig. 9.4 – MIMES model results for the scenario examining large-scale wind development with all human activities excluded from wind areas. The average monthly change in biomass (relative to baseline conditions of no wind development) is shown across the study area for two focal species: humpback whales (a) and one of their main prey sources, Atlantic herring (b). Wind development areas in (a) and (b) are designated by black polygons inside the study area. Panel (c) presents scatter plots for the average monthly change in Atlantic herring (x-axis) versus the average monthly change in humpback whales (y-axis) in wind areas (open circles) and outside wind areas (closed circles).

trast, all other fishing sectors exhibit either no tradeoff or very minimal losses in net profits (across all scenarios the greatest loss in annual net profits to any fishing sector was 0.02%; Fig. 9.5). Thus, for the small tradeoffs exhibited, gains in net profits exceed losses considerably. In other words, win-win opportunities appear to be stronger than the win-lose cases when considering relative changes in net profits.

Tradeoffs measured across the study area suggest how offshore wind energy developments affect productivity of human-use sectors. All sector members, however, may not experience gains and losses equally when access to marine resources may be spatially partitioned through formal or informal agreements (Acheson and Gardner 2004; Martin and Hall-Arber 2008). To understand the effects of spatial scale on tradeoffs, we describe changes in species populations and human-activity dynamics within wind areas, focusing our attention on changes associated with development at the largest spatial scale and with other human-use sectors granted unrestricted access to these areas. This scenario is reflected by points located at the maximum extent of the wind-development axis (far right along the x-axis) in Fig. 9.5.

Considering tradeoffs occurring just within wind areas, both lobster and gillnet fishing exhibit annual gains in net profits. For lobster fishing, the average annual

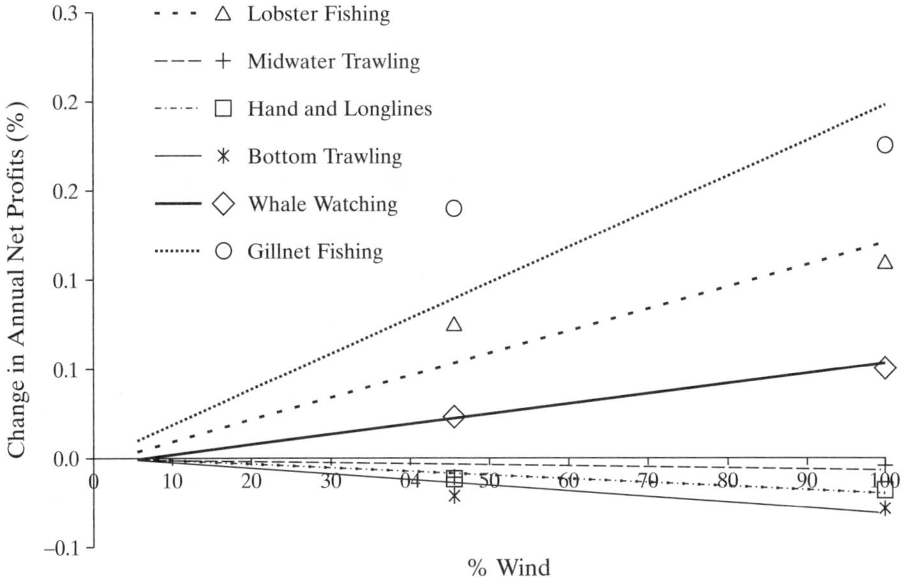

Fig. 9.5 – MIMES model results examining offshore wind developments in the Massachusetts Bay study area. Results demonstrate changes in profitability of different economic sectors for scenarios in which human activities are allowed full access to wind areas. The average annual change in net profit (compared to baseline conditions of no wind) is shown for five types of commercial fishing and whale watching. The changes in net profits are shown across four scales of offshore wind field development where the smallest development represents a single community wind project (15 turbines, ~5% wind development) and the largest represents the development of a two commercial-scale and two community-scale projects (278 turbines total, 100% wind development). Note, these results may be compared to tradeoffs presented in Fig. 9.3 in which human activities are restricted from operating in a 1 km^2 buffer around wind turbines.

increase in net profits is 3.7% (average across all years for this fishery in Fig. 9.6). Valued in dollars, this reflects an annual gain of more than $3,000 and represents a 30-fold increase compared to the same tradeoff across scale of the entirety of the study area. Lobster fishing is a territorial industry, in large part because it relies on a fairly sedentary species that is captured by stationary fishing gear (Acheson and Gardner 2005), and a limited number of fishers are likely to receive the benefits resulting from wind development. In contrast, the gillnet sector operates with gear that is frequently moved to different locations, and access to the living resources that support this sector are not spatially partitioned. Net profits to gillnet fishing within wind areas exhibit an average annual increase of 3.7% (across all years for this fishery Fig. 9.6), representing a dollar increase of $6,000 per year. In this case, the gains are more likely to be equally distributed among gillnet fishers.

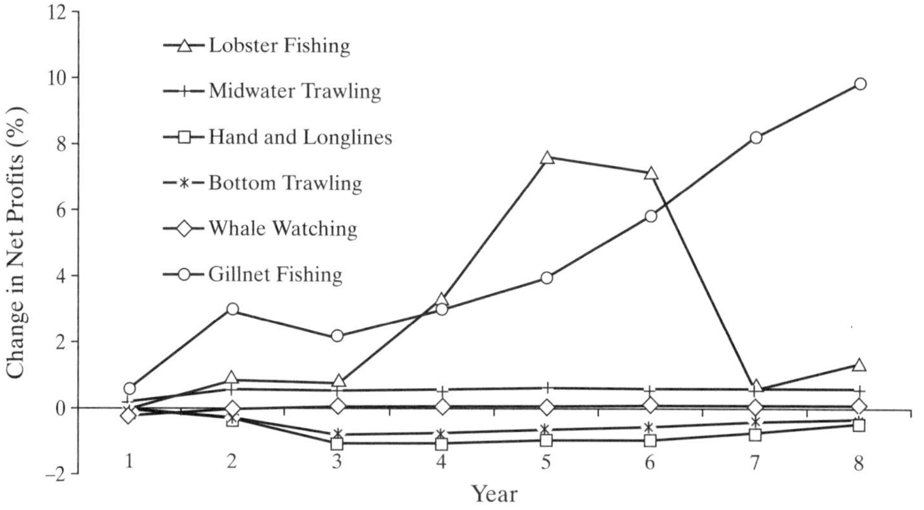

Fig. 9.6 – Percent change in net profits from the baseline for the wind scenario in which the scale of development is maximum and there are no restrictions to human uses within wind areas. Shown are the changes that occur within the wind areas only (*i.e.* not across the entire study as shown in previous figures). For the human activities displayed, each point is the average annual change in net profits from the baseline across the eight-year time period modeled.

Assessing average annual tradeoffs is useful for determining which sectors net the greatest gains and losses over a designated time horizon; calculating average tradeoffs, however, masks year-to-year dynamics and limits an understanding of long-term trends. Looking through time across the eight-year scenario run, we see the net profits to lobster fishing in wind areas exhibit a sharp increase in the middle years and then drop off at the end of the model run (Fig. 9.6). In contrast, net profits to gillnet fishing generally exhibit a steady increase. The bottom-trawling and hand- and long-line sectors exhibit slight negative impacts, with losses beginning in year three; by the end of the eight-year run, these sectors show very little effect from wind energy development (Fig. 9.7). These temporal patterns are the result of complex interactions among 40 species and seven human activities; we describe some of these dynamics below. Overall, the temporal patterns underscore

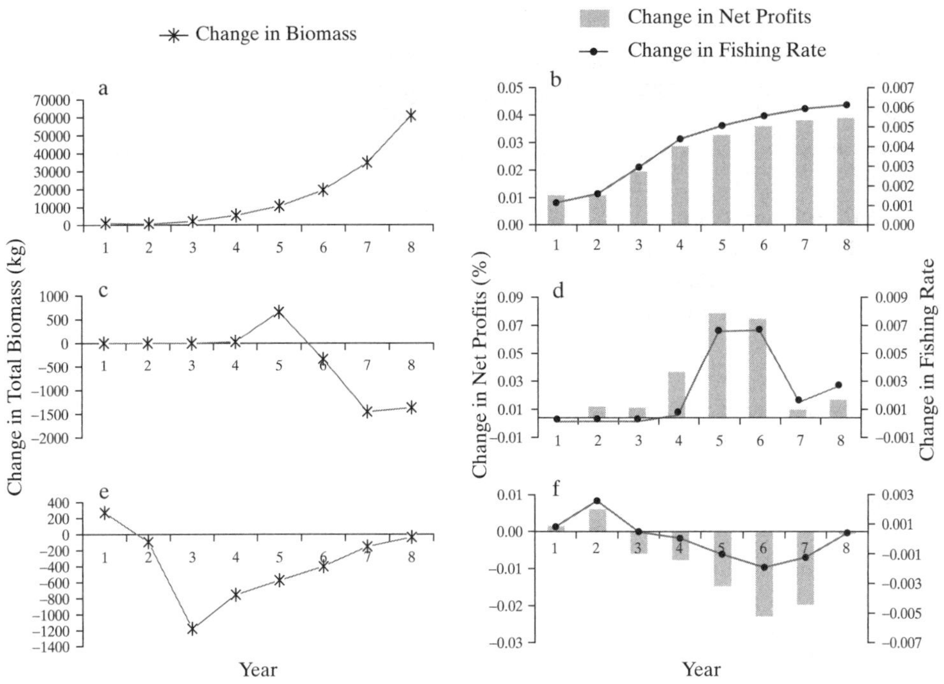

Fig. 9.7 – Annual changes from the baseline in natural and human metrics for the wind scenario with maximum development and no restrictions on human uses within wind areas. Shown are biomass, net profits, and fishing for three commercially targeted species. Left panels show change over the eight-year time horizon in average annual biomass (where total biomass = biomass present in the system + landings) for (a) Atlantic herring, (c) American lobster, and (e) winter flounder. Right panels show the change in annual net profits (hashed bars; primary y-axis) and change in average annual fishing rate (black circles; secondary y-axis) for same three species (b) Atlantic herring, (d) American lobster, and (f) winter flounder. Reading from left to right demonstrates the changes in the targeted population, economics, and fishing pressure within wind areas through time.

the importance of adopting a long-term view of ecosystem service tradeoffs, with appreciation that a tradeoff is not a single number, but a process that unfolds over time, with shifting benefits and costs that may eventually stabilize around a new dynamic.

Inside wind areas natural and human dynamics contribute to distinct patterns of economic tradeoffs. These are manifest in temporal trends in biomass, landings, and fishing rates for commercially targeted species (Fig 9.7). Changes in total biomass reflect the response of species to the presence of wind energy development either as a direct result of changing substratum conditions or via indirect effects of changes in community composition. The change in total biomass of Atlantic herring exhibits a sharp increase over time (Fig. 9.7a) as an indirect effect of small decreases in predator abundance that feed on this forage fish. Landings of the species also increase through time until the last year of the scenario run when the change in landings appears to reach a plateau. Importantly, increases in Atlantic herring landings within wind areas results from two factors: 1) increases in biomass and 2) increases in fishing rate relative to the baseline (Fig. 9.7b). Human activities respond dynamically to the additional resources by increasing fishing effort, yet unlike lobsters (described below), Atlantic herring is able to continue its positive biomass gains despite increasing fishing pressure.

For lobsters, the minimal increases in biomass exhibited in early years of the scenario run are likely the result of addition of hard substratum (preferred by lobsters over soft sediments), whereas the rapid increase exhibited in year five is more characteristic of a loss of predators (Fig. 9.7c). Increases in lobster landings during these early years reflect the small positive changes in the resource base with minimal changes in fishing pressure (Fig. 9.7d). At year five, however, total biomass of lobster rises more substantially and fishing pressure in this year exhibits a parallel increase. After this time fishing pressure is maintained at increased levels (Fig. 9.7d) even while the resource itself falls below the baseline. Thus, the overall positive tradeoff exhibited for the lobster fishing sector across the eight-year time period is driven in later years by a higher fishing rate even though the lobster population exhibits a loss in total biomass relative to baseline conditions. This tradeoff therefore appears to be unsustainable as a greater proportion of the lobster population is being removed from the system each year at the end of the scenario run.

A third species of commercial interest, winter flounder, provides a final example. The addition of hard substratum that results from wind development contributes to small losses of this soft-substratum-dependent species early in the model run (Fig. 9.7e). Changes in community structure, however, are also occurring. After the initial losses exhibited in first three years, total biomass of winter flounder begins a steady increase finally returning to baseline levels in the final year of the model run. Patterns of increasing biomass after year three are associated with decreasing fishing pressure beginning in the same year. Thus, reduced fishing pressure is one factor allowing the total biomass of this species to approach baseline conditions by the end of the eight-year model run. Also note that while fishing rates after year three remain below or at the baseline, temporal rates over the course of the model run are variable exhibiting a decreasing trend in years 3–6 followed by an increasing trend in the final years of the model (Fig. 9.7f).

3.4 Model limitations and future work

In most cases, tradeoffs calculated at the scale of the wind area run in the same direction as those found at the full study area only much stronger. These results suggest that ecosystem effects associated with harnessing offshore wind resources will be fairly contained to an area in close proximity to wind developments. It is possible, however, that effects from wind areas that act as *de facto* marine reserves (due to regulatory restrictions limiting human activities) could be an important effect of wind developments at larger scales (Inger et al. 2009). In the Massachusetts Bay MIMES model, spillover dynamics are not well developed given that only a portion of the modeled species are able to move from cell to cell (i.e., migrants can exhibit cell to cell interaction, but residents cannot). Thus, while the model currently accounts for interactions across 40 species and 7 human activities, and does so in the context of multiple spatially explicit habitat layers (e.g., bathymetry and sediment), technological constraints make it difficult to capture the full extent of layered complexity associated with this system. Developing the capabilities for all species to move through the modeled ecosystem will be a high priority of future work. Even so, we believe the breadth of dynamics modeled extend far beyond what many EBM tools have thus far been able to achieve.

While we focus on relative changes in net profits of human-use sectors, it should be recognized that tradeoffs can also be calculated in terms of their dollar value in the model. Moreover, the MIMES framework is designed to understand how monetary values may be distributed across the human landscape. One way to do this is to understand how economic productivity translates into the addition or loss of jobs and wages paid. While not described in detail here, this flexibility is a fundamental design point in MIMES and allows users to understand tradeoffs in the context of different value systems.

3.5 MIDAS decision support

The Massachusetts Bay MIMES model reflects a highly collaborative effort to integrate a wealth of observed information and expert knowledge about this system. Capturing the dynamic ecosystem relationships over time and space and understanding how they respond to relevant management decisions is an important achievement. The model itself, however, is limited in its ability to directly inform stakeholders since running it requires a high level of technical know-how as well as specific software and computing requirements. As described earlier in this chapter, we therefore rely on the user interface MIDAS to communicate the work to a broad stakeholder audience.

MIDAS contains the results of the two case studies developed in MIMES, including the wind scenarios described in detail here and a second case study exploring changes in fishing of forage fish species. In total, the interface captures results from 30 different resource management scenarios related to the two case studies. Because a fully functioning MIMES model is quite complicated, is computationally intensive, and can take some time to run, MIDAS currently uses only a limited number of specific probes to explore a library of pre-run scenarios. The interface, however, has built-in capabilities that encourage users to think beyond what is currently presented. For example, one feature encourages users to submit management scenarios of their own design to the MIDAS team (e.g., the establishment of a no-take marine reserve); another allows users to designate a polygon of interest within the study area from which scaled output can be requested. These features help capture users' interests and engage them in an ongoing dialogue about ecosystem-based planning in the marine environment.

Overall, MIDAS is designed to encourage users to explore model results through interactive features. For example, users can easily toggle between scenarios and compare different types of outputs (e.g., species biomass, human activities, and economic profits). Results are displayed using a variety methods including bar charts, line graphs, and movies that demonstrate modeled responses of species or human uses over space and time. Some MIDAS features were also described earlier including the role of user-selected motivations that provide an individualized backdrop for viewing MIMES model results.

There are additional features in development for MIDAS that could greatly aid our understanding of the behavior of human-use sectors and advance what has been called the second science of human-natural interactions (Stern 1993). Using appropriately designed surveys and consenting participants, MIDAS could be employed as a social science and economic data collection tool. Users could be polled to determine their stated preferences for a suite of ecosystem services. MIDAS could also be used to understand users' potential responses to changing ecosystem service flows. Empirical data obtained through MIDAS would provide deeper insight into socio-economic influences operating in the system and could be used to refine parameters governing human behavior dynamics in MIMES, providing increased realism and higher predictive capacity. There are some recent examples of models that incorporate aspects of human behavior into the modeled framework, including how perceptions affect decision-making (Wandersee et al. 2011) and how the integration of stakeholder knowledge can inform paths towards sustainability (Gray et al. 2012). Another way to enhance our understanding of coupled systems would be to develop MIDAS with gaming features that allow stakeholders to interact with the model and to explore the range of possible outcomes of alternative management actions in a virtual world (see also Fulton and Link, this volume).

4. Conclusions

In this chapter we have described how an ecosystem accounting approach and decision support tool can provide critical information to support EBM efforts in marine environments. Focusing on the MIMES model for accounting and MIDAS for visualizing and decision support, we presented the application of these methods to the Massachusetts Bay, detailing the changes in natural and human model features that occur in response to a set of management interventions. Below, we review the unique features of MIMES-MIDAS and discuss important future work to grow our understanding of the complex dynamics that govern social-ecological dynamical systems.

MIMES is distinguished from many other ecosystem service modeling approaches in its consideration of how suites of ecosystem services respond dynamically to changes in management decisions (or other influences like climate change) over space and through time. Both the spatial and temporal resolution of analysis is flexible, allowing users to address questions at the local, regional or global scale. In addition, MIMES returns results in terms of sustainability (tradeoffs between now and the future) and environmental equity, that is, losers and winners in environmental decision-making (Daily and Ehrlich 1996). The diverse streams of knowledge included in MIMES are necessary to ensure the model reflects a variety of stakeholder perspectives. Incorporating a diversity of information also allows for the discovery of unexpected outcomes. For example, in earlier sections we described how management scenarios that limit the area over which fishing activities and whale watching occur led, surprisingly, to an increase in the profitability of the latter. This occurs because humpback whales (the species targeted by whale watching activities) exhibit a positive response to increased prey availability in areas targeted for wind energy development that restrict fishing activities.

The MIDAS user interface is a highly interactive platform that allows stakeholders to readily visualize ecosystem service tradeoffs that result from alternative management actions. Design features of MIDAS support stakeholder engagement, including the presence of user-selected motivations that provide an individualized context for understanding ecosystem tradeoffs. In addition, users are encouraged to post comments on MIDAS and suggest additional management scenarios for exploration; these elements foster an ongoing dialogue among stakeholders and can promote collaborative decision-making. Future work on MIDAS will focus on using the interface to gather empirical data about how humans respond to changes in ecosystem services. Socio-economic dynamics are often poorly characterized in SES models; thus, obtaining empirical measures of human valuation and behavior will help bridge a critical gap in our current understanding.

The development of a MIMES-MIDAS case study depends on connecting different types of information about a system in order to understand the pathways that generate and maintain ecosystem service flows. This process encourages subject experts to collaborate in formulating a hierarchical, multiscale conception of ecosystem processes and interactions. In contrast to hypothesis testing in an isolated context, the accounting framework offers the possibility of doing so at the ecosystem scale. This kind of synthesis rarely arises spontaneously (McMichael et al. 2003) and can be one of the most significant scientific contributions of the current legal and practical demand for the development of EBM. Posing sets of hypotheses can also point the way towards gaining additional empirical data from the system to support or refute the behavior identified by the model. Continued reassessment of these relationships is essential and should be incorporated into the adaptive management cycle.

In parallel with developing the MIMES-MIDAS framework, we have been working with colleagues on several initiatives to increase our understanding of the temporal and spatial dynamics of this system; these include historical analysis of nearshore fish stocks, integrated data layers of benthic communities obtained through a mounted camera sled called HabCam, and nonlinear time series analysis of fish stock forecasting (Deyle and Sugihara 2011; Glaser et al. 2011; Liu et al. 2012). The empirical data and analytical results of these efforts will enhance realism of the model; however, care must be taken when adding complexity to modeled dynamics, since greater complexity also makes discerning the mechanisms that underlie outcomes more difficult. A tiered approach in which the model is constructed in stages, the sensitivity of parameters evaluated, and uncertainty formally considered is recommended.

The MIMES-MIDAS system for Massachusetts has already proven useful in its primary objective of synthesizing data, visualizing system processes and tradeoffs, sharing information with stakeholders and decision-makers, and facilitating discussion of policy alternatives. Both it and its sister project on wind farm configuration (White et al. 2012) generated considerable excitement when presented to regional ocean planners and managers.[3] Following completion of the pilot version of the model and decision support tool, we have shared the work with various stakeholder and potential user groups, including the Massachusetts Department of Coastal Zone Management and the organization SeaPlan (a public-private partnership that works to advance the governance and science of coastal EBM in the

[3] While we were immersed in considering knock-on consequences of wind farm development at a system level, colleagues from the Bren School for Environment (University of California at Santa Barbara) were working in parallel with a focus on achieving a detailed understanding of ecosystem service tradeoffs dependent upon the precise configuration of any hypothetical wind farm (White et al. 2012).

US Northeast). Improvements based on this initial feedback were made and we have begun to introduce the model to key decision-making agencies in the region including the Stellwagen Bank National Marine Sanctuary Program office. The assembly of this SES model for coastal Massachusetts was a demanding technical challenge; however, it is only the first step in moving toward a science of social-ecological systems that is both theoretically and empirically grounded. With urgency, we must now accomplish an even greater feat—to use this framework along with other complementary approaches to reveal the underlying laws that govern SESs. Only with this knowledge in hand can we at last aspire to illuminate the true meaning of sustainability.

Acknowledgments

We thank S. Moura, N. Napoli, and J. Swasey (SeaPlan) for extensive feedback throughout the development of this work and for help organizing and facilitating stakeholder enegagement meetings. MIDAS design and technical support were greatly aided by J. Pitt, G. Meyer, and E. Walker (Boston University). K. Lagueux and B. Wikgren (New England Aquarium) provided assistance in refining data layers to be used within the modeling framework. J. Link (Northeast Fisheries Science Center) provided biological output from NEUS ATLANTIS model. Data from fisheries-independent bottom trawl surveys conducted in Massachusetts state waters provided by J. King (Massachusetts Division of Marine Fisheries). Data from trawl surveys conducted in federal waters provided by M. Fogarty (Northeast Fisheries Science Center). L. Hatch, B. Haskell, C. MacDonald (Stellwagen Bank National Marine Sanctuary Office) and D. Simpson, T. Callaghan (Massachusetts Office of Coastal Zone Management) helped develop alternative management scenarios for MIMES and provided assistance in calibrating modeled dynamics. Expert scientists from throughout the region participated in workshops to characterize ecosystem service production flows and human impacts including L. Incze (National Science Foundation), A. Rosenberg (University of New Hampshire), L. Hatch (Stellwagen Bank National Marine Sanctuary Office), and P. Auster (University of Connecticut). B. Costanza (Portland State University) was a leader in the original conceptualization and development of MIMES. Two anonymous reviewers provided constructive editorial comments. Funding for this work provided by the Gordon and Betty Moore Foundation and the NSF-NOAA program Comparative Analysis for Marine Ecosystem System Organization.

References

Acheson, J. M. and R. J. Gardner. 2004. Strategies, conflict, and the emergence of territoriality: The case of the Maine lobster industry. *American Anthropologist* **106:**296–307.

Acheson, J. M. and R. J. Gardner. 2005. Spatial strategies and territoriality in the Maine lobster industry. *Rationality and Society* **17:**309–341.

Bailey, B. H. and J. M. Freedman. 2008. A regional assessment of the US offshore wind energy resource through the use of mesoscale modeling. *Marine Technology Society Journal* **42:**8–18.

Boumans, R., R. Costanza, J. Farley, M. A. Wilson, R. Portela, J. Rotmans, F. Villa, and M. Grasso. 2002. Modeling the dynamics of the integrated earth system and the value of global ecosystem services using the GUMBO model. *Ecological Economics* **41:**529–560.

Caddy, J. F. 1999. Fisheries management in the twenty-first century: will new paradigms apply? *Reviews in Fish Biology and Fisheries* **9:**1–43.

Costanza, R., A. Voinov, R. Boumans, T. Maxwell, F. Villa, L. Wainger, and H. Voinov. 2002. Integrated ecological economic modeling of the Patuxent River watershed, Maryland. *Ecological Monographs* **72:**203–231.

Daily, G. C. and P. R. Ehrlich. 1996. Socioeconomic equity, sustainability, and Earth's carrying capacity. *Ecological Applications* **6:**991–1001.

Daily, G. C., S. Polasky, J. Goldstein, P. M. Kareiva, H. A. Mooney, L. Pejchar, T. H. Ricketts, J. Salzman, and R. Shallenberger. 2009. Ecosystem services in decision making: time to deliver. *Frontiers in Ecology and the Environment* **7:**21–28.

Deyle, E. R. and G. Sugihara. 2011. Generalized Theorems for Nonlinear State Space Reconstruction. *Plos One* **6.**

Dietz, T., E. Ostrom, and P. C. Stern. 2003. The struggle to govern the commons. *Science* **302:**1907–1912.

European Environment Agency (EEA). 2010. Ecosystem accounting and the cost of biodiversity losses. Technical report No 3/2010. EEA: Copenhagen.

Gaudiosi, G. 1996. Offshore wind energy in the world context. *Renewable Energy* **9:**899–904.

Glaser, S. M., H. Ye, M. Maunder, A. MacCall, M. Fogarty, and G. Sugihara. 2011. Detecting and forecasting complex nonlinear dynamics in spatially structured catch-per-unit-effort time series for North Pacific albacore (Thunnus alalunga). *Canadian Journal of Fisheries and Aquatic Sciences* **68:**400–412.

Granek, E. F., S. Polasky, C. V. Kappel, D. J. Reed, D. M. Stoms, E. W. Koch, C. J. Kennedy, L. A. Cramer, S. D. Hacker, E. B. Barbier, S. Aswani, M. Ruckelshaus, G. M. E. Perillo, B. R. Silliman, N. Muthiga, D. Bael, and E. Wolanski. 2010. Ecosystem Services as a Common Language for Coastal Ecosystem-Based Management. *Conservation Biology* **24:**207–216.

Gray, S., A. Chan, D. Clark, and R. C. Jordan. 2012. Modeling the integration of stakeholder knowledge in social–ecological decision-making: benefits and limitations to knowledge diversity. *Ecological Modelling* **229:**88–96.

Gregr, E. J. and K. M. A. Chan. 2011. Making science relevant to marine ecosystem-based management. *Biological Conservation* **144:**670–671.

Halpern, B. S., S. Walbridge, K. A. Selkoe, C. V. Kappel, F. Micheli, C. D'Agrosa, J. F. Bruno, K. S. Casey, C. Ebert, H. E. Fox, R. Fujita, D. Heinemann, H. S. Lenihan, E. M. P. Madin, M. T. Perry, E. R. Selig, M. Spalding, R. Steneck, and R. Watson. 2008. A global map of human impact on marine ecosystems. *Science* **319:**948–952.

Inger, R., M. J. Attrill, S. Bearhop, A. C. Broderick, W. J. Grecian, D. J. Hodgson, C. Mills, E. Sheehan, S. C. Votier, M. J. Witt, and B. J. Godley. 2009. Marine renewable energy: potential benefits to biodiversity? An urgent call for research. *Journal of Applied Ecology* **46:**1145–1153.

Kortsch, S., R. Primicerio, F. Beuchel, P. E. Renaud, J. Rodrigues, O. J. Lonne, and B. Gulliksen. 2012. Climate-driven regime shifts in Arctic marine benthos. *Proceedings of the National Academy of Sciences of the United States of America* **109:**14052–14057.

Lester, S. E., K. L. McLeod, H. Tallis, M. Ruckelshaus, B. S. Halpern, P. S. Levin, F. P. Chavez, C. Pomeroy, B. J. McCay, C. Costello, S. D. Gaines, A. J. Mace, J. A. Barth, D. L. Fluharty, and J. K. Parrish. 2010. Science in support of ecosystem-based management for the US West Coast and beyond. *Biological Conservation* **143:**576–587.

Link, J. S., E. A. Fulton, and R. J. Gamble. 2010. The northeast US application of ATLANTIS: A full system model exploring marine ecosystem dynamics in a living marine resource management context. *Progress in Oceanography* **87:**214–234.

Liu, H., M. J. Fogarty, S. M. Glaser, I. Altman, C. H. Hsieh, L. Kaufman, A. A. Rosenberg, and G. Sugihara. 2012. Nonlinear dynamic features and co-predictability of the Georges Bank fish community. *Marine Ecology Progress Series* **464:**195-U228.

Liu, J., T. Dietz, S. R. Carpenter, M. Alberti, C. Folke, E. Moran, A. N. Pell, P. Deadman, T. Kratz, J. Lubchenco, E. Ostrom, Z. Ouyang, W. Provencher, C. L. Redman, S. H. Schneider, and W. W. Taylor. 2007. Complexity of coupled human and natural systems. *Science* **317:**1513–1516.

Lotze, H. K., H. S. Lenihan, B. J. Bourque, R. H. Bradbury, R. G. Cooke, M. C. Kay, S. M. Kidwell, M. X. Kirby, C. H. Peterson, and J. B. C. Jackson. 2006. Depletion, degradation, and recovery potential of estuaries and coastal seas. *Science* **312:**1806–1809.

Lowenthal, D. 2000. Nature and morality from George Perkins Marsh to the millennium. *Journal of Historical Geography* **26:**3–27.

Lubchenco, J. and N. Sutley. 2010. Proposed US Policy for Ocean, Coast, and Great Lakes Stewardship. *Science* **328:**1485–1486.

Maler, K. G., S. Aniyar, and A. Jansson. 2008. Accounting for ecosystem services as a way to understand the requirements for sustainable development. *Proceedings of the National Academy of Sciences of the United States of America* **105:**9501–9506.

Maler, K. G., S. Aniyar, and A. Jansson. 2009. Accounting for Ecosystems. *Environmental & Resource Economics* **42:**39–51.

Marsh, G. P. 1864. *Man and Nature; or, Physical Geography as Modified by Human Action.* Scribner, New York.

Martin, K. S. and M. Hall-Arber. 2008. Creating a place for "Community" in New England fisheries. *Human Ecology Review* **15:**161–170.

Massachusetts Executive Office of Energy and Environmental Affairs (EOEEA) 2009a. *Massachusetts Ocean Management Plan. Volume 1. Management and Administration.* http://www.mass.gov/eea/ocean-coastal-management/mass-ocean-plan/final-massachusetts-ocean-management-plan.html.

Massachusetts Executive Office of Energy and Environmental Affairs (EOEEA) 2009b. *Massachusetts Ocean Management Plan. Volume 2. Baseline Assessment and Science Framework.* http://www.mass.gov/eea/ocean-coastal-management/mass-ocean-plan/final-massachusetts-ocean-management-plan.html.

Massachusetts Oceans Act. 2008. General Court of the Commonwealth of Massachusetts. Acts 2008. Chapter 114. http://www.malegislature.gov/Laws/SessionLaws/Acts/2008/Chapter114.

McMichael, A. J., C. D. Butler, and C. Folke. 2003. New visions for addressing sustainability. *Science* **302:**1919–1920.

McPeak, J. G., D. R. Lee, and C. B. Barrett. 2006. Introduction: The dynamics of coupled human and natural systems. *Environment and Development Economics* **11:**9–13.

Morthorst, P. E., H. Auer, and A. Garrad. 2009. Part III The Economics of Wind Power. In *Wind Energy—The Facts: A Guide to the Technology, Economics, and Future of Wind Power.* European Wind Energy Association (EWEA), editor. Earthscan, London; Sterling, VA.

Mulder, K., A. Troy, and R. Boumans. 2007. The role of built, human, social, and natural capital in determining land values, and the influence of demographics upon this relationship. *Spatial Economic Analysis* **2:**135–156.

National Marine Fisheries Service (NMFS). 2011. National Bycatch Report. W. A. Karp, L. L. Desfrosse, S. G. Brooke, Editors. U.S. Dep. Commer., NOAA Tech. Memo. NMFS-F/SPO-117C, 508. p.

O'Reilly, T. 2005. http://oreilly.com/web2/archive/what-is-web-20.html.

Ostrom, E. 2007. A diagnostic approach for going beyond panaceas. *Proceedings of the National Academy of Sciences of the United States of America* **104:**15181–15187.

Ostrom, E., J. Burger, C. B. Field, R. B. Norgaard, and D. Policansky. 1999. Sustainability—Revisiting the commons: Local lessons, global challenges. *Science* **284:**278–282.

Patel, H., S. Gopal, L. Kaufman, M. Carleton, C. Holden, V. Pasquarella, M. Ribera, and B. Shank. 2011. MIDAS: A Spatial Decision Support System for Monitoring Marine Management Areas. *International Regional Science Review* **34:**191–214.

Pelc, R. and R. M. Fujita. 2002. Renewable energy from the ocean. *Marine Policy* **26:**471–479.

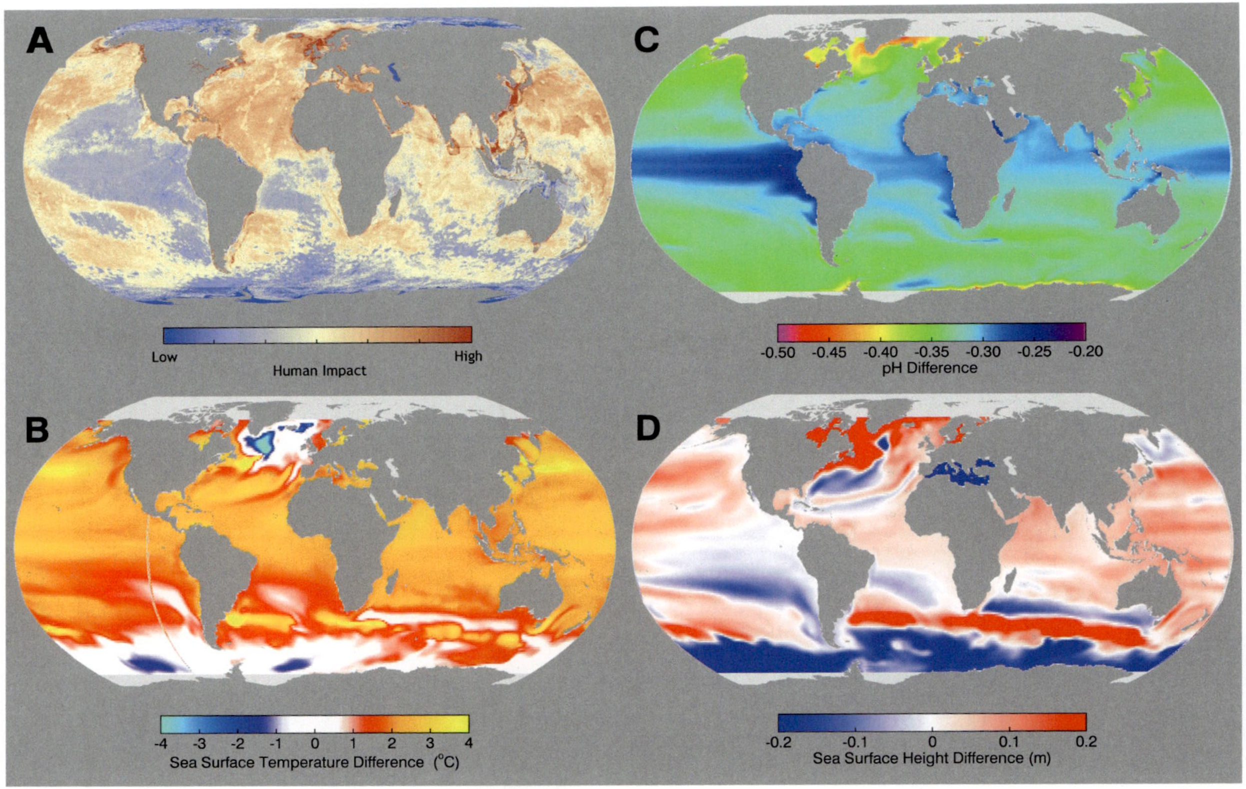

Fig. 1.1 – Maps of the global distribution of (a) cumulative anthropogenic impacts (data provided by Ben Halpern, UCSB and used with permission), (b) projected change in sea surface temperature by 2100 (V. Saba, NOAA), (c) projected change in pH by 2100, and (d) local difference relative to average global sea level rise by 2100". Data for mapping derived from the GFDL ESM2 Global Coupled Climate–Carbon Earth System Model (V. Saba, NOAA personal communication; for model details and caveats, see Dunne et al. 2012; Dunne et al. in press). Maps prepared by S. Lucey and Kim Hyde, NOAA. Note: no estimates for Arctic regions (grey) because of ice cover. Reference period for SST, pH, and SLR. 1996–2005.

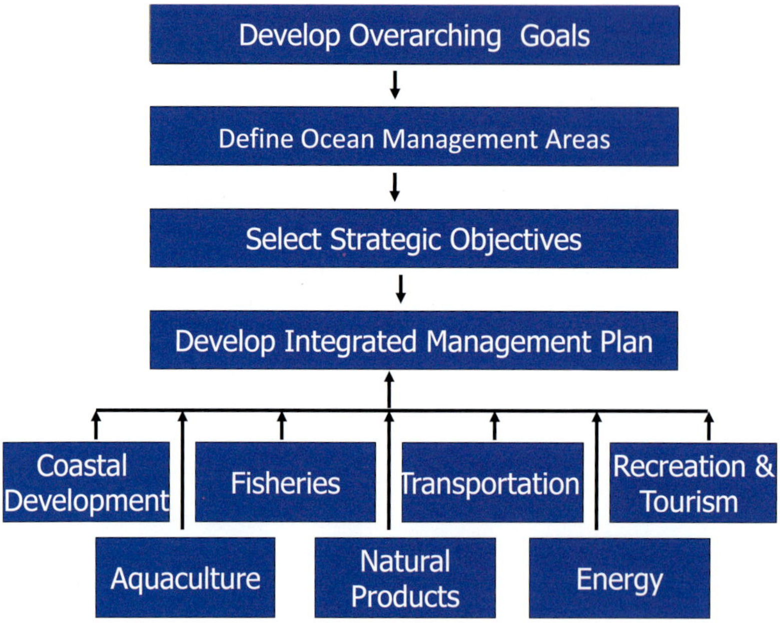

Fig. 1.2 – Roadmap to marine ecosystem-based management culminating in development of an integrated management plan incorporating sectoral consideration from a selection of ocean use communities.

Fig. 1.3 – Elements of marine ecosystem-based management engagement with stakeholders (scoping and setting objectives); selection of spatial management units, ecosystem research and monitoring, integrated ecosystem assessment (incorporating selection of indicators, management strategy evaluation, risk analysis and choice of tactical management tools [including spatial management strategies and input or output controls affecting ecosystem impacts]); and development of scientific advice for management.

Fig. 1.4 – Developmental cycle for an integrated ecosystem assessment. (Courtesy of the NOAA IEA Program Office)

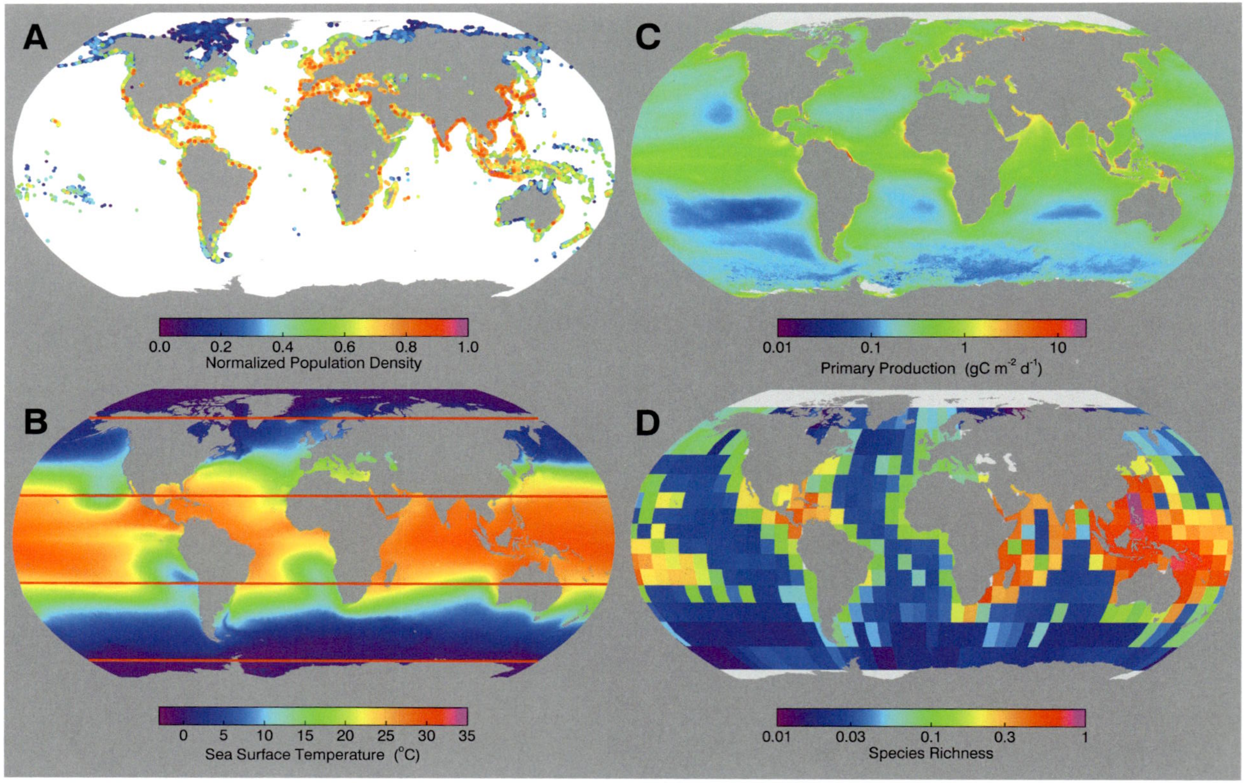

Fig. 1.5 – Maps of the global distribution of (a) human population density (data provided by Ben Halpern UCSB and used with permission), (b) sea surface temperature, (c) primary production, and (d) marine biodiversity (data provided by Derek Tittensor and Boris Worm, Dalhousie University and used with permission). (Maps prepared by S. Lucey and Kim Hyde, NOAA)

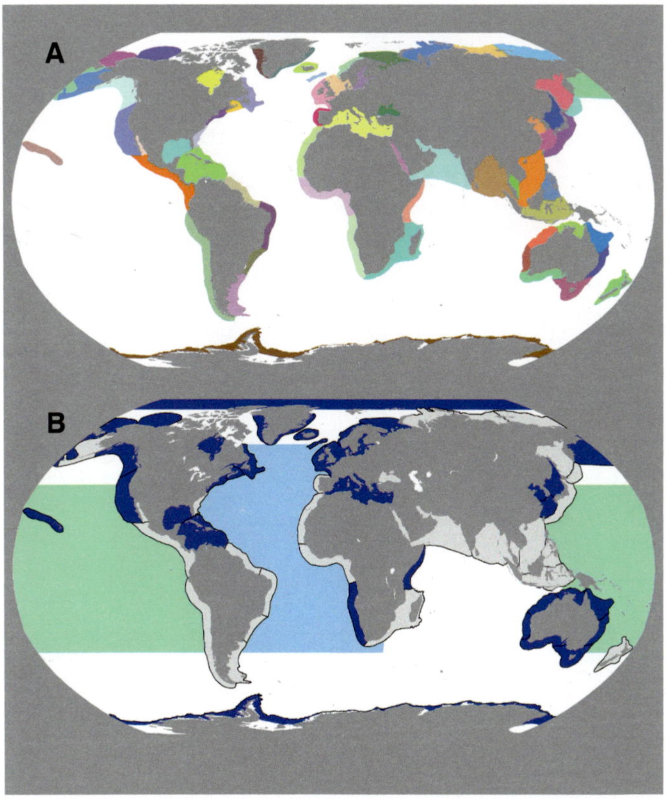

Fig. 1.6 – Location of (a) designated large marine ecosystems and (b) LMEs covered in part or in whole in this volume (dark blue). Open ocean areas in the Atlantic and Pacific Oceans covered here also shown in light blue and green, respectively. (Maps prepared by Kim Hyde, NOAA)

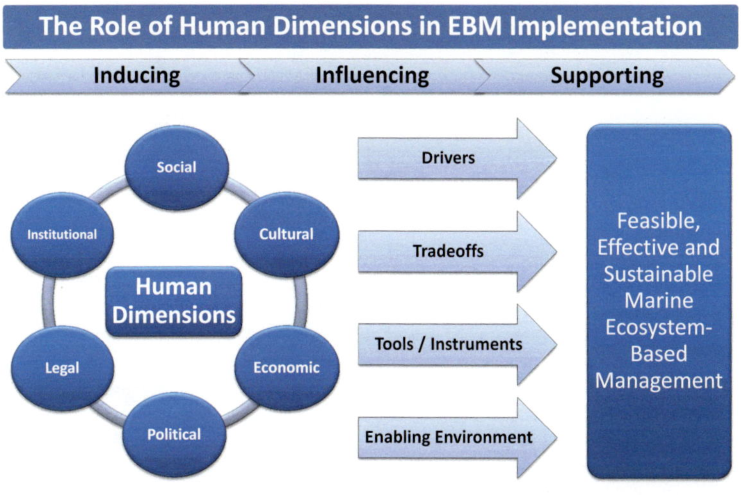

Fig. 3.1 – The human dimensions relevant to marine ecosystem-based management, and their role in inducing, influencing and supporting EBM implementation, acting through a set of *entry points:* drivers, tradeoffs (including costs and benefits of EBM), tools / instruments (as suitable for EBM), and the enabling environment (supporting and constraining factors within the corresponding marine system).

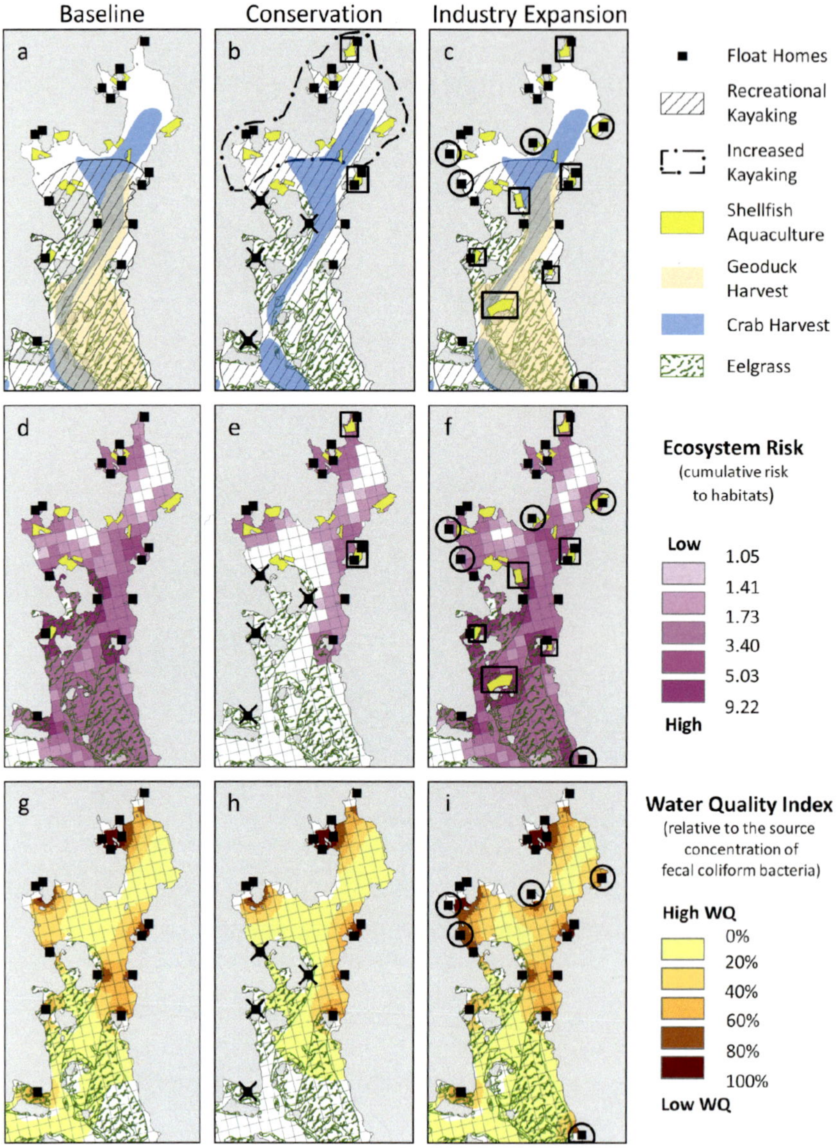

Fig. 8.3 – Three alternative management scenarios for Lemmens Inlet, BC identified by West Coast Aquatic (a–c) and some InVEST outputs (d–i). The three scenarios explored were: (a) Baseline (no changes to current uses or zones); (b) Conservation (zoning rules restrict float homes and aquaculture in areas near eelgrass beds—four float homes are removed from areas where they overlap with eelgrass (shown under black X's); two new oyster deepwater tenures are located outside of sensitive habitat zones (shown in black squares); kayaking routes are expand into previously unused areas (shown in dashed line); and geoduck harvest is prohibited throughout the Inlet; (c) Industry expansion (five new float home leases are added, as shown in black circles; five new oyster tenures are added, as shown in black squares; and wild geoduck harvest is allowed). InVEST model outputs are shown in the remaining 6 panels. (d–f) Ecosystem risk, which is the cumulative risk of human activities (float homes, aquaculture, geoduck harvest) to nearshore habitats (eelgrass, soft, and hard bottom) under the three alternative management scenarios. (g–i) Concentration of fecal coliform bacteria relative to concentrations at float home sources under the three scenarios. (Reprinted from Guerry et al. 2012b, by permission of the publisher, Tayor & Francis Ltd., http://www.tandf.co.uk/journals)

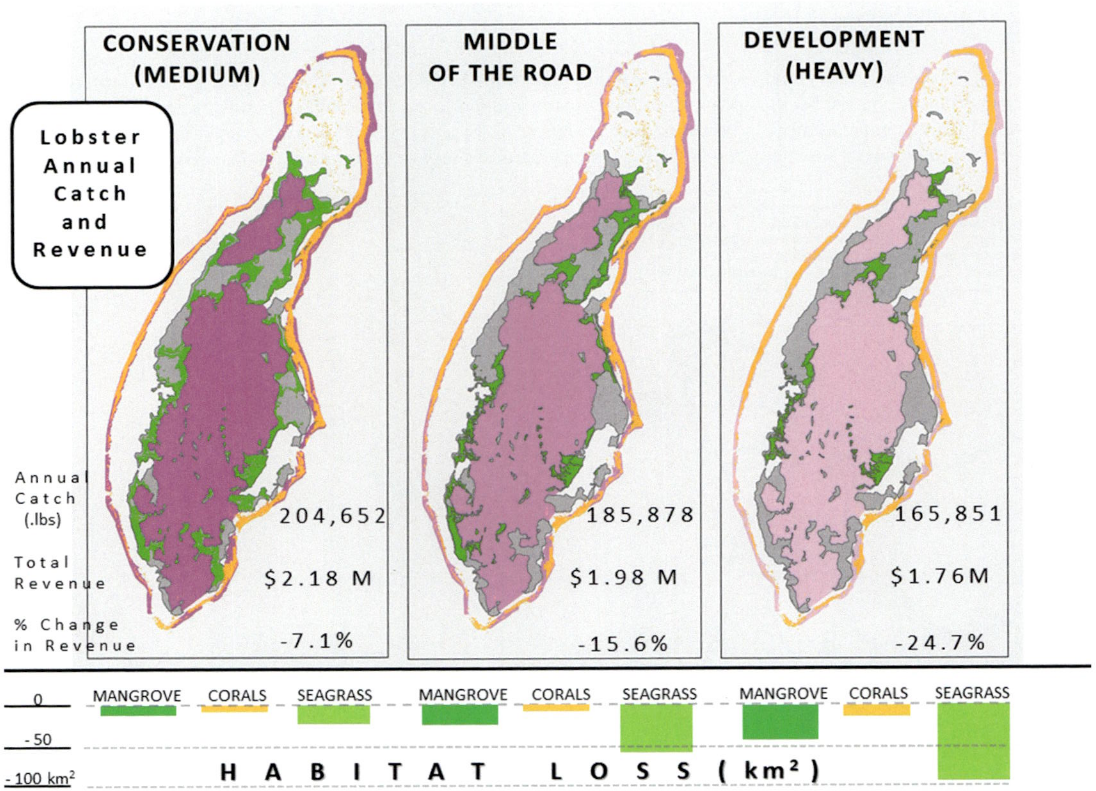

Fig. 8.4 – Outputs from the InVEST lobster fishery and habitat risk assessment model for Turneffe Atoll in Belize. The three panels show outputs from scenarios representing significant conservation efforts, a middle-of-the-road scenario, and a future with heavy coastal development. Darker pink colors indicate more harvest of lobster—with landings, total lobster revenue, and change from current revenue indicated. Colored bars along the bottom represent summarized losses, for each scenario, of each habitat from the habitat risk assessment model. These outputs from the habitat model served as inputs into the fishery model to produce differences in fishery yield.

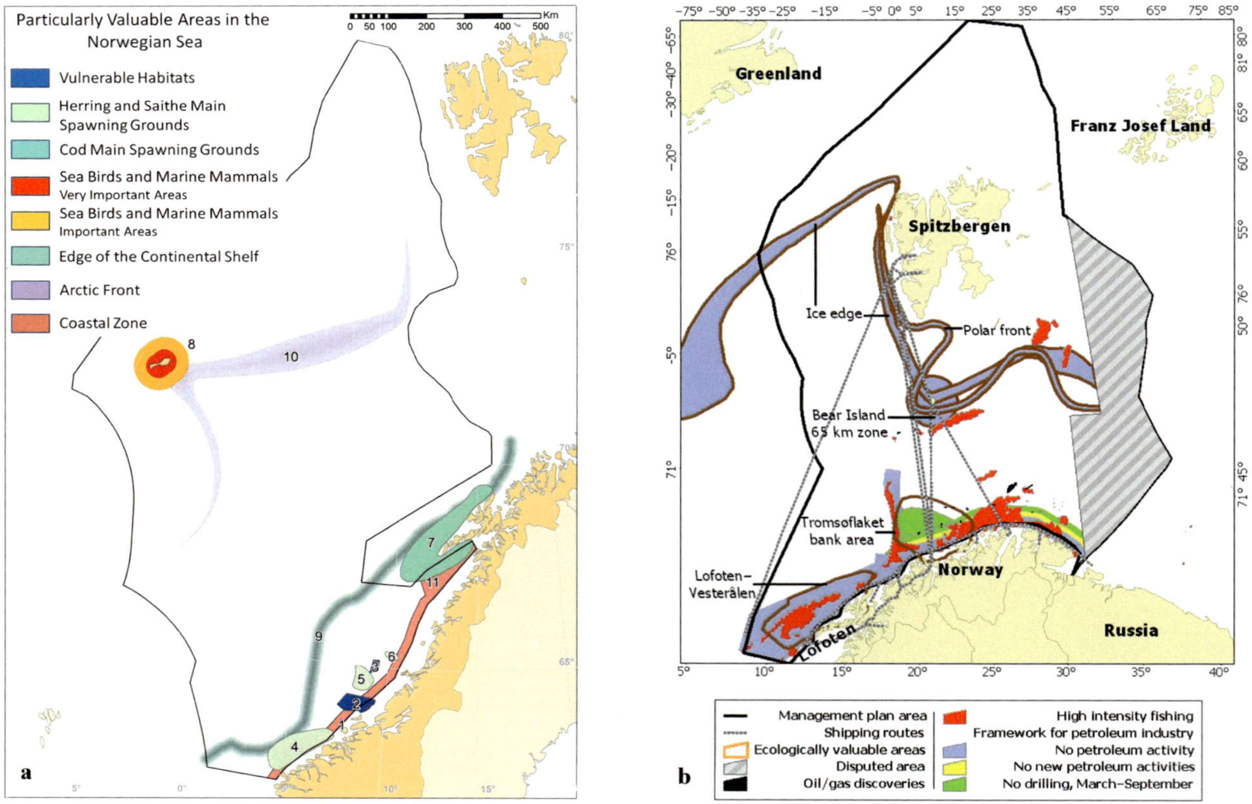

Fig 10.2 – Spatial aspects of Barents and Norwegian Sea Integrated Management Plans. (a) Particularly valuable areas identified in the Norwegian Sea [1 Remman Archipelago; 2 Froan Archipelego and Sula Reef; 3; Iverryggen Reef; 4; Mere Banks; 5; Halten Bank; 6; Sklinna Bank; 7 Vestfjorden/Vesterålen Islands; 8 Jan Mayen; 9 Shelf Edge; 10 Arctic Front; 11 Coastal Zone]. (Figure courtesy of Geir Ottersen, and reproduced with permission from Elsevier, from Ottersen et al. 2011); (b) Integrated management in the Barents Sea. (Figure reproduced with permission from Oxford University Press, from Olsen et al. 2007)

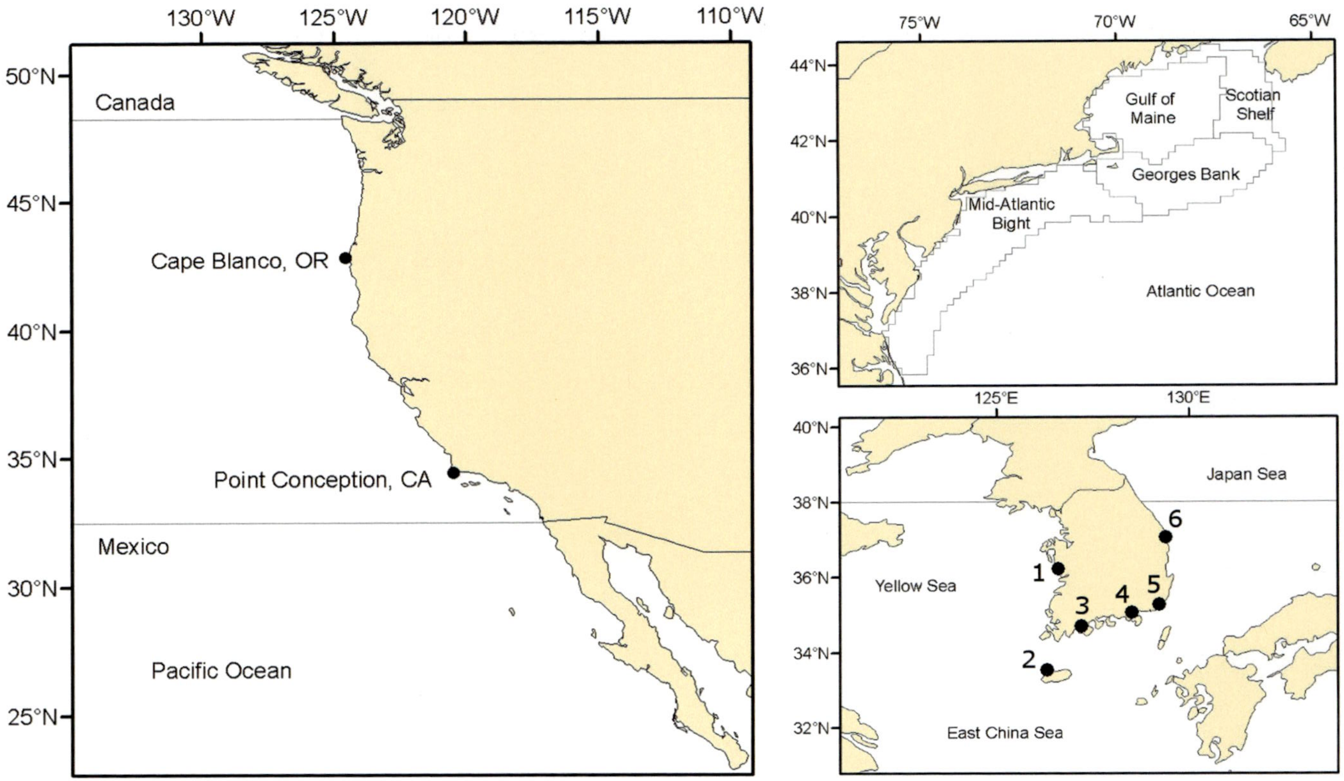

Fig. 11.1 – Maps and place names of three case study regions: California Current LME (left), Northeast Shelf LME (upper right), Korean Marine Ecosystem (lower right: the numbers denote marine ranch areas (1: Taean, 2: Jeju, 3: Jeonnnam, 4: Tongyeong, and 6: Uljin), and major container port (5: Busan).

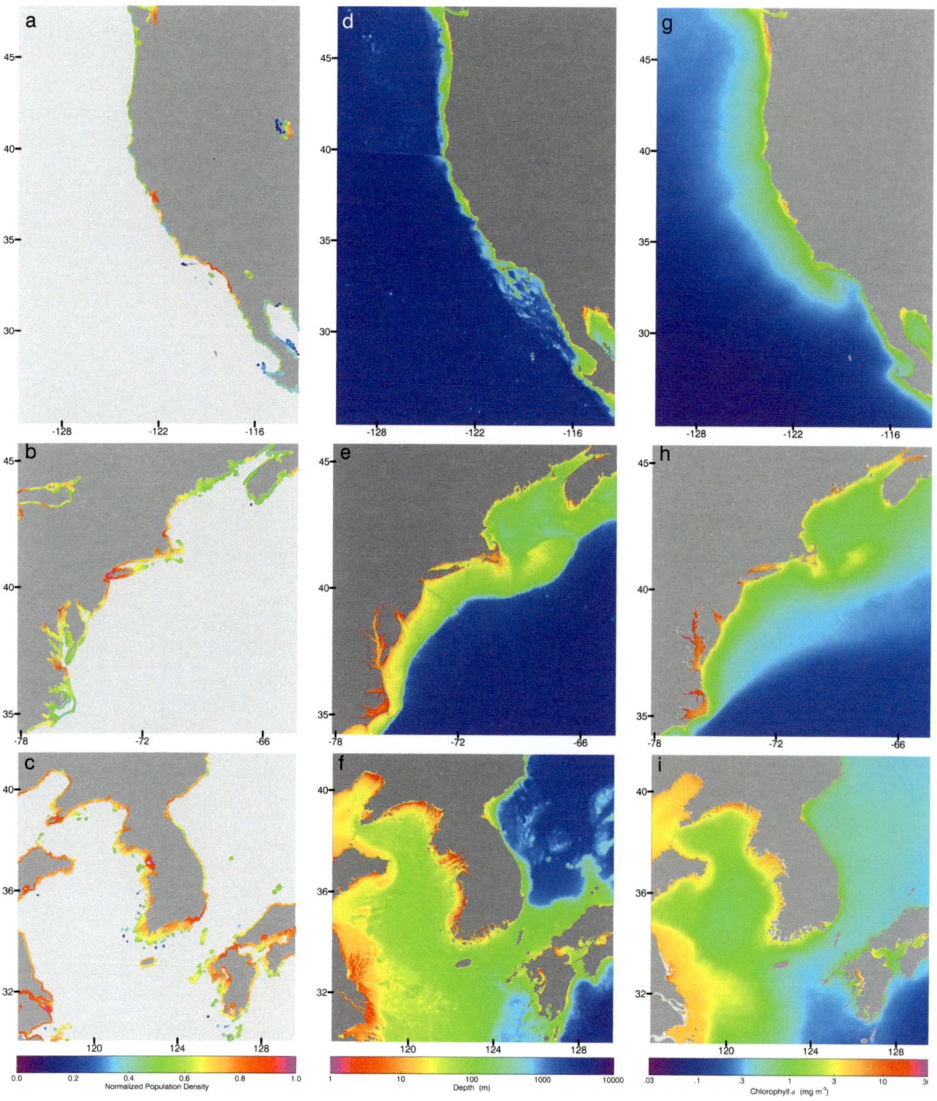

Fig. 11.2 – Maps of (a–c) human population density along the California Current Large Marine Ecosystem (CCLME), the Northeast Continental Shelf Large Marine Ecosystem (NESLME), and the Korean marine ecosystem (KME), (d–f) bathymetry for the CCLME, NESLME, and KME, (g–i) chlorophyll *a* concentrations for the CCLME, NESLME, and KME. (Kimberly Hyde, NOAA, personal communication)

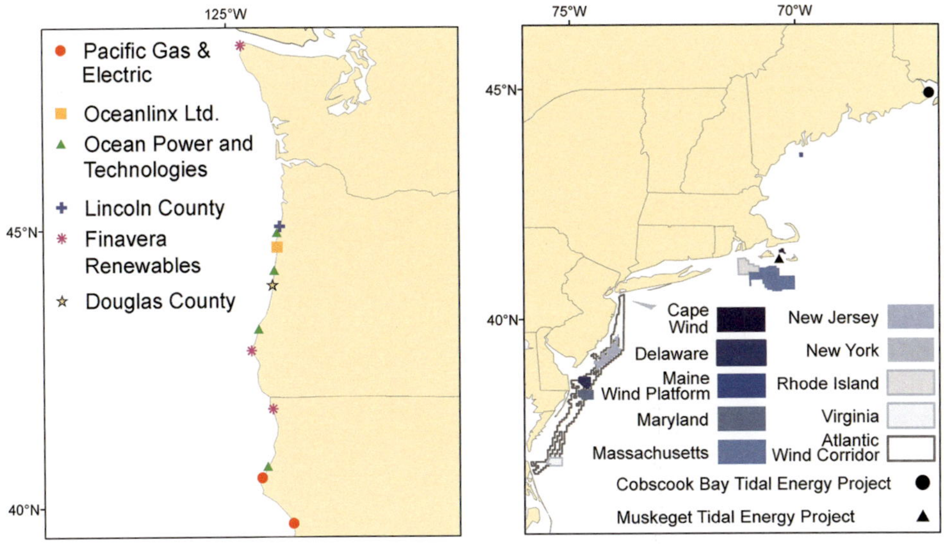

Figure 11.3 – Location of Wave energy projects in the California Current Large Marine Ecosystem (left). Proposed lease tracts for wind farm development, tidal power installations and an energy transmission corridor on the Northeast Continental Shelf (right). (Courtesy Sean Lucey, NOAA)

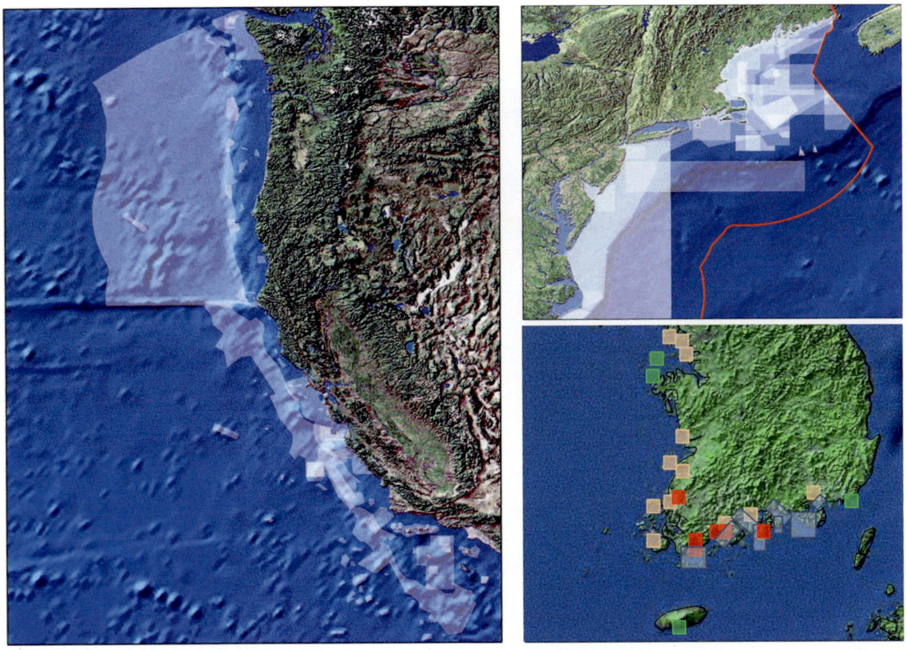

Fig. 11.4 – Locations of Marine Protected Areas on the U.S. Pacific (left), Northeast Coasts (upper right) and southern Korean peninsula (lower right). (Data sources for US locations: NOAA Marine Protected Areas Program. Korea: Chang Zhang [Blue: Fisheries Protected Area (PA); Orange: Wetland PA; Green: Marine Ecosystem PA; Red: Marine Conservation PA])

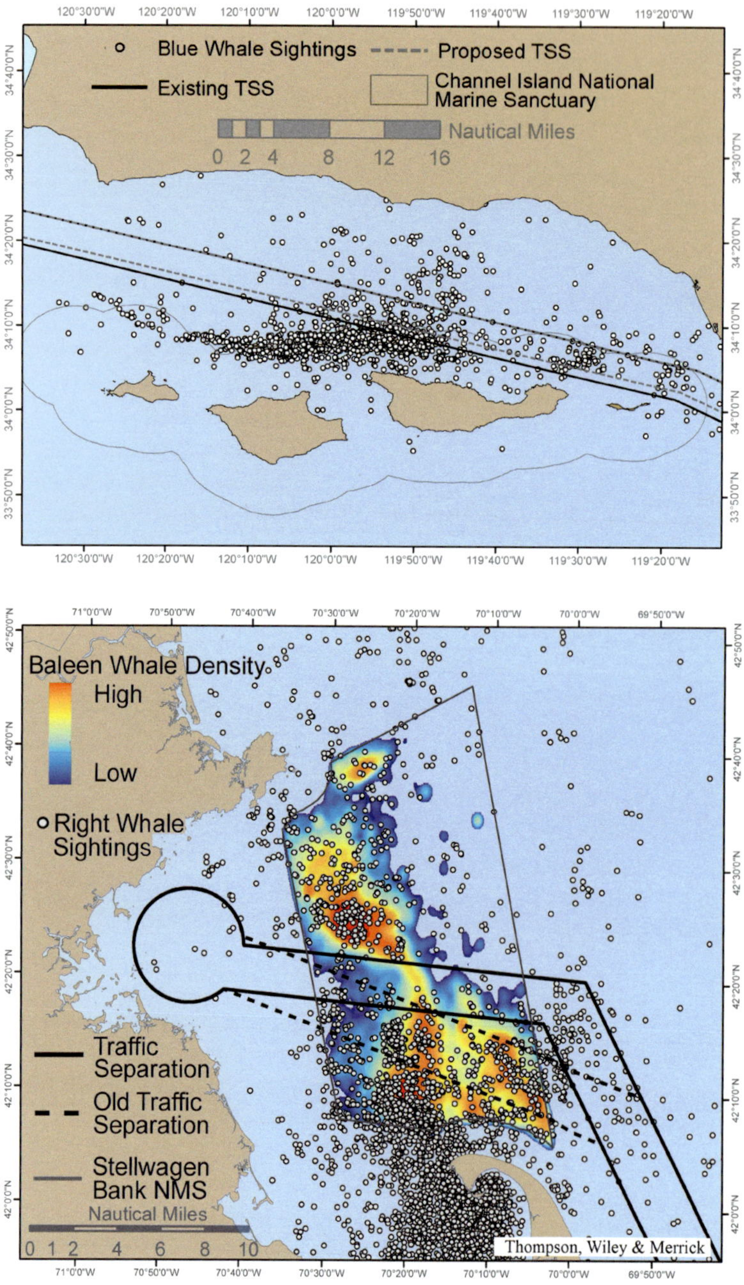

Fig. 11.5 – Proposed 1 nm shift northward of the southbound shipping lane near the Channel Islands National Marine Sanctuary (upper). Also shown is an overlay of blue whale sightings data collected by Channel Islands National Marine Sanctuary volunteers from 1999 through 2010. Realignment of the shipping lanes (TSS) into the Port of Boston, Gulf of Maine in 2007 by the International Maritime Organization to reduce the risk of ship strikes to baleen whales in the Stellwagen Bank National Marine Sanctuary (lower).

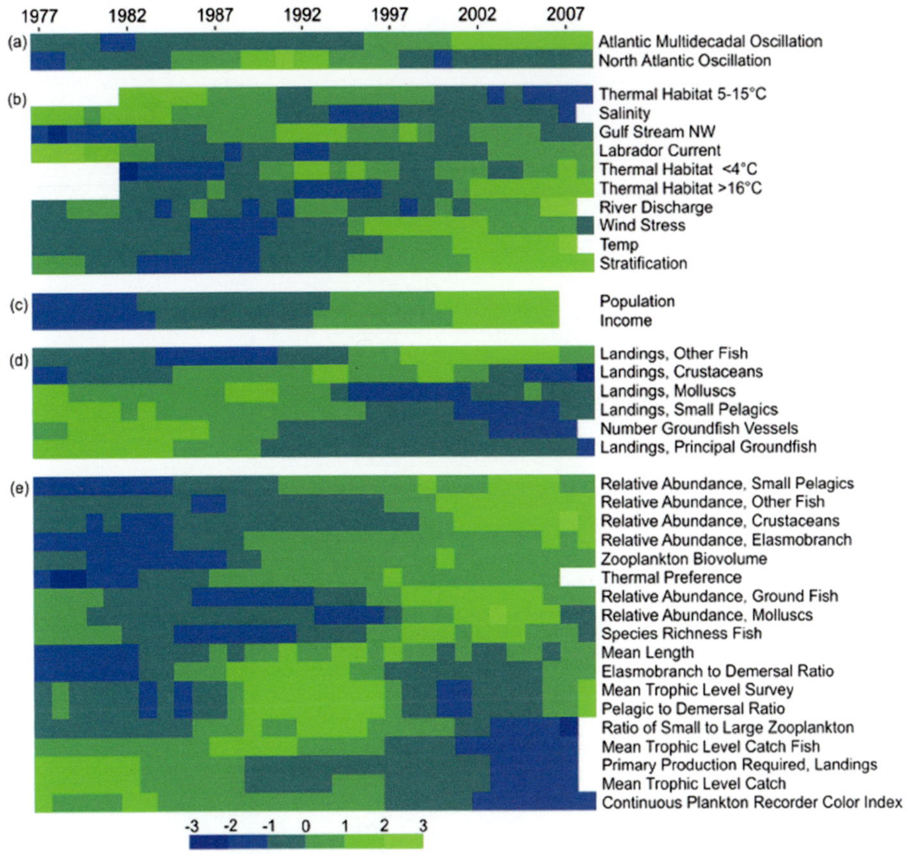

Fig. 11.6 – Normalized indictors for (a) physical drivers, (b) physical pressures, (c) anthropogenic drivers, (d) anthropogenic pressures, and (e) ecosystem state variables. Green colors indicate high values, blue colors represent low values. Blank rectangles indicate no data for time period. (Adapted from Fogarty et al. 2012)

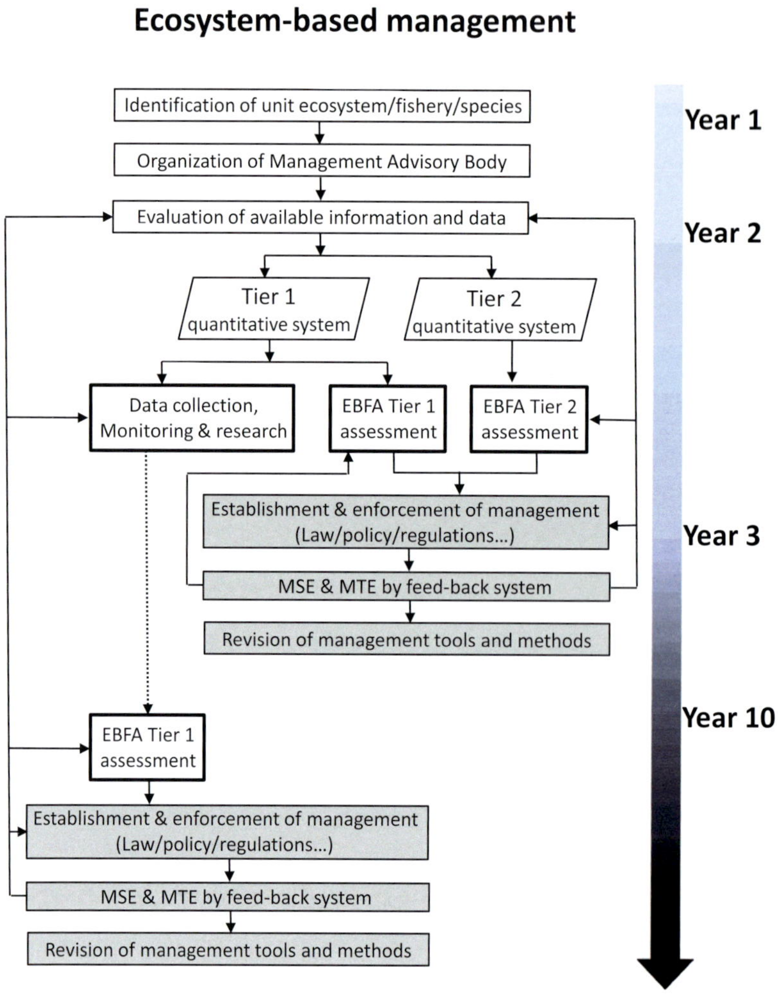

Fig. 11.8 – A flowchart showing the procedure for the ecosystem-based fishery management system in Korea. The overall process includes (a) identification of other user groups, such as aquaculture, sand mining, etc. (b) Spatial units include Large Marine Ecosystems: East Sea, Yellow Sea, East China Sea; Coastal Ecosystems: Various coastal fisheries; Inshore Ecosystems: Fishing village cooperatives, Marine ranches, Self-management units, (c) Consultations and decisions on assessment methods, plans for assessment, and setting management tools and methods, (d) Decision on appropriate analytical methods, (e) specification of Minor species of TAC, Minor species of marine ranches, Target species of stock-rebuilding program, Main species of self-management units and (f) Key species of TAC, Target species of marine ranches, Main species of stock-rebuilding program.

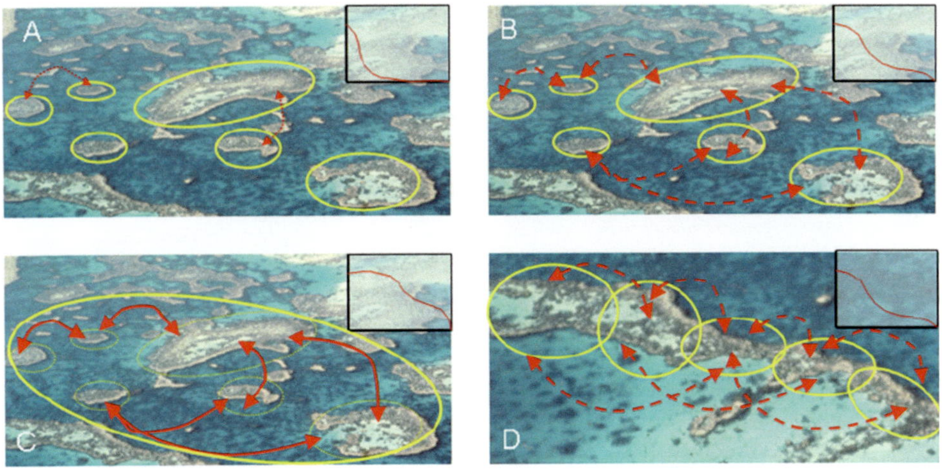

Fig. 12.2 – Each panel shows a patchy array of reef habitat, some of which is occupied by a reef fish species (ovals = local aggregations of fish). Dispersal (chiefly by larvae) among sites is shown by arrows, graded to show slight (A), moderate (B, D) or extensive exchange (C). Mean scale of dispersal is shown as a graph of proportion of larvae (y axis) against distance from source (x axis) in the upper right corner of each panel—mean dispersal distance is least in A, intermediate and identical in B and D, and greatest in C. Cases A, B, and C differ only in the scale of dispersal relative to the scale of patchiness of habitat, yet yield essentially independent local populations (A), a metapopulation (B) in which local populations are sufficiently connected by dispersal for some interaction, and a single, but subdivided, population (C), occupying a number of patches of habitat. Case D is typical of regions where coral reef habitat is more contiguous, yet the spatially explicit mating pattern and scale of larval dispersal still provide a functional metapopulation even though patch structure is primarily an analytical construct. (Reprinted from Sale and Kritzer 2008)

Fig. 12.3 – Fossil corals embedded in a matrix of compressed biogenic sand along the coast of Curacao, Netherlands Antilles (A). The bell tower of the Cathedral of Havana, Cuba (B), one of many colonial structures in Old Havana and around the Caribbean constructed of coral limestone. (Photos by Jake Kritzer)

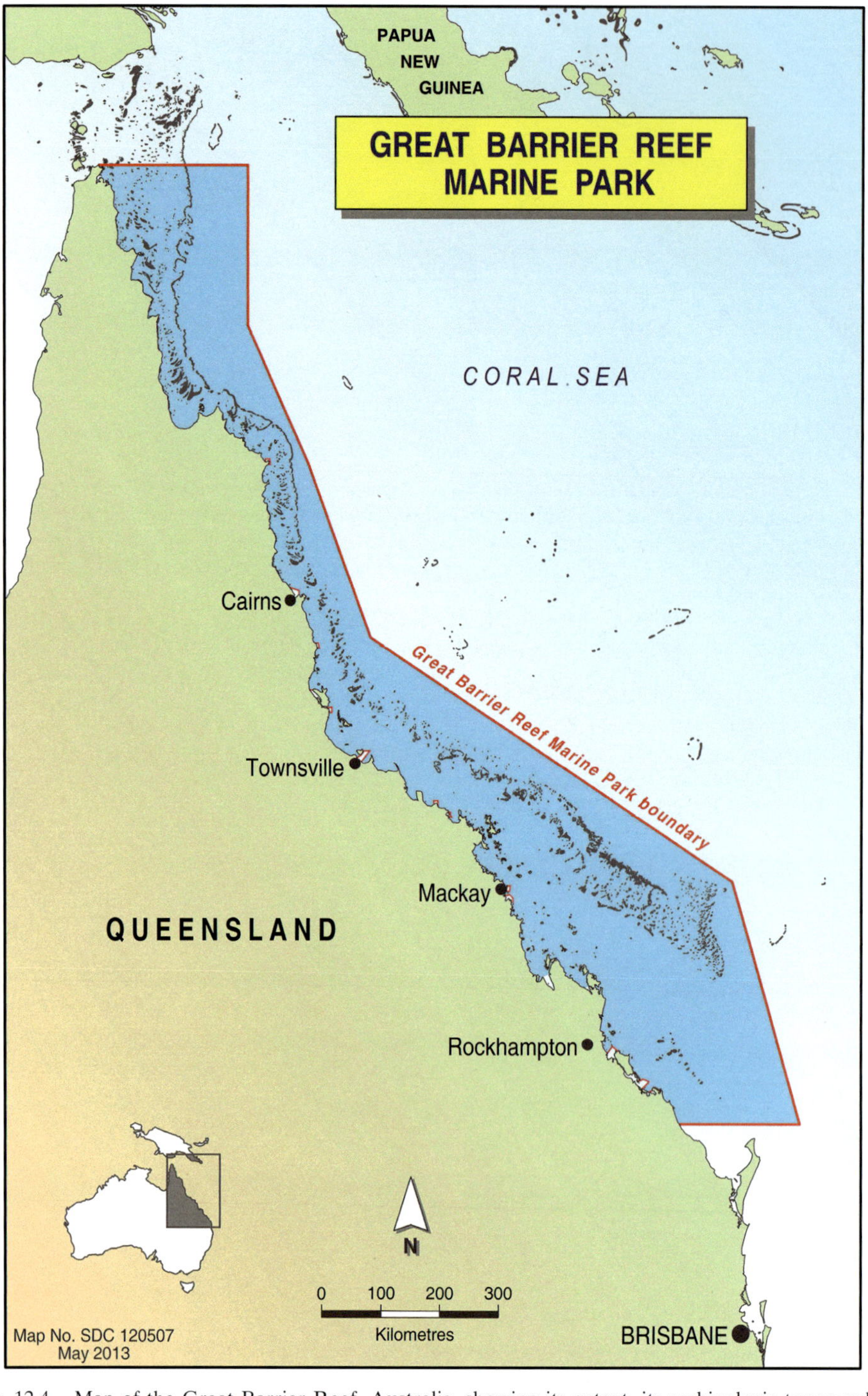

Fig. 12.4 – Map of the Great Barrier Reef, Australia, showing its extent, its archipelagic topography, and the boundaries of the Great Barrier Reef Marine Park. (Map courtesy of the Spatial Data Centre, Great Barrier Reef Marine Park Authority, © Commonwealth of Australia [GBRMPA] 2012)

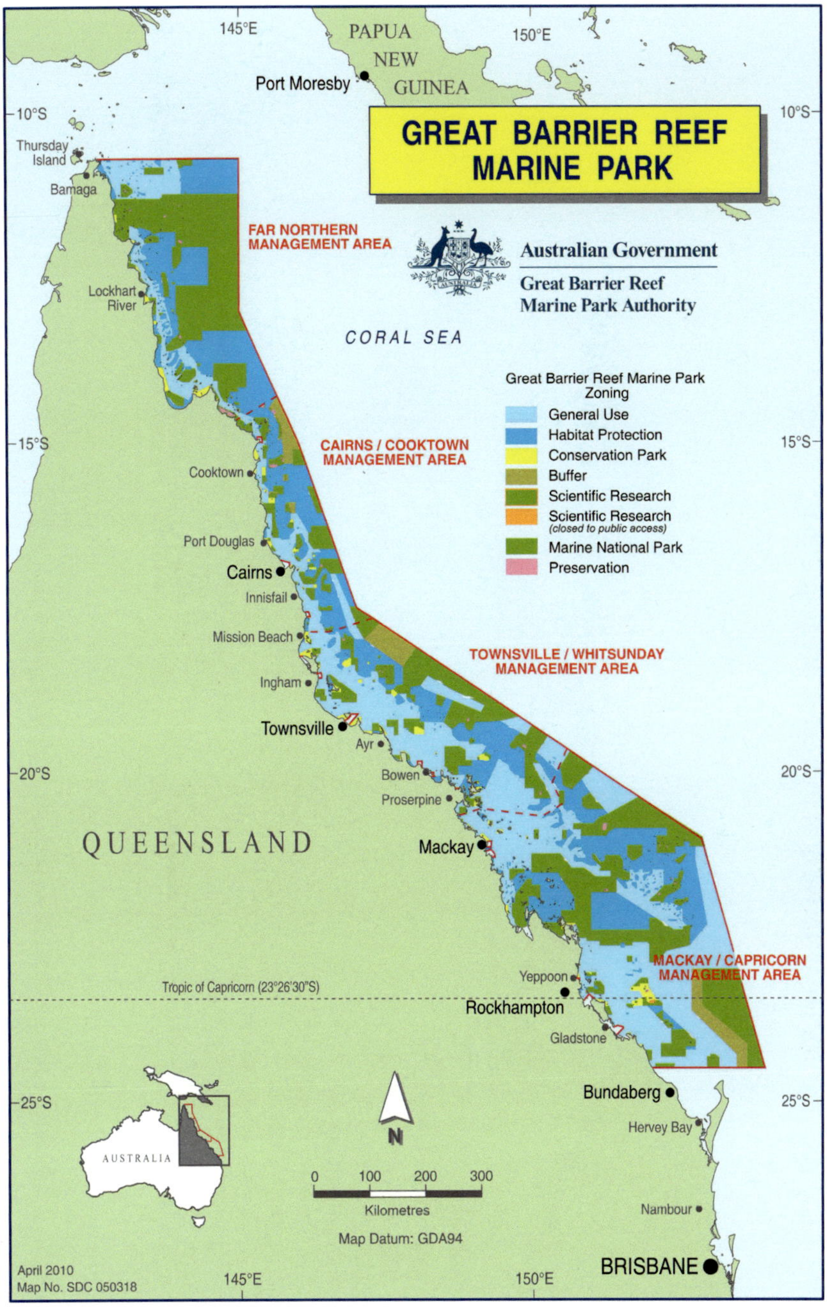

Fig. 12.5 – Map of management zones in the Great Barrier Reef Marine Park implemented in 2004 following extensive review of previous zoning provisions in place since the late 1980s. The revised zoning plan was based on identification of 70 reef and non-reef benthic bioregions within the Great Barrier Reef Maine Park, analyses of major uses of the park, and principles of comprehensiveness, adequacy, and representativeness that ensured that at least 20% of each bioregion was included in multiple 'no-take zones' distributed across the geographic range of each bioregion. (Map courtesy of the Spatial Data Centre, Great Barrier Reef Marine Park Authority, © Commonwealth of Australia [GBRMPA] 2012)

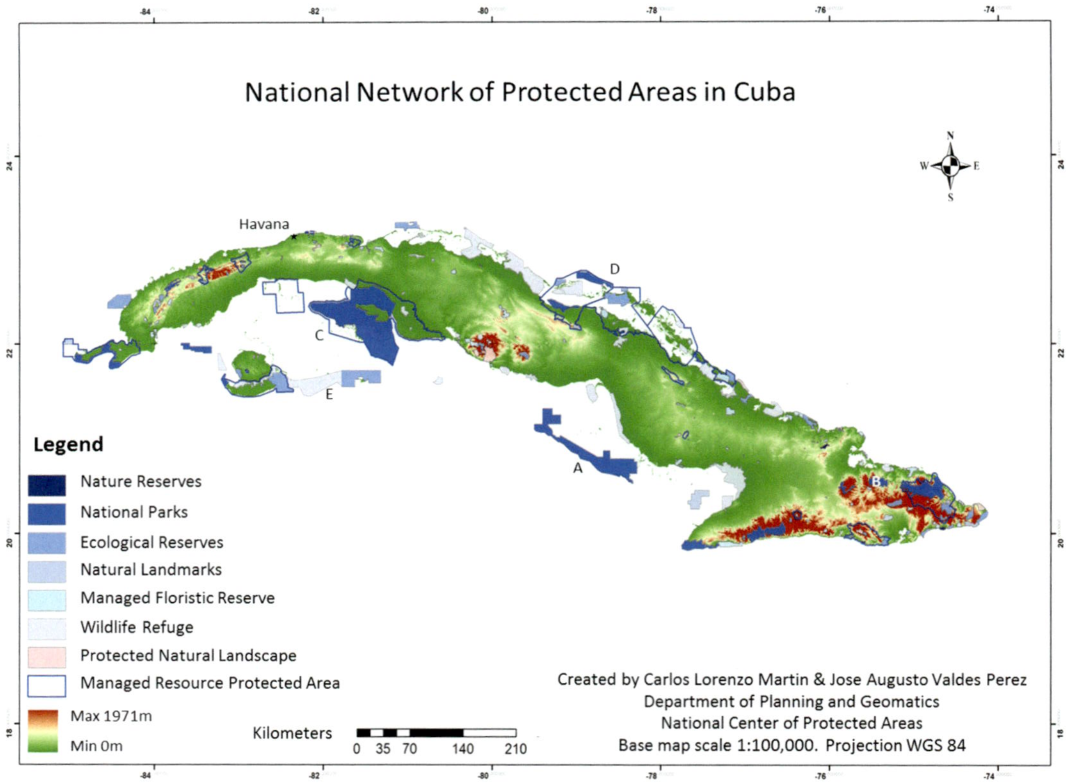

Fig. 12.6 – Marine, coastal, freshwater and terrestrial protected areas in Cuba. Areas discussed in the text include (in the order referenced): Jardines de la Reina (Gardens of the Queen) National Park (A), Sierra Cristal National Park (B), Cienéga de Zapata (Zapata Swamp) National Park (C), Sabana-Camagüey Archipelago (containing multiple protected areas of different designations; D), and Canarreos Archipelago (also containing multiple designations; E).

Fig. 12.7 – Habitat and species diversity in the Jardines de la Reina (Gardens of the Queen) National Park, Cuba, including elkhorn coral reef crest (A), red mangrove sapling fringed by eelgrass (B), Caribbean reef sharks (C), Goliath groupers (D), hawksbill turtle (E), and black coral (F). (Photos A–D by Noel Lopez, and E–F by Fausto de Nevi)

Fig. 12.11 – Differential responses to thermal stress and bleaching among coral species and morphologies in Kenya. Bleaching-sensitive corals include species of branching *Acropora* (A) and plating *Montipora* (B), while species of massive *Porites* (C) are often less affected by bleaching. (Photos by Emily Darling)

Fig. 12.13 – Fish trap typical of those used by Kenya's artisanal fishermen (A), with an escape gap that allows bycatch species to escape (B). (Photos by Tim McClanahan)

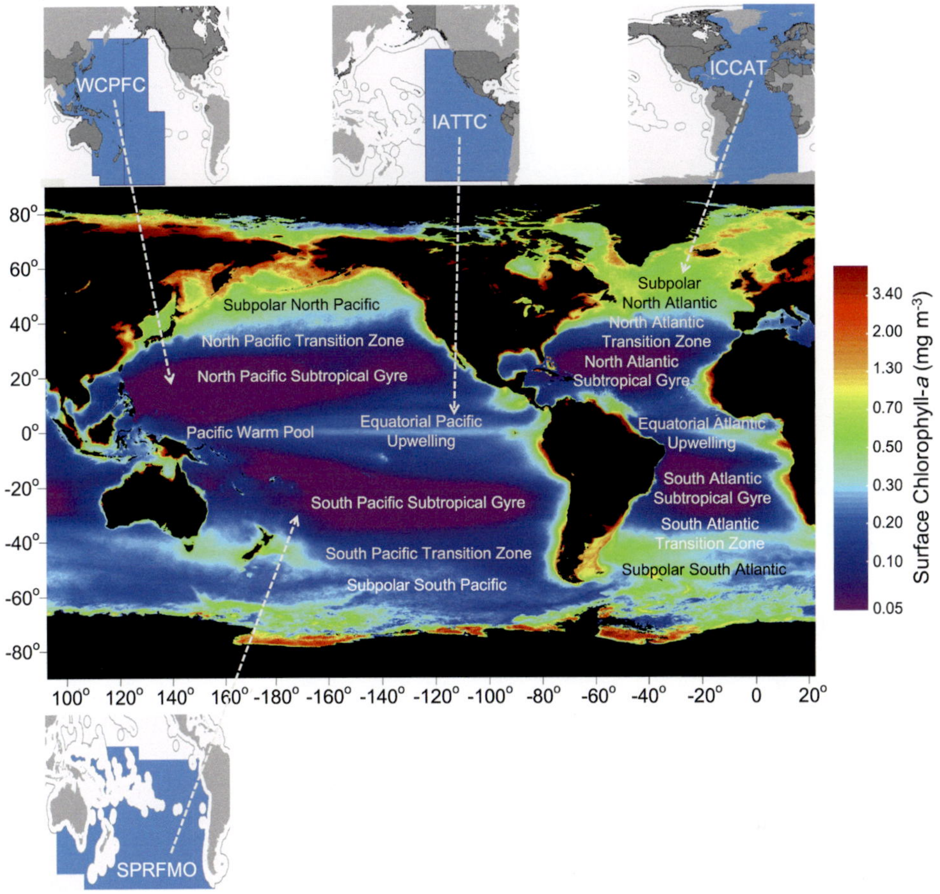

Fig. 13.1 – Surface chlorophyll-*a* estimates (a proxy for phytoplankton biomass) derived from the SeaWiFS ocean color satellite sensor and the major open ocean biomes of the Pacific and Atlantic Oceans. The image is a composite using monthly level-3 data (9 km resolution) from 1997 to 2010 (http://oceancolor.gsfc.nasa.gov/). The color bar for chlorophyll-*a* is on a log scale. The regions highlighted in blue in the perimeter maps are the regional fishery management organization (RFMO) boundaries (http://www.fao.org/fishery/rfb/en) for the major fisheries of the open ocean Pacific and Atlantic Oceans. These organizations are the Western and Central Pacific Fisheries Convention (WCPFC), the Inter-American Tropical Tuna Commission (IATTC), the South Pacific Regional Management Organization (SPRFMO), and the International Commission for the Conservation of Atlantic Tunas (ICCAT). The countries highlighted in gray are members of each respective RFMO.

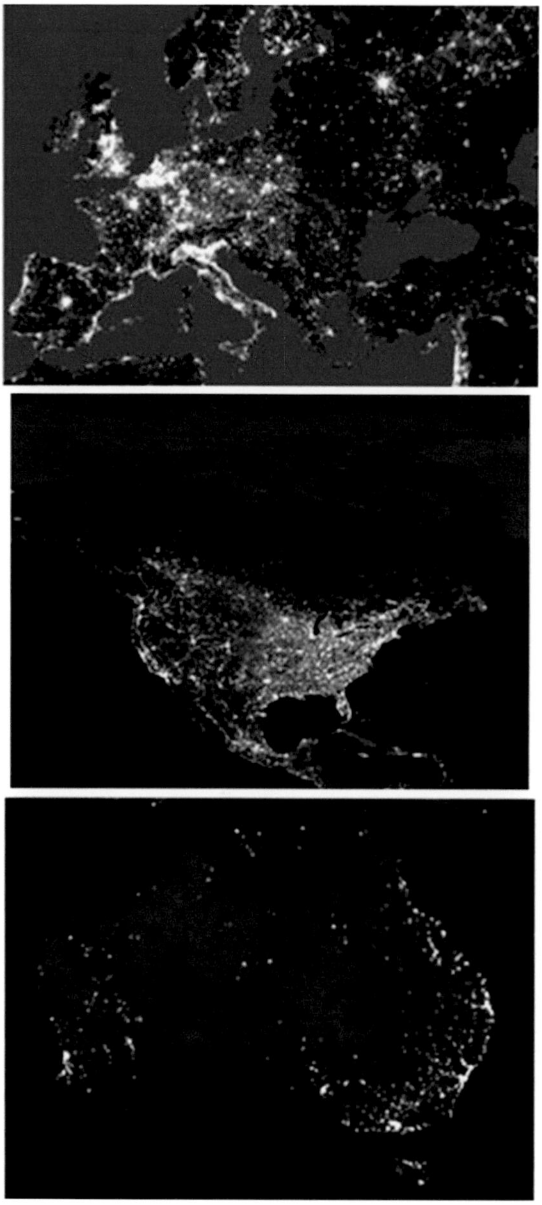

Fig. 14.1 – Satellite night images of Europe, North America, and Australia, revealing areas of urbanization along the coasts. (Google Earth)

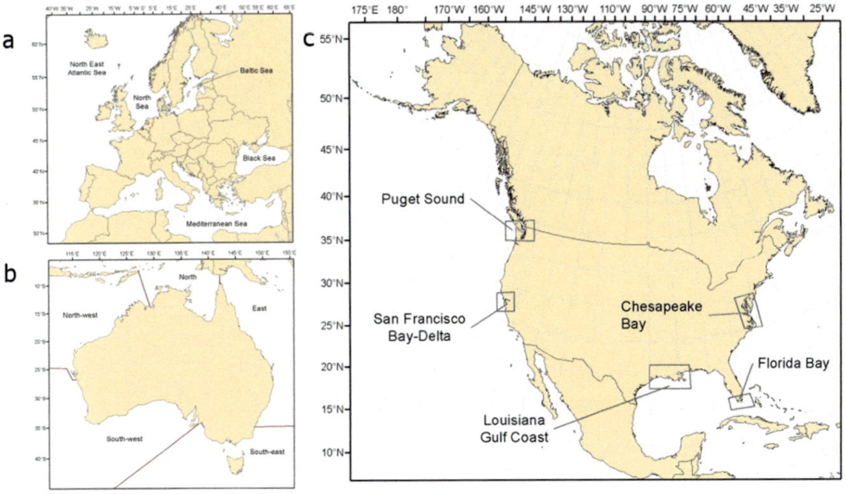

Fig. 14.2 – a. Europe. b. Australia. c. North America. The chapter focuses on EBM in highly impacted coastal areas of these three broad geographical regions. (Map courtesy of Sean Lucey, NEFSC)

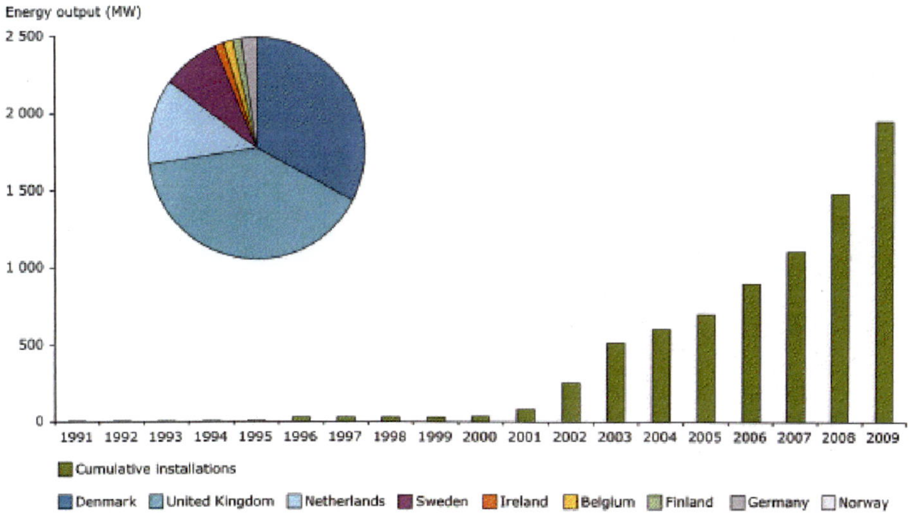

Fig. 14.3 – European wind energy production. (EEA, 2010)

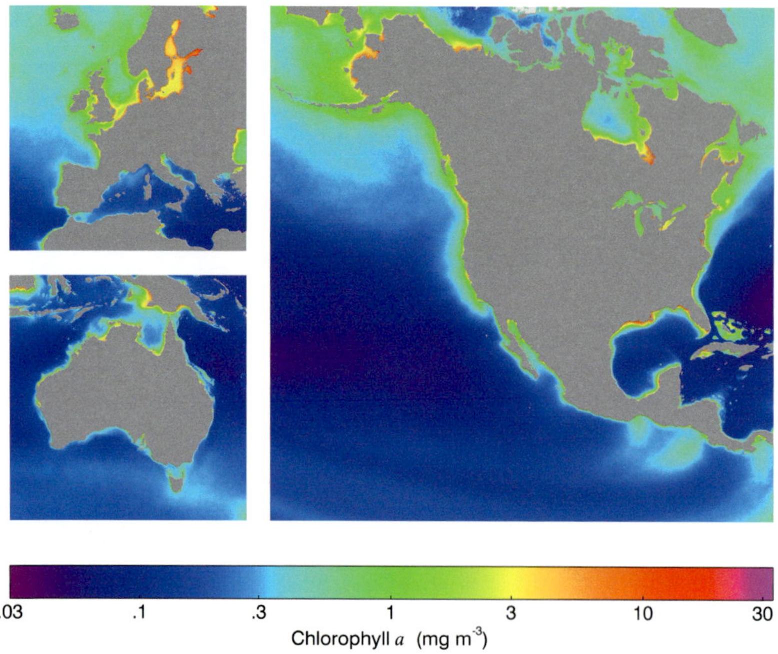

Fig. 14.4 – Chlorophyll distributions in waters of a) Europe, b) Australia, and c) North America, indicating the high primary productivity of near-shore coastal areas. (Map courtesy of Kim Hyde, NEFSC)

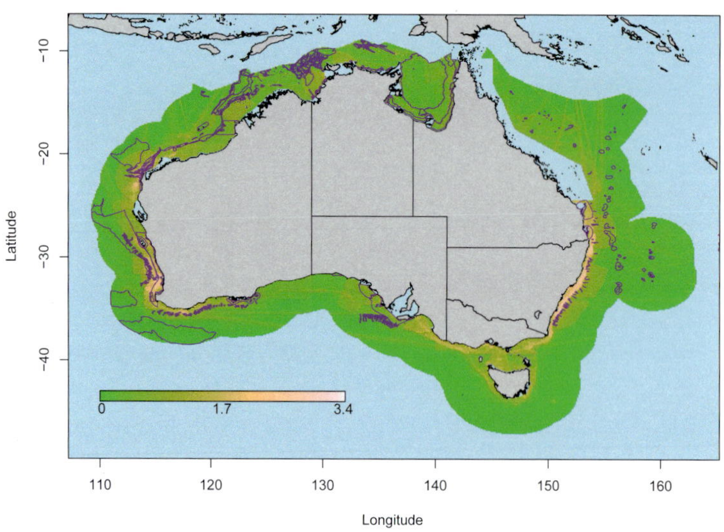

Fig. 14.5 – Standardized composite of human pressures in Australia's Exclusive Economic Zone. Features in purple are KEFs (see text). (Source: Hayes et al. 2011)

Raudsepp-Hearne, C., G. D. Peterson, and E. M. Bennett. 2010. Ecosystem service bundles for analyzing tradeoffs in diverse landscapes. *Proceedings of the National Academy of Sciences of the United States of America* **107:**5242–5247.

Reinhart, C. M. and K. S. Rogoff. 2011. From Financial Crash to Debt Crisis. American *Economic Review* **101:**1676–1706.

Rosenberg, A. A. 2006. Regional governance and ecosystem-based management of ocean and coastal resources: can we get there from here. *Duke Environmental Law and Policy Forum* **16:**179–185.

Ruckelshaus, M., T. Klinger, N. Knowlton, and D. R. Demaster. 2008. Marine ecosystem-based management in practice: Scientific, and governance challenges. *Bioscience* **58:**53–63.

Scheffer, M. and S. R. Carpenter. 2003. Catastrophic regime shifts in ecosystems: linking theory to observation. *Trends in Ecology & Evolution* **18:**648–656.

Sheridan, B., S. D. Baker, N. S. Pearre, J. Firestone, and W. Kempton. 2012. Calculating the offshore wind power resource: Robust assessment methods applied to the US Atlantic Coast. *Renewable Energy* **43:**224–233.

Stern, P. C. 1993. A Second Environmental Science: Human-Environmnetal Interactions. *Science* **260:**1897–1899.

Sugihara, G., J. Beddington, C.-h. Hsieh, E. Deyle, M. Fogarty, S. M. Glaser, R. Hewitt, A. Hollowed, R. M. May, S. B. Munch, C. Perretti, A. A. Rosenberg, S. Sandin, and H. Ye. 2011. Are exploited fish populations stable? *Proceedings of the National Academy of Sciences of the United States of America* **108:**E1224-E1225.

System of Environmental-Economic Accounting (SEEA). 2003. *Handbook of national accounting: Integrated environmental and economic accounting 2003.* United Nations, European Commission, International Monetary Fund, Organisation for Economic Co-operation and Development, World Bank. https://unstats.un.org/unsd/envaccounting/pubs.asp.

Tallis, H., P. Kareiva, M. Marvier, and A. Chang. 2008. An ecosystem services framework to support both practical conservation and economic development. *Proceedings of the National Academy of Sciences of the United States of America* **105:**9457–9464.

Thomas, W. L. 1956. *Man's Role in Changing the Face of the Earth.* University of Chicago Press, Chicago.

Turner, B. L. I., W. Clark, R. Kates, J. Richards, J. Mathews, and W. Meyer. 1990. *The Earth as Transformed by Human Action: Global and Regional Changes in the Biosphere over the Past 300 Years.* Cambridge University Press, Cambridge, United Kingdom.

United Nations Conference on Environment and Development (UNCED). Rio de Janeiro, Brazil, 3–14 June 1992.

Vervoort, J. M., K. Kok, R. van Lammeren, and T. Veldkamp. 2010. Stepping into futures: Exploring the po-tential of interactive media for participatory scenarios on social-ecological systems. *Futures* **42:**604–616.

Vitousek, P. M., H. A. Mooney, J. Lubchenco, and J. M. Melillo. 1997. Human domination of earth's ecosystems. *Science* **227:**494–499.

Voinov, A., C. Fitz, R. Boumans, and R. Costanza. 2004. Modular ecosystem modeling. *Environmental Modelling & Software* **19:**285–304.

Walker, B. and J. A. Meyers. 2004. Thresholds in ecological and social-ecological systems: a developing database. *Ecology and Society* **9:**3.

Wandersee, S. M., L. An, D. López-Carr, and Y. Yang. 2012. Perception and decisions in modeling coupled human and natural systems: a case study from Fanjingshan National Nature Reserve, China. *Ecological Modelling* **229:** 37–49.

White, C., B. S. Halpern, and C. V. Kappel. 2012. Ecosystem service tradeoff analysis reveals the value of marine spatial planning for multiple ocean uses. *Proceedings of the National Academy of Sciences of the United States of America* **109:**4696–4701.

Worm, B., E. B. Barbier, N. Beaumont, J. E. Duffy, C. Folke, B. S. Halpern, J. B. C. Jackson, H. K. Lotze, F. Micheli, S. R. Palumbi, E. Sala, K. A. Selkoe, J. J. Stachowicz, and R. Watson. 2006. Impacts of biodiversity loss on ocean ecosystem services. *Science* **314:**787–790.

Chapter 10. Ecosystem-Based Management in High Latitude Ecosystems

SARAH GAICHAS

Ecosystem Assessment Program
National Oceanic and Atmospheric Administration
Northeast Fisheries Science Center

CHRISTIAN REISS

Antarctic Ecosystem Research Division
National Oceanic and Atmospheric Administration
Southwest Fisheries Science Center

MARIANO KOEN-ALONSO

Fisheries and Oceans Canada
Northwest Atlantic Fisheries Centre

Contents

1. Introduction

In this chapter we compare and contrast progress towards Ecosystem-Based Management (EBM, defined in Part 1) in high latitude marine ecosystems. We consider EBM implementations and the challenges presented within both northern and southern hemisphere high latitude marine ecosystems (Fig. 10.1). Our geographic definition of high latitudes is above 45° in each hemisphere. In the southern hemisphere, we define the Antarctic marine ecosystem within the Southern Ocean as waters extending from the continent of Antarctica to the Antarctic Convergence, an oceanographic front which ranges from approximately 45–60°S latitude (Kock et al. 2007). In the northern hemisphere, we examine both "high Arctic" and "subarctic" (cool-temperate/boreal) marine ecosystems, which have some common and some distinct characteristics, if not fixed latitudinal boundaries. For example, although the Arctic is generally defined as the region north of the Arctic Circle (66.33°N latitude), subarctic ecological characteristics extend seasonally well above this latitude into the Barents Sea due to the influence of warmer North Atlantic

The Sea, Volume 16, edited by Michael J. Fogarty and James J. McCarthy
ISBN 978-0-674-07270-1

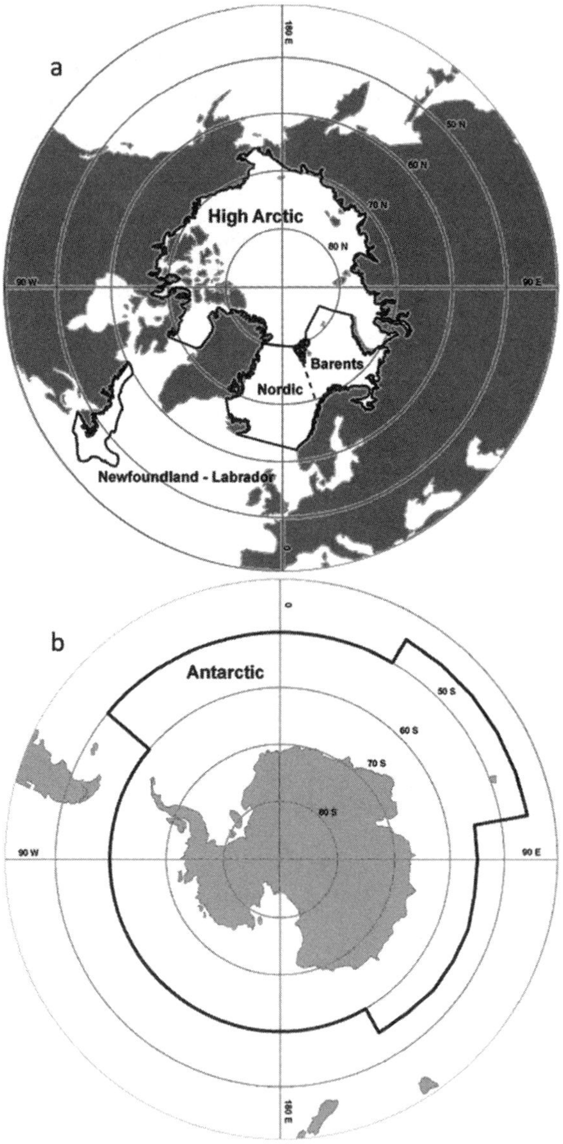

Fig. 10.1 – High latitude marine ecosystems of the (a) Northern, and (b) Southern hemisphere. (Figure by Scott Geis)

currents. Further, regions included in the Arctic Climate Impact Assessment (ACIA 2005) extend below the Arctic Circle into the subarctic waters of the Bering Sea and Aleutian Islands in the Pacific, and onto the Labrador shelf in the North Atlantic. We therefore roughly define the boundary between the high Arctic and subarctic as the Polar Front, such that the ecosystems of the high Arctic and Antarctic will be more comparable, and to distinguish the more productive subarctic systems as we examine progress towards EBM.

In the rest of the chapter, we first outline the general ecosystem services provided by and threats and impacts to these ecosystems. We then review EBM progress in high latitude ecosystems by examining four examples across a range of international, multinational, national, and regional implementations of EBM. This treatment is not intended to be exhaustive, but illustrative of a range of systems and governance structures, and to identify successful partial or full EBM examples as suggested by Leslie and McLeod (2007). We compare the international contexts of the Antarctic and high Arctic marine ecosystems, and the primarily multinational and/or national contexts of the European (Barents Sea), and the Canadian North Atlantic (Newfoundland and Labrador) subarctic continental shelves. For each example system, we review key ecosystem services, major human use patterns, and assessments of threats and impacts in each of the ecosystems (Table 10.1). The current EBM framework is described for each ecosystem, including a list of policy objectives and progress towards achieving them. Finally, we look for common issues for EBM across high latitude regions, and ask what challenges remain in achieving EBM, and what conclusions can be drawn regarding best practices and lessons learned.

2. High Latitude Marine Ecosystems—What Makes Them Distinctive?

Several physical characteristics distinguish high latitude marine ecosystems from all others worldwide: the extreme seasonality of temperature and daylight, and the presence of sea ice. Above 66.33°North or South latitude, seasonal daylight variation is most extreme (full sun 24 hours at the summer solstice, full darkness 24 hours at the winter solstice), with complete light or darkness lasting the full season at the poles. Winters are extremely cold (-40 to 0°C air temperature on average in marine regions, with a global extreme of -89°C recorded on the continent of Antarctica), and summer average temperatures can range from -10 to 10°C (and higher over land). Sea ice seasonally covers large proportions of the open water in these ecosystems, forcing all inhabitants to adapt to its annual cycle of advance and retreat. Human populations at high latitudes tend to have some of the lowest densities on earth, largely due to these conditions (and in the Antarctic, due to the isolation of the continent). However, there is increasing interest in the natural and mineral resources of these regions as human populations expand.

The seasonal extremes of temperature, light duration, and sea ice formation and retreat shape the major ecosystem services and affect nearly all human uses of high latitude seas. At the global scale, both terrestrial and marine habitats in high latitudes are important for climate regulation. Temperature gradients and ice extent influence temperature regulation, ocean circulation and carbon sequestration, the major global-scale supporting and regulating services of high latitude ecosystems (Millennium Ecosystem Assessment 2005; Anisimov et al. 2007). Large areas of ice and snow covered surfaces at high latitudes reflect rather than absorb incoming sunlight, helping to control global temperature, and sinking of cold high salinity ocean water at high latitudes drives global ocean circulation (Loeng et al. 2005; Rahmstorf 2006). Marine food webs export carbon from the atmosphere to the ocean floor (Millennium Ecosystem Assessment 2005). While each ecosystem interacts differently with global climate as described in the case studies below, it is important to note that high Arctic and Antarctic marine ecosystems provide supporting and regulating ecosystem services extending well beyond their boundaries.

TABLE 10.1.
Comparison of geographic, productive, and human attributes of four high latitude marine ecosystems.
World ocean total average catch 2000–2010 from FAO (2012).
Total proven world oil reserves from BP (2012).

	Antarctic	High Arctic	Barents Sea (Nordic Seas)	Newfoundland Labrador
Latitude range (degrees)	45–77 °S	66–90 °N	70–80 °N	42–60 °N
Area (km^2)	31,800,000	11,500,000 (CIA 2011)	1,400,000 (1,156,000)	785,000
Seasonal ice extent (average min- max, km^2, recent trends)	3–18 million Stable/increasing overall, WAP decreasing, Ross increasing	7–13 million 3.4 million 2012 low min, declining	150,000–865,000 401,000 2012 low max, declining	291,598 87,974 (2010)–477,885 (1990) Declining since 1990s
Primary production (gC/m^2y)	16–100 (Dayton et al. 1994)	20–70; Chukchi 20–400 (Carmack et al. 2006)	20–200 (Carmack et al. 2006)	150–300 (Hunt, Jr. and Drinkwater 2005)
Commercial fisheries catch				
2000–2010 average t/y	149,200	100–10,000	886,000 (3,242,000)	302,450
average t/km^2y	0.005	0.000	0.633	0.385
% world catch	0.2% (FAO 2012)	0.0–0.1% (Zeller et al. 2011, FAO 2012, ICES 2012)	1.1% (ICES 2011a, b, d, c, e, 2012)	0.4% (NAFO 2012)
Commercial marine mammal catch				
2000–2010 average n	490 whales	0‡	21,240 seals 560 whales (168,480 seals* 40 whales†)	232,000 seals
% world catch	0.1% (FAO 2012)	0% (FAO 2012, IWC 2012)	5% (41%) (ICES 2011f, IWC 2012)	48% (DFO 2010, Hammill et al. 2011)
Hydrocarbon potential				
oil, million barrels	Unknown	60,362	10,930	1,933
gas, trillion cu ft	Unknown	789.8	379.786	11.048
% of 2012 proven world reserves	Unknown	3.7% oil, 10.7% gas	0.7% oil, 5.2% gas	0.1% oil, 0.2% gas
Current offshore hydrocarbon production	0	0	9.33 million cu ft gas per year (2007–2010 average)	112 million barrels oil (2002–2011 average)

Human population (% indigenous)	<5,000 (0.0%)	2,070,000 (10%) 13,000 USA (15%) 40,000 Canada (50%) 55,000 Greenland (80%) 2,000,000 Russia (4%)	500,000 (Norway on Barents Sea) (15%) 1,160,000 (Russia on Barents Sea) (4%) 2,600,000 (OSPAR Arctic region) (OSPAR 2000, 2010)	500,000 (entire province) (3%) (Census 2011, Statistics Canada—www.statcan.gc.ca)
Governance	International Treaty	International Forum, National policies (overlapping)	International Treaty, Multinational agreements, National policies (overlapping)	National policy within EEZ, international management outside EEZ, and international agreements to for straddling stocks (Non-overlapping, complementary)

*Greenland Northeast and Northwest Atlantic seal landings reported to FAO; includes seals taken in subarctic Greenland.
†Whales taken in Iceland EEZ; includes whales taken in subarctic Iceland.
‡Subsistence (non-commercial) marine mammal takes in subarctic Alaska averaged 2,000 animals between 2000–2008 (Wolfe et al. 2009), and 50–100 bowhead whales are taken in the Alaskan high Arctic annually; other high latitude subsistence marine mammal catches are uncertain and not included here.

Primary and secondary production in these systems is seasonally intense, at times achieving levels of annual production comparable with mid-latitude systems in half or less the growing season (Loeng et al. 2005; Ducklow et al. 2007). Food chains are relatively short and animals have high lipid storage capacity and content at all trophic levels (Grebmeier et al. 2006; Weslawski et al. 2006; Falk-Petersen et al. 2009; Weslawski et al. 2009). Provisioning services based on these food webs and supported by sea ice cycles are strongly linked to cultural ecosystem services for human populations surrounding high latitude seas, particularly in the northern hemisphere where populations are higher (Table 10.1). Subsistence hunting and fishing have sustained humans for millennia in northern hemisphere high latitude ecosystems, with indigenous populations deriving socio-cultural, spiritual, and religious benefits from these systems as well (Nuttall et al. 2005; Huntington & Pungowiyi 2009).

Given the importance of ice to high latitude ecosystem services, climate change processes that alter the ice cycle lead the list of threats to the sustainability of these services (ACIA 2005; Millennium Ecosystem Assessment 2005; Anisimov et al. 2007). Potential threats to high latitude ecosystems include climate change, hydrocarbon and mineral industrial development, contaminants, fisheries, ultraviolet radiation and exotic species introductions, as summarized in the Millennium Ecosystem Assessment (2005). In the case studies below we highlight more specific threats to each ecosystem.

To this point we have described the similarities between northern and southern hemisphere high latitude systems. However, the systems have important physical differences and contrasting management frameworks addressing specific aspects of the above threats and impacts. Physically, northern and southern hemisphere high latitude marine ecosystems are nearly opposite in terms of land-sea-ice configurations. The high Arctic Ocean and subarctic seas are surrounded by continents with shallow wide shelves and substantial freshwater inputs, while Antarctica is a continent surrounded by ocean with deep narrow shelves and little freshwater influence. In the Southern Ocean, winter sea ice can extend to 60°S with a total ice area of 18 million km^2, exposing the ice edge habitat to daily light even during winter. In the Arctic, sea ice extent is constrained by continents, with the early spring ice edge in relative darkness above 70°N in the Barents Sea or confined to the better lit waters of the subarctic shelves. Geological history combined with these physical characteristics has led to different patterns in biodiversity and ecological production between the Arctic/subarctic and Antarctic seas (Dayton et al. 1994). Humans have inhabited lands surrounding northern hemisphere high latitude seas for millennia, but have only recently occupied Antarctica, and remain at extremely low densities there. Politically, the high Arctic Ocean includes the territorial seas of six adjacent nations as well as (ice-capped) international waters, thus falling under national and international governance structures, while the Antarctic ecosystem is mostly within international waters and is governed by an international treaty. Both areas have international maritime boundary disputes (CIA 2011). Individual subarctic marine systems are smaller in physical scale and in closer proximity to modern human population centers than the high Arctic and Antarctic, with EBM falling within multinational, national, and regional jurisdictions. The combination of physical, ecological, and political contrasts between these high latitude ecosystems drives some of the contrasts in progress towards EBM that we detail in each section below.

3. EBM in Four High Latitude Ecosystems

3.1 Antarctic: an international context for EBM

This section focuses on EBM in the Antarctic, within the management boundaries of the Commission for the Conservation of Antarctic Marine Living Resources (CCAMLR). CCAMLR is responsible for managing most living marine resources from the edge of the continent to its northern boundaries (Fig. 10.1). Overall, the Antarctic Treaty System (ATS), which includes the Convention establishing CCAMLR, provides for the protection and permitting of activities, including research, tourism and development on the continent and in the Southern Ocean.

3.1a Ecosystem services, human use patterns, threats and impacts
The Southern Ocean and Antarctic continent provide supportive and regulatory ecosystem services of global consequence. However, the lack of permanent human settlements and an existing governance structure (the ATS) minimize or reduce many threats and impacts. The most important global ecosystem services provided are climate regulation services, provisioning services (for krill and finfish), and cultural services (principally scientific study and tourism). Petrochemical and mining issues are inconsequential at present.

Global and regional climate regulation services are disproportionately evident in the Antarctic. Despite the relatively small size of the Southern Ocean, it is responsible for ~40% of the world's oceans CO_2 uptake (Sabine et al. 2004). This outsized effect on climate means that changes in the Southern Ocean are likely to have dramatic impacts globally. The Antarctic is experiencing some of the most rapid and dramatic climate-driven changes on the planet (Vaughan et al. 2003). Among the documented changes are increases in mid-depth ocean temperatures (Gille 2002), changes in wind direction and strengthening of the westerly winds (Meredith & King 2005), and changes to the sea-ice duration and extent (Stammerjohn et al. 2008; Stammerjohn et al. 2012). Due to strong ocean/atmospheric links, changes in wind stress will dramatically affect the meridional overturning circulation, deepwater formation and heat content, further impacting global ocean circulation and carbon dioxide uptake (Ito et al. 2010).

Commercial fishing is the primary provisioning service in the Southern Ocean. Although many regard the Southern Ocean as relatively pristine, impacts from fish and wildlife overexploitation exist. Considerable declines in abundance of higher predators and fish began as early as the late 1700s (Bonner 1968). Antarctic fur seals (*Arctocephalus gazellae*) were nearly extirpated in just a few years of sealing in the Antarctic Peninsula, and in other areas between about 1790 and 1825 (Stackpole 1955; O'Gorman 1963; Payne 1977; Hofmeyr et al. 1997). Large baleen whale harvest increased following World War II with severe impacts to some stocks. Japan continues a small scientific harvest despite the 1986 commercial whaling moratorium instituted by the International Whaling Commission (IWC). More recently, fish such as marbled rockcod (*Notothenia rossii*) and mackerel icefish (*Champsocephalus gunnari*) were over-harvested between the 1960s and the mid-1980s, and a number of these stocks have not recovered. Since the mid-1990s, new fisheries have developed for Patagonian toothfish (*Dissostichus eleginoides*) and Antarctic toothfish *(Dissostichus mawsonii*). Illegal, unregulated, and unreported (IUU) harvest of toothfish in the CCAMLR area initially exceeded 100 000 metric tons per year, but has declined recently with improved regulation. The highest volume fishery for Antarctic krill began in the 1960s, with peak annual harvest exceeding 500 000 metric tons (total catch is lower now; Table 10.1). As krill are the principal prey for many predators, regulation of catch was a major impetus for establishing CCAMLR to prevent ecosystem impacts from overfishing (http://www.ccamlr.org).

In addition to the provisioning services provided by the Southern Ocean, scientific study represents the major cultural service. A founding premise of the AT is the scientific study of the Antarctic, and major advances in evolution, cold-water physiology, paleoclimate, planetary science and basic physics have occurred here (Hightower & Renfro 1988; Clarke & Crane 1992; Leuenberger & Siegenthaler 1992; Eastman & McCune 2000; Anchordoqui & Halzen 2006; Taboada 2012). Antarctica has also inspired human exploration. Both the history and natural beauty in the Antarctic are internationally appreciated. Many of the huts and camps from various exploratory expeditions in the late 19th and early 20th centuries are designated historical protected sites through the ATS and are visited by tourists. Between 1997 and 2010, the number of visitors increased from ~9 000 to ~37 000 (IAATO 2012).

Owing to the remoteness from population centers and the distance from shipping routes, support for commercial shipping is less important in the Antarctic compared to other high latitude seas. Most vessels in the Antarctic carry tourists or supplies to scientific bases, or participate in scientific research or fishing. Shipping of commercial goods is generally within the EEZs of southern hemisphere nations like South Africa, Chile and Argentina, or within open ocean areas outside the jurisdiction of the ATS. Parties of the AT in recent years have asked the International Maritime Organization to place some restrictions on the fuel oils allowed to be used in the Antarctic (28th ATCM Meeting), and to consider the issues relating to vessel construction, engineering requirements, and ballast water discharge in the Antarctic (29th ATCM Meeting).

Climate change impacts are likely to alter ecosystem services provided by the Antarctic and Southern Ocean. In addition to the potential for changes in carbon sequestration and global ocean circulation detailed above, both declines and increases in sea ice extent and duration in the Antarctic Peninsula region are also well documented, varying by region (Stammerjohn et al. 2008; Stammerjohn et al. 2012). Changes in the magnitude and duration of sea ice will impact the length and duration of fishing seasons. It will also open areas previously inaccessible and may reallocate or concentrate fishing effort into other areas as ice increases in the Ross Sea. A recent review of likely changes in the Antarctic and consequences for CCAMLRs ecosystem-based management by Trathan and Agnew (2010) outlines other possible impacts of climate change including shifts in the distribution of populations, loss of habitat, changes in phenology impacting productivity, increased ocean acidification, and species invasions. Changes in sea–ice have been associated with changes in phytoplankton community structure (Montes-Hugo et al. 2009), and increases in water temperature (Gille 2002). Southern Ocean pH is declining as CO_2 is absorbed by seawater, and there is concern that pelagic zooplankton communities could be altered as calcareous species like *Limacina helicina* divert energy from reproduction to preserving calcified structures (Fabry et al. 2008). Additionally, ecologically destructive invasive species could alter the benthos in the future. Lithodid crabs, which had been absent from Antarctic shelves, have established populations in deeper, warmer areas of the Antarctic Peninsula shelf (Smith et al. 2012). If bottom waters continue to warm, these species could impact the benthic communities that have evolved in the absence of crushing predators. Such impacts may be greater if lower pH waters erode calcareous shells of invertebrates like clams.

These impacts challenge the common belief that the Antarctic ecosystem is pristine (but see Halpern et al. 2007). Developing responsive EBM strategies is therefore complicated as much by the ongoing recovery from large scale removals (Laws 1977) as by shifting ecosystem structure owing to rapid climate change and consequent population perturbations (Trivelpiece et al. 2011).

3.1b Ecosystem-Based Management in the Antarctic

The Antarctic Continent and Southern Ocean ecosystem is managed within the context of several Conventions and Treaties. The IWC, established in 1946 to regulate the harvest of baleen whales globally, continues to manage cetaceans in the Southern Ocean. However, the passage and ratification of the Antarctic Treaty (AT) in 1959 was the first major step in providing a framework for international cooperation and scientific investigation of the Continent. Here, we give a brief synopsis of governance structures leading to the CCAMLR to frame the context for EBM. Detailed AT accomplishments are reviewed elsewhere (c.f. Berkman et al. 2011).

The AT established a framework for cooperative international governance of approximately 10% of the world's land and ocean area. The twelve original signatories agreed to set aside sovereignty issues, facilitate and cooperate in scientific research, and make consensus decisions on management issues on the Continent through a Consultative Process (Antarctic Treaty Consultative Meetings; ATCMs). The *Agreed Measures on the Conservation of Antarctic Flora and Fauna,* approved through the ATCM Process (1959–1964, adopted 1972), formulated restrictions on the capture and molestation of animals, required control of other damaging activities especially during breeding seasons, dealt with non-indigenous species introduction to Antarctica, and allowed for the establishment of specially protected species (e.g., Ross seals (*Ommatophoca rossii*) and fur seals). Further, 15 (and eventually 50) areas around the Antarctic were designated as Specially Protected Areas, affording them increased protection from disturbance. The *Agreed Measures* were replaced by the Protocol on Environmental Protection to the Antarctic Treaty (The Environmental Protocols) that established a more complete environmental protection scheme for the Antarctic in 1991. A Committee on Environmental Protection (CEP) was established to provide advice to the ATCM. Other non-governmental groups like the Scientific Committee on Antarctic Research (SCAR), tour operators and other stakeholders also provide input at the ATCM. Including these user groups has allowed a more diverse and representative discussion of management issues among Parties to the AT.

Other Conventions have been ratified or developed to manage particular extractive uses of the Antarctic resources. *The Convention for the Conservation of Antarctic Seals* (CCAS, signed 1972, effective 1978), established provisional conservation measures and provides for the establishment of a regulatory body should a commercial sealing industry develop, in particular for the crabeater seal (*Lobodon carcinophagous*), whose population was thought to exceed 50 million animals (Erickson et al. 1971). (However, more recent estimates suggest that there are only around 5 to 10 million crabeater seals (Erickson & Hanson 1990; Southwell et al. 2008), and sealing never developed.) There have been similar attempts to develop a Convention for the commercial exploration and extraction of minerals from the Southern Ocean and Antarctic Continent. However, although a Convention framework exists (The *Convention on the Regulation of Antarctic Mineral Resource Activities* (CRAMRA)), it has not been ratified by any nation. Nevertheless, commercial exploration or extraction of minerals from the Antarctic is prohibited by Article 7 of the Environmental Protocols.

The *Convention for the Conservation of Antarctic Marine Living Resources* (the Convention), is often described as a mature and leading example of ecosystem-based fishery management (Lodge et al. 2007; McFadden & Barnes 2009). Between 1977 and 1982, interested parties developed and subsequently ratified the Convention in response to concern that the extraction of Antarctic krill in the Southern Ocean, widely recognized as a 'keystone' species, could disrupt the Antarctic ecosystem (El-Sayed 1994). The Convention extends beyond AT boundaries northward to the the Antarctic Polar Front, an ecologically rather than politically important boundary (Fig. 10.1). The Convention established the Commission (CCAMLR), whose purview is broad, including all harvested species except whales and seals. Because it must also account for fishing effects on krill-dependent predators, CCAMLR also regulates activities that could affect marine mammals. For the remainder of this section, we examine how the CCAMLR applies ecosystem principles and what has been accomplished since its inception in 1982.

The ecosystem-based, precautionary approach to management within CCAMLR is embodied in 3 paragraphs of Article II of the Convention:

ARTICLE II

1. The objective of this Convention is the conservation of Antarctic marine living resources.
2. For the purposes of this Convention, the term 'conservation' includes rational use.
3. Any harvesting and associated activities in the area to which this Convention applies shall be conducted in accordance with the provisions of this Convention and with the following principles of conservation:
   (a) prevention of the decrease in the size of any harvested population to levels below those which ensure its stable recruitment. For this purpose its size should not be allowed to fall below a level close to that which ensures the greatest net annual increment;
   (b) maintenance of the ecological relationships between harvested, dependent and related populations of Antarctic marine living resources and the restoration of depleted populations to the levels defined in sub-paragraph (a) above; and

(c) prevention of changes or minimization of the risk of changes in the marine ecosystem which are not potentially reversible over two or three decades, taking into account the state of available knowledge of the direct and indirect impact of harvesting, the effect of the introduction of alien species, the effects of associated activities on the marine ecosystem and of the effects of environmental changes, with the aim of making possible the sustained conservation of Antarctic marine living resources.

Article II of the Convention provides a wide scope for managing fishery impacts on targeted and non-targeted species of the Southern Ocean, and applies the precautionary approach at the ecosystem level. Other paragraphs, along with Article II, established a decision-making process that relies on scientific working groups (WG) to inform a Scientific Committee (SC) that provides advice to CCAMLR for the development of Conservation Measures. Conservation measures are developed by CCAMLR through consensus. Given the simplicity of these statements in Article II, and the extent, distance and geo-political considerations required in managing long-lining, pot fishing, pelagic and benthic trawling fisheries in an area comprising more than 10% of the entire world ocean, the framework that has been built and the success that has been achieved is compelling.

After CCAMLR was established, it was clear that in order to implement the Convention effectively, two critical issues needed resolution. The first issue was how the broad statements in Article II were to be operationalized. The second issue was how the necessary ecosystem information could be obtained and incorporated into management strategies. The requirements in Article II for rational use, reversibility, and for consideration of fished and unfished species, dependent species, and the environment implied an approach beyond single species management, most critically for the Antarctic krill fishery. However, the prior overharvested or depleted status of several finfish at the inception of CCAMLR required that initial Conservation Measures be developed and approved before effective ecosystem-based management strategies could be implemented. Therefore, two WGs were established between 1982 and 1986 and charged with evaluating information on 1) the finfish fisheries and 2) the krill fishery and the other dependent species.

The WG on Fish Stock Assessment approached their work in a traditional single species context to bring fisheries under control after decades of uncontrolled and excessive harvest, and to begin to develop the necessary data collection and monitoring protocols to determine stock status and trends for important fisheries around the Southern Ocean. The WG used closed areas combined with a two-tier classification of fisheries to accomplish this task. "New and Exploratory" fisheries were designed to generate, through the fishery, sufficient data to develop a stock assessment. Upon sufficient data collection, these fisheries would graduate to "Fully-assessed" fisheries, with different data collection requirements, including minimal sampling, tagging, and biological data collection on finfish vessels, as well as 100% observer coverage. Resolving traditional fishery issues like bycatch and IUU fishing have been the focus of considerable management effort. IUU toothfish fishing required CCAMLR to develop, through international cooperation, more complex management strategies including the implementation of Vessel Monitoring Systems and a catch documentation scheme (CDS) that follows fish as they are sold and traded on the commercial market. To address finfish bycatch, "move-on rules" have been established to ensure bycatch is minimized and that effort is distributed across the fishing grounds. WGs were also established to address the mitigation of marine mammal and seabird bycatch mortalities within CCAMLR managed waters.

Initial progress in managing the fisheries was stymied by nations seeking to continue fishing given the economics of the time, and the resistance to embracing the precautionary approach in the face of uncertainty regarding the status of stocks (Constable et al. 2000; Constable 2001; Kock et al. 2007). Over time, with the slow addition of Conservation Measures that would provide data on the fisheries, closures of areas that were overfished, and gear restrictions in some areas, assessments have become more complete. In addition to fishery independent surveys in some parts of the Southern ocean, scientific surveys in a number of areas (Jones et al. 2000; Kock & Jones 2005; Kock et al. 2007; Belchier & Collins 2008) illustrate CCAMLR member efforts to provide data to manage these fisheries. IUU and bycatch prevention measures have also been successful. From 1996 to 2008 the percentage of illegally caught toothfish declined from about 72% to 10% of the catch, from a high of >32 000 tons to less than 1200 tons, while the total catch in the fishery declined as conservation measures were put in place. In addition, between 1998 and 2003 catch rates for longline vessels in South Georgia declined from 0.032 birds per 1000 hooks to 0.0003 birds per 1000 hooks. However, IUU fishing continues to be the major source of bird mortality, with 2003 estimates ranging from 10 to 30 thousand birds.

The WG on Krill (now merged with the WG on CCAMLR Ecosystem Monitoring Program and renamed the WG on Ecosystem Monitoring and Management (WG-EMM)) employed more of an ecosystem approach from the start. Implementing a krill fishery quota based on highly variable historical catch data was problematic, so CCAMLR initially set a quota at the maximum observed catch until programs were in place to monitor the dependent species as required in Article II. Establishing a precautionary catch limit based on the biomass of krill rather than on the historical catch required accounting for the different aspects of Article II. The quota was determined with three decision rules providing the first direct ecosystem based feedback:

A) Choose a krill exploitation rate such that the probability of the median stock biomass dropping to less than 20% of pre-exploitation biomass is just 10% over a 20 year period (reflecting the reversibility required by Article II);
B) Choose a second exploitation rate such that spawning biomass over a 20 year period is 75% of the pre-exploitation biomass (allocating biomass to krill-dependent predators by ensuring that no more than 25% of the spawning biomass is allocated to the fishery);
C) Finally, select the lower (precautionary approach) of the two exploitation rates to calculate the acceptable yield.

The second goal for ecosystem-based fishery management was to better understand the ecosystem structure and function, in particular being able to separate the effects of fishing from natural environmentally driven variability. The CCAMLR

WG on Ecosystem Monitoring and Management established an ecosystem monitoring program (the CCAMLR Ecosystem Monitoring Program; CEMP) across the Southern Ocean to inform the SC as to the status and trends of populations of krill-dependent species, as they responded to change in krill abundance or availability. Work from CCAMLR members suggested that upper level predators (mammals and birds) could be important indicators of environmental conditions, as a number of antecedent studies had shown population responses to variability in environment and food (Reid & Croxall 2001; Mori & Boyd 2004). The combined WG-EMM has been effective in developing integrated, basin scale surveys for providing advice regarding the biomass of krill.

CCAMLR has also addressed spatial allocation of catches, impacts to benthic habitats, and most recently, Marine Protected Areas (MPAs). Through the WGs (SC-CCAMLR 2002, 2003) CCAMLR used the considerable data on the distribution and abundance of birds and mammals in their at-sea foraging areas to further subdivide the 4 sub-Areas into 17 smaller Small Scale Management Units (SSMUs) for fisheries. A number of options for allocating the catch among these 17 areas were proposed (Hewitt et al. 2004) and evaluated with modeling (Watters et al. 2005; Plagányi & Butterworth 2006a,b; Watters et al. 2006). CCAMLR has also devoted significant effort to assessing benthic impacts of fishing in different areas of the Southern Ocean, using a scheme that requires reporting on the number, taxa, and amount of by-catch species indicating vulnerable marine ecosystems (VME). In areas directly managed by CCAMLR, the VME indicator reporting process has documented and closed small areas on the slope of the Ross Sea in waters deeper than 550m (SC-CCAMLR 2008). Similar areas have been closed around the South Orkney Islands when VME indicator taxa have been discovered and documented (SC-CCAMLR 2010). The CCAMLR Commission and the Scientific Committee has also actively pursued the development of strategies to develop a network of MPAs. A large bio-regionalization effort was undertaken in 2008, and various tools have been used to understand the location of unique or highly important areas (SC-CCAMLR 2007, 2008). The first MPA established by CCAMLR was a relatively large area in the Northern Weddell Sea (SC-CCAMLR 2010). At the most recent CCAMLR meeting, however, none of three newly proposed MPAs were implemented, largely because some Members hesitated to proactively close areas without a direct and imminent threat. CCAMLR has been a world leader in implementing Ecosystem-Based Fisheries Management (EBFM) within the international context. Clearly, the future EBM challenge for CCAMLR will be to be appropriately precautionary and to balance the different values placed on conservation (including rational use) and preservation.

3.2 High Arctic: international, national, and regional contexts for EBM

This section focuses on EBM in the high Arctic marine regions above 66.33°N off Russia, the USA, Canada, and northern Greenland, and in international waters in the central Arctic Ocean (Fig. 10.1). Our geographic definition of the high Arctic Ocean is that used in the Arctic Climate Impact Assessment (McBean et al. 2005). This definition excludes the Barents and Nordic (Greenland, Iceland and Norwegian) Seas with mixed arctic and subarctic ecology, which we consider separately (Section 3.3).

3.2a Ecosystem services, human use patterns, threats and impacts

The services provided by high Arctic marine ecosystems both extend beyond the immediate Arctic ecosystems (e.g. climate regulation) and are related to the major human use patterns within the ecosystems, which include shipping, hydrocarbon extraction, subsistence fishing and hunting by local indigenous peoples, and tourism. Nearly all of the threats and impacts to high Arctic ecosystems relate to climate change-driven sea ice loss and the potential shifts in human use patterns associated with climate change: habitat loss, biodiversity loss, chemical spills, and introduced species (ACIA 2005; Anisimov et al. 2007; Arctic Council 2010).

The high Arctic region provides climate regulation services for the northern hemisphere as well as the globe (McBean et al. 2005; Budikova 2009). Changes in Arctic sea ice coverage and distribution combined with melting of the land-based ice sheet on Greenland alter the flow, temperature and salinity characteristics of water leaving the Arctic for the north Atlantic, changing its density, which determines whether and how far it sinks, thus influencing the formation of North Atlantic Deep Water and the "overturning" circulation (Loeng et al. 2005). The sinking and export of deep water drives the import of warmer surface water to the north Atlantic, maintaining temperate conditions over much of northern Europe (Broecker 1991). Loss of sea ice due to climate warming is generally predicted to increase flow out of the Arctic, reduce the formation of North Atlantic Deep Water, and therefore weaken the global thermohaline circulation, potentially resulting in cooling of the north Atlantic and northern Europe due to decreased warm surface water flow (Loeng et al. 2005; McBean et al. 2005).

Arctic sea ice also regulates and supports regional climate via the atmosphere. Both modeling and observational studies suggest that lower wintertime ice extent reduces wind speeds and storminess north of 45°N, but intensifies both further south as far as the tropics; higher wintertime ice extent reverses these patterns (Budikova 2009). Since the mid 1990s, more ocean surface has become ice-free in late summer, so the Arctic Ocean has stored more heat, locally warming the atmosphere and changing circulation patterns throughout the following winters (Francis et al. 2009; Overland & Wang 2010). In particular, a weakening in prevailing winds during the winters of 2009–2010 and 2010–2011 moved more cold Arctic air south over the continents (Overland et al. 2011). The result has been colder winters in the mid-latitudes of Europe, North America, and Russia where major population centers exist (Honda et al. 2009; Petoukhov & Semenov 2010). Therefore, climate change impacts on Arctic sea ice and the climate regulation services it provides are likely to be felt in marine and terrestrial ecosystems and across socio-political boundaries throughout the northern hemisphere.

Changes to Arctic sea ice extent and distribution will also alter supporting services for human activities within the ecosystem. The high Arctic Ocean could support the shortest possible shipping corridor between the Atlantic and Pacific Ocean basins if not for the ecosystem's characteristic sea ice, in particular that of the fabled "Northwest Passage" along the North American coast (Williams 2003, 2011). However, international and local shipping on the Northern Sea Route along Russia's Arctic Ocean coast has been increasing recently (Denisov & Mikhaylichencko 2009). Increased shipping in the Arctic seas may have positive economic effects while introducing further potential threats and impacts, including oil and hazardous materials spills, exotic species introductions, air pollution, port and other coastal infrastructure development, and disturbance or accidental mortality of marine animals (Arctic Council 2009a).

The most obvious marine ecosystem provisioning service is the production of harvestable living resources. In contrast with the other case studies in this chapter, high Arctic marine ecosystems do not support major commercial fisheries at present (Table 10.1). Even reconstructed catch histories for nearshore and estuarine fish species in arctic Russia, Canada, and the U.S. (Zeller et al. 2011) rank the high Arctic one of the least commercially fished ecosystems on earth. The only large scale commercial fishery that has taken place in the Pacific high Arctic was for bowhead whales off Alaska and Canada between the 1850s and early 1900s. Commercial high Arctic whaling came to an end as the expense of operating in this extreme environment exceeded the profits available from whale oil products (Bockstoce 1978).

Although commercial fishing is minimal, subsistence fishing and hunting of seabirds and marine mammals is critically important to humans inhabiting high Arctic ecosystems. Provisioning and cultural ecosystem services overlap extensively for the indigenous peoples of the Arctic (Huntington & Pungowiyi 2009). In particular, subsistence hunting of Arctic marine mammals is important to many indigenous Arctic people for maintaining nutritional provisions, economies, social structures, cultural character, and spiritual identity (Nuttall et al. 2005; Hovelsrud et al. 2008). Indigenous Arctic peoples have inhabited the Arctic and foraged in its marine ecosystems for thousands of years (Pitulko et al. 2004; Anderson 1984; Savinetsky et al. 2004). At least 17 groups of indigenous peoples now reside in the Arctic (Table 10.2), combining both commercial and subsistence economies within their communities. Nearly all indigenous communities (and some non-indigenous) harvest marine mammals, seabirds, seabird eggs, and fish, as well as caribou and other terrestrial animals and plants, primarily because these foods are both less expensive and more culturally valued than store-bought foods imported from lower latitudes (Nuttall et al. 2005; Merkel & Barry 2008).

Oil exploration represents the other major human activity on Arctic continental shelves, which brings both economic enrichment and the potential for contamination of ecosystems if there are spills or other industrial accidents. The high Arctic may contain 13% of the planet's undiscovered oil and 30% of its undiscovered natural gas, primarily offshore (Bird et al. 2008; Gautier et al. 2009). Most Arctic oil and gas development to date has occurred on land, but climate-driven sea ice loss may allow easier access to Arctic continental shelves for oil and gas exploration and extraction (Gautier et al. 2009). The high Arctic shelves off Alaska, Canada, Norway, Russia, and Greenland are being actively explored (Anonymous 2009a; Hoel et al. 2009; Siron et al. 2009), while development and extraction are in progress in the Barents Sea and off the Newfoundland Labrador subarctic shelf (Table 10.1 and see the following case studies). Oil and gas development is controversial in the Arctic, with some stakeholders envisioning economic gains and increasing world oil supply as others envision ecological damage from both the footprint of development in previously undisturbed areas and the potential for more catastrophic damage due to large accidental spills (e.g., Schiermeier 2012).

Climate change-driven effects are changing the high Arctic rapidly. Record low ozone was observed over the high Arctic in 2011, rivaling the annual Antarctic ozone hole (Bernhard et al. 2011). The most obvious change is the continuing

TABLE 10.2.
Indigenous peoples of the Arctic
(Nuttall et al. 2005)

Region	People
Alaska	Inupiat Yup'ik Alutiiq Aleuts Athapaskans
Canada	Inuit Inuvialuit Dene Athapaskans
Greenland	Kalaallit Inughuit
Fennoscandia	Saami
Russian Kola Peninsula	Saami
Russian far North and Siberia	Chukchi Even Evenk Nenets Yukaghir

decline in summer sea ice cover, which reached a new record minimum in 2012 (National Snow & Ice Data Center 2012). The previous record minimum ice extent was in 2007 (Overland et al. 2008; Parkinson & Cavalieri 2008), with 2007–2011 below all other records (Perovich et al. 2011; Richter-Menge et al. 2011). In 2010–2011, shipping routes along both the Russian and Canadian Arctic coasts were open (Perovich et al. 2011). Declining sea ice has resulted in the replacement, by annual ice, of nearly 30% of the perennial ice which existed in 1979 (Carmack et al. 2006). Since perennial ice is thicker than annual ice, this suggests that annual ice may be more prone to quicker melting in the summer, both continuing the trend and perhaps increasing the overall variance of ice cover relative to past conditions. The perennial sea ice is also reportedly getting younger and thinner overall (Maslanik et al. 2007; Maslanik et al. 2011). This reduction in ice cover is happening much faster than climate change models have predicted (Walsh 2008; Rampal et al. 2011). Overall, biodiversity of indigenous biota in the Arctic is threatened by climate-driven habitat loss, range extensions of non-endemic species, and transfer of exotic species by increased ship traffic as described above (Arctic Council 2009a; Bluhm et al. 2011).

Climate change and contaminants also threaten cultural and provisioning services. Human foragers in the Arctic are immediately affected by earlier melts, thinner ice, ice further from shore, and changes in animal migratory patterns (Mallory et al. 2006; Krupnik & Ray 2007). Sea ice loss endangers the traditionally hunted ice-dependent marine mammals such as hooded seals, narwhals and polar bears (Hovelsrud et al. 2008; Laidre et al. 2008). Furthermore, the extreme seasonality of production and short food chains, combined with the preferential atmospheric transport of some contaminants to the Arctic causes long-lived, lipid-rich marine mammals and birds to accumulate contaminants and toxins which may threaten human health (Anonymous 1991; Alexander 1995; Van Oostdam et al. 2005; Mallory et al. 2006).

Ocean acidification is another threat to marine ecosystems caused when excess atmospheric carbon dioxide dissolves in ocean waters, disrupting the acid-base chemistry (Feely et al. 2004; Sabine et al. 2004). As acidic ocean waters are less saturated with calcium carbonate, less is available for uptake by marine organisms to support physiological processes, including growing shells and other calcified structures. This in turn could lead to shifts in community structure where more acidification-sensitive organisms become rare compared with less sensitive organisms, changes which could be amplified by food web effects (Fabry et al. 2008). Ocean acidification and its consequences are of particular concern in the north Pacific and western high Arctic, where baseline conditions are naturally more acidic and less saturated with calcium carbonate (Bates & Mathis 2009; Fabry et al. 2009; Mathis 2011). As sea ice coverage is reduced, the Arctic Ocean may take up more atmospheric carbon dioxide, potentially increasing acidification in the basin (Bates & Mathis 2009).

However, not all changes are predicted to have negative impacts. As open water increases, primary productivity may increase (Arrigo et al. 2008), potentially benefitting Bowhead whales (Moore & Laidre 2006) and polar cod (Fortier et al. 2006). Likewise, earlier ice breakup and more open water may benefit some marine birds (Mallory et al. 2006). However, the pelagic food web interactions in the high Arctic may complicate the separate predictions for bowhead whales, marine birds, and polar cod, given that they may compete for any increased zooplankton production in open water high Arctic seas (Frost & Lowry 1984; Bluhm & Gradinger 2008).

These recent changes in Arctic ecosystems have also increased accessibility and interest in tourism to the region (Edes 2008). Tourist travel to the high Arctic has increased since the early 1990s, with some measures showing visitors doubling in this short time (Stewart et al. 2005; Fay & Karlsdóttir 2011). Many visitors to the Arctic arrive by cruise ship, some to the high Arctic in search of sea ice associated wildlife. The loss of sea ice in the Northwest Passage increased ship-based tourism from 1984 on (Stewart et al. 2007), while in Hudson Bay recent sea ice loss appears to have decreased cruise tourism (Stewart et al. 2010). With tourism now one of the dominant economic sectors in the global economy, Arctic nations are actively promoting tourism in order to reap the significant economic benefits it can provide (Edes 2008; Anonymous 2009a; UNWTO 2012). Despite the economic benefits, increased tourism potentially alters the Arctic ecosystems and human cultures that attract tourists and increases travel-related carbon emissions, which exacerbate climate warming and acidification (Edes 2008).

3.2b Ecosystem-Based Management in the high Arctic

As is clear from the list of human uses above, the diversity of physical and ecological characteristics in the Arctic are matched by a diversity of human social, cultural, and political interactions in the region. High Arctic EBM is developing as a multi-scale process; here we describe international, national and regional levels of high Arctic EBM (see also Cochrane et al., this volume).

In 1996, the Arctic nations[1] established the Arctic Council, the primary international governance structure in the region (Arctic Council 1996). The Arctic Council has permanent participation from six organizations representing Arctic indigenous peoples: the Arctic Athabaskan Council, the Aleut International Association, the Gwich'in Council International, the Inuit Circumpolar Council, the Russian Arctic Indigenous Peoples of the North, and the Saami Council. Observer status is granted to other parties with interests in the Arctic, including non-Arctic states and non-governmental organizations (NGOs). Consensus-based decision making is used to achieve "soft law" (not legally binding) international agreements within the Arctic Council.

The Arctic Council arose from the 1991 Arctic Environmental Protection Strategy (AEPS). The objectives and principles of the AEPS reflected many EBM tenets such as including humans within the description of the ecosystem, balancing environmental quality and sustainable use of natural resources, considering interdependence of components within ecosystems, and regularly reviewing environmental status and assessing cumulative impacts. One objective highlighted the importance of indigenous peoples in the Arctic: "to recognize and, to the extent possible, seek to accommodate the traditional and cultural needs, values, and practice of the indigenous peoples as determined by themselves, related to the protection of the Arctic environment (Anonymous 1991)." The AEPS identified environmental threats including persistent organic contaminants, oil pollution, heavy metals, noise, radioactivity, and acidification from terrestrial/industrial sources, and identified actions to address these threats. It also established the Arctic Monitoring and Assessment Program (AMAP) to monitor and assess the status and trends of pollutants and their impacts, the Protection of the Arctic Marine Environment (PAME) working group addressing environmental protection policy, the Conservation of Arctic Flora and Fauna (CAFF) working group to coordinate biodiversity research and exchange data, and the Emergency Prevention, Preparedness, and Response (EPPR) coordination program (Anonymous 1991). These programs and groups were designed to integrate and augment the work of existing national and multinational programs and agreements to the whole-Arctic level, and continue today as working groups of the Arctic Council, along with the Arctic Contaminants Action Program (ACAP) and the Sustainable Development Working Group (SDWG).

With this foundation, the Arctic Council has developed an international EBM framework through several of its working groups, and most recently with a new expert group. In 2004, the Arctic Council adopted PAME's Arctic Marine Strategic Plan for integrated management in the Arctic, with four objectives: "reduce and prevent pollution in the Arctic marine environment, conserve Arctic marine biodiversity and ecosystem functions, promote the health and prosperity of all Arctic inhabitants, [and] advance sustainable Arctic marine resource use (Arctic Council 2004)." The Plan identifies 29 strategic actions to improve and respond to knowledge about the marine ecosystem, comply with existing regional and interna-

[1] Six nations bordering the Arctic Ocean or Nordic Seas: Canada, Denmark (with respect to the Faroe Islands and Greenland), Iceland, Norway, the Russian Federation, and the United States, and two nations with terrestrial Arctic territory: Sweden and Finland.

tional agreements, apply an ecosystem approach, facilitate partnerships, build local capacity, and support communication, all to be implemented by Arctic Council working groups, member nations, and other agencies. Many of the strategic actions have been completed, including a comprehensive Arctic Marine Shipping Assessment (Arctic Council 2009a), an update to Offshore Oil and Gas Guidelines (Arctic Council 2009b), a report on Arctic Biodiversity Trends (Arctic Council 2010), part of a larger ongoing Biodiversity Assessment, and a legally binding Aeronautical and Maritime Search and Rescue agreement (Arctic Council 2011c).

Arctic Marine Strategic Plan actions related to the ecosystem approach are also in progress. PAME initiated a formal EBM framework for Arctic marine ecosystems using a five-module Large Marine Ecosystem (LME) approach assessing indicators of productivity, pollution and ecosystem health, fish and fisheries, socioeconomics, and governance (Siron et al. 2008). PAME and SDWG collaborated to identify "Best practices in ecosystem-based oceans management in the Arctic (Hoel 2009)," including flexible application of EBM, science-based integrated decision making, commitment to EBM, area-based but trans-boundary approaches, participation by stakeholders, and adaptive management (Arctic Council 2008). PAME has applied ecological criteria to redraw LME boundaries, and synthesized existing regional marine ecosystem status reports to the full Arctic level (Arctic Council 2011b). More recently, the Arctic Council established an Ecosystem-Based Management Experts Group (Arctic Council 2011a), which has reviewed monitoring and assessment efforts throughout the region, has agreed that LMEs are the appropriate scale for integrated assessment and EBM, and has recommended integration of AMAP, CAFF, PAME and SDWG activities as steps towards achieving integrated assessment in the LMEs (Arctic Council 2012).

A key strength of the Arctic Council in facilitating high Arctic EBM is stakeholder involvement. As noted above, indigenous peoples have been involved as permanent participants in the Arctic Council throughout its history, meaning they must be consulted on all proposals and actions, and they may propose initiatives as well. This is a more powerful position than the observer or NGO status that indigenous peoples generally have in international organizations (Kiovurova 2008). Because of this participation, each Arctic Council declaration and plan explicitly recognizes and supports indigenous communities' stake and formal involvement in international EBM, including incorporation of traditional knowledge (Arctic Council 1996, 2011a). Incorporating traditional knowledge into EBM is particularly important in the Arctic, where indigenous peoples have generations of experience subsisting within the ecosystem and observing recent changes, although the process for integrating traditional knowledge with academic science requires ongoing interaction and collaborative effort (Huntington & Pungowiyi 2009; Huntington 2011).

Given the above monitoring and assessment activities and products, the Arctic Council seems poised to provide an effective international forum to support high Arctic EBM. However, some stakeholders have been concerned that "soft law" governance may lead to ineffective management of human activities in the high Arctic (Nowlan 2001). The Arctic Council has no enforcement capability, so it currently "encourages" the adoption of its recommendations for meeting its objectives by other regulatory bodies such as the International Maritime Organization (IMO) or national governments (Kao et al. 2012). Another concern is that Arctic nations may withdraw funding or other support from the Arctic Council as participation is voluntary (Nowlan 2001; Lennon 2008). Proposals to develop a legally binding Arctic Treaty similar to the Antarctic Treaty have been raised throughout the history of the Arctic Council and its predecessor, the AEPS. However, there is also concern that environmental protection progress made to date by the Arctic Council has been politically possible because it has not been legally binding, and that contentious negotiations over a binding Arctic Treaty would stall this progress (Stokke 2007; Kiovurova 2008).

Outside the Arctic Council, the nations bordering the Arctic Ocean have all developed national policies and activities related to marine EBM to some degree. We focus on Norway's approach to EBM in Section 3.3. Here, we briefly outline the remaining arctic nations' EBM activities in the high Arctic (with a fuller discussion of Canadian national policy in Section 3.4).

Canada's integrated ocean management system has developed recently from more regionally and sectorally oriented management (Siron et al. 2008; Siron et al. 2009). In the Canadian Arctic, the Beaufort Sea Large Ocean Management Area (LOMA) is cooperatively managed by the Inuvialuit and the Canadian government. In 2008, an ecosystem overview report was compiled from scientific and traditional knowledge, and an assessment report was compiled from similar sources though interactive workshops with stakeholders (Cobb et al. 2008; Siron et al. 2008; Siron et al. 2009). Ecologically and biologically significant areas were identified within the Beaufort Sea LOMA, and an Integrated Ocean Management Plan for the Beaufort Sea was released by the Beaufort Sea partnership which outlines 24 management objectives for the ecosystem (Beaufort Sea Partnership 2009). While the Beaufort Sea LOMA Integrated Ocean Management plan was successfully completed, further progress is unclear. In 2010, Canada's Department of Fisheries and Oceans (DFO) produced a comprehensive ecosystem status and trends report for the Beaufort Sea and other high Arctic systems (Niemi et al. 2010), but no mention was made of the Beaufort Sea LOMA or its management objectives. DFO discontinued the LOMA program in 2011, as discussed in Section 3.4. This leaves considerable uncertainty surrounding the Beaufort Sea LOMA's objectives and EBM implementation in the region.

In the United States, a national policy framework for marine EBM is based on a 2010 Executive Order establishing the National Ocean Council, comprised of the heads of all agencies involved on managing ocean activities and resources (Obama 2010). This order built upon the previous Ocean Action Plan (2004), which had led to the ongoing development of an ecosystem approach to management within the National Oceanic and Atmospheric Administration (Anonymous 2009b). The United States interacts with the high Arctic primarily through the state of Alaska and Federal agencies responsible for resource management in that region, which remain sector-based. However, the Alaska Marine Ecosystem Forum provides a voluntary means for interagency cross-sector coordination and collaboration, and there are other government, Alaska native, and international bodies that might contribute to the development of EBM in the region (Mangerink et al. 2008). In addition, the fisheries sector is incorporating ecosystem approaches in its management. The North Pacific Fisheries Management Council recently preemptively closed federal waters in the U.S portions of the Chukchi and Beaufort Seas to commercial fishing until a scientifically sustainable basis for opening fisheries is established, citing both ecological and economic factors (Wilson & Ormseth 2009). Some local initiatives also incorporate aspects of EBM; for example, the community of Barrow has been active in seeking stricter environmental review of offshore oil exploration in order to preserve the offshore environment (Itta 2008). The U.S. approach to EBM is to provide a general national framework that allows for regional development of specifics, which appears to be happening in the U.S. high Arctic.

Similar to the U.S., the Russian Federation has a complex federal, regional, and local governance structure which is integrating aspects of ocean EBM within a sector-based system under the 2001 Marine Doctrine of the Russian Federation (Denisov & Mikhaylichencko 2009). Although ten agencies potentially manage ocean resources in Russia, a Marine Board has been established to coordinate federal and regional entities' activities in the marine environment. While this framework is very important to implementing coordinated policy, other EBM-related initiatives were negatively impacted by the economic downturn in Russia. Therefore, most management remains regional and sector based, with economic considerations generally taking precedence over environmental protection (Denisov & Mikhaylichencko 2009). This illustrates another real-world difficulty with EBM implementation: it may be easier to sell the idea and the extensive work it entails during economic expansions than during times of economic stress. However, the Russian Federation has made progress on EBM outside the high Arctic: Russia and Norway have bilateral agreements which contribute to EBM in the Barents Sea (Section 3.3).

Given the number and diversity of ecosystems, nations, and cultures in the high Arctic, along with complex and changing human uses and threats to the ecosystems, it is both impressive and unsurprising that EBM is developing here in a variety of ways. While some EBM frameworks are well-developed, difficult conflicts still exist. However, nations generally recognize a need for coordination across sectors and boundaries to achieve common goals.

3.3 Northeast Atlantic and Barents Sea: a multinational context for EBM

In this section, we focus primarily on EBM efforts in the Barents Sea (Fig. 10.1) under multinational conventions and advisory bodies, bilateral Norway-Russia agreements, and Norway's national policy. We also review the larger context of EBM within the wider northeast Atlantic south of Fram Strait in the Nordic (Greenland, Iceland, and Norwegian) Seas encompassed by the same conventions and advisory bodies.

3.3a Ecosystem services, human use patterns, threats and impacts

The regional climate regulation and global ocean circulation supporting services provided by northeast Atlantic high latitude ecosystems are directly linked to those described for the high Arctic (Section 3.2), as Fram Strait is the main region of North Atlantic Deep Water formation. The northeast Atlantic high latitude ecosystems also support shipping and provide hydrocarbon resources, with increased demand for both services likely in the future. The Nordic Seas already have important high traffic international shipping routes, and the Barents Sea region is likely to provide increased support for shipping as industrial uses (in particular, oil and gas exploration and extraction) increase and sea ice extent decreases (Norwegian MTI 2007; OSPAR 2010b). The Barents and Norwegian Seas link the high Arctic Northern Sea Route described in section 3.2 with busy European ports, reducing Europe-Asia shipping distance by a third compared with southern routes (Astill 2012). While increased shipping potentially brings economic benefits, and sea transport may have environmental advantages over road or air transport, risks associated with shipping include water and air pollution from exhaust emissions, intentional waste disposal, unintentional oil and chemical spills, and exotic species introductions from ballast water or ships' hulls (OSPAR 2010b).

Similarly, oil and gas production and development in the Nordic and Barents Seas bring economic benefits combined with environmental risks. Offshore petroleum production is the largest industry sector in Norway, accounting for over 20% of GDP, and Norway is currently among the top three exporters of natural gas and the top eight exporters of crude oil worldwide (CIA 2011; Hansen & Rasen 2012). The highly productive North Sea fields are in decline, but 18 fields in the Norwegian Sea and one field in the Barents Sea have been developed, and both Norway and Russia have considerable interest in developing the potentially rich Barents Sea resources further (Gautier et al. 2009; OSPAR 2010b; Hansen & Rasen 2012; Schiermeier 2012). The primary risks of offshore petroleum development include oil spills and routine contaminant releases, and seabed disturbance, all of which may be more difficult to remediate in cold, dark, icy high latitude seas. Further, development of coastal infrastructure for increased oil and gas development and cargo shipping may alter nearshore environments (OSPAR 2010b).

In contrast with high Arctic systems, commercial harvest of marine resources is a critically important ecosystem service in the northeast Atlantic, and represents the primary human use of these ecosystems along with petroleum exploration and production (OSPAR 2000). Fishing is a dominant industry in Greenland, Iceland, Russia, and Norway. These nations land over 4 million metric tons of fish annually (Table 10.1), and Norway is second only to China in worldwide fish export value (FAO 2012a, b). The Barents and Nordic Seas currently support the largest cod (*Gadus morhua*) and herring (*Cluepa harengus*) fisheries in the world (Gjøsæter 1995; Hjermann et al. 2004a; Loeng & Drinkwater 2007; Corkeron 2009; Johannesen et al. 2012), with many other fish and invertebrates supporting important Barents and Nordic Seas fisheries (OSPAR 2000; Denisov & Mikhaylichencko 2009; ICES 2012c). These stocks have been fished for over 500 years, but overfishing contributed to collapses of herring in the 1970s and capelin (*Mallotus villosus*) in the 1980s, and reduced productivity of cod and other stocks (Hjermann et al. 2004a; Vilhjalmsson et al. 2005). With both capelin and herring stocks at low levels, impacts spread to predators: Barents Sea cod productivity was further reduced, and both seabirds and seals had recruitment failures (Gjøsæter et al. 2009). Implementation of quotas, harvest moratoria, and other management measures to reduce fishing effort, as well as consideration of species interactions in fishery management (see below) have generally improved stock conditions throughout the region, and Barents Sea cod and Norwegian spring spawning herring stocks are currently at high levels (OSPAR 2010b; ICES 2012c). More recent collapses of Barents Sea capelin stocks under the current fishery management regime have been attributed to climate and species interactions, in particular predation by cod and the recovered herring stock, and have had less severe food web consequences (Hjermann et al. 2004a; Gjøsæter et al. 2009).

The northeast Atlantic high latitude seas also support a thriving finfish aquaculture industry, primarily in Norway (OSPAR 2010b). While the quantity of fish (mainly salmon) produced by aquaculture remains lower than the total quantity of wild-caught fish, in Norway the value of farmed fish has exceeded that of wild-caught fish since 2005, and was twice as valuable in 2010, with employment in the sector increasing (Statistics Norway 2012). Along with these economic and social benefits, the ecological impacts of aquaculture may include disease and parasite introduction to wild communities, interbreeding of cultured and wild stocks, introduction of non-native species, and local contamination, eutrophication, or other disturbance directly around fish farms (OSPAR 2010b; Taranger et al. 2011).

In addition to finfish, commercial harvest of marine mammals and birds has historically been an important provisioning service of the Nordic and Barents Seas, although commercial importance is reduced at present due to controversy over the practice and regulation-limited markets for products (European Commission 2010). Commercial harvest of seabirds is permitted in Greenland and Iceland for domestic sale, but only eider down (non-lethally collected from empty nests) is of international commercial importance (Merkel & Barry 2008). Right and bowhead whaling began in the 1500–1600s; both of these species are now either critically endangered or extinct in the northeast Atlantic as a result of overexploitation prior to the 20th century (Allen & Keay 2006; Reilly et al. 2012a, b). While many nations participated in high seas industrial whaling for the other great whales through the 20th century in the northeast Atlantic and worldwide (Ellis 1991), Norway and Iceland in particular had substantial whaling industries, and presently continue commercially hunting minke and some fin whales within their exclusive economic zones (Table 10.1) under objection to the 1984 IWC whaling moratorium (IWC 2012). Harp and hooded seals are also commercially hunted in the Barents and Greenland Seas by Norway and Russia (Skaug et al. 2007; ICES 2011), and Greenland has a mixed subsistence/commercial whale and seal hunt by indigenous Inuit (Department of Fisheries 2009, OSPAR 2010b). Determining the ecological interactions of marine mammal harvest and fisheries has been a major challenge for ecosystem modeling in the Barents Sea, where finfish stocks already strongly interact (Corkeron 2009; Lindstrøm et al. 2009; Howell & Bogstad 2010).

Other human uses of and impacts to the high latitude northeast Atlantic, such as tourism, contamination, and species introductions, are projected to increase as climate warming reduces seasonal sea ice and opens access to the region (OSPAR 2010b). Tourism is not new in these high latitudes, but it has increased since the late 1980s. While the Norwegian North Cape has been visited since 1845, cruise ship tourism to Greenland, Iceland, and Norway (Svalbard) has doubled from the early 1990s to 2010 (Fay & Karlsdóttir 2011). Increased tourism can benefit coastal high latitude communities with increased economic diversity, revenue, and employment, but costs may include disruption of rural lifestyles and subsistence cultures, as well as environmental impacts from coastal development and increased ship and land-based pollution (OSPAR 2010b; Fay & Karlsdóttir 2011).

Impacts from species introductions are also mixed: Pacific red king crabs were released into the Barents Sea in the 1960s by Russian scientists to encourage a new coastal fishery (Jorstad et al. 2002). By the 1990s, the crabs were established, and a lucrative quota-managed commercial fishery began in the early 2000s (Sundet & Hjelset 2002; Nilssen & Sundet 2006). Ecological impacts of the invasive crabs may include altered benthic communities and increased habitat for finfish disease vectors (Falk-Petersen et al. 2011). Climate warming and loss of sea ice may further complicate invasive and native species interactions in the Barents and Nordic seas. Along with the predicted loss of sea ice, increased freshwater inputs, and ocean acidification impacts from climate warming, plankton and fish communities are shifting, and the timing of production is changing, potentially altering prey availability throughout the food web while reducing biodiversity with the loss of ice dependent communities (OSPAR 2010b; Dalpadado et al. 2012).

3.3b EBM in the Northeast Atlantic and Barents Sea

The northeast Atlantic is among the most densely populated and commercially exploited of the high latitude systems we compare (Table 10.1). EBM in this region takes place within a framework of international and multinational treaties and agreements, all interacting with national policies. We first describe the wider context of multinational EBM for the region, and follow with descriptions of national policy (see also Cochrane et al.; Houde et al., this volume).

The 1992 Convention for the Protection of the Marine Environment of the North-East Atlantic (OSPAR) provides a basis for coordinated international (multilateral) marine EBM in the seas from 36°N to the North Pole, and from 42°W to the continental European coast. A combination of the previous Oslo (1972) and Paris (1974) Conventions designed to prevent marine pollution from airborne/waterborne and land-based sources, respectively, OSPAR's mandate was extended to include marine environmental assessment, and conservation of ecosystems and biodiversity (OSPAR 1992; Skjærseth 2006). OSPAR's North-East Atlantic Environment Strategy sets objectives for environmental protection using an explicit ecosystem approach, with seven strategic objectives supported by specific operational objectives and targets, guiding principles, and timelines for implementation within five themes: biological diversity and ecosystems, eutrophication, hazardous substances, offshore oil and gas industry, and radioactive substances (OSPAR 2010a). Climate change impacts are evaluated by OSPAR as well. Using these themes and objectives as a basis, two comprehensive assessments of ecosystem status for the entire region were completed (OSPAR 2000, 2010b). While OSPAR's assessments address all human uses of the ecosystem including fisheries and shipping, the Convention defers fisheries management advice to the International Council for the Exploration of the Sea (ICES) and other organizations, and shipping management to the IMO. OSPAR also coordinates with the European Union (EU)'s Marine Strategy Framework Directive, which requires EU members to attain "good environmental status" by 2020 (OSPAR 2010a). OSPAR's decisions are adopted with three-quarters majority vote by, and are legally binding on, its Contracting Parties[2] (OSPAR 1992). Thus OSPAR's "hard law" regulation contrasts with the Arctic Council's "soft law" framework (Section 3.2).

[2] Belgium, Denmark, the European Community, Finland, France, Germany, Iceland, Ireland, Luxembourg, the Netherlands, Norway, Portugal, Spain, Sweden, Switzerland, and the United Kingdom.

For monitoring and assessment, OSPAR divides the northeast Atlantic into five regions. Region I, Arctic Waters, encompasses the Nordic and Barents Seas, as well as a portion of the high Arctic Ocean. OSPAR's 2010 Quality Status Report (QSR; OSPAR 2010b) found that Region I is the least polluted OSPAR area and has generally good status. Qualitatively, there were no problems with meeting the stated objectives for eutrophication; some problems with hazardous and radioactive substances, pollution from offshore petroleum development, and fisheries; and many problems with biodiversity due to declining species and habitats. The status of noise, marine litter, and non-indigenous species was unknown. The 2010 QSR identified the major ongoing concerns for Region I ecosystems as climate change, contaminant transport from lower latitudes, ocean acidification, seabird population declines, and pollution from potentially increased offshore oil and gas development and shipping. However, improvements were noted in some fish stocks' status, aquaculture management, reduction in illegal Barents Sea cod catch, habitat protection through Marine Protected Areas (MPAs), and integrated management plan development by Norway for the Barents and Norwegian Seas (detailed below). OSPAR's recommendations for Region I were to develop integrated management plans for the remaining seas, continue monitoring for climate change, ocean acidification, and industrial impacts, and collaborate with Arctic Council working groups to achieve these goals (OSPAR 2010b).

In general, the 2010 QSR demonstrated that OSPAR and its Contracting Parties (CPs) have made progress in reducing nutrient inputs, chemical contaminants, radioactive substances, and pollution from offshore oil and gas development in the entire OSPAR region since the previous (2000) assessment, although some targets are not being met and stricter regulation is recommended (OSPAR 2010b). Further, the 2010 QSR provides a comprehensive list of human activities and expected future pressures on these ecosystems, emphasizing the need for assessing cumulative impacts and developing integrated management plans, both critical to improving EBM in the northeast Atlantic. The 2010 QSR also provided recommendations to reduce the ecosystem effects of fishing (which were identified as more severe outside Region I) (OSPAR 2010b). The fisheries management regime in the OSPAR region is complex, with five international and multiple bilateral agreements for the separate management of marine mammals, tunas, salmon, and various individual fish species which range throughout the territorial waters of several nations (OSPAR 2010b). While management is by many agencies, advice comes primarily from one: ICES provides independent, precautionary advice on sustainable yields for commercially targeted species in the northeast Atlantic to its 20 member nations,[3] and supports the ecosystem approach developing under the EU's Marine Strategy Framework Directive (ICES 2012a,b).

The Barents Sea provides an example of developing EBFM. While individual species remain the basis for fishery management, several stock assessments (and therefore catch quotas) address multispecies and ecosystem interactions. The capelin stock assessment formally models cod predation, and managers close cap-

[3] Belgium, Canada, Denmark (including Greenland and Faroe Islands), Estonia, Finland, France, Germany, Iceland, Ireland, Latvia, Lithuania, the Netherlands, Norway, Poland, Portugal, Russia, Spain, Sweden, the United Kingdom, and the United States of America.

elin fisheries when the stock biomass falls below a threshold level to prevent capelin recruitment failure and its potential impacts throughout the food web (Gjøsæter et al. 2002; Gjøsæter et al. 2009). Recent assessments for northeast Arctic cod and haddock have included predation mortality as well, and environmental factors influencing northeast Arctic cod recruitment are also formally incorporated within the assessment and resulting advice (ICES 2012c). The Joint Norwegian Russian Fisheries Commission has made efforts to reduce illegal and unreported fishing, discards, bycatch of nontarget species, and other ecosystem impacts of fishing in the Barents Sea (Norwegian MFCA 2007; OSPAR 2010b; ICES 2012c). Further integration of species, fisheries, and climate interactions into the Barents Sea EBFM could be based on current research efforts incorporating whale and seal predation and harvest, climate driven shifts in species overlap and predation pressure, and multispecies yield optimization (Hjermann et al. 2004b; Filin & Tjelmeland 2005; Tjelmeland 2005; Lindstrøm et al. 2009; Howell & Bogstad 2010; ICES 2012c).

Barents Sea EBFM is developing in concert with Norway's national EBM policy for the region. The Barents Sea Integrated Management Plan (Norwegian ME 2006), its update (Norwegian ME 2011), and the Norwegian Sea Integrated Management Plan (Norwegian ME 2009) implement EBM in high latitude waters off Norway (Olsen et al. 2007; Hoel et al. 2009; Anonymous 2011; Ottersen et al. 2011). The plans coordinate fisheries, shipping, hydrocarbon, and other industry management under a single science-based framework applying the ecosystem and precautionary approaches, primarily to reduce the conflicts between and to evaluate the cumulative impacts of these human uses on the ecosystem. The existing regulatory regimes and ministries for each industry were left in place, and initial sectoral environmental impact assessments were conducted by these ministries as the basis for cumulative assessment in the Barents and Norwegian Seas. The Integrated Plans add marine spatial planning (Fig. 10.2) and identify ecosystem quality objectives (EcoQOs) as cross-sectoral management tools to address overall strategic objectives and cumulative environmental impacts (Olsen et al. 2007; Ottersen et al. 2011). By design, the plans are periodically revisited and adjusted for new scientific knowledge in the ecosystem (Hoel et al. 2009; von Quillfeldt et al. 2009). To facilitate integrated management under the plans, three new working groups were established to report to the Ministry implementing the plans: one for coordinating monitoring to regularly assess system state based on the EcoQOs outlined in the plan, one to assess pollution and other environmental risks, and one to coordinate and implement the science necessary for EBM (Olsen et al. 2007; von Quillfeldt et al. 2009); a fourth group representing stakeholders was added later in the process (Ottersen et al. 2011).

The Norwegian Integrated Management Plans have been successful in integrating the work of previously sector-based scientists and agencies to evaluate cumulative effects of human activity in Norwegian waters. Although national policies for EBM in Norway and Russia are developing at different rates, Russian scientists participated in the Norwegian planning process (Olsen et al. 2007), and scientific institutions in the Barents Sea region have conducted bilateral ecosystem assessments, resulting in a "Barents Portal" website and joint Norwegian Russian environmental status report (Arneberg et al. 2009; Stiansen et al. 2009), and an ecosystem considerations summary for ICES fishery management advice (ICES

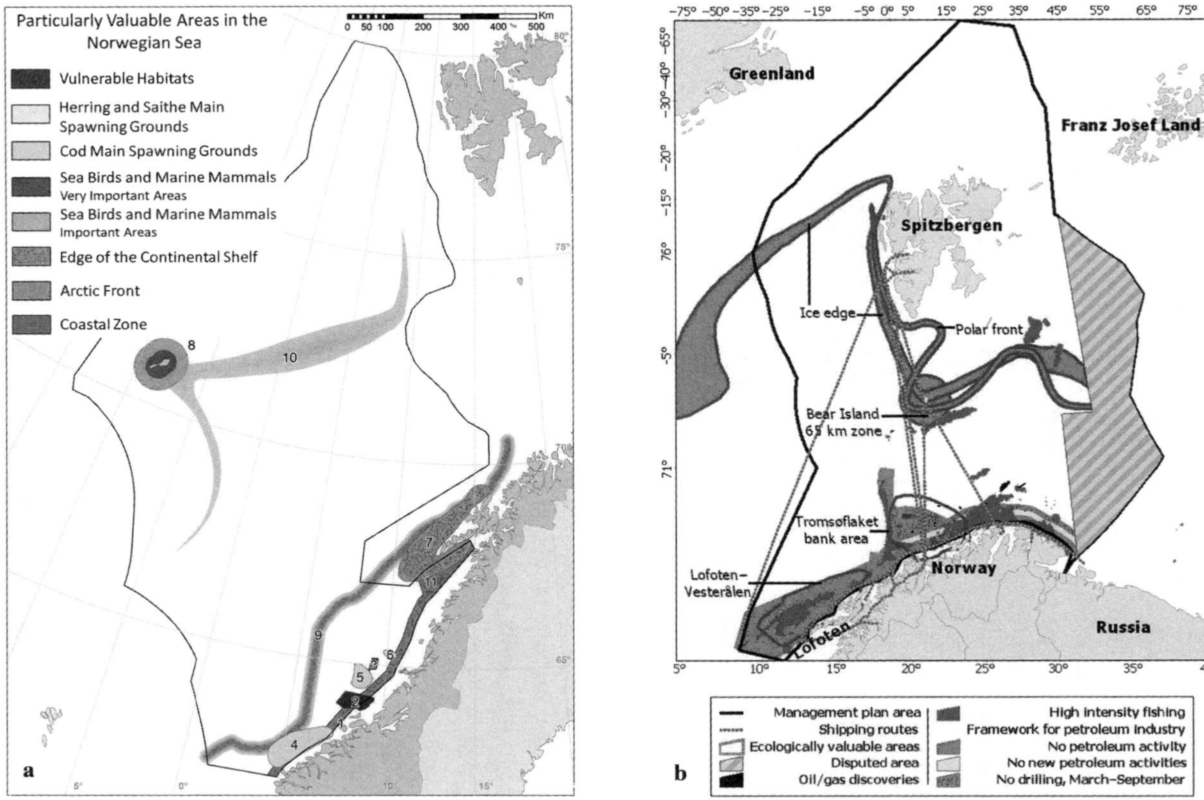

Fig 10.2 – (See color insert) Spatial aspects of Barents and Norwegian Sea Integrated Management Plans. (a) Particularly valuable areas identified in the Norwegian Sea [1 Remman Archipelago; 2 Froan Archipelego and Sula Reef; 3 Iverryggen Reef; 4 Mere Banks; 5 Halten Bank; 6 Sklinna Bank; 7 Vestfjorden/Vesterålen Islands; 8 Jan Mayen; 9 Shelf Edge; 10 Arctic Front; 11 Coastal Zone]. (Figure courtesy of Geir Ottersen, and reproduced with permission from Elsevier, from Ottersen et al. 2011); (b) Integrated management in the Barents Sea. (Figure reproduced with permission from Oxford University Press, from Olsen et al. 2007)

2012c). While this represents clear scientific progress, aspects of the planning process were criticized. For example, the initial Barents Sea Plan was developed without consultation of the resident indigenous Saami Parliament (although they now have some level of participation in implementation (Hoel et al. 2009)). Stakeholders groups provided "written feedback" on sectoral environmental impact assessments for both plans and attended a conference reviewing the Norwegian Sea Plan (Knol 2010; Ottersen et al. 2011), but it is unclear whether there were other opportunities for involvement in initial plan development. Finally, there are still questions about political motivations during ongoing implementation, with increasing pressure for oil and gas development in this potentially rich region. It will remain challenging to balance the often conflicting needs of the existing fishing industry and the expanding oil industry (Knol 2010). Nevertheless, planned updates encourage continuous improvement in both knowledge and management, and the Norwegian Integrated Management Plans have been cited as good examples of applied EBM (OSPAR 2010b).

In summary, EBM is in progress in high latitude waters of the northeast Atlantic, primarily through the multinational OSPAR Convention, bilateral and multinational fisheries agreements and increasingly EBFM oriented ICES advice, and the national EBM policy of Norway. Given the many and increasing human uses of these ecosystems, further development of EBM may help manage potential resource use conflicts, as well as challenges that arise due to climate change.

3.4 Newfoundland and Labrador: a national and multinational approach to EBM

The Newfoundland and Labrador (NL) shelves ecosystem in the Canadian Northwest Atlantic encompasses subarctic and temperate waters off the coast of Labrador and the island of Newfoundland (Fig. 10.1). This region is one of the three major biogeographic units currently being considered by Fisheries and Oceans Canada (DFO) for the purposes of i) assessing and reporting on ecosystem status and trends, and ii) spatial planning for the conservation of ecosystem properties and management of human activities (DFO 2009).

3.4a Ecosystem services, human uses, threats, and impacts

The primary ecosystem service in this region has been provisioning from commercial exploitation of living resources. For most of its recorded history, Newfoundland fisheries meant essentially Atlantic cod fisheries, which in turn structured the human immigrant communities. Fishing cod was the main economic incentive that attracted European fleets to the Grand Banks in the 1500s, and drove the settlement of permanent European colonies in the region. The cod fishery became a core economic factor, as well as an iconic and culturally defining activity for the people in Newfoundland for the next 500 years (Rose 2007). Together with the cod fisheries, other activities that developed early on included salmon fisheries, whaling, and sealing (Rose 2007). Although sealing has historically targeted mostly harp seals and to a lesser extent hooded seals *Cystophora cristata,* other pinnipeds were also significantly impacted; Atlantic walrus *Odobenus rosmarus* was extirpated from this ecosystem by the 1800s. The sealing industry reached its maximum impact and development during the 1800s and early 1900s when annual catches frequently surpassed 300 thousand individuals and reached peak landings above

600 thousand seals. This exploitation drastically reduced seal populations in this ecosystem; harp seals reached its minimal historical numbers in the early 1970s (Healey & Stenson 2000).

There are two clear collapses in the Newfoundland fisheries in the 20th century. The first one was associated with overfishing after the initial development of large scale offshore fishing fleets in the 1950s and 1960s (Rice 2002). During the 1960s and early 1970s, total annual catches from this ecosystem typically surpassed 600 thousand tons, and peaked over 1.2 million tons in the late 1960s. With the extension of Canadian jurisdiction to 200 nm in 1977, most commercial fish stocks were in poor shape and showing clear indications of overfishing. After fishing pressure was reduced and many commercial stocks showed signals of rebuilding, fishing mortality started to rise again (Rice 2002). In contrast with the 1950s and 1960s, when warm ocean climate conditions predominated, periods of colder conditions started in the early 1970s and became typical during the late 1980s and early 1990s, when the coldest ocean conditions on record were experienced (Drinkwater 2002). Although many commercial stocks had dramatically declined by the early 1990s, the second and most iconic collapse involved Atlantic cod, driven by significantly in-creased fishing pressure during the 1980s and early 1990s combined with harsh environmental conditions (Rice 2002; Lilly 2008). In 1992 the moratorium on the fishing of Northern cod was put in place by DFO, followed in 1994 by a Northwest Atlantic Fisheries Organization (NAFO) moratorium on Grand Bank cod.

The cod moratorium marked a major shift in the Newfoundland and Labrador fisheries; with no or very reduced access to most of the traditional groundfish species, strong snow crab and northern shrimp fisheries developed taking advantage of the increases in shellfish stocks. During the 2000s, most of the catches from this ecosystem were of snow crab and northern shrimp, with only a minor fraction composed by groundfishes. Although total annual landings in the 2000s were around 300 thousand tons (Table 10.1), well below landings in the years prior to the cod moratorium, the higher market value of shellfish maintained the value of the landings at or above record levels. Still, the newly developed shellfish fisheries did not provide the same level of employment nor achieve the large geographical reach into rural Newfoundland and Labrador as the traditional cod fishery.

While traditional fisheries were collapsing in the 1990s, the oil and gas industry developed in the Grand Banks. Oil and gas exploration in the region started in the 1960s, and a large number of fields were discovered during the 1970s and 1980s. By the late 1990s, the first production project, *Hibernia,* came on line, and several others have followed since. By the end of 2011, there were four production fields on the Grand Banks (*Hibernia, Terra Nova, White Rose,* and *North Amethyst*) which had produced in excess of 1.25 million barrels of oil (Table 10.1). In the 2005–2010 period the oil and gas industry represented around 30% (~8 billion dollars) of the Gross Domestic Product (GDP) of the province of Newfoundland and Labrador while fishing and processing was less than 4% of the GDP; however, these energy and fishing sectors were at par from a jobs perspective, representing approximately 4% of the employment in the province by 2010.

Although in comparison with oil and gas any other economic activity appears dwarfed, another industry that rapidly developed in the region after the collapse of the groundfish fisheries was aquaculture. Aquaculture sites are all located on the island of Newfoundland, usually no further than 5 nm from shore. Aquaculture production, initially based on blue mussels and later dominated by salmonids, was less than 2 thousand tons in the mid 1990s and by 2011 it reached 17 thousand tons, and represented a total market value of 120 million dollars. Many other activities use the marine environment in the NL ecosystem, ranging from aboriginal subsistence fishing and nearshore operations like tourism focused on marine wildlife (e.g. whale watching) to offshore ones like marine transportation involving both cargo and tanker vessels. Marine traffic also includes ferries connecting communities, as well as smaller vessels like recreational and fishing boats. This intense maritime transport impacts large marine species like whales and sea turtles through noise pollution and injuries or deaths from collisions, but also represents a potentially important vector for invasive species. Mining and major hydroelectric projects are also planned or in development in Newfoundland and Labrador. Many of these activities will have important impacts on watersheds and coastal environments. Overall, it was estimated that between 2001 and 2005, ocean related activities contributed to nearly 45% of the GDP and 25% of the employment in the province of Newfoundland and Labrador (Templeman 2010).

In summary, the NL marine ecosystem is exposed to impacts and subject to threats from a diverse set of human activities (Park et al. 2010; Templeman 2010). Climate change is also expected to modify temperature levels and patterns, raise sea levels, and impact ice distribution and dynamics. Many of these changes would likely affect the ranges and distribution of species, but given the emerging understanding of the importance of seasonal sea ice dynamics in the regulation of bottom-up driven production processes, climate change could significantly alter the basic structure and productivity of this ecosystem.

3.4b Ecosystem-based management in the NL ecosystem

The NL ecosystem extends beyond the 200 nm Economic Exclusive Zone (EEZ) boundary. This implies that management of its resources has both a Canadian and an international component. Within Canadian waters, Fisheries and Oceans Canada (DFO) has the lead federal role in managing Canada's fisheries. In international waters, also including several straddling stocks, fisheries management is the responsibility of NAFO.

There are three key pieces of legislation that govern fisheries in Canada. The *Fisheries Act* provides the framework for the management of fisheries, fish habitat and aquaculture; the *Species at Risk Act* tasks DFO with the management of aquatic species at risk; and the *Oceans Act*. The 1997 *Oceans Act* establishes Canada's commitment to the integrated management of oceans and marine resources, holding that conservation, based on an ecosystem approach, is of fundamental importance to maintaining biological diversity and productivity in the marine environment. It promotes wide application of the precautionary approach to conservation, management and exploitation of marine resources, and tasks DFO to lead and coordinate the development and implementation of a national system of marine protected areas (MPAs). The *Oceans Act* set the stage for the development of several ocean-related initiatives and policies (*Canada's Ocean Strategy* in 2002, *Canada's Ocean Action Plan* in 2005, and the *Health of the Oceans Initiatives* in 2007). As part of these initiatives several programs and activities were carried out. Two inshore MPAs were designated in 2005 under the *Ocean Act:* the Gilbert Bay and Eastport MPAs. Further, collaborative management was encouraged in marine regions through the creation of Large Ocean Management Areas (LOMAs), where all levels of government, aboriginal groups, industry organizations, environmental and community groups, and academia worked together to develop a strategic, long-term plan for sustainable management of resources. Within the southern subunit of the NL ecosystem, corresponding to NAFO Divisions 3LNOP, the Placentia Bay—Grand Banks (PBGB) LOMA was established.

As part of the Integrated Management Planning for the PBGB LOMA, conservation objectives were developed to guide the planning process by setting the bounds within which social, cultural and economic objectives could be established to ensure ecosystem sustainability to human impacts. Following DFO national guidelines (DFO 2004, 2006a,b, 2007a), LOMA-specific Ecologically and Biologically Significant Areas (EBSAs) were identified (DFO 2007b). Follow-up work involved the characterization and analysis of risk to ecosystem components, and their prioritization based on the estimated risk of harm from human activities within the PBGB LOMA (Park et al. 2010; Park et al. 2011). Although DFO discontinued the LOMAs program in 2011, work towards the development of integrated management in the NL ecosystem has continued; a process directed to identifying EBSAs in the NL ecosystem north of the former PBGB LOMA boundaries is currently under way. More recently, a National Framework for Canada's Network of Marine Protected Areas was initiated (Government of Canada 2011). Within this framework, the Laurentian Channel along the southern boundary of the NL ecosystem (previously identified as an EBSA by the PBGB LOMA planning process) was identified as an area of interest for an offshore MPA.

The LOMA program provided an envelope for developing integrated planning, but it did not have direct management authority; LOMA committees had a purely advisory role. Fisheries management authority always remained with DFO, with management guided by DFO's Sustainable Fisheries Framework (SFF), which employs an ecosystem-based and precautionary approach to fisheries management in Canada. This framework includes specific policies on i) the application of the precautionary approach for fishery decision making, ii) impacts of fishing on benthic habitats, communities, and species, and iii) new fisheries for forage species. The SFF is applied through Integrated Fisheries Management Plans (IFMPs), which are developed for individual fisheries and include short and long-term sustainable fisheries objectives, information on the fisheries, status of the stock, management measures, and considerations of other issues such as by-catch, species interactions, and habitat.

At present there are no formal mechanisms linking individual IFMPs across species, habitats, or communities; this makes addressing trade-offs among fisheries operating within the same ecosystem difficult. Also, the lack of an overarching and comprehensive national policy on how DFO will develop and implement ecosystem-based management beyond the more limited scope covered by the SFF adds uncertainty to regional planning processes. The need for such overarching and comprehensive policy framework, which could provide linkages to existing policies as well as guide the development of new ones, has already been identified (DFO 2012). Only some fisheries in the NL ecosystem have IFMPs (e.g. harp seals, northern shrimp, snow crab). Ecosystem information and advice currently feeds into these IFMPs, but it is also included in more traditional management processes in the form of "ecosystem considerations" sections in stock-assessment reports (e.g., DFO 2010).

Many important fish stocks in this ecosystem are straddling stocks, and therefore managed under NAFO. NAFO regulatory authority is limited to the area outside the EEZ; therefore, cooperation and consistency between Canada and NAFO approaches are essential for an effective implementation of any ecosystem-based management strategy in the NL ecosystem. In 2007, NAFO adopted an amendment to its convention where it commits to *apply an ecosystem approach to fisheries management in the Northwest Atlantic that includes safeguarding the marine environment, conserving its marine biodiversity, minimizing the risk of long term or irreversible adverse effects of fishing activities, and taking account of the relationship between all components of the ecosystem.* Although this amendment is still in the process of being ratified by contracting parties, NAFO has already advanced in the process of developing an ecosystem approach to fisheries management.

Scientific advice is provided by the NAFO Scientific Council (SC), which established a Working Group on Ecosystem Approaches to Fisheries Management (WGEAFM). The NAFO Fisheries Commission (FC) establishes regulations and makes management decisions; it created a Working Group of Fisheries Managers and Scientists on Vulnerable Marine Ecosystems (WGFMS). Since 2008 these working groups have been providing analyses and advice to their respective parent bodies on ecosystem-related issues (NAFO 2008b, 2009b, 2010b,c, 2012). During the last five years, NAFO SC has adopted a "Roadmap for the development of an Ecosystem Approach to Fisheries for NAFO" (NAFO 2010a), while providing advice on distribution of cold-water corals and sponges, threshold values for encounter protocols, and other VME issues (NAFO 2008a, 2009a, 2010a). NAFO FC has defined the fisheries footprint in the NRA and put in place an exploratory fishing protocol for any new fishery outside this footprint, closed many locations and seamounts to bottom fishing for the protection of VMEs, and implemented encounter protocols for cold-water corals and sponges (NAFO 2012). In 2012, FC and SC replaced the WGFMS with a new joint SC-FC working group on Ecosystem Approaches to Fisheries Management with broader scope, recognizing the need for close dialogue and collaboration between managers and scientists in the development and implementation of EBFM.

The NAFO SC Roadmap to EAF states that an ecosystem approach to fisheries should be: a) objective driven, b) focused on long-term ecosystem and stock sustainability, c) place-based, and d) addressing trade-offs among human activities explicitly. It is developed around the concept of Integrated Ecosystem Assessments, defined as *a synthesis and quantitative analysis of information on relevant physical, chemical, ecological, and human processes in relation to specified ecosystem management objectives* (Levin et al. 2009). The implementation of this roadmap is being developed around three practical sets of activities: a) definition of geographical management units, b) determination of ecosystem state and function, and c) development of management tools. In terms of ecosystem state and function, the Roadmap to EAF outlines a three-tier hierarchical process, going from overall ecosystem fishery production to single-species yields. These three tiers involve the determination of fisheries production potential at the ecosystem level, the allocation of the fisheries production among target species (i.e. multispecies trade-offs and other ecosystem considerations like protection of biodiversity), and the validation/verification of sustainability at the single-stock level (NAFO 2010c).

Overall, the challenges to implementing ecosystem-based management in the NL ecosystem are still numerous and diverse; both from the science (e.g. multispecies interactions, climate change) and management (e.g. trade-offs between human activities, coordination and consistency between Canadian and NAFO regulations) sides. Nonetheless, significant progress has been made in the last 10 years; we are developing a better understanding on how this ecosystem functions, and management frameworks in Canada and NAFO are increasingly being developed with ecosystem-based and precautionary approaches. Limited resources and frequently changing political priorities can certainly impact the pace at which ecosystem-based management is developed and the precise form it takes, but the path towards achieving it seems firmly set.

4. Conclusions: EBM in High Latitude Seas

The unique physical attributes of high latitude seas present particular challenges for EBM, but have also provided opportunities for EBM development. In this final section, we first summarize the key differences between high latitude seas and other marine ecosystems with respect to EBM, and note where progress is most advanced in addressing these key features. Finally, we discuss whether and where the challenges proposed by Leslie and McLeod (2007) are being met in high latitude seas: defining common EBM objectives, developing governance frameworks promoting EBM, and determining whether EBM goals are achieved by management practices.

4.1 High latitude ecosystem attributes most important to EBM

What do we gain from examining high latitude seas? Here, we highlight seven characteristics of high latitude systems drawn from our case studies which pose challenges for EBM development.

1. Human populations, and therefore local impacts to ecosystem services, are relatively small in high latitude systems at present, but impacts from human activities outside the systems can be large. Compared with lower latitude marine ecosystems, high latitude seas have fewer problems with point source pollution, agricultural runoff, shoreline development, and other impacts associated with locally dense human populations. This is an advantage in that EBM may focus on sustaining, rather than restoring, many services. However, wind- and water-borne contaminants accumulate at high latitudes in substrates and animals. EBM in the high Arctic began with Arctic nations addressing human health impacts of contamination from distant global sources. Similarly, global use of aerosols depleted ozone layers at the poles, and increasing global production of CO_2 is changing climate conditions rapidly in these systems.

2. High latitude seas, with their characteristic sea ice and ice-driven production cycles, are particularly vulnerable to climate change impacts on local and global ecosystem services, and are already changing rapidly. Climate impacts in the high

Arctic and Antarctic will have multi-scale repercussions through alteration of provisioning services, biodiversity, climate regulation and ocean circulation. These services extend beyond the boundaries of the ecosystems, and therefore beyond the jurisdiction of organizations attempting to implement EBM within them. While CCAMLR, the Arctic Council, OSPAR, ICES, NAFO and other regional organizations clearly acknowledge climate warming impacts to their respective ecosystems, systems are changing faster than predicted, and global action to reduce atmospheric CO_2 has been limited. The biggest challenge faced by regional EBM in high latitude seas is that global effort is required to sustain the global supporting and regulating services provided locally.

3. Provisioning services in high latitude seas have been compromised historically by commercial overexploitation, and recovery is complex. In these systems, the great whales were systematically removed, and other apex predators including fish were depleted, fundamentally changing community composition with unknown impacts on food web structure. Some stocks, like the great whales, have never recovered to pre-exploitation levels, while some fish stocks have recovered to levels where sustainable harvest is maintained. Recovery of overexploited species in these systems has been primarily achieved through "good single species management" where fishing mortality rates on individual species are constrained. As illustrated in the Barents Sea example for a small subset of species, the tradeoffs associated with attempting to recover multiple interacting stocks (e.g. both predators and their prey) need to be addressed by EBM; so far, tradeoffs are addressed for only a subset of exploited fish in the Barents Sea.

4. Today, high latitude seas are the only places where EBM must consider the commercial catch of marine mammals. Over 90% of the worldwide catch of marine mammals (primarily seals) reported in the FAO database is taken from the high latitude north Atlantic systems, and the only remaining commercial catch of whales comes from this same region and the Southern Ocean (Table 10.1). Commercial exploitation of whales and seals is controversial due to competing social concerns over animal welfare, indigenous cultural heritage, and sovereign national rights. While marine mammal harvest is considered sustainably managed through single species catch quotas in these systems, there remains a substantial clash between differing worldviews on the appropriateness of hunting marine mammals internationally. Therefore, EBM in these regions will need to address tradeoffs between cultural and provisioning services provided by marine mammals, in addition to interactions between marine mammals and other exploited species.

5. High latitude marine ecosystems also provision the world from lower on the food web, harvesting the some of the world's largest volumes of key forage species, including Antarctic krill in the Southern Ocean, Atlantic herring and capelin in the Nordic and Barents Seas, and shrimp off Newfoundland and Labrador. EBM (or EBFM) must address the tradeoffs between the provisioning from these fisheries and the energy forage species supply supporting ecosystem biodiversity (especially culturally significant marine mammals and birds) and other commercially important fish. EBFM efforts under CCAMLR in the Southern Ocean and ICES, Norway, and Russia in the Barents Sea account for the importance of some forage species to the wider ecosystem or to other key commercial species. CCAMLR's

EBFM may be further tested by increasing efficiency and interest in the krill fishery (Ito et al. 2010). As worldwide demand for products from these key forage species grows and climate change alters production cycles in high latitude seas, EBFM will be further challenged to balance provisioning and cultural services throughout the food web while sustaining these fisheries.

6. Cultural ecosystem services for indigenous peoples are closely linked to provisioning services in the northern hemisphere high latitude seas. Subsistence fishing and hunting are critically important to indigenous peoples throughout the high Arctic, and on a smaller scale in the north Atlantic. Therefore, EBM will need to involve indigenous peoples and subsistence users, and this has explicitly been incorporated into Arctic EBM under the Arctic Council. Continued indigenous stakeholder participation is vital to monitoring and maintaining subsistence provisioning and cultural services as other human uses increase in these ecosystems.

7. Conflicts between human uses in high latitude systems are projected to increase rapidly as climate change increases access to previously unavailable resources. EBM in these systems will need to accommodate previously unobserved and changing system states, uses, and user groups. Sustaining ecosystem services under these conditions will require flexibility and decision making under considerable uncertainty. Among the high latitude seas examined here, the Barents Sea has the most potentially conflicting human uses (oil and gas development, fishing, shipping, and tourism) and also has the best developed EBM policies to manage conflicts between these uses, mainly through spatial segregation.

4.2 Evaluating EBM development in high latitude seas

Despite the progress noted in each region, none of the case studies for high latitude seas can be considered fully implemented EBM. However, current management regimes incorporate many EBM attributes. In all of the case studies above, resource management is based on science, and the general objective of management is maintaining or restoring ecosystem functions which provide sustainable goods and services. Scientific ecosystem assessment capacity is high in these regions, with comprehensive ecosystem status reports prepared in every region reviewed. One missing component of EBM in several high latitude seas reviewed above is integration or management across all resource use sectors at the scale of the full ecosystem. This is not surprising given the scale of the regions and the complexity of governance within them. Nor is it unique to high latitude seas given the challenges associated with integrating management across sectors with potentially conflicting objectives. Nevertheless, clear progress is evident on cross sector integration in a few high latitude seas, in particular the Barents Sea.

Leslie and McLeod (2007) identify the first challenge for EBM as defining clear objectives. All of the high latitude management plans reviewed had strategic EBM objectives. However, clearly defined quantitative objectives and timelines for achieving them have yet to be developed in most systems, with the notable exceptions of seas covered by OSPAR, and fisheries management sectors in all regions. In the Southern Ocean under CCAMLR and in the Barents Sea under ICES and Norway, quantitative fishery management objectives for key forage species also consider their importance to the wider ecosystem, a clear EBFM objective.

CCAMLR's great success has been transforming the general objectives given in the articles of the treaty into operational EBFM objectives. EBFM in these high latitude seas also incorporates objectives for reducing bycatch of non-commercial species and impacts to habitats. However, fisheries management remains focused on individual population sustainability and so has not defined clear objectives where tradeoffs between species are ecologically inevitable (outside the Barents Sea and Southern Ocean). Nor has it addressed how to resolve tradeoffs between preferred social states (e.g. high profitability and high employment) and preferred ecological states (e.g. abundant, accessible, and desirable food fish species and charismatic megafauna). These tradeoffs would be better balanced under management with expanded focus; even high quality, scientifically based fisheries management is insufficient to sustain both ecosystem provisioning and cultural services.

Within cross-sector EBM, OSPAR's objectives for environmental quality were developed through multinational negotiation, and are mostly quantitative with clear timelines for achieving them. Norway's Barents Sea Integrated Ecosystem Plan also establishes clear ecosystem objectives with quantitative baselines and some action thresholds, with others in development, and a general timeline for reducing impacts by 2020. OSPAR and Norway's plans are binding, leaving implementation to member nations and sector-specific ministries, respectively. In contrast, the integrated plans developed by Canada for both the high Arctic Beaufort Sea LOMA and the subarctic Newfoundland Labrador (PBGB) LOMA stated clear but qualitative objectives for maintaining the ecological function of existing ecosystem components, without quantitative action thresholds (or management authority). The Arctic Council's Arctic Marine Strategic Plan identifies objectives for providing an effective EBM governance framework in the high Arctic. This is an essential first step towards EBM, but development of operational management objectives and authority to achieve them is left to individual governments and agencies.

In all of the high latitude seas, statements of priority for different ecosystem services or resource use sectors are absent; rather the goal is to prevent or reduce conflict between uses and maintain all services. Formal plans for tradeoffs between sectors have yet to be developed outside of Norway, where fishermen are financially compensated for losses due to offshore energy development activities (Norwegian ME 2011), implying that energy has priority over fishing as a use. In high latitude seas with relatively low human populations and historically limited human uses, it may be possible to defer tradeoffs, or prevent them with spatial management. However, human uses are increasing in high latitude systems, and are projected to increase further even as systems are altered by climate change. Well-designed EBM is the best framework to deal with these politically difficult tradeoffs as long as all affected stakeholders are involved in the process (Leslie & McLeod 2007).

We found at least partial examples of successful EBM across all of the governance frameworks from international through regional scales and with both hard and soft law structures. Therefore, no single optimal governance structure led to EBM success; however, successful tactics were apparent. One successful EBM tactic possible under hard law governance is pre-emptive restriction of access to resources before industrial uses develop. Because mineral development is not allowed under the ATS, Antarctic EBM can focus almost entirely on fisheries through CCAMLR. Similarly, fisheries managers closed Arctic waters off Alaska to fishing prior to industry development, restricting use of these waters to indigenous and petroleum industry uses. This can be an effective short term EBM policy while research and planning for more extensive integrated management progresses. This tactic has likely been successful because isolation and low population have historically restricted human uses in these ecosystems.

Another successful approach for EBM in high latitude seas arose under nearly opposite conditions in the northeast Atlantic, where diverse human uses and heavy exploitation began centuries ago. The multinational objectives set under OSPAR to improve environmental quality in the far more impacted North Sea have benefitted higher latitude seas in OSPAR's Region I. The advantage of OSPAR's wide geographic reach has been to apply these management measures in the neighboring high latitude systems prior to similar large scale industrial development. Now, as oil and gas development proceeds in the Barents Sea, much stricter environmental rules apply than in the 1970s when the industry developed in the North Sea.

Other partial EBM success stories in high latitude seas can arise from soft-law governance frameworks. The Arctic Council provides a forum for stakeholder interaction, scientific assessment, and regional management advice; this soft-law institution could be the effective governance structure which promotes EBM for this large, politically complex region. Interactions between the Arctic Council and hard law frameworks under OSPAR and other international management organizations including Norway, with its national EBM policy, could further develop EBM within the wider Arctic. During the 1990s, OSPAR's then limited effectiveness was enhanced by interaction with soft-law institutions such as the International North Sea Conferences, which led to non-binding agreements among a subset of states. This ultimately improved OSPAR's decision-making process, leading eventually to more effective environmental laws for the entire region through both OSPAR and the EU (Skjærseth 2006).

The ultimate EBM challenge is to determine whether goals have been achieved by the management frameworks in place (Leslie & McLeod 2007). Even in the best developed EBM example for the Barents Sea, where both OSPAR and Norway have made considerable progress in meeting objectives for environmental quality and reduced user conflicts, EBM is still evolving, and some objectives (e.g., stabilizing seabird populations) will be more difficult to achieve. Proactive restriction of human use can prevent resource damage, but this seems unlikely to be a good long term strategy in fast-changing high latitude systems as global human needs for ecosystem services increase. As high latitude seas change rapidly, continued development and implementation of EBM will be essential, and EBM programs must be capable of iterative re-evaluation to respond to change.

References

ACIA (2005) *Arctic Climate Impact Assessment—Scientific Report.* Cambridge UP, New York. 1042p.

Alexander V (1995) The influence of the structure and function of the marine food web on the dynamics of contaminants in Arctic Ocean ecosystems. *Sci Total Environ* 160/161:593–603

Allen RC, Keay I (2006) Bowhead Whales in the Eastern Arctic, 1611–1911: Population Reconstruction with Historical Whaling Records. *Environment and History* 12:89–113

Anchordoqui L, Halzen F (2006) IceHEP: high energy physics at the South Pole. *Annals of Physics* 321:2660–2716

Anderson DD (1984) Prehistory of north Alaska. In: Damas D (ed) Volume 5, *Arctic.* Smithsonian Institution, Washington D.C.

Anisimov OA, Vaughan DG, Callaghan TV, Furgal C, Marchant H, Prowse TD, Vilhjálmsson H, Walsh JE (2007) Polar regions (Arctic and Antarctic). In: Parry ML, Canziani OF, Palutikof JP, van der Linden PJ, C. E. H (eds) *Climate Change 2007: Impacts, Adaptation and Vulnerability Contribution of Working Group II to the Fourth Assessment Report of the Intergovernmental Panel on Climate Change.* Cambridge UP, Cambridge

Anonymous (1991) Arctic Environmental Protection Strategy. arctic_environment.pdf from http://arctic-council.org

Anonymous (2009a) Greenland. In: Hoel AH (ed) *Best practices in ecosystem-based oceans management in the Arctic.* Report Series 129. Norwegian Polar Institute, Tromso

Anonymous (2009b) USA: An integrated approach to ecosystem-based management. In: Hoel AH (ed) *Best practices in ecosystem-based oceans management in the Arctic,* Report Series 129. Norwegian Polar Institute, Tromso

Anonymous (2011) Updated version of the integrated management plan for the Barents Sea–Lofoten area. Office of the Prime Minister (Norway)

Arctic Council (1996) Declaration on the Establishment of the Arctic Council. Ottowa, Canada. http://www.arctic-council.org/index.php/en/about/documents/category/5-declarations

Arctic Council (2004) Arctic Marine Strategic Plan. http://www.pame.is/index.php/arctic-marine-strategic-plan2

Arctic Council (2008) Observed Best Practices in Ecosystem-based Oceans Management in the Arctic Countries. ACSAO-NO04_5_3BrPOMAr.pdf from http://arctic-council.org

Arctic Council (2009a) Arctic Marine Shipping Assessment. 2009 Report. http://www.pame.is/amsa-2009-report

Arctic Council (2009b) Arctic Offshore Oil and Gas Guidelines. http://www.pame.is/offshore-oil-and-gas/77-arctic-offshore-oil-and-gas-guidelines-2009

Arctic Council (2010) Arctic Biodiversity Trends 2010: Selected Indicators of Change. http://www.arcticbiodiversity.is/

Arctic Council (2011a) Outcomes from Ecosystem-based Management Experts Group October 18–19, 2011. http://www.arctic-council.org/index.php/en/about-us/expert-groups/275-expert-groups

Arctic Council (2011b) Report from the PAME workshop on ecosystem approach to management. http://www.pame.is/reports

Arctic Council (2011c) Search and rescue in the Arctic. http://www.arctic-council.org/index.php/en/oceans/search-and-rescue

Arctic Council (2012) Report from the PAME workshop on ecosystem approach to management. http://www.pame.is/reports

Arneberg P, Korneev O, Stiansen JE, Titov O (2009) Joint Norwegian-Russian environmental status 2008. Report on the Barents Sea Ecosystem http://www.barentsportal.com/barentsportal09/index.php?option=com_content&view=article&id=184%3Ajoint-norwegian-russian-environmental-status-2008-report-on-the-barents-sea-ecosystem&catid=84%3Aintroduction&Itemid=201&lang=en

Arrigo KR, van Dijken G, Pabi S (2008) Impact of a shrinking Arctic ice cover on marine primary production. *Geophys Res Lett* 35: L19603, doi:10.1029/2008GL035028

Astill J (2012) The Melting North. Special Report: The Arctic. *The Economist* June 16, 2012

Bates NR, Mathis JT (2009) The Arctic Ocean marine carbon cycle: evaluation of air-sea CO_2 exchanges, ocean acidification impacts and potential feedbacks. *Biogeosciences* 6:2433–2459

Beaufort Sea Partnership (2009) Integrated Ocean Management Plan for the Beaufort Sea: 2009 and beyond. http://www.beaufortseapartnership.ca/bsimp.html

Belchier M, Collins MA (2008) Recruitment and body size in relation to temperature in juvenile Patagonian toothfish (*Dissostichus eleginoides*) at South Georgia. *J Mar Biol* 155:493–503

Berkman PA, Lang MA, Walton DWH, Young OR (2011) *Science Diplomacy: Antarctica, Science and the Governance of International Spaces.* Smithsonian Institution. Scholarly Press, Inc.

Bernhard G, Manney G, Fioletov V, Grooß J-U, Heikkilä A, Johnsen B, Koskela T, Lakkala K, Müller R, Lund Myhre C, Rex M (2011) Ozone and UV radiation. In: Arctic Report Card 2011. http://www.arctic.noaa.gov/report11/ozone_uv.html

Bird KJ, Charpentier RR, Gautier DL, Houseknecht DW, Klett TR, Pitman JK, Moore TE, Schenk CJ, Tennyson ME, Wandrey CJ (2008) Circum-Arctic resource appraisal; estimates of undiscovered oil and gas north of the Arctic Circle: U.S. Geological Survey Fact Sheet 2008–3049. U.S. Geological Survey. http://pubs.usgs.gov/fs/2008/3049/

Bluhm BA, Gebruk AV, Gradinger R, Hopcroft RR, Huettmann F, Kosobokova KN, Sirenko BI, Weslawski JM (2011) Arctic marine biodiversity: An update of species richness and examples of biodiversity change. *Oceanography* 24:232–248

Bluhm BA, Gradinger R (2008) Regional variability in food available for Arctic marine mammals. *Ecological Applications* 18:S77-S96

Bockstoce J (1978) History of commercial whaling in Arctic Alaska. *Alaska Geographic* 5:17–25

Bonner WN (1968) The fur seal of South Georgia. *British Antarctic Survey Report* 56:1–95

BP (2012) BP Statistical review of world energy. London

Broecker WS (1991) The great ocean conveyor. *Oceanography* 4:79–89

Budikova D (2009) Role of Arctic sea ice in global atmospheric circulation: A review. *Global and Planetary Change* 68:149–163

Carmack E, Barber D, Christensen J, Macdonald R, Rudels B, Sakshaug E (2006) Climate variability and physical forcing of the food webs and the carbon budget on panarctic shelves. *Progress In Oceanography* 71:145–181

CIA (2011) The World Factbook 2011. Accessed December 12, 2011. https://www.cia.gov/library/publications/the-world-factbook/geos/xq.html

Clarke A, Crane JA (1992) The Southern Ocean benthic fauna and climate change: an historical perspective. *Phil Trans R Soc Lond B* 338:299–309

CNLOPB (2012) Canada-Newfoundland Labrador Offshore Petroleum Board Annual Report.

Cobb D, Fast H, Papst MH, Rosenberg D, Rutherford R, Sareault JE (eds) (2008) Beaufort Sea Large Ocean Management Area: Ecosystem overview and assessment report. *Can. Tech. Rep. Fish. Aquat. Sci.* 2780: ii-ix + 188 p.

Constable AJ (2001) CCAMLR ecosystem monitoring and management: future work. *CCAMLR Science* 9:233–253

Constable AJ, de la Mare WM, Agnew DJ, Iverson I, Miller D (2000) Managing fisheries to conserve the Antarctic marine ecosystem: practical implementation of the Convention on the Conservation of Antarctic Marine Living Resources (CCAMLR). *ICES J Mar Sci* 57:778–791

Corkeron PJ (2009) Marine mammals' influence on ecosystem processes affecting fisheries in the Barents Sea is trivial. *Biology letters* 5:204–206

Dalpadado P, Ingvaldsen RB, Stige LC, Bogstad B, Knutsen T, Ottersen G, Ellertsen B (2012) Climate effects on Barents Sea ecosystem dynamics. *ICES J Mar Sci* 69:1303–1316

Dayton PK, Mordida BJ, Bacon F (1994) Polar marine communities. *American Zoologist* 34:90–99

Denisov VV, Mikhaylichencko YG (2009) Management of the Russian Arctic seas. In: Hoel AH (ed) Best practices in ecosystem-based oceans management in the Arctic. Report Series 129. Norwegian Polar Institute, Tromso

Department of Fisheries, Hunting, and Agriculture (2009) Management and utilization of seals in Greenland. The Greenland Home Rule

DFO (2004) Identification of Ecologically and Biologically Significant Areas. Ecosystem Status Report 2004/006. DFO Canadian Science Advisory Secretariat

DFO (2006a) Final Report of the Fisheries Oceanography Committee 2006 Annual Meeting; 29–31 March 2006. Proceedings Series 2006/033. DFO Canadian Science Advisory Secretariat

DFO (2006b) Identification of Ecologically Significant Species and Community Properties. Science Advisory Report 2006/041. DFO Canadian Science Advisory Secretariat

DFO (2007a) Guidance Document on Identifying Conservation Priorities and Phrasing Conservation Objectives for Large Ocean Management Areas. Advisory Report 2007/010. DFO Canadian Science Advisory Secretariat

DFO (2007b) Placentia Bay-Grand Banks Large Ocean Management Area Conservation Objectives. Science Advisory Report 2007/042. DFO Canadian Science Advisory Secretariat

DFO (2009) Development of a framework and principles for the biogeographic classification of Canadian marine areas. Science Advisory Report 2009/056. DFO Canadian Science Advisory Secretariat

DFO (2010) Assessment of capelin in SA2 + Div. 3KL in 2010. Science Advisory Report 2010/090. DFO Canadian Science Advisory Secretariat

DFO (2012) Results and recommendations from the Ecosystem Research Initiative—Newfoundland and Labrador's Expanded Research on Ecosystem Relevant but Under-Surveyed Splicers. Science Advisory Report 2012/058. DFO Canadian Science Advisory Secretariat

Drinkwater K (2002) A review of the role of climate variability in the decline of northern cod. *American Fisheries Society Symposium* 32:113–130

Ducklow HW, Baker K, Martinson DG, Quetin LB, Ross RM, Smith RC, Stammerjohn SE, Vernet M, Fraser W (2007) Marine pelagic ecosystems: the West Antarctic Peninsula. *Phil Trans R Soc Lond B* 362:67–94

Eastman JT, McCune AR (2000) Fishes on the Antarctic continental shelf: evolution of a marine species flock? *Journal of Fish Biology* 57 (Supplement A):84–102

Edes ME (2008) Ecotourism in the Arctic Circle: regional regulation is necessary to prevent concerned environmentalists from further contributing to climate change. *Global Business and Development Law Journal* 21:251–283

El-Sayed SZ (ed) (1994) *Southern Ocean Ecology: the BIOMASS perspective.* Cambridge UP

Ellis R (1991) *Men and Whales.* Knopf, New York

Erickson AW, Hanson MB (1990) Continental estimates and population trends of antarctic ice seals. In: Kerry KR, Hempel G (eds) *Antarctic Ecosystems Ecological change and conservation.* Springer-Verlag, Heidelberg, Germany

Erickson AW, Siniff DB, Cline DR, Hofman RJ (1971) Distributional ecology of Antarctic seals. In: Deacon G (ed) *Symposium on Antarctic Ice and Water Masses.* Comm. Antarct Res., Cambridge, UK

European Commission (2010) Commission Regulation (EU) No 737/2010 of 10 August 2010 laying down detailed rules for hte implementaiton of Regulation (EC) No 1007/2009 of the European Parliament and of hte Council on trade in seal products. European Commission, Brussels. http://eur-lex.europa.eu/JOHtml.do?uri=OJ:L:2010:216:SOM:EN:HTML

Fabry VJ, McClintock JB, Mathis JT, Grebmeier JM (2009) Ocean acidification at high latitudes: the bellweather. *Oceanography* 22:160–171

Fabry VJ, Seibel BA, Feely RA, Orr JC (2008) Impacts of ocean acidification on marine fauna and ecosystem processes. *ICES J Mar Sci* 65:414–432

Falk-Petersen J, Renaud P, Anisimova N (2011) Establishment and ecosystem effects of the alien invasive red king crab (*Paralithodes camtschaticus*) in the Barents Sea—a review. *ICES J Mar Sci* 68:479–488

Falk-Petersen S, Haug T, Hop H, Nilssen KT, Wold A (2009) Transfer of lipids from plankton to blubber of harp and hooded seals off East Greenland. *Deep Sea Research Part II: Topical Studies in Oceanography* 56:2080–2086

FAO (2012a) Fisheries and Aquaculture Information and Statistics Service. http://www.fao.org/fishery/statistics/global-capture-production/query/en

FAO (2012b) The State of World Fisheries and Aquaculture 2012. FAO, Rome

Fay G, Karlsdóttir A (2011) Social indicators for arctic tourism: Observing trends and assessing data. *Polar Geography* 34:63–86

Feely RA, Sabine CL, Lee K, Berelson W, Keypas J, Fabry VJ, Millero FJ (2004) Impact of anthropogenic CO_2 on the CO_3 system in the oceans. *Science* 305:362–366

Filin A, Tjelmeland S (2005) Ecosystem approach to estimation of long-term yield of cod in the Barents Sea. In: Shibanov V (ed) *Proceedings of the 11th Joint Russian-Norwegian Symposium: Ecosystem dynamics and optimal long-term harvest in the Barents Sea fisheries.* Murmansk, Russia 15–17 August 2005. IMR/PINRO Report series 2/2005. Institute of Marine Research, Bergen

Fortier L, Sirois P, Michaud J, Barber D (2006) Survival of Arctic cod larvae (*Boreogadus saida*) in relation to sea ice and temperature in the Northeast Water Polynya (Greenland Sea). *Canadian Journal of Fisheries and Aquatic Sciences* 63:1608–1616

Francis JA, Chan W, Leathers DJ, Miller JR, Veron DE (2009) Winter Northern Hemisphere weather patterns remember summer Arctic sea-ice extent. *Geophys Res Lett* L07503, doi:10.1029/2009GL037274

Frost KJ, Lowry LF (1984) Trophic relationships of vertebrate consumers in the Alaskan Beaufort Sea. In: Barnes PW, Schell DM, Reimnitz E (eds) *The Alaskan Beaufort Sea, Ecosystems and Environments.* Academic Press Inc., Orlando, FL

Gautier DL, Bird KJ, Charpentier RR, Grantz A, Houseknecht DW, Klett TR, Moore TE, Pitman JK, Schenk CJ, Schuenemeyer JH, Sorensen K, Tennyson ME, Valin ZC, Wandrey CJ (2009) Assessment of undiscovered oil and gas in the Arctic. *Science* 324:1175–1179

Gille ST (2002) Warming of the Southern Ocean since the 1950s. *Science* 295:1275–1277

Gjøsæter H (1995) Pelagic Fish and the Ecological Impact of the Modern Fishing Industry in the Barents Sea. *Arctic* 48:267–278

Gjøsæter H, Bogstad B, Tjelmeland S (2002) Assessment methodology for Barents Sea capelin, *Mallotus villosus* (Müller). *ICES J Mar Sci* 59:1086–1095

Gjøsæter H, Bogstad B, Tjelmeland S (2009) Ecosystem effects of the three capelin stock collapses in the Barents Sea Published in collaboration with the University of Bergen and the Institute of Marine Research, Norway, and the Marine Biological Laboratory, University of Copenhagen, Denmark. *Marine Biology Research* 5:40–53

Government of Canada (2011) National Framework for Canada's Network of Marine Protected Areas. Fisheries and Oceans Canada, Ottowa. 31p.

Grebmeier JM, Cooper LW, Feder HM, Sirenko BI (2006) Ecosystem dynamics of the Pacific-influenced Northern Bering and Chukchi Seas in the Amerasian Arctic. *Progress In Oceanography* 71:331–361

Hammill, MO, Stenson GB (2011) Estimating abundance of Northwest Atlantic harp seals, examining the impact of density dependence. DFO Canadian Science Advisory Secretariat

Halpern BS, Selkoe KA, Micheli F, Kappel CV (2007) Evaluating and ranking the vulnerability of global marine ecoystems to anthropogenic threats. *Conserv Biol* 21:1301–1315

Hansen JO, Rasen B (2012) *Facts 2012: The Norwegian Petroleum Sector.* Ministry of Petroleum and Energy and Norwegian Petroleum Directorate

Healey BP, Stenson GB (2000) Estimating pup production and population size of the northwest Atlantic harp seal (*Phoca groenlandica*). Research Document 2000/081. DFO Canadian Science Advisory Secretariat

Hewitt RP, Watters G, Trathan PN, Croxall JP, Goebel ME, Ramm D, Reid K, Trivelpiece WZ, Watkins JL (2004) Options for allocating the precautionary catch limit of krill among small-scale management units in the Scotia Sea. CCAMLR *Science* 11:81–97

Hightower LE, Renfro JL (1988) Recent applications of fish cell culture to biomedical research. *J Exp Zool* 248:290–302

Hjermann DO, Ottersen G, Stenseth NC (2004a) Competition among fishermen and fish causes the collapse of Barents Sea capelin. *Proc Natl Acad Sci* USA 101:11679–11684

Hjermann DO, Stenseth NC, Ottersen G (2004b) Indirect climatic forcing of the Barents Sea capelin: a cohort effect. *Marine Ecology Progress Series* 273:229–238

Hoel AH (ed) (2009). *Best practices in ecosystem-based oceans management in the Arctic.* Report Series 129. Norwegian Polar Institute, Tromso

Hoel AH, von Quillfeldt CH, Olsen E (2009) Norway and integrated oceans management—the case of the Barents Sea. In: Hoel AH (ed) *Best practices in ecosystem-based oceans management in the Arctic.* Report Series 129. Norwegian Polar Institute, Tromso

Hofmeyr GJG, Bester MN, Jonker FC (1997) Changes in population sizes and distribution of fur seals at Marion Island. *Polar Biology* 17:150–158

Honda M, Inoue J, Yamane S (2009) Influence of low Arctic sea-ice minima on anomalously cold Eurasian winters. *Geophys Res Lett* 36:L08707, doi:10.1029/2008GL037079

Hovelsrud GK, McKenna M, Huntingont HP (2008) Marine mammal harvests and other interactions with humans. *Ecological Applications* 18:S135–S147

Howell D, Bogstad B (2010) A combined Gadget/FLR model for management strategy evaluations of the Barents Sea fisheries. *ICES J Mar Sci* 67:1998–2004

Hunt Jr. GL, Drinkwater K (eds) (2005) Background on the climatology, physical oceanography and ecosystems of the Sub-Arctic Seas. Appendix to the ESSAS Science Plan. GLOBEC Report No. 20. GLOBEC International Project Office, Plymouth, UK

Huntington HP (2011) The local perspective. *Nature* 478:182–183

Huntington HP, Pungowiyi C (2009) Indigenous perspectives. In: Hoel AH (ed) *Best practices in ecosystem-based oceans management in the Arctic.* Report Series 129. Norwegian Polar Institute, Tromso

IAATO (2012) Report of the International Association of Antarctica Tour Operators 2011–12. Antarctic Treaty Consultative Meeting XXXV, Hobart, Tasmania

ICES (2011) Status and harvest potential of the Harp seal stocks in the Greenland Sea and the White Sea/Barents Sea, and of the hooded seal stock in the Greenland Sea. In: ICES (ed) *ICES Advice 2011,* Book 3.

ICES (2012a) General context of *ICES Advice,* Book 1

ICES (2012b) ICES scientific and advisory services of relevance to the EU Marine Strategy Framework Directive. www.ices.dk/iceswork/bulletin/20120521_ICES_Strategy_MSFD.pdf

ICES (2012c) Report of the Arctic Fisheries Working Group 2012 (AFWG), 20–26 April 2012, ICES Headquarters, Copenhagen. ICES CM 2012/ACOM:05

Ito T, Woloszyn M, Mazloff M (2010) Anthropogenic carbon dioxide transport in the Southern Ocean driven by Ekman flow. *Nature* 463:80–83

Itta ES (2008) Shell abandons 2008 Beaufort drilling plans. *North Slope News,* July 2008, Issue 1, Barrow, AK. http://www.co.north-slope.ak.us/departments/mayorsoffice/NSnews/July08_no1.pdf

IWC (2012) Catches taken: Under Objection. http://iwcoffice.org/table_objection

Johannesen E, Ingvaldsen RB, Bogstad B, Dalpadado P, Eriksen E, Gjosaeter H, Knutsen T, Skern-Mauritzen M, Stiansen JE (2012) Changes in Barents Sea ecosystem state, 1970–2009: climate fluctuations, human impact, and trophic interactions. *ICES J Mar Sci* 69:880–889

Jones CD, Kock K-H, Balguerias E (2000) Changes in biomass of eight species of finfish around the South Orkney Islands (subarea 48.2) from three bottom trawl surveys. *CCAMLR Science* 7:53–74

Jorstad KE, Farestveit E, Rudra H, Agnalt AL, Olson S (2002) Studies on red king crab (Paralithodes camtschaticus) introduced to the Barents Sea. In: Paul AJ, Dawe EG, Elner R, Jamieson GS, Kruse GH, Otto RS, Sainte-Marie B; Shirley TC, Woodby D (eds) *Crabs in Cold Water Regions: Biology, Management, and Economics.* University of Alaska Sea Grant, Fairbanks, AK. 425–438

Kao S-M, Pearre NS, Firestone J (2012) Adoption of the arctic search and rescue agreement: A shift of the arctic regime toward a hard law basis? *Marine Policy* 36:832–838

Kiovurova T (2008) Alternatives for an Arctic Treaty—evaluation and a new proposal. *RECIEL* 17:14–26

Knol M (2010) Scientific advice in integrated ocean management: The process towards the Barents Sea plan. *Marine Policy* 34:252–260

Kock K-H, Jones CD (2005) Fish Stocks in the southern Scotia Arc region—a review and prospectus for future research. *Reviews in Fisheries Science* 13:75–108

Kock K-H, Reid K, Croxall JP, Nicol S (2007) Fisheries in the Southern Ocean: an ecosystem approach. *Phil Trans R Soc Lond B* 362:2333–2349

Krupnik I, Ray GC (2007) Pacific walruses, indigenous hunters, and climate change: Bridging scientific and indigenous knowledge. *Deep Sea Research Part II: Topical Studies in Oceanography* 54:2946–2957

Laidre KL, Stirling I, Lowry LF, Wiig O, Heide-Jorgensen MP, Ferguson SH (2008) Quantifying the sensitivity of Arctic marime mammals to climate-induced habitat change. *Ecological Applications* 18:S97-S125

Laws RM (1977) The significance of vertebrates in the Antarctic marine ecosystem. In: Llano GA (ed) *Adaptations within Antarctic ecosystems.* Gulf Publishing Co. for the Smithsonian Institution, Washington, DC., USA

Lennon E (2008) A tale of two poles: a comparative look at the legal regimes of the Arctic and the Antarctic. *Sustainable Development Law & Policy* 8:32–66

Leslie HM, McLeod KL (2007) Confronting the challenges of implementing marine ecosystem-based management. *Frontiers in Ecology and the Environment* 5:540–548

Leuenberger M, Siegenthaler U (1992) Ice-age atmospheric concentration of nitrous oxide from an Antarctic ice core. *Nature* 360:449–451

Levin PS, Fogarty MJ, Murawski SA, Fluharty D (2009) Integrated ecosystem assessments: developing the scientific basis for ecosystem-based management of the ocean. *PLoS Biology* 7:23–28

Lilly G (2008) The decline, recovery, and collapse of Atlantic cod (*Gadus morhua*) off Labrador and eastern Newfoundland. In: Kruse GH, Drinkwater K, Ianelli JN, Link JS, Stram DL, Wespestad VG, Woodby D (eds) *Resilience of gadid stocks to fishing and climate change.* University of Alaska Sea Grant, Fairbanks, AK

Lindstrøm U, Smout S, Howell D, Bogstad B (2009) Modelling multi-species interactions in the Barents Sea ecosystem with special emphasis on minke whales and their interactions with cod, herring and capelin. *Deep Sea Research Part II: Topical Studies in Oceanography* 56:2068–2079

Lodge MW, Anderson D, Løbach T, Munro G, Sainsbury K, Willock A (2007) Recommended Best Practices for Regional Fisheries Management Organizations: Report of an independent panel to develop a model for improved governance by Regional Fisheries Management Organizations. Chatham House (The Royal Institute of International Affairs)

Loeng H, Brander K, Carmack EC, Denisenko SG, Drinkwater KF, Hansen B, Kovacs K, Livingston P, McLaughlin F, Sakshaug E (2005) Marine systems. In: *Arctic Climate Impact Assessment—Scientific Report.* Cambridge UP

Loeng H, Drinkwater K (2007) An overview of the ecosystems of the Barents and Norwegian Seas and their response to climate variability. *Deep Sea Research Part II: Topical Studies in Oceanography* 54:2478–2500

Mallory ML, Gilchrist HG, Braune BM, Gaston AJ (2006) Marine birds as indicators of Arctic marine ecosystem health: linking the Northern Ecosystem Initiative to long-term studies. *Environmental monitoring and assessment* 113:31–48

Mangerink K, Altman S, Hunter K (2008) *Integrated ecosystem-based management of the U.S. Arctic marine environment: assessing the feasibility of program development and implementation.* Environmental Law Institute, Washington DC

Maslanik J, Fowler C, Stroeve J, Drobot S, Zwally J, Yi D, Emery W (2007) A younger, thinner Arctic ice cover: increased potential for rapid, extensive sea-ice loss. *Geophys Res Lett* 34, L24501, doi:10.1029/2007GL032043

Maslanik J, Stroeve J, Fowler C, Emery W (2011) Distribution and trends in Arctic sea ice age through spring 2011. *Geophys Res Lett* 38, L13502, doi:10.1029/2011GL047735

Mathis JT (2011) The extent and controls on ocean acidification in the western Arctic Ocean and adjacent continental shelf areas. In: Arctic Report Card 2011. http://www.arctic.noaa.gov/report11/ocean_acidification.html

McBean G, Alekseev G, Chen D, Forland E, Fyfe J, Groisman PY, King R, Melling H, Vose R, Whitfiled PH (2005) Arctic climate: past and present. In: *Arctic Climate Impact Assessment—Scientific Report.* Cambridge UP

McFadden KW, Barnes C (2009) The implementation of an ecosystem approach to management within a federal government agency. *Marine Policy* 33:156–163

Meredith MP, King JC (2005) Rapid climate change in the ocean west of the Antarctic Peninsula during the second half of the 20th century. *Geophys Res Lett* 32, L19604, doi:10.1029/2005GL024042

Merkel F, Barry T (eds) (2008) Seabird harvest in the Arctic. CAFF International Secretariat, Circumpolar Seabird Group (CBird), CAFF Technical Report No. 16

Millennium Ecosystem Assessment (2005) Polar Systems. Chapter 25. In: Hassan R, Scholes R, Ash N (eds) *Ecosystems and human well-being: current state and trends,* Vol. 1. Island Press, Washington DC

Montes-Hugo M, Doney SC, Ducklow HW, Fraser W, Martinson D, Stammerjohn SE, O. S (2009) Recent Changes in Phytoplankton Communities Associated with Rapid Regional Climate Change Along the Western Antarctic Peninsula. *Science* 323:1470–1473

Moore SE, Laidre KL (2006) Trends in sea ice cover within habitats used by bowhead whales in the western Arctic. *Ecological Applications* 16:932–944

Mori Y, Boyd IL (2004) The Behavioral Basis for Nonlinear Functional Responses and Optimal Foraging in Antarctic Fur Seals. *Ecology* 85:398–410

NAFO (2008a) Report of Scientific Council meeting, 5–19 June 2008. SCS Document 08/19. NAFO

NAFO (2008b) Report of the NAFO Scientific Council Working Group on Ecosystem Approach to Fisheries Management (WGEAFM). SCS Document 08/10. NAFO

NAFO (2009a) Report of Scientific Council Meeting, 4–18 June 2009. SCS Document 09/23. NAFO

NAFO (2009b) Report of the NAFO Scientific Council Working Group on Ecosystem Approach to Fisheries Management (WGEAFM). SCS Document 09/06. NAFO

NAFO (2010a) Report of Scientific Council meeting, 3–16 June 2010. SCS Document 10/18. NAFO

NAFO (2010b) Report of the 3rd Meeting of the NAFO Scientific Council Working Group on Ecosystem Approaches to Fisheries Management (WGEAFM). I SCS Document 10/24. NAFO

NAFO (2010c) Report of the NAFO Scientific Council Working Group on Ecosystem Approach to Fisheries Management (WGEAFM). SCS Document 10/19. NAFO

NAFO (2012) Northwest Atlantic Fisheries Organization conservation and enforcement measures. FC Document 12/01. NAFO

National Snow & Ice Data Center (2012) Arctic sea ice extent settles at record seasonal minimum. http://nsidc.org/arcticseaicenews/2012/09/arctic-sea-ice-extent-settles-at-record-seasonal-minimum/

Niemi A, Paulic J, Cobb D (2010) Ecosystem status and trends report: Arctic Marine Ecozones. In: Department of Fisheries and Oceans Canada (ed), Science Advisory Secretariat. Res. Doc. 2010/066. Department of Fisheries and Oceans, Canada

Nilssen EM, Sundet JH (2006) The introduced species red king crab (*Paralithodes camtschaticus*) in the Barents Sea. *Fisheries Research* 82:319–326

Norwegian ME (2006) Integrated Management Plan of the Marine Environment of the Barents Sea and the Sea Areas off the Lofoten Islands. Book Report No. 8 (2005–2006) to the Storting. Ministry of the Environment, Oslo

Norwegian ME (2009) Integrated Management of the Marine Environment of the Norwegian Sea. Report No. 37 (2008–2009) to the Storting. Ministry of the Environment, Oslo

Norwegian ME (2011) First update of the Integrated Management Plan for the Marine Environment of the Barents Sea-Lofoten Area. Meld. St. 10 (2010–2011) Report to the Storting (white paper). Ministry of the Environment, Oslo

Norwegian MFCA (2007) Norwegian Fisheries Management. Norwegian Ministry of Fisheries and Coastal Affairs, Oslo

Norwegian MTI (2007) The Norwegian Maritime Strategy 2007. Steady as she goes: The Government's strategy for environmentally friendly growth in maritime industry. Norwegian Ministry of Trade and Industry, Oslo

Nowlan L (2001) Arctic legal regime for environmental protection. International Union for the Conservation of Nature (IUCN), Gland, Switzerland and Cambridge, UK, and International Council of Environmental Law (ICEL), Bonn, Germany

Nuttall M, Berkes F, Forbes B, Kofinas GP, Vlassova T, Wenzel G (2005) Hunting, herding, fishing, and gathering: indigenous peoples and renewable resource use in the Arctic. *Arctic Climate Impacts Assessment, Scientific Report.* Cambridge UP

O'Gorman FA (1963) The return of the Antarctic fur seal. *New Scientist* 20:374–376

Obama B (2010) Executive Order 13547—Stewardship of the Ocean, Our Coasts, and the Great Lakes. The White House, Office of the Press Secretary, Washington, D.C.

Olsen E, Gjøsæter H, Rottingen I, Dommasnes A, Fossum P, Sandberg P (2007) The Norwegian ecosystem-based management plan for the Barents Sea. *ICES J Mar Sci* 64:599–602

OSPAR (1992) Convention for the Protection of the Marine Environment of the North-East Atlantic. OSPAR Commission, London

OSPAR (2000) Quality Status Report 2000, Region I—Arctic waters, Vol. OSPAR Commission, London

OSPAR (2010a) The North-East Atlantic Environment Strategy: Strategy of the OSPAR Commission for the Protection of the Marine Environment of the North-East Atlantic 2010–2020 (OSPAR Agreement 2010–3). OSPAR Commission, London

OSPAR (2010b) Quality Status Report 2010. OSPAR Commission, London

Ottersen G, Olsen E, van der Meeren GI, Dommasnes A, Loeng H (2011) The Norwegian plan for integrated ecosystem-based management of the marine environment in the Norwegian Sea. *Marine Policy* 35:389–398

Overland J, Bhatt U, Key J, Liu Y, Walsh J, Wang J (2011) Temperature and Clouds. In: Arctic Report Card 2011. http://www.arctic.noaa.gov/report11/temperature_clouds.html

Overland JE, Wang M (2010) Large-scale atmospheric circulation changes are associated with the recent loss of Arctic sea ice. *Tellus A* 62:1–9

Overland JE, Wang M, Salo S (2008) The recent Arctic warm period. *Tellus A* 60:589–597

Park LE, Beresford LA, Anderson MR (2010) Characterization and analysis of risks to key ecosystem components and properties. In: Oceans, Habitat and Species at Risk Publication Series, Newfoundland and Labrador Region, No. 0003. Fisheries and Oceans Canada, Newfoundland and Labrador Region

Park LE, Beresford LA, Kissler E (2011) Prioritization of key ecosystem components based on risk of harm from human activities within the Placentia Bay/Grand Banks Large Ocean Management Area.

In: Oceans, Habitat and Species at Risk Publication Series, Newfoundland and Labrador Region, No. 0004. Fisheries and Oceans Canada, Newfoundland and Labrador Region

Parkinson CL, Cavalieri DJ (2008) Arctic sea ice variability and trends, 1979–2006. *J of Geophys Res* 113: C07003, doi:10.1029/2007JC004558

Payne MR (1977) Growth of a fur seal population. *Phil Trans R Soc Lond B* 279:67–79

Perovich DK, Meier W, Maslanik J, Richter-Menge J (2011) Sea Ice. In: Arctic Report Card: 2011. http://www.arctic.noaa.gov/report11/sea_ice.html

Petoukhov V, Semenov VA (2010) A link between reduced Barents-Kara sea ice and cold winter extremes over northern continents. *J of Geophys Res* 115: D21111, doi:10.1029/2009JD013568

Pitulko VV, Nikolsky PA, Girya EY, Basilyan AE, Tumskoy VE, Koulakov SA, Astakhov SN, Pavolva EY, Anisimov MA (2004) The Yana RHS site: humans in the arctic before the last glacial maximum. *Science* 303:52–56

Plagányi ÉE, Butterworth DS (2006a) A spatial multi-species operating model (SMOM) of krill-predator interactions in small-scale management units in the Scotia Sea. Workshop document presented to WG-EMM subgroup of CCAMLR (Commission for the Conservation of Antarctic Marine Living Resources), WG-EMM-06/12

Plagányi ÉE, Butterworth DS (2006b) An illustrative management procedure for exploring dynamic feedback in krill catch limit allocations among small-scale management units. Workshop document presented to WG-EMM subgroup of CCAMLR (Commission for the Conservation of Antarctic Marine Living Resources), WG-EMM-06/28

Rahmstorf S (2006) Thermohaline ocean circulation. In: Elias SA (ed) *Encyclopedia of Quaternary Sciences.* Elsevier, Amsterdam

Rampal P, Weiss J, Dubois C, Campin JM (2011) IPCC climate models do not capture Arctic sea ice drift acceleration: Consequences in terms of projected sea ice thinning and decline. *J Geophys Res* 116:C00D07, doi:10.1029/2011JC007110

Reid K, Croxall JP (2001) Environmental response of upper trophic-level predators reveals a system change in an Antarctic marine ecosystem. *Proc R Soc Lond B* 268:377–384

Reilly SB, Bannister JL, Best PB, Brown M, Brownell Jr. RL, Butterworth DS, Clapham PJ, Cooke J, Donovan G, Urbán J, Zerbini AN (2012a) *Balaena mysticetus* (Svalbard-Barents Sea (Spitsbergen) subpopulation). In: IUCN 2012. IUCN Red List of Threatened Species. Version 2012.2. http://www.iucnredlist.org/details/2472/0

Reilly SB, Bannister JL, Best PB, Brown M, Brownell Jr. RL, Butterworth DS, Clapham PJ, Cooke J, Donovan G, Urbán J, Zerbini AN (2012b) Eubalaena glacialis. In: IUCN 2012. IUCN Red List of Threatened Species. Version 2012.2. http://www.iucnredlist.org/details/41712/0

Rice J (2002) Changes to the Large Marine Ecosystem of the Newfoundland-Labrador Shelf. In: Sherman K, Skjoldal HR (eds) *Large Marine Ecosystems of the North Atlantic: changing states and sustainability.* Elsevier, Amsterdam

Richter-Menge J, Jeffries MO, Overland JE (2011) Arctic Report Card 2011. http://www.arctic.noaa.gov/reportcard

Rose GA (2007) *Cod: The ecological history of the North Atlantic fisheries.* Breakwater Books, St. John's

Sabine CL, Feely RA, Gruber N, Key RM, Lee K, Bullister JL, Wanninkhof R, Wong CS, Wallace DWR, Tilbrook B, Millero FJ, Peng T-H, Kozyr A, Ono T, Rios AF (2004) The Oceanic Sink for Anthropogenic CO2. *Science* 305:367–371

Savinetsky AB, Kiseleva NK, Khassanov BF (2004) Dynamics of sea mammal and bird populations of the Bering Sea region over the last several millennia. *Palaeogeography, Palaeoclimatology, Palaeoecology* 209:335–352

SC-CCAMLR (2002) Report of the Twenty-first Meeting of the Scientific Committee. Commission for the Conservation of Antarctic Marine Living Resources (CCAMLR), Hobart, Australia

SC-CCAMLR (2003) Report of the Twenty-second Meeting of the Scientific Committee. Commission for the Conservation of Antarctic Marine Living Resources (CCAMLR), Hobart, Australia

SC-CCAMLR (2007) Report of the Twenty-sixth Meeting of the Scientific Committee. Commission for the Conservation of Antarctic Marine Living Resources (CCAMLR), Hobart, Australia

SC-CCAMLR (2008) Report of the Twenty-seventh Meeting of the Scientific Committee. Commission for the Conservation of Antarctic Marine Living Resources (CCAMLR), Hobart, Australia

SC-CCAMLR (2010) Report of the Twenty-ninth Meeting of the Scientific Committee. Commission for the Conservation of Antarctic Marine Living Resources (CCAMLR), Hobart, Australia

Schiermeier Q (2012) The great Arctic oil race begins. *Nature* 482:13–14

Siron R, Sherman K, Skjoldal HR, Hiltz E (2008) Ecosystem-Based Management in the Arctic Ocean: A Multi-Level Spatial Approach. *Arctic* 61, Suppl. 1:86–102

Siron R, VanderZwaag D, Fast H (2009) Ecosystem-based ocean management in the Canadian Arctic. In: Hoel AH (ed) *Best practices in ecosystem-based oceans management in the Arctic,* Report Series 129. Norwegian Polar Institute, Tromso

Skaug HJ, Frimannslund L, Øien NI (2007) Historical population assessment of Barents Sea harp seals (*Pagophilus groenlandicus*). *ICES J Mar Sci* 64:1356–1365

Skjærseth JB (2006) Protecting the North-East Atlantic: enhancing synergies by institutional interplay. *Marine Policy* 30:157–166

Smith CR, Grange LJ, Honig DL, Naudts L, Huber B, Guidi L, Domack E (2012) A large population of king crabs in Palmer Deep on the west Antarctic Peninsula shelf and potential invasive impacts. *Proc R Soc B* 279:1017–1026

Southwell C, Paxton CGM, Borchers D, Boveng P, de la Mare W (2008) Taking account of dependent species in management of the Southern Ocean krill fishery: estimating crabeater seal abundance off east Antarctica. *Journal of Applied Ecology* 45:622–631

Stackpole EA (1955) The voyage of the Huron and the Huntress: the American sealers and the discovery of the continent of Antarctic. Marine Historical Association, Mystic, Conn. Pub. no. 29

Stammerjohn SE, Martinson DG, Smith RC, Iannuzzi RA (2008) Sea ice in the western Antarctic Peninsula region: spatio-temporal variability from ecological and climate change perspectives. *Deep Sea Research Part II: Topical Studies in Oceanography* 55:2041–2058

Stammerjohn SE, Massom R, Rind D, Martinson D (2012) Regions of rapid sea ice change: An inter-hemispheric seasonal comparison. *Geophys Res Lett* 39:L06501, doi:10.1029/2012GL050874

Statistics Norway (2012) Focus on Fishing and fish farming. http://www.ssb.no/english/subjects/10/05/fiskeri_havbruk_en/

Stewart EJ, Draper D, Johnston ME (2005) A review of tourism research in the polar regions. *Arctic* 58:383–394

Stewart EJ, Howell SEL, Dawson J, Draper D (2010) Cruise Tourism and Sea Ice in Canada's Hudson Bay Region. *Arctic* 63:57–66

Stewart EJ, Howell SEL, Draper D, Yackel J, Tivy A (2007) Sea ice in Canada's Arctic: implications for cruise tourism. *Arctic* 60:370–380

Stiansen JE, Korneev O, Titov O, Arneberg P, Filin A, Hansen JR, Høines Å, Marasaev S (eds) (2009) Joint Norwegian-Russian environmental status 2008. Report on the Barents Sea Ecosystem. Part II—Complete report. IMR/PINRO Joint Report Series, 2009(3), 375 p.

Stokke OS (2007) A legal regime for the Arctic? *Marine Policy* 31:402–408

Sundet JH, Hjelset M (2002) The Norwegian red king crab (*Paralithodes camtschaticus*) fishery: Management and bycatch issues. In: Paul AJ, Dawe EG, Elner R, Jamieson GS, Kruse GH, Otto RS, Sainte-Marie B; Shirley TC, Woodby D (eds) *Crabs in Cold Water Regions: Biology, Management, and Economics.* University of Alaska Sea Grant, Fairbanks, AK. 681–692

Taboada I (2012) A review of particle astrophysics with ICECUBE. *Mod Phys Lett* A 27: 1230042. doi:10.1142/S021773231230042X

Taranger GL, Boxaspen KK, Madhun AS, Svasand T (eds) (2011) *Risk assessment—environmental impacts of Norwegian aquaculture* (Extracts from "Fisken og havet, særnummer 3–2010"—printed August 2011), Institute of Marine Research, Bergen

Templeman ND (2010) Ecosystem status and trends report for the Newfoundland and Labrador Shelf. Research Document 2010/026. DFO Canadian Advisory Science Secretariat

Tjelmeland S (2005) Evaluation of long-term optimal exploitation of cod and capelin in the Barents Sea using the Bifrost model. In: Shibanov V (ed) Proceedings of the 11th Joint Russian-Norwegian Symposium: Ecosystem dynamics and optimal long-term harvest in the Barents Sea fisheries Murmansk, Russia 15–17 August 2005, IMR/PINRO Report series 2/2005. Institute of Marine Research, Bergen

Trathan PN, Agnew D (2010) Climate change and the Antarctic marine ecosystem: an essay on management implications. *Antarctic Science* 22:387–398

Trivelpiece WZ, Hinke JT, Miller AK, Reiss CS, Trivelpiece SG, Watters GM (2011) Variability in krill biomass links harvesting and climate warming to penguin population changes in Antartica. Proceedings of the National Academy of Sciences, doi:10.1073/pnas.1016560108

UNWTO (2012) UNWTO Tourism Highlights, 2012 Edition, http://mkt.unwto.org/en/publication/unwto-tourism-highlights-2012-edition

Van Oostdam J, Donaldson SG, Feeley M, Arnold D, Ayotte P, Bondy G, Chan L, Dewaily E, Furgal CM, Kuhnlein H, Loring E, Muckle G, Myles E, Receveur O, Tracy B, Gill U, Kalhok S (2005) Human health implications of environmental contaminants in Arctic Canada: A review. *Sci Total Environ* 351–352:165–246

Vaughan DG, Marshall GJ, Connolley WM, Parkinson CL, Mulvaney R, Hodgson DA, King JC, Pudsey CJ, Turner J (2003) Recent rapid regional climate change on the Antarctic Peninsula. *Climate Change* 60:243–274

Vilhjalmsson H, Hoel AH, Agnarsson S, Arnason R, Carscadden JE, Eide A, Fluharty D, Honneland G, Hvingel C, Jakobsson J, Lilly G, Nakken O, Radchenko V, Ramstad S, Schrank W, Vestergaard N, Wilderbuer T (2005) FIsheries and aquaculture. In: *Arctic Climate Impacts Assessment, Scientific Report.* Cambridge UP

von Quillfeldt CH, Olsen E, Dommasnes A, Vongraven D (2009) Integrated ecosystem-based management of the Barents Sea-Lofoten area. In: Sakshaug E, Johnsen GH, Kovacs KM (eds) *Ecosystem Barents Sea.* Tapir Academic, Trondheim, Norway

Walsh JE (2008) Climate of the Arctic environment. *Ecological Applications* 18:S3-S22

Watters GM, Hinke JT, Reid K, Hill S (2005) A krill-predator-fishery model for evaluating candidate management procedures. CCAMLR WG-EMM-05/13

Watters GM, Hinke JT, Reid K, Hill S (2006) KPFM2, be careful what you ask for—you just might get it. CCAMLR WG-EMM-06/22

Weslawski JM, Kwasniewski S, Stempniewicz L (2009) Warming in the Arctic May Result in the Negative Effects of Increased Biodiversity. *Polarforschung* 78:105–108

Weslawski JM, Kwasniewski S, Stempniewicz L, Blachpwiak-Samolyk K (2006) Biodiversity and energy transfer to top trophic levels in two contrasting Arctic fjords. *Polish Polar Research* 27:259–278

Williams G (2003) *Voyages of delusion: the quest for the Northwest Passage.* Yale UP, New Haven

Williams G (2011) *Arctic labyrinth: the quest for the Northwest Passage.* University of California Press

Wilson WJ, Ormseth OA (2009) A new management plan for the Arctic Waters of the United States. *Fisheries* 34:555–558

Wolfe RJ., Fall JA, Riedel M (2009) The subsistence harvest of harbor seals and sea lions by Alaska Natives in 2008. Alaska Native Harbor Seal Commission and Alaska Department of Fish and Game Division of Subsistence, Anchorage, AK

Zeller D, Booth S, Pakhomov E, Swartz W, Pauly D (2011) Arctic fisheries catches in Russia, USA, and Canada: baselines for neglected ecosystems. *Polar Biology* 34:955–973

Chapter 11. Marine Ecosystem-Based Management in Temperate Systems

JAMEAL F. SAMHOURI

National Oceanic and Atmospheric Administration, Northwest Fisheries Science Center

SUAM KIM

Pukyong National University

CHANG-IK ZHANG

Pukyong National University

MICHAEL J. FOGARTY

National Oceanic and Atmospheric Administration, Northeast Fisheries Science Center

Contents

1. Introduction

The global footprint of human activity in the ocean is perhaps nowhere as evident as in temperate seas, particularly in the northern hemisphere. The temperate zones host the highest concentration of industrialized activity on the planet. This activity and the high coastal population densities with which it is associated place high demands on ecosystem services. As a result, there is an increasingly urgent need for integrated management of human activities to ensure the sustainable delivery of these services.

In this chapter, we describe progress in implementing Marine Ecosystem-Based Management (mEBM) in selected temperate-water systems with a focus on drawing contrasts among regions with distinct properties and characteristics. We highlight the disparate challenges in two U.S. continental shelf systems sharing a common national regulatory framework but facing distinctive pressures and impacts related to geography, history and local and regional interests. We contrast

The Sea, Volume 16, edited by Michael J. Fogarty and James J. McCarthy
ISBN 978-0-674-07270-1

these observations to ecosystem-based management off the Korean Peninsula in a unique management system characterized by active artificial enhancement strategies for habitat and biota. The strong influence of human activity in the immediate coastal and estuarine zones of the world is accorded separate treatment by Houde et al. (this volume), including selected temperate coastal and estuarine systems. Although the proximity of population centers to coastal and estuarine systems results in immediate localized effects, the reach of human intervention extends to the continental shelves and beyond. We accordingly attempt to complement the nearshore perspective provided by Houde et al. by extending the area of consideration more broadly within temperate waters.

Here, we seek to develop themes common to temperate seas in which ecosystem-based management efforts are growing worldwide. Many temperate marine systems have strong existing regulatory mechanisms for fisheries and other marine uses in place. High human population densities along their coasts have characterized most of them for centuries or even millennia (Houde et al. this volume). This long history of peopled oceans means that fisheries in many temperate systems have been overexploited and coastlines have been altered dramatically. Integrated management is complicated not only by multi-sector conflicts but also by state, provincial, and national jurisdictional boundaries. These characteristics augment the need for integrated management of the marine environment, but also make the hurdles to achieving EBM higher. The case studies we have chosen aim to highlight these issues, while also remaining cognizant of the inherent social and ecological contrasts among temperate shelf ecosystems. Where appropriate, we have interwoven our discussion of these three case studies within the narrative for each major theme area to facilitate comparisons among them.

Our evaluation of temperate continental shelf systems in the United States focuses on two Large Marine Ecosystems: the California Current (CCLME) in the Northeastern Pacific Ocean, and the Northeast Continental Shelf (NESLME) in the Northwestern Atlantic Ocean. Both systems share a border with Canada and the CCLME is also shared with Mexico (Figure 11.1). In the following, we concentrate principally on the elements of the CCLME and NESLME that fall within US waters. Our area of interest off the Korean Peninsula resides within the Northwestern Pacific Ocean, and its waters meet with those of North Korea, China, and Japan (Figure 11.1). In contrast to our US examples, the Korean Marine Ecosystem (KME) is characterized by semi-closed regional seas. Additionally, the three systems differ markedly in other characteristics including their history of human settlement, coastal development, and the distribution of human populations in coastal areas (Figure 11.2a,b,c), the physiography of their continental shelves (Figure 11.2d,e,f) and their basic ecological characteristics including chlorophyll concentrations (Figure 11.2g,h,i). All these factors shape the scope for different ecosystem services and approaches taken to mEBM in these areas.

Population densities along the coasts (within 100km) of each study region vary widely, and speak to the degree of human influence on, and reliance upon, these marine ecosystems. Densities of ~458 people/km^2 in the Korean Peninsula exceed those along the Northeast Shelf (~220 people/km^2), which exceed population densities along the coast of the California Current (~140 people/km^2). Cultural differences also hold important implications for resource demand in our study areas. For example, annual per capita sea food consumption in the United States is 17 kg but

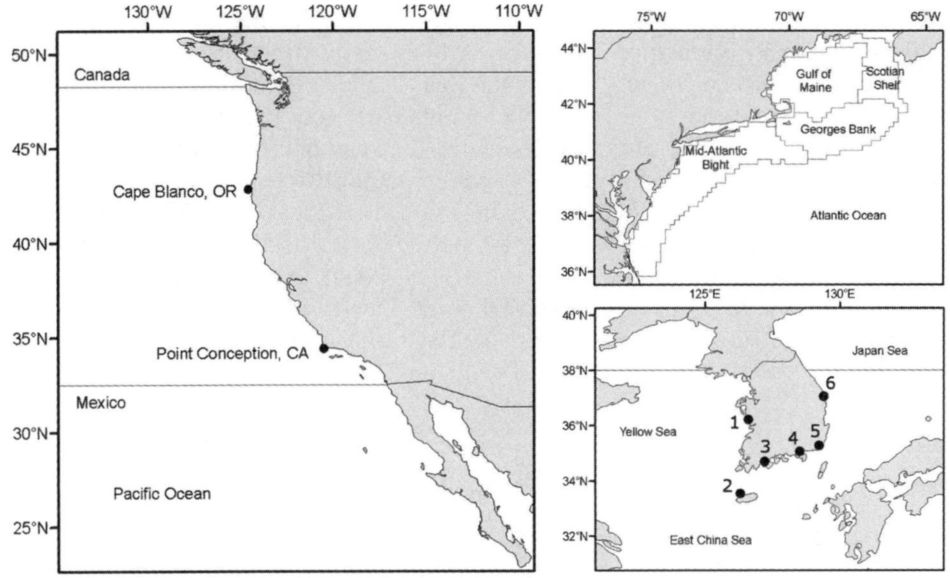

Fig. 11.1 – (See color insert) Maps and place names of three case study regions: California Current LME (left), Northeast Shelf LME (upper right), Korean Marine Ecosystem (lower right: the numbers denote marine ranch areas (1: Taean, 2: Jeju, 3: Jeonnnam, 4: Tongyeong, and 6: Uljin), and a major container port (5: Busan).

is 50–60 kg in Korea (Kim and Low 2011), resulting in very different demands for food provisioning services from the sea.

Approximately half of the population of the conterminous United States resides within coastal counties (Crosset et al. 2004). The California Current LME includes the shorelines of several of North America's most populous and most visited cities—San Diego, Los Angeles, San Francisco, Seattle, and Vancouver (Figure 11.2a)—and at the same time contains vast expanses of undeveloped coastline. Six of the ten fastest growing coastal counties in the U.S. are in California (Crosset et al. 2004). Its beauty, variety of marine-derived goods and services, and progressive environmental policies have made this region a natural test-bed for ecosystem-based management policies and practice in the ocean. These characteristics have also presented challenges for developing practical, scalable, comprehensive approaches to making mEBM a reality.

The coastline of the Northeast Shelf LME is more densely populated throughout the region than the CCLME (Figure 11.2b) and is more broadly industrialized. Nonetheless, the intrinsic beauty of the coastline retains its allure in attracting visitors seeking recreational opportunities. The Northeast has historically supported, and continues to supply, a diverse and economically important spectrum of ecosystem goods and services. Implementing marine ecosystem-based management in the Northeast U.S. entails greater jurisdictional challenges than on the Pacific Coast, with the interests of eleven coastal states and the federal government to balance and reconcile.

Northeast Asia is one of the most densely populated regions in the world, responsible for a dominant share of fishery production and consumption, and contributing significantly to the international trade of these products from the northwestern Pacific Ocean (Kim and Low 2011). Over the course of more than two millennia, human populations have increasingly concentrated in coastal areas of the Korean Peninsula (Figure 11.2c). Due to a long history of high fishing pressure and environmental changes caused by rapid industrialization and urbanization, the KME has been severely altered throughout the region, with most marine living resources over-utilized and some depleted altogether. Moreover, industrial development and intensive aquaculture activities in coastal areas, especially of the Yellow Sea, have strongly modified ecosystem structure and function resulting in serious impacts in the region (Kim and Khang 2000). Also, the majority of the population in Korea resides near coastal cities. Due to the rapid industrialization in South Korea since the 1960s, the percentage of the population living in urban areas has increased rapidly (74.4% in 1990; Kim and Oever 1992).

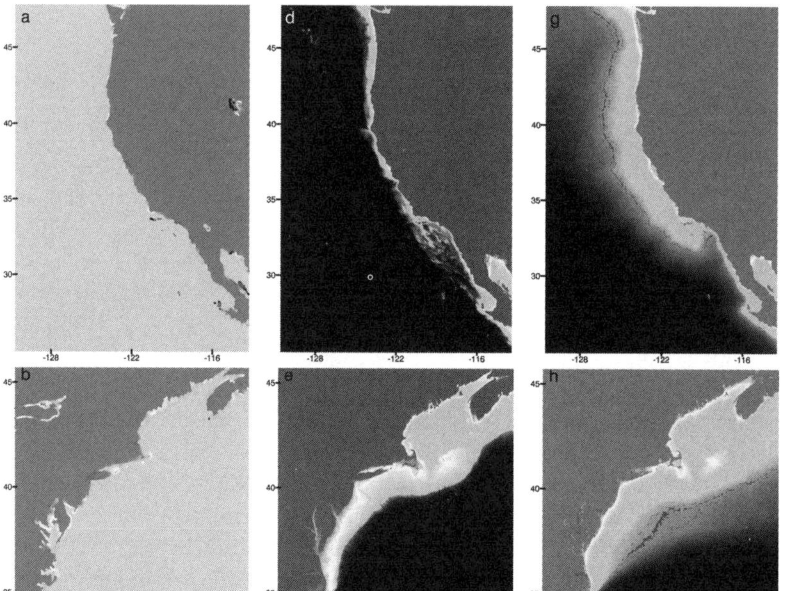

Fig. 11.2 – (See color insert) Maps of (a–c) human population density along the California Current Large Marine Ecosystem (CCLME), the Northeast Continental Shelf Large Marine Ecosystem (NESLME), and the Korean marine ecosystem (KME), (d–f) bathymetry for the CCLME, NESLME, and KME, (g–i) chlorophyll *a* concentrations for the CCLME, NESLME, and KME. (Kimberly Hyde, NOAA, personal communication)

1.1 Physical setting

In all three systems, physical characteristics strongly shape the range of possibilities for the development of different ecosystem service sectors. For example, wind fields, water depth, substrate type and proximity to population centers strongly influence decisions concerning siting of renewable energy installations. The geological history of an area sets the stage for the extraction of non-renewable resources, and the oceanography and hydrography of the region controls important aspects of ecosystem and fishery productivity.

The California Current Large Marine Ecosystem (CCLME) spans nearly 3,000 km of coastline off the west coast of North America, from Vancouver Island, British Columbia, Canada, to Baja California, Mexico, with a distinctively narrow continental shelf area (Figure 11.2d). The CCLME is a highly productive system, due to the transport of cold water from higher to lower latitudes by the eastern boundary current, and the strong winds that, in concert, drive high levels of coastal upwelling. Upwelling brings cold, nutrient-rich water to the surface and fuels system-wide productivity (Hickey 1989). Chlorophyll concentrations are highest near the coast (Figure 11.2g). Other periodic physical oceanographic processes cause occasional basin-scale subarctic and subtropical water mass intrusions (e.g., El Niño) or shifts in sea surface temperature and productivity regimes (e.g., decadal-scale climate forcing) (Bakun 1993; Aquarone and Adams 2008). The west coast of North America, notable for its tectonic activity, is an actively rising margin with much of the sedimentary material deposited on the coast coming from inland areas. Rocky reefs, sandy beaches, wave-swept shores, seamounts, and deep-sea canyons are part of a diverse geomorphology that has shaped the region's ecological communities and human development patterns. Additionally, the three major estuaries—San Francisco Bay, Columbia River, and Puget Sound—contribute significantly to local economies and connect ecological processes on land with those in the sea.

The NESLME, encompassing the area between Cape Hatteras and the Gulf of Maine (Figure 11.2e), spans approximately 1,600 km of coastline and covers approximately 260,000 km^2 within the 200m isobath. This boundary ecosystem rests between subtropical and boreal ocean provinces and is highly dynamic. The shelf is comparatively broad relative to the California Current LME (Figure 11.2d) and can be subdivided into four major ecological production units including the southwestern Scotian Shelf, the Gulf of Maine, Georges Bank, and the Mid-Atlantic Bight on the basis of physiographic considerations, oceanography and hydrography, and productivity (Fogarty et al. 2011). The oceanography of the region is dominated by the interplay of the Gulf Stream, a wind-driven western boundary current, and buoyancy driven flow from major current systems in the region, including the Labrador Current (Drinkwater and Mountain 1997). Climate and weather patterns in the region are dominated by the North Atlantic Oscillation (Drinkwater and Mountain 1997; Townsend et al. 2006). As with the CCLME, chlorophyll concentrations are highest nearshore (Figure 11.1h), reflecting nutrient inputs from land-based and other sources. High chlorophyll concentrations and levels of primary production, however, are also found offshore, on Georges Bank, an historically important fishing ground.

The Korean marine ecosystem consists of a shelf area of variable width (Fig 11.2f), with coasts that border the Yellow Sea, East China Sea, Korea Strait, and the East/Japan Sea. Coastal areas of the Korean Peninsula have different characteristics. On the west coast, sandy-mud tidal flats with high tidal range (5–10 m) form fertile benthic habitats. On the east coast, the continental shelf is narrow with a steep slope. The warm and saline Kuroshio Current flows over the continental slope along the Asian coast, but its speed is much reduced on the continental shelf of the East China Sea (Lie et al. 1998). A branch of the Kuroshio Current, called the Tsushima Warm Current, enters into the East/Japan Sea through the Korea Strait, and flows out through Tsugaru and Soya Straits in Japan. The major hydrographic features and ecological phenomena in surface waters of the East China Sea and the southern East/Japan Sea are mostly influenced by variability of the Tsushima Warm Current.

2. Ecosystem Services, Threats, and Impacts

In the following, we focus on ecosystem services that can be readily quantified in the CCLME, NESLME, and KME in terms of their present or projected levels and/or their market or non-market values to allow comparisons between these regions. We use three ecosystem service categories proposed in the Millennium Assessment—provisioning, cultural, and regulating services—and leave a conceptual overview of how natural and social systems are linked through production functions to other chapters in this volume (see Plummer and Levin; Guerry and Tallis; Altman et al.; this volume).

Ocean-related ecosystem services contribute directly to the national economies in the U.S. and Korea in important ways. In 2009, industries dependent on the ocean contributed over $40 billion to the combined Gross Domestic Product (GDP) of California, Oregon, and Washington [National Ocean Economics Program (NOEP) data available at http://www.oceaneconomics.org/Market]. Ocean-related industries provided over $50 billion to the GDP in the Northeast. In the southern Korean Peninsula, it is estimated that marine sectors account for at least $50 billion in revenues each year (Table 11.1).

The multi-billion dollar contribution of maritime industries to the market economy does not provide insight into the non-market values associated with other components of these systems (see Guerry and Tallis, this volume; Altman et al. this volume). An appreciation of the importance of the full suite of ecosystem services will necessarily encompass intrinsic, market, and non-market values.

2.1 Provisioning services

Provisioning services of current or potential importance in our temperate case study regions include production of seafood from capture fisheries and aquaculture, energy from renewable (wind, waves, tides) and non-renewable (oil, gas) sources, natural products such as corals, shells, seaweeds, minerals, and fish for the aquarium trade; and, transportation corridors for shipping. Though not yet in high demand, desalinization of marine waters for use as drinking water and in irrigation promises to hold significant value in the face of continued population growth and climate change, most notably in the southern portions of the CCLME.

TABLE 11.1.

Estimates of revenues and/or magnitudes for selected ecosystem services in the California Current Large Marine Ecosystem (CCLME), the Northeast Shelf Large Marine Ecosystem (NESLME), and the Korean Marine Ecosystem (KME)

Type of ecosystem service	Ecosystem component underlying service	Importance to CCLME	Importance to NESLME	Importance to KME
Provisioning				
seafood from wild fisheries	fishery species	Landings of ~600,000 mt[1], worth ~$500 million[2]	Landings of ~500,000 mt[1], worth ~$1.2 billion[2]	Landings of ~1 million mt[5], worth ~$15 billion[6]
seafood from aquaculture	fishery species	Harvest worth ~$110 million[2] on ~conducted on 9000 ha[3]	Harvest worth ~$70 million[2] on ~conducted on 33,000 ha[3]	Harvest worth ~$1.8 billion[6]
energy from renewable sources	habitats	Wind potential 33.4 GW[4]	Wind potential 186.9 GW[4]	Not available
energy from non-renewable sources	habitats	Reserves of 10.5 billion barrels of oil[4]	Market value of $11 billion[2]	Market value of ~$1.3 billion[7]
transportation corridors	physical setting	Market value of $14 billion[2]	Market value of $12 billion[2]	Average yearly per capita wage for shipping and transportation, ~$16,000[8]
Cultural and Aesthetic				
recreational activities	all ecosystem components	23 million participants; beach visitation: 185 million activity days; Wildlife viewing 96 million activity days. Total estimated value $18 billion[2]	9 million participants; beach visitation: 233 million activity days; Wildlife viewing 207 million activity days. Total estimated value $32 billion. Non-market value of ~$32 billion[2]	Total estimated value of tourism, ~$56 billion[9]
education and research	all ecosystem components	more than 50 marine research institutions are located in the CCLME	more than 30 marine research institutions are located in the NESLME	two major research institutes (KIOST, NFRDI), plus ~20 marine-related Departments in universities that carry out ocean/fishery research independently or jointly with those two institutes.

[1]NMFS, [2]NOEP, [3]USDA (2005), [4]MMS (2009), [5]MLTM (2009), [6]FAO http://www.fao.org/fishery/statistics/software/fishstat/en, 2007 values in 2010 US dollars, [7]US Energy Information Administration International Energy Statistics, Baker Hughes international rig information, and BP Statistical Review of World Energy (2011); 2008 values in 2010 US dollars, [8]http://www.nber.org/oww/, 2001 values in 2010 US dollars, [9]http://www.wttc.org/, 2010 values for total contribution of travel and tourism to GDP data (not specific to marine tourism) in 2010 US dollars

2.1a Capture fisheries

Capture fisheries have provided critical ecosystem services in temperate systems for centuries. The CCLME, NESLME, and KME are characterized by high and roughly comparable productivity, which in turn has supported important fisheries. The ecological structure of these systems, however, has been altered radically by fishing and other human activities. The decimation of populations of large whales and other marine mammals, sea turtles and some seabirds, the sequential depletion of fish populations, and the decline of diadromous species through obstruction of river systems have fundamentally changed ecosystem structure across all three regions. Because many of these species are high trophic level predators, reductions in their abundance have had cascading consequences for the ecological communities of which they are a part (Estes et al. 2011).

Species targeted by fisheries in the CCLME have tremendous historical and contemporary cultural, economic, and ecological importance. Harvesting of marine mammals, notably several groups of pinnipeds, fueled the early economic development of the Pacific coast. Most of the major fisheries recognized today in this region trace their origins to the turn of the 20th century (Field and Francis 2006), helping to fuel the colonization and development of the American West by European-Americans.

On the Northeast U.S. shelf, rights to the lucrative salt-cod trade were established as early as 1620. Fisheries in the Northeast steadily increased in economic importance over the next three centuries, dominated by groundfish including cod, haddock and flounders, and pelagic fishes (principally herring and mackerel; Lear 1998; Bolster 2008; Rosenberg and Claesson 2008). The whaling industry, ultimately attaining a global reach, was a dominant economic force in New England towns and cities from the mid 18th century until its demise in the mid-late 19th century as stocks were depleted and alternative oil-based products became widely available.

High primary and secondary production support various kinds of mid- and high-trophic level fisheries in the southern Korean Peninsula. Short-lived small pelagic fishes are dominant in the surface layer of the Tsushima Warm Current ecosystem and coastal areas (Kim et al. 2007). Marked changes in species composition of the offshore fish community have been observed during the last three decades. Ten major small pelagic fishes and common squid (*Todarodes pacificus*) accounted for about 30% of total Korean commercial catch in the 1970s–1980s, but increased to about 60% through the 1990s (Kim 2003).

Recent landings from the Northeast Shelf LME are comparable in magnitude to those of the California Current but the combined dockside revenues from oceanic and estuarine commercial fisheries on the Northeast Shelf in 2010 were substantially higher, reflecting the importance of high-value invertebrate resources (Table 11.1). In contrast, fisheries from Korean waters yield more than twice as much by weight than the CCLME and NESLME combined—over 1 million metric tons (MT) of marine fishes and invertebrates annually (MLTM 2009). The northwestern Pacific has produced the largest fishery yields among FAO fishery statistical areas since the 1970s (FAO 2010). The catch from this area was at least about one quarter, i.e., more than 20 million MT, of the world total in 2007, though this area occupies only 6.1% of the entire ocean.

Angling provides both seafood and an important source of recreational activity. It therefore underpins both provisioning and cultural services to society. Recreational fishing provides important second-order stimuli to local economies related to sale of boats, fuel, travel expenses, fees for charter and party boats, tackle, bait and other items (Kaplan and Leonard, 2012).

2.1b Aquaculture

Globally, the production of farmed fish from marine and freshwater sources now exceeds that of wild fish (Naylor et al. 2009). In the United States, mariculture is steadily increasing in importance to local food supplies and economies but currently still provides a relatively small overall contribution to marine-based food provisioning service relative to that of many other nations, including South Korea.

Marine aquaculture in the United States is dominated by the production of high value, low trophic level species, such as bivalve mollusks in coastal and estuarine waters. However, interest in offshore mariculture in the United States is rapidly increasing (NOAA 2007; Corbin 2010). The most recent nationwide estimate of aquaculture in the United States indicates a higher harvest value of mariculture production on the Pacific Coast despite substantially higher allocation of space given over to mariculture in the Northeast (Table 11.1).

The Korean aquaculture industry produced about 1.2 million MT of seafood in 2007. Seaweeds (mostly kelp and laver) and shellfish (oyster, mussel) accounted for 56% and 29% of the total aquaculture production, respectively. In contrast, production of marine finfish aquaculture (flounder, rockfish, sea bream, and sea bass) occupied only 15% of the total production, while its economic value exceeds 50% of the total value (MLTM 2009). So, in contrast to the US marine ecosystems, aquaculture in South Korea is dominated by the production of low value, low trophic level species, but most of the financial value from cultured stocks come from higher trophic level species.

2.1c Energy

Interest in the development of offshore renewable energy sources as part of an overall national strategy for energy independence has dramatically changed the future outlook for ocean use patterns in the United States. Significantly, it has also generated considerable interest in development of comprehensive marine spatial planning to deal with competing uses of ocean space (see section 3.2 below).

National assessments of the energy production potential for offshore wind resources in the U.S. indicate that up to 20% of the electricity needs of coastal states could be met (MMS 2009). Current technologies and engineering constraints restrict consideration to water depths less than 30m. Prospects for floating installations and other approaches, however, are currently under active investigation to remove these constraints.

Wind energy production potential is estimated to be greater in the Northeast Shelf LME than in the CCLME. Consistently strong wind fields and proximity to major urban centers with steadily increasing energy requirements and well developed energy transmission facilities have focused attention on the Northeast for further wind energy development. A number of potential offshore sites are currently under active consideration (Figure 11.2a). Although strong offshore wind resources exist off California, Oregon, and Washington, the highest potential resides over deep water that is currently prohibitive for active development (MMS 2009). The estimated wind potential in water depths less than 60m in the Northeast exceeds that of the Pacific Coast by an order of magnitude (Table 11.1).

The first offshore windfarm in the United States (the Cape Wind project; see Figure 11.3) was proposed in 2001 (see also Altman et al., this volume). The Cape Wind project has passed several levels of regulatory review (Rodgers and Olmsted 2010) and test towers are now slated for installation. The proposal entails placement of 130 wind turbines distributed over approximately 7,500 ha in federal waters off Massachusetts.

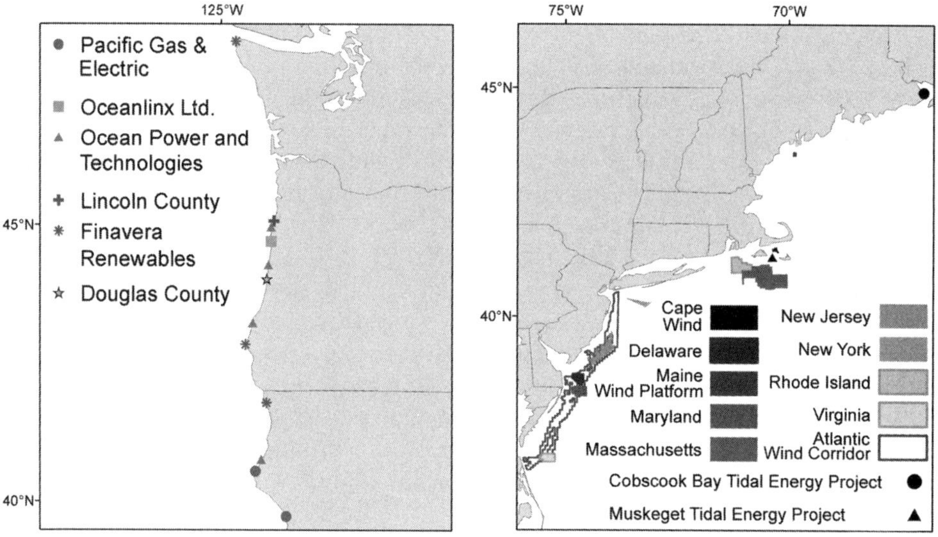

Figure 11.3 – (See color insert) Location of Wave energy projects in the California Current Large Marine Ecosystem (left). Proposed lease tracts for wind farm development, tidal power installations and an energy transmission corridor on the Northeast Continental Shelf (right). (Courtesy S. Lucey, NOAA)

Beyond wind, other renewable marine energy sources include wave, ocean current, and tidal energy. Options for harvesting these untapped resources are in varying but still nascent degrees of development in the United States. The highest estimated potential for wave energy is in the Pacific region (including Alaska and Hawaii; MMS 2009). Several wave energy projects are emerging along the coasts of California, Oregon, and Washington (Figure 11.3). High tidal amplitudes in the Gulf of Maine in proximity to the Bay of Fundy have focused interest in potential tidal power installations in the Gulf of Maine. In Cobscook Bay, ME, the first grid-connected ocean energy project in the nation has recently begun operation using tidal energy (Figure 11.3). A 5 MW pilot installation (the Muskeget Tidal Energy Project) is currently in development off Martha's Vineyard, MA (Figure 11.3).

Evaluation of potential future production of offshore oil and natural gas resources is again under active review. Currently on the U.S. Pacific and Northeast

Coasts, extraction of oil and gas reserves is confined to California. The Outer Continental Shelf (OCS) area of the United States contributed 27% of domestic oil production and 14% of domestic natural gas production in 2007 (MMS 2009; Table 11.1).

In Korea, the development of marine energy resources is regarded as one of 11 promising fields in ocean science and engineering (KORDI 2007). Tidal currents are very strong, reaching 0.4–1.0 m/sec, on the west coast (Kim and Khang 2000), so that several pilot projects on tidal plants have been carried out via feasibility studies since the 1970s, and one 245 MW tidal power plant is under construction at Siwha Lake on the northwest coast. Investigation of underwater mineral resources including offshore oil and gas and hydrocarbon is also very important for the sustainability of Korean economy. However, economically valuable mining is not available in Korean waters currently, although small amounts of natural gas have been produced from the coastal areas of the East/Japan Sea.

2.1d Transportation

Seaports are critical lifelines for ocean and coastal economies. The estimated combined value of the shipping and marine transportation sector on the U.S. Pacific Coast was over $14 billion dollars, representing about 34% of the GDP generated nationwide by this sector (Table 11.1). The estimate for shipping activity in the Northeast U.S. was over $12 billion, accounting for 28% of GDP attributable to transportation and shipping throughout the nation (Table 11.1). Because enhancement of international export is regarded as the principal driving force for economic development in Korea, the availability of maritime transport route and harbor facilities are essential components of the Korean economy. As the international trade of Korea has increased in recent years, activities at major harbors have expanded accordingly. Busan Container Port handled about 13,000,000 Twenty-foot Equivalent Units (TEUs) in 2008, so that it was ranked as the 5th busiest global container port worldwide. However, due to the global economic crisis in 2008, there was a sharp reduction at this port in 2009 (Ko et al. 2010).

2.2 Cultural services

Cultural services encompass a rich set of considerations involving important market and non-market values. Included among contemporary cultural ecosystem services are contributions of the marine environment to: (a) a sense of place, including aesthetic enjoyment, existence values, and identity; (b) recreational activities; (c) spiritual sites and other places of special significance; and, (d) education and research (Chan and Ruckelshaus 2010).

In the CCLME there is a rich cultural heritage of indigenous peoples relating to the marine environment, reflected in their art, religious beliefs, and oral traditions. For example, stories, traditions, and livelihoods in First Nation and Native American communities revolve around annual salmon migrations; the intrinsic value of salmon in the CCLME extends well beyond the price they fetch in seafood markets. Colonization of the Pacific Coast also resulted in strong traditions and cultures associated with the coast and sea among settlers. Today, domestic tourism and recreation, another type of cultural service, on the US west coast draws over

60% of residents to participate in activities such as visiting beaches, SCUBA diving, snorkeling, and surfing (Leeworthy et al., 2005; http://coastalsocioeconomics.noaa.gov/core/nsre/nsreforecast.pdf) (Table 11.1).

In the Northeast Shelf LME, the longstanding importance of coastal and marine resources to native peoples is no less evident. The remains of ancient middens stand testament to the reliance of native cultures on the bounty of nearshore ecosystems over the millennia (e.g. Bourque 2004). From the inception of European colonization, maritime trades were critically important to the success of the entire enterprise. These traditions remain deeply interwoven into the cultural fabric of coastal communities of the Northeast. Indeed, tourism and recreation is a contemporary cultural service related to these traditions, and as in the CCLME, ~60% of residents participate in activities such as visiting beaches, SCUBA diving, snorkeling, and surfing on the New England coast (Leeworthy et al., 2005) (Table 11.1).

In Korea, tourism is a multi-billion dollar industry (Table 11.1), and ocean sports and leisure activities such as yachting and eco-tourism are regarded as the most promising future industries in the archipelago areas along the south coast, as income level surpasses an annual per capita GDP of $20,000 (Schwarzbach et al. 2008). There are many marine-related cultural and historic relics in coastal areas that entice visitors. Ocean-related folk festivals with local fishery products are increasingly popular in fishing towns around the Korean Peninsula. Traditional seafoods are also an important attraction and economic driver within the coastal areas.

Religious and spiritual values associated with marine ecosystems are no less important in the southern Korean Peninsula. Fishery-related shrines dedicated to shaman gods and mythical stories are important cultural landmarks throughout coastal areas. In small fishing villages in Korea, female shamans conduct spiritual ceremonies for the success and safety of fishermen before a big fishing trip. Songs and dances embodying and encoding Traditional Ecological Knowledge of fishermen are common.

Though it is possibly the most daunting challenge for the science of ecosystem services, the development of new methods to document and quantify the value provided by these cultural services is an essential step to ensuring that they weather storms caused by economic downturns and are not underserved as future management and conservation plans are implemented.

2.3 *Regulating services*

Regulating services provided by marine ecosystems include but are not limited to climate and weather regulation; carbon storage and sequestration in coastal habitats; coastal protection; and, bioremediation. Because the full role and import of certain regulatory services are difficult to quantify in terms of their production value, we do not attempt to assign monetary values. Houde et al. (this volume) describe many of these considerations for estuarine and coastal areas of the U.S. In addition, as many regulating services are supplied in part by living habitats, we address them to some extent in section 2.4 below by characterizing gross changes in habitat areal extent over time.

The weather and climate in the coastal regions of the NESLME and the KME are strongly influenced by the Gulf Stream and the Tsushima Warm Current, respectively. In the southern Korean Peninsula, for instance, mild winters and cool summers in coastal areas are a typical result of characteristic oceanographic dynamics. However, heavy rains accompanied by strong storms in summer result in frequent floods.

Carbon storage and sequestration by burial of sediments and via biomass production have recently gained attention as a key service due to climate change. Indeed, the areal extent of corrosive waters in the CCLME in particular is climbing, at least in part due to ocean acidification processes resulting from the ocean's absorption of carbon emitted as a byproduct of a variety of human activities (Feely et al. 2008; Gruber et al. 2012).

Coastal protection from flooding and erosion is essential to maintaining the values of coastal properties. Part of the production function underlying coastal protection services is determined by the amount and characteristics of living habitats (e.g., kelps, seagrasses), in addition to bathymetry, exposure, and geomorphology (see Guerry and Tallis, this volume). Finally, bioremediation by living and non-living nearshore habitats is a key service for degrading chemicals and eliminating them as safety concerns for people and seafood in the marine environment (Singh and Tripathi 2006). The agricultural productivity of the terrestrial environment bordering all three of our case study regions is due in part to the heavy use of pesticides, herbicides, and fertilizers, highlighting the importance of marine bioremediation via seagrasses, shellfish, and other organisms.

2.4 Pressures and threats

The delivery of ecosystem services in temperate regions is threatened by a range of human activities occurring at an unprecedented scale and magnitude. This scaling effect can render activities like fishing, coastal development, agriculture, industry, and shipping unsustainable, turning them into impacts such as overexploitation, habitat degradation and loss, pollution, species invasions, and anthropogenic climate change. We note that viewed through the lens of tradeoff analysis, certain impacts on some ecosystem services are to be expected and may confer benefits to other services. The key issue is to determine societal choices on the relative balance of goods and services desired and to hold impacts to sustainable levels that do not impair ecosystem resilience.

2.4a Effects of human activities on seafood provision

In U.S. waters of the CCLME six species are currently listed as overfished (NOAA 2012). All are either salmon species (2 of 6) or low productivity rockfish species. Salmon in the Pacific Coast region are subject to the interplay of threats during both the freshwater and marine phases of their life history. Several subpopulations of salmon in the CCLME are also protected under the Endangered Species Act (ESA). Rockfishes are inherently vulnerable to overharvesting because of slow growth and delayed reproduction resulting in low productivity and resilience to exploitation. Three species in Puget Sound, WA, were protected under the ESA in 2010. In the Northeast, 14 stocks are currently classified as overfished (NOAA

2012). Of these, several are species or stocks at the southern extent of their range and vulnerable to the interplay of exploitation and climate-related shifts in productivity.

Although fishing effects by commercial fleets are often implicated most heavily in overexploitation, recreational fisheries can also strongly contribute to population declines. Coleman et al. (2004) reported that although the overall contribution of recreational landings to total fish landings in the US was low (4%) in 2002, landings of species of concern in recreational fisheries comprised 23% of the total landings. Important regional differences were also noted, with recreational landings accounting for 59% of landings of species of concern on the Pacific Coast but only 12% in the Northeast in 2002. Recreational fisheries therefore hold the potential to have a disproportionate effect on species that are considered to be at risk through overfishing.

In Korean waters, a history of fishing pressure dating back millennia has resulted in overexploitation, with current commercial catches dominated by younger, smaller fish. The coastal and offshore commercial fisheries peaked in the mid 1980s, producing over 1.5 million MT, but decreased to below 1.0 million MT in the early 2000s, and landings have been relatively stable since that time (MLTM 2009). Due to the decrease in higher trophic level species such as small yellow croaker and the increase in common squid and anchovy, the mean trophic level of the catch decreased continuously in the Japan/East Sea, the Yellow Sea and the East China Sea during the 1967~2002 period (Zhang and Lee, 2004). In addition to declines in mean trophic level of the catch, population biomass, annual catch, and catch per unit of effort for major fisheries resources all have decreased, indicating changes in quantity as well as quality of fisheries resources (MLTM 2009). Stable yields in the last decade are likely attributable to persistence of an overfished ecosystem state, rather than to strong management actions curtailing catches. In artisanal fisheries, which target a different set of species than the commercial fleets, illegal and unreported fishing has severely depleted coastal fish resources relative to historical abundances.

Korean fishermen have also suffered from the increased occurrence of harmful algal blooms since the 1980s. The formation of a hypoxic water mass in coastal bays was caused by the discharge of organic materials and pollutants from nearby metropolitan areas (Hong and Lee 1983; Seo et al. 2012). Flow patterns of coastal currents along the southern coast are complicated, and some semi-closed bays with limited circulation patterns are characterized by high concentrations of pollutants. High concentrations of pollutants including heavy metals, PAHs, and other contaminants from the coastal industrial complex have accumulated in coastal ecosystems. Abnormal sex ratios and reproductive properties of benthic animals are often reported in and around Korean waters (Shim et al., 2000; Vaschenko et al., 1997).

Recently, the frequency and intensity of red tides has greatly increased in Korean coastal areas. Toxic red tides have occurred along the southern Korean Peninsula since the 1990s, leading to massive deaths of fish in caged aquaculture farming. Huge blooms of green algae, *Entermorpha prolifera,* occurred in Chinese waters in May and July 2008 (Sun et al. 2008). Sporadic outbursts of giant jellyfish (*Nemopilema nomurai*) have been reported off the Chinese, Korean, and Japanese coasts, causing serious losses and damage in Korean and Japanese fisheries.

2.4b Habitat loss

Radical transformation of coastal, estuarine and nearshore landscapes over the last three centuries has had profound implications for key regulating and supporting services in the United States (Houde et al. this volume). As discussed in section 2.3, wetland areas play essential roles as important nursery areas of many species; in carbon storage and sequestration; in shoreline protection through attenuation of wave energy; and for bioremediation through filtering and uptake of nutrients and contaminants. During the period 1790~1985, it is estimated that nearly 70% of combined freshwater and coastal wetlands were lost in the Pacific Coast states (see Table 4.5 in Mitsch and Gosselink 2000). Estimated wetland losses in California were the most pronounced, with a reduction of approximately 90%. In the Northeast, the estimated loss in total wetland area was 43% (range 9–74% for the states of Maine to North Carolina). The historical loss in potential buffering capacity against rising CO_2 levels in the atmosphere and protection against storm surge is immense.

No net loss provisions in U.S. and Canadian national policies (National Wetlands Policy Forum, 1988, http://www.dfo-mpo.gc.ca/habitat/role/141/1415/14155/fhm-policy/page03-eng.asp) provide a vehicle for the protection of wetlands in the CCLME and NESLME, but other marine habitats are less well represented at the federal level (MB 1994). Individual states and provinces do, however, include levels of protection for marine habitats more broadly (e.g., no net loss of shoreline ecological functions, WA Shoreline Management Act; SMP 2010) and specifically designated areas such as National Marine Sanctuaries have a variety of restrictions on use. For the period 2004–2009, the loss in all intertidal wetlands including marine and estuarine locations was 1.4% (Dahl 2011). The largest losses in this category were in estuarine intertidal vegetated habitats (2.4%). An earlier estimate for the entire Atlantic coast of the U.S. (Maine to Florida) indicated a loss of 1.4% of intertidal wetlands from 1998–2004 (Stedman and Dahl 2008).

Active coastal development on the west coast of Korea during the last three decades has led to the destruction of benthic habitats and nursery grounds of coastal fishes. Many seaweed and seagrass communities with associated fish species are disappearing as a consequence. Coastal erosion threatens river deltas, human communities, ecotourism, and aquaculture facilities. Dredging for safe navigation, sand mining and port construction also could greatly damage habitats of benthos. Trans-shipment and traffic accidents often result in oil spills. Two big collision accidents in 1995 and 2007 in Korea have profoundly affected the benthic ecosystem in coastal areas.

2.4c Climate change

Houde et al. (this volume) describe potential impacts of climate change in coastal and estuarine waters with consideration of sea level rise, shoreline erosion, and other impacts. Below we focus on selected issues affecting not only coasts, but the continental shelf regions.

Ocean acidification. The uptake of anthropogenically released CO_2 in the ocean is influencing marine ecosystems significantly. Decreasing pH may interfere with the ability of organisms to form calcium carbonate structures: tests, shells, and otoliths. Also, changes in pH alter the fundamental chemical balances in the ocean, potentially affecting the chemical speciation of phosphorus, nitrogen, and other elements that are critical to ocean life. Fundamental differences in oceanography and physical processes on the Pacific Coast relative to Northeast Shelf result in different responses and vulnerabilities to increased atmospheric CO_2 levels. Upwelling waters in the California Current are characterized by naturally occurring high CO_2 levels, corresponding low pH, low oxygen, and high nutrient contents (Feeley et al. 2009). The capacity of the system to absorb atmospheric CO_2 is accordingly low. On the Northeast Shelf, important latitudinal differences can be anticipated due to the strong temperature-dependence of solubility of CO_2 in seawater; the cold waters off New England are expected to decline more rapidly in buffering capacity than those of the Mid-Atlantic region.

Cooley and Doney (2009) provided estimates of potential economic losses to the fishery industry in the United States under different CO_2 emission scenarios. First order estimates of decline in net present value for New England fisheries under the high emission scenario ranged from \$0.47–1.19 billion by 2060 and \$0.12–0.30 billion for the Pacific Coast (2007 dollars at a 2% discount rate). The greater projected impact in the Northeast was attributed to a greater reliance on high value invertebrate species that are more vulnerable to acidification.

Ocean warming. Temperature directly and indirectly affects virtually every aspect of the biology and ecology of marine organisms with important implications for ecosystem services. Clear trends in global warming of the oceans have been reported (e.g., Levitus et al. 2000) and these are manifest in the CCLME (McGowan et al. 1998), the NESLME (EcoAP 2012), and the KME. In fact, the warming rate of SST in the northwest Pacific was one of the world's highest during the 1980s–2000s (Belkin 2009). Oceanographic observations for the past 30 years indicate annual and winter SST warming rates of 0.019°C/yr and 0.035°C/yr in the KME, respectively (MLTM 2009). Potentially important impacts on provisioning services include reorganization of marine communities, overall changes in productivity, and shifts in distribution of economically important species (Ainsworth et al. 2012; Sorte et al. 2010; Cheung et al. 2009; 2013).

Climate change effects on temperature and stratification are predicted to result in changes in phytoplankton community composition with an increase in dinoflagellates over diatoms (e.g., Bouman et al. 2003; Morán et al. 2009). Diatoms are more efficiently converted into biomass of higher trophic level species than dinoflagellates and a switch to a system dominated by dinoflagellates could result in reduced overall productivity at higher trophic levels (e.g., Mann 1921). Recent declines in diatoms relative to dinoflagellates has been documented on the Northeast Continental Shelf based on Continuous Plankton Recorder observations (EcoAP 2012).

Shifts in distribution have been observed for a number of species in the three case study systems. Interestingly, Pinsky et al. (in review) found a slight southward change in the center of mass of fish populations in the CCLME based on research vessel survey observations, reflecting an overall cooling related to oceanographic processes over the last several decades. On the Northeast Continental Shelf, an evaluation of three dozen species indicated that three quarters shifted either poleward, to deeper water, or both, over the last several decades (Nye et al. 2009). Updated and expanded observations by Pinsky et al. (in review) confirm these results using changes in climate velocity as an explanatory variable. Pinksy and Fogarty (2012) tracked changes in the distribution of both abundance and commercial landings for four economically important species and detected greater inertia in shifts in landings than in abundance as revealed by research vessel surveys. Social and regulatory constraints in fishing communities were invoked to explain these results. Due to the continuous warming of ocean waters off of the southern Korean Peninsula, the fishing areas of chub mackerel, one of the representative purse-seine fisheries, were located further north in the early 2000s than in the 1980s (Kim et al., 2012).

Deoxygenation. Ocean warming and increased stratification are expected to reduce dissolved oxygen in the global ocean between 1–7% by 2100 and such declines could continue for millennia (Sarmiento et al. 1998; Keeling et al. 2010). The North Pacific Ocean stands out as an area that has already experienced rapid deoxygenation during the past 50 years (i.e. 22% decrease; Whitney et al. 2007; Keeling et al. 2010). Expansion of oxygen minimum zones (OMZs) is predicted wherever they occur throughout the world's oceans (Whitney et al., 2007), and shoaling of OMZs up and onto the continental shelf is becoming a conspicuous new phenomenon along the west coast of North America (Grantham et al. 2004; Whitney et al. 2007). In the East/Japan Sea, the dissolved oxygen (DO) minimum layer deepened during the 1960s–1990s (Kim et al. 1996). The NESLME is not invulnerable to deoxygenation associated with climate change either (Levin et al. 2009). The majority of hypoxic estuaries are reported from the North Atlantic Ocean and adjacent seas, though many of these occurrences are attributed to eutrophication (Diaz and Rosenberg 2008).

3. EBM in Action: Similar Principles—Variable Implementation

Management of temperate seas is undergoing a fundamental transformation worldwide, from the US, Europe, and East Asia, to Australia, South Africa, and South America. Examples of efforts to integrate decisions about how people can use and benefit from marine systems are increasingly easy to identify (Cicin-Sain and Knecht 1998; WSSD 2002; Rosenberg et al. 2009). This transition toward marine Ecosystem-Based Management is playing out against a complicated backdrop of a changing climate, new societal demands, and existing legal and policy infrastructure that vary from place to place. As a consequence, marine EBM has "many faces" (Tallis et al. 2010), a feature exemplified clearly by our three case study regions.

In the United States, certain elements of marine EBM have been developed and implemented over the last several decades. However, interest and commitment to this issue were revitalized with the recommendations of two high level policy panels, the Pew Oceans Commission (POC 2003) and The U.S. Commission on Ocean Policy (USCOP 2004). USCOP noted:

> U.S. ocean and coastal resources should be managed to reflect the relationships among all ecosystem components, including human and nonhuman species and the environments in which they live. Applying this principle will require defining relevant geographic management areas based on ecosystem, rather than political, boundaries.

In 2010, an Executive Order establishing a new National Policy for the Stewardship of the Oceans, Coasts, and Great Lakes was implemented (see Cochrane et al. this volume; Houde et al. this volume). This 'National Ocean Policy (NOP)' established nine priorities to guide management activities (Ocean Policy Task Force 2010). The policy designates Ecosystem-Based Management as the foundation, and regional ocean planning as a critical tool, for meeting the goal of integrated management for the oceans, coasts, and Great Lakes.

In South Korea, marine EBM has taken shape around efforts to ensure the sustainability of provisioning services, primarily those from capture fisheries and aquaculture. To this end, the Korean government has been developing a comprehensive ecosystem-based marine ranching program, pursuing four major elements: (1) sustainability of production, (2) maintenance of biodiversity, (3) protection of habitat, and (4) socio-economic benefits. This approach differs from that described above for the US in that it is focused more on sustainability of extractive, renewable resources, and secondarily on other types of ocean uses and benefits. As such, it is best characterized as ecosystem-based fisheries management (EBFM) rather than EBM. We develop this distinction more fully in section 3.4.

Despite its differences, the Korean ecosystem-based marine ranching program has many similarities with EBFM efforts in the US. As it is designed to carry out the enhancement and efficient management of fisheries resources, it requires an understanding of ecological interactions among major species with respect to predation, competition for prey species, effects of climate on fish ecology, interactions between fishes and their habitats, and the effects of fishing on fish stocks and their ecosystems. By developing this knowledge and understanding, it is hoped that fisheries management can avoid significant risks and potentially irreversible changes in marine ecosystems caused by fishing or marine ranching.

Beyond the ecosystem-based marine ranching program, Korean ocean policy has provisions for the management of threatened and protected marine species and communities. The Minister of Food, Agriculture, Forest and Fisheries (MIFAFF) is responsible for taking conservation action when an animal species is categorized as endangered. The designation of endangered animals requires consideration of all of the following: (1) fisheries animals which are regulated by CITES (the Convention on International Trade in Endangered Species of Wild Fauna and Flora), (2) endangered wild fauna and flora, specified by the Law of Natural Environment Conservation, Article No. 2 Clauses 6 and 7, and (3) fish species that are valuable for protection scientifically, and are requested for protection for research by research institutes endorsed by the Minister.

3.1 Multiscale initiatives

Ecosystem-Based Management in the California Current and Northeast Continental Shelf LMEs has been developing organically on a range of spatial scales, spanning the full spectrum from local to international initiatives. Within the United States, national, regional, state, and local EBM efforts have now gained traction. These multiscale initiatives typically reflect a broad range of objectives but share certain common elements, including a focus on the need for holistic and cross-sectoral management approaches and an emphasis on spatial management strategies as a major tool for mEBM (see Table 11.2). As noted above, the rapidly emerging requirement for offshore renewable energy sources has served as a catalyst for marine spatial planning in the United States, especially in the Northeast and this is strongly represented in the east coast initiatives reported in Table 11.2.

Coordination among these diverse initiatives is currently evolving and its ultimate form will reflect the interplay among federal, state, and local authorities and responsibilities. The federal government has responsibility from 3 to 200 miles from shore with state and local authority residing within the 3 mile limit on the Pacific and Northeast coasts. Federal management authority maps to a number of legislative mandates including the Sustainable Fisheries Act, the Endangered Species Act, the Marine Mammal Protection Act, the Clean Water Act, the National Environmental Protection Act, and the Coastal Zone Management Act (see also Houde et al. this volume). The National Ocean Policy does not supersede these legal mandates but rather provides a vehicle for coordination among management authorities responsible for implementing the provisions of these acts. Under NOP, a National Ocean Council with high level representatives from 27 federal agencies and departments has been created to provide this level of coordination.

A total of nine Regional Planning Bodies (RPBs) are called for under the NOP. These planning bodies hold the responsibility of developing management strategies with a focus on spatial planning. The first of the RPBs has now been formed in the Northeast. In each area, regional ocean councils have been formed. In some cases, these councils have built on existing interstate agreements and structures. The councils representing the Pacific and Northeast states, which share similar goals (Table 11.2), are the West Coast Governors' Alliance (WCGA), the Mid-Atlantic Regional Council on the Ocean (MARCO), and the Northeast Regional Ocean Council (NROC). All of these ocean councils were formed in the last decade, and are intended to promote cooperation that results in management strategies that benefit each region in addition to individual member states. The regional ocean councils are well positioned to support the RPBs.

Movement toward mEBM in South Korean waters also requires coordination between federal and regional authorities. Various laws currently enforced by MIFAFF, MLTM (Ministry of Land, Transport and Maritime Affairs), and ME (Ministry of Environment) cause inter-agency conflicts in policy making. Thus, there is an apparent lack of integrated, holistic management. Although a variety of management projects mentioned above are under way to manage fish resources by MIFAFF, these projects are not systematically, nor consistently, operated. Despite a number of ongoing stock rebuilding activities, the projects have not been effective; fisheries catch and catch per unit effort (CPUE) have not improved. Several difficulties are also associated with traditional control devices in Korean fisheries management. Fortunately, the MIFAFF has recently established the 5-Year Basic Plan of the Korean Fisheries Resources Management (2011–2015), started development of ecosystem-based fisheries assessment methodology as one of the major targets of the project and made a plan for the establishment of an EBFM system by 2013 (MIFAFF 2011). Currently there are five marine ranching programs requiring coordination in the following areas: in Tongyoung, Uljin, Taean, Jeonnam, and Jeju (MIFAFF 2009) (Fig. 11.1). The Tongyoung Marine Ranching Program is the oldest, and has been conducted since 1998 as a pilot program.

3.2 State and Local EBM and CMSP Initiatives

A number of initiatives to establish elements of mEBM and/or spatial planning have developed at local to regional scales. Houde et al. (this volume) cover several local initiatives (the West Coast EBM Network identified in Table 11.2 comprises a consortium of local-scale initiatives). These range from efforts encompassing local estuaries and embayments to state or province-wide programs. The challenge is to nest these activities within regional and national-scale efforts to ensure effective coordination and cooperation. These finer scale programs can ensure that local needs, objectives and knowledge can be fully represented and that stakeholders at a grassroots level can be effectively engaged, meeting a critical need for mEBM (Charles, this volume; Houde et al. this volume).

Consideration of a sample of programs at state and regional (interstate) levels indicates a set of focal interests and objectives tailored to local conditions and needs but holding certain elements in common. We identified eight major recurring themes (Table 11.2). We note that in a number of instances, theme areas were implicit in some of the program overviews but were not explicitly stated and are not listed here. Additional examples of formal Marine Spatial Planning efforts in the U.S. are described in Rice et al. (this volume).

3.3 Marine protected areas

Spatial management strategies in the marine environment have an extensive history in the United States, particularly for fishery management purposes. These objectives include protection of habitat, spawning and/or juvenile fish and shellfish, and protection of threatened or endangered species.

The passage of the California Marine Life Protection Act in 1999 provided the first effort in the United States to enact a comprehensive spatial management strategy for the marine environment. In 2000 an Executive Order calling for the establishment of a national system of marine protected areas/MPAs in the United States quickly followed. Large swaths of the Pacific Coast and Northeast Shelf are covered by a number of types of MPAs and Marine Managed Areas (Figure 11.4) with widely varying restrictions and degrees of control exerted within each. Both systems are covered by a complex mosaic of managed spaces with substantial overlap among designated MPAs designed to achieve multiple objectives. Nearly 380 MPAs are part of the Pacific Coast MPA system and more than 400 are listed in the Northeast Shelf as part of this voluntary system; however, the area set aside for MPAs along the Pacific Coast is greater by a factor of 10 (Table 11.3). By far the largest proportion of the MPAs on both coasts are designated for Uniform Multiple Use and only a very small fraction are designated for more complete protection (Table 11.3).

The network of MPAs put in place along the coast of California has led to rebuilding of recreationally valuable species within MPA borders, which may benefit local dive tourism and sportfishing industries (Hamilton et al. 2011). Increases in biomass within fishery MPAs on the Northeast Shelf have been documented (e.g., Fogarty and Murawski 2005) and beneficial effects to local fisheries resulting from spill-over effects have been demonstrated (Murawski et al. 2004; 2005).

Table 11.2.

Focal considerations for mEBM by location/management authority at provincial, state, and local levels: (A) use resources sustainably, (B) protect water quality/quantity, (C) maintain ecosystem health [coastal habitats, biodiversity, species abundance, food webs, invasive species], (D) adapt to local climate change, (E) preserve marine economies/working water fronts/access to marine resources, (F) ensure healthy human populations and vibrant communities, (G) develop effective spatial planning and conflict resolution, (H) foster effective participatory management, scientific knowledge and communication.

Geographic location	Project/Authority	Foci	Initiated	Website
California	California Ocean Protection Act/Ocean Protection Council	A,C,H	2004	opc.ca.gov/
West Coast	West Coast Governors Alliance for Ocean Health	A,B,C,E,H	2006	westcoastoceans.org
West Coast	West Coast Ecosystem-Based Management Network	A,B,C,D,E,F,G,H	2008	westcoastebm.org
Massachusetts	Massachusetts Ocean Management Plan	A,C,G,H	2008	mass.gov/eea/ocean-coastal-management/mass-ocean-plan
Rhode Island	Rhode Island Special Area Management Plan	A,C,E,H	2010	seagrant.gso.uri.edu/oceansam
New York	New York Offshore Spatial Plan	A,C,G,H	2006	estuary.cog.ny.us
Delaware	Delaware Marine Spatial Project	G	2012	deseagrant.org/research/development/SCD-1
Maryland	Maryland Coast Atlas	G,H	2009	dnr.state.md.us/ccp/coastalatlas/
Virginia	Virginia Special Area Management Planning	A,G,H	2008	deq.state.va.us/Programs/CoastalZoneManagement/CZMIssuesInitiatives
Tongyoung	Korea Fisheries Resources Agency (FIRA)	A,C,E,H	1998	fira.or.kr/marinerc
Jeonnam	Korea Fisheries Resources Agency (FIRA)	A,C,E,H	2001	fira.or.kr/marinerc
Uljin	Korea Fisheries Resources Agency (FIRA)	A,C,E,H	2002	fira.or.kr/marinerc
Taean	Korea Fisheries Resources Agency (FIRA)	A,C,E,H	2002	fira.or.kr/marinerc
Jeju	Korea Fisheries Resources Agency (FIRA)	A,C,E,H	2002	fira.or.kr/marinerc

Korea has a total of 16 different types of MPAs designated by 13 MPA related laws operated by three ministries (MIFAFF, MI, and MLTM) and one administration (Cultural Heritage Administration). The total marine area protected is larger than 6,000 km^2. A total of 10 Fisheries Resources Protected Areas were designated to protect spawning and nursery grounds of fishery resources. These areas, which total 3,866 km^2, are managed by the MIFAFF. Four Marine Conservation Areas, which total 1,882 km^2, were designated to control marine environmental quality for protecting habitats of marine biological resources from pollution and are managed by the ME. A total of 12 Wetland Protected Areas and four Marine Ecosystem Protected Areas were designated to protect marine biodiversity and ecosystems and are managed by the MLTM. Their areas total 220.4 km^2 and 70.4 km^2, respectively (Figure 11.4).

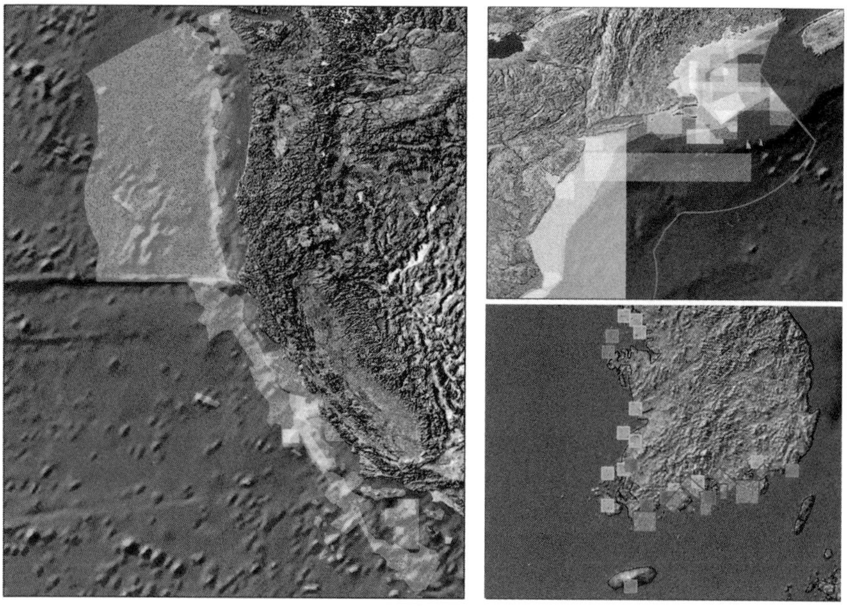

Fig. 11.4 – (See color insert) Locations of Marine Protected Areas on the U.S. Pacific (left), Northeast Coasts (upper right) and southern Korean peninsula (lower right). (Data sources for US locations: NOAA Marine Protected Areas Program. Korea: Chang Zhang [Blue: Fisheries Protected Area (PA); Orange: Wetland PA; Green: Marine Ecosystem PA; Red: Marine Conservation PA])

In addition to formally designated MPAs, de facto reserves and protected areas are often created through choices made in siting competing ocean use activities (White et al. 2012). It is worth noting that specialized forms of spatial management constraints have been implemented to reduce impacts on marine mammals associated with ship strikes. For example, concerns about ship strikes affecting blue (*Balaenoptera musculus*) and humpback (*Megaptera novaeangliae*) whales in the southern California Current, between the Channel Islands and the mainland, have led to proposals for shifts in the location of shipping lanes (Figure 11.5). Modest changes in shipping lanes have the potential for important reductions in injury and mortality to these whale species (Redfern et al. 2013). On the Northeast shelf, a

TABLE 11.3.
Marine Protected Areas classified by level of protection maintained within each MPA for the NESLME and CCLME. Data provided by the NOAA Marine Protected Area Center.

	Pacific Coast		Northeast Coast	
Level of Protection	Number	Area (km^2)	Number	Area (km^2)
No Access	23	12,012	6	428
No Impact	1	< 1	1	0.00
No Take	13	364,188	4	<1
Uniform Multiple Use	326	6,375,891	362	415,005
Zoned Multiple Use	15	22,064	33	19,023
Zoned with No Take Areas			2	58
Total	378	6,446,155	408	434,515

risk and remediation effort in response to concerns over ship strikes affecting the northern right whale (*Eubalaenus glacialis*) off Massachusetts led to changes in shipping lanes and recommended vessel speeds in the approach to Boston Harbor (Figure 11.5). Predictions indicated that transit times into the harbor would increase by only 9–22 minutes but the risk of ship strikes to large whales would decrease by 81% (Wiley et al. 2006). Thus, the economic impacts are small relative to the potential conservation benefits for endangered species.

3.4 Ecosystem-based fishery management

Sectoral management in an ecosystem context is arguably most fully developed in fisheries applications. Indeed, strategies for the development of Ecosystem-Based Fishery Management, in which the objective is to sustain healthy marine ecosystems along with the fisheries they support (Pikitch et al. 2004), are now under way in all three of our case study regions.

As noted above, under the provisions of the U.S. National Ocean Policy, the ultimate authority for each ocean use sector remains with existing management bodies. The NOP is structured to ensure coordination among these management agencies and institutions. A total of nine federal fishery management councils are responsible for regional management decisions in federal waters. State and Federal officials and appointed representatives from different stakeholder groups comprise the membership of the councils. Individual states retain control within their territorial waters. The Magnuson-Stevens Fishery Management and Conservation Act as amended (Sustainable Fisheries Act 2006) calls for explicit consideration of ecosystem structure, including protection of essential fish habitat, minimizing incidental catch of non-target species, and management of inter-related stocks of fish. It further recognizes the critical human dimension in fishery management and explicitly calls for measures that consider the full effects on fishing communities of management actions to achieve conservation goals.

The Pacific Fishery Management Council (PMFC) holds responsibility for fishery management in US federal waters in the California Current LME. PMFC has developed a draft Fishery Ecosystem Plan designed to build on the Council's four existing fishery management plans (PFMC 2010). The purpose of the FEP is to

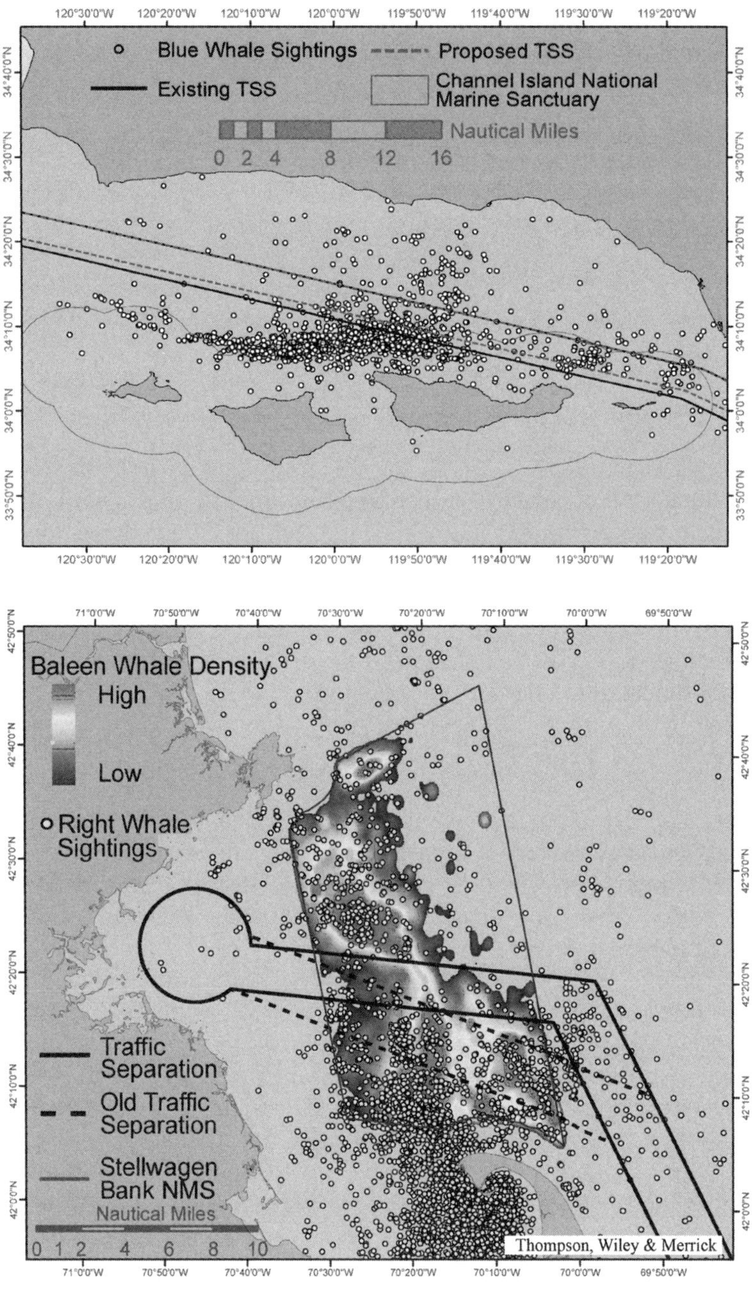

Fig. 11.5 – (See color insert) Proposed 1 nm shift northward of the southbound shipping lane near the Channel Islands National Marine Sanctuary (upper). Also shown is an overlay of blue whale sightings data collected by Channel Islands National Marine Sanctuary volunteers from 1999 through 2010. Realignment of the shipping lanes (TSS) into the Port of Boston, Gulf of Maine in 2007 by the International Maritime Organization to reduce the risk of ship strikes to baleen whales in the Stellwagen Bank National Marine Sanctuary (lower).

enhance the Council's species-specific management programs with more ecosystem science, broader ecosystem considerations and management policies that coordinate Council management across its Fishery Management Plans (FMPs) and the CCLME in order to inform policy choices and trade-offs.

The PFMC has called for the development of annual State of the Ecosystem Reports for the CCLME to inform ongoing management decisions. A preliminary discussion document of this type was delivered to the PFMC in 2011 (Levin and Wells 2011), and focused on an evaluation of ecosystem conditions (climate, predators, prey, and non-fisheries threats) for several example stocks in the CCLME. The FEP notes the importance of the ongoing effort to develop an Integrated Ecosystem Assessment for the California Current to inform policy choices (see below). In addition to consideration of the ecosystem requirements established under the Sustainable Fisheries Act described above, PFMC has instituted requirements to protect forage species at the nexus of the CCLME food web. For example, in recognition of the role of krill as key forage for many fishery and protected species, PMFC preemptively closed any possibility of developing a krill fishery in the US portion of the CCLME in 2006.

Both the New England and the Mid-Atlantic Fishery Management Councils have signaled their interest in moving toward Ecosystem-Based Fishery Management. The NEFMC has prepared a strategy document on options for implementation of EBFM within their jurisdiction (NEFMC 2010; O'Boyle et al. 2012). The approach entails the development of a transition strategy in which the 9 existing fishery management plans administered by the council are initially retained and evaluated for consistency with ecosystem objectives. The interdependencies among the plans (including technical (by-catch) interactions and biological interactions among species) are to be considered along with explicit consideration of environmental change that might alter overall productivity patterns. These transitional plans are to be ultimately replaced by fully integrated plans within defined ecological management units (NEFMC 2010; Fogarty et al. 2012).

The MAFMC has initiated the development of an EBFM Guidance Document. The objective of the guidance document is ". . . to enhance the Council's species-specific management programs with more ecosystem science, broader ecosystem considerations and management policies that coordinate Council management across its Fishery Management Plans (FMPs) and the relevant ecosystems. The document is intended to provide a framework for considering policy choices and trade-offs as they affect FMP species and the broader ecosystems.

Contemporary Korean fisheries have been managed by a variety of tools, such as input and output controls and technical measurements. Current fisheries management in Korea includes the establishment of precautionary total allowable catch (TAC)-based fishery management. The TAC-based management applies to eleven species and several related gear types. In addition to TACs, some technical measures have been implemented including closed fishing seasons/areas, fish size- and sex-controls, and fishing gear restrictions. A variety of projects are currently under way to recover depleted fish resources in Korea, but these projects operate separately, not systematically, nor consistently. Conventional management with its focus on managing target species, rather than on managing for true sustainability in the use of each species within its ecological context, has not been effective.

Therefore, there is much interest in using an ecosystem-based approach in assessment and management.

To sustain marine productivity and economic profitability and to maintain healthy ecosystems, therefore, the Korean Government laid the foundation for a comprehensive Ecosystem-based Integrated Fishery Management Framework (EBIFM; MOMAF 2007). The framework establishes three major components: (1) the offshore large ecosystem management system, (2) the coastal and inshore self-regulatory ecosystem management system and (3) the marine ranching ecosystem management system.

The offshore component is intended to manage the activities of medium-large fishing vessels operating in open waters under the control of the central government or relevant international bodies. This sector fishes on transboundary resources.

The coastal and inshore component manages a diverse group comprising small boat fisheries, mariculture operations for shellfish and seaweeds, village cooperative fisheries, and diver fisheries. This sector is specifically designed to be self-regulatory and under local control using a precautionary ecosystem-based approach. The system is intended to prevent illegal fishing and to enhance and protect fishing grounds.

The overarching EBIFM Framework has seen a variety of methods of implementation, including programs for marine ranching, resource enhancement, marine forest enhancement, and self-management. The Korean government is currently developing a comprehensive ecosystem-based marine ranching program. The marine ecosystem ranch system is nested within the inshore and coastal sector. Currently, there are five designated marine ranches in the country (see section 3.1). The marine ranches limit fishing activities to hook and line fishing and actively engage in ecosystem enhancement activities including placement of artificial reefs, reconstruction of seaweed beds to restore fish habitats, and the culture and release of larvae and juveniles of some species (see Zhang and Kim 2011). Management is under control of local ranching communities using an ecosystem-based precautionary system.

Artificial habitat and fishery enhancement programs are a cornerstone of Ecosystem-Based Integrated Fishery Management in Korea. Construction of artificial reefs is aimed at improving productivity of devastated fishing grounds by providing fish resources with habitats, and spawning and nursery grounds. In Korea, a total of 55% of the area with artificial reefs is utilized as fishing grounds and the other 45% is preserved solely for fisheries. Since 1998, the National Fisheries Research and Development Institute (NFRDI) has developed seed production technology to release strong juveniles of rockfish and sea bream. Seed production has successfully enhanced fishery resources and increased the incomes of fishermen (Jeon 2004).

Similarly, the Korean marine 'forest' enhancement program began in 1997, with the aim of artificial enhancement of marine plant and seaweed habitats. The program is intended to construct seaweed beds to recover whitening-affected marine forest ecosystems in an effort to boost fish production and fishermen's incomes. It also aims to produce seaweed-based green energy and to combat global warming with CO_2 reduction technology. The program has a target to establish massive marine forest areas covering 35,000 ha in Korean waters by 2020 (MIFAFF 2010).

The Korean style fisheries co-management system has become a unique way to activate social processes and to encourage the involvement of fishermen in coastal fisheries management (Lee and Shin 2004). The self-management approach in Korea was designed to enhance fisheries resources, to protect fishing grounds of self-regulatory communities, and to manage their fisheries resources by their own regulations and knowledge. Since the pilot projects started in 2001 (Park and Zhang 2008), a total of 879 fishing communities have participated in this self-management system (as of 2010).

4. Science to Support Ecosystem-Based Management

Development of management strategies to meet defined EBM goals requires an analytical evaluation of the status of the system in relation to established reference points and an assessment of the probable outcomes of alternative management actions. The US and Korean systems have met this need with related but distinct initiatives. In the US, a protocol for the development of Integrated Ecosystem Assessments (IEAs) has been established to assess ecosystem condition and health in relation to specified management objectives (Levin et al. 2008, 2009). An earlier effort established in 2001 periodically provides a Coastal Condition report prepared by the US Environmental Protection Agency in cooperation with other US federal and state agencies [http://water.epa.gov/type/oceb/assessmonitor/nccr]—effectively a report card on the status of coastal ecosystems in the United States arrayed along five major dimensions. Although very different in intent from an IEA, the Coastal Condition Report is complementary to the integrated assessment format. The Nature Conservancy has also produced extremely valuable Ecoregional Assessments in terrestrial and aquatic ecosystems including parts of both the CCLME and the NESLME (http:// http://east.tnc.org/reports/all_assessment _docs). In South Korea, Zhang et al. (2009, 2010, 2011) have developed an integrated ecosystem-based fisheries assessment (EBFA) method for determining risk to, and risk to the ecosystem from, Korean fisheries.

4.1 Integrated Ecosystem Assessments

An IEA is the synthesis and analysis of all available information on relevant physical, chemical, ecological and human processes in relation to specified ecosystem management objectives. The principal elements of an IEA comprise (1) Scoping, (2) Indicator Selection and Ecosystem Status Assessment, (3) Risk Analysis, and (4) Management Strategy Evaluation (Levin et al. 2008, 2009; Plummer and Levin, this volume). Following implementation of the selected management strategy from step 4, the effectiveness of monitoring and the management program as whole is evaluated and mid-course corrections undertaken as needed. The overall approach is framed within the context of adaptive management and monitoring (Levin et al. 2008, 2009).

IEAs are now being developed throughout the country in a phased implementation strategy reflecting funding priorities and constraints. The California Current LME has been selected for the first full IEA development; work on this project was initiated in 2009. The NESLME is slated for full development following completion of the IEAs for the CCLME and the Gulf of Mexico.

Scoping is the key to creating an informative and useful IEA. To succeed, it requires scientists to successfully navigate the science-policy interface in order to identify stakeholder goals and objectives. It must also decompose the ecosystem into sub-systems thought to be most influential to those goals and objectives, including identification of key drivers and pressures. In the CCLME, scientists worked with regional managers to scope stakeholder goals and objectives, and selected a limited set of EBM drivers, pressures, and components to use in the initial phase of a California Current IEA (Levin and Schwing 2011). In the NESLME, initial scoping sessions were undertaken in 2005 in 26 coastal cities and towns to assess stakeholder perspectives on integrating ecosystem principles into fishery management. Plans are now under way to broaden the remit of scoping to include other ocean use sectors.

4.1a Ecosystem indicators and status

The process of identifying a concise and informative set of ecosystem indicators is a formidable challenge (Kershner et al. 2011). In the CCLME, the process of vetting indicators winnowed an initial list of >500 indicators to a much more manageable and reliable set of ~20 (Levin and Schwing 2011). It also distinguished between indicators that could be locally measured but were regionally informative versus those that could be regionally estimated but locally informative. IEA reports in the CCLME are currently produced on a two-year cycle, with the latest version expected to be published in a web-based format in 2013.

In the NESLME, the Driver-Pressure-State-Impact Response (DPSIR) framework has been adopted to characterize indicators drawn from an initial suite of more than 80 human-related, climate, physical, and ecological metrics (Fogarty et al. 2012b, Figure 11.6). Based on analyses of the relationships between drivers and pressures and between pressures and state metrics, a more restricted set of indicators has been identified to inform management decisions. These indicators are responsive to change and can be connected to specific drivers and pressures (Fogarty et al. 2012b). Ecosystem Status Reports have also been established to provide periodic updates on changes in system health and condition for this system (EcoAP 2012).

4.1b Risk assessment

In addition to using indicators to report on the status of the ecosystem, the IEA includes an assessment of risk, a first cut approach for identifying problems that may require management intervention. A preliminary risk assessment focused exclusively on Puget Sound, WA, suggested that fishing was much less of a concern than pollution associated with residential land use and industry, with little exception. These types of insights can be used to develop scenarios that inform the final step of the IEA, management strategy evaluation (MSE) (Samhouri and Levin 2012).

4.1c Management strategy evaluation

Management Strategy Evaluation calls for the use of one or more operating models that provide a virtual representation of the ecosystem from climate and physical processes through human social structures and within which different management approaches can be tested (Fulton and Link, this volume). The system embeds a representation of the assessment process itself to allow an evaluation of

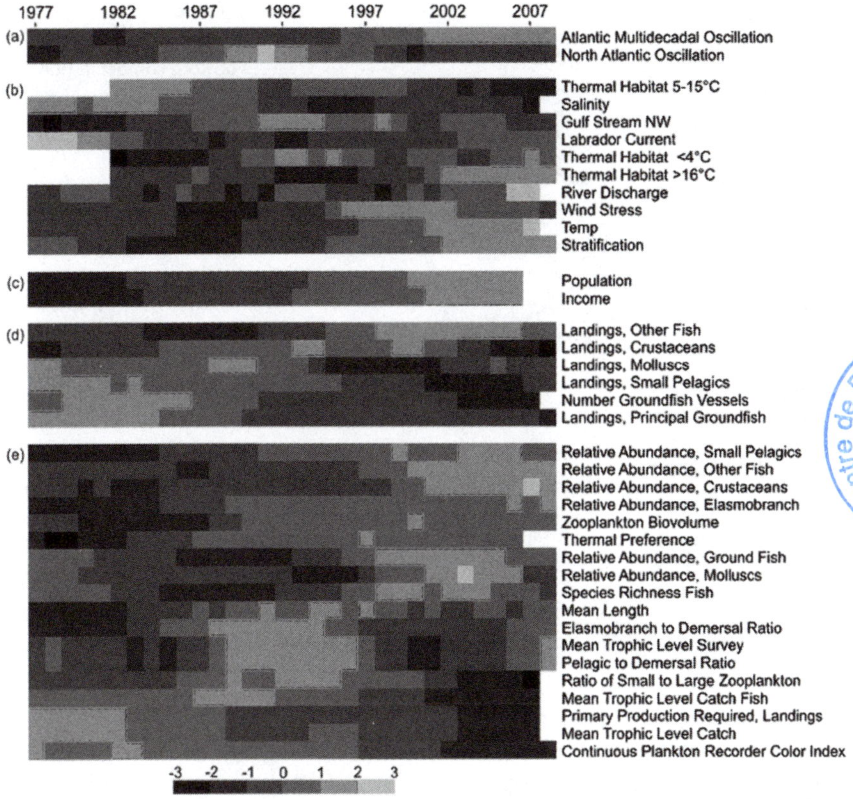

Fig. 11.6 – (See color insert) Normalized indictors for (a) physical drivers, (b) physical pressures, (c) anthropogenic drivers, (d) anthropogenic pressures, and (e) ecosystem state variables. Green colors indicate high values, blue colors represent low values. Blank rectangles indicate no data for time period. (Adapted from Fogarty et al. 2012)

its efficacy. In both the CCLME and NESLME, the comprehensive end-to-end model Atlantis is being used as the principal operating model (see Fulton et al. 2011; Fulton and Link, this volume). Scientists in the CCLME have also used a suite of simulation and statistical models to examine the influence of alternative management scenarios on a variety of ecosystem components not tracked with strong fidelity by Atlantis (e.g., salmon; Kaplan et al. in review). This approach is based on the Millennium Ecosystem Assessment in which integrated scenarios were developed in a qualitative sense, but the work of quantitatively assessing their influence is distributed amongst a variety of tools and experts.

Existing MSEs for the California Current have focused on potential impacts of climate change on food webs and fisheries (Ainsworth et al. 2012), consequences of forage fish depletion for yields to commercial fisheries (Smith et al. 2011), and comparisons of alternative types of spatial and non-spatial management for trade-offs between different ecosystem goals (Figure 11.7). It will be interesting to see how these novel scientific products inform and guide management decisions along the west coast of the US. At the moment, unless forced via litigation by environmental NGOs, such decisions tend to be made with a single sector focus.

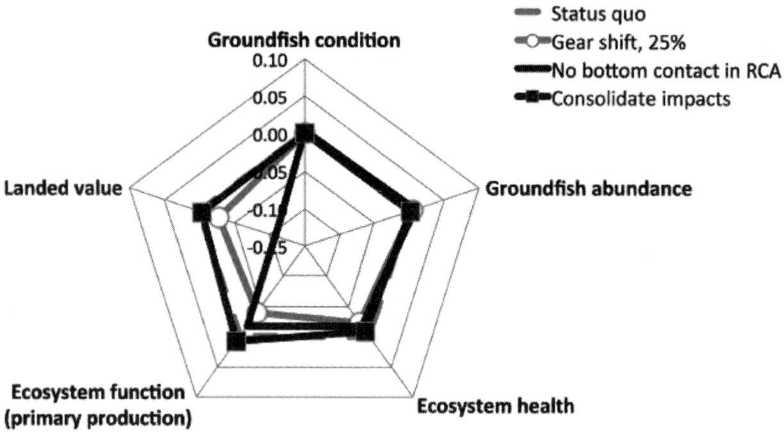

Fig. 11.7 – Performance of alternative management scenarios in terms of ecosystem components relevant to the IEA in the California Current. Note that values here represent proportional difference from a status quo (business-as-usual) projection. The scenarios include a gear shift from trawl to longline and pot, eliminating bottom-contact gear in the Rockfish Conservation Area (RCA), and consolidating the footprint of bottom-contact impacts. Here the management action occurs coast wide and the indicators are calculated on a coast-wide basis.

4.1d Ecosystem-based Fisheries Assessment (EBFA)

In Korea, Ecosystem-based fisheries assessment uses a two-tier analytical method for determining risk: Tier 1 is designed for data-rich situations, while Tier 2 applies to data-poor situations (Zhang et al. 2009, 2010 2011). This assessment method is composed of the following steps (see Figure 11.8). The first step involves setion of the target management unit, i.e., ecosystem, fishery, or individual species. Once the management unit is selected, a management advisory body composed of stakeholders is organized to consult and decide on assessment methods and set management tools and methods. For Korean fisheries, four objectives were identified: sustainability, biodiversity, habitat quality and socio-economic benefits.

Next, all available data and information are identified and evaluated, a decision is made as to the appropriate analysis (either Tier 1 or Tier 2), and assessment of the system is undertaken. Principal fishery/species with sufficient data to permit quantitative assessment are assigned to Tier 1, while other fishery/species are assigned to Tier 2. For major fishery/species lacking sufficient data and information for quantitative assessment, efforts to enhance data collection, monitoring and (if necessary) proper research are undertaken to permit scientific evaluation under Tier 1 guidelines (Figure 11.8).

Where data are sufficient for a Tier 1 or 2 analysis, ecosystem indicators, along with target and limit reference points, are identified for each selected objective. Three considerations are used in the indicator selection process: 1) ease of understanding by users, 2) responsiveness to management of human activities, and 3) measurability, using existing data or currently monitored information.

Relative weights for each indicator were obtained by conducting a series of

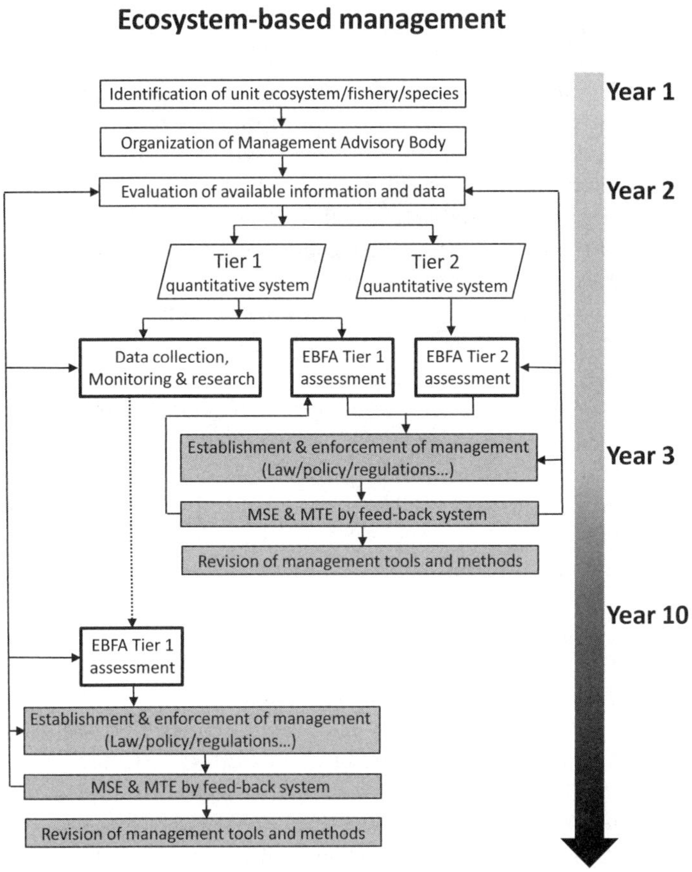

Fig. 11.8 – (See color insert) A flowchart showing the procedure for the ecosystem-based fishery management system in Korea. The overall process includes (a) identification of other user groups, such as aquaculture, sand mining, etc. (b) Spatial units include Large Marine Ecosystems: East Sea, Yellow Sea, East China Sea; Coastal Ecosystems: Various coastal fisheries; Inshore Ecosystems: Fishing village cooperatives, Marine ranches, Self-management units, (c) Consultations and decisions on assessment methods, plans for assessment, and setting management tools and methods, (d) Decision on appropriate analytical methods, (e) specification of Minor species of TAC, Minor species of marine ranches, Target species of stock-rebuilding program, Main species of self-management units and (f) Key species of TAC, Target species of marine ranches, Main species of stock-rebuilding program.

expert workshops, considering: 1) the importance of an indicator for achieving management objectives, 2) scientific basis for estimating indicators and reference points, and 3) availability of data and information. The same indicators across Tier 1 and Tier 2 assessments can be weighted differently, depending on the situation. Target reference points in this application correspond to a state of each indicator that is considered desirable, while the limit reference point is defined as the limit beyond which the state of each indicator is not considered desirable. Species were assigned a status for each indicator to denote risk (for details on the scoring process, see Zhang et al. 2009).

Information is produced that allows the construction of four different but nested indices useful for characterizing the status of the system. These indices are the objective risk index (ORI), species risk index (SRI), fishery risk index (FRI), and ecosystem risk index (ERI). A risk assessment diagram is created to facilitate consideration of the constructed indices. The four axes represent each ORI. Detailed methods and data used for scoring Tier 1 and Tier 2 indicators and calculating risk indices are explained in Zhang et al. (2009, 2010, and 2011).

Following this process, the subject or subjects which resulted in high risk scores during the EBFA are enumerated. This step is followed by the identification and implementation of appropriate management measures in accordance with existing laws, policies and regulations. In Korea, these processes are initiated under the authority of MIFAFF, and there is little opportunity for coordination with sectors outside of fisheries. The final step in the process is the development of procedures capable of evaluating the effectiveness of management measures and revising the management system as needed using management strategy evaluation (MSE) or management tactic evaluation (MTE). In each step the involvement of stakeholders is desirable to achieve consensus. This approach to ecosystem-based fisheries management (EBFM) reduces potential fishing impacts while at the same time allowing for the extraction of fish resources at levels sustainable for the ecosystem.

Forecasting the results of any management action is very important, but difficult because the dynamics of ecosystems are complex and poorly understood. Complementing the assessment method described above will be research to develop appropriate forecasting methods prior to implementing actual management measures (Zhang and Kim 2011). Zhang et al. (2011) extended the EBFA approach to create an Integrated Fisheries Risk Analysis Method for Ecosystems (IFRAME). Figure 11.9 shows the structure of the IFRAME framework. The first panel is the assessment component, which constructs ecosystem structure and assesses risks. The middle panel centers on forecasting ecosystem structure and risks based on various scenarios. The lower panel is the management component, which evaluates and implements management approaches. The IFRAME is explained in detail in Zhang et al. (2011), which is still in the developing stages for application. Recently, Hollowed et al. (2012) selected seven representative classes of models to illustrate the utility of the approach for assessing climate change impacts on higher trophic level species, and concluded that the IFRAME modeling approach was best suited to evaluate the performance of the mitigation strategies relative to the principles of ecosystem-based management.

5. Discussion

The California Current, Northeast Continental Shelf, and Korean Marine Ecosystems are foci in an evolving quest to define and implement effective strategies for Ecosystem-Based Management. Formal approaches to defining national visions for marine EBM are comparatively recent, however. The diversity of ongoing activities, each of which operates on a broad range of spatial and temporal scales, presents unique challenges to melding the parts into an organic whole in each region.

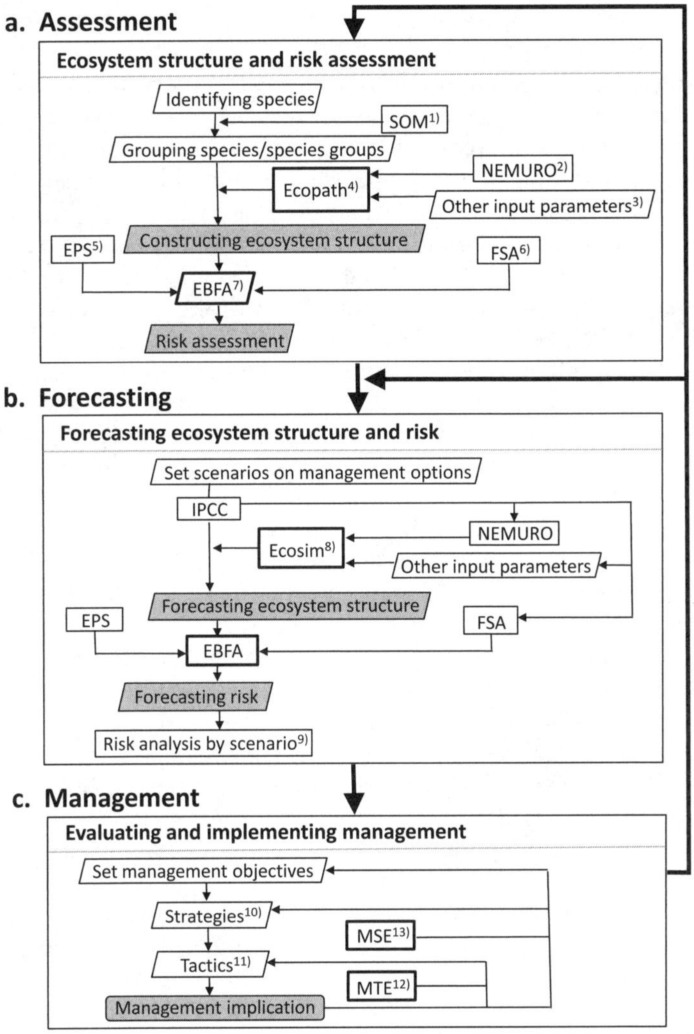

Fig. 11.9 – A flowchart showing the procedure for the ecosystem-based fishery management system in Korea.

Stark differences on several key issues necessitate a tailored approach to EBM in each case study region:

- the history of settlement and development;
- the natural supply of different ecosystem services;
- scientific understanding;
- jurisdictional and transboundary complexities;
- governance structures; and,
- culture.

We believe the same will be true in most of the world's temperate regions as they embark on EBM implementation, which is why it is encouraging to see flexible, quantitative scientific assessment frameworks emerging to support such transitions.

In the United States, colonization by European settlers in significant numbers on the east coast pre-dated that on the west by nearly two centuries. The availability of abundant natural resources in the Americas was a critical catalyst for colonization. The appropriation of ecosystem services by native peoples considerably preceded these events in North America, but differed from the present day situation in large part because coastal population densities were so much lower. In Korea and in Asia more generally, dense coastal populations dependent on near-shore resources have existed for millennia. The critical importance of protein from the sea to human well-being in Korea, and indeed throughout Asia, highlights the necessity of dealing effectively with threats to mariculture and fisheries through integrated management. As elsewhere throughout the shores of much of the world ocean, impacts on ecosystem services have escalated with the burgeoning human population. The trajectory of impact and change can be mapped to these historical events (Lotze et al. 2006).

In an evaluation of the state of implementation of EBM throughout the North Pacific, Jamieson and Zhang (2005) traced differences in the needs and objectives of holistic management in different countries to variation in coastal population densities and history of resource use. China, Japan and Korea are all characterized by high population densities on the coasts and an extensive history of full exploitation of renewable resources. These features have led to a focus on minimizing widespread overexploitation, rebuilding depleted stocks, and minimizing the impacts of eutrophication and pollution principally related to runoff from land-based sources. In contrast, Russia and western North America, all characterized by lower coastal population densities relative to Asian countries, have identified concerns centered on fishing impacts, oil and gas development, and ocean transportation.

Regulatory and cultural environments are no less consequential in the development and implementation of marine EBM in the three case studies examined in this chapter. In this context, the U.S. has been characterized as a weak federal system with important roles played by the coastal states in devising policies for EBM and Coastal Marine Planning (CMP) (Rosenberg et al. 2009). Indeed, it is clear that state and local authorities in the United States have been highly influential in moving the process forward (see also Altman et al., Houde et al., this volume). This progress is perhaps an important lesson for considering EBM implementation in the KME, where coordination must occur at subregional levels but also at the international level with neighboring China, Japan, and North Korea. The US experience suggests that efforts at multiple scales can integrate and converge as the policy environment allows.

An essential challenge in the U.S. is the need for effective coordination among a diverse set of regulators and stakeholders so that the whole is ultimately greater than the sum of the parts. Prior to the establishment of the National Ocean Policy, Turnipseed et al. (2009) called for designation of ocean resources as a public trust to provide a legal foundation for coordination among local, state, and federal governments to meet management objectives. In part this call was motivated by the overwhelming complexity of spatial management practices in the Northeast (see Figure 1 of Turnipseed et al. 2009 and Figure 11.3). Crowder et al. (2006) highlighted the fragmentation of management of human activities in the southern California region in their recommendation for an integrated strategy of coastal and marine spatial planning to resolve mismatches in spatial governance mechanisms in the US. Effective coordination of multi-scale efforts in moving toward marine EBM is clearly one of the most important challenges to confront not just in the U.S. but in temperate seas worldwide.

Ecosystem-based fishery management is one of many inroads that have been made in recent decades, taking temperate regions several steps closer to mEBM. However, on its own, EBFM represents a piecemeal approach to achieving mEBM that could result in undue emphasis on parts (fisheries) rather than the whole. Without full consideration and coordination of management actions for other ocean use sectors, EBFM is necessarily incomplete as an implementation strategy for EBM. Accordingly, plans for EBFM will be most effective at helping to achieve EBM with explicit consideration of how to integrate them with other ocean sector management strategies. For instance, in the U.S. integration could be accomplished under the National Ocean Policy. In this context, EBFM must ultimately accommodate the transition from a perspective in which fisheries are the focal point to one in which they are viewed as one of many valued ecosystem and human-use components.

The importance of the effective use of co-management strategies with shared responsibilities among governments and stakeholders is recognized but still in an early stage of development in the United States. In contrast, the devolution of management authority and responsibility in the Korean coastal and inshore ecosystem units and the marine ranch ecosystem unit to stakeholders is quite highly developed. While the central government retains the lead role in management of the offshore large ecosystem component, a strategy of self-regulation in the remaining two ecosystem components, with the objective of achieving cooperation and buy-in for the management programs, has been adopted.

Indeed, the Korean ranch ecosystem concept provides a novel example of how to implement marine EBFM. Its focus not only on conservation and protection, but also on active remediation as a central theme, sets it apart from many other ecosystem-based fishery management strategies now being employed throughout the world. In particular, the emphasis on restoring natural habitat, the use of artificial reefs, and supplementing some natural populations through artificial enhancement along with important constraints on modes of fishing are important elements of an overall strategy designed to recover the production potential at these sites. The nesting of the five current ranch ecosystems within the broader coastal and inshore ecosystem structure will also afford the possibility of interesting contrasts and comparisons to help guide adaptive management approaches in Korea.

The marine ranch ecosystem concept also serves to highlight a lesson common to many of the anecdotes reviewed in this chapter: EBM has not often been put into practice in the holistic, cross-sectoral, integrated manner to which it aspires. Instead, it is frequently the case that individual ocean use sectors operate within their own management spheres, and different sectors do not collaborate at the outset to achieve solutions that minimize conflict and maximize benefit to all parties. Indeed, in many cases management activities have been largely reactive. Several of the outcomes we have reviewed do not exist because of proactive decision making, but instead are a result of interventions that occurred after problems arose.

While it will not always be possible to anticipate intra- and inter-sectoral conflict, rather than having sectoral activities limited by one another, why not challenge the current system to identify management strategies that can augment the benefits accrued to each sector and all sectors taken together? Recent scientific advances suggest that a priori planning for efficient allocation of different uses in seascapes can produce more efficient and effective outcomes than de facto single sector management (Lester et al. 2010, White et al. 2012, Allnutt et al. 2012). If one goal of EBM is to reduce conflicts and enhance cooperation among ocean user groups, what is needed is the management and policy infrastructure capable of ingesting emerging scientific analyses and applying their insights at appropriate spatial scales.

Fortunately, in the three examples of EBM in temperate seas reviewed here, scientific assessments provide increasingly hopeful and content-rich information about what is possible with integrated marine management. In fact, in the U.S. and Korea, approaches to assessment and evaluation of ecosystem status as a guide to management action have undergone a form of convergent evolution, albeit strongly influenced by the deeply interconnected world we now inhabit. The Integrated Ecosystem Assessments now under way in the United States and the Ecosystem-Based Fisheries Assessment (EBFA) approach in Korea share many common elements and approaches including the recognition of the importance of developing informative indicators and the development of reference points to guide management decisions based on these indicators; the central role of risk analysis to account for uncertainty; the value of management strategy evaluation to test potential management choices first in a virtual world; and finally, recognition of the importance of adaptive management that can learn from past experience and evolve in the face of changing conditions.

As in so much of the world ocean, the state of play in implementing marine EBM in both countries is best characterized as an evolving process. In the United States, an attempt is being made to develop management strategies that encompass the broad range of ecosystem services, threats to sustainable delivery of these services, and the tradeoffs that must be confronted in dealing with potentially conflicting goals and objectives. Developments in EBFM are steadily progressing but ultimately require integration with other sectoral management concerns to meet their full promise. Spatial management strategies are relatively well developed and formal methods of marine spatial planning are quickly being adopted, often at the state and regional levels. In Korea, sophisticated approaches to EBFM are being developed and implemented. The zoned approach to delineation of ecosystem management units with open ocean, coastal and inshore, and marine ranch components is a unique spatial management strategy. Collectively, the progress toward implementing aspects of marine EBM is clear. The challenges remaining for full implementation are daunting but can be met with continued commitment and innovation.

Acknowledgments

We are grateful to Kimberly Hyde and Sean Lucey of the Northeast Fisheries Science Center for assistance in mapping, graphics, and analysis; Jordan Gass of the NOAA MPA Center for assistance with MPA mapping and analysis, Sarah Lester of the Sustainable Fisheries Group at UC Santa Barbara, Natalie Semyk and Steve Katz at the Channel Islands National Marine Sanctuary and Isaac Kaplan and Blake Feist at the Northwest Fisheries Science Center for analytical assistance. We are grateful for the thoughtful and provocative comments of Phil Levin, Jodie Toft, and Elliott Hazen.

References

Ainsworth, C. H., J. F. Samhouri, D. S. Busch, W. L. Cheung, J. Dunne, and T. A. Okey. 2012. Potential impacts of climate change on northeast Pacific marine fisheries and food webs. *ICES J. Mar. Sci.* In press.

Allnutt, T. F., T. R. McClanahan, S. Andréfouët, M. Baker, E. Lagabrielle, C. McClennen, A. J. M. Rakotomanjaka, T. F. Tianarisoa, R. Watson, and C. Kremen. 2012. Comparison of Marine Spatial Planning Methods in Madagascar Demonstrates Value of Alternative Approaches. *PLoS ONE* 7:e28969.

Aquarone, M. C., and S. Adams. 2008. XIV-44 California Current: LME No.3. In K. Sherman and G. Hempel (eds.), The UNEP large marine ecosystem report: A perspective on changing conditions in LMEs of the world's regional seas, p. 593–603. *UNEP Regional Seas Report and Studies No.182.* United Nations Environment Programme, Nairobi, Kenya.

Bakun, A. 1993. The California Current, Benguela Current, and Southwestern Atlantic Shelf ecosystems: A comparative approach to identifying factors regulating biomass yields. In K. Sherman, L. M. Alexander, and B. Golds (eds.), *Large marine ecosystems—Stress, mitigation, and sustainablity,* p. 199–224. Am. Assoc. Adv. Sci. Publ., Washington, DC.

Beck, M. W., K. L. Heck, J., K. W. Able, D. L. Childers, D. B. Eggleston, B. M. Gillanders, B. Halpern, C. G. Hays, K. Hoshino, T. J. Minello, R. J. Orth, P. F. Sheridan, and M. P. Weinstein. 2001. The identification, conservation, and management of estuarine and marine nurseries for fish and invertebrates. *Biosci.* 51(8):633–641.

Belkin, I. M. 2009. Rapid warming of large marine ecosystems. *Progr.Oceanogr.* 81: 207–213.

Bolster, W. J. 2008. Putting the oceans in Atlantic history: maritime communities and marine ecology in the Northwest Atlantic, 1500–1800. *Am. Hist. Rev.* 113:19–47.

Bouman, H. A., T. Platt, S. Sathyendranath, W. K. W. Li, V. Stuart, C. Fuentes-Yaco, and H. Maass. 2003. Temperature as indicator of optical properties and community structure of marine phytoplankton: Implications for remote sensing. *Mar. Ecol. Prog. Ser.* 258:19–30.

Bourque, B. J. 2004. *Twelve Thousand Years. American Indians in Maine.* Univ. Nebraska Press, Lincoln 368pp.

Chan, K. M., and M. Ruckelshaus. 2010. Characterizing changes in marine ecosystem services. F1000 Biology Reports, 2.

Chen, C-T. A. 2000. The impact of dams on fisheries: Case of the Three Gorges dam. *Glob. Change News Lett.* 44: 4–5.

Cheung, W. W. L., V. W. Y. Lam, J. L. Sarmiento, K. Kearney, R. Watson, and D. Pauly. 2009. Projecting global marine biodiversity impacts under climate change scenarios. Fish and Fisheries 10:235–251.

Cheung, W. W. L., R. Watson, and D. Pauly. 2013. Signature of ocean warming in global fisheries catch. Nature 497:365–368.

Cicin-Sain B. and R. Knecht. 1998. *Integrated Coastal and Ocean Management: Concepts And Practices.* Washington, D.C.: Island Press.

Coleman, F. C., W. F. Figueira, J. S. Ueland, and L. B. Crowder. 2004. The impact of United States Recreational Fisheries on Marine Fish Populations. *Science.* 305:1958–1960.

Cooley, S. R. and S. C. Doney. 2009. Anticipating ocean acidification's economic consequences for commercial fisheries. *Env. Res. Lett.* 4:024007 doi:10.1088/1748–9326/4/2/024007.

Corbin, J. S. 2010. Sustainable U.S. marine aquaculture in the 21st century. *Mar. Tech. Soc. J.* 44:3:5.

Crossett, K. M., T. J. Culliton, P. C. Wiley, and T. R. Goodspeed. 2004. Population trends along the coastal United States: 1980–2008. Nat. Ocean. Atmos. Admin. Washington D.C.

Crowder, L. B., G. Osherenko, O. R. Young, S. Airame, E. A. Norse, N. Baron, J., C. Dat, F. Douvere, C. N. Ehler, B. S. Halpern, S. J. Langdon, K. L. McLeod, J. C. Ogden, R. E. Peach, A. A. Rosenberg, and J. A. Wilson. 2006. Resolving mismatches in U.S. ocean governance. *Science* 313:617–618.

Dahl, T. E. 2011. Status and trends of wetlands in the conterminous United States 2004–2009. U.S. Fish. Wildl. Serv. Washington, D.C.

Diaz, R. J. and R. Rosenberg. 2008. Spreading dead zones and consequences for marine ecosystems. *Science* 321:926–929.

Drinkwater, K. F. and D. G. Mountain. 1997. Climate and oceanography . . . In: J. Boreman, B. S. Nakashima, J. A. Wilson, and R. L. Kendall (eds.), *Northwest Atlantic Groundfish: Perspectives on a Fishery Collapse.* American Fisheries Society, Bethesda, Maryland, USA.

EcoAP. 2012. Ecosystem Status Report for the Northeast U.S. Continental Shelf Large Marine Ecosystem—2011. Ecosystem Assessment Program NEFSC Center Ref. Doc. 12–07, Northeast Fisheries Science Center, Woods Hole, MA.

Estes, J. A., J. Terborgh, J. S. Brashares, M. E. Power, J. Berger, W. J. Bond, S. R. Carpenter, T. E. Essington, R. D. Holt, J. B. C. Jackson, R. J. Marquis, L. Oksanen, T. Oksanen, R. T. Paine, E. K. Pikitch, W. J. Ripple, S. A. Sandin, M. Scheffer, T. W. Schoener, J. B. Shurin, A. R. E. Sinclair, M. E. Soule, R. Virtanen, and D. A. Wardle. 2011. Trophic downgrading of planet earth. *Science* 333:301–306.

FAO 2010. State of World Fisheries and Aquaculture. Food and Agriculture Organization of the United Nations, Rome.

Feely, R. A., C. L. Sabine, J. M. Hernandez-Ayon, D. Ianson, and B. Hales. 2008. Evidence for upwelling of corrosive "acidified" water onto the continental shelf. *Science* 320:1490–1492.

Feely, R. A., S. C. Doney, and S. R. Cooley. 2009. Ocean acidification: present conditions and future changes in a high-CO_2 world. *Oceanogr.* 22:37–47.

Field, J. C. and R. C. Francis. 2006. Considering ecosystem-based fisheries management in the California Current. *Mar. Pol.* 30:552–569.

Fogarty, M. J. and S. A. Murawski. 1998. Large-scale disturbance and the structure of marine systems: fishery impacts on Georges Bank. *Ecol. Appl.* 8:6–22.

Fogarty, M. J. and S. A. Murawski. 2005. Do marine protected areas really work? *Oceanus* 43:42–44.

Fogarty, M. J., W. J. Overholtz, and J. S. Link. 2012a. Aggregate-species production models for the Demersal Fish Complex of the Gulf of Maine. *Mar. Ecol. Prog. Ser.* In press.

Fogarty, M. J., R. Gamble, K. Hyde, S. Lucey, and C. Keith. 2011. Spatial considerations for ecosystem-based fishery management on the Northeast U.S. continental shelf. In (D. Packer, Ed.) Proceedings of the Mid-Atlantic Management Council's Habitat-Ecosystem Workshop, *NOAA Tech. Mem.* NMFS-F/SPO-115 pp 31–33.

Fogarty, M. J., K. Friedland, L. Col, R. Gamble, J. Hare, K. Hyde, J. Link, S. Lucey, H. Liu, J. Nye, W. Overholtz, D. Richardson, B. Rountree, and M. Taylor. 2012. Status of the northeast continental shelf large marine ecosystem: an indicator-based approach 2012. In (R. Stephenson et al., Eds.) *Advancing Ecosystem-Based Management in the Gulf of Maine.* In press.

Frank, K. T., B. Petrie, N. L. Shackell, and J. S. Choi. 2006. Reconciling differences in trophic control in mid-latitude marine ecosystems. *Ecol. Lett.* 9:1096–1105.

Fulton, E. A., J. S. Link, I. C. Kaplan, M. Savina-Rolland, P. Johnson, C. Ainsworth, P. Horne, R. Gorton, R. J. Gamble, A. D. M. Smith, and D. C. Smith. 2011. Lessons in modeling and management of marine ecosystems: the Atlantis experience. *Fish. Fisher.* 12:171–188.

Gaichas, S., R. Gamble, M. Fogarty, H. Benoît, T. Essington, C. Fu, M. Koen-Alonso, and Jason Link. 2012. Assembly rules for aggregate-species production models: Simulations in support of management strategy evaluation. *Mar. Ecol. Prog. Ser.* 459:275–292 .

Grantham, B. A., F. Chan, K. J. Nielsen, D. S. Fox, J. A. Barth, A. Huyer, J. Lubchenco, et al. 2004. Upwelling-driven nearshore hypoxia signals ecosystem and oceanographic changes in the northeast Pacific. *Nature* 429:749–754.

Gruber, N., C. Hauri, Z. Lachkar, D. Loher, T. L. Frölicher, and G.-K. Plattner. 2012. Rapid progression of ocean acidification in the California current system. *Science* 337:220–223.

Hamilton, S. L., J. E. Caselle, D. P. Malone, and M. H. Carr. 2011. Incorporating biogeography into evaluations of the Channel Islands marine reserve network. *Proceedings of the National Academy of Sciences USA* 107:18272–18277.

Hare, S. R. and N. J. Mantua. 2000. Empirical evidence for North Pacific regime shifts in 1977 and 1989. *Prog. Oceanogr.* 47:104–145.

Hickey, B. M. 1989. Patterns and processes of circulation over the Washington continental shelf and slope. In M. R. Landry and B. M. Hickey (eds.), *Coastal Oceanography of Washington and Oregon.* Elsevier, Maryland Heights, MD. 41–115.

Hollowed, A. B., E. N. Curchitser, C. A. Stock, and C. I. Zhang. 2012. Trade-offs associated with different modeling approaches for assessment of fish and shellfish responses to climate change. *Climate Change,* DOI 10.1007/s10584–012–0641-z.

Hong, J.-S., and J. H. Lee. 1983. Effects of the pollution on the benthic macrofauna in Masan Bay, Korea. *J. Oceanol. Soc. Korea.* 18:169–179.

Jamieson, G. and C.-I Zhang. 2005. Report of the study group on ecosystem-based management science and its application to the North Pacific. *PICES Scientific Report* No. 29.

Jeon, I. G. 2004. Status and problems of stock enhancement. Symposium on the stock rebuilding in Korean waters. NFRDI, Busan, Korea. 95–123.

Keeling, R. F., A. Kortzinger, and N. Gruber. 2010. Ocean deoxygenation in a warming world. *Annual Review of Marine Science* 2:199-229.

Kershner, J., J. F. Samhouri, C. A. James, and P. S. Levin. 2011. Selecting Indicator Portfolios for Marine Species and Food Webs: A Puget Sound Case Study. *PLoS ONE* 6:e25248.

Kaplan, I. C. and J. Leonardo. 2012. From krill to convenience stores: Forecasting the economic and ecological effects of fisheries management on the US West Coast. *Marine Policy* 36:947–954.

Kim, S. 2003. Changes in fisheries resources in relation to variability of oceanic environments. *J. Korean Soc. Fish. Res.* 6:11–20. (written in Korean with English abstract)

Kim, S., S. Kang, C-I. Zhang, H. Seo, M. Kang, and J. J. Kim. 2012. Comparison of fisheries yield and oceanographic features at the southern boundaries of the western and eastern Subarctic Pacific Ocean. *ICES J. Mar. Sci.,* 69:1141–1147.

Kim, S. and S. H. Khang, 2000. Yellow Sea. In: C. R. C. Sheppard (Ed.), *Seas at the millennium: an environmental evaluation,* Vol. II, Regional Chapters: The Indian Ocean to the Pacific. 487–498.

Kim, K., K.-R. Kim, Y.-G. Kim, Y.-K. Cho, J.-Y. Chung, B.-H. Choi, S.-K. Byun, G.-H. Hong, M. Takematsu, J.-H. Yoon, Y. Volkov, and M. Danchenkov. 1996. New findings from CREAMS observation: water masses and eddies in the East Sea. *J. Kor. Soc. Oceanogr.* 31:155–163.

Kim, S. and L.-L. Low. 2011. Emerging issues of East Asian fisheries in conjunction with changes in climate and social systems in the 21st century. *Env. Pol.* 10:73–91.

Kim, O. K. and P. Oever. 1992. Demographic transition and patterns of natural resources use in the Republic of Korea. *Ambio* 21: 56–62.

Kim, S., C.-I. Zhang, J.-Y. Kim, J.-H. Oh, S. Kang, and J. B. Lee. 2007. Climate variability and its effects on major fisheries in Korea. *Ocean Sci. J.* 42:179–192.

Kimbrough, K. L., W. E. Johnson, G. G. Lauenstein, J. D. Christensen, and D. A. Apeti. 2008. An assessment of two decades of contaminant monitoring of the nation's coastal zone, *NOAA Tech, Mem. NOS NCOS* 74, 105 pp.

Ko, B.-W., E.-S. Kim, and K. S. Kil. 2010. The positive externality of port calling: A case study of Busan Container Port. KMI *Int. J. Mar. Affairs Fish.* 2:63–77.

KORDI. 2007. Report on Nation's promising marine technology toward the future society. Korea Ocean Research and Development Institute, Ansan, Korea. 1185 p.

Lear, W. H. 1998. History of the fisheries of the Northwest Atlantic: the 500-year perspective. *J. Northw. Atl. Fish. Sci.* 23:41–73.

Lee, S. K. and Y. M. Shin. 2004. A study on the self regulatory management model of coastal fisheries in Korea. *The Journal of Fisheries Business Administration.* 35(1): 87–115 (in Korean with English abstract).

Lester, S. E., K. L. McLeod, H. Tallis, M. Ruckelshaus, B. S. Halpern, P. S. Levin, F. P. Chavez, C. Pomeroy, B. J. McCay, C. Costello, S. D. Gaines, A. J. Mace, J. A. Barth, D. L. Fluharty, and J. K. Parrish. 2010. Science in support of ecosystem-based management for the US West Coast and beyond. *Biological Conservation* 143:576–587.

Levin, P. S. and F. B. Schwing (Eds.). 2011. Technical background for an integrated ecosystem assessment of the California Current: Groundfish, salmon, green sturgeon, and ecosystem health. *U.S. Dept. of Commerce, NOAA Tech. Memo.*, NMFS-NWFSC-109, 330 p.

Levin, P. S., M. J. Fogarty, G. C. Matlock and M. Ernst. 2008. Integrated ecosystem assessments. *U.S. Dept. of Commerce, NOAA Tech. Memo., NMFS-NWFSC-92, 20.*

Levin, P. S, M. J. Fogarty, S. A. Murawski, and. D. Fluharty. 2009. Integrated Ecosystem Assessments: developing the scientific basis for ecosystem-based management of the ocean. *PLoS Biol.* 7(1) e1000014.

Levin, P., C. Ainsworth, Y. deReynier, R. Dunsmore, M. Fogarty, K. Holsman, E. Howell, C. Kelble, M. Monaco, S. Oakes, R. Shuford, and C. Werner. 2012. Integrated Ecosystem Assessment: Guidance for Implementation. Report Prepared for NOAA Science Advisory Board.

Levitus, S., J. I. Antonov, T. P. Boyer, and C. Stephens. 2000. Warming of the World Ocean. *Science* 287:2225–2229.

Lie, H.-J., C.-H. Cho, J.-H. Lee, P. Niiler, and J.-H. Hu. 1998. Separation of the Kuroshio water and its penetration onto the continental shelf west of Kyushu. *J. Geophys. Res.* 103, 2963–2976.

Link, J. S., R. Gamble, and M. Fogarty. 2011. The NEFSC Ecosystem Modeling Enterprise in Support of Ecosystem-Based Fishery Management. NEFSC Center Res. Doc. 11–23.

Link, J., W. Overholtz, J. O'Reilly, J. Green, D. Dow, D. Palka, C. Legault, J. Vitaliano, V. Guida, M. Fogarty, J. Brodziak, E. Methratta, W. Stockhausen, Col, L., G. Waring, and C. Griswold, 2008. An Overview of EMAX: The Northeast U.S. Continental Shelf Ecological Network. *J. Mar Sys.* 74:453–474.

Link, J., J. O'Reilly, M. Fogarty, D. Dow, J. Vitaliano, C. Legault, W. Overholtz, J. Green, D. Palka, V. Guida, and J. Brodziak. 2008. Energy flow on Georges Bank revisited: the Energy Modeling and Analysis Exercise in historical context. *J. Northwest Atl. Fish. Sci.* 39:83–101.

Lotze, H. K., H. S. Lenihan, B. J. Bourque, R. H. Bradbury, R. G. Cooke, M. C. Kay, S. M. Kidwell, M. X. Kirby, C. H. Peterson, and J. B. C. Jackson. 2006. Depletion, degradation, and recovery potential of estuaries and coastal seas. *Science* 312:1806–1809.

Mann, A. 1921. The dependence of the fishes on the diatoms. Ecol. 2:79–83

McGowan, J. A., D. R. Cayan, and L. M. Dorman. 1998. Climate-Ocean Variability and Ecosystem Response in the Northeast Pacific. *Science* 281:210–217.

MB. 1994. *Restoring and Protecting Marine Habitat: The Role of Engineering and Technology.* Marine Board, The National Academies Press.

McLeod, K. and H. Leslie. 2009. *Ecosystem-Based Management for the Oceans . . .* Island Press, Washington. 368 pp.

MIFAFF (Ministry for Food, Agriculture, Forestry and Fisheries). 2009. Green growth and fishing village of hope: Design of future green marine ranching. 105 pp (in Korean).

MIFAFF (Ministry for Food, Agriculture, Forestry and Fisheries). 2010. Final report of marine forest enhancement program in 2009. 628 pp (in Korean).

MIFAFF (Ministry for Food, Agriculture, Forestry and Fisheries). 2011. The Basic plan of the Korean fisheries resources management (2011–2015). 89 pp (in Korean).

Mitsch, W. J. and J. G. Gosselink. 2000. *Wetlands.* Third Edition. J. Wiley and Sons, New York. 920 pp.

MLTM. 2009. State report of the Seas of Republic of Korea. Korea Ocean Research & Development Institute, Ministry of Land, Transport and Maritime Affairs. 86 p.

MLTM. 2011. National Investigation of Marine Ecosystem. Ministry of Land, Transport and Maritime Affairs. 927 p.

MMS. 2009. Report to the secretary U.S. Department of the Interior. Survey of available data on OCS resources and identification of data gaps. *OCS Rept. MMS* 2009–015.

MOMAF. 2007. Studies on the ecosystem based fishery management system. Seoul (in Korean).

Morán, X. A., A. López-Urrutia, A. Calvo-Díaz, and W. W. Li. 2009. Increasing importance of small phytoplankton in a warmer ocean. *Glob. Change Biol.* 16(3):1137–1144.

Murawski, S. A., P. Rago, and M. Fogarty 2004. Spillover effects from temperate marine protected areas. *AFS Symp. Vol.* 42:167–184.

Murawski, S. A., S. E. Wigley, M. J. Fogarty, P. J. Rago, and D. G. Mountain, 2005. Effort distribution and catch patterns adjacent to temperate MPAs. *ICES J. Mar. Sci.* 62:1150–1167.

National Ocean Economics Program (NOEP). 2012. Ocean Economy Data, 2009, generated by the NOAA Coastal Services Center. Available at: http://www.oceaneconomics.org/Market/ocean/ocean Econ.asp.

Naylor, R. L., R. W. Hardy, D. P. Bureau, A. Chiu, M. Elliott, A. P. Farrell, I. Forster, D. M. Gatlin, R. J. Goldburg, K. Hua, and P. D. Nichols. 2009. Feeding aquaculture in an era of finite resources. *Proceedings of the National Academy of Sciences USA* 106:15103–15110.

NEFMC. 2010. White Paper on Ecosystem-Based Fishery Management. New England Fishery Management Council, Newburyport, MA. 25 pp.

NMFS. 2009. Our living oceans. Report on the status of U.S. living marine resources, 6th edition. U.S. Dept. Commer., NOAA Tech. Memo. NMFS-F/SPO-80. 369p.

NOAA. 2007. NOAA 10-year plan for marine aquaculture, October 2007. www.aquaculture.noaa.gov

NOAA. 2012. Status of Stocks: Report on the Status of U.S. Fisheries for 2011. NOAA Fisheries, Washington D.C. 20 pp.

Nye, J. A., J. S. Link, J. A. Hare, and W. J. Overholtz. 2009. Changing spatial distribution of fish stocks in relation to climate and population size on the Northeast United States continental shelf. *Mar. Ecol. Progr. Ser.* 393:111–129.

O'Boyle, R., S. Cadrin, D. Georgianna, J. Kritzer, M. Sissenwine, M. Fogarty, C. Kellogg, and P. Fiorelli. Ecosystem-Based Fishery Management for the New England Fishery Management Council. 2012. In G. H. Kruse, H. I. Browman, K. L. Cochrane, D. Evans. G. S. Jamieson, P. A. Livingston, D. Woodby, and C. I. Zhang (Eds.), *Global Progress in Ecosystem-Based Fishery Management.* Alaska Sea Grant, University of Alaska, Fairbanks. Doi:10.4027/gpebfm. 2012.05:87–98.

Ocean Policy Task Force. 2010. Final recommendations of the Interagency Ocean Policy Task force. The White House Council on Environmental Quality, Washington, D.C. (*http://www.whitehouse .gov/files/documents/OPTF_FinalRecs.pdf*)

Park, H. W. and C. I. Zhang. 2008. A study on the ecosystem-based resource management system of self-regulatory community fisheries. *J. Kor. Soc. Fish. Tech.* 44(4):345–352 (in Korean with English abstract).

Parrish, R. H., C. S. Nelson, and A. Bakun. 1981. Transport mechanisms and reproductive success of fishes in the California Current. *Biol. Oceanogr.* 1:175–203.

Pauly, D., V. Christensen, J. Dalsgaard, R. Froese, and F. Torres. 1998. Fishing down marine food webs. *Science* 279:860–863.

Pendleton, L. H. 2010. The economic and market value of coasts and estuaries: What's at stake. Restore America's Estuaries.

PFMC. 2010. Pacific coast fishery ecosystem plan for the U.S. portion of the California Current Large Marine Ecosystem. Pacific Fishery Management Council, Portland, OR. 82p.

PFMC. 2012. Pacific coast fishery ecosystem plan for the U.S. portion of the California Current Large Marine Ecosystem. Pacific Fishery Management Council, Portland, OR. 83p.

PICES. 2007. Report of WG 21 on non-indigenous aquatic species. PICES Annual Report 2007, pp. 183–191. http://www.pices.int/publications/annual_reports/Ann_Rpt_07/2007%20WG21_f.pdf.

Pikitch, E. K., C. Santora, E. A. Babcock, A. Bakun, R. Bonfil, D. O. Conover, P. Dayton, P. Doukakis, D. Fluharty, B. Heneman, E. D. Houde, J. Link, P. Livingston, M. Mangel, M. K. McAllister, J. Pope, and K. J. Sainsbury. 2004. Ecosystem-based fishery management. *Science* 305:346–347.

Pinsky, M. L. and M. J. Fogarty 2012. Lagged social-ecological responses to climate and range shifts in fisheries. *Climatic Change* 115:883-891.

POC. 2003. America's living ocean: charting a course for sea change. A report to the nation. Pew Oceans Commission, Pew Trusts, Washington, DC.

Redfern, J. V., M. F. McKenna, T. J. Moore, J. Calambokidis, M. L. DeAngelis, E. A. Becker, J. Barlow, K. A. Forney, P. C. Fiedler, and S. J. Chivers. 2013. Assessing the risk of ships striking large whales in marine spatial planning. *Conservation Biology* 27:292–302.

Rodgers, M. and C. Olmsted. 2007. Engineering and regulatory challenges facing the development of viable offshore wind projects. *Mar. Tech. Soc. J.* 42:44–50.

Rosenberg, A. A. and S. H. Claesson. 2008. Stellwagen Bank Marine Historical Ecology Final Report. Gulf of Maine Cod Project, Univ. New Hampshire. Durham, N.H. 132 pp.

Rosenberg, A. A. and P. Sandifer. 2009. What do managers need? In K. McLeod and H. Leslie (eds.), *Ecosystem-Based Management for the Oceans.* Island Press, Washington, 13–30.

Rosenberg A. A., M. L. Mooney-Seus, I. Kiessling, C. B. Mogensen, R. O'Boyle, and J. Peacey. 2009. Lessons from National-Level Implementation Across the World. In: K. McLeod, H. Leslie (eds.), *Ecosystem-Based Management for the Oceans.* Washington DC: Island Press. 294–313.

Sarmiento, J. L., T. M. C. Hughes, R. J. Stouffer, and S. Manabe. 1998. Simulated response of the ocean carbon cycle to anthropogenic climate warming. *Nature* 393:245-249.

Samhouri, J. F. and P. S. Levin. 2012. Linking land- and sea-based activities to risk in coastal ecosystems. *Biological Conservation* 145:118–129.

Schwarzbach, H., S.-G. Kim, S.-W. Lee, J.-h. Lee, J. Hong, and Y.-J. Lee. 2008. A Study on Marine Tourism Policies to Upgrade the Marina and Pleasure Boat Industry of Korea: Relevance of the U.S. Experience to Korea. Korea Maritime Institute, 2008 Korea-America Joint Marine Policy Research Center, Cooperative Research 01–01, 157 p.

Seo, J.-Y., S.-H. Park, J.-H. Lee, and J.-W. Choi. 2012. Structural changes in macrozoobenthic communities due to summer hypoxia in Gamak Bay, Korea. *Ocean Science Journal.* 47:27–40.

Shim, W. J., S. H. Kahng, S. H. Hong, N. S. Kim, S. K. Kim, and J. H. Shim. 2000. Imposex in the rock shell, Thais clavigera, as evidence of organotin contamination in the marine environment of Korea. *Mar Environ Res.* 49:435–451.

Singh, S. N. and R. D. Tripathi. 2006. *Environmental bioremediation technologies.* Springer.

Smith, A. D. M., C. J. Brown, C. M. Bulman, E. A. Fulton, P. Johnson, I. C. Kaplan, H. Lozano-Montes, et al. 2011. Impacts of Fishing Low-Trophic Level Species on Marine Ecosystems. *Science* 333:1147–1150.

Sorte, C. J. B., S. L. Williams, and J. T. Carlton. 2010. Marine range shifts and species introductions: comparative spread rates and community impacts. *Global Ecology and Biogeography* 19:303–316.

Stedman, S. and T. E. Dahl. 2008. Status and trends of wetlands in the coastal watersheds of the eastern United States. NOAA, NMFS and U.S. Dept. Int. 32 pp.

Sun, S., F. Wang, C. Li, S. Qin, M. Zhou, L. Ding, S. Pang, D. Duan, G. Wang, B. Yin, R. Yu, P. Jiang, Z. Liu, G. Zhang, X. Fei, and M. Zhou. 2008. Emerging challenges: Massive green algae blooms in the Yellow Sea. *Nature Proc.* dl:10101/npre.2008.2266.1.

Tallis H., P. S. Levin, D. L. Fluharty, and B. S. Halpern. 2010. The many faces of ecosystem-based management: Making the process work today in real places. *Marine Policy* 34:340–348.

Townsend, D. W., A. C. Thomas, L. M. Mayer, M. Thomas, and J. Quinlan. 2006. Oceanography of the Northwest Atlantic Continental Shelf. pp. 119–168. In: A. R. Robinson, and K. H. Brink (eds.), *The Sea,* Volume 14, Harvard University Press.

Turnipseed, M., L. B. Crowder, R. D. Sagarin, and S. E. Roady. 2009. Legal bedrock for rebuilding America's ocean ecosystems. *Science* 324:183–184.

USCOP. 2004. An ocean blueprint for the twenty-first century. United States Commission on Ocean Policy, Washington, DC.

U.S. Department of Agriculture (USDA), National Agricultural Statistics Service (NASS). 2005. Census of Aquaculture. USDA 2002 Census of Agriculture.

USEPA. 2012. National Coastal Condition Report IV. EPA-842-R-10–003. USEPA Off. Res. Dev./Off. Water. Washington, DC. 20460.

Vaschenko, M. A., I. G. Syasina, P. M. Zhadan, and L. A. Medvedeva. 1997. Reproductive function state of the scallop Mizuhopecten yessoensis Jay from polluted areas of Peter the Great Bay, Sea of Japan. *Hydrobiologia.* 352:231–240.

Ware, D. M. and R. E. Thomson. 2005. Bottom-up ecosystem trophic dynamics determine fish production in the Northeast Pacific. *Science* 308:1280–1284.

White, C., B. S. Halpern, and C. V. Kappel. 2012. Ecosystem service tradeoff analysis reveals the value of marine spatial planning for multiple ocean uses. *Proceedings of the National Academy of Sciences USA* 109:4696–4701.

Whitney, F. A., H. J. Freeland, and M. Robert. 2007. Persistently declining oxygen levels in the interior waters of the eastern Subarctic Pacific. *Progress in Oceanography* 75:179–199.

Wiley, D. N., M. A. Thompson, and R. Merrick. 2006. Realigning the Boston Traffic Separation Scheme to Reduce the Risk of Ship Strike to Right and Other Baleen Whales. In K. Börner and E. F. Hardy (eds.), *Science Maps for Science Policy Makers, Places and Spaces: Mapping Science.*

WSSD. 2002. World Summit on Sustainable Development Plan of Implementation. Johannesburg, South Africa. 62 p.

Zhang, C. I. and S. K. Lee. 2004. Trophic levels and fishing intensities in Korean marine ecosystems. *J. Korean. Soc. Fish. Res.* 6(2):140–152 (In Korean).

Zhang, C. I. and S. Kim. 2011. A pragmatic approach for ecosystem-based fisheries assessment and management: a Korean marine ranch ecosystem. In A. Belgrano and C. Fowler (eds.), *Ecosystem-based Management for Marine Fisheries.* Cambridge University Press, Cambridge, UK. 153–180.

Zhang, C. I., A. B. Hollowed, J. B. Lee, and D. H. Kim. 2011. An IFRAME approach for assessing impacts of climate change on fisheries. *ICES J. Mar. Sci.* 68(6):1318–1328.

Zhang, C. I., S. Kim, D. Gunderson, R. Marasco, J. B. Lee, H. W. Park, and J. H. Lee. 2009. An ecosystem-based fisheries assessment approach for Korean fisheries. *Fish. Res.* 100:26–41.

Zhang, C. I., H. W. Park, J. H. Lim, H. C. Kwon, and D. H. Kim. 2010. A study on indicators and reference points for the ecosystem-based resource assessment. *J. Kor. Soc. Fish. Tech.* 46(1):32–49 (in Korean).

Zhang, C. I., Y.-J. Kwon, J. H. Lim, and Y.-I. Seo. 2012. Fisheries assessment and management of Korean coastal ecosystems. Prepared to submit to *Ocean and Coastal Management.* 24 p.

Chapter 12. Ecosystem-Based Management of Coral Reefs and Interconnected Nearshore Tropical Habitats

JACOB P. KRITZER

Environmental Defense Fund

CHRISTINA C. HICKS

James Cook University

BRUCE D. MAPSTONE

Commonwealth Science and Industrial Research Organization
Marine & Atmospheric Research

FABIAN PINA-AMARGÓS

Center for Coastal Ecosystems Research, Cuba

PETER F. SALE

United Nations University
Institute for Water, Environment, and Health

Contents

1. Introduction

The shallow, nearshore and intertidal areas of tropical regions across the globe harbor some of the most productive and diverse, but also most heavily impacted, marine ecosystems in the world (Wilkinson 2008; De'ath et al. 2012). By virtue of their proximity, coral reefs are inextricably linked to mangrove forests and seagrass beds, in terms of both facing common threats and supporting common species. These interconnected ecosystems provide a wide range of goods and services to human communities, often in otherwise impoverished areas. Indeed, developing

The Sea, Volume 16, edited by Michael J. Fogarty and James J. McCarthy
ISBN 978-0-674-07270-1

nations are disproportionately distributed within tropical latitudes, in part because the hot climate at low latitudes is conducive to drought, famine, disease and natural disasters (Harrison 1979). Yet, coral reefs and other coastal marine ecosystems in the tropics provide rich reservoirs of natural resources. However, terrestrial and climate impacts, and scientific and governance limitations, have resulted in severe degradation of many nearshore tropical ecosystems, and continued threats to many more. In fact, coral reefs might become the first marine ecosystem to be driven to extinction by anthropogenic activity, perhaps within the next century (Sale 2011).

This chapter will primarily focus on ecosystem-based management of coral reefs, as they are the most diverse marine ecosystem in the tropics capable of producing perhaps the widest array of goods and services (at least on a per-unit-area basis), yet are also the most vulnerable. By necessity we will also examine ecosystem-based management of adjacent seagrass beds and especially mangroves, given the ecological and, at least in some cases, governance linkages between these habitats and coral reefs.

2. Ecological Overview

2.1 An improbable equilibrium

Coral reefs are wondrously improbable things, wondrous because of their sheer biological diversity and intricate inter-relationships among resident species, and improbable because the conditions necessary for their existence are both precise and rare. Coral reefs are always living on the edge, and in past geological periods they have been absent for periods of tens of millions of years.

Coral reefs are the only totally biogenic ecosystem, although deep-sea vents and rainforests approach this condition. With the exception of basaltic intrusions in locations such as Hawaii, terrigenous sediments in some reefs near continental coastlines, and exposed underlying bedrock in subtropical locations where coral growth is scant, everything on a coral reef is produced by living organisms. Many Pacific reefs perch on top of thousands of meters of biogenic limestone deposited on a slowly subsiding base over thousands of years. Coral reef formation is tightly constrained by the narrow tolerances of corals and other calcifiers such as calcareous reef-building algae. While there are differences among species, corals generally require a firm substratum, depths less than 50m (unless the water is exceptionally clear), oceanic salinity and pH, temperatures between about 18°C and 35°C, and minimal pollution or sediment load (Wells 1957; Brown 1997).

Every reef is a dynamic equilibrium between accretive and erosional forces. Accretive forces are the active growth of corals, coralline algae, and other taxa that form skeletal structures out of calcium carbonate. These are augmented by physical forces building sand cays and rubble banks, and by chemical processes leading to production of 'beach rock', or consolidated carbonate sands. The erosional forces include waves, tidal scour, and storm damage, and the actions of a diverse group of bioeroders that dissolve, burrow into, or scrape the surfaces of live and dead coral formations (Glynn 1997).

The dynamic equilibrium between accretion and erosion has played out in a world of more or less stable sea level for the last 5,000 years. Prior to that, reefs experienced rising sea levels back to 20,000 years BP. Eustatic forces can cause local sea level to change, and reefs can be drowned or desiccated by these changes, as well as by changes in global sea level. If sea level rises or reefs subside more quickly than they can grow, calcification slows or may stop completely, in which case the calcifying species die and the reef exists as a deep-water, flat-topped guyot. If sea level falls or the substratum rises, reefs are elevated above low tide level and exposed corals die. We are now entering a new period of global sea level rise, and coral reefs will need to maintain a rate of calcification sufficient to counter erosional forces and permit progressive upward growth to keep the reef in shallow water.

2.2 *Geological history of coral reefs*

2.2a *Rugose, tabulate and early scleractinian reefs*

As sketched by Veron (1986), coral reefs first appear in the Ordovician, about 490 x 10^6 years BP, with the evolution of rugose, and subsequently tabulate corals. Reefs dominated by tabulate corals persisted through the Devonian, but dwindled and disappeared by the close of the Permian, about 245 x 10^6 years BP. Reef-building did not reoccur until the mid-Triassic, about 230 x 10^6 years BP, with the appearance of early scleractinian corals. Reef-building prospered and Scleractinia evolved through the Triassic and into the Jurassic so that by mid-Jurassic, 175 x 10^6 years BP, most modern orders of corals were present. In the early Cretaceous reef-building was uncommon, but it increased until reef development was as strong by the end of the Cretaceous, 65 x 10^6 years BP, as it is today. The end-Cretaceous mass extinction eliminated nearly all reefs, which were globally absent for about 10 x 10^6 years. Reef-building commenced again and the approximately modern coral fauna was present by the Eocene, 54 x 10^6 years BP. There were further fluctuations in abundance until the widespread reef-building episode coinciding with the end-Pleistocene sea-level rise commencing 10–20 x 10^3 years BP (Veron et al. 1996).

2.2b *History of modern coral reefs*

Stationary sea levels of the past 5,000 years (Tornqvist and Hijma 2012) have permitted a considerable diversification of habitats in shallow waters as reefs at sea level are forced to grow laterally by some minimal accretion on the seaward face and deposition of eroded sediments on the leeward. Some of these leeward sediments become secondarily cemented and colonized by coral. The result is a considerable variation in habitat type from the leeward to windward edge of a reef and distinctively different windward and leeward slopes. At times when sea level is rising rapidly, reef growth is primarily vertical and the diversity of habitats may be less. However, a variety of differing habitat types represented over distances of 1 km or less is a feature of all reefs. The considerable drop in sea level during the Pleistocene means that most reefs are relatively young structures that have grown on top of a substratum that was dry land as little as 10,000 years BP. That substratum may be limestone of earlier reefal origin, but it can also be limestone of other origin or rock of other type.

2.3 *Intricate linkages in a complex ecological structure*

2.3a *High biodiversity*

An obvious ecological feature of coral reefs is high biodiversity. A widely cited estimate is that coral reefs currently support 25% of all marine species (Knowlton et al. 2010). A number of studies have considered how that high biodiversity is accommodated. Considerable emphasis was initially placed on the evident segregation of similar species among habitats, such that between-habitat diversity is particularly high. This partitioning can be over-emphasized, however. Even on relatively less diverse Caribbean reefs there are abundant examples of groups of ecologically similar taxa occurring together in a single habitat but not occupying other habitats that are in turn occupied by other groups of species. Moreover, high temporal variability of assemblage structure within habitats suggests more dynamic conditions than would generate strong niche partitioning (Fig. 12.1). The simplest explanation for their high diversity may be the correct one: species have evolved over a relatively long time in an environment not interrupted by glaciation, and with a substantial area of habitat to be occupied (Sale 1991).

2.3b *Sedentary organisms*

Most reef species are sedentary, if not sessile, due to high habitat diversity and typically high habitat specificity. Despite dividing almost equally into species that spawn in mid-water with no egg care and species that care for the eggs until hatching, only a handful of fish species extend care beyond hatching. For the overwhelming majority, larvae are pelagic. The larval phase ranges from about 10 days in a few small Gobiidae and the pomacentrid genus *Amphiprion,* to about 2–3 months in the Acanthuridae and some other taxa. A typical larval duration for reef fish species is about 1 month, and the larvae develop substantial, though still not well documented, sensory and behavioral capabilities. At the end of larval life, fish settle to reef or other nursery habitat, and while many change habitat with age, the initial selection appears to be a choice, not a process of elimination of individuals that settle to the 'wrong' habitat. Numerous studies that have censused fish in reef habitats immediately following settlement makes untenable the idea that precise habitat 'preferences' are actually due to post-settlement winnowing.

Behavioral studies have revealed that most fishes move about over small home ranges within preferred habitat (Meyer et al. 2010, Marshall et al. 2011), and range size scales with fish size (Sale 1978). During post-settlement life stages, reef fishes exhibit a range of fascinating and complex social, communication and other behavioral systems. In fact, before reef fish ecologists began addressing population, metapopulation, and ecosystem-level questions, the earliest research breakthroughs were mostly in the realm of behavior and life history (reviewed by Kritzer and Sale 2006), due in large part to the clear, shallow waters typical of most coral reefs that make them highly amenable to observation and experimentation (Sale 1991).

The tendency toward habitat specificity and a sedentary nature is furthered by the patchiness of reef systems. This patchiness is revealed on scales from meters to kilometers, and the result is that even over contiguous habitat, individuals are quite restricted to a site. Species are then represented as numerous, small populations within patches of suitable habitat. Indeed, reef fish species are best viewed as metapopulations, with dynamics both at the local population and metapopulation

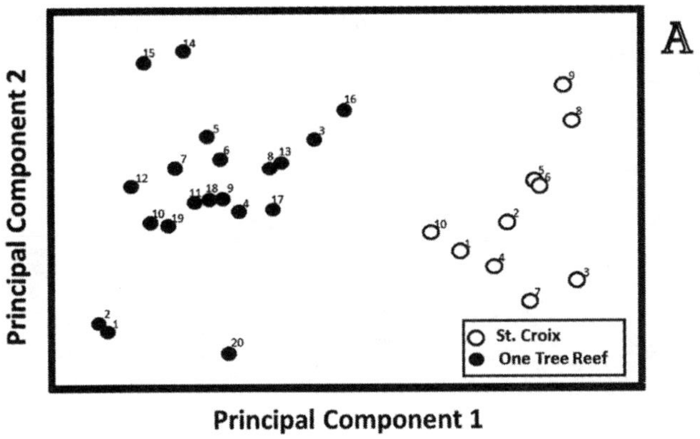

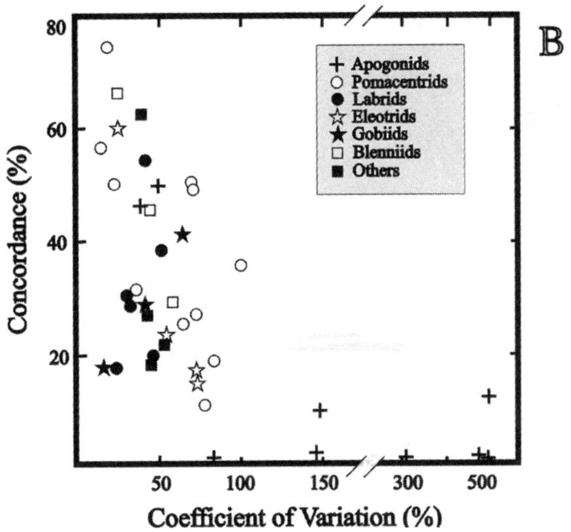

Fig. 12.1 – The high temporal variability in structure of reef fish assemblages is illustrated in two different ways. Principal components analysis displays changing structure of assemblages across 10 successive censuses over 5 years at St. Croix, U.S.V.I., and across 20 successive censuses over 10 years at One Tree Reef, Australia (A). Data are numbers of fish in each of 11 trophic guilds. There is a clear separation (different structure) between the two sites, but equally substantial changes through time within each site. Underlying reasons for the varying structure using the same One Tree Reef data are revealed by a plot including the 40 most common species across the 20 censuses (B). Each species is located in a space defined by concordance in pattern of distribution across patch reefs as measured by Kendall's W, and the extent to which they varied in abundance across censuses as measured by the coefficient of variation. Few species are located in the upper left quadrant representing those that are consistently present and abundant. (A redrawn from Chittaro and Sale 2003; B redrawn from Sale 1991)

scale functioning accordingly (Kritzer and Sale 2004, 2006; Fig. 12.2). Such patchiness means that localized impacts can rapidly deplete species that remain abundant only kilometers away.

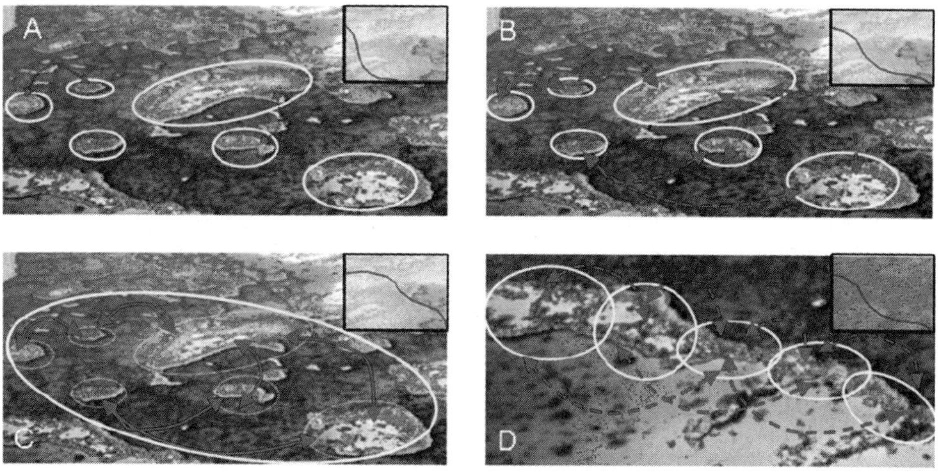

Fig. 12.2 – (See color insert) Each panel shows a patchy array of reef habitat, some of which is occupied by a reef fish species (ovals = local aggregations of fish). Dispersal (chiefly by larvae) among sites is shown by arrows, graded to show slight (A), moderate (B, D) or extensive exchange (C). Mean scale of dispersal is shown as a graph of proportion of larvae (y axis) against distance from source (x axis) in the upper right corner of each panel—mean dispersal distance is least in A, intermediate and identical in B and D, and greatest in C. Cases A, B, and C differ only in the scale of dispersal relative to the scale of patchiness of habitat, yet yield essentially independent local populations (A), a metapopulation (B) in which local populations are sufficiently connected by dispersal for some interaction, and a single, but subdivided, population (C), occupying a number of patches of habitat. Case D is typical of regions where coral reef habitat is more contiguous, yet the spatially explicit mating pattern and scale of larval dispersal still provide a functional metapopulation even though patch structure is primarily an analytical construct. (Reprinted from Sale and Kritzer 2008)

2.3c Efficient nutrient capture and recycling

Another characteristic of coral reefs is efficient nutrient cycling. Most reefs occur in waters low in nutrients, but achieve exceptionally high levels of productivity nonetheless. Nutrients are efficiently captured and recycled among organisms on the reef, a process initially described by Odum and Odum (1955), and refined by much subsequent work. Nutrients arrive in the form of plankton that are filtered by fish, corals, sponges, and other planktivores. Some also arrive as dissolved organic carbon (DOC) shed from nearby seagrass beds (Ziegler and Brenner 1999), or as feces of fish that fed over seagrass beds at night, returning to the reef to rest during the day (Meyer et al. 1983). Coral reef food webs are rich with pathways. The mucus of corals and other species are stripped off by water movement and support a rich microbial flora that in turn is fed on by other planktivores or filter-feeders, or accumulates on reef sediments (D'Elia and Wiebe 1990). Feces of fish swimming above the reef are typically consumed multiple times before reaching the substratum (Robertson 1982). Microbial flora contained in reef sediments, distributed as a film over non-living rocky surfaces, or living as endosymbionts of many reef species is rich in cyanobacteria that are highly effective nitrogen fixers (Larkum et al. 1988; D'Elia and Wiebe 1990; Fiore et al. 2010). The result is that water streaming off the leeward side of a reef is notably deficient in the plankton and nutrients found on the seaward side, while primary production on the reef is the highest for a natural ecosystem and rivals that of intensively farmed crops.

2.3d Linkages to other systems

In addition to linkages with the pelagic realm through larval dispersal, coral reefs are linked to other coastal ecosystems. The most important links are with mangrove forests and back-reef lagoons supporting seagrass beds, sponge habitats and sandy, shallow substrata. All of these environments serve as important nursery habitats for reef species, while seagrass beds also serve as important foraging habitat (McFarland et al. 1985). The trophic links between seagrass beds and coral reefs are mediated by fish families including grunts (Haemulidae), goatfishes (Mullidae), emperors (Lethrinidae), and cardinalfishes (Apogonidae). Reef corals have been shown to benefit both from DOC released by seagrasses (Ziegler and Brenner 1999) and from transfer of nutrients by fishes (Meyer et al. 1983).

The ontogenetic links well demonstrated by fish species are also demonstrated by *Panulirus* lobsters, the most important fishery species on coral reefs. Depending on the species, the larval stages settle inshore into precise microhabitats within mangroves, seagrass beds, sponge beds, or other back-reef habitat. They spend days to months in these habitats, perhaps moving to other non-reef habitats as they grow, until they move out to the reef late in juvenile life. On the reef they may occupy a second juvenile habitat before moving into typical adult habitat (Butler et al. 1997). Among fishes, commercially important families such as the groupers (Serranidae), snappers (Lutjanidae), emperors (Lethrinidae) and grunts (Haemulidae) show such behavior, as do numerous families of smaller species.

2.3e Ecological resilience

A final ecological feature worthy of mention is the high resilience of coral reefs not degraded by human activities. High diversity, rich trophic pathways, and intricate relationships among species and habitats that ensure effective transfer of energy and nutrients together ensure that a coral reef is capable of withstanding impacts of storms, outbreaks of disease and pest species such as the crown-of-thorns starfish, *Acanthaster* sp., or fluctuations in recruitment success due to variable larval survival and delivery. Recovery tends to be rapid when the reef is not degraded (Hughes et al. 2010), and the frequency of impacts is not excessive relative to patterns over its evolutionary history (De'ath et al. 2012).

3. Ecosystem Services

Coral reefs are distributed among a diverse array of socio-economic and governance contexts, in which opportunities for food and income from terrestrial resources vary just as widely. Consequently, overall reliance upon reef resources and the specific needs and uses are difficult to generalize. However, coral reefs have several attributes discussed above that promote economic importance regardless of location and, despite variability in their value among locales, nowhere is their importance negligible. These attributes include proximity to shore and shallow depth, which together promote easy access. Structurally complex habitat creates an effective coastal barrier and abundant living space. Those habit attributes, combined with efficient capture and conversion of nutrients, generate biomass and diversity that provide cultural, recreational and food resources. However, their nearshore location also creates conflicts with ecosystem services that rely more on the space occupied by reefs, seagrass beds and mangrove forests, rather than their ecological attributes.

This section aims to convey the spectrum of services provided by coral reefs and related nearshore habitats, and provide examples of the scales, importance and challenges associated with those services, but will not attempt a comprehensive review of all services in all contexts (for an earlier, and excellent, complementary review, see Moberg and Folke 1999). The focus is on services that can more readily be measured in quantifiable ecological (e.g., biomass, biodiversity) and economic (e.g., revenue, employment) terms. Therefore cultural, aesthetic and existence values are not covered in depth, although these ecosystems certainly provide those values in varying ways across the globe, and their importance should not be disregarded.

3.1 Shipping

At least since Captain James Cook's *Endeavour* ran aground on Endeavour Reef within the Great Barrier Reef complex in 1770, shipping lanes have generally tried to avoid coral reefs. Unlike most of the high seas, coral reefs occupy shallow water and present a significant navigational hazard. Shipping lanes often run parallel with reef systems, and by necessity must sometimes pass through gaps in reef architecture to reach the open ocean. In many places, development and maintenance of infrastructure needed for cargo ships has meant direct removal, displacement or persistent impacts upon coral, seagrass and mangrove habitats, to the point where these habitats have been severely diminished or lost outright. Therefore, shipping is generally not a key service provided by contemporary coral reefs, seagrass beds and mangroves, and in fact is largely mutually incompatible. One important exception is the cruise ship industry, often a significant component of the tourism sector that is considered below.

3.2 Mining and mineral extraction

Industrial-scale mining of minerals and fuels beneath coral reefs has not been extensive, despite concerns about potentially significant reserves within ancient limestone motivating exploration and extraction (Moberg and Folke 1999). However, mining of emergent and near-surface limestone (Fig. 12.3a) as a building material is practiced in multiple locations such as the Maldives, Tanzania, Sri Lanka, India, Indonesia and the Philippines, where alternatives for construction are limited (Dulvy et al. 1995). Limestone can be mined from fossil deposits that form the foundation of many tropical islands, as well as the skeletal foundation underlying living corals. The presence of fossil corals in many colonial structures across the Caribbean illustrates that this practice has persisted for at least several centuries, if not longer (Fig. 12.3b).

Fig. 12.3 – (See color insert) Fossil corals embedded in a matrix of compressed biogenic sand along the coast of Curacao, Netherlands Antilles (A). The bell tower of the Cathedral of Havana, Cuba (B), one of many colonial structures in Old Havana and around the Caribbean constructed of coral limestone. (Photos by Jake Kritzer)

Fossil reef limestone provides the primary architectural foundation for most coral reefs, with living tissue comprising only a very thin surface layer. Therefore, "dead" reef is structurally important, yet takes decades or centuries to accumulate. Furthermore, as sea-level rise drives reefs to migrate landward as well as toward the surface, limestone foundation in shallow areas currently lacking extensive reef development might become more important. In fact, even though structural complexity in shallow back reef areas is generally less than high rugosity fore reef areas, interstitial space in otherwise flat limestone "pavement" provides important shelter for small individuals. Therefore, limestone mining is a service that needs to be managed carefully to promote ecological function and resilience.

3.3 Renewable energy

The most extensive source of renewable offshore energy exploited to date is wind (Inger et al. 2009). Most wind turbines are sited on soft sediments because foundations can be anchored more easily than on consolidated substrates, which require drilling, and the perceived ecological impacts are less (Snyder and Kaiser 2009). Therefore, construction of wind turbines directly on coral reefs seems unlikely, although construction on adjacent soft sediments and seagrass beds that interact with reefs ecologically might be possible. However, wind turbines are also more commonly constructed in developed countries outside the tropical zone, and further offshore where wind fields are greater and conflicts with coastal users are less (Snyder and Kaiser 2009). Since other forms of offshore renewable energy (wave, tidal) are much less developed (Inger et al. 2009), industrial-scale renewable energy is not likely to be a significant service provided by nearshore tropical ecosystems in the near future.

Coastal ecosystems in the tropics do provide renewable energy at an artisanal or subsistence scale through firewood collected from mangrove forests. Like limestone mining, this service is generally utilized to meet local needs where other resources are lacking, and harvest is technologically simple. However, the aggregate importance of this service is significant with mangroves providing wood for fuel and construction, and food and medicinal products from fruits and leaves across the globe (e.g., Dahdouh-Guebas et al. 2000). Timber production is sustainable if managed properly, but harvest in many places is done at rates resulting in outright loss of mangrove forests. Valiela et al. (2001) estimate that one-quarter of the 35% loss of mangroves over the last two decades of the 20th century, or 9% of the overall loss, was due to timber harvest. Because mangroves are important nursery habitats for reef-associated species, and provide sediment trapping and nutrient processing services, timber harvest has the potential to compromise other valuable ecosystem services.

3.4 Communications

By virtue of their nearshore location, coral reefs, mangroves and seagrass beds seem likely locations for laying of undersea telecommunications and energy cables. However, this use has received little discussion in the scientific and management communities relative to others. This may be due in part to the fact that coral reefs are predominantly located in the developing world, where technology lags developed nations. As tropical nations increase coastal infrastructure and standards of living, communications might become a service extracted more frequently from these ecosystems. This service has been provided more extensively by coral reefs in South Florida, U.S.A., one of the developed nations with considerable coral reef area, and analysis and debate about the costs and benefits has ensued (Sultzman et al. 2002; Spurgeon 2003). The Florida experience might foretell similar uses and debates to come in currently less developed areas, although increasing importance of wireless communications might reduce the frequency of such conflicts.

3.5 Tourism

Worldwide, tourism is estimated to be the most valuable service provided by coral reefs in terms of revenue generated. This high value is due to the fact that reefs are readily accessible from shore, in shallow water, often clear and sheltered, have diverse and complex habitats, and harbor abundant and diverse fish and invertebrate fauna, not to mention marine mammals, reptiles and seabirds. Costanza et al. (1997) estimate this value to be approximately US$187 billion, or around half of the total value of coral reefs. However, their methodology might over-extrapolate location-specific values resulting in a gross overestimate, especially in light of the considerable variability in tourism value across the globe (Dixon 1998; Brander et al. 2007). Cesar et al. (2003) estimated that coral reefs could generate US$9.6 billion in global revenue annually if managed sustainably, a more modest but not inconsequential figure.

Notably, tourism accounted for approximately one-half of the overall value of coral reefs estimated by Costanza et al. (1997) and one-third of the value estimated by Cesar et al. (2003). So, the relative importance of tourism is comparable by either approach, and it is the most economically important service provided by coral reefs. Using the analysis of Costanza et al. (1997) for a relative comparison if not absolute estimate, the direct tourism value of mangroves and seagrass beds is much less than coral reefs. However, the importance of these habitats in supporting coral reef fish communities makes them an important contributor to the high value of reef-based tourism that might not be captured.

The nature and importance of reef-based tourism varies substantially across the globe. Scuba diving, snorkeling and forms of eco-tourism are by far the most common activities of visitors to coral reefs. The extent to which the tourism industry is developed in a given locale is determined by a combination of factors including inherent natural attributes (water clarity, biodiversity), impacts on those attributes, remoteness (potentially an asset or barrier), transportation and lodging infrastructure, marketing, availability of other activities, and socio-economic attributes (e.g., safety, or perceptions thereof). Consequently, while tourism on coral reefs has been estimated to generate an average of US$184 per visitor day based on data from locales across the globe, the median value is only US$17 (Brander et al. 2007), indicating a distribution with a long tail toward areas with the right combination of the attributes.

Another important determinant of the realized economic benefits of tourism is the management of and reliance upon other services that affect attributes sought by visitors. Burke et al. (2011) compiled region-specific estimates of the value of tourism relative to fisheries, and found that the value of tourism exceeds that of fisheries in all regions considered by a factor of two to one hundred, with the exception of the Philippine and Indonesian region. There, food security is a much larger concern and access to lucrative Asian seafood markets is more direct, so the estimated present value of tourism is approximately one-tenth that of fishing.

3.6 *Fishing*

Fish provide an average of 15% of daily protein needs for one-third of the world's population, and more than 50% in some African countries (Smith et al. 2010). The majority of those dependent upon fish are in the coastal tropics, where fishing helps meet income and food security needs. The potential aggregate economic value of fishing on the world's coral reefs under sustainable management is estimated to be US$5.7 billion, just over half of the value attainable from tourism or coastal protection services (Cesar et al. 2003). Newton et al. (2007) estimate the aggregate sustainable yield from coral reef fisheries in 49 island nations to be approximately 589,000 mt, although current yields are approximately 64% higher than this level. Like tourism, the absolute and relative importance of fishing varies considerably across the globe (Burke et al. 2011). Moreover, because mangroves are important nurseries for snappers and groupers, species often targeted by reef fisheries, the fisheries production value of mangroves should not be overlooked. In Mexico's Gulf of California, mangroves are estimated to generate US$37,500·$ha^{-1}$·$yr^{-1}$ in fisheries production (Aburto-Oropeza et al. 2008).

Sport fishing plays a significant and growing role in many regions, especially where development of the broader tourism sector results in greater visitation by anglers. According to one estimate, the value generated by sport fishing on coral reefs approximates that of snorkeling and other forms of wildlife viewing, other than scuba diving (Brander et al. 2007). Arguably, sport fishing is a service more similar to tourism than other types of fishing because the value derives less from the product caught than from all of the expenses necessary for a successful trip (airfares, lodging, meals, etc.), and the catch is often released. However, in some locales (e.g., Belize) what is often referred to as 'recreational' fishing by local anglers is actually a form of subsistence fishing.

Commercial, subsistence or recreational fishing can be incompatible with diving, snorkeling and wildlife viewing because each use compromises the other. Non-fishermen tourists can occupy space sought by fishermen and affect the behavior of quarry so that catchability is reduced. Fishermen can likewise occupy space sought by tourists and lower abundance of species tourists hope to see either by dispersing or removing them, or removing attractors such as forage fish. Maximizing value of both services therefore requires carefully managing, and most likely separating, these activities. Also, use of reefs for fishing and tourism must not degrade habitat integrity lest these and other services be compromised and value lost.

3.7 Aquaculture

Aquaculture development on coral reefs is more limited than elsewhere. The only aquaculture production to take place directly on the primary reef architecture might be rearing of giant clams, *Tridacna gigas* (e.g., Rubec et al. 2001), whereby young clams are spawned shoreside and moved to rearing beds on or near the reef. In contrast, broodstock of commercially important finfish (e.g., groupers; Tucker 1999) or coral fragments (Delbeek 2001) are removed from the reef for grow-out in facilities on land, in deeper water off the reef, or in back reef locations. If managed well, these limited removals and subsequent culture have the potential to reduce impacts of wild harvest for food and the aquarium trade (Pomeroy et al. 2006). Propagation of coral fragments can also help reef restoration, although the scale of degradation relative to the current scale of production means this is likely to primarily be a very localized benefit at present (Hoegh-Guldberg et al. 2007). Coral propagation as a restoration strategy is almost certain to be ineffective, and is therefore not advisable, if the conditions that led to reef decline in the first place are not reversed or at least reduced (Edwards 2010).

Although *in situ* culture of marine flora and fauna in off-reef or back reef areas can reduce effects on wild populations, such operations can still compromise ecosystem function. For example, algal farming in coastal lagoons in Tanzania has been shown to affect abundance, species richness, trophic identity and community composition of nearby fish assemblages (Bergman et al. 2001). Seaweed farming in Tanzania also results in finer sediment, lower levels of organic matter in sediments, declines in sea grass and macroalgae, and reduced abundance and biomass of macrofauna, particularly suspension-feeding bivalves. These attributes are more similar to sand banks than more productive biotic habitats (Eklof et al. 2005). Additionally, some of these ecosystem changes have been linked to health problems among farmers (Frocklin et al. 2012).

The most significant aquaculture development in the coastal tropics takes place not on reefs, but rather in lagoons and the intertidal zone often occupied by mangroves. There, extensive facilities for production of shrimp and other species have been constructed around the world. Systems for integrating aquaculture facilities with mangroves have been piloted in Vietnam (Binh et al. 1997), but these systems have not been widely adopted. Much more commonly, shrimp aquaculture involves removal of mangroves, which means that this is less an ecosystem service and more a displacement of the ecosystem. This displacement is often accompanied by significant pollution of nearby reef and seagrass systems (Naylor et al. 2000). The overall scale of mangrove displacement worldwide is on the order of 25–50% or more in some regions (Ellison 2008), and more than 50% of that total loss is due to aquaculture (Valiela et al. 2001).

3.8 Coastal protection and buffering

Given their proximity to coastal settlements, coral reefs and mangroves might be expected to serve important buffering roles. These benefits are apparent, but should not be overstated since even healthy, intact habitats might do little to lower effects of the strongest events (e.g., the 2004 Indian Ocean tsunami; Alongi 2008). Still, estimates suggest that, despite the much larger literature devoted to fisheries benefits, coral reefs and mangroves collectively generate greater value through protection and buffering services. Like estimates of the value of tourism services, figures from Costanza et al. (1997) and Cesar et al. (2003) differ by at least an order of magnitude (US$170 billion versus US$9 billion, respectively). However, both studies again agree that the value of buffering and protection services approximates that of tourism. These benefits are derived both from protection from storms and day-to-day sheltering that facilitates access to the sea for fishing, tourism and other uses.

The relative difference between buffering and protection services provided by coral reefs and vegetated intertidal habitats, including mangroves, is much less than the difference in tourism value (Costanza et al. 1997), although the amount attributable to mangroves alone cannot be readily determined since they are combined with non-tropical intertidal vegetation in the analysis. Other global estimates of the value of protection and buffering services of mangroves, akin to those for coral reefs provided by Cesar et al. (2003), are not readily available, although regional estimates are available and generally corroborate the comparative value relative to coral reefs (reviewed by Wells et al. 2006; Conservation International 2008). In fact, many estimates are not for one habitat type or the other, but rather for the combined coral reef/mangrove complex, further illustrating the importance of habitat mosaics in delivering ecosystem services.

4. Threats and Impacts

Like ecosystem services, the nature and magnitude of threats to coral reefs, mangroves and seagrasses vary considerably among regions. However, there are attributes common to all coral reefs and associated habitats making them especially vulnerable to certain threats and impacts. Many of these attributes also allow them to provide such diverse ecosystem services. The long-term evolution of reef systems that has generated such high ecological complexity and diversity also makes them vulnerable to rapid and irregular changes relative to those experienced over their evolution. Their shallow and nearshore location that allows easy accessibility also places these systems in direct contact with marine, coastal and land-based anthropogenic activity relative to ecosystems in deeper water and further offshore. We address a broad range of threats and impacts, but again without attempting a comprehensive review. For a thorough review of the long-term decline of a reef system and assessment of the primary drivers, see De'ath et al. (2012).

4.1 Habitat loss

Habitat loss might be the most severe impact compromising the ability of coral reefs, mangroves and seagrasses to deliver goods and services, especially the most valuable ones (tourism, fishing, coastal protection). Three-dimensional structural complexity extending into the water column, and above the sea surface in the case of mangroves and some emergent corals, absorbs and dissipates swells and surge. Structural complexity also creates abundant space for spawning, sheltering and feeding by fishes and invertebrates sought by both fishermen and tourists. Moreover, in addition to resident fauna, these habitats are an attraction in their own right. Unfortunately, all of the habitats in coastal tropical ecosystems are highly vulnerable, and therefore have experienced significant declines (Valieila et al. 2001; Hoegh-Guldberg et al. 2007; Waycott et al. 2009). Direct removal of mangrove wood for fuel, limestone for construction or coral for the aquarium trade, displacement of habitat for development, and incidental damage by ships, divers and fishermen all exacerbate habitat loss. Physiological stresses linked to land-based inputs of sediment, fertilizers, pesticides and other pollutants further enhance losses. Prospects for corals in particular might be dim in light of increasing temperatures, sea level and acidity brought on by climate change and acidification (Hoegh-Guldberg et al. 2007), although seagrasses and mangroves are far from immune to climate-induced stresses.

Shipping has interactions with and impacts upon nearshore habitats in two important ways. First, poor charts, navigational errors or simple carelessness can lead to ships running aground, like Cook's *Endeavour* more than two centuries ago. Such collisions cause localized damage to reef architecture, which can persist for many years due to slow growth rates (Dennis and Bright 1988). Second, contamination of contacted sediments by anti-foulants can inhibit coral recruitment (Negri et al. 2002) and further delay recovery. In addition to groundings, development and use of shipping infrastructure can have impacts. Most shipping infrastructure is at coastal population centers, where reefs are either naturally absent or have been lost or severely degraded by urban and industrial impacts. However, in contrast to cargo vessels, cruise ships are increasingly using more remote and pristine locations that appeal to tourists. Cruise ship anchors can cause significant localized damage (Allen 1992). Of greater concern, construction and maintenance of cruise ship terminals in otherwise less developed areas can lead to more widespread and persistent effects. These include damage to submerged and intertidal vegetation by herbicidal antifoulants (Gardinali et al. 2004), and increased turbidity and sedimentation due to dredging (Goreau et al. 2008).

Anchors, chains and lines from dive, fishing and other vessels can damage reefs, although many dive destinations now use fixed moorings. However, scuba divers themselves damage reefs through intentional or incidental contact. These impacts are each minor compared with anchor damage, but the large number of divers and other users means aggregate effects can be more significant. Areas with high diver traffic have lower coverage of live coral and higher coverage of dead coral and rubble (Hawkins et al. 1999; Tratalos and Austin 2001), although the magnitude of effects depends upon topography and other attributes (Rouphael and Inglis 1997). Fortunately, successful strategies for mitigating diver impacts exist, including limiting the number of dives at a site (Hawkins et al. 1999), briefings on responsible practices (Medio et al. 1997), and targeted intervention of high-risk user groups (Rouphael and Inglis 2001). Similarly, destructive fishing practices can significantly reduce habitat cover and complexity (e.g., Mangi and Roberts 2006), but cooperative management, education and enforcement can mitigate this impact (see 6.3b below).

4.2 Harmful algal blooms

In contrast to more enclosed systems such as estuaries and lakes, harmful blooms of planktonic algae (i.e., harmful algal blooms, or HABs) seem to be less of a concern on coral reefs, the waters of which are more open and well-mixed. Indeed, the frequency, magnitude and effects of HABs have received comparatively little attention on coral reefs (Bauman et al. 2010). However, HABs are a threat to coral reefs, and one that has been recognized since at least the 1980s, when severe dinoflagellate outbreaks caused up to 100% coral mortality in near-surface waters of Costa Rica and Panama (Guzmán et al. 1990). More recently, a 2008 dinoflagellate outbreak in the Gulf of Oman caused complete loss of two branching corals, and substantial reductions in the abundance, richness, and trophic diversity of fish communities (Bauman et al. 2010). Documenting such effects is motivating research into mechanisms underlying HABs.

A more immediate and significant threat to coral reefs are blooms of fleshy or turf macroalgae that can overrun settlement space for coral and other sessile fauna. Littler et al. (2006) developed a trophic model showing the complex interplay between top-down herbivory and bottom-up nutrient loading in driving algal dynamics on coral reefs. This interplay illustrates the potential severe synergistic effects of multiple stressors, here terrestrial run-off increasing nutrient load and loss of vegetation that processes nutrients, alongside overfishing of parrotfishes and other herbivores that reduces algal grazing. Climate-induced effects on coral reproduction, growth and recruitment can also exacerbate ecosystem shifts by creating additional imbalances between algae and corals. Mumby and Steneck (2009) provide a comprehensive overview of these complex dynamics that are increasingly becoming better understood (as they unfold and are documented, unfortunately), changing paradigms about the stability of coral reef systems.

4.3 Introduced species

Like any ecosystem, coral reefs, mangroves and seagrass beds are not immune to introduction of non-native species through accidental or intentional transport.

Like most threats, the risks and severity of introduced species vary among systems based on proximity to sources of introduction, such as shipping terminals (ballast water), shoreside aquaculture facilities raising non-native species (e.g., Silva-Oliveira et al. 2011), and residential areas where aquarium species are prevalent (Semmens et al. 2004). As high diversity ecosystems, coral reefs might have greater inherent resilience to effects of introduced species, as the reef community is evolved to cope with a greater and more variable array of residents. Of course, the impact of any introduction will depend upon the characteristics of both the intruder and the system it joins. Two well-known case studies illustrate the range of potential effects: blueline snapper *Lutjanus kasmira* in Hawaii and lionfishes *Pterois volitans* and *P. miles* in the Caribbean.

Blueline snapper were intentionally introduced through documented efforts in the late 1950s and early 1960s by the State of Hawaii to enhance local fisheries, and had spread across the archipelago by 1992 (Gaither et al. 2010). That same year, the first reported release of lionfish into the Caribbean and western Atlantic took place when Hurricane Andrew destroyed a large aquarium at a private residence in Florida, releasing six animals into Biscayne Bay (Ruiz-Carus et al. 2006). This invasion may have been enhanced by subsequent inadvertent, or at least unsanctioned, releases that were not documented. Despite the smaller scale and unintentional nature of the lionfish introduction, its spread was even more rapid, with fish present across the wider Caribbean, along the U.S. Atlantic coast and into the Gulf of Mexico by 2010 (Schofield 2010). Lionfish abundance remains comparatively low in most places it has invaded (Ruiz-Carus et al. 2006), whereas blueline snapper are one of the most abundant species on Hawaiian reefs (Friedlander et al. 2002).

Despite their abundance, Friedlander et al. (2002) detected no strong ecological effects of blueline snapper on the fish community, which perhaps reflects the system having adapted to their presence over four decades. It is too early to determine whether lionfish will have severe ecological effects, although they have already served as a vector for a non-native parasite (Ruiz-Carus et al. 2006). Furthermore, blueline snapper do not represent a fundamentally different species morphologically, behaviorally, taxonomically or toxicologically in their new home. In contrast, lionfish are completely unique among Caribbean fishes. Therefore, predation by or predation on lionfish could have more severe consequences if their abundance dramatically increases. Finally, although introduction of fishes is more visible and garners more attention, introductions of corals, algae and other benthos have been documented and, as the structural foundation for the system, might have greater effects (e.g., Silva et al. 2011).

4.4 Noise pollution

Coral reefs are incredibly noisy places. Butterflyfish emit a variety of sounds to establish territories and warn off intruders (Tricas et al. 2006). Success in conflicts between snapping shrimp depends heavily upon both producing and interpreting auditory cues (Herbholz and Schmidt 1998). The result of sound production by these and other species is a complex auditory environment, with "pop", "trumpet", "drumming" and "banging" calls emitted by multiple species with different daily, lunar, spatial, and seasonal patterns (McCauley and Cato 2000). Beyond the intraspecific interactions driven by these sounds, the collective noise seems to be important in driving larger scale population and community dynamics. Specifically, free-swimming larvae of many fishes use sound as a cue in locating a reef on which to settle (Vermeij et al. 2010). Therefore, sound is a key determinant of inter-reef connectivity and metapopulation dynamics.

Despite the importance of sound, understanding the severity of noise pollution for most marine ecosystems is limited and anecdotal (Popper and Hastings 2009). McCauley et al. (2002) used an air gun to generate high intensity sound that simulated seismic soundings, and showed severe damage to the ears of pink snapper. As their name implies, sound is important for communication in many snappers. Effects of noise pollution on marine mammals have received the most attention of any taxonomic group in the ocean (see Weilgart 2007). Marine mammals are rarely resident on coral reefs, but many interact with reefs seasonally or opportunistically, primarily for feeding. Therefore, effects of noise pollution on marine mammals can have top-down effects on reef ecosystems, whereas effects on prey species can have bottom-up effects on marine mammals.

Noise pollution in the ocean is increasing across the globe. Because sound travels so effectively through water, this can affect vast areas. However, the significance of this impact for any given reef will be dependent on its proximity to urban areas, shipping centers, energy or mineral extraction, military installations and activity, and nearshore industrial development.

4.5 Chemical spills

Pollution impacts have long been a focus of marine ecologists working on coastal habitats. Run-off of sediment, fertilizer and pesticides from agricultural activity and coastal development can substantially alter the physical and chemical profile of waters in which coral reefs, mangroves and seagrasses grow (reviewed by Rawlins et al. 1998). Research has shown that habitats are especially vulnerable in sheltered areas with lower wave energy and flushing, which increases residence time of the contaminant (Gundlach and Hayes 1978).

Beyond localized effects, ship groundings can have widespread and long-lasting effects when oil or other contaminants are dispersed. Persistent petrochemical pollution through leakage or accidents at extraction, processing and distribution facilities is a more severe problem in some locales. This is most notable in the highly enclosed Arabian (Persian) Gulf (Banat et al. 1998; El-Sammak 2001), where isolation, high temperatures and high salinity have shaped reefs with lower diversity but higher relative endemism (Coles and Tarr 1990). While they persist, petroleum products can increase mortality of vegetation and corals (Jackson et al. 1989), and have longer term impacts by compromising recruitment, growth and viability of eggs and larvae (Loya and Rinkevich 1980).

4.6 Biodiversity loss

The high diversity of coral reefs means species loss has two opposing implications. On the one hand, loss of even a small proportion of species means a greater loss of marine species globally than comparable proportional loss in other ecosystems. On the other hand, high diversity ecosystems often have high functional redundancy, and are buffered against species loss to a degree. Of course, some species are more abundant, more unique, play more important ecological roles, or have greater social, cultural or economic importance than others, so the implications depend upon which species are lost. For example, herbivorous parrotfishes play an especially important role regulating algal cover on coral reefs, particularly where sea urchin die-offs have been severe (Mumby et al. 2006). Furthermore, even high diversity systems will eventually suffer functional breakdown if species loss becomes too high.

On coral reefs, overall diversity is strongly tied to coral cover and habitat diversity. An 8-year study of coral cover and fish diversity in Papua New Guinea by Jones et al. (2004) documented declines in coral cover from 66% down to 7%, with associated decline in fish diversity of 22%. Pratchett et al. (2011) reviewed a range of studies like Jones et al. (2004), and found consistent results in many cases. Substantial decreases in fish diversity were evident whenever coral cover decreases by 60% or more. Loss of fish diversity is not directly proportional to loss of coral cover because the rate of decline will vary among species depending upon their dependence on coral. Many species derive refuge, spawning or feeding benefits from dead coral structures, even if demographic rates suffer, until severe structural degradation eventually reduces habitat complexity too much. Corallivores will be affected most quickly due to the loss of an important food source.

The drivers of geological cycles of reef loss and recovery outlined above illustrate the strong potential for climate change to significantly reduce reef biodiversity. Veron (2008) concludes that acidification is the most significant cause of mass coral extinctions. However, acidification also has synergistic effects with sea level rise and temperature regimes, including extreme highs, extreme lows, and rapid fluctuations. Importantly, Veron (2008) also notes that responses of corals to climate-driven stressors have varied through time due to the precise conditions present during each episode of extinction and recovery and the largely unknown, but undoubtedly varying, aspects of coral biology that affect susceptibility and resilience. Under ongoing climate change, modern corals might have more capacity to withstand thermal stress than once thought due to ability to alter the composition of symbiotic zooxanthellae in their tissues such that more heat-resistant strains dominate (Berkelmans and van Oppen 2006). However, if acidification remains the major factor in reef decline (Veron 2008), then such adaptations might have limited benefit.

In addition to global loss of corals, Valiela et al. (2001) highlight the fact that global loss of mangrove forest has exceeded that of coral reefs or rainforests with at least 35% of the historical area worldwide lost by the end of the 20th century. This trend will also exacerbate biodiversity loss on coral reefs (e.g. Mumby et al. 2004). Seagrass loss has been nearly as severe, with estimated 29% global loss (Waycott et al. 2009). The combined losses of mangroves, seagrasses and corals within these habitat mosaics might result in nonlinear loss of biodiversity at a more rapid rate than that due to loss of any single habitat type.

4.7 Yield foregone

Mismanagement of harvest, consequent overfishing and other impacts have resulted in tremendous declines in fish stocks and foregone yield in most other marine ecosystems. Worldwide, lost value due to unsustainable fishing in all marine ecosystems is estimated to be US$50 billion annually (World Bank 2009). Lost value attributable to fisheries dependent upon coral reefs, mangroves and seagrasses is unknown, but management effectiveness varies considerably among coral reef fisheries. Newton et al. (2007) synthesized data from coral reef fisheries from 49 island nations. They found harvest levels and resource status relative to estimates of sustainable yield ranged from considerable overfishing and even collapse, to potential for increased yields in some nations.

However, Newton et al. (2007) estimate overall yield to be 64% greater than the aggregate sustainable level. Excess yield does not yet represent yield foregone at the global scale, but these fisheries are accruing an ecological debt that will eventually result in substantially reduced harvests. This has already been observed in nations with fully collapsed fisheries, representing around 20% of those studied by Newton et al. The authors predict that offsetting this growing ecological debt would require nearly 200,000 km^2 of additional reef area by 2050, which represents a near doubling of the 255,000 km^2 of reef area worldwide estimated by Spalding and Grenfell (1997).

5. Context and Approaches for Ecosystem-Based Management of Coral Reefs

Principles, tools (science and policy) and processes for ecosystem-based management are outlined in Part I of this volume, building on an extensive literature preceding (e.g., Pikitch et al. 2004; Levin et al. 2009; Worm et al. 2009; Thrush and Dayton 2010). Most proposals for the practice of ecosystem-based management call for attention to a range of ecosystem components, including species, benthic habitats, oceanography, climate, and pollutants, as well as the comparably complex socio-economic systems that affect, extract value from, and are themselves part of the ecosystem (see, e.g., Kritzer 2004 for an example of the consequences of failing to account for human behavior). Some or all of these components can be combined into complex models that illustrate how different uses and impacts affect other services through physical, chemical or biological linkages. This means that ecosystem-based management in its idealized form can be data-hungry and scientifically sophisticated. Moreover, governance systems required to implement ecosystem-based management are complex because of the larger number of economic sectors and stakeholders involved.

This complexity should not be prohibitive and the costs are likely worth bearing, for the alternative is often greater uncertainty in decisions due to including only a limited number of drivers while excluding those that fall within a different scientific discipline, governance entity, or both. For example, the scientific basis for managing fisheries is based primarily on estimates of demographic traits for target species, which are typically time-averaged and not linked to extra-population factors. However, when factors such as temperature (e.g., MacKenzie et al. 2008) or predation (e.g., Moustahfid et al. 2009) are taken into account, estimates of reference points, stock status or uncertainty can change drastically. Errors introduced by such omissions can lead to overexploitation and lost value, as well as increased costs stemming from scarcity, economic hardship and legal disputes. Even if the science informing management of a single service is sound, failing to address interactions among uses is likely to lead to inefficiencies, lost value, greater conflicts and higher costs. In other words, complexities and costs associated with ecosystem-based management are likely to be offset by the unplanned costs associated with failing to adopt such an approach.

Unfortunately, it is difficult to test this conclusion because ecosystem-based management has not been adopted comprehensively in any setting, and at best has been implemented in incomplete or incremental ways where it has been tried at all. All nations face the scientific and governance challenges associated with implementing ecosystem-based management to some degree. However, these challenges are often much greater in nations in the developing tropics harboring coral reefs. There, capacity for research, analysis and governance are often insufficient for the demands of ecosystem-based management in its idealized form.

The ecological attributes outlined above make it even more difficult to develop the scientific foundation called for by many ecosystem-based management proponents. The spatially heterogeneous environment (Fig. 12.2) and tendency for organisms to be sedentary and habitat specialists combine to facilitate, and perhaps demand, management at a much smaller spatial scale. The physical structure of reefs and their nearshore and near-surface location force management to be differently structured than in other marine ecosystems. This need for finer scale management does not, however, negate the need for management at larger scales, as climactic and oceanographic processes create ecological linkages and synchronies through dispersal and recruitment processes operating at larger scales (i.e., 10s to 1000s of km). Therefore, ecosystem-based management of coral reefs and associated habitats must be inherently multi-scale in nature. This is true of most marine ecosystems, but the range of scales that must be accommodated is arguably much greater for coral reefs. In addition to spatial complexity, high biodiversity and spatio-temporal variability in assemblage structure (Fig. 12.1) make modeling coral reef food webs and assessment of trade-offs much more complicated than systems with lower diversity and turnover.

In most developing countries, coral reef management usually strives to achieve two goals: sustainable catch for local consumption and trade, and conservation of the structurally and biologically diverse reef environment to deliver other services, especially coastal protection and tourism. While these goals are compatible, limited management capacity in many developing countries means that a rather complex task is not done successfully, and most coral reefs are poorly managed if at all. Management limitations are exacerbated by scientific limitations. Even in developed nations harboring coral reefs (Australia, United States), detailed understanding of more than a few target stocks is rare (see Cochrane et al., this volume; Houde et al., this volume). Understanding more complex ecosystem-level dynamics to a level that allows informed, quantitative decisions concerning trade-offs between uses is even rarer.

Nevertheless, it is eminently feasible, and quite effective, for coral reef management to be spatially structured, on scales from less than a kilometer to hundreds or even thousands of kilometers. The use of spatially explicit management strategies, in particular the establishment of marine protected areas (MPAs) of various types, is readily implemented and commonly used. MPAs can take a variety of forms, from fully no-take marine reserves offering full protection from extractive and possibly other uses as well, to those that restrict a more limited range of uses (Rice et al. this volume). Where management of coral reefs has shown success, MPAs are often a central, if not sole, strategy, and one that is in many ways inherently ecosystem-based, even if that is not explicit. This is because the underlying rationale for MPAs is to preserve a more unimpacted ecological state, including population structure, behavior, trophic structure, and habitat complexity. To be most effective, this must happen at the scale of individual MPAs and networks that capture processes and spread risk at larger scales (Sale et al. 2005). Also, because coral reefs are inextricably linked to mangroves, seagrass beds, adjacent soft sediment areas (Jones et al. 1992), and inter-reef areas containing oases of "stepping stone" structure (Kritzer and Sale 2006), spatial management should not focus solely on primary reef architecture.

Despite the importance of MPAs, management outside MPA borders cannot be neglected. Fisheries management on coral reefs is usually complicated by the fact that these are inevitably multi-gear, multi-species fisheries. These fisheries are seldom capital-intensive, can develop rapidly in new locations, and operate in a relatively unregulated way in many locations, imposing significant effort on target stocks. Unfortunately, limited capacity means that management of water quality, coastal pollution, and terrestrial run-off is frequently neglected, with pronounced consequences for reefs.

6. Case Studies

We review three case studies where management of coral reef systems has shown some measure of success and is ecosystem-based to some degree. The case studies span a range of capacity in terms of science, policy, governance, monitoring and enforcement, and all have developed systems that work within the capacity at hand. All adopt spatiallyexplicit strategies, including management at comparatively small scales. Still, although these case studies show effectiveness, all continue to confront remaining challenges in terms of emerging threats, competing uses, knowledge gaps, and implementation effectiveness.

6.1 High capacity: Great Barrier Reef

The Great Barrier Reef (GBR) sits within a well-established, though complex, governance context that is relatively well resourced and well informed by research (see Houde et al. this volume). Ecological sustainability, ecosystem-based management, and precautionary decision-making are common principles in most of the legislation affecting anthropogenic activities on the GBR. Explicit reference to these principles is relatively recent, however, reflecting amendments since the 1990s to older legislation that predated the rise of ecosystem-based management. Hence, ecosystem-based management has not been based on assessments of ecosystem responses to threats, but rather has emerged as a principle shaping the revision of management strategies in the 21st century. The combination of extensive spatial regulation of allowable uses and use-specific strategies, such as conventional fisheries management, grew out of a desire for conservation of ecosystem function in the context of multiple uses in establishment legislation in the 1970s and 1980s. This means the GBR is likely to be managed relatively conservatively in the interests of protecting ecosystem function against anthropogenic activities.

Nevertheless, challenges remain in managing activities in the coastal hinterland that affect key processes. Adequately integrating catchment management with reef management will be essential for true ecosystem-based management of the GBR.

6.1a The setting

The GBR is the Earth's largest coral reef system. It is a biologically diverse system including over 4,000 reefs, shoals, and islands across 70 bioregions spread over 15 degrees of latitude along the western margin of the Coral Sea off north-east Australia (Fig. 12.4). It extends from the south coast of Papua New Guinea at around 9°S to around 24.5°S, or about a third of the way down Australian's east coast, and from the coast to over 250km offshore at its widest point, with the outer margin sitting at the continental shelf margin for most of its length.

Most of this vast system was declared a multiple use marine park (Fig. 12.4) through the Great Barrier Reef Marine Park Act in 1975, motivated mainly by conservation concerns and the desire to prevent oil exploration and mining. The Great Barrier Reef Marine Park (GBRMP) includes 344,400 km^2 of the GBR, comprising over 3,000 reefs, shoals, and islands and all habitats from low water to just beyond the continental shelf break between 10.7°S and 24.5°S. A total of 348,000 km^2 of the GBR, including all of the GBRMP, was included in the World Heritage Register in 1981, meeting all four natural heritage criteria of geological, ecological, aesthetic, and biological diversity significance.

The GBRMP coastal hinterland is extremely sparsely populated over its northern half (to around 17°S), mainly by Australian indigenous people, and moderately populated over the remainder, mostly by widely spaced urban centers of up to 200,000 people. Coastal areas paralleling the GBR generally have higher population density than the national average of 3 people·km^{-2}, with densities upwards of 100 people·km^{-2} in the population centers (Queensland Treasury and Trade 2012). These population centers support a flourishing tourism industry attracting an estimated 6 million visitors annually. Diverse recreational and commercial fishing activities extend across the GBR from coastal towns. They also are commercial and transportation gateways to mining and grazing industries further inland, and sugar cane and fruit growing along the intervening coastal lowlands. Several major ports along the GBR coast are centers for export of ores or livestock and some have ore processing industries, usually close to the coast. The GBR lagoon is also a major shipping route, with an estimated 6,000 transits per year by vessels over 50m long.

The northern extremity beyond the Marine Park and along the eastern Torres Strait to near Papua New Guinea is populated by 19 traditional communities of Torres Strait Islanders, who have inhabited the region for thousands of years. Major human uses of this part of the GBR are commercial fishing, subsistence fishing and hunting, relatively light tourism, and as the northern extent of the alongshore shipping route.

The GBR is a well researched system. The region is home to six island research stations that been established for over three decades, spanning the length of the GBR at 9, 14, 16, 18, 23, and 23.5°S. The Australian Institute of Marine Science and James Cook University have been established in Townsville (19°S) for over 30 years, each with very strong research programs focused on the GBR covering biology, ecology, hydrodynamics, geology, water quality, fisheries, and social sci-

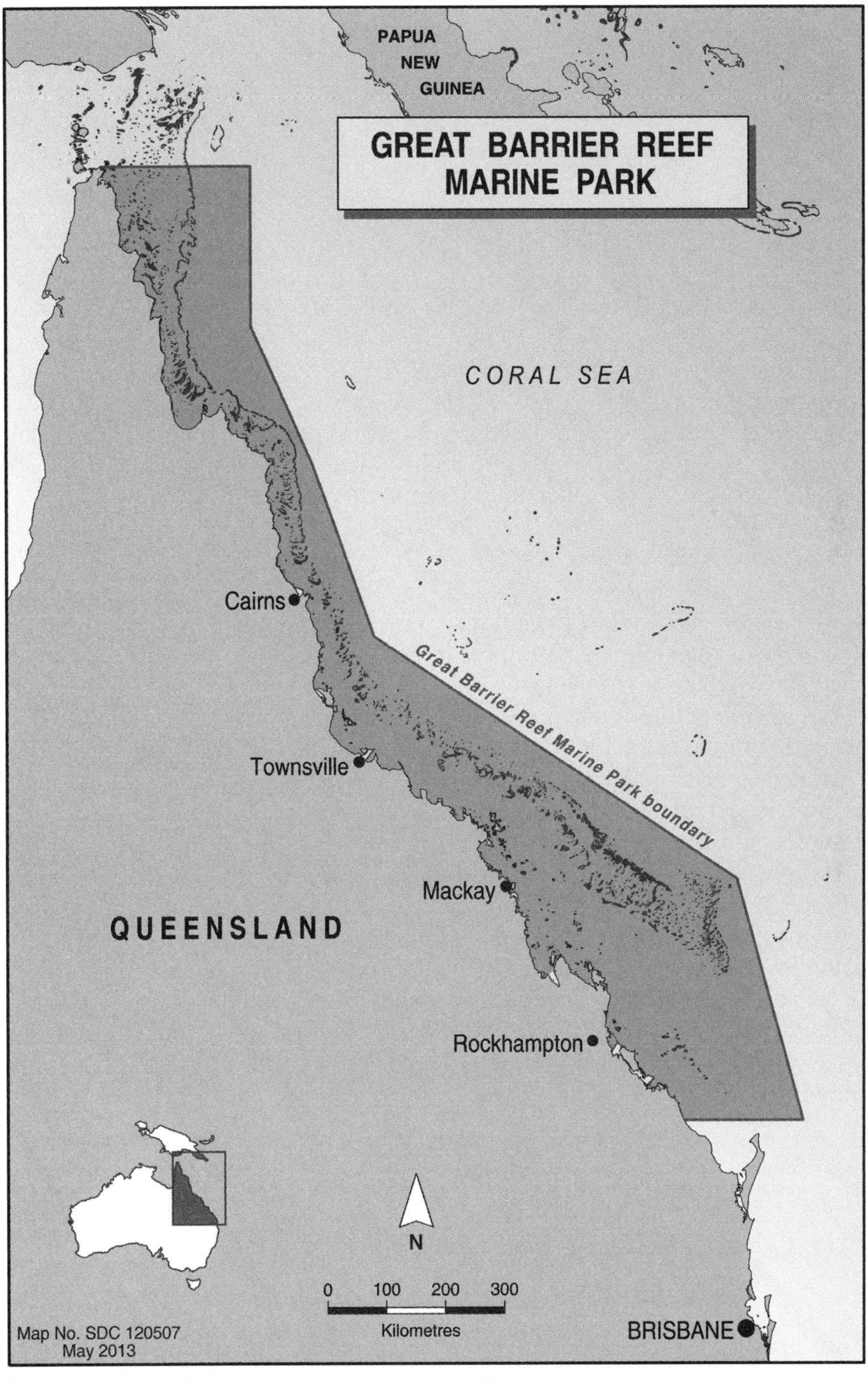

Fig. 12.4 – (See color insert) Map of the Great Barrier Reef, Australia, showing its extent, its archipelagic topography, and the boundaries of the Great Barrier Reef Marine Park. (Map courtesy of the Spatial Data Centre, Great Barrier Reef Marine Park Authority, © Commonwealth of Australia [GBRMPA] 2012)

ences. The Universities of Queensland and Sydney and the Australian Commonwealth Scientific and Industrial Research Organisation (CSIRO) also have longstanding research on the GBR. World Heritage Area status and iconic standing have meant the Australian government has invested heavily in research and many international researchers work on projects with collaborators based in Australia.

6.1b Governance

Governance of the GBR is complex and multi-jurisdictional but generally well established in legislation. The GBRMP Act (1975) declared the GBRMP and established the Great Barrier Reef Marine Park Authority (GBRMPA) to manage the park, originally with the goal, '*to provide for the protection, wise use, understanding and enjoyment of the Great Barrier Reef in perpetuity through the care and development of the Great Barrier Reef Marine Park*'. The Act was amended in 2008 to align better with ecosystem-based management principles and other legislation including the Environment Protection and Biodiversity Conservation (EPBC) Act (1999) and Australia's Oceans Policy statement (1998). The Act now requires the GBRMPA to recommend management strategies that, "*. . . provide for the long term protection and conservation of the environment, biodiversity and heritage values of the Great Barrier Reef Region*" and, "*. . . allow ecologically sustainable use of the Great Barrier Reef Region . . .*". The Act is specific in making '*reasonable use*' subject to maintaining ecological integrity and heritage values and invokes explicitly ecosystem-based management, intergenerational equity, and the precautionary principle. Other Commonwealth Acts apply to activities on the GBR, and their application to the GBRMP is expected to be consistent with the GBRMP Act and management strategies in place. In most respects the specific requirement for ecosystem-based management is relatively recent even though the original intent and subsequent management of the GBR have been consistent with principles underpinning ecosystem-based management.

The continental islands of the GBR and the coastal hinterland are governed under Queensland law and local regulations. A plethora of State legislation applies to activities in or around the GBR, including Acts to manage conservation, coastal development, water quality, fishing, ports, and boating, among others. Much of the State legislation dates from the 1990s or earlier but has been updated to give primacy to ecological sustainability, and be consistent with the GBRMP Act. For example, the objectives of the Queensland Fisheries Act (1994) when drafted were, ". . .

(a) ensuring fisheries resources are used in an ecologically sustainable way; and
(b) achieving the optimum community, economic and other benefits obtainable from fisheries resources; and
(c) ensuring access to fisheries resources is fair".

The "*primary purpose*" of the Act was amended in 2002 to, "*. . . provide for the use, conservation and enhancement of the community's fisheries resources and fish habitats in a way that seeks to—*

(a) apply and balance the principles of ecologically sustainable development; and
(b) promote ecologically sustainable development",

increasing the emphasis on ecological sustainability and introducing requirements for habitat protection and application of the precautionary principle and intergenerational equity. There is no explicit requirement at this time for ecosystem-based management.

Hence, overarching management of the GBRMP for conservation purposes vests with the GBRMPA but management of many specific activities including fishing, shipping, and coastal development sits with multiple Commonwealth or State agencies outside the explicit jurisdiction of the GBRMPA. Day-to-day management and enforcement across the GBR are shared between the GBRMPA and the State of Queensland, through relevant State government departments including those responsible for parks and wildlife, boating, and fisheries, under "The Emerald Agreement" reached between the Commonwealth and the State in 1977. Enforcement includes structured, randomized aerial surveillance and on-water vessel patrols over the entire region, various regular compliance checks, such as catch and vessel inspections, at multiple ports and boat ramps, water quality monitoring around key ports, and strong penalties for discharge or spillage of oils and other pollutants.

6.1c Conservation and biodiversity management

The GBRMPA relies on spatial regulation as its primary management strategy. Spatial management is implemented via legislated Zoning Plans that classify all areas of the GBR into zones in which various activities are either allowed or prohibited. All forms of mining or drilling are prohibited over the whole GBR. International shipping is restricted to defined routes through the GBRMP and all ships are required to have locally qualified pilots in charge during transit. Zoning for regulation of all other activities ranges from General Use Zones in which most activities, including demersal prawn trawling, are allowed to Preservation Zones, to which all access is prohibited unless for specifically permitted exceptions such as non-destructive research.

Slightly over 33% of the GBRMP is zoned Marine National Park (effectively 'look but don't touch', no extraction) or more restrictive, equivalent to IUCN Categories IA-II, whilst the remainder of the park is accessible for commercial and recreational fishing or collecting to some degree, equivalent to IUCN Categories IV or VI. These figures are the result of a comprehensive revision of zoning in 2004, informed by a detailed bio-regionalization based on a wide array of ecological, geomorphological, sedimentary, and bathymetric datasets (Fig. 12.5). That review resulted in each of the 70 bioregions, ranging from deep muddy habitats to emergent coral reefs, having at least 20% of their area zoned Marine National Park or a more protected designation, and several bioregions having more than 40% of their area similarly protected (Fernandes et al. 2005a,b). All zones are represented in most bioregions in at least two, and often more, areas that are distributed across the latitudinal and cross-shelf extent of each bioregion. No similar regime of area management exists in the Torres Strait.

The principle underlying use of zoning as the primary conservation management instrument is that protecting ecological systems from direct interference by human activities will enable the maximum amount of ecological resilience to cope with periodic perturbations. Hence, spatial management could be considered as a coarse instrument for ecosystem-based management. In most cases, spatial man-

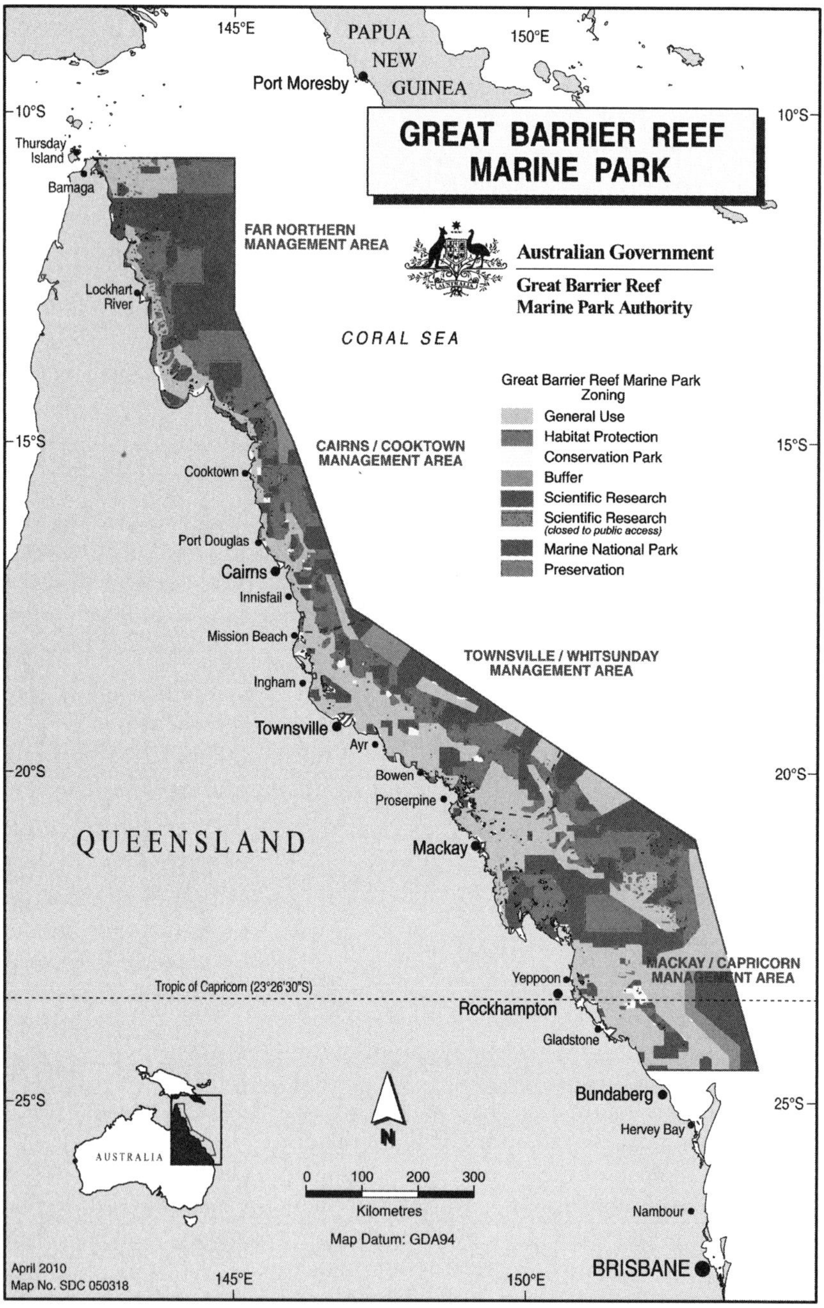

Fig. 12.5 – (See color insert) Map of management zones in the Great Barrier Reef Marine Park implemented in 2004 following extensive review of previous zoning provisions in place since the late 1980s. The revised zoning plan was based on identification of 70 reef and non-reef benthic bioregions within the Great Barrier Reef Marine Park, analyses of major uses of the park, and priciples of comprehensiveness, adequacy, and representativeness that ensured that at least 20% of each bioregion was included in multiple 'no-take zones' distributed across the geographic range of each bioregion. (Map courtesy of the Spatial Data Centre, Great Barrier Reef Marine Park Authority, © Commonwealth of Australia [GBRMPA] 2012)

agement is not accompanied by ecosystem-based assessments of performance, although many single or multi-species abundance-based assessments of the effect of zoning have been made since the 1980s (see McCook et al. 2010 for review). The zoning approach on the GBR also recognizes that explicit and direct control of many activities is beyond the jurisdiction of GBRMPA and so protecting selected areas from those activities might effectively provide refuges, or buffers, against overall adverse effects of those activities. For example, exclusion of fishing from 20–67% (depending on bioregion) of available habitat effectively means that approximately the same proportion of a harvestable stock will be protected from harvest.

Many commercial and all research activities in the GBRMP are regulated by the issue of permits or licensing under Commonwealth or Queensland State regulation in order to identify and manage explicitly the commercial and social values that are dependent on the ecological well-being of the GBR. Eco-tourism and local recreational users of the GBR arguably are those most directly benefitting from effective conservation (spatial) management and close regulation of permitted uses, but most other activities (e.g. commercial fishing) also depend on a healthy ecosystem to provide essential economic benefits that flow through the local and national communities.

Other threatening processes, however, are less likely to be moderated directly by spatial management. Ocean warming from climate change, ocean acidification, and the distribution of flood plumes from GBR hinterland catchments, for example, will not respect zoning boundaries and so will affect protected and 'exploited' (e.g., through fishing) areas of the GBRMP equally. A recent assessment of significant decline in live coral cover since the 1980s, for example, attributed the main causes of that decline to be repeated outbreaks of the coral-eating crown of thorns starfish (COTS; *Acanthaster planci*), repeated physical damage from severe tropical storms, and coral bleaching associated with rising ocean temperatures (De'ath et al. 2012). COTS outbreaks have been linked, at least theoretically, to increased nutrient loading through terrestrial runoff from agricultural areas resulting in periodically increased survival of the larval stages leading to population booms. The latter two phenomena have been linked to anthropogenic climate change, though it remains unclear whether the frequency or intensity of tropical cyclones can be attributed to climate change. It remains unknown whether a protected system will be more resilient to such effects than a system already affected by human interventions. A recent review of diverse measures of effectiveness of the GBRMP zoning plans (McCook et al. 2010) demonstrated that areas closed to access or extractive use had significantly better ecological status by a variety of criteria. That review also provided initial evidence that protected communities may be less vulnerable to pan-system perturbations such as COTS outbreaks, terrestrial runoff or ocean warming, though spatial protection from local threats is likely to provide at best an attenuation of global impacts rather than lasting resilience.

6.1d Threats

The GBRMPA in 2009 published a major review of the status, pressures, and threats of the GBR and an outlook of the future challenges for management, including challenges associated with hinterland catchment processes affecting the

GBR (Anonymous 2009). The main threats were identified as: climate change, including ocean warming and ocean acidification; catchment inputs of sediment, nutrients, and pollutants; coastal development, especially port and major tourism and marina developments; and potential overfishing and the ecological effects of demersal prawn trawling. The aforementioned study by De'ath et al. (2012) adds to that list increases in the frequency and severity of tropical cyclones and COTS outbreaks as the most important drivers of declining coral cover in particular. Cyclone patterns are linked to climate change, and COTS outbreaks might be linked to altered nutrient profiles due to watershed impacts, so the findings are consistent with those of the GBRMPA assessment.

Climate change is driven by factors largely beyond the GBR and responses tend to be based on management that is believed to facilitate resilience. Coastal development and terrestrial runoff are regulated mainly by Queensland State agencies outside the jurisdictional control of the GBRMPA. Prospects of a boom in coastal tourism and port development recently prompted a 'reactive monitoring mission' from World Heritage Commission of UNESCO to review the status of the GBR World Heritage Area (WHA) that concluded that the 'Outstanding Universal Value' of the GBR remained but also included adverse statements on prospective threats to World Heritage status, especially from coastal development and terrestrial runoff (Douvere and Badman 2012). Considerable effort is put into regulation of these activities in the interests of minimizing adverse effects, assisted by sympathetic legislation in many areas that increasingly requires explicit attention to ecological sustainability.

Fishing is perhaps the human use with the most direct effects on components of GBR ecosystems *per se.* Area closures to fishing will provide protection from direct disturbance of included habitats but the extent to which area protection from fishing allows the maintenance of unexploited stock structure will depend on movement of individuals between protected and fished zones at all stages of a species' life cycle, with highly sedentary, low dispersal species being most likely to develop unexploited stock characteristics in protected zones over time (Little et al. 2009a). The likelihood of some exchange between protected and fished zones at some life history stage, however, makes the coexistence of adequate regulations in areas open to fishing an important adjunct to spatial protection. We therefore outline below the interaction between management of one commercial fishery and conservation objectives for the GBRMP as an example of the benefits of complementary spatial and use management.

6.1e Managing fishing with conservation

Fishing within the GBRMP is managed under the Queensland Fisheries Act (1994) through a variety of input controls (e.g., licensing, effort regulation, gear restrictions), output controls (e.g., catch quotas, bag limits, size limits), or access controls (seasonal closures) (Houde et al., this volume). Commonwealth marine park legislation, including zoning plans, is superior to State legislation applicable in the GBRMP, so fishery management strategies apply only to areas where fishing is allowed. However, the GBRMP and Fishery Acts largely are sympathetic to one another in that both emphasize ecological sustainability, caution, and intergenerational equity.

The line fishery on the GBR is regulated via the 2003 Fisheries (Coral Reef Fin Fish) Management Plan under the Queensland Fishery Act. The fishery is comprised of a commercial line sector, a charter sector, and a recreational sector. The charter fishery caters mainly to paying recreational fishermen whereas the recreational fishery mainly involves people operating under their own control. Harvest by all sectors is highly diverse, with over 125 species taken, though over 80% of the catch is comprised of just two species: common coral trout (*Plectropomus leopardus*) and red throat emperor (*Lethrinus miniatus*).

Commercial and charter fisheries are regulated by limited entry licensing, but there is no recreational licensing. The commercial sector underwent a significant effort reduction in 2003–04 concurrent with the introduction of the Management Plan and revised GBRMP Zoning Plan, as did most GBR fisheries, accompanied by a major structural adjustment package funded by the Commonwealth government to offset the effects of the major increase in areas closed to fishing. Commercial fishing is subject to gear restrictions, effort controls, and annual individual transferable quotas for each of the two primary target species and collectively for other species. These quotas were introduced in 2003–04 based on historical catches and are not based on formal stock assessment and adjustment procedures, although a stock assessment has been conducted for *L. miniatus*. Management Strategy Evaluations for coral trout (Mapstone et al. 2004, 2008) and red throat emperor (Little et al. 2009b) have predicted that quotas are unlikely to be realized given current effort controls and associated fishing power, and that has been so to date, though the same analyses indicate that the underlying stocks will remain robust. Recreational and charter harvest is limited by 'in possession' bag limits for day and extended trips.

Minimum legal sizes apply to take of all species by all sectors and some species also have maximum allowable sizes. Extensive biological information exists for the main target species and for many of the minor species and minimum legal sizes generally are set conservatively at sizes likely to have allowed 100% of individuals to spawn in at least one year prior to recruitment to the fishery. Most species are long lived (>10 years), mature at young ages, and have 'table top' asymptotic growth by which they reach near-maximum size typically at less than 20% of expected longevity. These characteristics mean that minimum size limits for several secondary harvest species are set close to maximum size, effectively almost precluding harvest.

The combination of the GBRMP zoning plans that protect over 30% of the stocks from harvest and the conservative off-reserve fishery management strategies are likely to satisfy the objectives of both the GBRMP and Queensland Fishery acts, at least for line fishing, though arguably neither conservation nor fishery management strategies have been designed specifically via assessments of their viability as instruments for ecosystem-based management.

6.1f Summary

The GBR is recognized for high biodiversity value that underpins multiple social and economic values for the Australian and international community. Most of the GBR is included in a highly regulated, strongly managed marine park with well-informed and comprehensive spatial management for biodiversity conservation complemented by generally conservative activity-specific management of human uses. The combination of specific legislative requirements for ecological sustainability and intergenerational equity, stable and well-resourced multi-jurisdiction regulation, and a relatively good understanding of the biophysical system is likely to result by default in successful ecosystem management, even though the current management strategies reflect the evolution of strategies put in place before ecosystem-based management was an explicit policy driver. A future challenge will be to better integrate regulatory frameworks for the GBR and the coastal hinterland such that future ecosystem-based management planning can include comprehensive assessment of all processes ultimately determining ecological sustainability. An associated challenge is to develop robust ecosystem assessments necessary to verify that the system is operating explicitly, rather than implicitly and serendipitously, under ecosystem-based management.

6.2 Intermediate capacity: Cuba

Cuba is the most important island in the Western Hemisphere in terms of biological diversity. Its nearly 110,000 km^2 terrestrial surface area (Fig. 12.6) comprises about one-third of the land area of the Caribbean, but the nation boasts nearly four times as many plant species as Jamaica and almost 12 times as many as Puerto Rico. Recent biological surveys show that 40% of macrofauna species are endemic. More than 50% of the most important ecosystems and 55% of the endemic species of Caribbean islands are found in Cuba. Cuban marine and coastal ecosystems are particularly diverse and productive, as more than 95% of the 3,800 km shelf is fringed by coral reefs (Fig. 12.7a), seagrass beds cover half of the Cuban shelf and mangroves represent one-fifth of Cuban forests (Fig. 12.7b) (CITMA 2010). Although Cuba remains a developing nation, ambitious and evolving policy goals, and a strong and growing scientific infrastructure, are moving the nation in the direction of ecosystem-based management. Spatial management through MPAs and zoning systems is the foundation of Cuba's efforts to more effectively manage its rich ecological diversity for both conservation and human use.

6.2a Principal uses and threats

Cuban coastal and marine ecosystems support human settlements, commercial, subsistence and recreational fishing, shipping, tourism, aquaculture, oil and gas development, and other industries (e.g., sugar mills and nickel mining). Nearly four million people, or about one-third of the Cuban population, live in approximately 70 coastal cities and towns that depend on those activities as the cornerstone of residents' livelihoods. Most tourism infrastructure and activity are located in the coastal and marine areas of Cuba. Cuba's average population density is approximately 106 people·km^{-2} (World Bank 2009), with much higher densities in the six cities with 200,000 people or more. Because Cuba is a fairly narrow island (i.e., <200 km wide over its entire 1,250 km length), land-based activities anywhere on the island can have significant impact on coastal and marine systems.

Although most of Cuba's shoreline, marine waters and small islands have suffered less degradation than other Caribbean and coastal tropical locales, several threats to living and non-living resources are of concern. In its National Environmental Strategy for 2011–2015, the Cuban government identified the major threats

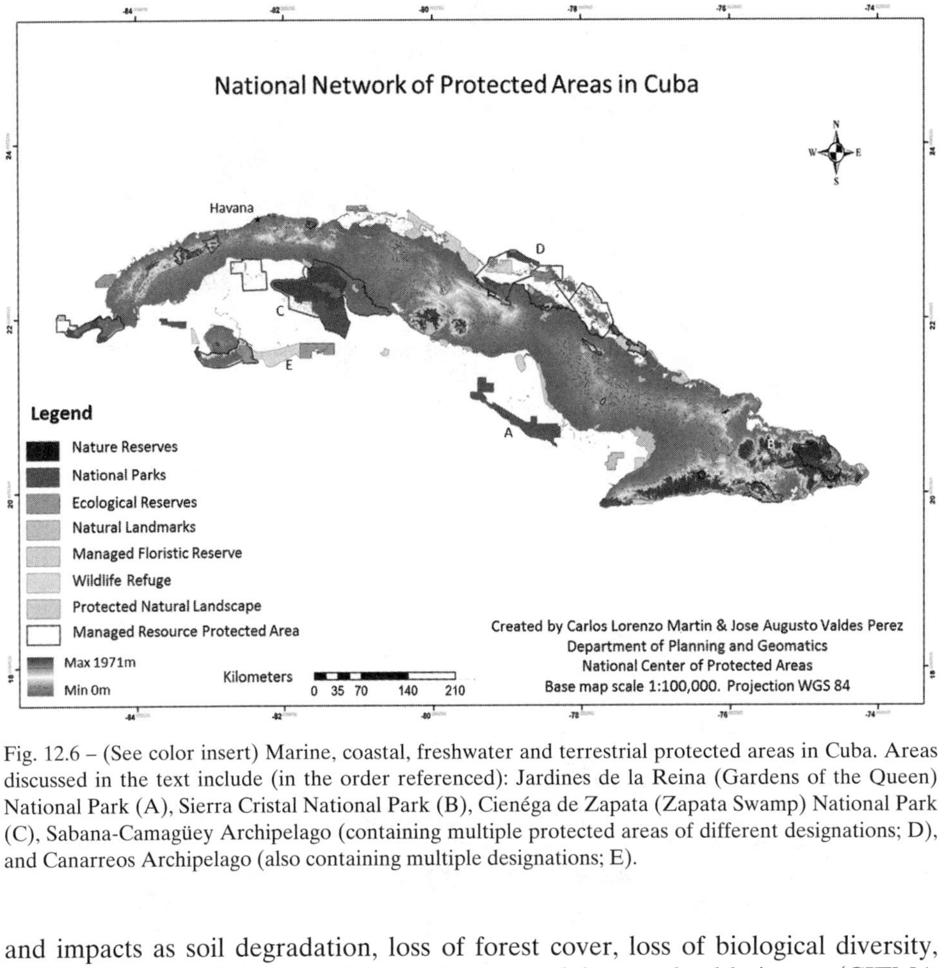

Fig. 12.6 – (See color insert) Marine, coastal, freshwater and terrestrial protected areas in Cuba. Areas discussed in the text include (in the order referenced): Jardines de la Reina (Gardens of the Queen) National Park (A), Sierra Cristal National Park (B), Cienéga de Zapata (Zapata Swamp) National Park (C), Sabana-Camagüey Archipelago (containing multiple protected areas of different designations; D), and Canarreos Archipelago (also containing multiple designations; E).

and impacts as soil degradation, loss of forest cover, loss of biological diversity, pollution, water shortages, climate change, and human health issues (CITMA 2010).

6.2b Policy and governance context

Cuba has a diverse set of legal, policy, scientific, and governance tools to deal with threats to its natural heritage (CITMA 2010). In 1994, Cuba created the Ministry of Science, Technology and the Environment (CITMA), elevating environmental issues to the ministerial level. With the creation of CITMA, the National Center for Protected Areas (CNAP) was born in 1995. CNAP has played a critical role in creating the National System of Protected Areas (SNAP) and achieving important conservation benefits. The Executive Committee of the Council of Ministers (CECM), one of the highest decision-making bodies in Cuba, has the authority to declare protected areas. This provides important government support and a vehicle for potential rapid implementation, although scientific limitations, stakeholder conflicts, and other barriers still present delays. Decisions are also supported by a SNAP Coordination Board, a collaborative body directed by CNAP and including five other key decision-making institutions: Fishery Regulations Office, State Forestry Service, National Enterprise for Flora and Fauna, Corps of Forest Wardens,

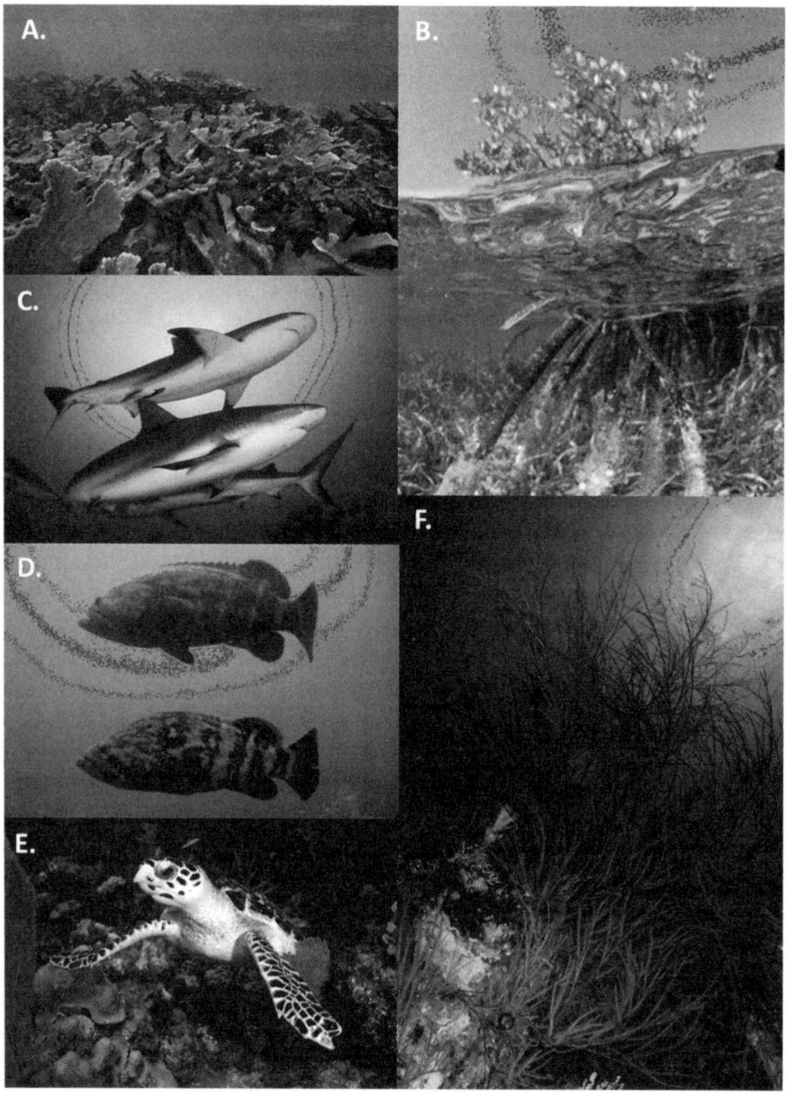

Fig. 12.7 – (See color insert) Habitat and species diversity in the Jardines de la Reina (Gardens of the Queen) National Park, Cuba, including elkhorn coral reef crest (A), red mangrove sapling fringed by eelgrass (B), Caribbean reef sharks (C), Goliath groupers (D), hawksbill turtle (E), and black coral (F). (Photos A–D by Noel Lopez, and E–F by Fausto de Nevi)

and Directorate of the Environment. The approval and planning of protected areas occurs through a participatory process, including national and provincial agencies, local communities, and relevant socio-economic groups (e.g., fishermen, tourism workers) through formal and informal meetings, negotiations, planning workshops, conflict resolution, and education.

The three main legal tools governing management of shallow coastal tropical habitats in Cuba include Environmental Law 81, which is the framework law for environmental management, Decree Law 201 outlining operations of the SNAP, and Decree Law 212 for Coastal Zone Management (CITMA 2010). Law 81, adopted in 1997, defines the SNAP as an integrated marine-terrestrial system and establishes its objectives and basic principles. Decree Law 201, adopted in 1999, is the primary legal instrument for the SNAP. It defines protected area categories (which are the same for terrestrial and marine areas), administrative issues, mechanisms for proposals and approvals, and guidance for participatory area planning. Decree Law 212, adopted in 2000, defines the coastal zone and different sub-zones within it based on shoreline type (sandy, rocky, mangrove, etc.). Decree Law 212 also outlines procedures for uses of the coastal zone.

Beginning in 2003, MPAs established under the SNAP include 5-year management plans (CNAP 2008a), supported by methodologies for preparation and operations (CNAP 2008b), and for determining carrying capacity of visitors in MPAs (CNAP 2006). In addition to MPAs created primarily for conservation, Cuba has been implementing different types of multi-use management zones. These include spatial management strategies established by fisheries institutions known as Zones under Special Regimes of Use and Protection (ZBREUP). These zones are effectively fishery reserves that act as additional protected areas.

Building upon lessons learned within the fishing sector through implementation of ZBREUPs, a similar and relatively new type of zone is being developed to solve conflicts among a more diverse array of uses and users. These Zones under Regimes of Integrated Coastal Management (ZBRMIC) are intended to coordinate government, community, science and the economy in the pursuit of conservation and socio-economic development in coastal and marine zones. In 2007, CITMA established a nationwide procedure to assess and approve processes of integrated coastal management throughout the country. This procedure verifies that each proposal has guaranteed minimal organizational, functional, technical and methodological requirements to start a process of integrated coastal management, and oversees the implementation and functioning of the ZBRMICs.

6.2c Scientific basis for spatial management

Cuban MPAs are proposed, selected and designed based on high biological diversity (species richness), presence of coral reef, presence of endangered, rare, charismatic and depleted species, high tourism interest or potential, and low conflicts among users. Among developing countries, Cuba has a relatively strong, and growing, scientific foundation with which to measure sites against these criteria, although considerably more information is still needed to more effectively manage most existing MPAs and to design new ones. However, research emerging from Jardines de la Reina (Gardens of the Queen) National Park is an especially good example of an information base being built to guide design and management of Cuban MPAs.

A comprehensive checklist of fish species has been developed for the Jardines de la Reina (Pina-Amargós et al. 2007), and several reports of species new to science on both global (Ortea and Espinosa 1998; Espinosa and Ortea 1998a,b,c, 1999) and national (Abreu y del Valle 1998; Ibarzábal 1998; Claro et al. 2000, 2001; Hernández-Fernández and Varela 2007; Varela et al. 2008; Guimarais et al. 2009; Acevedo et al. 2010; Parada-Isada et al. 2012) scales have been made. More detailed information on life histories, behavior, connectivity and population dynamics has been compiled for many of the larger species with greatest tourism value and conservation concern, including sharks (Graham et al. 2007; Fig. 12.7c), goliath grouper (*Epinephelus itajara;* Pina-Amargós and González-Sansón 2009; Fig. 12.7d), sea turtles (Moncada 1998, 1999, 2006, 2010; Medina et al. 2010; Fig. 12.7e), and other fish species (Pina-Amargós et al. 2008a; Castellanos-Gell et al. 2012).

Other research has focused on community- and ecosystem-level processes. Characteristics of coral reef habitats have been described (Pina-Amargós et al. 2008b), with particular attention to community composition of octocorals and stony corals and patterns of recovery from bleaching (Hernández-Fernández et al. 2011). Similarly, overall trophic structure of coral reefs in the Jardines de la Reina has been described (Newman et al. 2006), with more detailed investigation into demography of longspine black sea urchin (*Diadema antillarum;* Martín-Blanco et al. 2010) and their effects on algal communities (Martín-Blanco et al. 2011). Impacts of hurricanes on marine ecosystems have been studied to understand implications of severe and large-scale disturbances (Pina-Amargós et al. 2008c). Effects of management actions have been studied in terms of spillover of fish from MPAs (Pina-Amargós et al. 2010). Finally, to supplement this body of biological and ecological research with socio-economic information, gamefishing (Figueredo-Martín 2010a) and diving (Figueredo-Martín, 2010b) activities have been characterized.

Beyond efforts to build scientific foundations for individual sites, an important strength of the SNAP is that it has taken into account complex macro-scale issues. These include the interconnections between terrestrial and marine ecosystems, and issues specific to marine habitats. SNAP also considers connectivity among marine populations and intends to build a network of interconnected MPAs through knowledge of oceanographic processes, larval dispersal (Paris et al. 2005), and spawning aggregations (Claro and Lindeman 2003). Resilience in the face of climate change is also considered and has been taken into account within several components of SNAP to produce a system that more effectively manages shallow coastal habitats and their biological diversity.

6.2d Progress toward spatial management

Cuba began efforts to declare protected areas in the early 20th century. Sierra Cristal National Park was established in 1930 under Presidential Decree No. 487 (CNAP 2008a; Fig. 12.6, 12.8). By the century's end, many of Cuba's ecosystems were represented in protected areas, and Cuba's progress toward ecosystem-based management is founded upon this spatially explicit framework. The ambitious policy related to protected areas is novel in that it integrates marine, coastal and terrestrial systems. Cuba's SNAP has 253 protected areas planned, which will help to protect 20% of the national territory, 17% of terrestrial area and 25% of marine area, including 26% of Cuban wetlands and 57% of coral reefs (CNAP 2008a).

Progress toward these ambitious goals has been dramatic in the 21st century. In 2001, the Executive Committee of the Council of Ministers (CECM) declared the first 18 Cuban MPAs (CNAP 2008a; Fig. 12.8). In 2004, Cuba had 45 protected areas approved, including 21 with coastal and marine components, which represented 28% of the SNAP and covered 6% of Cuba area (CNAP 2008a). By 2012, Cuba had almost doubled these figures: 103 protected areas approved, including 56 with coastal and marine components representing 41% of the SNAP, covering 17% of Cuba's area and 20% of marine waters (CNAP 2012; Fig. 12.8).

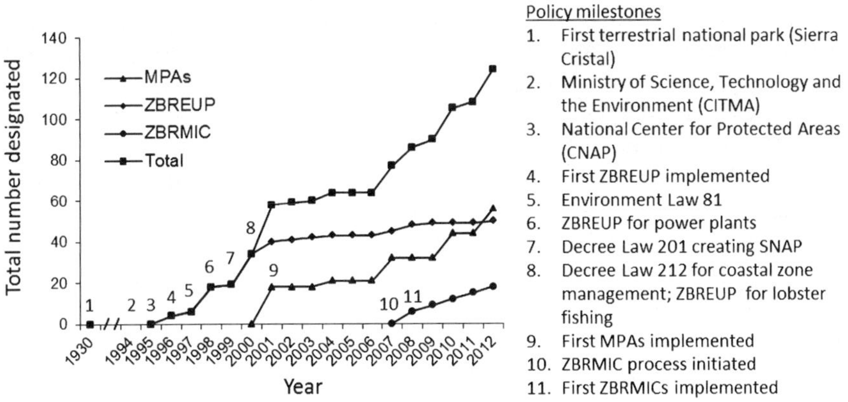

Fig. 12.8 – History of implementation of Marine Protected Areas (MPAs), Zones under Special Regimes of Use and Protection (ZBREUP), and Zones under Regimes of Integrated Coastal Management (ZBRMIC), with timing of major policy milestones. See text for descriptions of the different types of managed areas and policies.

The Ministry of the Fishing Industry (now the Ministry of Food Industry) through the Office of Fishing Regulations began declaring ZBREUPs in 1994. Initially, most did not allow harvest of finfish, with the exception of catch and release gamefishing, but did allow lobster fishing. Currently, ZBREUP are moving towards a more diverse set of regulations and zoning types, similar to protected areas zoning (Fig. 12.6), although a finfishing ban remains. ZBREUP have been the initial phase for many Cuban MPAs, such as Jardines de la Reina National Park (Fig. 12.6), which was declared in 2010. Jardines de la Reina Marine Reserve, established in 1996, now sits within the National Park, and is the largest marine reserve in the Caribbean (Appeldoorn and Lindeman 2003).

The biodiversity value, size, and importance of Jardines de la Reina National Park and Ciénaga de Zapata (Zapata Swamp) National Park (Fig. 12.6) warrant special mention. Jardines de la Reina has been effective in protecting large, economically important fish such as sharks (Fig. 12.7c), snappers and groupers (Fig. 12.7d) inside the marine reserve (Pina-Amargós 2008), and also supplying biomass to surrounding fishing grounds (Pina-Amargós et al. 2010). A special protected area classification, related to IUCN category VI, is the Special Regions of Sustainable Development (SRSD) category. This type of protected area includes large areas of high economic and conservation importance. Iconic Cuban SRSD are the two largest archipelagos in Cuba (Sabana-Camagüey and Canarreos) and Ciénaga de Zapata, the largest wetland in the insular Caribbean (CNAP 2008a; Fig. 12.6).

In the direction of integrated management of multiple uses, Cuba now has in place 15 ZBRMIC since the first six were approved in 2008 (Fig. 12.8). The largest are Península de Guanahacabibes, Golfo de Guacanayabo, and Ciénaga de Zapata. Thirteen ZBRMIC are accomplishing their objectives and eleven have protected areas within. There is strong synergy between ZBRMIC and protected areas, especially MPAs, given that protected areas complement and make feasible ZBRMIC objectives including conservation and reducing user conflicts. Conservation actions and protected area planning often become part of ZBRMIC management plans. Some existing MPAs have become part of a larger ZBRMIC. Mechanisms developed in creating the SNAP have served as a basis to effectively implement integrated coastal management programs. One CITMA objective is to approve at least three ZBRMIC annually. Thus, by 2015, Cuba is expected to have 28 ZBRMIC comprising 63% of its coastline.

ZBRMIC have made important contributions to management in Cuba. They provide space for conflict resolution, while promoting integration among management plans inside the ZBRMIC (forestry, fisheries, territorial planning, protected areas). Participating in the ZBRMIC process has facilitated qualification and environmental education of institutions and communities, and created room for citizen participation through representation on management bodies. An especially important result has been greater participation, commitment and integration of local governments in management. ZBRMIC have created opportunities to carry out specific environmental actions with an ecosystem-based and multi-user approach.

Ecosystem-based management through MPAs, ZBREUP and ZBRMIC has been applied in projects such as the Cuba/GEF/UNDP Sabana-Camagüey Project from 1993 to the present (Proyecto Sabana-Camagüey 1999) and Cuba/GEF/UNDP Archipiélagos del Sur Project from 2010 to the present (Proyecto Archipiélagos del Sur 2009). Integration of these tools has increased protection and sustainable use of natural resources with seven ZBRMIC and eight MPAs already declared in Sabana-Camagüey archipelago (Proyecto Sabana-Camagüey 1999), and eight ZBRMIC, five extended MPAs, six new MPAs and several ZBREUP proposed within Cuba's southern archipelagos (Proyecto Archipiélagos del Sur 2009).

By 2014, the Archipiélagos del Sur Project will increase MPA coverage from 28% to 35%, MPA area inside ZBRMIC from 0% to 68%, total ZBRMIC coverage from 0% to 47%, and total ZBREUP coverage from 12% to at least 20% in Cuba's southern archipelagos (Proyecto Archipiélagos del Sur 2009). Habitat representation in Cuba's southern archipelagos will increase substantially: coral reefs in MPAs will increase from 12% to 20% and in ZBRMICs from 0% to 48%, seagrass beds in MPAs will increase from 11% to 19% and in ZBRMICs from 0% to 36%, and mangrove forests in MPAs will increase from 66% to 74% and in ZBRMICs from 0% to 73% (Proyecto Archipiélagos del Sur 2009).

6.2e Remaining challenges

Its progress notwithstanding, Cuba faces important challenges in implementation of its multi-faceted spatial management system, many of which are discussed in the National Environmental Strategy (CITMA 2010). Operationally, enforcement to prevent and punish illegal fishing, hunting and trade of endangered species (Fig. 12.7e, f) is weak, which limits effectiveness within and outside managed areas. Procedures for protected area approval are often too time-consuming and complicated, unnecessarily delaying implementation. Also, environmental impact assessment procedures and resulting spatial planning have not always sufficiently considered intrinsic and use value of biodiversity. In terms of geographic coverage, ZBRMIC are underrepresented in southern Cuba, accounting for just 27% of the national total.

The three components of Cuba's spatial management system (MPAs,

ZBREUP, ZBRMIC) are still mainly seen by decision makers and the public as separate tools. A key challenge is articulating, designing and implementing these components in a way that not only allows each to achieve its objectives, but also to perform synergistically toward common goals. Although integrating these tools in the sea will be a significant challenge in its own right, integration with other structures such as Watershed Management Units will be necessary as well. Better educating decision makers and stakeholders about ecosystem-based management concepts and practices, and then conducting pilot projects to demonstrate effectiveness, are critical steps toward more effective integration. To maximize success, greater capacity for adaptive management than available at present is necessary so deficiencies are quickly identified and corrected. For example, many MPAs boundaries are not well placed to facilitate enforcement, and adaptive management strategies could help adjust boundaries as lessons are learned after implementation.

In addition to improved coordination, education, and enforcement, overcoming scientific deficiencies will be critical. Baseline studies to determine trends of biodiversity loss and recovery after impacts and mitigation are lacking. Knowledge gaps exist for several taxonomic groups, primarily invertebrates and microorganisms, limiting the ability to determine status and threats, and adopt conservation measures. Effects of climate change, ocean acidification, increased frequency and strength of storms, sea level rise and invasive species are poorly understood in Cuba, as is the value of ecosystem services needed to understand the full economic impact of those processes. Individual managed areas and networks should be designed taking into account oceanographic phenomena, larval dispersal, location of spawning aggregations, juvenile and adult movement, and nursery grounds.

Building the needed science is compromised by funding. Although all MPAs and ZBRMIC are financially supported through national and international projects, funds are insufficient for thorough management, enforcement, research, and monitoring. Reliance upon international sources is especially high for MPA management, which are far from being self-sufficient. Additional economic and social science research could better estimate values of ecosystem goods and services, allowing more appropriate allocation of resources and development of financing mechanisms to cover costs.

6.2f Fisheries management

Although Cuba faces important challenges in management of MPAs, ZBREUP and ZBRMIC, far greater challenges exist in the majority of national waters outside of managed areas (Fig. 12.6). A variety of uses take place in these areas, including transportation, tourism and growing interest in energy production (Piñon and Muse 2010), fishing is the most widespread activity involving the largest number of people. The Cuban fishing industry experienced drastic changes following the break-up of the Soviet Union in the early 1990s, and with it the loss of Soviet markets and subsidies. The fleet transitioned from one comprised of fewer and larger vessels targeting high volume and low value species to one comprised of more numerous smaller vessels targeting a greater diversity of low volume but high value species (Adams 2000). Application of Schaefer models to lobster, shrimp, and tuna suggests management of those stocks has been reasonably successful since this transition (Baisre Hernandez 2006). However, concerns persist about sustainable harvest of other species, particularly reef-associated finfish (Claro et al. 2009). Scientific deficiencies are again a barrier, and research is needed on many harvested species, as well as impacts of multi-species and multi-gear fisheries on an ecosystem scale.

To address the changed character of Cuba's fisheries, management has been moving toward more decentralized approaches (Adams 2000), which might lead to adoption of cooperatives, territorial user rights for fishing (TURFs), and other rights-based approaches that have been demonstrated to be more effective in lower capacity settings (Gutiérrez et al. 2011). The Cuban government passed a new decree law in 2012 authorizing creation of cooperatives in a variety of industries, including fisheries. This policy has the potential to fundamentally shift the nature of incentives, responsibilities, and interactions among fishermen, scientists and decision makers. Given the large number of small-scale fishermen spread among many small and widely dispersed ports, those changes hold great potential for more effective management (Prince 2003). However, carefully planned pilot projects designed to determine necessary elements of these governance systems in the Cuban context will be essential to building workable models.

6.2g Summary

Cuba has made important progress toward ecosystem-based management, even without explicitly defining it as such, through an ambitious spatial management program. Sustaining marine and coastal ecosystems will be essential to preserving Cuba's natural heritage for its socio-economic well-being. Replicating successes, and especially increased attention to scientific and management deficiencies, will continue moving Cuba toward ecosystem-based management.

Of course, Cuba's efforts are relatively young, so it is too early to fully evaluate the outcomes. However, two case studies show potential benefits to local communities of MPAs that are well designed, well managed, and supported by science. After declaration of Punta Francés National Park, protection of fish and coral, combined with improved administration, are predicted to generate more than US$100 million annually for the nation (Angulo-Valdés 2005). JRMR has already fueled a thriving tourist economy (Figueredo Martín et al. 2009) through protection of large charismatic species such as sharks, snappers and groupers (Pina-Amargós 2008), high quality dive sites (Figueredo Martín et al. 2010a), and opportunities to fish for bonefish, tarpon and permit (Figueredo Martín et al., 2010b). Additionally, JRMR benefits coastal communities dependent on commercial fishing (Figueredo Martín et al. 2009) through spillover of biomass to surrounding fishing grounds (Pina-Amargós et al. 2010).

6.3 Low capacity: Kenya

Kenya's 600km coastline is situated just south of the equator (Fig. 12.9). Extensive coral reefs, mangroves, and nearshore islands are found in the far south, and a fringing reef, approximately 3 km from the shore, extends for much of the coast as far north as Malindi. This fringing reef creates a wide lagoon where the majority of marine resource extraction occurs. The coast has a high population density (~40 people·km^{-2}) but below the national average (~66 people·km^{-2}) (Kenya Bureau of Statistics 2012). Activities along the coast are a major contributor to the national economy, although the region remains geographically, historically, and socially distinct.

6.3a Ecosystem services

Coral reefs in the western Indian Ocean provide multiple ecosystem services (Fig. 12.10). Services that provide economic benefits are most commonly quantified, but social and cultural services such as education (i.e., local ecological knowledge) and bequest (i.e., knowledge that future generations will benefit) are recognized, and in some areas their importance exceeds economic benefits (Rönnbäck et al. 2007; Hicks et al. 2009). Appreciation of these services can contribute to adaptive man-

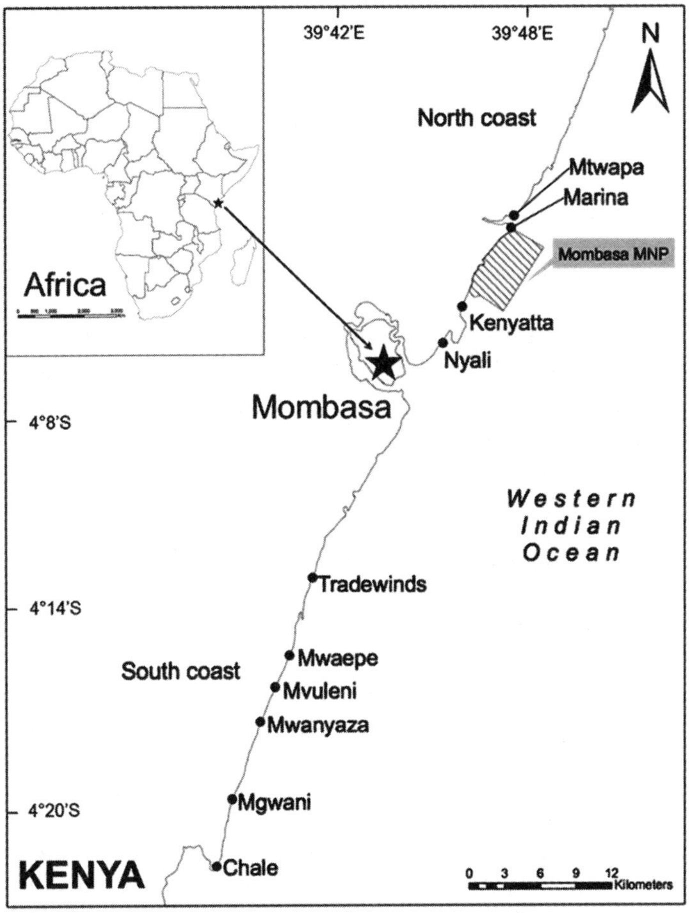

Fig. 12.9 – Detail of the Kenyan coast centered on the port city of Mombasa. Shipping is an important ecosystem service for Mombasa and the nation as a whole. Outside of the well-protected harbor, fishing, tourism and shoreline protection are important services delivered by the fringing reef adjacent to the many coastal towns. Marine protected areas, such as Mombasa Marine National Park, have been an important component of management in this developing nation.

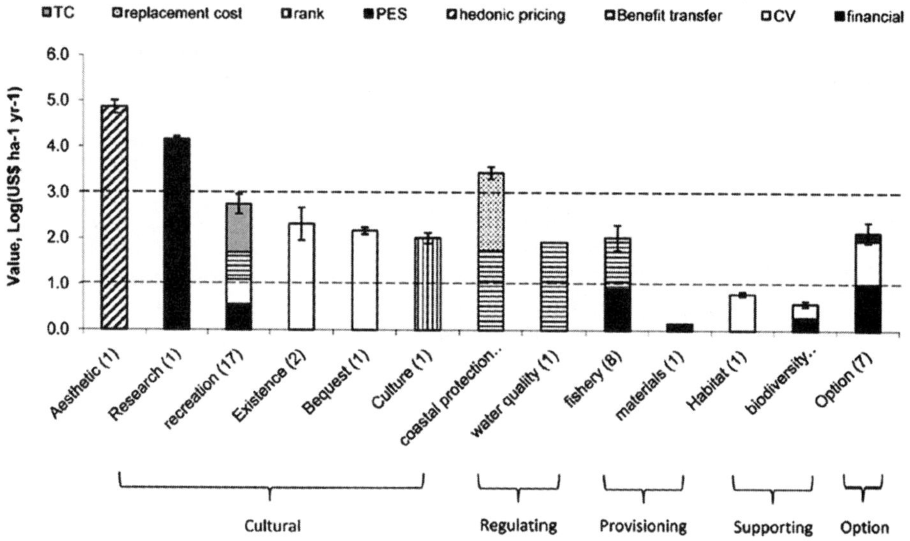

Fig. 12.10 – Estimated value of ecosystem services in the western Indian Ocean. Services are ordered according to categories from the Millennium Ecosystem Assessment. (Reprinted from Hicks 2011, by permission of the publisher, Taylor & Francis Ltd., http://www.tandf.co.uk/journals)

agement, enabling learning from and adapting to changes in the coupled social-ecological system.

The port of Mombasa, in Kenya's principal coastal town (Fig. 12.9), was originally established as part of the spice trade route in the sixteenth century. The port has the best natural harbor in East Africa, and has become the largest port in the region (Foeken et al. 2000). Mombasa port handles more than 15 million tons of cargo each year, contributing significantly to the region's economy (Kenya Ports Authority 2012). Along the coast, erosion is a critical issue, as many hotels are built within 30m of the high tide mark (legal limit) and have had to invest in coastal defenses (i.e., sea walls). The reef provides some protection and for most of the year calm conditions exist in the lagoon, making access to the coral, seagrass, and sand habitats by swimming or dugout canoe possible.

This reef system, and the easy access to it, sustains livelihoods and nutritional needs of many coastal communities. The majority of fishing in Kenya is conducted in the lagoon. Fishing tends to be small scale, artisanal, or subsistence, and fishermen use a diversity of gears to land a diversity of species (McClanahan et al. 2008b). Approximately 30% of coastal households are involved in fishing (Cinner et al. 2009a) predominantly as a primary occupation. Subsistence farming, fish trading, or the informal sector (i.e., economic activity not regulated by the government) provide secondary occupations (Cinner and Bodin 2010). This suggests that, although fishing may not generate a great deal of revenue, it is vital to coastal livelihoods. Limited industrial-scale fishing also takes place, and the government receives limited revenue in international trawler license fees.

Tourism is a major industry in Kenya contributing 12% to GDP. Sandy beaches and coral reefs attract many tourists. Although tourism is important to the national economy, and its value dwarfs that of other coastal ecosystem services (Hicks et al. 2009), local and regional benefits are minimal (Sindiga 1995). The informal sector is responsible for local benefits accruing from tourism, and only ~7% of households along the coast are involved in tourism (Cinner et al. 2009a).

A cement quarry set back from the coast excavates fossil reef for construction. This cement factory generates income and employment. Associated with the quarry are ecosystem restoration projects and visitor education centers providing recreational and educational benefits. Mangroves bordering the coast provide local provisioning, regulating, cultural and other services (Rönnbäck et al. 2007). Perhaps most significant at a global scale is the carbon sequestration service mangroves provide.

6.3b Threats and impacts

Climate change and fishing are the two greatest threats along the Kenyan coast. These threats influence different components of the ecosystem with impacts potentially reinforcing one another. Fishing targets large-bodied fish and climate change induces habitat loss and declines of small species (Graham et al. 2011). Other common climate effects are bleaching and mortality of corals. Kenya's worst coral bleaching event occurred in 1998 when the El-Niño Southern Oscillation (ENSO) coincided with the Indian Ocean Dipole (IOD), resulting in the warmest ocean temperatures on record and up to 90% coral mortality across the western Indian Ocean (McClanahan et al. 2007a; Graham et al. 2008). Loss of coral cover following bleaching (Fig. 12.11) opens up space for recolonization by fast growing algae or weedy coral species (Green et al. 2008; Darling et al. 2010). Changes in benthic composition have substantial impacts on fish assemblages, leading to reductions in species richness and abundance of small-bodied and strongly coral-dependent species (Wilson et al. 2006; Pratchett et al. 2008). Substantial changes in fish communities occurred across the western Indian Ocean following the 1998 bleaching event (Graham et al. 2008). However, changes in Kenya were relatively small, likely because the reefs and fish assemblages were already altered by fishing.

Destructive and unsustainable fishing is the main local threat to Kenya's coastal systems. Dense human populations lead to heavy fishing pressure, and population density at Kenya's coast is predicted to increase. Furthermore, natural population growth is likely to be exacerbated by climate-induced coastward migration. Population density increases demand for fish and the number of fishermen pursuing a dwindling resource, creating a Malthusian tendency for fishermen to use increasingly destructive methods (Pauly 1989; McClanahan et al. 2008b). These three factors—growing population, climate-induced coastward migration, and destructive fishing—combine to intensify fishing impacts. Local threats are more amenable to management, so there is greater potential for more effective management of fishing effort and destructive methods.

Fishing has the greatest impact on species with slow demographic rates, which are typically at higher trophic levels (Jennings et al. 1999; Rochet et al. 2000; Cheung et al. 2005) and fetch higher prices (McClanahan 2010). Fishing in Kenya has consequently caused fast-growing herbivores to dominate the assemblage (e.g., green parrotfish *Leptoscarus vaigensis* and white-spotted rabbitfish *Siganus sutor;* McClanahan et al. 2008b; McClanahan and Hicks 2011; Hicks and McClanahan

Fig. 12.11 – (See color insert) Differential responses to thermal stress and bleaching among coral species and morphologies in Kenya. Bleaching-sensitive corals include species of branching *Acropora* (A) and plating *Montipora* (B), while species of massive *Porites* (C) are often less affected by bleaching. (Photos by Emily Darling)

2012). The result is that where fishing is unregulated, effort is higher, destructive gears dominate, and incomes are lower (McClanahan 2010). Dragging gears or trampling whilst fishing can also directly damage corals and other benthos (Mangi and Roberts 2006). Further indirect effects of fishing influence competition, settlement, and predation by altering the balance of functional groups. For example, removing higher trophic level species may reduce predation, resulting in prey release (Sandin et al. 2010). This has been most noticeable in Kenya through removal of red-lined trigger fish *Balistipus undulates,* the dominant predator of sea urchins. In areas where fishing pressure is high, such as outside protected areas, densities of red-lined trigger fish are low and urchin populations have exploded (McClanahan 2000). The implications of community shifts are unclear; however, system productivity is clearly declining and climate-driven loss of charismatic fish and corals is likely to impact tourism (Cesar et al. 2002). Other stressors include sedimentation and eutrophication, which are more episodic and spatially restricted due to strong currents and tidal flushing (Obura 2001).

6.3c Management and governance

At the beginning of the last century, social norms and traditional ecological knowledge determined which rules governed resource use (Ogwang et al. 2006). Traditionally, an elder leader of a fish landing site, provided advice on seasons, who could fish and which species were prohibited, and dealt with accidents and conflicts. However, the colonial and post-colonial era drive for economic development reduced local authority resulting in an unregulated open access approach to fisheries management. The exception was a parks system established independently of fisheries management to meet the demands of a growing tourism market

(McClanahan 2007; Hicks et al. 2009). The lack of fisheries regulation resulted in severe depletion of artisanal fishing stocks (McClanahan and Muthiga 1988).

Decline in fish stocks was the impetus for Kenya's fisheries management to forge a relationship between research and management, promoting research that is solution-oriented. Three institutional developments characterize these changes. First, in 1986 the coral reef conservation project (CRCP), a research non-governmental organization (NGO), was established to understand effects of human influences on Kenya's coral reefs. At its core was assessment and understanding of ecological and biophysical processes, and an appreciation of the link to exploitation and management (McClanahan and Muthiga 1988; McClanahan et al. 2005a). Secondly, in a similar trend, the early 1990s saw a change toward integrated coastal management (ICM), catalyzed by the first meeting between sectors where the interdependency of ecology and economics was recognized (McClanahan et al. 2005a).

Finally, in 2006, the fisheries department introduced legislation that devolved management to local landing sites again. This legislation enables re-emergence of the role played by elders, recognizing the importance of robust socio-ecological systems that include co-management and build upon existing traditions and institutions, enabling users to develop and enforce rules governing resource use. Recent developments under this system include proliferation of community-led, no-take, locally managed marine areas (LMMAs), similar to those developed in the Philippines (Alcala et al. 2006).

The establishment of Kenya's marine parks was through a top-down command-and-control approach. Although independent of fisheries management, it has resulted in an effectively enforced system of no-take areas along the coast, which in turn has enabled deep understanding of reef recovery dynamics. Ecosystems within protected areas have responded substantially, with fish biomass recovering to asymptotes at 1200 $kg \cdot ha^{-1}$ over periods of ~20 years (McClanahan et al. 2007b), and benthos tending away from fleshy algae toward calcifiers (McClanahan and Graham 2005; O'Leary et al. 2010). It should be noted, however, that a diversity of factors, including human population density, amount of area closed to fishing and perceived compliance levels, influence the likelihood of recovery (Pollnac et al. 2010; Daw et al. 2011).

There has been progress towards understanding complex social-ecological drivers. Investigations have found that economic development is related to reef health (Cinner et al. 2009b), and that poverty can result in fishermen being unable to reduce pressure on dwindling resources (Cinner et al. 2009a). This work has led to development of management frameworks that help fit social and environmental contexts (McClanahan et al. 2008a; Cinner et al. 2012). For example, where environmental susceptibility is high and adaptive capacity is low, efforts should be made to reduce reliance on vulnerable resources (McClanahan et al. 2008a).

Kenyan fishermen use several gears that can be broadly categorized into five types: spear gun, beach seine, gill net, trap and line. Gear use and attitudes toward each differ along the coast. Beach seine and spear guns damage habitat and overlap in selectivity with most other gears (McClanahan and Mangi 2004; Mangi and Roberts 2006). Beach seines have been effectively discouraged by traditional elders in many sites for more than 20 years (McClanahan 2007). Consequently, the

Kenyan fisheries department took advantage of natural differences, backed by evidence of their destructive nature, to support local restrictions and ban use of beach seines and spears in 2001 (Kenya Gazette (2001)). These developments represent co-management between government, local communities, and research organizations. Although enforcement has been spatially variable (McClanahan et al. 2005b), when effective, both area- and gear-based approaches have increased catch per unit effort of seagrass and coral reef fisheries (McClanahan et al. 2008b; McClanahan and Hicks 2010; Fig. 12.12).

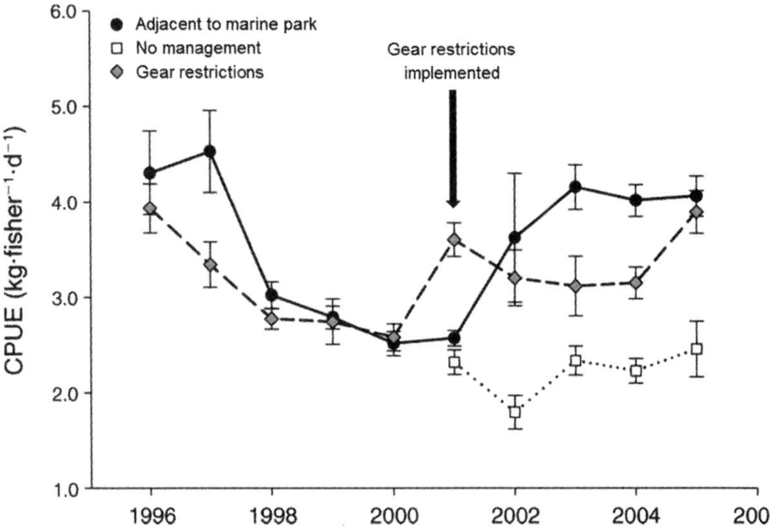

Fig. 12.12 – Trends in catch per unit effort (CPUE) from Kenya's artisanal fishery. Areas with gear restrictions achieved greater CPUE relative to areas without management. The highest CPUE was achieved in areas adjacent to a no-take marine protected area, but only after gear restrictions were also introduced. (Reprinted from McClanahan et al. 2008)

Gear-based approaches can be further designed to improve yields, protect vulnerable life stages, and enable adaptive management in the face of climate change (Cinner et al. 2009c; Hicks and McClanahan 2012). These approaches are favorable in fisheries that support a growing population as they adapt a fishery rather than reducing it (Hilborn et al. 2004). This adaptive quality enables gear restrictions to be tailored to local socio-economic and ecological contexts (McClanahan and Cinner 2008), facilitating design that is more acceptable and less intrusive. Gears selectively target specific life history traits. Modifying the portfolio of gears can protect vulnerable traits, and mesh restrictions can protect key species past vulnerable life stages (Hicks and McClanahan 2012). Information on catch composition by different gears has recently been combined with information on vulnerability of species to population declines following coral mortality and importance of fish species in facilitating ecosystem recovery (Cinner et al. 2009c). This work has shown that spear fishers target species that are vulnerable to population crashes following extensive habitat loss. Furthermore, spear guns and traps also capture the highest proportion of species providing key ecological functions, such as herbivores and macro-invertebrate feeders (Cinner et al. 2009c). Such assessments may provide the basis for adaptive gear regulations following habitat disturbances, to both sustain catches and allow ecosystem recovery. Recently, Kenya has been experimenting with escape gaps in fish traps that allow vulnerable species to escape (Fig. 12.13), but still maintain catch rates and profits. Similar work in the Caribbean has shown success with these approaches (Johnson 2010).

Fig. 12.13 – (See color insert) Fish trap typical of those used by Kenya's artisanal fishermen (A), with an escape gap that allows bycatch species to escape (B). (Photos by Tim McClanahan)

A recent study across the western Indian Ocean made the first attempt to identify non-linear thresholds in ecological processes and state variables associated with fishable biomass (McClanahan et al. 2011). Fishable biomass was used as the driver of these ecosystem dynamics as it is tightly linked to fishing pressure, the dominant local stressor in the region. Tipping points were identified, whereby substantial changes in variables occurred as fishable biomass declined, such as the ratio of macro algae to corals, the proportion of herbivores, or the extent of predation. Critically, this analysis enabled identification of reference points and targets for ecosystem-based management (McClanahan et al. 2011). This threshold analysis can form the basis for determining multi-species maximum sustainable yield. If fishable biomass can be maintained between 300 and 600 $kg{\cdot}ha^{-1}$, dual objectives of sustaining fisheries and maintaining the ecosystem above key tipping points should be achieved (McClanahan et al. 2011).

In Kenya, marine National Parks were maintaining fishable biomass well above these ecosystem-based targets, areas with gear restrictions had fishable biomass within the window, and areas with no restrictions had fishable biomass substantially below the window and below all ecosystem threshold points (McClanahan et al. 2011). This work suggests that gear-based management along Kenya's coast is proving an effective tool for ecosystem-based management and should incorporate morc of the coastline.

Management developments are also evident beyond reef fisheries. Recognizing the carbon sequestration value of mangroves, a longstanding rehabilitation program run by the Kenya Marine and Fisheries Research Institute (KEMFRI) became a recent candidate for a trial in 'Blue carbon'. Blue carbon attempts to finance conservation and rehabilitation initiatives through global emissions trade and REDD+ schemes (Grimsditch et al. 2012). Elsewhere, and following a different model, Lafarge ecosystems has a restoration program in its quarries based on biodiversity objectives, and capable of responding to evaluations that use neighboring natural habitats as references (Kahumba 2009). These approaches contribute to climate change adaptation and mitigation (Kitha and Lyth 2011).

6.3d Remaining challenges

The magnitude and persistence of threats mean significant challenges must still be addressed to move Kenya further toward ecosystem-based management of coral reefs and associated coastal systems. Although there has been coordination of Kenya's fisheries management, many successes have been ad hoc and many unregulated areas remain. Consequently, science and management need to scale these efforts up, capitalizing on successes and filling both social and ecological gaps in Kenya's management mosaic. This approach should feed localized, community-specific successes into a regional adaptive strategy. It is critical that this approach equally emphasize social and ecological goals, enabling mechanisms for learning, communication, and conflict resolution. It is often users who are intimately linked to resources that are most likely to detect and adapt to changes first. Such adaptability needs to be fostered. Adaptive mechanisms are necessary as unexpected threats emerge, presenting new challenges requiring flexible solutions. For example, increased Somali pirate activity is likely to create new social-ecological challenges, potentially undermining effective management strategies.

7. Summary and Conclusions

According to many, the future of coral reefs and associated habitats is grim. These systems occupy very precise locales defined by depth, temperature, substrate, salinity, acidity, turbidity and other factors. Even subtle changes in these attributes can lead to significant degradation or outright loss. Capacity for habitats to migrate elsewhere when conditions in existing locations become unsuitable is limited as development and other impacts reduce colonization sites, and increasing rates of environmental change outpace rates of dispersal, recruitment, growth and evolution. A growing human population and its demands for food and income exacerbate the strain.

Nevertheless, coral reefs are visible and valuable ecosystems, continually motivating improved management, even if the pace of action too often is slow. The case studies reviewed herein illustrate some of the more ambitious efforts toward managing reef ecosystems, importantly in three very different contexts. The case studies span three major ocean basins (Pacific, Atlantic/Caribbean, Indian). Top-down management by national government is a necessary, but not sufficient, characteristic of all three, but high level policy development and ability to effectively implement policy vary due to a complex combination of history, culture, political model, wealth and associated scientific, monitoring and enforcement capacity, competing uses, and other factors. Despite those differences, all three rely, or are beginning to rely, on smaller scale governance and incentives to complement high level authority and offset its limitations. Spatial management aimed at balancing competing uses, restoring ecosystem function, improving science, and buffering against uncertainty is a central strategy in all three cases. However, spatial management will not be fully effective in isolation.

The case studies show how integration of MPAs with other management tools can increase effectiveness. Australia has paired sophisticated zoning with incentive-based fisheries management in the form of individual transferable quotas (ITQs) to more effectively manage the most heavily exploited stocks. This represents a form of management devolved to the individual level, as ITQs empower fishermen to make more decisions regarding when and how to fish than under inefficiencies forced by input controls (Costello et al. 2008; Worm et al. 2009; Grimm et al. 2012). Similarly, Kenya has devolved important aspects of management to smaller scales, although the key units are local villages and their traditional social structures. Collective management can be effective, especially where it matches finer ecological scales, and builds upon existing social and cultural systems (Prince 2003; Ostrom 2010; Gutiérrez et al. 2011). Certain outcomes of Kenya's management experience illustrate that effectiveness, although key challenges remain. In contrast to Australia and Kenya, Cuba has historically relied more on top-down than smaller scale approaches. However, Cuba's planning processes for multi-user management zones and recent authorization of fishing cooperatives illustrate how it too is increasingly turning to management at smaller scales.

One important difference between Australia and the other case studies is use of output controls alongside spatial management and multi-scale governance and incentive structures. Output-based management, especially when combined with incentive-based allocation systems, has been shown to be more effective at achieving target biomass, fishing within quotas, and reducing variance in biomass, exploitation rate and catch (Worm et al. 2009; Essington 2010; Grimm et al. 2012; Melnychuk et al. 2011). Of course, scientific, monitoring and enforcement capacity needed to effectively set and implement catch limits are often lacking, which, combined with lack of acceptance for output controls by key stakeholders, can be prohibitive. In such cases, input controls might remain a necessary and reasonably effective strategy, as evidenced by Kenya, until increased capacity and outreach change the logistical and political conditions in favor of output controls. Notably, the work of McClanahan et al. (2011) has made assessment and management of reef fisheries on a multi-species ecosystem level possible. Should Kenya formalize those metrics as part of its management strategy, it will be a rare instance of such an approach pioneered by a low capacity developing nation.

Fishing has been a focus of this discussion because it is one of the most significant uses, and impacts. Also, of the three most valuable services provided by coral reefs (i.e., alongside tourism and coastal protection), it is arguably the least effectively managed in most places, which has implications for other services. However, other uses and impacts still need increased attention in the cases examined here, and just about every other reef ecosystem. Although results have not been adequately measured, and benefits have not yet been fully realized due to the efforts beginning in earnest only within the last decade, Cuba has perhaps demonstrated the most ambitious policies toward integrated management of uses of our three case studies. This is done through two primary vehicles. First, Cuba is establishing marine protected areas in concert with those along the coast, on land, in wetlands, and in freshwater in a rare example of a truly integrated protected area network. This not only combines management of coral reefs with associated coastal habitats (mangroves, seagrass beds) but also with freshwater and terrestrial systems. Second, within marine waters, Cuba's multi-user planning zones (ZBRMIC) are now attempting to plan and coordinate ocean uses in ways that promote equity and maximize effectiveness. More detailed scientific and policy analyses of Cuba's approaches should be a priority to determine their effectiveness and value as a model for other locales.

Daunting challenges continue to confront coral reefs, mangroves and seagrasses. Pessimism is pervasive among many, and not unjustifiably. However, encouraging developments continue to emerge in both developed nations, such as the United States' creation of the 362,073 km^2 Papahānaumokuākea Marine National Monument in the Northwest Hawaiian Islands, and smaller developing nations, such as Palau's policy to place 20% of its land and 30% of its marine waters within protected areas. Increasing awareness and implementation of ecosystem-based management continues as science progresses and degradation unfolds. Whether and to what extent science and management will keep pace with ecosystem change is unclear. We are likely to see reduced area and productivity of coral reefs, mangroves and seagrasses. However, these systems are able to deliver a wide range of goods and services, even in a diminished state. The biggest question is how much value will be lost due to uncertain future trends in the ecosystem and management responses.

Acknowledgments

We thank the editors for inviting this chapter and for constructive comments that greatly improved its quality. Valuable comments were also provided by two anonymous reviewers that likewise resulted in a greatly improved chapter. Owen Liu provided invaluable assistance in the final preparation, without which this chapter would not have been completed. CCH thanks Tim McClanahan for valuable discussions that shaped the Kenya case study. FPA was supported by a Pew Fellowship in Marine Conservation during preparation of this chapter. Field work by JPK that informed the Cuba case study was supported by the Waitt Foundation and the MacArthur Foundation.

References

Abreu, M. and del Valle, R. 1998. Presencia de *Lytechinus williamsi* Chesher, 1969 (Echinodermata: Echinoidea) en aguas del Mar Caribe de Cuba. *Avicennia* 8/9:163.

Aburto-Oropeza, O., Ezcurra, E., Danemann, G., Valdez, V. Murray, J. and Sala, E. 2008. Mangroves in the Gulf of California increase fishery yields. *Proceedings of the National Academy of Sciences* 105: 10456–10459.

Acevedo, C. J., López, D. and Parada, A. 2010. Se localiza *Copernicia macroglosa* (Arecaceae) en Cayo Caguama, archipiélago Jardines de la Reina, Camagüey, Cuba. *Bissea* 4.

Adams, C. 2000. Recent changes in management structure and strategies of the Cuban fishing industry. In *Cuba in Transition, Volume 10,* 121–127. Miami, Florida: Association for the Study of the Cuban Economy.

Alcala, A. C. and Russ, G. R. 2006. No-take marine reserves and reef fisheries management in the Philippines: a new people power revolution. *Ambio* 35: 245–254.

Allen, W. H. 1992. Increased dangers to Caribbean marine ecosystems: cruise ship anchors and intensified tourism threaten reefs. *BioScience* 42: 330–335.

Alongi, D. M. 2008. Mangrove forests: resilience, protection from tsunamis, and responses to global climate change. *Estuarine, Coastal, and Shelf Science* 76: 1–13.

Angulo-Valdés, J. 2005. Effectiveness of a Cuban Marine Protected Area in Meeting Multiple Management Objectives. Ph.D. diss., Dalhousie University Halifax, Nova Scotia, Canada.

Anonymous. 2009. The Great Barrier Reef Outlook Report. Townsville, Australia: Great Barrier Reef Marine Park Authority.

Appeldoorn, R. S. and Lindeman, K. C. 2003. A Caribbean-wide survey of marine reserves: spatial coverage and attributes of effectiveness. *Gulf and Caribbean Research* 14: 139–154.

Baisre Hernández, J. A. 2006. Cuban fisheries management regime: current state and future prospects. Reykjavik: United Nations Fisheries Training Programme.

Banat, I. M., Hassan, E. S., El-Shahawi, M. S. and Abu-Hilal, A. H. 1998. Post-Gulf War assessment of nutrients, heavy metal ions, hydrocarbons, and bacterial pollution levels in the United Arab Emirates coastal waters. *Environment International* 24: 109–116.

Bauman, A. G., Burt, J. A., Feary, D. A., Marquis, E. and Usseglio, P. 2010. Tropical harmful algal blooms: an emerging threat to coral reef communities? *Marine Pollution Bulletin* 60: 2117–2122.

Bergman, K. C., Svensson, S. and Öhman, M. C. 2001. Influence of algal farming on fish assemblages. *Marine Pollution Bulletin* 42: 1379–1389.

Berkelmans, R. and Van Oppen, M. J. 2006. The role of zooxanthellae in the thermal tolerance of corals: a 'nugget of hope' for coral reefs in an era of climate change. *Proceedings of the Royal Society B: Biological Sciences* 273: 2305–2312.

Binh, C. T., Phillips, M. J. and Demaine, H. 1997. Integrated shrimp-mangrove farming systems in the Mekong delta of Vietnam. *Aquaculture Research* 28: 599–610.

Brander, L. M., Van Beukering, P. J. H. and Cesar, H. J. S. 2007. The recreational value of coral reefs: a meta-analysis. *Ecological Economics* 63: 209–218.

Brown, B. E. 1997. Disturbances to reefs in recent times. In C. Birkeland, ed., *Life and Death of Coral Reefs,* 354–379. New York: Chapman and Hall.

Burke, L., Reytar, K., Spalding, M. and Perry, A. 2011. Reefs at Risk Revisited. Washington, DC: World Resources Institute.

Butler, M. J. IV, Herrnkind, W. F. and Hunt, J. H. 1997. Factors affecting the recruitment of juvenile Caribbean spiny lobsters dwelling in macroalgae. *Bulletin of Marine Science* 61: 3–19.

Castellanos-Gell, J., Robainas-Barcia, A., Casane, D., Chevalier-Monteagudo, P., Pina-Amargós, F. and García-Machado, E. 2012. The surgeonfish, *Acanthurus bahianus,* has crossed the Amazon-Orinoco outflow barrier. *Marine Biology* 159:1561–1565.

Cesar H., Pet-Soede, L, Westmacott, S. and Mangi, S. 2002. Economic analysis of coral bleaching in the Indian Ocean—Phase II. In Linden O., D. Souter, D. Wilhelmsson, and D. Obura, eds., *Coral Reef Degradation in the Indian Ocean: Status Report* 2002, 251–262. Kalmar, CORDIO.

Cesar, H. J. S., Burke, L. and Pet-Soade, L. 2003. The economics of worldwide coral reef degradation. Cesar Environmental Economics Consulting, Arnhem, Netherlands.

Cheung, W. W. L., Pitcher, T. J. and Pauly, D. 2005. A fuzzy logic expert system to estimate intrinsic extinction vulnerabilities of marine fishes to fishing. *Biological Conservation* 124: 97–111.

Chittaro, P. M. and Sale, P. F. 2003. Structure of patch-reef fish assemblages at St. Croix, U.S. Virgin Islands, and at One Tree Reef, Australia. *Marine Ecology Progress Series* 249: 277–287.

Cinner, J. E. and Bodin, Ö. 2010. Livelihood diversification in tropical coastal communities: a network-based approach to analyzing 'livelihood landscapes'. *PloS One* 5: e11999.

Cinner, J. E., Daw., T. and McClanahan, T. R. 2009a. Socioeconomic factors that affect artisanal fishers' readiness to exit a declining fishery. *Conservation Biology* 23: 124–130.

Cinner, J. E., McClanahan, T. R., Daw, T. M., Graham, N. A. J., Maina, J., Wilson, S. K. and Hughes, T. P. 2009b. Linking social and ecological systems to sustain coral reef fisheries. *Current Biology* 19: 206–212.

Cinner, J. E., McClanahan, T. R., Graham, N. A. J., Pratchett, M. S., Wilson, S. K. and Raina, J-B. 2009c. Gear-based fisheries management as a potential adaptive response to climate change and coral mortality. *Journal of Applied Ecology* 46: 724–732.

Cinner, J. E., McClanahan, T. R., Graham, N. A. J., Daw, T. M., Maina, J., Stead, S. M., Wamukota, A., Brown, K. and Bodin, O. 2012. Vulnerability of coastal communities to key impacts of climate change on coral reef fisheries. *Global Environmental Change* 22: 12–20.

CITMA (Cuban Ministry of Science, Technology and the Environment). 2010. Estrategia Ambiental Nacional Cuba 2011–2014. 71 pp.

Claro, R. and Lindeman, K. C. 2003. Spawning aggregation sites of snapper and grouper species (Lutjanidae and Serranidae) on the insular shelf of Cuba. *Gulf and Caribbean Research* 14: 91–106.

Claro, R., Gilmore, R. G., Robins, C. R. and McCosker, J. E. 2000. Nuevos registros para la ictiofauna marina de Cuba. *Avicennia* 12/13:19–24.

Claro, R., García-Arteaga, J. P. and Pina-Amargós, F. 2001. Nuevos registros de peces marinos para Cuba. *Avicennia* 14:125–127.

Claro, R., de Mitcheson, Y. S., Lindeman, K. C. and García-Cagide, A. R. 2009. Historical analysis of Cuban commercial fishing effort and the effects of management interventions on important reef fishes from 1960–2005. *Fisheries Research* 99: 7–16.

CNAP (Cuban National Center for Protected Areas). 2006. Metodología para Capacidad de Carga de Visitantes en las Áreas Marinas Protegidas de Cuba. 33 pp.

CNAP (Cuban National Center for Protected Areas). 2008a. Sistema Nacional de Área Protegidas 2009–2014. 222 pp.

CNAP (Cuban National Center for Protected Areas). 2008b. Metodología para la Elaboración de Planes de Manejo en Áreas Protegidas de Cuba. 89 pp.

CNAP (Cuban National Center for Protected Areas). 2012. Base de Datos del Sistema Nacional de Área Protegidas. Copia electrónica.

Coles, S. L. and Tarr, B. A. 1990. Reef fish assemblages in the Western Arabian Gulf: A geographically isolated population in an extreme environment. *Bulletin of Marine Science* 47: 696–720.

Conservation International. 2008. Economic Values of Coral Reefs, Mangroves, and Seagrasses: A Global Compilation. Arlington, VA, USA: Center for Applied Biodiversity Science, Conservation International.

Costanza R., d'Arge, R., de Groot, R., Farber, S., Grasso, M., Hannon, B., Limburg, K., Naeem, S., O'Neill, R. V., Paruelo, J., Raskin, R. G., Sutton, P. and van den Belt, M. 1997. The value of the world's ecosystem services and natural capital. *Nature* 387: 253–260.

Costello, C., Gaines, S. D. and Lynham, J. 2008. Can catch shares prevent fisheries collapse? *Science* 321: 1678–1681.

Dahdouh-Guebas, F., Mathenge, C., Kairo, J. G. and Koedam, N. 2000. Utilization of mangrove wood products around Mida Creek (Kenya) amongst subsistence and commercial users. *Economic Botany,* 54: 513–527.

Darling, E. S., McClanahan, T. R. and Côté, I. M. 2010. Combined effects of two stressors on Kenyan coral reefs are additive or antagonistic, not synergistic. *Conservation Letters* 3: 122–130.

Daw, T. M., Cinner, J. E., McClanahan, T. R., Graham, N. A. J., and Wilson, S. K. 2011. Design factors and socioeconomic variables associated with ecological responses to fishery closures in the western Indian Ocean. *Coastal Management* 39: 412–424.

De'ath, G., Fabricius, K. E., Sweatman, H., and Puotinen, M. 2012. The 27-year decline of coral cover on the Great Barrier Reef and its causes. *PNAS* 109: 17995–17999.

Delbeek, J. C. 2001. Coral farming: past, present and future trends. *Aquarium Sciences and Conservation* 3: 171–181.

D'Elia, C. F. and Wiebe, W. J. 1990. Biogeochemical nutrient cycles in coral reef ecosystems. In Dubinsky, Z., ed. *Coral Reefs: Ecosystems of the World,* 25: 49–74. Amsterdam: Elsevier.

Dennis, G. D. and Bright, T. J. 1988. The impact of ship grounding on the reef fish assemblage at Molasses Reef, Key Largo National Marine Sanctuary, Florida. *Proceedings of the 6th International Coral Reef Symposium, Australia* 2: 213–218.

Dixon, J. A. 1998. Economic values of coral reefs: what are the issues? In *Coral reefs: Challenges and Opportunities for Sustainable Management,* edited by M. E. Hatziolos, A. J. Hooten, and M. Fodor, 157–179. Washington, D.C.: World Bank Publications.

Douvere, F. and Badman, T. 2012. Mission Report: Reactive Monitoring Mission to Great Barrier Reef (Australia), 6th to 14th March 2012. UNESCO World Heritage Centre—IUCN. 99 pp.

Dulvy, N. K., Stanwell-Smith, D., Darwall, W. R. T. and Horrill, C. J. 1995. Coral mining at Mafia Island, Tanzania: a management dilemma. *Ambio* 24: 358–365.

Edwards, A. J., ed. 2010. *Reef Rehabilitation Manual.* Coral Reef Targeted Research & Capacity Building for Management Program: St. Lucia, Australia. 166 pp.

Ehrlich, P. R. 1975. The population biology of coral reef fishes. *Annual Review of Ecology and Systematics* 6: 211–247.

Eklöf, J. S., Castro, M. T., Adelsköld, L., Jiddawi, N. S. and Kautsky, N. 2005. Differences in macrofaunal and seagrass assemblages in seagrass beds with and without seaweed farms. *Estuarine, Coastal and Shelf Science* 63: 385–396.

Ellison, A. M. 2008. Managing mangroves with benthic biodiversity in mind: moving beyond roving banditry. *Journal of Sea Research* 59: 2–15.

El-Sammak, A. 2001. Heavy metal pollution in bottom sediment, Dubai, United Arab Emirates. *Bulletin of Environmental Contamination and Toxicology* 67: 295–302.

Espinosa, J. and Ortea, J. 1998a. Nuevo género y nueva especie de Molusco Gastrópodo Margineliforme (Mollusca: Gastropoda) con rádula taenioglossa. *Avicennia* 8/9: 113–116.

Espinosa, J. and Ortea, J. 1998b. Nuevas especies de la Familia Marginellidae (Mollusca: Neogastropoda) de Cuba y los Cayos de la Florida. *Avicennia* 8/9: 117–134.

Espinosa, J. and Ortea, J. 1998c. Una nueva especie del género Olivella (Mollusca: Neogastropoda) del Mar Caribe cubano. *Avicennia* 8/9: 155–158.

Espinosa, J. and Ortea, J. 1999. Descripción de nuevas marginelas (Mollusca: Marginellidae) de Cuba y del Caribe de Costa Rica y Panamá. *Avicennia* 10/11: 165–176.

Essington, T. E. 2010. Ecological indicators display reduced variation in North American catch share fisheries. *PNAS* 107: 754–759.

Fernandes, L., Day, J., Kerrigan, B., Breen, D., Mapstone, B. D., Coles, R., De'ath, G., Marsh, H., Poiner, I., Ward, T., Williams, D. and Kenchington, R. 2005a. Biophysical principles to design a network of no-take areas: The Great Barrier Reef case study. In *Proceedings of the 1st International Marine Protected Areas Congress, October 2005, Geelong, Australia,* 4–7. Townsville, Australia: Great Barrier Reef Marine Park Authority.

Fernandes, L., Day, J., Lewis, A., Slegers, S., Kerrigan, B., Breen, D., Cameron, D., Jago, B., Hall, J., Lowe, D., Innes, J., Tanzer, J., Chadwick, V., Thompson, L., Gorman, K., Simmons, M., Barnett, B., Sampson, K., De'ath, G., Mapstone, B. D., Marsh, H., Possingham, H., Ball, I., Ward, T., Dobbs, K., Aumend, J., Slater, D. and Stapleton, K. 2005b. Establishing representative no-take areas over 1/3 of the Great Barrier Reef: large-scale implementation of marine protected area theory with lessons for global application. *Conservation Biology* 19: 1733–1744.

Figueredo-Martín, T. 2009. Factibilidad económica del área protegida marina de Jardines de la Reina. Tesis presentada en opción al Título Académico de Master en Manejo integrado de Zonas Costeras. Habana, Cuba: Universidad de la Habana.

Figueredo-Martín, T., Pina-Amargós, F., Angulo Valdés, J. and Gómez Fernández, R. 2010a. Buceo contemplativo en Jardines de la Reina: caracterización y percepción sobre el estado de conservación en el área. *Revista de Investigaciones Marinas* 31: 23–32.

Figueredo-Martín, T., Pina-Amargós, F., Angulo Valdés, J. and Gómez Fernández, R. 2010b. Pesca recreativa en Jardines de la Reina, Cuba: caracterización y percepción sobre el estado de conservación del área. *Revista de Investigaciones Marinas.* 31: 141–148.

Fiore, C. L., Jarett, J. K., Olson, N. D. and Lesser, M. P. 2010. Nitrogen fixation and nitrogen transformations in marine symbioses. *Trends in Microbiology* 18: 455–463.

Foeken, D., Hoorweg, J. and Obudho, R. A. 2000. The Kenya coast: a regional study. In Hoorweg, J., D. Foeken, and R. A. Obudho, eds., *Kenya Coast Handbook. Culture, Resources and Development in the East African Littoral,* 1–10. Münster: LIT Verlag.

Friedlander, M., Parrish, J. D. and DeFelice, R. C. 2002. Ecology of the introduced snapper *Lutjanus kasmiva* (Forsskal) in the reef fish assemblage of a Hawaiian bay. *Journal of Fish Biology* 60: 28–48.

Fröcklin, S., de la Torre-Castro, M., Lindström, L., Jiddawi, N. S. and Msuya, F. E. 2012. Seaweed mariculture as a development project in Zanzibar, East Africa: a price too high to pay? *Aquaculture* 356: 30–39.

Gaither, M. R., Bowen, B. W., Toonen, R. J., Planes, S., Messmer, V., Earle, J. and Robertson, D. R. 2010. Genetic consequences of introducing allopatric lineages of bluestriped snapper (*Lutjanus kasmira*) to Hawaii. *Molecular Ecology* 19: 1107–1121.

Gardinali, P. R., Plasencia, M. D. and Maxey, C. 2004. Occurrence and transport of Irgarol 1051 and its major metabolite in coastal waters from South Florida. *Marine Pollution Bulletin* 49: 1072–1083.

Glynn, P. W. 1997. Bioerosion and coral reef growth: a dynamic balance. In Birkeland, C. (Ed.), *Life and Death of Coral Reefs,* 69–98. New York: Chapman and Hall.

Goreau, T. J., Fisher, T., Perez, F., Lockhart, K., Hibbert, M. and Lewin, A. 2008. Turks and Caicos Islands 2006 coral reef assessment: Large-scale environmental and ecological interactions and their management implications. *Revista de Biologia Tropical* 56: 25–49.

Graham, N. A. J., McClanahan, T. R., MacNeil, M. A., Wilson, S. K., Polunin, N. V. C., Jennings, S., Chabanet, P., Clark, S., Spalding, M. D., Letourneur, Y., Bigot, L., Galzin, R., Öhman, M. C., Garpe, K. C., Edwards, A. J. and Sheppard, C. R. C. 2008. Climate warming, marine protected areas and the ocean-scale integrity of coral reef ecosystems. *PLoS ONE* 3: e3039.

Graham, N. A. J., Chabanet, P., Evans, R. D., Jennings, S., Letourneur, Y., MacNeil, M. A., McClanahan, T. R., Öhman, M. C., Polunin, N. V. C. and Wilson, S. K. 2011. Extinction vulnerability of coral reef fishes. *Ecology Letters* 14: 341–348.

Graham, R.T, Roberts, C. M. and Pina-Amargós, F. 2007. Patterns of movement of large migratory fish and the design of marine protected areas networks: whale sharks in the Western Caribbean. In J. C. Day, J. Senior, S. Monk and W. Neal, eds., *First International Marine Protected Areas Congress, October 23–27 2005, conference proceedings:* IMPAC 1 2005, Geelong, Victoria, Australia, 152–153.

Green, D. H., Edmunds, P. J. and Carpenter, R. C. 2008. Increasing relative abundance of *Porites astreoides* on Caribbean reefs mediated by an overall decline in coral cover. *Marine Ecology Progress Series* 359: 1–10.

Grimm, D., Barkhorn, I., Festa, D., Bonzon, K., Boomhower, J., Hovland, V. and Blau, J. 2012. Assessing catch shares' effects evidence from federal United States and associated British Columbian fisheries. *Marine Policy* 36: 644–657.

Grimsditch, G., Alder, J., Nakamura, T., Kenchington, R. and Tamelander, J. 2012. The blue carbon special edition—introduction and overview. *Ocean & Coastal Management.* In press.

Guimarais, M., Suárez, A. M. and Cabrera, R. 2009. Especies de *Halimeda,* sección Micronesica (Briopsidales: Halimedacea) en el archipiélago Jardines de la Reina, SE de Cuba. *Revista de Investigaciones Marinas* 30: 169–172.

Gutiérrez, N. L., Hilborn, R. and Defeo, O. 2011. Leadership, social capital and incentives promote successful fisheries. *Nature* 470: 386–389.

Gundlach, E. R. and Hayes, M. O. 1978. Vulnerability of coastal environments to oil spill impacts. *Marine Technology Society Journal* 12: 18–27.

Guzmán, H. M., Cortés, J., Glynn, P. W. and Richmond, R. H. 1990. Coral mortality associated with dinoflagellate blooms in the Eastern Pacific (Costa Rica and Panama). *Marine Ecology Progress Series* 60: 299–303.

Harrison, P. 1979. *Inside the Third World: the Anatomy of Poverty.* London: Penguin Books.

Hawkins, J. P., Roberts, C. M., Van'T Hof, T., Meyer, K. D., Tratalos, J. and Aldam, C. 1999. Effects of recreational SCUBA diving on Caribbean coral and fish communities. *Conservation Biology* 13: 888–897.

Herberholz, J. and Schmitz, B. 1998. Role of mechanosensory stimuli in intraspecific agonistic encounters of the snapping shrimp (*Alpheus heterochaelis*). *Biological Bulletin* 195.2: 156–167.

Hernández-Fernández, L. and Varela, C. 2007. Primer registro de *Carijoa riseii* (Octocorallia: Telestacea), para aguas cubanas. *Cocuyo* 16: 5.

Hernández-Fernández, L., Guimarais-Bermejo, M., Arias-Barreto, R. and Clero-Alonso, L. 2011. Composición de las comunidades de octocorales y corales pétreos y la incidencia del blanqueamiento del 2005 en Jardines de la Reina, Cuba. *Rev. Mar. Cost* 3: 77–90.

Hicks, C. C. and McClanahan, T. R. 2012. Assessing gear modifications needed to optimize yields in a heavily exploited, multi-species, seagrass and coral reef fishery. *PloS ONE* 7: e36022.

Hicks C. C., McClanahan, T. R., Cinner, J. E. and Hills, J. M. 2009. Trade-offs in values assigned to ecological goods and services associated with different coral reef management strategies. *Ecology and Society* 14: 10.

Hilborn, R., Punt, A. E. and Orensanz, J. 2004. Beyond band-aids in fisheries management: fixing world fisheries. *Bulletin of Marine Science* 74: 493–507.

Hoegh-Guldberg, O., Mumby, P. J., Hooten, A. J., Steneck, R. S., Greenfield, P., Gomez, E., Harvell, C. D., Sale, P. F., Edwards, A. J., Caldeira, K., Knowlton, N., Eakin, C. M., Iglesias-Prieto, R., Muthiga, N. Bradbury, R. H., Dubi, A. and Hatziolos, M. E. 2007. Coral reefs under rapid climate change and ocean acidification. *Science* 318: 1737–1742.

Hughes, T. P., Graham, N. A. J., Jackson, J. B. C., Mumby, P. J. and Steneck, R. S. 2010. Rising to the challenge of sustaining coral reef resilience. *Trends in Ecology and Evolution* 25: 633–642.

Ibarzábal, D. 2001. Eunícidos (Annelida: Polychaeta) del Archipiélago Jardines de la Reina, plataforma suroriental de Cuba. *Avicennia* 14:75–83.

Inger, R., Attrill, M. J., Bearhop, S., Broderick, A. C., James Grecian, W., Hodgson, D. J., Mills, C., Sheehan, E., Votier, S. C., Witt, M. J. and Godley, B. J. 2009. Marine renewable energy: potential benefits to biodiversity? An urgent call for research. *Journal of Applied Ecology* 46: 1145–1153.

Jackson, J. B. C., Cubit, J. D., Keller, B. D., Batista, V., Burns, K., Caffey, H. M., Caldwell, R. L., Garrity, S. D., Getter, C. D., Gonzalez, C., Guzman, H. M., Kaufmann, K. W., Knap, A. H., Levings, S. C., Marshall, M. J., Steger, R., Thompson, R. C. and Weil, E. 1989. Ecological effects of a major oil spill on Panamanian coastal marine communities. *Science* 243: 37–44.

Jennings, S., Greenstreet, S. P. R. and Reynolds, J. D. 1999. Structural change in an exploited fish community: a consequence of differential fishing effects on species with contrasting life histories. *Journal of Animal Ecology* 68: 617–627.

Johnson, A. E. 2010. Reducing bycatch in coral reef trap fisheries: escape gaps as a step towards sustainability. *Marine Ecology Progress Series* 415: 201–209.

Jones, G. P., Ferrell, D. J. and Sale, P. F. 1992. Fish feeding and dynamics of soft-sediment mollusc populations in a coral reef lagoon. *Marine Ecology Progress Series* 80: 175–190.

Jones, G. P., McCormick, M. I., Srinivasan, M. and Eagle, J. V. 2004. Coral decline threatens fish biodiversity in marine reserves. *Proceedings of the National Academy of Sciences* 101: 8251–8253.

Kahumba, P. 2009. Planning and evaluating performance of ecosystem restoration projects—the case of Bamburi limestone quarry. In Hoorweg, J. and N. Muthiga, eds., *Advances in Coastal Ecology: People, Processes and Ecosystems in Kenya,* 209–219. Leiden: African Studies Centre.

Kenya Bureau of Statistics. http://www.knbs.or.ke/Population%20Statistics.php accessed 5 August 2012.

Kenya Gazette. 2001. Notice No. 7565. Nairobi.

Kenya Ports Authority. http://www.kpa.co.ke accessed 5 August 2012.

Kitha, J. and Lyth, A. 2011. Urban wildscapes and green spaces in Mombasa and their potential contribution to climate change adaptation and mitigation. *Environment and Urbanization* 23: 251–265.

Knowlton, N., Brainard, R. E., Fisher, R., Moews, M., Plaisance, L. and Caley, M. J. 2010. Coral reef biodiversity. In: McIntyre, A. D. (ed.), *Life in the World's Oceans. Diversity, Distribution and Abundance,* 65–77. Oxford: Wiley-Blackwell.

Kritzer, J. P. 2004. Effects of noncompliance on the success of alternative designs of marine protected-area networks for conservation and fisheries management. *Conservation Biology* 18: 1021–1031.

Kritzer, J. P. and Sale, P. F. 2004. Metapopulation ecology in the sea: from Levins' model to marine ecology and fisheries science. *Fish and Fisheries* 5: 131–140.

Kritzer, J. P. and Sale, P. F. 2006. The metapopulation ecology of coral reef fishes. In Kritzer, J. P. and P. F. Sale (eds.), *Marine Metapopulations,* 31–67. San Diego: Academic Press.

Larkum, A. W. D., Kennedy, I. R. and Muller, W. J. 1988. Nitrogen fixation on a coral reef. *Marine Biology* 98: 143–155.

Levin, P. S., Fogarty, M. J., Murawski, S. A. and Fluharty, D. 2009. Integrated ecosystem assessments: developing the scientific basis for ecosystem-based management of the ocean. *PLoS Biology* 7: e1000014.

Little, L. R., Punt, A. E., Mapstone, B. D., Begg, G. A., Goldman, B. and Ellis, N. 2009a. Different responses to area closures and effort controls for sedentary and migratory harvested species in a multi-species coral reef line fishery. *ICES Journal of Marine Science* 66: 1931–1941.

Little, L. R., Punt, A. E., Mapstone, B. D., Begg, G. A., Goldman, B. and Williams, A. J. 2009b. An agent-based model for simulating trading of multi-species fisheries quota. *Ecological Modelling* 220: 3404–3412.

Littler, M. M., Littler, D. S. and Brooks, B. L. 2006. Harmful algae on tropical coral reefs: bottom-up eutrophication and top-down herbivory. *Harmful Algae* 5: 565–585.

Loya, Y. and Rinkevich, B. 1980. Effects of oil pollution on coral reef communities. *Marine Ecology Progress Series* 3: 167–180.

MacKenzie, B. R., Horbowy, J. and Köster, F. W. 2008. Incorporating environmental variability in stock assessment: predicting recruitment, spawner biomass, and landings of sprat (*Sprattus sprattus*) in the Baltic Sea. *Canadian Journal of Fisheries and Aquatic Science* 65: 1334–1341.

Mangi, S. C. and Roberts, C. M. 2006. Quantifying the environmental impacts of artisanal fishing gear on Kenya's coral reef ecosystems. *Marine Pollution Bulletin* 52: 1646–1660.

Mapstone, B. D., Davies, C. R., Little, L. R., Punt, A. E., Smith, A. D. M., Pantus, F., Lou, D. C., Williams, A. J., Jones, A., Russ, G. R. and MacDonald, A. D. 2004. The effects of line fishing on the Great Barrier Reef and evaluation of alternative potential management strategies. CRC Reef Research Centre Technical Report No 52. Townsville. Australia: CRC Reef Research Centre.

Mapstone, B. D., Little, L. R., Punt, A. E., Davies, C. R., Smith, A. D. M., Pantus, F., McDonald, A. D., Williams, A. J. and Jones, A. 2008. Management strategy evaluations for line fishing in the Great Barrier Reef: balancing conservation and multi-sector fishery objectives. *Fisheries Research* 94: 315–329.

Marshall, A., Mills, J. S. and Rhodes, K. L. 2011. Passive acoustic telemetry reveals highly variable home range and movement patterns among unicornfish within a marine reserve. *Coral Reefs* 30: 631–642.

Martín Blanco, F., González Sansón, G., Pina Amargós, F. and Clero Alonso, L. 2010. Abundance, distribution and size structure of *Diadema antillarum* (Echinodermata: Diademantidae) in South Eastern Cuban coral reefs. *International Journal of Tropical Biology and Conservation* 58: 663–675.

Martín Blanco, F., Clero Alonso, L., González Sansón, G. and Pina-Amargós, F. 2011. Influence of *Diadema antillarum* populations (Echinodermata: Diadematidae) on algal community structure in Jardines de la Reina, Cuba. *Revista de Biología Tropical* 59.

McCauley, R. D. and Cato, D. H. 2000. Patterns of fish calling in a nearshore environment in the Great Barrier Reef. *Philosophical Transactions of the Royal Society of London B* 355: 1289–1293.

McCauley, R. D., Fewtrell, J. and Popper, A. N. 2002. High intensity anthropogenic sound damages fish ears. *Journal of the Acoustic Society of America* 113: 638–642.

McClanahan, T. R. 2000. Recovery of a coral reef keystone predator, *Balistapus undulates,* in East African marine parks. *Biological Conservation* 94: 191–198.

McClanahan, T. R. 2007. Management of area and gear in Kenyan coral reefs. In McClanahan T. R. and J. C. Castilla, eds., *Fisheries Management: Progress Towards Sustainability,* 166–185. London: Blackwell Press.

McClanahan, T. R. 2010. Effects of fisheries closures and gear restrictions on fishing income in a Kenyan coral reef. *Conservation Biology* 24: 1519–1528.

McClanahan, T. R. and Cinner, J. E. 2008. A framework for adaptive gear and ecosystem-based management in the artisanal coral reef fishery of Papua New Guinea. *Aquatic Conservation: Marine and Freshwater Ecosystems* 18: 493–507.

McClanahan, T. R. and Graham, N. A. J. 2005. Recovery trajectories of coral reef fish assemblages within Kenyan marine protected areas. *Marine Ecology Progress Series* 294: 241–248.

McClanahan T. R. and Hicks, C. C. 2011. Changes in life history and ecological characteristics of the fish catch community with increasing fishery management. *Fisheries Management and Ecology* 18: 50–60.

McClanahan, T. R. and Mangi, S. C. 2004. Gear-based management of a tropical artisanal fishery based on species selectivity and capture size. *Fisheries Management and Ecology* 11: 51–60.

McClanahan, T. R. and Muthiga, N. A. 1988. Changes in Kenyan coral reef community structure and function due to exploitation. *Hydrobiologia* 166: 269–276.

McClanahan, T. R., Mwaguni, S. and Muthiga, N. A. 2005a. Management of the Kenyan coast. *Ocean & Coastal Management* 48: 901–931.

McClanahan, T., Davies, J. and Maina, J. 2005b. Factors influencing resource users and managers' perceptions towards marine protected area management in Kenya. *Environmental Conservation* 32: 42–49.

McClanahan, T. R., Ateweberhan, M., Graham, N. A. J., Wilson, S. K., Ruiz Sebastian, C., Bruggemann, J. H. and Guillaume, M. M. M. 2007a. Western Indian Ocean coral communities: bleaching responses and susceptibility to extinction. *Marine Ecology Progress Series* 337: 1–13.

McClanahan, T. R., Graham, N. A. J., Calnan, J. and MacNeil, M. A. 2007b. Towards pristine biomass: reef fish recovery in coral reef marine protected areas in Kenya. *Ecological Applications* 17: 1055–1067.

McClanahan, T. R., Cinner, J. E., Maina, J., Graham, N. A. J., Daw, T. M., Stead, S. M., Wamukota, A., Brown, K., Ateweberhan, M., Venus, V. and Polunin, V. C. 2008a. Conservation action in a changing climate. *Conservation Letters* 1: 53–59.

McClanahan, T. R., Hicks, C. C. and Darling, E. S. 2008b. Malthusian overexploitation and efforts to overcome it on Kenyan coral reefs. *Ecological Applications* 18: 1516–1529.

McClanahan, T. R., Graham, N. A. J., MacNeil, M. A., Muthiga, N. A., Cinner, J. E., Bruggemann, J. H. and Wilson, S. K. 2011. Critical thresholds and tangible targets for ecosystem-based management of coral reef fisheries. *PNAS* 108: 17230–17233.

McCook, L. J., Ayling, T., Cappo, M., Choat, J. H., Evans, R. D., De Freitas, D. M., Heupel, M., Hughes, T. P., Jones, G. P., Mapstone, B. D., Marsh, H., Mills, M., Molloya, F. J., Pitcher, C. R., Pressey, R. L., Russ, G. R., Sutton, S., Sweatman, H., Tobin, R., Wachenfeld, D. R. and Williamson, D. H. 2010. Adaptive management of the Great Barrier Reef: a globally significant demonstration of the benefits of networks of marine reserves. *PNAS* 107: 18278–18285.

McFarland, W. N., Brothers, E. B., Ogden, J. C., Shulman, M. J., Bermingham, E. L. and Kotchian-Prentiss, N. M. 1985. Recruitment patterns in young french grunts, *Haemulon flavolineatum* (Family: Haemulidae) at St. Croix, U.S.V.I. *Fishery Bulletin* 83: 413–426.

Medina, Y., Moncada, F. and Nodarse, G. 2010. Selección del sitio de anidación y éxito de eclosión en nidos de tortuga carey (*Eretmochelys imbricata*). *Revista Cubana de Investigaciones Pesqueras* 27: 60–65.

Medio, D., Ormond, R. F. G. and Pearson, M. 1997. Effect of briefings on rates of damage to corals by SCUBA divers. *Biological Conservation* 79: 91–95.

Melnychuk, M. C., Essington, T. E., Branch, T. A., Heppell, S. S., Jensen, O. P., Link, J. S., Martell, S. J. D., Parma, A. M., Pope, J. G. and Smith, A. D. 2011. Can catch share fisheries better track management targets? *Fish and Fisheries* 13: 267–290.

Meyer, C. G., Papastamatiou, Y. P. and Clark, T. B. 2010. Differential movement patterns and site fidelity among trophic groups of reef fishes in a Hawaiian marine protected area. *Marine Biology* 157: 1499–1511.

Meyer, J. L., Schultz, E. T. and Helfman, G. S. 1983. Fish schools: an asset to corals. *Science* 220: 1047–1049.

Moberg, F. and Folke, C. 1999. Ecological goods and services of coral reef ecosystems. *Ecological Economics* 29: 215–233.

Moncada, F., Koike, H., Espinosa, G., Manolis, Ch., Pérez, C., Nodarse, G., Tanabe, S., Sakai, H., Webb, G., Carrillo, E., Díaz, R. and Tsubouchi, T. 1998. Movimiento e integridad de la población. *Revista Cubana de Investigaciones Pesqueras* 22: 135–142.

Moncada, F., Carrillo, E., Saenz, A. and Nodarse, G. 1999. Reproduction and nesting of the hawksbill turtle, *Eretmochelys imbricada,* in the Cuban Archipelago. *Chelonian Conservation and Biology* 3: 257–263.

Moncada, F., Abreu-Grobois, A., Muhlia-Melo, A., Bell, C., Troeng, S., Bjorndal, K., Bolten, A., Meylan, A., Zurita, J., Espinosa, G., Nodarse, G., Márquez-Millán, R., Foley, A. and Ehrhart, L. 2006. Movement patterns of green turtles (*Chelonia mydas*) in Cuba and adjacent Caribbean waters inferred from flipper tag recaptures. *Journal of Herpetology* 40: 22–34.

Moncada, F., Nodarse, G., Medina, Y. and Escobar, E. 2010. Twelve years of monitoring hawksbill turtle (*Eretmochelys imbricate*) nesting at Doce Leguas Keys and Labyrinth, Jardines de la Reina archipelago, Cuba. *Marine Turtle Newsletter* 127: 6–8.

Moustahfid, H., Link, J. S., Overholtz, W. J. and Tyrrell, M. C. 2009. The advantage of explicitly incorporating predation mortality into age-structured stock assessment models: an application for Atlantic mackerel. *ICES Journal of Marine Science* 66: 445–454.

Mumby, P. J. and Steneck, R. S. 2008. Coral reef management and conservation in light of rapidly evolving ecological paradigms. *Trends in Ecology & Evolution* 23: 555–563.

Mumby, P. J., Edwards, A. J., Arias-Gonzalez, J. E., Lindeman, K. C., Blackwell, P. G., Gall, A., Gorczynska, M. I., Harborne, A. R., Pescod, C. L., Renken, H., Wabnitz, C. C. C. and Llewellyn, G. 2004. Mangroves enhance the biomass of coral reef fish communities in the Caribbean. *Science* 427: 533–536.

Mumby, P. J., Dahlgren, C. P., Harborne, A. R., Kappel, C. V., Micheli, F., Brumbaugh, D. R., Holmes, K. E., Mendes, J. M., Broad, K., Sanchirico, J. N., Buch, K., Box, S., Stoffle, R. W. and Gill, A. B. 2006. Fishing, trophic cascades, and the process of grazing on coral reefs. *Science* 311: 98–101.

Naylor, R. L., Goldburg, R. J., Primavera, J. H., Kautsky, N., Beveridge, M. C., Clay, J., Folke, C., Lubchenco, J., Mooney, H. and Troell, M. 2000. Effect of aquaculture on world fish supplies. *Nature* 405: 1017–1024.

Negri, A. P., Smith, L. D., Webster, N. S. and Heyward, A. J. 2002. Understanding ship-grounding impacts on a coral reef: potential effects of anti-fouling paint contamination on coral recruitment. *Marine Pollution Bulletin* 44: 111–117.

Newman, M. J. H., Paredes, G. A., Sala, E. and Jackson, J. B. C. 2006. Structure of Caribbean coral reef communities across a large gradient of fish biomass. *Ecology Letters* 9:1216–1227.

Newton, K., Côté, I. M., Pilling, G. M., Jennings, S. and Dulvy, N. K. 2007. Current and future sustainability of island coral reef fisheries. *Current Biology* 17: 655–658.

Obura, D. O. 2001. Kenya. *Marine Pollution Bulletin* 42: 1264–1278.

Odum, H. T. and Odum, E. P. 1955. Trophic structure and productivity of a windward coral reef community on Eniwetok Atoll. *Ecological Monographs* 25: 291–320.

Ogwang,V. O., Odende, T. and Okwach, R. 2006. Guidelines for beach management units for Kenya. Edf project no. 8, Acp Ror 029. Nairobi: Kenya Department of Fisheries.

O'Leary, J. and McClanahan, T. R. 2010. Trophic cascades result in large-scale coralline algae loss through differential grazer effects. *Ecology* 91: 3584–3597.

Ortea, J. and Espinosa, J. 1998. Dos nuevas especies de Moluscos marinos (Mollusca: Gastropoda) recolectadas en los subarchiélagos Jardines del Rey y Jardines de la Reina, descritas en honor de los Reyes de España por su primera visita a Cuba. *Avicennia* 8/9: 1–6.

Ostrom, E. 2009. A general framework for analyzing sustainability of social-ecological systems. *Science* 325: 419–422.

Parada-Isada, A., Socarrás-Torres, E., Primelles-Rivero, J. and Hernández-Álvarez, D. 2012. New bird species and distributional records for Jardines de la Reina archipelago, Cuba, during autumn and spring migrations 2009–10. *Cotinga* 34: 55–60.

Paris, C. B., Cowen, R. K., Claro, R. and Lindeman, K. C. 2005. Larval transport pathways from Cuban snapper (Lutjanidae) spawning aggregations based on biophysical modelling. *Marine Ecology Progress Series* 296: 93–106.

Pauly, D., Silvestre, G. and Smith, I. 1989. On development, fisheries and dynamite: a brief review of tropical fisheries management. *Natural Resource Modeling* 3: 307–329.

Pikitch, E. K., Santora, C., Babcock, E. A., Bakun, A., Bonfil, R., Conover, D. O., Dayton, P., Doukakis, P., Fluharty, D., Heneman, B., Houde, E. D., Link, J., Livingston, P. A., Mangel, M., McAllister, M. K., Pope, J. and Sainsbury, K. J. 2004. Ecosystem-based fishery management. *Science* 305: 346–347.

Pina-Amargós, F. 2008. Efectividad de la Reserva Marina de Jardines de la Reina para la conservación de la ictiofauna. Tesis doctoral, Universidad de La Habana, Cuba. 92 pp.

Pina-Amargós, F. and González-Sansón, G. 2009. Movement patterns of goliath grouper *Epinephelus itajara* around southeast Cuba: implication for conservation. *Endangered Species Research* 7: 243–247.

Pina-Amargós, F., Claro, R., García-Arteaga, J. P., López-Fernández, N. and González-Sansón, G. 2007. Ictiofauna del archipiélago Jardines de la Reina, Cuba. *Revista de Investigaciones Marinas* 28: 217–224.

Pina-Amargós, F., González-Sansón, G., Jiménez del Castillo, A., Zayas Fernández, A., Martín Blanco, F. and Acosta de la Red. W. 2008a. Movement of adult fish in the Jardines de la Reina Marine Reserve, Cuba. *Proceedings of the Symposium on Caribbean Connectivity* 127–144.

Pina-Amargós, F., Hernández Fernández, L., Clero Alonso, L. and González Sansón, G. 2008b. Características de hábitats coralinos en Jardines de la Reina, Cuba. *Revista de Investigaciones Marinas.* 29: 225–237.

Pina-Amargós, F., González Sansón, G., del Castillo, A. J., Fernández, A. Z., Blanco, F. M. and de la Red, W. A. 2008c. Effects of Hurricane Dennis on Jardines de la Reina, Cuba. In Datta, R. K., ed.,

Coastal Ecosystems: Hazards, Management and Rehabilitation, 215–233. New Delhi: Daya Publishing House.

Pina-Amargós, F., González Sansón, G., del Castillo, A. J., Fernández, A. Z., Blanco, F. M., and de la Red, W. A. 2010. An experiment of fish spillover from a marine reserve in Cuba. *Environmental Biology of Fishes* 87: 363–372.

Piñon, J. and Muse, R. 2010. Coping with the Next Oil Spill: Why US-Cuba Environmental Cooperation is Critical. Issue Brief No. 2. Washington, D.C.: Brookings Press.

Pollnac, R., Christie, P., Cinner, J. E., Dalton, T., Daw, T. M., Forrester, G. E., Graham, N. A. J. and McClanahan, T. R. 2010. Marine reserves as linked social-ecological systems. *PNAS* 107: 18262–18265.

Pomeroy, R. S., Parks, J. E. and Balboa, C. M. 2006. Farming the reef: is aquaculture a solution for reducing fishing pressure on coral reefs? *Marine Policy* 30: 111–130.

Popper, A. N. and Hastings, M. C. 2009. The effects of human-generated sound on fish. *Integrative Zoology* 4: 43–52.

Pratchett, M.S, Munday, P. L., Wilson, S. K., Graham, N. A. J., Cinner, J. E., Bellwood, D. R., Jones, G. P., Polunin, N. V. C. and McClanahan, T. R. 2008. Effects of climate-induced coral bleaching on coral-reef fishes: ecological and economic consequences. *Oceanography and Marine Biology: An Annual Review* 46: 251–296.

Pratchett, M. S., Hoey, A. S., Wilson, S. K., Messmer, V. and Graham, N. A. J. 2011. Changes in biodiversity and functioning of reef fish assemblages following coral bleaching and coral loss. *Diversity* 3: 424–452.

Prince, J. D. 2003. The barefoot ecologist goes fishing. *Fish and Fisheries* 4: 359–371.

Proyecto Sabana-Camagüey. 1999. Protecting Biodiversity and Establishing Sustainable Development in the Sabana-Camagüey Ecosystem. 145 pp.

Proyecto Archipiélagos del Sur. 2009. Aplicación de un enfoque regional al manejo de las áreas marino-costeras protegidas en la Región Archipiélagos del Sur de Cuba. 89 pp.

Queensland Treasury and Trade. 2012. Population growth highlights and trends 2012. Office of Economic and Statistical Research. QLD Australia

Rawlins, B. G., Ferguson, A. J., Chilton, P. J., Arthurton, R. S., Rees, J. G. and Baldock, J. W. 1998. Review of agricultural pollution in the Caribbean with particular emphasis on small island developing states. *Marine Pollution Bulletin* 36: 658–668.

Robertson, D. R. 1982. Fish feces as fish food on a Pacific coral reef. *Marine Ecology Progress Series* 7: 253–265.

Rochet M. J., Cornillon, P. A., Sabatier, R. and Pontier, D. 2000. Comparative analysis of phylogenetic and fishing effects in life history patterns of Teleost fishes. *Oikos* 91: 255–270.

Rönnbäck, P., Crona, B. and Ingwall, L. 2007. The return of ecosystem goods and services in replanted mangrove forests: perspectives from local communities in Kenya. *Environmental Conservation* 34: 313–324.

Rouphael, A. B. and Inglis, G. J. 1997. Impacts of recreational SCUBA diving at sites with different reef topographies. *Biological Conservation* 82: 329–336.

Rouphael, A. B. and Inglis, G. J. 2001. "Take only photographs and leave only footprints"?: an experimental study of the impacts of underwater photographers on coral reef dive sites. *Biological Conservation* 100: 281–287.

Rubec, P. J., Pratt, V. R. and Cruz, F. 2001. Territorial use rights in fisheries to manage areas for farming coral reef fish and invertebrates for the aquarium trade. *Aquarium Sciences and Conservation* 3: 119–134.

Ruiz-Carus, R., Matheson, R. E. Jr., Roberts, D. E. Jr. and Whitfield, P. E. 2006. The western Pacific red lionfish, *Pterois volitans* (Scorpaenidae), in Florida: evidence for reproduction and parasitism in the first exotic marine fish established in state waters. *Biological Conservation* 128: 384–390.

Sale, P. F. 1978. Reef fishes and other vertebrates: A comparison of social structures. In Reese, E. S., and F. J. Lighter (eds.), *Contrasts in Behavior: Adaptations to the Aquatic and Terrestrial Environments.* 313–346. New York: Wiley Interscience.

Sale, P. F. 1991. Reef fish communities: open, nonequilibrial systems. In Sale, P. F. *The Ecology of Fishes on Coral Reefs.* 564–598. San Diego, Academic Press.

Sale, P. F. 2011. *Our Dying Planet: An Ecologist's View of the Crisis We Face.* Berkeley: University of California Press.

Sale, P. F. and J. P. Kritzer. 2008. Connectivity: what it is, how it is measured, and why it is important for management of reef fishes? In Grober-Dunsmore, R. and B. D. Keller (eds.), *Caribbean Connectivity: Implications for Marine Protected Area Management,* 12–26. Silver Spring, Maryland, USA: National Oceanic and Atmospheric Administration, Office of National Marine Sanctuaries.

Sale, P. F., Cowen, R. K., Danilowicz, B. S., Jones, G. P., Kritzer, J. P., Lindeman, K. C., Planes, S., Polunin, N. V. C., Russ, G. R., Sadovy, Y. J. and Steneck, R. S. 2005. Critical science gaps impede use of no-take fishery reserves. *Trends in Ecology & Evolution* 20: 74–80.

Sandin, S. A., Walsh, S. M. and Jackson, J. B. C. 2010. Prey release, trophic cascades, and phase shifts in tropical nearshore marine ecosystems. In Estes, J. and J. Terborgh, eds., *Trophic Cascades: Predators, Prey, and the Changing Dynamics of Nature,* 71–90. Island Press.

Schofield, P. J. 2010. Update on geographic spread of invasive lionfishes (*Pterois volitans* [Linnaeus, 1758] and *P. miles* [Bennet, 1828]) in the Western North Atlantic Ocean, Caribbean Sea and Gulf of Mexico. *Aquatic Invasions* 5: S117–S122.

Semmens, B. X., Buhle, E. R., Salomon, A. K. and Pattengill-Semmens, C. V. 2004. A hotspot of non-native marine fishes: evidence for the aquarium trade as an invasion pathway. *Marine Ecology Progress Series* 226: 239–244.

Silva, A. G., Lima, R. P., Gomes, A. N., Fleury, B. G. and Creed, J. C. 2011. Expansion of the invasive corals *Tubastraea coccinea* and *Tubastraea tagusensis* into the Tamoios Ecological Station Marine Protected Area, Brazil. *Aquatic Invasions* 6: S105–S110.

Silva-Oliveira, G. C., Ready, J. C., Iketani, G., Bastos, S., Gomes, S., Sampaio, I. and Maciel, C. 2011. The invasive status of *Macrobrachium rosenbergii* (De Man, 1879) in Northern Brazil, with an estimation of areas at risk globally. *Aquatic Invasions* 6: 319–328.

Sindiga, I. 1995. Wildlife-based tourism in Kenya: land use conflicts and government compensation policies over protected areas. *The Journal of Tourism Studies* 6: 45–55.

Smith, M. D., Roheim, C. A., Crowder, L. B., Halpern, B. S., Turnipseed, M., Anderson, J. L., Asche, F., Bourillón, L., Guttormsen, A. G., Khan, A., Liguori, L. A., McNevin, A., O'Connor, M. I., Squires, D., Tyedmers, P., Brownstein, C., Carden, K., Klinger, D. H., Sagarin, R. and Selkoe, K. A. 2010. Sustainability and global seafood. *Science* 327: 784–786.

Snyder, B. and Kaiser, M. J. 2009. Ecological and economic cost-benefit analysis of offshore wind energy. *Renewable Energy* 34: 1567–1578.

Spalding, M. D. and Grenfell, A. M. 1997. New estimates of global and regional coral reef areas. *Coral Reefs* 16: 225–230.

Spurgeon, J. 2003. Economic issues relating to coral reef damages from submarine fiber optic cables permitted across the sovereign submerged lands of the State of Florida. Jacobs, Reading, UK: Public Employees for Environmental Responsibility.

Sultzman, C., Halter, H. A., Craig, R. K., Meyer, D., Ruggieri, J. A. and Spurgeon, J. 2002. A professional jury report on the biological impacts of submarine fiber optic cables on shallow reefs off Hollywood, Florida. Vero Beach: Public Employees for Environmental Responsibility Coral Reef/Fiber Optic Cable Jury.

Thrush, S. F. and Dayton, P. K. 2010. What can ecology contribute to ecosystem-based management? *Annual Review of Marine Science* 2: 419–441.

Tornqvist, T. E. and Hijma, M. P. 2012. Links between early Holocene ice-sheet decay, sea-level rise and abrupt climate change. *Nature Geoscience* doi: 10.1038/NGEO1536.

Tratalos, J. A. and Austin, T. J. 2001. Impacts of recreational SCUBA diving on coral communities of the Caribbean island of Grand Cayman. *Biological Conservation* 102: 67–75.

Tricas, T., Kajiua, S. M. and Kosaki, R. 2006. Acoustic communication in territorial butterflyfish: test of the sound production hypothesis. *Journal of Experimental Biology* 209: 4994–5004.

Tucker, J. W. 1999. Species profile grouper aquaculture. SRAC Publication no. 721. Stoneville, Mississippi: Southern Regional Aquaculture Center, Mississippi State University.

Valiela, I., Bowen, J. L. and York, J. K. 2001. Mangrove forests: one of the world's threatened major tropical environments. *Bioscience* 51: 807–815.

Varela, C., Castellanos, S. and Hernández, L. 2008. Registros nuevos de invertebrados (Cnidaria y Crustacea) para Cuba. *Cocuyo* 17: 12–14.

Vermeij, M. J. A., Marhaver, K.L, Huijbers, C. M., Nagelkerken, I. and Simpson, S. D. 2010. Coral larvae move toward reef sounds. *PLoS ONE* 5.5: e10660.

Veron, J. E. N. 1986. *Corals of Australia and the Indo-Pacific.* Sydney: Angus and Robertson.

Veron, J. E. N. 2008. Mass extinctions and ocean acidification: biological constraints on geological dilemmas. *Coral Reefs* 27: 459–472.

Veron, J. E. N., Odorico, D. M., Chen, C. A. and Miller, D. J. 1996. Reassessing evolutionary relationships of scleractinian corals. *Coral Reefs* 15: 1–9.

Waycott, M., Duarte, C. M., Carruthers, T. J. B., Orth, R. J., Dennison, W. C., Olyarnik, S., Calladine, A., Fourqurean, J. W., Heck, K. L. Jr., Hughes, A. R., Kendrick, G. A., Judson Kenworthy, W., Short, F. T. and Williams, S. L. 2009. Accelerating loss of seagrasses across the globe threatens coastal ecosystems. *Proceedings of the National Academy of Sciences* 106.30: 12377–12381.

Weilgart, L. S. 2007. The impacts of anthropogenic ocean noise on cetaceans and implications for management. *Canadian Journal of Zoology* 85: 1091–1116.

Wells, J. 1957. Coral reefs. *Geological Society of America Memoir* 67: 609–631.

Wells, S., Ravilous, C. and Corcoran, E. 2006. In the Front Line: Shoreline Protection and Other Ecosystem Services from Mangroves and Coral Reefs. Cambridge, UK: United Nations Environment Programme World Conservation Monitoring Centre.

Wilkinson, C. 2008. *Status of coral reefs of the world: 2008.* Townsville, Australia: Global Coral Reef Monitoring Network and Reef and Rainforest Research Centre.

Wilson, S. K., Graham, N. A. J., Pratchett, M. S., Jones, G. P. and Polunin, N. V. C. 2006. Multiple disturbances and the global degradation of coral reefs: are reef fishes at risk or resilient? *Global Change Biology* 12: 2220–2234.

World Bank. 2009. *The Sunken Billions. The Economic Justification for Fisheries Reform: Examples of Economic Studies on Specific Fisheries.* Washington, D.C.

Worm, B., Hilborn, R., Baum, J. K., Branch, T. A., Collie, J. S., Costello, C., Fogarty, M. J., Fulton, E. A., Hutchings, J. A., Jennings, S., Jensen, O. P., Lotze, H. K., Mace, P. M., McClanahan, T. T., Minto, C., Palumbi, S. R., Parma, A. M., Ricard, D., Rosenberg, A. A., Watson, R. and Zeller, D. 2009. Rebuilding global fisheries. *Science* 325: 578–585.

Ziegler, S. and Brenner, R. 1999. Dissolved carbon cycling in a seagrass dominated lagoon. *Marine Ecology Progress Series* 180: 149–160.

Chapter 13. Open Ocean Systems

JEFFREY J. POLOVINA

National Oceanic and Atmospheric Administration
Pacific Islands Fisheries Science Center

ALISTAIR J. HOBDAY

Commonwealth Scientific and Industrial Research Organization
Marine and Atmospheric Research

J. ANTHONY KOSLOW

Scripps Institution of Oceanography

VINCENT S. SABA

National Oceanic and Atmospheric Administration
Northeast Fisheries Science Center

Contents

1. Introduction

Open ocean systems are marine ecosystems located away from the influences of continental shelves or coasts. They are dominated by a pelagic food web, with the only benthic influences coming from oceanic seamounts, oceanic ridges, or islands. Spatially, most of the global ocean is composed of open ocean systems. These systems include a diverse set of biomes including equatorial upwelling zones, and subtropical and subarctic gyres (Longhurst et al. 1995). In this chapter we examine the practice of ecosystem-based management (EBM), largely associated with fisheries, for the Pacific and Atlantic Oceans tropical and subtropical ecosystems. We select these regions based on their relatively long history of exploitation and the range of EBM approaches being implemented. Seamounts and oceanic ridges represent a unique subset of open ocean ecosystems and are covered as a separate section at the end of the chapter.

The Sea, Volume 16, edited by Michael J. Fogarty and James J. McCarthy
ISBN 978-0-674-07270-1

2. Geography, Oceanography, Ecosystem, and Management Jurisdictions

2.1 Pacific and Atlantic tropical and subtropical Ocean

The North Pacific tropical and subtropical Ocean is broadly composed of two biomes, the North Pacific Subtropical Gyre (NPSG) and the Pacific equatorial upwelling biome (EU). The North Pacific Subtropical Gyre is one of the largest biomes on the planet, about 20 million square kilometers in area, covering much of the North Pacific. Spatially, it rotates in an anticyclone (clockwise) direction with boundaries defined by the Kuroshio Extension Current and the North Pacific Current on the north, the California Current on the east, the North Equatorial Current on the south and the Kuroshio Current on the west. Vertically, it is characterized by warm, vertically stratified surface water with low surface nutrients, a seasonally variable mixed layer, and low surface chlorophyll with a subsurface chlorophyll maximum (Fig. 13.1; Karl 1999).

The Pacific equatorial upwelling (EU), also referred to as the cold tongue, spans the eastern portion of the equator, bounded by a salinity front located, on average, approximately at the dateline on the west and the North Pacific and South Pacific Subtropical Gyres on the north and south respectively (Fig. 13.1). It is characterized by equatorial upwelling resulting in strong horizontal and vertical gradients in temperature, salinity, and phytoplankton density with relatively high primary production compared to the NPSG. The size and productivity of the EU increases (decreases) substantially with La Niña (El Niño) conditions.

The South Pacific Subtropical Gyre (SPSG) extends from Australia in the west to Chile on the east, and is bounded to the south by the eastward flowing Antarctic circumpolar current, and to the north by the westward flowing South Equatorial Current (SEC) (Fig. 13.1). When the broad SEC meets the islands in the west one portion flows around northern Papua New Guinea and on into the Indian Ocean, while the southern portion forms the East Australian Current (EAC) flowing south along the Australian coast at around 18°S (Ridgway and Dunn 2003). The EAC transports heat poleward and separates from the east Australian coast at around 32°S to form the eastward flowing Tasman Front. However, EAC filaments and eddies extend farther south than the mean separation point, often extending south to Tasmania (Ridgway 2007; Holbrook 2010). The South Pacific is dotted with many small island nations, and as a result territorial control of open ocean waters via Exclusive Economic Zones (EEZ) is much greater in this world region than most others. In this chapter, when considering the South Pacific, we focus on the western portion of the South Pacific, where ecosystem-based management has been implemented in a number of ways.

The North Atlantic Subtropical Gyre circulates clockwise and is bounded by the Gulf Stream Current on the west, the North Atlantic Drift Current on the north, the Canary Current on the east, and the Atlantic Northern Equatorial Current on the south (13.1). The South Atlantic Subtropical Gyre circulates counter-clockwise and is bounded by the Brazil Current on the west, the Atlantic Southern Equatorial Current on the north, the Benguela Current on the east, and the South Atlantic Current on the south (13.1). Physically and biologically the Atlantic subtropical gyres are very similar to the Pacific subtropical gyres with nutrient input to the euphotic zone limited by vertical stratification resulting in low surface phytoplankton biomass. Mid-ocean eddies are also common in the subtropical Atlantic and

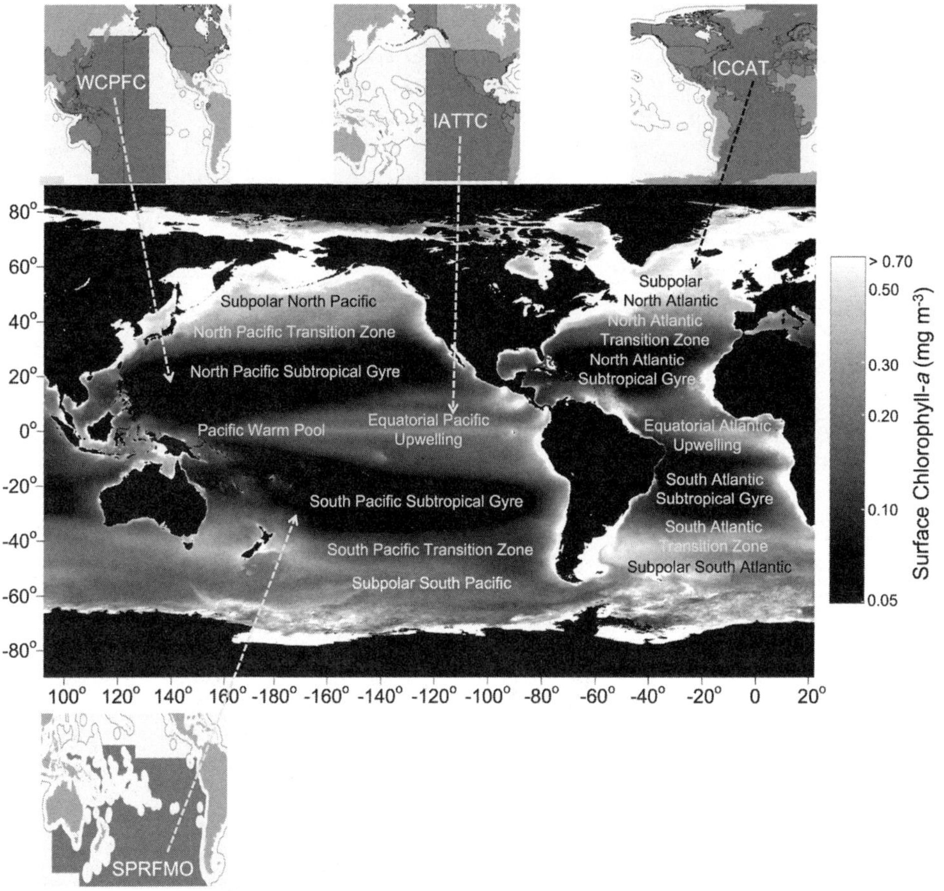

Fig. 13.1 – (See color insert) Surface chlorophyll-*a* estimates (a proxy for phytoplankton biomass) derived from the SeaWiFS ocean color satellite sensor and the major open ocean biomes of the Pacific and Atlantic Oceans. The image is a composite using monthly level-3 data (9 km resolution) from 1997 to 2010 (http://oceancolor.gsfc.nasa.gov/). The color bar for chlorophyll-*a* is on a log scale. The regions highlighted in blue in the perimeter maps are the regional fishery management organization (RFMO) boundaries (http://www.fao.org/fishery/rfb/en) for the major fisheries of the open ocean Pacific and Atlantic Oceans. These organizations are the Western and Central Pacific Fisheries Convention (WCPFC), the Inter-American Tropical Tuna Commission (IATTC), the South Pacific Regional Management Organization (SPRFMO), and the International Commission for the Conservation of Atlantic Tunas (ICCAT). The countries highlighted in gray are members of each respective RFMO.

Pacific and disrupt the vertical stratification to supply a substantial portion of nutrients to phytoplankton (McGillicuddy et al. 2007). Eddies are a particularly important feature in the southeast portion of the South Atlantic subtropical gyre where the Agulhas Current carries more eddy kinetic energy than any other region in the Southern Hemisphere such that large (300 km) eddies can cross the entire South Atlantic and reach the coast of Brazil (Longhurst 2007).

Wedged between the Atlantic subtropical gyres, spanning the equator between Africa and South America, lies the Equatorial Atlantic Upwelling biome. This biome is characterized by higher phytoplankton biomass (in the central and east sections) relative to the subtropical gyres in the north and south (Fig. 13.1). This is due to zonal wind stress causing the equatorial Atlantic thermocline to shoal in the east and deepen in the west (similar to the equatorial Pacific) and thus enhancing the vertical advection of nutrients to the euphotic zone.

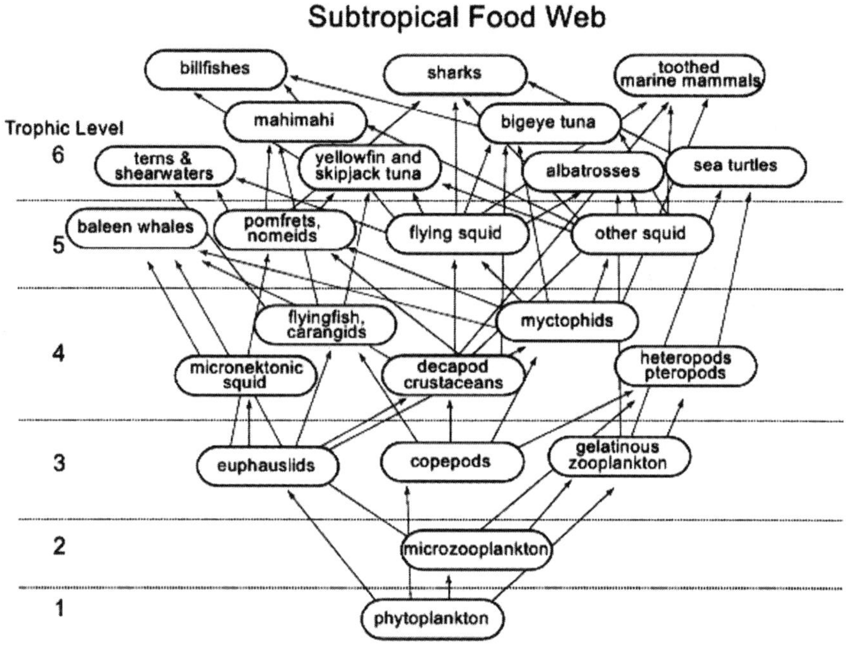

Fig. 13.2 – A generalized subtropical food web. (Seki and Polovina 2001)

2.2 *Pacific and Atlantic tropical and subtropical ecosystems*

Small plankton (picoplankton and microzooplankton) dominate the base of subtropical foodwebs, yet a long food chain with six or more trophic levels supports a highly diverse community populated at the top trophic levels by large pelagics including billfishes, tunas, sharks, and toothed marine mammals (Fig. 13.2, Seki and Polovina 2001). At the low to midtrophic levels organisms can be partitioned by depth into three broad groups: epipelagic species (0 to 200 m), mesopelagic species (200 m to 1,000 m), and bathypelagic species (below 1,000 m), although many organisms of the latter two groups make vertical migrations to shallow depths at night to feed. Large pelagic species have well defined and distinct foraging depths with some such as yellowfin and skipjack tuna using the epipelagic zone, while others such as bigeye tuna and swordfish feeding in the mesopelagic zone (Young et al. 2010). Food webs in the more productive equatorial upwelling zones are less nitrate limited, more limited by iron, and have higher primary production and a greater abundance of large plankton than the subtropical gyres but include the same large pelagic species found in the subtropical gyres.

Pelagic habitat in the tropical and subtropical biomes consists of dynamic oceanographic features including major current boundaries, fronts, and eddies and locations with interactions between currents and bathymetric features including islands, seamounts, and ridges. At the poleward boundary of the subtropical gyres, between the subtropical and subarctic gyres, are the productive transition zones that have been shown to be critical migration and foraging habitat for many upper trophic level marine organisms including sea turtles, sea birds, tunas, and sharks (Polovina et al. 2001; Block et al. 2011; Mansfield et al. 2009).

Both pelagic and demersal fisheries take place in the open ocean. Commonly caught pelagic species in the Atlantic and Pacific tropical and subtropical tuna fisheries include a diverse community of targeted species of tunas, billfishes, sharks and other pelagic fishes. However, a unique characteristic of the tropical and subtropical regions is their high biodiversity resulting in a large number of non-targeted species captured as bycatch in the longline and purse seine tuna fisheries. For example, in the east coast Australian pelagic longline fishery, some 100 species have been recorded as bycatch (Hobday et al. 2011).

2.3 Regional fisheries management organizations

At the global level there are several legal instruments that provide a framework for EBM in the open ocean. The United Nations Convention on the Law of the Sea of 10 December 1982 includes recognition of the need to balance conservation objectives and in particular the effects of fishing on the ecosystem with the need for food from the sea (Cochrane et al. this volume). The 1992 Convention of Biological Diversity highlights EBM objectives including the need for conservation of biological biodiversity, establishment of protected areas, protection of ecosystems and natural habitats, maintenance of viable species populations, and restoration of degraded ecosystems (Cochrane et al. this volume).

Under this global framework, the management of tuna and billfish fisheries in the tropical and subtropical Pacific occurs at several levels. The tropical and subtropical Pacific is covered by two international regional fishery management organizations (RFMO). The Western and Central Pacific Fisheries Management Commission (WCPFC) manages pelagic fisheries west of 150° W in the North Pacific and west of 130° W in the South Pacific. The Inter-American Tropical Tuna Commission (IATTC) manages pelagic fisheries from the WCPFC boundary eastward in both the North and South Pacific (Fig. 13.1). In addition to international management, individual countries may impose regulations on their fishing fleets or in their Exclusive Economic Zone (EEZ). For example, in the case of the Hawaii-based longline fishery which targets bigeye tunas and swordfish over a broad region in the central NPSG, the Western Pacific Regional Fishery Management Council (WPRFMC) under the United States Department of Commerce has responsibility to ensure the fishery conforms to applicable U.S. laws, such as the Magnuson-Stevens Fisheries Conservation and Management Act and the Endangered Species Act (ESA).

While the WCPFC and IATTC cover both the North and South Pacific (Fig. 13.1) their focus is primarily highly migratory pelagic resources. In 2009 the South Pacific Regional Fisheries Management Organization (SPRFMO) was formed to close a gap in the international conservation and management of non-highly migratory fisheries including pelagic and demersal species and protection of biodiversity in the marine environment extending from the most eastern part of the South

Indian Ocean through the Pacific towards the Exclusive Economic Zones (EEZs) of South America. There is an effort to establish a similar regional fisheries management organization for the North Pacific to address management of non-highly migratory species.

The predominant intergovernmental organization responsible for the management of Atlantic tunas, swordfish, billfish, small tuna, and pelagic sharks is the International Commission for the Conservation of Atlantic Tunas (ICCAT). The ICCAT management area covers the entire Atlantic basin from pole to pole (Fig. 13.1). The major responsibilities of ICCAT consist of field research, stock assessment, and science-based management advice, and it provides a forum where government bodies from various countries can come to a consensus on management strategies.

Open ocean Atlantic fisheries targeting species not regulated by ICCAT (i.e. deep-sea fishes, squids, crabs) are regulated by other regional fishery bodies. The Northeast Atlantic Fisheries Commission (NEAFC) manages the deep-sea fisheries of the northeast Atlantic. The regulatory area of the NEAFC ranges from the waters outside the EEZ of Spain to the mid-Atlantic ridge and extends in the north to the Barents Sea and south to about 38° N. The Northwest Atlantic Fisheries Organization (NAFO) manages fisheries outside the EEZs of the northwest Atlantic. The fisheries managed by NAFO are predominately bottom trawl fisheries targeting demersal fishes such as cod but also catch many true deep-sea fishes. The Southeast Atlantic Fisheries Organization (SEAFO) is the managing body for deep-sea fisheries in the southeast Atlantic with a convention area ranging from 6° S to 50° S and from outside of Africa's EEZs to about 20° W. The Western Central Atlantic Fisheries Commission (WEFCAFC) manages the area from 35° N to 10° S and is bounded by 40° W and outside the EEZs of the United States, Central America, and South America. Herrings, sardines, and anchovies make up the majority of the catch in the WEFCAFC management area.

3. Ecosystem Services

The value of services provided by oceanic species and ecosystems can be expressed in economic terms, with varying ease depending on the service (Costanza et al. 1997; Levin and Lubchenco 2008). Most difficult is the valuation of "ecosystem function" services, such as filtering water or regulating climate (e.g. Costanza et al. 1997). For example, while the oceans are crucial in absorbing anthropogenic carbon dioxide, and phytoplankton are responsible for about one third of world primary production (Chavez et al. 2011), it is very difficult to calculate an economic value for these services. More easily valued are the extractive industries, such as fishing, or activities that generate significant employment, such as tourism. The primary ecosystem services in the tropical and subtropical Pacific, at least in pure economic terms, are those associated with the fisheries of large pelagic fishes, principally tunas and billfishes. These revenues include not only the value of the catch but license fees paid to fish within another country's EEZ, revenues generated from canneries, fishing gear sales, boat building and repair, and infrastructure provision (e.g. wharves and roads).

At the basin-scale, in 2009, tuna fisheries for the entire Pacific caught about 3.0

million metric tons (mt) of tuna representing 72% of the global tuna landings (WCPFC 2009). In the Pacific, about 80% of the tuna landings came from the WCPFC region and 20% from the IATTC region (WCPFC 2009). In the WCPFC region the tuna catches by weight were 72% skipjack, 18% yellowfin, 5% bigeye, and 5% albacore tunas (WCPFC 2009). The tuna stocks are fished by both distant water nations as well as vessels from Pacific Island nations (also known as 'coastal states'). The value of the tuna catches in the WCPFC region in 2009 was US$4.1 billion (FFA 2010). An important ecosystem service is the benefit of the fisheries to the small Pacific Island nations. In 2009 it was estimated that the industry provided employment to 12,000 islanders, US$167 million to the island economies, and food security for island populations (FFA 2010). Even within the US, for the State of Hawaii, American Samoa, Guam, and the Commonwealth of the Marianas, the value of the pelagic fishery is significant. For example, in 2009, the domestic Hawaii pelagic fisheries landed 12.3 thousand mt of pelagic fishes valued at $67 million, largely from the Hawaii-based longline fishery (WPRFMC 2010).

As in the Pacific Ocean, the major economic ecosystem services of the open ocean Atlantic Ocean are the landings of tuna, billfish, and shark species (Fig. 13.3). Fishing effort and catches increased from the 1950s to the 1990s followed by a modest decline to the present as a result of catch limits on target species (Fig. 13.3, ICCAT SC-RS 2011). In 2010, the total landings of tuna in the entire Atlantic Ocean and Mediterranean Sea from all gear types was approximately 420 thousand mt (ICCAT SC-RS 2011), substantially less than the total catch in the Pacific Ocean from 2009. The breakdown by species of these landing weights for 2010 was 43% skipjack, 26% yellowfin, 18% bigeye, 10% albacore and 3% northern bluefin (ICCAT SC-RS 2011). Northern bluefin tuna is the most valuable species and those landed in the eastern Atlantic Ocean and Mediterranean Sea in 2010 had a market value of US$300 million (Bonini et al. 2011). Swordfish represents an important component of the fishery catches, accounting for about 40 thousand mt in 2010 (Fig. 13.3). Sharks are caught as bycatch in the tuna and swordfish longline fishery, but they are often not retained so catch and retention rate data are not well reported. However, the reported catch data indicates substantial numbers, particularly of blue shark, are caught (Fig. 13.3).

Although this chapter is focused on pelagic fisheries and ecosystems, it is worth discussing the deep-sea fisheries, even though the knowledge base on these fisheries, ecosystems, and fish is relatively data poor. The decline in coastal fishery resources combined with advances in gear technology has led to the gradual increase in fishing effort in deep-sea ecosystems. For example, in the northeast Atlantic, there are deep-sea fisheries employing bottom-trawls and highly mechanized longlines between the shelf areas and west of the mid-Atlantic ridge (Gordon et al. 2003). The year-round target species are not true deep-sea species (as defined by the International Council for the Exploration of the Sea) and consist primarily of cod (*Gadus morhua*), redfish (*Sebastes mentella*), and Greenland halibut (*Reinhardtius hippoglossoides*). In 2009, the largest landings of true deep-sea species in the northeast Atlantic were ling (*Molva molva*), argentines (*Argentina silus*), tusk (or cusk) (*Brosme brosme*), blue ling (*Molva dypterygia*), and black scabbardfish (*Aphanopus carbo*) (NEAFC 2009). Other highly landed species in

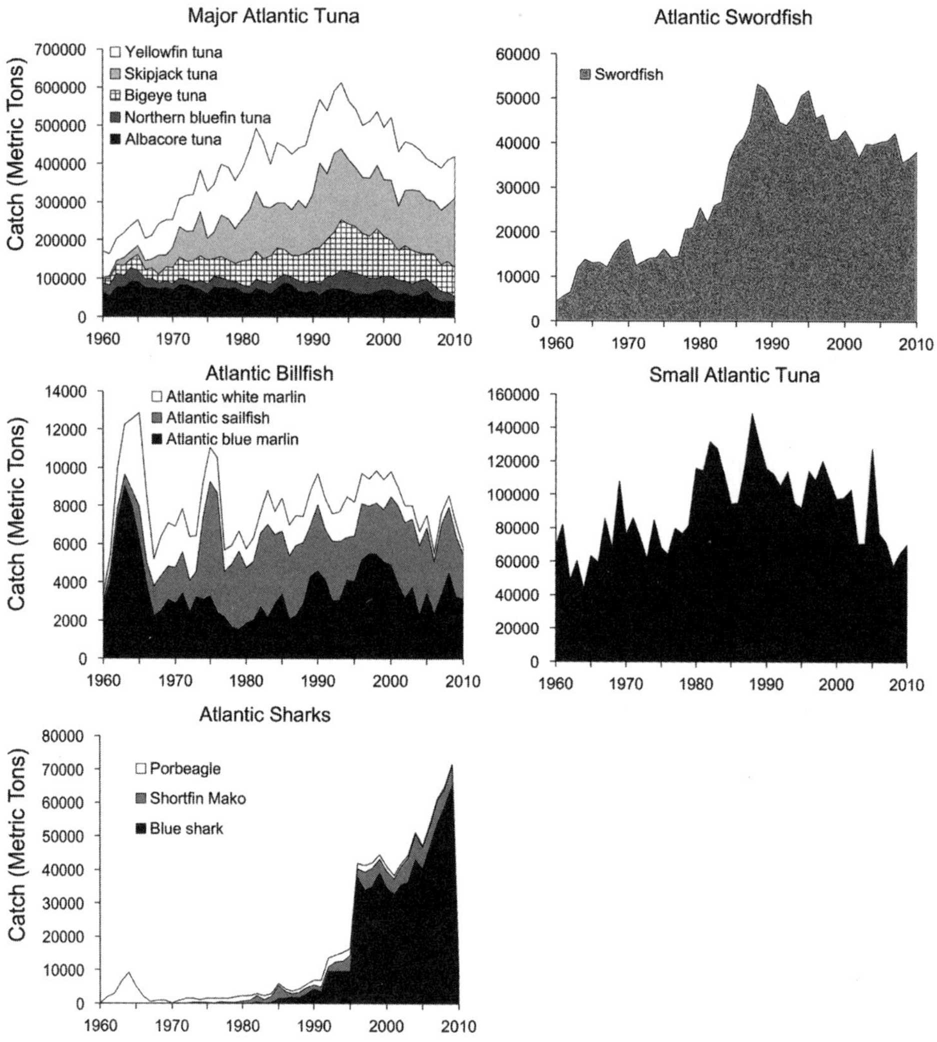

Fig. 13.3 – Time-series (1960–2010) of nominal catch of the various major and small tuna, swordfish, billfish, and sharks in the Atlantic Ocean and Mediterranean Sea (http://www.iccat.es/en/accesingdb.htm). Note the scale differences on each panel's y-axis. Major tuna are yellowfin (*Thunnus albacares*), skipjack (*Katsuwonus pelamis*), bigeye (*Thunnus obesus*), northern bluefin (*Thunnus thynnus*), and albacore (*Thunnus alalunga*). Swordfish and billfish are Atlantic swordfish (*Xiphias gladius*), white marlin (*Tetrapturus albidus*), sailfish (*Istiophorus albicans*), and blue marlin (*Makaira nigricans*). Small tuna are blackfin tuna (*Thunnus atlanticus*), plain bonito (*Orcynopsis unicolor*), frigate tuna (*Auxis thazard*), little tunny (*Euthynnus alletteratus*), wahoo (*Acanthocybium solandri*), bullet tuna (*Auxis rochei*), serra Spanish mackerel (*Scomberomorus brasiliensis*), king mackerel (*Scomberomorus cavalla*), West African Spanish mackerel (*Scomberomorus tritor*), Atlantic bonito (*Sarda sarda*), cero (*Scomberomorus regalis*), seerfishes nei (*Scomberomorus* spp.), and Atlantic Spanish mackerel (*Scomberomorus maculatus*). Major sharks are blue shark (*Prionace glauca*), shortfin mako (*Isurus oxyrinchus*), and porbeagle (*Lamna nasus*). (ICCAT SC-RS 2011)

previous years include Baird's slickhead (*Alepocephalus bairdii*), European conger (*Conger conger*), and roundnose grenadier (*Coryphaenoides rupestris*) (NEAFC 2007). In some areas of the northeast Atlantic, the extremely long-lived orange roughy (*Hoplostethus atlanticus*) and various species of Squaliformes (dogfish sharks) represent a significant portion of the catch (Gordon et al. 2003). In the southeast Atlantic, highly landed species comprise orange roughy, deep-sea crab (*Chaceon* spp.), and Patagonian toothfish (*Dissostichus eleginoides*) (www.seafo.org).

In the Pacific Ocean, the majority of deep-sea catch derives from the South Pacific, primarily near Australia, New Zealand, and Chile (FAO 2007). In 2004, the highest landings in the Southwest Pacific were blue grenadier (*Macruronus novaezelandiae*), southern blue whiting (*Micromesistius australis*), pink cusk-eel (*Genypterus blacodes*), and oreo dory (*Neocyttus rhomboidalis*). Patagonian grenadier (*Macruronus magellanicus*), southern blue whiting, and Patagonian toothfish (*Dissostichus eleginoides*) were the predominant catch in the Southeast Pacific (FAO 2007). In the North Pacific, the majority of the deep-sea catch was hairtail (*Trichiurus* spp.).

Another ecosystem service includes ecotourism benefits from sports fishing and whale, dolphin, and shark tours offered by commercial operators in many locations. Whale watching is an increasingly important economic and recreational activity. In 1998, whale watching occurred in 492 communities in 87 countries (Hoyt and Hyenegaard 2002). More than 9 million tourists participated in whale-watching excursions, which resulted in an estimated $299.5 million in direct expenditures and $1,049 million in total expenditures. Since 1991, global whale-watching numbers have increased by 12.1% per year (Hoyt and Hyenegaard 2002). For example, in Hawaii in 1999, 52 vessels offered whale-watching trips for 370,000 passengers (Utech 2000). Direct revenues from these trips for 1999 are estimated at $11–16 million and the total economic impact to Hawaii was estimated at $19–27 million supporting 280–390 full-time jobs (Utech 2000). Coastal whale watching in Australia focuses on two migratory species, humpback whales (*Megaptera novaeangliae*) and southern right whales (*Eubalaena australis*) that are seasonally present, with some million participants per year, and considerable value for regional economies. More recently, Minke whale-based tourism has appeared, with swim-with-the-whale experiences available from Cairns (Great Barrier Reef) (Valentine et al. 2004). Many south Pacific nations have declared their EEZ as whale sanctuaries—indicating the value of these ecosystem services.

Similarly, Pacific Island nations are finding the value of sharks to their dive industry greatly exceeds their fishery value. A study in Palau documented that the opportunity to dive with sharks was the primary reason about 21% of the divers (about 8,600 annually) visited Palau (Vianna 2010). The annual economic benefit from these divers was estimated to be US$18 million to the economy representing about 8% of the islands gross domestic product. This income represented US$1.2 million in salaries paid to the local community and US$1.5 million in tax revenues. It was thus estimated that the economic value of an individual reef shark to the Palauan economy is US$179,000 annually, representing a lifetime value of US$1.9 million to the tourism industry compared to an estimated value of US$108 for a shark caught by the fishery (Vianna et al. 2010).

4. Threats and Impacts

There are a range of threatening activities that can impact marine ecosystems in open ocean regions, including overfishing, climate change, pollution, invasive species, mining, shipping, and oil and gas extraction, most of which are concentrated in the upper 200 meters (Halpern et al. 2008; Game et al. 2009). However, there is also a growing awareness of human-induced impacts to deep ocean and deep benthic habitats and these threats will likely increase in the future, especially as mining and oil and gas extraction move into deeper waters (Ramirez-Llodra et al. 2011). In the open ocean, fishing is considered the most widespread and dramatic impact for biological systems (Halpern et al. 2008; Game et al. 2009).

4.1 Fishing impacts on target fish populations

By one estimate, since 1950, fisheries have removed 50 million tons of tunas, billfishes, and other pelagic fishes from the Pacific Ocean (Sibert et al. 2006). There are several perspectives on the current status of Pacific tunas and billfishes as a result of fishing. One analysis using Japanese longline catch per unit effort (CPUE) data as an index of population abundance concluded that the aggregated biomass of large pelagic fishes, principally tunas and billfishes, has declined to 10% of its pre-fishery level (Myers and Worm 2005). However, many fisheries researchers argue that longline catch per unit of effort is not an accurate representation of population abundance (Hampton et al. 2005; Maunder et al. 2006). A subsequent analysis based on stock assessment models finds fishing has had a more moderate impact, at least on six target tuna species, estimating that their current biomasses are 36 to 91% of their estimated biomasses in the absence of fishing (Sibert et al. 2006). However, fish larger than 175 centimeters fork length have decreased to about 20% of their pre-exploitation abundance (Sibert et al. 2006).

The long-lived nature of large tunas and billfishes along with the restricted geographical range and short spawning periods of some of these species (i.e. bluefin tuna) make them more susceptible to collapse when fishing pressure is high. In particular, the extremely high economic value of bluefin tuna renders their stocks more likely to be overexploited and fished well beyond maximum sustainable yield (MSY) (Collette et al. 2011a). For example, in the Atlantic Ocean and Mediterranean Sea, the International Union for Conservation of Nature (IUCN) has listed northern bluefin tuna as endangered while bigeye, blue marlin, and white marlin are listed as vulnerable (Collette et al. 2011b–e). In the South Atlantic and throughout its range in the South Pacific and South Indian Oceans, the southern bluefin tuna (*Thunnus maccoyii*) is listed as critically endangered (Collette et al. 2011e). The present-day biomass of southern bluefin tuna is estimated to be 5% of its virgin biomass (Collette et al. 2011a). It is important to note that successful management and recovery of some tuna and billfish have occurred. Atlantic swordfish and populations of Spanish mackerel in the southeastern United States have been rebuilt after long periods of overfishing (Collette et al. 2011a), and an international agreement now exists for rebuilding of southern bluefin tuna stocks.

4.2 *Fishing impacts on bycatch species and biodiversity*

Longline and purse seine fishing for tunas and billfishes catch non-target species including both protected species (many sea birds, sea turtles, sharks, and cetaceans) and a diverse group of pelagic fishes and sharks. The main oceanic sharks caught in pelagic longline fisheries include the blue shark (*Prionace glauca*), the common thresher shark (*Alopias vulpinus*), the bigeye thresher shark (*A. superciliousus*), the shortfin and longfin mako sharks (*Isurus oxyrinchus* and *Isurus paucus*), the silky shark (*Carcharhinus falciformis*) and the oceanic whitetip shark (*Carcharhinus longimanus*). Often when sharks are caught their fins are removed and retained (finning) while the rest of the body is discarded but some sharks, typically silky, oceanic whitetip, and shortfin mako sharks, are often fully retained (Clarke 2011). A recent review of the status of key shark species in the central and western Pacific noted that the status of most shark species is worsening over time due to finning and directed fishing, with the greatest declines occurring for the oceanic whitetip shark (Clarke 2011).

Interactions between sea turtles and longlines are rare events and often the turtle is released alive. However, given the large longline effort in the Pacific, the fact that longlines catch older, reproductive age classes of endangered leatherback (*Dermochelys coriacea*) and loggerhead (*Caretta caretta*) sea turtles, the cumulative negative population impact of pelagic longline bycatch mortalities on these species is thought to be significant (Lewison and Crowder 2007). Sea birds are caught by longline gear as they eat bait when the gear is at the surface during setting or retrieval. For example, in 2000 in the Hawaii-based longline fishery an estimated 2433 sea birds were hooked (Gilman et al. 2008). One of the best studied marine mammal interactions with pelagic fishing gear occurred in the eastern tropical Pacific when purse seine fishing gear was set on combined schools of tunas and dolphins that in the 1960s and 1970s resulted in hundreds of thousands of dolphin mortalities (Lewison et al. 2004). In the central Pacific cetaceans including pilot whales, false killer whales (*Pseudorca crassidens*) and others are occasionally caught in pelagic longlines as they take fish or bait from the gear.

The diversity of tunas and billfishes has been examined based on Japanese longline catch data revealing distinct subtropical "hotspots" with maximum diversity often occurring at sea surface temperature (SST) of 25°C, at oceanographic fronts, and at shelf breaks (Worm et al. 2005). Seamounts are also habitats that support a zone of high pelagic biodiversity extending up to 40 km from their summits (Morato et al. 2010). One study examining tuna and billfish diversity found it has declined globally concurrent with an increase in fishing effort in the Japanese longline data (Worm et al. 2005). The value of maintaining biodiversity is highlighted in a study examining trends and impacts of biodiversity in coastal large marine ecosystems which found that marine biodiversity loss is increasingly impairing the ocean's capacity to provide food, maintain water quality, and recover from perturbations (Worm et al. 2006). It was estimated that rates of resource collapse increased and recovery potential, stability, and water quality decreased exponentially with declining diversity (Worm et al. 2006). The restoration of biodiversity, by contrast, increased productivity, on average, fourfold and decreased variability by 21% (Worm et al. 2006). It is likely that fishing negatively impacts open ocean biodiversity given that longlines and purse seines also catch and kill a diverse assemblage of fishes (Hobday et al. 2011), but a challenge in documenting fishery impacts on pelagic fish bycatch and biodiversity is that reliable estimates require a high level of fishery observer coverage which is lacking in many open ocean fisheries.

4.3 Fishing impacts on the ecosystem

A number of ecosystem model simulations, using Ecopath with Ecosim, have explored the role of top down controls on tropical and subtropical ecosystems. In one study a model for the central North Pacific subtropical system was used to investigate top down ecosystem impacts and concluded that there was not any single species group of apex predators that functioned as keystone predators but a broad removal of apex predators due to fishing could result in an increased abundance at the next lower trophic level (Kitchell et al. 2002). Thus, in effect the fishing fleet is the keystone predator in these systems.

An empirical study from the Hawaii-based longline fishery using a decade-long time series of observer and logbook data shows an increase in relative abundance of mid-trophic level species concurrent with a decline in apex predators over the period 1996–2006 (Polovina et al. 2009). Specifically, catch rates for five apex predators, blue shark (*Prionace glauca*), bigeye (*Thunnus obesus*) and albacore (*Thunnus alalunga*) tunas, shortbill spearfish (*Tetrapturus angustirostris*), and striped marlin (*Tetrapturus audax*), declined by 3% to 9% per year while catch rates for four mid-trophic species, mahimahi (*Coryphaena hippurus*), sickle pomfret (*Taractichthys steindachneri*), escolar (*Lepidocybium flavobrunneum*), and snake mackerel (*Gempylus serpens*), increased by 6% to 18% per year (Fig. 13.4, Polovina et al. 2009). The mean trophic level of the catch for these 13 species declined 5% and the proportion of apex predators in the catch declined from 70% to 40%. Thus, both model and empirical studies suggest that if fishing broadly reduces apex predator abundance then fishing can alter the structure of the ecosystem by reducing the abundance at the highest trophic level that reduces predation and competition resulting in an increase in the abundance at lower trophic levels. However if only a subset of the apex predators are impacted there appears to be sufficient redundancy in the diets of the remaining predators to maintain predation pressure on lower trophic levels and the ecosystem structure is not significantly altered (Kitchell et al. 2002; Griffiths et al. 2010).

In the Atlantic Ocean and Mediterranean Sea, the landings of major tuna and swordfish have substantially increased since the 1960s (Fig. 13.3) and thus there is concern over the ecosystem impacts from the increased harvest of these top predators (Collette et al. 2011a). A meta-analysis of fisheries and ecosystem model simulations including examples from the oceanic Atlantic and Pacific found that a reduction in the abundance of top predators appeared to trigger an increase in abundance in lower trophic levels that was widespread (Baum and Worm 2009). However, yet to be answered is whether changes in ecosystem structure could be avoided if predators were maintained above some threshold level (Baum and Worm 2009). Also, it was noted that due to our inability to adequately sample the entire ecosystem, especially in the open ocean, it was not known how far the ecosystem changes triggered by a reduction of top predators moved down through the trophic levels (Baum and Worm 2009). Presently, most reports of top-down con-

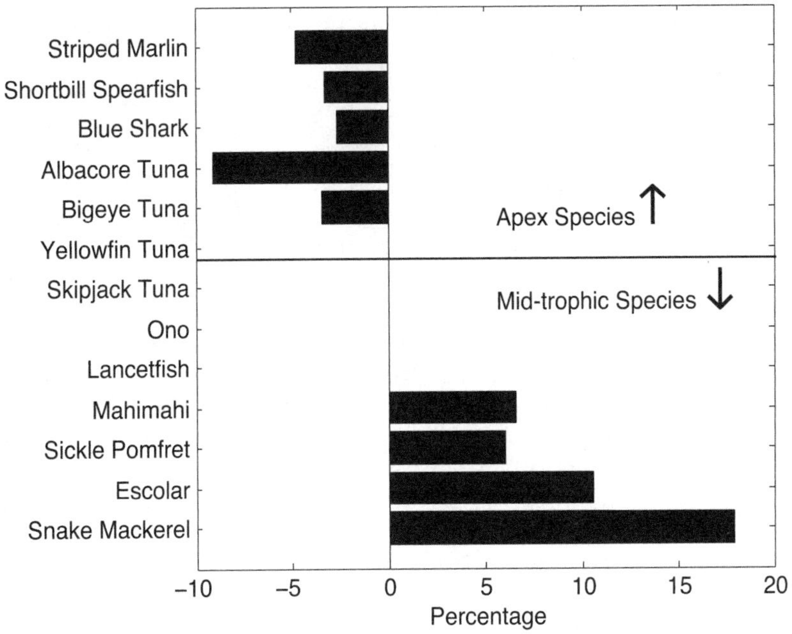

Fig. 13.4 – Annual percent change in catch rate by species in the Hawaii-based longline fishery, 1996–2007. (Polovina et al. 2009)

trol in marine ecosystems only involve predators at two adjacent trophic levels rather than through the entire food web (Baum and Worm 2009). Simultaneous removal of both predators and prey can inhibit the top-down response while increased predator diversity may minimize top-down forcing (Baum and Worm 2009). While there are reports that have used fishery data to infer top-down effects of top predator removal, there are also conflicting reports suggesting that fishery landings of mean trophic level are not an indicator of the impact of fishing on marine ecosystems (Branch et al. 2010).

4.4 Climate impacts

The physical oceanography, chemistry, and biology of the tropical and subtropical oceans are impacted by climate variation on interannual, decadal, and longer time periods. At the interannual time period, during El Niño events, the distribution of purse seine catch in the western and central equatorial Pacific shifts eastwards as much as 1,000 km, suggesting a spatial shift in the distribution of skipjack tuna with positive and negative impacts on their availability to island nations depending on their locations (Lehodey et al. 1997). For longline fisheries in the equatorial Pacific, the change in the vertical thermal structure especially along the equator during El Niño and La Niña events changes the depth distribution of the preferred thermal habitat of yellowfin and bigeye tunas, with impacts on catchability of the species. Recruitment in tuna populations is also influenced by El Niño Southern Oscillation (ENSO) variability, although the conditions favoring recruitment vary between species. El Niño events appear to result in higher recruitment for skipjack and yellowfin, while South Pacific albacore recruitment is higher under La Niña conditions (Lehodey et al. 2003).

On longer time scales, a 20-year time series of satellite-observed surface chlorophyll suggests that decadal changes in surface chlorophyll in the North Pacific vary with changes in the atmospheric patterns resulting in inferred ecosystem impacts (Martinez et al. 2009). A decade-long time series of satellite-derived surface chlorophyll data found the most oligotrophic region of the North and South Pacific and Atlantic subtropical oceans had expanded between 1 and 4% per year over the period 1998–2006 coherent with an increase in SST (Polovina et al. 2008). Further, over the period 1998–2007, phytoplankton median cell diameter exhibited statistically significant linear declines of about 2% in the North and South Pacific, and 4% in the North Atlantic (Polovina and Woodworth, 2012).

In the North Atlantic Ocean, the major decadal climate index that is typically used to investigate the relationship between climate and fishery landings is the North Atlantic Oscillation (NAO). The NAO is an index of sea level atmospheric pressure differences between the Azores high and Icelandic low. A large-scale study that investigated the link between NAO and tuna and billfish CPUE found that the NAO may affect many of these Atlantic fishery time-series but that this relationship was strongly spatially structured such that there were stronger NAO impacts in the North Atlantic than in the South Atlantic (Rouyer et al. 2008). However, the complex dynamics among fish, catchability, space, and oceanographic conditions are still poorly understood (Rouyer et al. 2008). Ultimately, the biology and life history traits of tuna and billfish combined with geographical location and changes in fishing gear and effort may result in nonlinear relationships between the environment and ecosystem responses (Fig. 13.5; Rouyer et al. 2008).

A somewhat new and less investigated climate index, the Atlantic Multidecadal Oscillation (AMO) is a measure of sea surface temperature variability in the North Atlantic Ocean and may also be linked with decadal variability in fishery landings (Condron et al. 2005). Climate variability could have significant impacts on the catchability of tunas, especially tropical tuna species (Miller 2007). Indeed, the nominal CPUE of yellowfin tuna in the eastern equatorial Atlantic Ocean was highest in subsurface water temperatures above 24°C and when the thermocline was deepest (Lan et al. 2011).

There has been debate over the association between the NAO and northern bluefin tuna in the northeastern Atlantic Ocean and Mediterranean Sea (Ravier and Fromentin 2004). One study found that albacore and northern bluefin tuna migrations were affected by the 1988–1989 regime shift in the northeast Atlantic Ocean, which was reported to have affected all trophic levels of the ecosystem (Dufour et al. 2010). In the Bay of Biscay, the arrival time for these tuna was earlier (8 to 14 days) by the year 2005 due to increasingly warmer ocean temperatures (Dufour et al. 2010). Although no correlations were found between coastal trap fishery catches of northern bluefin and the NAO (lagged or not), long-term variability in trap catches was negatively correlated with long-term temperature trends (Ravier and Fromentin 2004). This study postulated multiple hypotheses to explain the association between temperature and catch but ultimately concluded that colder (warmer) periods caused a restriction (expansion) in northern bluefin reproductive areas (Ravier and Fromentin 2004).

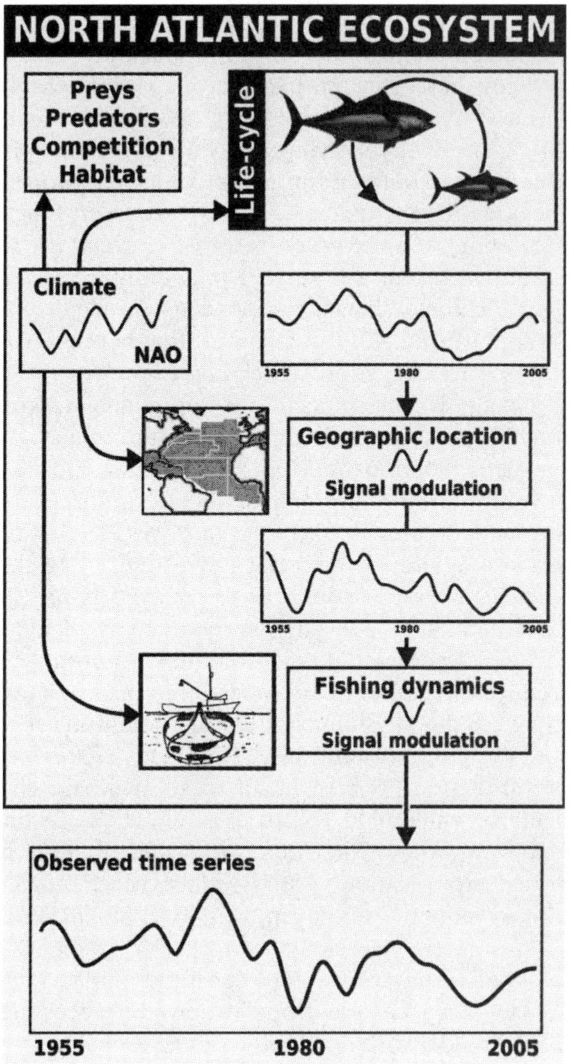

Fig. 13.5 – Representation of the successive signal modulation that can shape the time-series of fishery landings and CPUE in the North Atlantic. The dynamics induced at the population level is first influenced by the ecosystem and climate. The signal is then modulated depending on the geographic location and on the local properties, also influenced by the climate. Finally, the different fishing gears and dynamics also constitute a source of modulation. (Reproduced with permission from Rouyer et al. 2008; © 2008 National Academy of Sciences, U.S.A.)

Sea turtle populations also appear to respond to interannual and multidecadal climate variation. A statistical comparison between loggerhead sea turtle (*Caretta caretta*) nesting abundance anomalies and climatic indices in both the North Atlantic and Pacific found that nesting variability was regionally synchronized and strongly correlated to ocean conditions, with climate models alone explaining up to 88% of the observed changes over the past several decades (Van Houtan and

Halley 2011). Loggerheads in the South Pacific also appear to respond to climate variability as reflected in their annual nesting numbers (Chaloupka et al. 2008). The reproductive frequency of leatherback turtles (*Dermochelys coriacea*) in the eastern Pacific was associated with ENSO such that mature females had a higher (lower) probability of nesting following La Niña (El Niño) phases (Saba et al. 2007). This was due to climate-forced primary productivity transitions in the eastern tropical Pacific affecting the turtles' foraging resources. Interannual variability in the number of nesting Atlantic green turtles (*Chelonia mydas*) in Costa Rica was largely associated with local sea surface temperature variability (Solow et al. 2002). Anthropogenic climate change is also likely to negatively affect sea turtle populations via a rapid warming of the major nesting beaches, which could lead to an increase in the mortality of eggs and hatchlings (Saba et al. 2012).

Global climate change is also projected to impact the subtropical and tropical oceans. While much remains to be learned about the impacts from rising atmospheric anthropogenic carbon dioxide (CO_2) on the subtropical pelagic ecosystem, impacts may alter food web structure, functions such as energy and material flows, biogeochemical cycles, and ultimately ecosystem services (Doney et al. 2012). The oceanic uptake of CO_2 is altering the seawater chemistry of the world's oceans, making it more acidic, with consequences for marine biota (Fabry et al. 2008; Doney et al. 2012). The ability of certain marine animals, including pteropod molluscs, foraminifera, and some benthic invertebrates, to produce calcareous skeletal structures is reduced as seawater becomes more acidic (Fabry et al. 2008). In addition, CO_2 influences the physiology of marine organisms as well through acid-base imbalance and reduced oxygen transport capacity (Fabry et al. 2008). Behavioral changes may also occur as a result of these acid-base changes, with tunas considered particularly vulnerable (Nilsson et al. 2012). Along with changing ocean chemistry, the warming of the oceans may shift thermal habitats poleward and increase vertical stratification, which reduces nutrients in the subtropical oceans, and ultimately reduces overall productivity (Polovina et al. 2011; Doney et al. 2012).

Several model studies have examined possible impacts in the subtropical oceans. A simulation of an eastern tropical upwelling ecosystem based on the Ecopath with Ecosim model with a global warming scenario found that if current fishing levels were unchanged, the abundances at all trophic levels declined but reductions in fishing could enhance or mitigate these declines depending on the trophic level due to top-down impacts (Watters et al. 2003).

In another study, a climate model that includes a phytoplankton model suggested that the North Pacific subtropical gyre (NPSG) will increase in area by about 30% over the 21st century in response to a weakening and poleward shift of northern hemisphere westerlies while the EU will decrease in area by 28% (Polovina et al. 2011). The northern boundary of the NPSG is projected to shift northward by as much as 100 km per decade while the southwestern boundary at the equator is estimated to expand eastward by as much as 200 km per decade (Polovina et al. 2011). Fish catch in the NPSG is projected to decline based on two different ecosystem models by about 20–30% in the gyre interior and 50% at the northern boundary over this century but this impact could be mediated with a reduction of fishing effort (Woodworth-Jefcoats et al. 2012; Howell et al., 2012). Further, a new pelagic thermal habitat is projected to emerge just north of the

North Pacific equatorial zone, defined by SST permanently exceeding 31°C, and which is projected to cover over 25 million km^2 by the end of the century (Polovina et al. 2011). This new thermal habitat may exceed the thermal tolerance of many animals currently inhabiting the region and hence provide habitat for a new community of animals with high thermal tolerances.

Climate impacts on the Pacific bigeye tuna population over the 21st century were projected based on a Spatial Ecosystem And Population Dynamics Model (SEAPODYM) (Lehodey et al. 2010). The model forecasted complex spatial and temporal changes in bigeye tuna spawning and foraging habitats resulting in an eastward population shift and an initial increase in abundance followed by a population decline by 2100 (Lehodey et al. 2010).

In the Southern Hemisphere, since the late 1970s, climate change has been marked by a strengthening of the circumpolar westerly winds extending from the stratosphere to the troposphere and Earth's surface (e.g. Gillett and Thompson 2003). Modelling studies suggest that these positive changes in the Southern Annular Mode (SAM) may result from greenhouse warming and ozone depletion and may be responsible for the southward shift and decadal spin-up of the South Pacific subtropical gyre since the 1990s, including a strengthening and southward penetration of the East Australian Current (EAC) through the Tasman Sea (Cai 2006), which would warm the region, consistent with observations. Regional warming some four times the average rate has also been observed in the Tasman Sea (Ridgway 2007), with observed changes in the distribution of many coastal marine species (e.g. Last et al. 2011) and projected changes in distribution for pelagic species (Hobday 2010). An Ecopath with Ecosim model study for the pelagic ecosystem off eastern Australia found that climate change that impacted phytoplankton biomass had a moderate ecosystem response. For example, a simulated gradual increase of 68% in phytoplankton biomass increased biomass at all higher trophic levels but at modest levels of 9–11% (Griffiths et al. 2010). However, if climate change altered vertical habitat that directly impacted mid-trophic organisms, specifically micronekton fishes and cephalopods, then there were strong ecosystem impacts at both higher and lower trophic levels (Griffiths et al. 2010). This sensitivity of the ecosystem to the abundance of mid-trophic species is thought to be because these species have high biomasses and high production and consumption rates and are important as both prey and predators and hence regulate ecosystem dynamics (Griffiths et al. 2010).

Climate change may also impact the fishery yields of open ocean Atlantic fisheries through both biological and physical changes in the ecosystem. The projected warming of the Atlantic Ocean and potential changes in primary productivity may alter food webs in terms of bottom-up forcing, physiological habitat constraints (i.e. temperature, oxygen), and circulation changes (larval settlement). Vertical expansion and shoaling of the oxygen minimum zone (OMZ) in the tropical Atlantic Ocean (0 to 25 °N, 12 to 30 °W) has been reported for the period of 1960 to 2010 (water layer exceeding 3.5 ml O_2l has shoaled by ~ 1 m/year) equivalent to an oxygen habitat loss of 15% (Stramma et al. 2012). The use of an optimum oxygen vertical habitat zone has been documented for many large pelagic animals from fisheries and tagging studies (e.g. Stramma et al. 2012). The future implication is that the OMZ expansion in the Atlantic and Pacific Ocean is projected to intensify with climate change and thus may increase the catchability of tunas and billfishes as their preferred oxygen habitat shrinks closer to the surface layer (Stramma et al. 2012).

4.5 Anthropogenic ocean noise

Ocean noise is generated from various human activities including shipping, naval sonar operations, and seismic exploration for oil and gas industries. Large areas of the ocean can be affected by a single noise source and anthropogenic noise levels are increasing. Ocean noise can be a serious issue for cetaceans but the full scope of the problem is difficult to ascertain (Weilgart 2007). Naval sonar has been linked to cetacean strandings and mortalities with local population level impacts likely for beaked whales (Weilgart 2007). Recent reports have shown that ocean noise levels are increasing (e.g. Hildebrand 2009), decreasing the distance over which sound can be used by marine species for communication, and resulting in louder whale calls to compensate (Parks et al. 2010). The role of these species in the ecosystem, such as feeding and hence nutrient cycling, may be disrupted as a result, although the implications are still unclear.

4.6 Pollution

Plastics and other marine debris, including discarded or lost fishing gear, is distributed unevenly across the open oceans. The surface convergence zone along the northern and eastern NPSG, sometimes referred to as the North Pacific garbage patch, aggregates plastics and other debris (Howell et al. 2011). A survey of the North Pacific garbage patch in 1999 estimated the density of plastic pieces at 334,271 pieces/km^2 with a mean weight of 5.1 kg/ km^2 (Moore et al. 2001). A similar surface convergence zone is located in the SPSG. Plastics are found in stomachs of organisms throughout the food web including pelagic fishes of all sizes, sea turtles, sea birds, and marine mammals resulting in mortalities in some cases but the overall impact while thought to be adverse is difficult to quantify (Derraik 2002).

4.7 Oil and gas, deep sea mining, shipping (oil spills)

Extractive industries such as mining and oil and gas are generally restricted by depth limitations to coastal regions. Technology developments have allowed drilling from large platforms in deeper and deeper waters, and while the individual footprints may be small, accidents such as the 2010 Deepwater Horizon spill in the Gulf of Mexico (drilling at a depth of 1500 m) illustrate the consequences of accidents far from land in deep water. Control and recovery following accidents in the open sea are more difficult than for coastal incidents; however, the long-term effects on pelagic ecosystems are contested. Shipping traffic and seismic surveys in open seas also impact whales and other marine mammals and contribute to ocean noise pollution, as discussed previously.

5. Ecosystems Approaches to Management

Here we discuss the main Ecosystem approaches that have been proposed or employed to reduce the impacts of many of the threats to open ocean ecosystems presented in the previous section. These approaches should also provide ocean ecosystems with increased resilience to cope with the looming threat of climate change.

5.1 Reducing and preventing overfishing of target species

Fishing is recognized as the main ecosystem threat in open ocean systems (Halpern et al. 2008; Game et al. 2009). The first step in an EBM is to insure that target stocks are not overfished or if they are that rebuilding plans are in place. There are many management tools to regulate fisheries but as described in this section regulating the total catch appears to be the approach that works "best" in the Atlantic and Pacific. This move to total catch management follows a time when effort was not regulated, or input controls, which do not limit total catch, were used.

In the Pacific Ocean annual catch quotas have been imposed for yellowfin and bigeye tunas and striped marlin. For bigeye tuna and striped marlin the catch quotas will decrease annually for several years to bring the fishing mortality in line with fishing mortality (F) at maximum sustainable yield (MSY) (F_{MSY}) (WCPFC 2008a; WCPFC 2010a).

In the Atlantic, estimates of the northern bluefin tuna stock size suggest a biomass decline by a minimum of 51% over the last 40 years (Collette et al. 2011b). As of 2011, fishing mortality of the eastern Atlantic stock of northern bluefin tuna was substantially above maximum sustainable yield (MSY) and the estimates of spawning stock biomass were well below MSY(Collette et al. 2011b). The eastern Atlantic stock has been under increased pressure from demand for larger individuals used for tuna farming (Collette et al. 2011b). The 2011 total allowable catch (TAC) for the eastern stock of northern bluefin tuna was set to 12,900 metric tons (mt) which was the first step in a plan to rebuild the stock by 2022 with a 60% probability (ICCAT SC-RS 2011). For the western stock of northern Bluefin, ICCAT's 2011 TAC was 1,750 t and the latest assessments suggest that this stock may be slowly rebuilding (ICCAT SC-RS 2011). Atlantic bigeye tuna biomass was estimated to have declined by 40% from 1992–2007 while blue marlin was estimated to have declined by 60% to 64% from 1990 to 2006 (Collette et al. 2011c). The biomass of white marlin was estimated have declined between 9% and 37% from 1990 to 2006 (Collette et al. 2011d). The latest projections for bigeye tuna suggest that a TAC of 85,000 mt or less will promote stock rebuilding. Restrictions on the TAC of blue and white marlin were established by ICCAT such that the annual catch of these two species from longlines and purse seines must be no more than 33% (for white marlin) and 50% (for blue marlin) of the 1996 or 1999 landings, whichever is greater (ICCAT SC-RS 2011). The rebuilding of the Atlantic swordfish stocks (particularly the northern stock) is indicative of successful ICCAT management measures. Whereas minimum size regulations had little effect on the overfishing rate of the northern stock, the implementation of constraints on TAC combined with fishery closures that were geared to reduce the mortality of juveniles appeared to be the key regulations that increased the stock's biomass (Neilson et al. 2013). As of 2012, the northern stock has increased since 2000 and is at or above B_{MSY} (ICCAT SC-RS 2011). There is much more uncertainty regarding the southern stock production model and the estimated relative biomass in 2009 (B_{2009}/B_{MSY}) was 1.04, suggesting the stock is currently not overexploited (ICCAT SC-RS 2011).

In many coastal fisheries, fishery independent surveys provide information on population size independent of data from commercial fisheries. Unfortunately, such scientific survey methods are not applicable to tunas and other large pelagic fishes due to the lack of effective survey technology, low density of the target species dispersed over a large spatial region, and the large geographic area covered by the fishery. However, large-scale tuna tagging is currently being conducted in the western Pacific by the Secretariat of the Pacific Community (SPC) to estimate important biological information for tunas including fish movement, natural and fishing mortality, and growth. Also, genetic close kin population estimates are being used for southern bluefin tuna.

5.2 Reducing bycatch

Bycatch in the broadest sense includes both the incidental catch of protected species, which includes sea turtles, sea birds, and cetaceans, and catch of non-target species, the majority of which have commercial value and are retained, although some bycatch species have no commercial value and are discarded. If non-protected species bycatch species are not being overfished and are landed, then bycatch should not be a management concern. Since stock assessments are typically not conducted for most bycatch species, it generally isn't known if most bycatch are overfished. However, bycatch is a concern when it involves protected species and charismatic species such as sharks that appear in many instances to be overfished.

The bycatch of protected species around the world is typically regulated by setting limits in terms of absolute numbers captured. Several approaches have been used to reduce bycatch, especially for protected species, and gear modification has been one of the most successful. Gear modifications have proven very effective in reducing sea bird and sea turtle interactions in the Hawaii-based longline fishery. A regulated gear and bait change in the fishery targeting swordfish from j-shaped hook with squid bait to a wider circle shaped hook with fish bait resulted in a 83% and 90% reduction in capture rate of leatherback and loggerhead sea turtles, respectively, and an 18% increase in swordfish catch rate (Gilman et al. 2007). In addition to gear regulations, limits on the number of sea turtle interactions by species, monitored with 100% observer coverage, are imposed in the Hawaii-based longline fishery for swordfish. Circle hooks and fish bait have also been regulated in the WCPFC region for longline fishing targeting swordfish with shallow sets (WCPFC 2008b).

Sea bird interactions with longline gear can also be substantially reduced with gear modifications. Total sea bird captures in the Hawaii-based longline fishery decreased by 96%, from an estimated 2433 in 2000 to 88 in 2006, the first year regulations were in place (Gilman et al. 2008). Regulations designed to make hooks sink faster, more difficult to see at the surface, or keep birds away from hooks near the surface involved either use of a weight near the hook and thawed blue-dyed bait or setting gear from the side of the vessel, use of a weight close to the hook, and use of a 3m long bird curtain with streamers (Gilman et al. 2008).

Similar regulations have been implemented in the larger WCPFC region (WCPFC 2007) and in the open ocean EEZ of other nations (e.g. Australia, New Zealand).

Spatial management has also been used to reduce bycatch. In eastern Australia, to reduce incidental catches of southern bluefin tuna, spatial management zones are created that regulate fisher access to the portions of the fishing grounds that represent good bluefin tuna habitat in a multi-species longline fishery (Hobday and Hartmann 2006; Hobday et al. 2010). These zones are updated every two weeks during the fishing season to ensure that only operators holding quota for this species are allowed to fish where capture is likely. This real-time management has been effective in reducing bluefin catches by fishers not holding bluefin quota (Hobday et al. 2009). In the Hawaii-based longline fishery, fishing is prohibited in a zone of 80–120 km around the archipelago to reduce gear interactions with the endangered Hawaii Monk Seal and minimize overlap with the local recreational troll fishery. A third example of spatial management to further reduce longline interactions with loggerhead sea turtles is a map product called TurtleWatch which is provided as an experimental service to the Hawaii-based longline fishery targeting swordfish. The swordfish fishery has mandatory observer coverage and closes when 17 loggerheads have been caught (Howell et al. 2008). TurtleWatch is produced from satellite remotely sensed SST and identifies a band around the temporally dynamic 18° C SST front within the fishing ground, which represents the region with the highest probability of interactions between longlines and loggerhead sea turtles (Fig. 13.6, Howell et al. 2008). Even such voluntary schemes can drive behavior change that can reduce negative ecosystem interactions, particularly when penalties can apply for interactions with particular species.

Gear modification and closed areas are proposed management actions to reduce the mortality of false killer whales in the Hawaii-based longline fishery. The use of "weak" hooks which still retain large tunas but straighten out and release the much larger false killer whales along with an expansion of the closed area around the core habitat of the insular false killer whale are proposed to reduce interactions. In the case of dolphin mortalities associated with purse seine fishing in the eastern tropical Pacific, the imposition of trade restrictions that prohibited the sale of tuna captured in association with dolphin schools resulted in the development of innovative operational procedures by the industry so now in most tuna sets on dolphins, the dolphins are released alive (Lewison et al. 2004).

Sharks are typically not targeted except in a few very specialized fisheries; they typically represent a bycatch species where the fins are removed and retained while the carcass is discarded. The practice of shark finning is a primary source of shark mortality from longline and purse seine fisheries. The WCPFC has implemented a regulation that only allows finning if the entire carcass is retained (WPCFC 2010b). The motivation for the regulation is that sharks apart from their fins have low market value so few would be retained. Current US regulations require sharks be landed with fins attached and some states and territories have banned the possession of shark fins. However, the WCPFC regulation on finning does not appear to have been entirely effective in reducing mortality on sharks and it was suggested that a policy of no retention with prompt release of sharks unharmed might be needed to provide sufficient protection (Clarke 2011).

The Sub-Committee on Ecosystems for ICCAT also handles bycatch mitigation

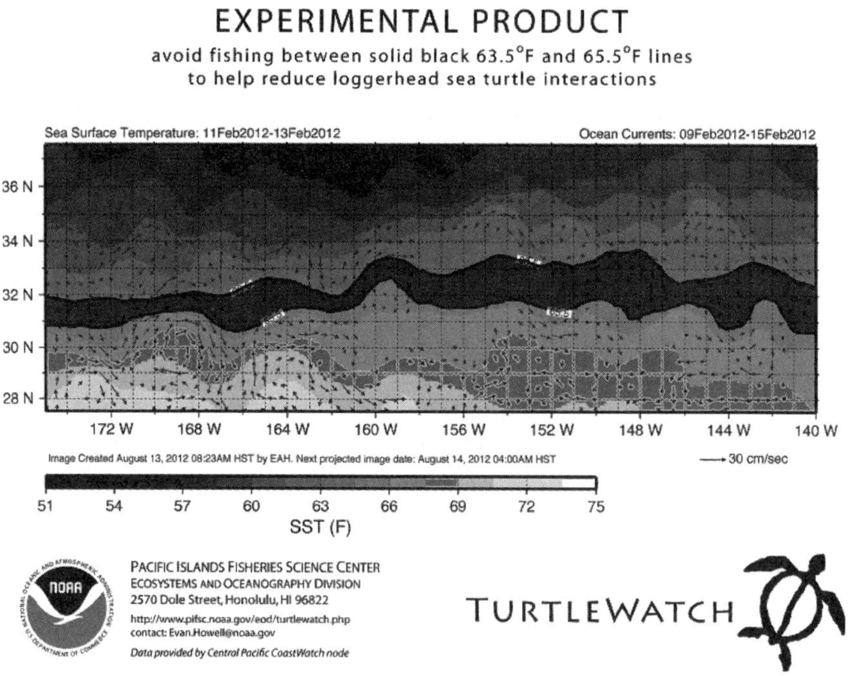

Fig. 13.6 – TurtleWatch, a dynamic map indicating the area with highest forecasted interaction between longline gear and loggerhead sea turtles. (Howell et al. 2008)

measures for sea birds and sea turtles. Across the ICCAT management area, the Japanese observer program has provided an extensive dataset from 1997–2009. High CPUE of albatrosses (Diomedeidae) and petrels (Procellariidae) were more common south of 30 °S from April to September and thus management strategies to mitigate bycatch should be targeted to this region (ICCAT SC-ECO 2011).

5.3 *Spatial management and marine protected areas (MPAs)*

Spatial management of pelagic fisheries such as those for tunas spans a range of options, from single species management aimed at reducing unwanted capture of target (Hobday et al. 2010) or bycatch species (e.g. Howell et al. 2008) to full ecosystem protection and exclusion of all extractive activities (MPAs; e.g. Hobday et al. 2011). Of the eight major threats to pelagic ecosystems described by Game et al. (2009), some threats are more easily obviated through spatial management than others. For entirely anthropogenic threats such as harvesting, mining or non-extractive use, spatial protection can result in direct localized abatement. Through a reduction in cumulative impact, MPAs can mitigate the severity of threats where direct abatement is not possible, such that recovery following some impact is greater. For example, spatial management might not protect against an oil spill, or climate related extreme events, but a less perturbed system can recover more quickly following such impacts (Worm et al. 2006). The large size of MPAs that would be needed to protect pelagic ecosystems, and particularly some of the wide-ranging pelagic species, has been considered a barrier to effective implementation (e.g. Martell et al. 2005; Kaplan et al. 2010), but some analyses have suggested that focussing on key habitat features or hotspots of distribution can still provide effective protection for a portion of the lifespan of many open ocean species (Game et al. 2009). For example, given the higher pelagic biodiversity around oceanic seamounts (Morato et al. 2010), MPAs around seamounts might be a tool to protect that biodiversity. Use of a network of MPAs has been suggested to protect high use habitats of mobile marine predators (Hooker et al. 2011). Even small protected areas can increase marine mammal survival rates, as evidenced for Hector's dolphin in New Zealand (Gormley et al. 2012). Regarding noise pollution, a concern in Hawaii is the impact of intense Navy sonar on some cetaceans, especially beaked whales. Mortalities of beaked whales in close proximity to Navy sonar activity have been documented (Weilgart 2007). Using pelagic protected areas to separate naval sonar activities from seamounts and islands with local populations of beaked whales could be beneficial.

There are some large marine protected areas in the Pacific with varying degrees of protection. The Papahanaumokuakea is a U.S. National Monument and World Heritage Site that covers 360,000 km^2 around the Northwestern Hawaiian Islands, extending 80 km offshore, and is closed to all fishing. The Phoenix Island Protected Area covers 408,250 km^2 of the waters within the EEZ of the Republic of Kiribati. It contains some insular closed areas but much of the oceanic waters are not closed to fishing although distant water fishing fleets fishing in these waters are required to have 100% observer coverage. The Mariana Trench U.S. National Monument and the Pacific Remote Islands U.S. National Monuments cover 246,608 and 210,000 km^2 respectively and create commercial fishing closures from the coast to 80 km offshore around three northern islands in the Marianas Archipelago and the Pacific Remote Islands of Kingman Reef, Palmyra Atoll, Howland Island, Baker Island, Jarvis Islands, Johnston Atoll, and Wake Island. The Cook Islands, in 2012, announced plans for a 1 million km^2 marine protected area. A recent move by the Australian Government to protect the Coral Sea in the southwest Pacific from fishing. While the size of these oceanic marine protected areas is impressive, surveillance and enforcement costs associated with the large areas are also substantial and often underfunded.

5.4 Risk-based approaches for fisheries management

Given the expense and data requirements associated with conventional fishery assessment models and the number of species that must be assessed for some fisheries, less intensive assessment methods are being implemented. As ecosystem-based management also requires attention to non-target species, habitats, and ecological communities, a range of assessment options is needed and risk-based approaches offer great potential (Smith et al. 2007). One approach developed in the southwest Pacific for fisheries and applied in a range of countries is based on an ecological risk assessment framework applied to fisheries, termed Ecological Risk Assessment for the Effects of Fishing (ERAEF) (Hobday et al. 2011b). The ERAEF consists of a hierarchy of tools, ranging from qualitative to fully quantitative approaches. For example, the semi-quantitative productivity-susceptibility analysis (PSA) scores species (or habitats or ecological communities) with regard to a set of productivity attributes, such as maximum size, and susceptibility attributes, such as overlap with the fishery, resulting in relative risk scores for each element species. The high risk species can be analyzed in more detail with a more quantitative tool, or managed to reduce risk implemented on the basis of the PSA results. Pelagic and demersal fisheries in the Australian region of the southwest Pacific are now assessed using this framework, complementing single species stock assessments and management, or substituting for these more complex approaches when data are limiting. Elements of the ERAEF approach have been modified for the U.S. (Patrick et al. 2010), ICCAT (Arrizabalaga et al. 2011), the Marine Stewardship Council and a range of smaller regions (Hobday et al. 2011b). Overall, the ERAEF approach offers a realistic method to assess ecological risk in an EBM context, and has applicability in a wide range of fisheries. The interactive and inclusive nature of the approach also has the advantage of bringing stakeholders, scientists and managers together to develop management solutions—important considerations in the EBM approach (Pikitch et al. 2004).

6. Seamount Ecosystems

Seamounts represent distinct ocean environments, whose topographically induced physical circulation can result in characteristic benthic and pelagic communities on the seafloor and in the waters in their vicinity. Many organisms in these communities exhibit distinct physiological, ecological, and life-history characteristics; these have led to particularly pronounced boom-and-bust fisheries that pose severe management challenges.

6.1 The physical and biological oceanography of seamounts

Seamounts are submerged mountains. Generally, they are of volcanic origin and conical in shape. Geologists classically defined them as being at least 1000 m in height, but features as small as 50–100 m in height are today generally recognized as being seamounts; many of the iconic seamount fisheries for orange roughy were conducted on seamounts rising from the continental slope only several hundred meters.

Seamounts were first encountered in the 19th century in the course of laying undersea cables and in early oceanographic expeditions, such as the *Challenger* expedition (1872–1876). However, they were not clearly recognized until the advent of ship's echo sounders. The first feature so-named was the Davidson Seamount off California in 1933.

The number of seamounts in the world's oceans is still known only approximately. Seamounts larger than about 1.5 km in height create gravity anomalies that can be detected from satellite altimetry data, but smaller seamounts are known only from ship's bathymetric surveys, so global estimates are mostly based on extrapolations. There are now 33,452 seamounts greater than 1 km in height and 138,412 knolls (500–1000 m in height) recorded (Yesson et al. 2011). However, only 6.5% of the seafloor has been surveyed, and estimates for the global number of seamount features > 100 m in height range into the tens of millions (Wessel et al. 2010). Overall, Yesson et al. (2011) estimated that seamounts > 1 km cover 4.7% of the seafloor and knolls cover a further 16.3%. Seamounts are associated with island arc systems, volcanic hot spots, spreading ridges and fracture zones.

They are most numerous in the Pacific, noted for its Ring of Fire and volcanic hot spots, but high densities of small seamounts are also associated with the Mid-Atlantic Ridge (Smith and Cann 1990).

The abrupt topography of seamounts interacts with the ocean's ambient current field. The interactions are complex, depending on the magnitude of the currents, the size of the seamount and its height above the seafloor, water column stratification, and latitude. Essentially, seamounts accelerate the typically weak (several cm/sec) open ocean ambient current flows by about an order of magnitude, particularly near the seamount summit and its steep slopes, sufficient to winnow away the sediment and foster benthic communities dominated by suspension feeders. If the seamount is of sufficient magnitude, a quasi-stationary eddy may be generated over the seamount known as a Taylor column, or Taylor cone if it is attenuated due to stratification. This eddy structure generates upwelling and downwelling, which may influence primary productivity over the seamount. However, evidence of sustained higher primary production over seamounts, such as higher concentrations of phytoplankton (chlorophyll), has been equivocal (e.g. Genin and Boehlert 1985). It is unlikely that the Taylor columns over seamounts are sufficiently stable, except in a few instances, to maintain a food web sustained by autochthonous plankton production (Genin and Dower 2007).

Nonetheless, a broad range of pelagic organisms, including tunas, squids, sharks, seabirds, turtles, and marine mammals, are noted to aggregate around seamounts with peaks within about 1000 m of the sea surface (Pitcher et al. 2007). It thus appears that mechanisms other than locally enhanced primary production must lead to so many higher predators aggregating at seamounts (Tseitlin 1985; Koslow 1997; Clark et al. 2010).

The enhanced currents in the vicinity of seamounts clearly play a major role in structuring seamount communities, and this is most apparent with respect to the benthos. Where currents are substantial, seamount benthic communities are dominated by hard and soft corals, sponges, brisingid starfishes, crinoids, ophiuroids, and other suspension feeders (Genin et al. 1986). Six species of scleractinian cold-water coral, such as *Lophelia pertussa* in the North Atlantic and *Solenosmilia variabilis* in the South Pacific, may form deepwater reefs. A range of species typically live in association with the corals and sponges, such that these stands are often highly diverse, dense, and in marked contrast with the soft bottom communities found on the surrounding seafloor (Richer de Forges et al. 2000; Koslow et al. 2001; Freiwald et al. 2004; Rowden et al. 2010).

Few seamounts have been intensively sampled biologically. The hard, rugged substrate and steep slopes, often covered with coral, renders them difficult to sample with conventional trawl gear. As a result, seamounts were largely avoided by biologists until relatively recently. Wilson and Kaufmann (1987) reported 597 invertebrate species from seamounts worldwide since the *Challenger* expedition of 1872, with 12% of fishes and 15% of invertebrates apparently representing endemics. However, most of these species (72%) were reported from just five seamounts. A little more than a decade later, Richer de Forges et al. (2000) reported ~850 species from a few clusters of seamounts on the continental slope off Tasmania (Australia) and the Norfolk Ridge and Lord Howe Rise in the Coral and Tasman Seas. About a third of these species were new to science. Parin et al. (1997) reported comparable levels of diversity based on Soviet studies of 22 seamounts on the Nazca and Sala y Gomez ridges in the southeast Pacific: 192 invertebrate species and 171 fish species, of which 51% and 44% were new to science and potentially endemic to seamounts.

Although seamounts are clearly 'hotspots' of benthic productivity and diversity, the issue of endemism remains unresolved (Stocks and Hart 2007). Recent studies of the galatheid crabs and brittle stars from the region around New Zealand, New Caledonia, and eastern Australia found that virtually all species on seamounts were also observed on the continental slope (Samadi et al. 2006; O'Hara 2007). McClain et al. (2009) observed that at most about 12% of the species on well-studied Davidson Seamount off the coast of California were restricted to seamounts; however, Davidson Seamount is only 120 km off the coast of California. They also noted that the community structure on Davidson Seamount differed markedly from that on the continental margin: the dominant seamount species were generally rare on the margin and vice-versa. Not surprisingly, levels of endemism are generally lower for fishes than invertebrates on seamounts relative to slope environments, given their mobility as adults (Tracey et al. 2004; Stocks and Hart 2007).

Several additional features of seamount benthic communities are worth noting. A significant number of sponge, crinoids, and brachiopod species appear to be 'living fossils,' relic species from archaic groups known primarily from the Mesozoic. Components of the fauna also exhibit exceptional longevity, with certain sponges, stalked crinoids, and gorgonians aged to approximately 300–500 years (Samadi et al. 2007). Such longevity suggests extremely low predation and natural disturbance rates, as well as, potentially, infrequent successful colonization. These characteristics have significant conservation implications.

The water column over seamounts is often dominated by aggregations of benthopelagic fishes in far higher densities than is found over the surrounding seafloor. How are these aggregations supported? Simple trophodynamic models for pelagic armorhead (Tseitlin 1985) and orange roughy (Koslow 1997) indicate that the in situ productivity over seamounts is too low by an order of magnitude to support these aggregations. Two mechanisms have been proposed to account for the maintenance of high-density aggregations in the deep sea. First, the amplification of currents over seamount summits and steep slopes enhances the flux of zooplankton and micronekton past seamounts. The aggregations may maintain themselves in regions of relative quiescence, much like trout hanging out in the back-eddies of a stream (Koslow 1997). Secondly, seamounts with shallow summits (< 1000 m depth) intercept vertical migrators, which are preyed upon by seamount-associated fishes (Isaacs and Schwartzlose 1965). Reductions in the density of migrators have been observed over seamounts (Genin et al. 1988, 1994), but although dietary studies have indicated feeding on migratory organisms (Fock et al. 2002), the pattern is not universal (Hirsch and Christiansen 2010).

The seamount-associated fish fauna is characterized by a suite of physiological, ecological, evolutionary, and life-history characteristics linked to features of the seamount environment that distinguish them from fishes that dominate pelagic or soft-bottom deep-sea environments. Andriashev (1953) first pointed out that there appeared to be two main evolutionary lineages in the deep sea: an 'ancient' lineage comprising most mesopelagic fishes (e.g. myctophids, gonostomatids and related families) and a 'secondary,' more highly evolved lineage comprising the dominant deep-sea demersal families in the Gadiformes and Ophidiiformes. However, the dominant seamount-associated fishes appear to be derived from a third, more modern Percomorph lineage, which includes the Beryciformes, Zeiformes, Scorpaeniformes, and Perciformes (Rosen 1973; Nelson 2006).

The biogeography of seamount-associated fishes is also distinct from that of other deep-sea fishes. The dominant families and key genera of deepwater fishes are generally cosmopolitan. Thus, myctophids and such gonostomatids as *Cyclothone* spp. dominate the meso- and bathypelagic fish fauna worldwide, and certain macrourids (e.g. *Coryphaenoides* spp.), morid cods, and ophidiid eels dominate the seafloor. However, on seamounts, different genera, families and even orders dominate in different climate zones and ocean basins. Orange roughy (*Hoplostethus atlanticus,* Trachichthyidae, Beryciformes) and to a lesser extent, several oreosomatids (Zeiformes) follow the distribution of Antarctic Intermediate Water (Koslow et al. 1994), dominating temperate zone seamounts in the South Pacific and Indian Oceans and extending into the North Atlantic. Pelagic armorhead (*Pseudopentaceros wheeleri,* Pentacerotidae, Perciformes) dominates seamounts in the central North Pacific, while species of alfonsino (*Beryx* spp., Berycidae, Beryciformes) dominate in the tropics and sub-tropics. Various rockfishes (*Sebastes* spp., Sebastidae, Scorpaeniformes) aggregate over seamounts in the eastern North Pacific, and Patagonian toothfish (*Dissostichus eleginoides,* Nototheniidae, Perciformes) and notothenid cods (*Lepidonotothen squamifrons, Notothenia rossii*) are dominant in the Southern Ocean. The more restricted distributions of seamount-associated fishes and their more advanced, more diverse evolutionary lineages suggest that these groups colonized the deep sea more recently and from more localized shallow-water groups. Thus, seamount fish faunas appear to have evolved in relative isolation from each other at the scale of large biogeographic provinces and water masses.

However, despite their diverse evolutionary lineages, seamount-associated fishes exhibit strong evolutionary convergence toward a deep-bodied body plan with wide caudal fins, adaptations for strong burst swimming and navigating the strong currents present around seamounts, in marked contrast with the slender body with tapering caudal region characteristic of demersal fishes over soft bottoms, such as the grenadier (*Coryphaenoides* spp.). There is also evidence of convergence in life history characteristics among these dominant seamount-associated fishes. Many exhibit remarkable longevity, with orange roughy, oreosomatids, and various species of *Sebastes* associated with seamounts (e.g. *S. alutus, S. borealis, S. ruberriumus, S. borealis,* and *S. aleutianus*) all reported to live more than 100 years (Fenton et al. 1991; Stewart et al. 1995; Caillet et al. 2001).

There have been few physiological studies of seamount-associated fishes. However, orange roughy food consumption and growth stand in marked contrast to that of demersal deepwater fishes living over soft bottoms. *Coryphaenoides* spp. are highly adapted to life in a food-poor environment, with rates of feeding and metabolism approximately an order of magnitude lower than for demersal species living on the shelf, such as cod (Smith 1978; Koslow 1996; Drazen 2002). The combination of low feeding and low metabolic costs can lead to highly efficient growth and reproduction, with estimates for *C. acrolepis* ranging from 52–66%, though only 16% for *C. armatus* (Koslow 1996; Drazen 2002). On the other hand, orange roughy feed at relatively high rates, but most of its food consumption is required for metabolism to live in the high-energy seamount environment (Koslow 1996). As a result, its growth efficiency is low, approximately 5%.

Seamount-associated fishes, such as orange roughy, *Sebastes* spp., oreos, pelagic armorhead, and Patagonian toothfish have firm, well-developed white muscle, consistent with their need for strong, burst-swimming activity. Their lipid and protein content are relatively high and water content low, unlike many species living over soft-bottom deep-sea environments (Koslow 1996). This, combined with their occurrence in aggregations over seamounts has rendered them attractive for exploitation.

6.2 Seamount fisheries

Exploitation of seamount-associated fishes predates the modern period. Polynesians traditionally fished the oilfish (*Ruvettus pretiosus*), a gempylid, at depths of 150–750 m around Pacific islands and seamounts using ingenious V-shaped wooden hooks (Gudger 1927). In the Atlantic, black scabbardfish (*Aphanopus carbo*), a trichiurid, was fished for centuries over seamounts around Madeira, and artisanal fisheries around the Azores targeted alfonsino, red seabream (*Pagellus bogaraveo*), and other species (Marques da Silva and Pinho 2007). Small-scale fisheries continue to be pursued for these species and others associated with islands and seamounts in the Pacific, Atlantic, and Indian Oceans, including tunas, rockfishes (*Sebastes* spp.), and groupers (*Epinephelus* spp.), snappers (*Etelis* spp., *Pristipomoides* spp.), and jacks (*Caranx* spp.). It is difficult to distinguish catches from seamounts from those on the slopes of islands, but it is estimated that annual landings from small-scale seamount fisheries are approximately 150,000–250,000 mt (Marques da Silva and Pinho 2007).

The modern era of seamount fisheries begins with Japanese pelagic fishermen, who exploited tunas with longlines over the Emperor seamounts in the central North Pacific before World War II and resumed operations not long after the war's end (Yasui 1986). In 1967, a Soviet trawler discovered large aggregations of pelagic armorhead (*Pseudopentaceros wheeleri*) there, and Soviet and Japanese trawl operations expanded rapidly across the Emperor and Hawaiian Island seamount chains. Within 10 years, these fleets landed almost a million mt of this species, fishing it to the point of commercial extinction. The fishery was conducted mostly in international waters and was unmanaged. It has never recovered (Fig. 13.7) (Koslow et al. 2000), although there are reports of sporadic fishing (Clark 2009).

Following the demise of the pelagic armorhead fishery, Soviet trawlers exploring elsewhere in the Pacific for demersal fishery resources were the first to encounter the orange roughy off New Zealand. Again, the orange roughy was an obscure fish and was initially identified by the Soviets as simply a 'berycoid fish.' However, before they discovered the large spawning aggregations, New Zealand took control of the deepwater fisheries within its Exclusive Economic Zone (EEZ).

The orange roughy resides predominantly at mid-slope depths at approximately 700–1500 m. Its distribution largely follows that of Antarctic Intermediate Water at temperate latitudes north of the Southern Ocean and extending into the North Atlantic (Koslow et al. 1994). Its center of abundance is on the mid-slope banks, plateaus, and seamounts around New Zealand, where the fishery began in the

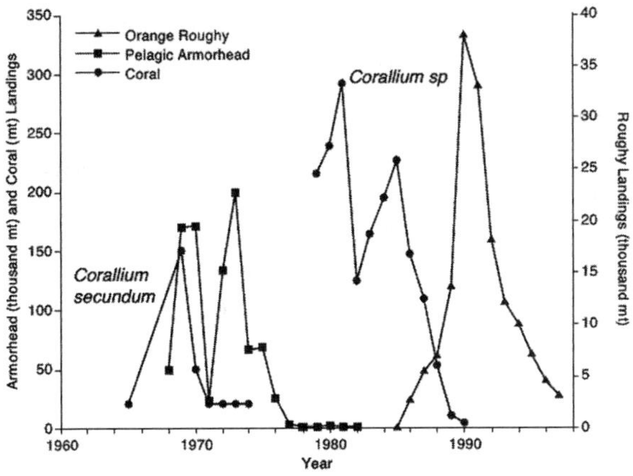

Fig. 13.7 – Time series of landings from the North Pacific seamount fisheries for pink coral (*Corallium* spp.) and pelagic armorhead and the Tasmanian orange roughy fishery. The coral fishery was for two species: *Corallium secundum* at ~400 m depth from 1965–mid-1970s, then the newly discovered *Corallium* sp. nov. at 900–1500 m from 1978–1990. (From Koslow 2007; used by permission.)

early 1980s. Orange roughy fisheries typically exhibit a 5–10-year boom and bust cycle (Fig. 13.7), and serial depletion generally characterized the spread of these fisheries from New Zealand to southeastern Australia at the end of the 1980s, and from there to seamounts and ridges in the southern Indian Ocean, off Namibia, Chile, and the North Atlantic (Branch 2001). Only the very largest orange roughy fishery on the Chatham Rise off New Zealand has lasted longer than a decade, mostly because of its greater size. However, it, too, is now assessed at about 21.5% of its virgin biomass, well below the target level of 30% B_0 for sustainability and estimated fishing mortality (F=0.63) remains 40% higher than is sustainable (F=0.45) (New Zealand Ministry of Fisheries 2009). To adopt a more precautionary approach, New Zealand fishery management recently changed from TAC targets derived from stock assessment models to a more adaptive method where catch is based on $C=M * B_{curr}$, where M is natural mortality (~4.5% for orange roughy) and B_{curr} is the current estimate of stock biomass (Clark & Dunn 2012).

Several factors contribute to this depletion pattern. Initially, little was known about the species, and commercial fishery development was allowed to press ahead without a strong scientific base. The species was initially believed to live approximately 20–25 years, based on misreading the otoliths. Quotas had to be cut dramatically, when it was realized that the species lived for more than 100 years (Fenton et al. 1991). Subsequently, orange roughy fisheries, such as off Tasmania (Australia), expanded unsustainably before scientific assessments of stock size were obtained, and the fishing industry pressed to maintain unsustainable quotas to maximize profits quickly from a resource with extremely low sustainable productivity (Koslow 2007). Other roughy fisheries, such as in the Indian Ocean and North Atlantic, were in international waters or deemed too small to be managed. Off Namibia, where care was taken to expand the fishery slowly, the fishery still collapsed in about five years (1995–99) due to the small size of the stock and its aggregation on small seamounts (Branch 2001).

Other seamount fisheries have followed this boom and bust pattern. One of the first seamount fisheries was for the pink coral, *Corallium secundum,* which Japanese fishermen discovered in 1965 at about 400 m depth on the Emperor Seamount chain. For millennia, related species of *Corallium* from the Mediterranean had been used for jewelry. The Pacific seamount pink coral fishery peaked around 1970 and then declined precipitously, until a second species of *Corallium* was discovered on the Emperor Seamounts in 1978 at 900–1500 m depth. Over 100 boats from Taiwan and Japan pursued this fishery in its peak year, 1981, but by the end of the decade this fishery, too, was depleted (Fig. 13.7). The fishery was mostly pursued in international waters and was unregulated. It was carried out with tangle-net gear, a particularly indiscriminate, wasteful, and destructive gear that consists of netting attached to heavy weights that is dragged up seamount slopes to break up the corals and recover what becomes entangled (Grigg 1993).

Many species are primarily fished on ridges, banks, and the continental slope and to a lesser extent on seamounts (Watson et al. 2007). Many of these species are also long-lived and their fisheries follow the pattern of rapid depletion. Examples include the fisheries for Pacific Ocean perch (*Sebastes alutus*) and other rockfishes from Northeast Pacific seamounts and Patagonian toothfish (*Dissostichus eleginoides*) from the sub-Antarctic Southern Ocean. The Patagonian toothfish fishery, unlike most seamount fisheries, was carried out predominantly with demersal-set longlines rather than trawls. Damage to benthic habitats was therefore less severe, but the bycatch of seabirds, particularly albatross and petrels, which dive for the baited hooks, was initially a severe problem. The toothfish fishery also experienced a high incidence of illegal, unreported, and unregulated (IUU) fishing, because it was carried out predominantly in the vicinity of isolated sub-Antarctic islands (e.g. Kerguelen, South Georgia, Macquarie, and Heard Islands), too remote to be effectively policed. As a result, many toothfish stocks were depleted. However, the Convention on the Conservation of Antarctic Marine Living Resources (CCAMLR), which manages the living marine resources over much of the Southern Ocean, explicitly adopts an ecosystem-based approach to management. Since 2004, it has applied age-structured stock assessments to many of the toothfish fisheries in its area (Collins et al. 2010).

It is estimated that global landings from large-scale commercial seamount fisheries total about 2.25 million mt since their inception in the late 1960s (Clark et al. 2007). They are thus a small fraction of global marine fish landings, currently on the order of 80 million tons annually.

6.3 Ecosystem-based management of seamounts

Ecosystem-based management of seamounts is complicated by several factors. First, sustainability issues for seamount fisheries include both impacts on the targeted stock and collateral damage to habitat or bycatch species (Clark and Dunn 2012). Second, many seamounts and their associated fisheries are in international waters.

Most fisheries are carried out either in the water column or over soft bottoms. However, fisheries on seamounts or in the vicinity of deepwater reefs (e.g. the *Lophelia* reefs of the North Atlantic or the coral and sponge 'gardens' of the Northeast Pacific) have had substantial impacts on the benthic habitat. Most of these fisheries are conducted with trawls, which remove the seamount benthos—the corals, sponges and associated benthic fauna—along with the targeted fish (Probert 1997; Anderson and Clark 2003). Coral rubble and bare substrate are typically left in the wake of these fisheries (Koslow et al. 2001). Given the longevity of much of this fauna, recovery is highly uncertain and is likely to be on century to millennial time scales. Seamounts heavily impacted by trawl fishing show little sign of recovery after several decades (Williams et al. 2010).

The impact of seamount fisheries on their target populations has also generally resulted in overfishing. As Clark and Koslow (2007) noted, "overexploitation is the norm, even where there have been active management regimes and research activities." There are several contributing factors. Fish aggregated at fixed topographic locations are highly vulnerable, and declining catch per unit effort, one of the clearest signs of over-fishing, may not occur until the population is depleted. Many seamount-associated populations are in international waters without rigorous management regimes or are too small to be actively managed. Most aggregations represent the accumulated biomass of an unproductive population, characterized by extreme longevity, low growth and highly episodic recruitment. Stocks are typically depleted rapidly, and recovery is slow and uncertain (Clark et al. 2007).

The short-term nature of most seamount fisheries, combined with their long-term environmental impacts, has led to increased efforts to protect seamount faunas and other deepwater coral environments. Deepwater marine protected areas or other provisions to limit trawling in vulnerable deepwater habitats have been established off southeastern Australia, New Zealand, the Hawaiian and Aleutian Islands, the west coast of Canada and the USA, Norway, the Azores, and elsewhere (Probert et al. 2007; Morato et al. 2010).

Given the poor record for the management of seamount fisheries, Clark and Dunn (2012) have called for a more highly precautionary approach based upon spatial management and very low fishing effort until credible biomass estimates are available. They point to the need to set aside seamounts in reserves before the onset of commercial fishing and to better balance conservation and fishing interests.

However, many seamount and ridge fisheries are in international waters, which pose a distinct challenge for their management and conservation. The International Seabed Authority (ISA), the UN agency created under the UN Convention on the Law of the Sea, has authority to manage mineral resources but not living marine resources on the high seas. However, in 2004 the UN General Assembly passed a resolution calling on high seas fishing nations and regional fisheries management organizations (RFMOs) to protect vulnerable marine ecosystems from destructive fishing practices in areas beyond national jurisdiction. Due to limited progress, the UN General Assembly passed further resolutions in 2006 and 2009 (Gianni et al. 2011). To recount progress to date, CCAMLR has restricted bottom trawling, sought to establish a representative system of MPAs, and worked to manage the toothfish and other seamount-associated fisheries within its jurisdiction in the Southern Ocean. However, of the more than 30 RFMO agreements, only CCAMLR explicitly adopts an ecosystem-based approach to management. In the Mediterranean, coastal states have adopted a Convention for the Protection of the Marine Environment, and steps have been taken to limit deepwater trawling.

The OSPAR Convention for the Protection of the Marine Environment of the Northeast Atlantic and the North East Atlantic Fisheries Commission have taken measures to protect coldwater corals, seamounts, ridge areas, deepwater sponge grounds and species such as the orange roughy in the waters off Europe extending out to the mid-Atlantic Ridge (Probert et al. 2007). In 2009, the NEAFC invoked management measures to conserve vulnerable marine ecosystems in the Northeast Atlantic, including the mid-Atlantic Ridge, Charlie-Gibbs Fracture Zone, and Reykjanes Ridge. These measures included fishery closures encompassing 330,000 km^2 for the years 2009 to 2015. In the southern Indian Ocean, where there has been fishing for orange roughy on seamounts and ridges in international waters, the Southern Indian Ocean Fishers Association together with IUCN in 2006 closed an area of about 300,000 km^2 to bottom trawling. A recent international program has examined seamounts and governance in this region (Garcia et al. 2013).

Taranto et al. (2012) have proposed a framework for the conservation of seamount biodiversity based on risk of human impact and criteria for ecologically or biologically significant areas (EBSA). In 2008, the Convention on Biological Diversity (CBD) established seven scientific criteria to identify EBSAs for open ocean and deep-sea habitats: uniqueness, rarity or endemism; special importance for the early life history stages of species; sensitivity to disturbance or slow recovery; productivity; diversity; and the condition of being pristine and undisturbed by human impacts. Indicators were based on benthic, benthopelagic, and pelagic components of seamount ecosystems, and human impacts from fishing and seabed mining were ranked on a five-point scale. The authors tested their method on simulated and actual seamount ecosystems, placing them into four portfolios based on high and low likelihood of being an EBSA and high and low threat likelihood. The framework is proposed as a means toward making progress on meeting the CBD conservation goal of protecting 10–30% of marine biomes by 2012.

Fulton et al. (2007) examined the use of ecosystem models to assess ecosystem-based management options for seamount fisheries based on a parameterization for a seamount in the Azorean region (Morato and Pitcher 2005). Not surprisingly, they found strong tradeoffs between economic and ecological objectives in the management of these fisheries.

On the horizon, management of mining as well as fishing will need to be considered in open-ocean ecosystems. Seamounts may be subject to mining for their ferromanganese crusts, which contain enriched concentrations of minerals, such as tellurium and cobalt, used in the manufacture of high-tech metals (Hein et al. 2010). Deepwater mining of hydrothermal vent massive sulfide deposits is imminent, with leases to Nautilus Minerals, a Canadian enterprise, already granted off Papua New Guinea in the Bismarck Sea. In what has been called a 'gold rush to the abyss,' China and Russia filed claims in 2011 to explore massive sulfide deposits on ridges in the Indian and Atlantic Oceans with the International Seabed Authority (ISA), the United Nations agency with jurisdiction over mineral resources on the high seas. In 2012, France and Korea filed claims with the ISA in as yet undisclosed areas. Private mining interests are in the process of staking claims with island nations in the western Pacific (Broad 2012). Massive sulfides contain significant concentrations of gold, silver, copper, and other metals. However, hydrothermal vent communities contain unique biological communities that are highly localized and with high levels of endemism. Until now, their conservation has not been an issue (Van Dover 2011). Ironically, it may be easier to obtain clearance for the environmental impacts of seabed mining within waters under national jurisdiction than in international waters where the ISA has jurisdiction. The ISA, in consultation with the scientific community, has adopted a 'Mining Code': protocols for managing the prospecting and development of high-seas mineral deposits, such as manganese nodules and massive sulfide deposits, which includes setting aside Reserve areas in parallel with mining sites, which may serve as a template for development of offshore mineral resources within nations' EEZs (ISA 1999, 2007; Koslow 2007).

7. Future Directions for Ecosystem Approaches for Open Oceans

In a recent evaluation of the approach and performance of 18 regional fisheries management organizations (RFMOs), the WCPFC was scored among the top tier of RFMOs in both categories, while both IATTC and ICCAT scored above average in approach but below average in performance; however, all RFMOs had room for improvement (Cullis-Suzuki and Pauly 2010). Greater transparency in process and decision making is required (Polacheck 2011), including access to finer resolution data by third parties (catch data is often only available to third parties on 5 degree resolution for commercial reasons) and independent auditing.

Given that one of the main threats to oceanic ecosystems is from fishing, effective fishery management is a critical element in EBM. Currently, management of pelagic fisheries in both the Atlantic and Pacific use single species quotas based on stock assessments of key target species. This work depends on comprehensive and accurate fisheries data, robust assessment models, and an understanding of key biological processes and parameters. For example, the WCPFC has identified monitoring of fishing activities, monitoring and assessment of stocks, and biological studies to support stock assessment models as continued high priorities. They also give priority to monitoring and assessing the ecosystem. This topic includes fisheries data on all catch not just target or landed species and finer spatial resolution of the fishery coverage, more work on the status of shark populations, use of ecological risk assessments, establishment of ecosystem indicators, identification of habitats of special concern, identification of oceanographic features, processes, and fishing practices that influence the distribution and abundance of fish stocks and their vulnerability to fishing gear, approaches for bycatch mitigation, investigating key trophic relationships, use of ecological and ecosystem-based models, and estimation of the maximum aggregate yield of all species that can be safely removed from an ecosystem without disrupting ecosystem structure and function (WCPFC-SC7 2011).

Regarding this last objective, several management bodies take an ecosystem approach to restricting yields at a level less than the sum of single species assessments. The North Pacific Regional Fisheries Management Council, responsible for managing the Bering Sea and Aleutian Island multispecies groundfish fishery, uses an approach based on the simultaneous application of quotas at the level of individual stocks, together with a combined total catch quota, lower than the sum of the individual quotas, along with incidental take quotas for non-target species, and, finally, incidental take limits for protected species, all to reduce the likelihood of long-term impairment of the ecosystem from fishing (Ruckelshaus et al. 2008).

Management of Antarctic fisheries by the Commission for the Conservation of Antarctic Marine Living Resources (CCAMLR) sets limits on the incidental mortality of non-target species and once these limits are met, the target fishery can be closed, even if the quota for the target species has not been harvested (Ruckelshaus et al. 2008). CCAMLR also attempts to preserve predator–prey relationships by setting its harvest quota for krill at 25% below the single species MSY to account for the key roll krill play as prey in the ecosystem (Ruckelshaus et al. 2008).

A recent meta-analysis of global fisheries explores tradeoffs between multispecies maximum sustainable yield (MMSY) and the collapse of individual stocks (Worm et al. 2009). Their model finds that for a wide range of exploitation rates ranging from 0.25 to 0.60 the resultant catches equal or exceed 90% of the MMSY, but with an exploitation rate of 0.60 almost half the species in the ecosystem are expected to collapse, while with an exploitation rate of 0.25 less than 10% of the species are expected to collapse (Worm et al. 2009). Thus in a multispecies context, taking into consideration aspects of ecosystem structure and function, the exploitation rate that achieves maximum sustainable yield should be considered an upper limit rather than a management target (Worm et al. 2009). More work in the Pacific and Atlantic is needed to expand single species management to identify multispecies yield and fishing mortality that are both close to MMSY and reduce the risk of individual stock collapse.

As mentioned previously, ICCAT has not yet incorporated ecosystems or multispecies interactions into their official fishery regulations. However, ICCAT has formed a Sub-Committee on Ecosystems and this group is the primary body that deals with the consideration of an Ecosystem Approach to Fisheries (EAF). This sub-committee also deals with oceanography, climate, and bycatch. As of 2011, the sub-committee recognized that moving forward towards EBM may help incorporate the complexity of the Atlantic ecosystem into future ICCAT stock assessments (ICCAT SC-ECO 2011). A spatial surplus production model called SEMIPRO was submitted to ICCAT to undertake multi-species, multi-area stock assessments (ICCAT SC-ECO 2011). This software package considers stock mixing among various areas and has shown that consideration of spatial catch and effort can provide a framework for simple yet reliable estimates of the spatial distribution and migration of stocks. The goal is to determine a spatial mosaic of fishing effort across all gear types that meet ICCAT management goals for multiple stocks (ICCAT SC-ECO 2011). Moreover, a multi-stock spatial model was designed for northern bluefin tuna to assess population biomass using conventional and electronic tag data, historic catch-at-age reconstructions, and otolith microchemistry stock-composition data (Taylor et al. 2011). This study concluded that if fishing in the northeastern Atlantic and Mediterranean continues at the same unregulated levels of 2007, both the eastern and western stocks will continue to decline due to stock mixing in their North Atlantic foraging areas.

Ecosystem indicators are being considered by working groups in both the WCPFC and IATTC and generally are either functions computed from catch or observer data such as mean trophic level of the landings or functions of environmental data such as the depth of the OMZ (ICCAT SC-ECO 2011). Two challenges arise when computing ecosystem indicators from fisheries data. First, of course, they only at best capture changes in the portion of the ecosystem that is exploitable by the gear and the operational characteristics of the fishery. This issue may be partially resolved by using simulations from an ecosystem model to help provide a more complete ecosystem perspective. Secondly, changes in species and trophic levels of the fishery catch may be the result of gear and operational changes in the fishery rather than ecosystem changes (Branch et al. 2010). Even subtle changes, for example, switching from use of a J-hook to a circle hook in a longline fishery can change the catchability of a species by as much as one third (Curran and Bigelow 2011). However, carefully constructed indicators from observer data which record all species caught together with gear and operational characteristics and used in a statistical model such as a generalized additive model by people with knowledge of the fishery can be informative of ecosystem changes. For tropical and subtropical fisheries, fisheries independent surveys are generally not feasible due to the large area over which the fisheries operate and low density of animals so catch-derived indicators are often the only tool available. However, if well-designed diet studies were instituted to complement the observer sampling, information on lower trophic levels could also be collected to identify or confirm major shifts in species composition in lower trophic levels (Allain et al. 2011).

Environmental indicators are equally challenging. The approach here is to provide an index linked to the catchability of the targeted fishes and/or one that tracks how productive the habitat is for the targeted fishes. The depth of the oxygen minimum layer averaged over the fishing ground could be an example of the former while surface chlorophyll averaged over the fishing ground could be an example of the latter. Environmental data come from a range of sources. Satellite sensed oceanographic data can provide surface temperature, winds, currents, and chlorophyll, operational ocean models can provide data for many more oceanographic parameters and especially important subsurface information, but they are model data, while oceanographic cruises and oceanographic data sets provide in-situ data, although these are often limited in space and time.

In theory, single species population dynamics assessment models can detect climate impacts on populations and provide managers with target levels that reflect changing conditions. However, it is not clear how well current stock assessment models perform in detecting population impacts arising from climate change to provide managers with timely advice under long-term climate change. The evaluation of the performance of current assessment models in response to climate change should be undertaken along with the use of ecosystem models driven by climate model output.

In summary, open ocean tropical and subtropical systems cover a large portion of the earth's surface and provide substantial economic and food security benefits. They face a range of threats including overfishing of target and bycatch species, climate change, and plastic and noise pollution. Seamounts with high densities of slow growing and long-lived species are ecosystems that are especially sensitive to overfishing. Regional Fisheries Management Organizations are making progress in addressing some of these threats, particularly overfishing on highly migratory species and reducing interactions with protected species. However, more needs to be done; for example, ecosystem impacts from climate change and fishery-induced ecosystem changes are not currently considered in management actions in these systems. Ecosystem status reports developed from observer and logbook data might provide the starting point for a qualitative ecosystem assessment that should accompany single species stock assessments. However, there is also a lack of theory on ecosystem dynamics that limits our ability to provide managers with specific ecosystem guidelines, indicators, and target levels. Large oceanic MPAs have been created but their value in managing oceanic ecosystems remains to be demonstrated. Ecosystem impacts from climate change and pollution could be substantial but our understanding of these impacts needs to be advanced.

References

Allain, V., S. Nicol, J. Polovina, M. Coll, R. Olson, S., Griffiths, J. Dambacher, J. Young, J. Jurado Molina, S. Hoyle, T. Lawson. 2011. International workshop on opportunities for ecosystem approaches to fisheries management in the Pacific Ocean tuna fisheries. *Rev Fish Biol Fisheries* **22**(1), 29–33.

Anderson, O. F., M. R. Clark. 2003. Analysis of bycatch in the fishery for orange roughy, *Hoplostethus atlanticus,* on the South Tasman Rise. *Marine Freshwater Research* **54,** 643–652.

Andriashev, A. P. 1953. Ancient deep-water and secondary deep-water fishes and their importance in a zoogeographical analysis. Akad. Nauk SSSR.

Arrizabalaga, H., P. de Bruyn, G. A. Diaz, H. Murua, P. Chavance, A. Delgado de Molina, D. Gaertner, J. Ariz, J. Ruiz, L. T. Kell. 2011. Productivity and susceptibility analysis for species caught in Atlantic tuna fisheries. *Aquatic Living Resources* **24,** 1–12.

Baum, J. K., B. Worm. 2009. Cascading top-down effects of changing oceanic predator abundances. *J Anim Ecol* **78,** 699–714.

Beerkircher, L. R., E. Cortes, M. Shivji. 2004. Characteristics of shark bycatch observed on pelagic longlines off the southeastern United States, 1992–2000. *Mar. Fish. Rev.* **64,** 40–49.

Behrenfeld, M. J. 2010. Abandoning Sverdrup's Critical Depth Hypothesis on phytoplankton blooms. *Ecology* **91,** 977–989.

Behrenfeld, M. J., P. G. Falkowski 1979. Photosynthetic rates derived from satellite-based chlorophyll concentration. *Limnology and Oceanograph.* **42,** 1–20.

Block, B. A., I. D. Jonsen, et al. 2011. Tracking apex marine predator movements in a dynamic ocean. *Nature* **475,** 86–90.

Bonini, S., N. Saran, L. Stein. 2011. Design for Sustainable Fisheries—Modeling Fishery Economics. McKinsey and Company.

Branch, T. A. 2001. A review of orange roughy *Hoplostethus atlanticus* fisheries, estimation methods, biology and stock structure. *S. Afr. J. Mar. Sci.* **23**(A Decade of Namibian Fisheries Science), 181–203.

Branch, T. A., R. Watson, et al. 2010. The trophic fingerprint of marine fisheries. *Nature* **468,** 431–435.

Broad, W. J. 2012. A gold rush in the abyss. New York Times, July 10.

Cai, W. 2006. Antarctic ozone depletion causes an intensification of the Southern Ocean supergyre circulation. *Geophysical Research Letters* **33,** L03712, doi:10.1029/2005GL024911.

Caillet, G. M., A. H. Andrews, E. J. Burton, D. L. Watters, D. E. Kline, L. A. Ferry-Graham. 2001. Age determination and validation studies of marine fishes: do deep-dwellers live longer? *Experimental Gerontology* **36,** 739–764.

Chavez, F. P., M. Messie, J. T. Pennington. 2011. Marine primary production in relation to climate variability and change. *Annu. Rev. Mar. Sci.* **3,** 227–260.

Chaloupka, M., N. Kamezaki, C. Limpus. 2008. Is climate change affecting the population dynamics of the endangered Pacific loggerhead sea turtle? *Journal of Experimental Marine Biology and Ecology* **356,** 136–143.

Clark, M. R. 2009. Deep-sea seamount fisheries: a review of global status and future prospects. *Latin American Journal of Aquatic Research* **37**(3): 501–512.

Clark, M. R., J. A. Koslow. 2007. Impacts of fisheries on seamounts. In: T. J. Pitcher, T. Morato, P. J. B. Hart, M. R. Clark, N. Haggan, R. S. Santos (Eds). *Seamounts: Ecology, Fisheries & Conservation.* Blackwell Publishing, Oxford: 413–441.

Clark, M. R., M. R. Dunn. 2012. Spatial management of deep-sea seamount fisheries: balancing exploitation and habitat conservation. *Environmental Conservation* **39**(2): doi:10.1017/S0376892912000021.

Clark, M. R., A. A. Rowden, T. Schlacher, A. Williams, M. Consalvey, K. I. Stocks, A. D. Rogers, T. D. O'Hara, M. White, T. M. Shank, J. M. Hall-Spencer. 2010. The ecology of seamounts: structure, function, and human impacts. *Annual Review of Marine Science* **2,** 253–278.

Clark, M. R., V. I. Vinnichenko, J. D. M. Gordon, G. Z. Beck-Bulat, N. N. Kukharev, A. F. Kakora. 2007. Large-scale distant-water trawl fisheries on seamounts. In: T. J. Pitcher, T. Morato, P. J. B. Hart, M. R. Clark, N. Haggan, R. S. Santos (Eds). *Seamounts: Ecology, Fisheries & Conservation.* Blackwell Publishing, Oxford: 361–399.

Clarke, S. 2011. A Status Snapshot of Key Shark Species in the Western and Central Pacific and Potential Management Options. Western and Central Pacific Fisheries Commission, Scientific Committee Paper. WCPFC-SC7–2011/EB-WP-04.

Collette, B. B., K. E. Carpenter et al. 2011a. High value and long life—double jeopardy for tunas and billfishes. *Science* **333,** 291–292.

Collette, B., A. F. Amorim et al. 2011b. *Thunnus thynnus.* In: IUCN 2011. IUCN Red List of Threatened Species. Version 2011.2.

Collette, B., A. Acero et al. 2011c. *Thunnus obesus.* In: IUCN 2011. IUCN Red List of Threatened Species. Version 2011.2.

Collette, B., A. Acero et al. 2011d. *Makaira nigricans.* In: IUCN 2011. IUCN Red List of Threatened Species. Version 2011.2.

Collette, B., A. F. Amorim et al. 2011e. *Kajikia albida.* In: IUCN 2011. IUCN Red List of Threatened Species. Version 2011.2.

Collins, M. A., P. Brickle, J. Brown, M. Belchier. 2010. The Patagonian toothfish: biology, ecology and fishery. *Advances in Marine Biology* **58,** 229–289.

Condron, A., R. DeConto et al. 2005. Multidecadal North Atlantic climate variability and its effect on North American salmon abundance. *Geophysical Research Letters* **32.**

Costanza, R., R. dArge, R. deGroot, S. Farber, M. Grasso, B. Hannon, K. Limburg, S. Naeem, R. V. O'Neill, J. Paruelo, R. G. Raskin, P. Sutton, M. van den Belt. 1997. The value of the world's ecosystem services and natural capital. *Nature* **387,** 253–260.

Cullis-Suzuki, S., D. Pauly. 2010. Failing the high seas: A global evaluation of regional fisheries management organizations. *Marine Policy* **34,** 1036–1042.

Curran, D., K. Bigelow. 2011. Effects of circle hooks on pelagic catches in the Hawaii-based tuna longline fishery. *Fish. Res.* **109,** 265–275.

Derraik, J. G. B. 2002. The pollution of the marine environment by plastic debris: a review. *Mar. Pollut. Bull.* **44,** 842–852.

Doney, S. C., M. Ruckelshaus, J. E. Duffy, J. P Barry, F. Chan, C. A. English, H. M. Galindo, J. M. Grebmeier, A. B. Hollowed, N. Knowlton, J. Polovina, N. N. Rabalais, W. J. Sydeman, and L. D. Talley. 2012. Climate Change Impacts on Marine Ecosystems. *Annu. Rev. Mar. Sci.* **4,** 11–37.

Drazen, J. C. 2002. Energy budgets and feeding rates of *Coryphaenoides acrolepis* and *C. armatus. Marine Biology* **140**(4), 677–686.

Dufour, F., H. Arrizabalaga et al. 2010. Climate impacts on albacore and bluefin tunas migrations phenology and spatial distribution. *Progress In Oceanography* **86,** 283–290.

Fabry, V. J., B. A. Seibel, R. A. Feely, J. C. Orr. 2008. Impacts of ocean acidification on marine fauna and ecosystem processes. *ICES Journal of Marine Science,* **65,** 414–432.

FAO Fisheries Report. No. 838. Rome, FAO. 2007. 203p.

Fenton, G. E., S. A. Short, D. A. Ritz. 1991. Age determination of orange roughy, *Hoplostethus atlanticus* (Pisces:Trachichthyidae) using $^{210}Pb/^{226}Ra$ disequilibria. *Marine Biology* **109,** 197–202.

FFA 2010. Economic indicators update 2010. Forum FisheriesAgency. http://www.ffa.int.

Fock, H. O., B. Matthiessen, H. Zidowitz, H. von Westernhagen. 2002. Diel and habitat-dependent resource utilisation of deep-sea fishes at the Great Meteor seamount (subtropical NE Atlantic): niche overlap and support for the sound-scattering layer-interception hypothesis. *Marine Ecology Progress Series* **244,** 219–233.

Fournier, D. A., J. Hampton, J. R. Sibert. 1998. MULTIFAN-CL: a length-based, age-structured model for fisheries stock assessment, with application to South Pacific albacore, *Thunnus alalunga. Can. Journ. Fish. And Aquat. Sci.* **55**(9), 2105–2116.

Friedland, K. D., C. A. Stock et al. 2012. Pathways between primary production and fisheries yields of Large Marine Ecosystems. *PLoS One* 7, e28945.

Freiwald, A., J. H. Fossa, A. Grehan, T. Koslow, J. M. Roberts. 2004. Cold-water coral reefs: out of sight—no longer out of mind. UNEP-WCMC, Cambridge, UK: 84 pp.

Fromentin, J., B. Planque. 1996. *Calanus* and environment in the eastern North Atlantic. 2. Role of the North Atlantic Oscillation on *Calanus finmarchicus* and *C. helgolandicus. Mar. Ecol. Prog. Ser.* **134,** 11–118.

Fulton, E. A., T. Morato, T. J. Pitcher. 2007. Modelling seamount ecosystems and their fisheries. In: T. J. Pitcher, T. Morato, P. J. B. Hart et al. (Eds). *Seamounts.* Blackwell, Oxford: 296–332.

Game E. T., H. S. Grantham, A. J. Hobday, R. L. Pressey, A. T. Lombard, L. E. Beckley, K. Gjerde, R. H. Bustamante, H. P. Possingham, A. J. Richardson. 2009. Pelagic protected areas: the missing dimension in ocean conservation. *Trends in Ecology and Evolution* **24,** 360–369 doi:10.1016/j.tree.2009.01.011.

Garcia, S. M., H. Cohen, D. Freestone, Carole Martinez, N. Oral, A. Rogers, P. A. Verlaan, D. Vousden. 2013. *An Ecosystem Approach to Management of Seamounts in the Southern Indian Ocean.* Vol. 4: A Road Map towards sustainable use and conservation of biodiversity in the Southern Indian Ocean. IUCN, Gland, Switzerland: 32 pp.

Genin, A., G. W. Boehlert. 1985. Dynamics of temperature and chlorophyll structures above a seamount: an oceanic experiment. *Journal of Marine Research* **43,** 907–924.

Genin, A., P. K. Dayton, P. F. Lonsdale, F. N. Spiess. 1986. Corals on seamounts provide evidence of current acceleration over deep sea topography. *Nature* **322,** 59–61.

Genin, A., C. Greene, L. Haury, P. Wiebe, G. Gal, S. Kaartvedt, E. Meir, C. Fey, J. Dawson. 1994. Zooplankton patch dynamics: daily gap formation over abrupt topography. *Deep-Sea Research* **41,** 941–951.

Genin, A., L. Haury, P. Greenblatt. 1988. Interactions of migrating zooplankton with shallow topography: predation by rockfishes and intensification of patchiness. *Deep-Sea Research* **35,** 151–175.

Genin, A., J. F. Dower. 2007. Seamount plankton dynamics. In: T. J. Pitcher, T. Morato, P. J. B. Hart, M. R. Clark, N. Haggan, R. S. Santos (Eds). *Seamounts: Ecology, Fisheries & Conservation.* Blackwell Publishing, Oxford: 85–100.

Gianni, M., D. E. J. Currie, S. Fuller, L. Speer, J. Ardron, B. Weeber, M. Gibson, G. Roberts, K. Sack, S. Owen, A. Kavanagh. 2011. Unfinished business: a review of the implementation of the provisions of UNGA resolutions 61/105 and 64/72 related to the management of bottom fisheries in areas beyond national jurisdiction. Deep Sea Conservation Coalition.

Gillett, N. P., D. W. J. Thompson. 2003. Simulation of Recent Southern Hemisphere Climate Change. *Science* **302,** 273–275.

Gilman, E., Kobayashi, D., Swenarton, T., Brothers, N., Dalzell, P., Kinan-Kelly, I. 2007. Reducing sea turtle interactions in the Hawaii-based longline swordfish fishery. *Biological Conservation* **139,** 19–28.

Gilman, E., Kobayashi, D., Chapoupka, M. 2008. Reducing seabird bycatch in the Hawaii longline tuna fishery. *Endang. Species Res.* **5** (2–3), 309–323.

Gordon, J. D., O. A. Bergstad, I. Figueiredo, and G. Menezes. 2003. Deep-water fisheries of the Northeast Atlantic: I. Description and Current Trends. *J. Northw. Atl. Fish. Sci.* **31,** 137–150.

Griffiths, S. P., J. W.Young, M. J. Lansdell, R. A. Campbell, J. Hampton, S. D. Hoyle, A. Langley, D. Bromhead, M. G. Hinton. 2010. Ecological effects of longline fishing and climate change on the pelagic ecosystem off eastern Australia. *Reviews in Fish Biology and Fisheries.* DOI 10.1007/s11160–009–9157–7.

Grigg, R. W. 1993. Precious coral fisheries of Hawaii and the U.S. Pacific islands. *Marine Fisheries Review* **55,** 50–60.

Gormley, A. M., Slooten, W., Dawson, S., Barker, R. J., Rayment, W., du Fresne, S. & Brager, S. 2012. First evidence that marine protected areas can work for marine mammals. *Journal of Applied Ecology,* DOI: 10.1111/j.1365–2664.2012.02121.x.

Gudger, E. 1927. Wooden hooks used for catching sharks and *Ruvettus* in the South Seas: a study of their variation and distribution. *Papers of the American Museum of Natural History* **28,** 199–355.

Halpern, B. S., S. Walbridge, K. A. Selkoe, C. V. Kappel, F. Micheli, C. D'Agrosa, J. F. Bruno, K. S. Casey, C. Ebert, H. E. Fox, R. Fujita, D. Heinemann, H. S. Lenihan, E. M. P. Madin, M. T. Perry, E. R. Selig, M. Spaulding, R. S. Steneck, R. Watson. 2008. A Global Map of Human Impact on Marine Ecosystems. *Science* **319,** 948–952.

Hampton, J., J. R. Sibert, P. Kleiber, M. N. Maunder, and S. J. Harley. 2005. Decline of Pacific tuna populations exaggerated? *Nature* **434,** E1–E2.

Hein, J. R., T. A. Conrad, H. Staudigel. 2010. Seamount mineral deposits: a source of rare metals for high-technology industries. *Oceanography* **23**(1), 184–189.

Hildebrand, J. 2009. Anthropogenic and natural sources of ambient noise in the ocean. *Marine Ecology Progress Series* **395,** 5–20.

Hirch, S., B. Christiansen. 2010. The trophic blockage hypothesis is not supported by the diets of fishes on Seine Seamount. *Mar. Ecol.* Suppl. 1 **31,** 107–120.

Hobday, A. J., K. Hartmann. 2006. Near real-time spatial management based on habitat predictions for a longline bycatch species. *Fisheries Management & Ecology* **13,** 365–380.

Hobday, A. J., N. Flint, T. Stone, J. S. Gunn. 2009. Electronic tagging data supporting flexible spatial management in an Australian longline fishery. In 'Tagging and Tracking of Marine Animals with Electronic Devices II. Reviews: Methods and Technologies in Fish Biology and Fisheries'. (Eds. J. Nielsen, J. R. Sibert, A. J. Hobday, M. E. Lutcavage, H. Arrizabalaga, N. Fragosa). (Springer: Netherlands), 381–403.

Hobday, A. J, J. W. Young, J. M. Dambacher, C. Moeseneder. 2011. Defining dynamic pelagic habitats on the east coast of Australia. *Deep Sea Research II* **58,** 734–745.

Hobday A. J., A. D. M. Smith, I. Stobutzki, C. Bulman, R. Daley, J. Dambacher, R. Deng, J. Dowdney, M. Fuller, D. Furlani, S. P. Griffiths, D. Johnson, R. Kenyon, I. A. Knuckey, S. D. Ling, R. Pitcher, K. J. Sainsbury, M. Sporcic, T. Smith, T. Walker, S. Wayte, H. Webb, A. Williams, B. S. Wise, S. Zhou. 2011b. Ecological Risk Assessment for the Effects of Fishing. *Fisheries Research* **108,** 372–384.

Holbrook N. L. 2010. South Pacific Ocean dynamics: potential for enhancing sea level and climate forecasts. Chapter 10 in: 'Climate Alert: Climate Change Monitoring and Strategy', eds. Y. You, A. Henderson-Sellers, Sydney University Press: 313–342.

Hooker, S. A., A. Canadas, K. D. Hyrenbach, C. Corrigan, J. J. Polovina, R. R. Reeves. 2011. How to make protected area networks effective for top predators. *Endangered Sp. Res.* **13,** 203–218.

Howell, E. A., D. R. Kobayashi, D. M. Parker, G. H. Balazs, and J. J. Polovina. 2008. TurtleWatch: A tool to aid in the bycatch reduction of loggerhead turtles (*Caretta caretta*) in the Hawaii-based pelagic longline fishery. *Endangered. Species Res.* **5,** 267–278.

Howell, E. A., S. J. Bograd, C. Morishige, M. P. Seki, J. J Polovina. 2011. On North Pacific circulation and associated marine debris concentrations. *Mar. Pollut. Bull.* doi:10.1016/j.marpolbul.2011.04.034.

Howell, E. A., C. C. C. Wabnitz, J. P. Dunne, and J. J. Polovina. 2012. Climate change and fishing impacts on the central North Pacific ecosystem and Hawaii-based pelagic longline fishery. *Climatic Change,* DOI: 10.1007/s10584–012–0597.

Hoyt, E., G. T. Hvenegaard. 2002. A Review of Whale-Watching and Whaling with Applications for the Caribbean. *Coastal Management* **30,** 381–399.

ICCAT. 2011. Statistical Bulletin Volume 40 (1960–2009). Madrid, Spain.

ICCAT SC-RS. 2011. Report for biennial period, 2010–11 Part II (2011)–Volume 2. Madrid, Spain.

ICCAT SC-ECO. 2011. Inter-Sessional Meeting of the Sub-Committee on Ecosystems. Miami, Florida, USA, May, 2011.

ISA. 1999. Deep-Seabed Polymetallic Nodule Exploration: Development of Environmental Guidelines, Kingston, Jamaica: International Seabed Authority. 289 pp.

ISA. 2007. Workshop on Mining of Cobalt Rich Ferromanganese Crusts and Polymetallic Sulphides—Technological and Economic Considerations. Kingston, Jamaica: International Seabed Authority. 203 pp.

Isaacs, J. D., R. A. Schwartzlose. 1965. Migrant sound scatterers: interaction with the sea floor. *Science* **150,** 1810–1813.

Kaplan, D. M., E. Chassot, A. Gruss, A. Fonteneau. 2010. Pelagic MPAs: The devil is in the details. *Trends in Ecology and Evolution,* DOI: 10.1016/j.tree.2009.09.003.

Karl, D. 1999. A sea of change: Biogeochemical variability in the North Pacific Subtropical Grye. *Ecosystems* **2,** 181–214.

Kitchell, J. F., T. E. Essington, C. H. Boggs, D. E. Schindler, C. J. Walters. 2002. The role of sharks and longline fisheries in a pelagic ecosystem of the central Pacific. *Ecosystems* **5,** 202–216.

Koslow, J. A. 1996. Energetic and life-history patterns of deep-sea benthic, benthopelagic and seamount-associated fish. *Journal of Fish Biology* 49 Supplement A, 54–74.

Koslow, J. A. 1997. Seamounts and the ecology of deep-sea fisheries. *American Scientist* **85,** 168–176.

Koslow, J. A., G. W. Boehlert, J. D. M. Gordon, R. L. Haedrich, P. Lorance, N. Parin. 2000. The impact of fishing on continental slope and deep-sea ecosystems. *ICES Journal of Marine Science* **57,** 548–557.

Koslow, J. A., C. M. Bulman, J. M. Lyle. 1994. The mid-slope demersal fish community off southeastern Australia. *Deep-Sea Research* **41,** 113–141.

Koslow, J. A., K. Gowlett-Holmes, J. Lowry, T. O'Hara, G. Poore, A. Williams. 2001. The seamount benthic macrofauna off southern Tasmania: community structure and impacts of trawling. *Marine Ecology Progress Series* **213,** 111–125.

Koslow, T. 2007. *The Silent Deep.* University of Chicago Press, Chicago.

Lan, K. W., M. A. Lee, et al. 2011. Ocean variations associated with fishing conditions for yellowfin tuna (*Thunnus albacares*) in the equatorial Atlantic Ocean. *ICES Journal of Marine Science* **68,** 1063–1071.

Last, P.R., W. T. White, D. C. Gledhill, A. J. Hobday, R. Brown, G. J. Edgar, G. T. Pecl. 2011. Long-term shifts in abundance and distribution of a temperate fish fauna: a response to climate change and fishing practices. *Global Ecology and Biogeography* **20,** 58–72, DOI: 10.1111/j.1466–8238.2010.00575.x.

Lehodey, P., M. Bertignac, J. Hampton, A. Lewis, J. Picaut. 1997. El Niño Southern Oscillation and tuna in the western Pacific. *Nature* **389,** 715–718.

Lehodey, P., F. Chai, J. Hampton. 2003. Modelling climate-related variability of tuna populations from a coupled ocean–biogeochemical-populations dynamics model. *Fish. Oceanogr.* **12,** 483–494.

Lehodey, P., Senina, I., Sibert, J., Bopp, L., Calmettes, B., Hampton, J., Murtugudde, R. 2010. Preliminary forecasts of Pacific bigeye tuna population trends under the A2 IPCC scenario. *Progress in Oceanography,* **86,** 302–315.

Levin, S. A., J. Lubchenco. 2008. Resilience, Robustness, and Marine Ecosystem-based Management. *BioScience* **58,** 27–32.

Lewiston, R. L., L. B. Crowder. 2007. Putting longline bycatch of sea turtles into perspective. *Conser. Bio.* **21,** 79–86.

Lewison, R. L., L. B. Crowder, A. J. Read, S. A. Freeman. 2004. Understanding impacts of bycatch on marine megafauna. *Trends in Ecology and Evolution* **19** (11), 598–604.

Longhurst, A. 2007. *Ecological Geography of the Sea.* Elsevier, London.

Longhurst A., S. Sathyendranath, T. Platt, C. Caverhill. 1995. An estimate of global primary production in the ocean from satellite radiometer data. *Journal of Plankton Research* **17,** 1245–1271.

Mansfield, K. L., V. S. Saba, J. A. Keinath, J. A. Musick. 2009. Satellite tracking reveals a dichotomy in migration strategies among juvenile loggerhead turtles in the Northwest Atlantic. *Marine Biology* **156,** 2555–2570.

Martell, S. J. D., T. E. Essington, B. Lessard, J. F. Kitchell, C. J. Walters, C. H. Boggs. 2005. Interactions of productivity, predation risk, and fishing effort in the efficacy of marine protected areas for the central Pacific. *Canadian Journal of Fisheries and Aquatic Sciences* **62,** 1320–1336.

Marques da Silva, H., M. R. Pinho. 2007. Small-scale fishing on seamounts. In: T. J. Pitcher, T. Morato, P. J. B. Hart, M. R. Clark, N. Haggan, R. S. Santos (Eds). *Seamounts: Ecology, Fisheries & Conservation.* Blackwell Publishing, Oxford: 335–360.

Martinez, E., D. Antoine et al. 2011. Phytoplankton spring and fall blooms in the North Atlantic in the 1980s and 2000s. *J. Geophys. Res.* 116, C11029, DOI: 10.1029/2010JC006836.

Martinez, E., D. Antoine, F. D'Ortenzio, B. Gentili. 2009. Climate-driven basin-scale decadal oscillations of oceanic phytoplankton. *Science* **326,** 1253–1256.

Maunder, M. N., J. R. Sibert, A. Fonteneau, J. Hampton, P. Keiber, S. J. Harley. 2006. Interpreting catch per unit effort data to assess the status of individual stocks and communities. *ICES Journ. Mar. Sci.* **63,** 1373–1385.

McClain, C. R., L. Lundsten, M. Ream, J. Barry, A. DeVogelaere. 2009. Endemicity, biogeography, composition, and community structure on a Northeast Pacific seamount. *PLOS Biology* **4** (1), e4141, DOI: 10.1371/journal.pone.0004141.

McGillicuddy, D. J., L. A. Anderson et al. 2007. Eddy/wind interactions stimulate extraordinary mid-ocean plankton blooms. *Science* **316,** 1021–1026.

Miller, K. A. 2007. Climate variability and tropical tuna: Management challenges for highly migratory fish stocks. *Marine Policy* **31,** 56–70.

Moore, C., S. Moore, M. Leecaster, S. Weisberg. 2001. A comparison of plastic and plankton in the North Pacific subtropical gyre. *Mar. Pollut. Bull.* **42,** 1297–1300.

Morato, T., Hoyle, S. D., Allain, V., Nicol, S. J. 2010. Seamounts are hotspots of pelagic biodiversity in the open ocean. *PNAS* **107** (21), 9707–9711.

Morato, T., T. J. Pitcher, M. R. Clark, G. Menezes, F. Tempera, F. Porteiro, E. Giacomello, R. S. Santos. 2010. Can we protect seamounts for research? A call for conservation. *Oceanography* **23**(1), 190–199.

Morato, T., T. J. Pitcher. 2005. Ecosystem simulatioins in support of management of data-liminted seamount fisheries. In: G. H. Kruse, V. F. Gallucci, D. E. Hay et al. (Eds). *Fisheries Assessment and Management in Data-Limited Situations.* Alaska Sea Grant, Fairbanks: 467–486.

Myers, R. A., B. Worm. 2005. Extinction, survival or recovery of large predatory fishes. *Phil. Trans. R. Soc. B.* **360,** 13–20.

NEAFC. 2009. Fisheries Status Report 1998–2007.

Neilson, J. D., F. Arocha, S., S. Cass-Calay, J. Mejuto, M. Ortiz, G. Scott, C. Smith, P. Travassos, I. Andrushchenko. 2013. The recovery of Atlantic swordfish: The comparative roles of the regional fisheries management organization and species biology. Reviews in Fisheries Science, 21(2), 59–97.

Nelson, J. S. 2006. *Fishes of the World.* John Wiley & Sons, NY.

New Zealand Ministry of Fisheries. 2009. Report from the Fisheries Assessment Plenary, May 2009: stock assessments and yield estimates. New Zealand Ministry of Fisheries, Wellington, New Zealand: 1036 pp.

Nilsson, G. E., D. L. Dixson, P. Domenici, M. I. McCormick, C. Sørensen, S.-A. Watson, P. L. Munday. 2012. Near-future carbon dioxide levels alter fish behaviour by interfering with neurotransmitter function. *Nature Climate Change,* DOI: 10.1038/NCLIMATE1352.

O'Hara, T. D. 2007. Seamounts: centres of endemism or species richness for ophiuroids? *Global Ecology and Biogeography* **16,** 720–732.

Palter, J. B., M. S. Lozier, R. T. Barber. 2005. The effect of advection on the nutrient reservoir of the North Atlantic Subtropical Gyre. *Nature* **437,** 687–692.

Parin, N. V., A. N. Mironov, K. N. Nesis. 1997. Biology of the Nazca and Sala y Gomez submarine ridges, an outpost of the Indo-West Pacific fauna in the Eastern Pacific Ocean: composition and distribution of the fauna, its communities and history. *Advances in Marine Biology* **32,** 145–242.

Parks, S. E., M. Johnson, D. Nowacek, P. L. Tyack. 2010. Individual right whales call louder in increased environmental noise. *Biology Letters* **7,** 33–35.

Patrick, W. S., P. Spencer, J. S. Link, J. Cope, J. Field, D. R. Kobayashi, P. Lawson, T. Gedamke, E. Cortes, O. Ormseth, K. A. Bigelow, W. J. Overholtz. 2010. Using productivity and susceptibility indices to assess the vulnerability of United States fish stocks to overfishing. *Fishery Bulletin* **108,** 305–322.

Pikitch, E. K., C. Santora, E. A. Babcock, A. Bakun, R. Bonfil, D. O. Conover, P. Dayton, P. Doukakis, D. Fluharty, B. Heneman, E. D. Houde, J. Link, P. A. Livingston, M. Mangel, M. K. McAllister, J. Pope, K. J. Sainsbury. 2004. Ecosystem-based fishery management. *Science* **305,** 346–347.

Pitcher, T. J., T. Morato, P. J. B. Hart, M. R. Clark, N. Haggan, R. S. Santos. 2007. *Seamounts: Ecology, Fisheries & Conservation.* Blackwell Publishing, Oxford.

Polacheck, T. 2011. Politics and independent scientific advice in RFMO processes: A case study of crossing boundaries. *Marine Policy,* DOI: 10.1016/j.marpol.2011.04.006.

Polovina, J. J., M. Abecassis, E. A. Howell, P. Woodworth. 2009. Increases in the relative abundance of mid-trophic level fishes concurrent with declines in apex predators in the central North Pacific subtropical gyre, 1996–2006. *Fish. Bull.* **107**(4), 523–531.

Polovina, J. J., J. P. Dunne, P. A. Woodworth, E. A. Howell. 2011. Projected expansion of the subtropical biome and contraction of the temperate and equatorial upwelling biomes in the North Pacific under global warming. *ICES J. Mar. Sci.* DOI: 10.1093/icesjms/fsq198.

Polovina J. J., E. Howell, D. R. Kobayashi, M. P. Seki. 2001. The transition zone chlorophyll front, a dynamic global feature defining migration and forage habitat for marine resources. *Progress in Oceanography* **49,** 469–483.

Polovina J. J., E. A. Howell, M. Abecassis. 2008. The ocean's least productive waters are expanding. *Geophys. Res. Lett.* **35,** L03618, DOI: 10.1029/2007GL031745.

Polovina, J. J. P. A. Woodworth. 2012. Declines in phytoplankton cell size in the subtropical oceans estimated from satellite remotely-sensed temperature and chlorophyll, 1998–2007. *Deep Sea Res.* II 77–80: 82–88.

Probert, P. K., S. Christiansen, K. M. Gjerde, S. Gubbay, R. S. Santos. 2007. Management and conservation of seamounts. In: T. J. Pitcher, T. Morato, P. J. Hart, M. R. Clark, N. Haggan, R. S. Santos (Eds). *Seamounts: Ecology, Fisheries & Conservation.* Blackwell Publishing, Oxford: 442–475.

Probert, P. K., D. G. McKnight, S. L. Grove. 1997. Benthic invertebrate bycatch from a deep-water trawl fishery, Chatham Rise, New Zealand. *Aquatic Conservation: Marine and Freshwater Ecosystems* **7,** 27–40.

Ramirez-Llodra, E., P. A. Tyler, M. C. Baker, O. A. Bergstad, M. R. Clark et al. 2011. Man and the Last Great Wilderness: Human Impact on the Deep Sea. *PLoS ONE* **6**(7): e22588. DOI: 10.1371/journal.pone.0022588.

Ravier, C., J. Fromentin. 2004. Are the long-term fluctuations in Atlantic bluefin tuna (*Thunnus thynnus*) population related to environmental changes? *Fisheries Oceanography* **13,** 145–160.

Richer de Forges, B., J. A. Koslow, G. C. B. Poore. 2000. Diversity and endemism of the benthic seamount fauna in the southwest Pacific. *Nature* **405,** 944–947.

Ridgway, K. R. 2007. Long-term trend and decadal variability of the southward penetration of the East Australian Current. *Geophysical Research Letters* **34,** L13613, DOI: 10.1029/2007GL030393.

Ridgway, K. R., J. R. Dunn 2003. Mesoscale structure of the mean East Australian Current System and its relationship with topography. *Progress in Oceanography* **56,** 189–222.

Rosen, D. E. 1973. Interrelationships of higher euteleostean fishes. In: P. H. Greenwood, R. S. Miles, C. Patterson (Eds). *Interrelationships of fishes.* Academic, New York: 397–513.

Rouyer, T., J. M. Fromentin, et al. 2008. Complex interplays among population dynamics, environmental forcing, and exploitation in fisheries. *Proc Natl Acad Sci USA* **105,** 5420–5425.

Rowden, A. A., T. A. Schlacher, A. Williams, M. R. Clark, R. Stewart, F. Althaus, D. A. Bowden, M. Consalvey, W. Robinson, J. Dowdney. 2010. A test of the seamount oasis hypothesis: seamounts support higher epibenthic megafaunal biomass than adjacent slopes. *Marine Ecology* **31,** 95–106.

Ruckelshaus, M., Klinger, T., Knowlton, N., Demaster, D., 2008. Marine ecosystem-based management in practice: Scientific and Governance challenges. *BioScience* **58,** 53–63.

Saba, V. S., P. Santidrián-Tomillo, R. D. Reina, J. R. Spotila, J. A. Musick, D. A. Evans, F. V. Paladino. 2007. The effect of the El Niño Southern Oscillation on the reproductive frequency of eastern Pacific leatherback turtles. *Journal of Applied Ecology* **44,** 395–404.

Saba, V. S., C. A. Stock, J. R. Spotila, F. P. Paladino, P. Santidrián-Tomillo. 2012. Projected response of an endangered marine turtle population to climate change. *Nature Climate Change,* DOI: 10.1038/NCLIMATE1582.

Samadi, S., L. Bottan, E. Macpherson, B. Richer de Forges, M.-C. Boisselier. 2006. Seamount endemism questioned by the geographic distribution and population genetic structure of marine invertebrates. *Marine Biology* **149,** 1463–1475.

Samadi, S., T. A. Schlacher, B. Richer de Forges. 2007. Seamount benthos. In: T. J. Pitcher, T. Morato, P. J. Hart, M. R. Clark, N. Haggan, R. S. Santos (Eds). *Seamounts: Ecology, Fisheries & Conservation.* Blackwell Publishing, Oxford: 119–140.

Seki, M. P., J. J. Polovina. 2001. Ocean gyre ecosystems. Volume 4 pp: 1959–1964. In: *The Encyclopedia of Ocean Sciences.* Steele, J. H. et al. (Eds.). Academic Press, San Diego, 3399 p.

Sibert, J., J. Hampton, P. Kleiber, M. Maunder. 2006. Biomass, size, and trophic status of top predators in the Pacific Ocean. *Science* **314,** 1773–1775.

Smith, D. K., J. R. Cann. 1990. Hundreds of small volcanoes on the median valley floor of the Mid-Atlantic Ridge at 24–30°N. *Nature* **348,** 152–155.

Smith, K. L. J. 1978. Metabolism of the abyssopelagic rattail *Coryphaenoides armatus* measured *in situ. Nature* **274,** 362–364.

Smith, A. D. M., E. A. Fulton, A. J. Hobday, D. C. Smith, P. Shoulder. 2007. Scientific tools to support practical implementation of ecosystem based fisheries management. *ICES Journal of Marine Science* **64,** 633–639.

Solow, A. R., K. A. Bjorndal, A. B. Bolten. 2002. Annual variation in nesting numbers of marine turtles: the effect of sea surface temperature on re-migration intervals. *Ecol. Lett.* **5,** 742–746.

Stocks, K. I., P. J. Hart. 2007. Biogeography and diversity of seamounts. In: T. J. Pitcher, T. Morato, P. J. Hart et al. (Eds.). *Seamounts: Ecology, Fisheries & Conservation.* Blackwell, Oxford: 255–281.

Stramma, L., E. D. Prince, S. Schmidtko, J. Luo, J. P. Hoolihan, M. Visbeck, D. W. R. Wallace, P. Brandt, A. Körtzinger. 2012. Expansion of oxygen minimum zones may reduce available habitat for tropical pelagic fishes. *Nature Climate Change* **2,** 33–37.

Stewart, B. D., G. E. Fenton, D. C. Smith, S. A. Short. 1995. Validation of otolith increment age estimates for a deepwater fish species, warty oreo, *Allocyttus verrucosus,* by radiometric analysis. *Marine Biology* **123,** 29–38.

Sverdrup, H. U. 1953. On conditions for the vernal blooming of phytoplankton. *Journal du Conseil International pour l'Exploration de la Mer* **18,** 287–295.

Taranto, G. H., K. Ø. Kvile, T. J. Pitcher, T. Morato. 2012. An ecosystem evaluation framework for global seamount conservation and management. *PLoS ONE* **7**(8), e42950, doi:10.1371/journal.pone.0042950.

Taylor, N. G., M. K. McAllister, et al. 2011. Atlantic bluefin tuna: a novel multistock spatial model for assessing population biomass. *PLoS ONE* **6,** e27693.

Tracey, D. M., B. Bull, M. R. Clark, K. A. Mackay. 2004. Fish species composition on seamounts and adjacent slope in New Zealand waters. *New Zealand Journal of Marine and Freshwater Research* **38,** 163–182.

Tseitlin, V. B. 1985. The energetics of fish populations inhabiting seamounts. *Oceanology* **25,** 237–239.

Vianna, G. M. S, M. G. Meekan, D. Pannell, S. Marsh S, N. S. J. Meeuwig. 2010. Wanted Dead or Alive? The relative value of reef sharks as a fishery and an ecotourism asset in Palau. Australian Institute of Marine Science and University of Western Australia, Perth.

Utech, D. 2000. "Valuing Hawaii's Humpback Whales: The Economic Impact of Humpbacks on Hawaii's Ocean Tour Boat Industry." In: *The Economic Contribution of Whalewatching to Regional Economies: Perspectives From Two National Marine Sanctuaries.* Marine Sanctuaries Conservation Series MSD-00–2. U.S. Department of Commerce, National Oceanic and Atmospheric Administration, Marine Sanctuaries Division, Silver Spring, MD.

Valentine, P. S., A. Birtles, M. Curnock, P. Arnold, A. Dunstan. 2004. Getting closer to whales—passenger expectations and experiences, and the management of swim with dwarf minke whale interactions in the Great Barrier Reef. *Tourism Management* **25,** 647–655.

Van Dover, C. L. 2011. Mining seafloor massive sulphides and biodiversity: what is at risk? *ICES Journal of Marine Science* **68,** 341–348.

Van Houtan, K. S, J. M. Halley. 2011. Long-Term Climate Forcing in Loggerhead Sea Turtle Nesting. *PLoS ONE* **6**(4), e19043. DOI: 10.1371/journal.pone.0019043.

Wallace, B. P., R. L. Lewison et al. 2010. Global patterns of marine turtle bycatch. *Conservation Letters* **3,** 131–142.

Watson, R., A. Kitchingman, W. W. L. Cheung. 2007. Catches from world seamount fisheries. In: T. J. Pitcher, T. Morato, P. J. B. Hart, M. R. Clark, N. Haggan, R. S. Santos (Eds.). *Seamounts: Ecology, Fisheries & Conservation.* Blackwell Publishing, Oxford: 400–412.

Watters, G. M., R. J. Olson, R. C. Francis, P. C. Fiedler, J. J. Polovina, S. B. Reilly, K. Y. Aydin, C. H. Boggs, T. E. Essington, C. J. Walters, J. F. Kitchell. 2003. Physical forcing and the dynamics of the pelagic ecosystem in the eastern tropical Pacific: simulations with ENSO-scale and global-warming climate drivers. *Canadian Journal of Fisheries and Aquatic Sciences* **60,** 1161–1175.

WCPFC 2007. Conservation and management measure to mitigate the impact of fishing for highly migratory fish stocks on seabirds. Western and Central Pacific Fisheries Convention. CMM 2007–04. http://www.wcpfc.int. 6 p.

WCPFC 2008a. Conservation and management measure for bigeye and yellowfin tunas in the Western and Central Pacific Ocean. Western and Central Pacific Fisheries Convention. CMM 2008–01. http://www.wcpfc.int. 42 p.

WCPFC 2008b. Conservation and management of sea turtles. Western and Central Pacific Fisheries Convention. CMM 2008–03. http://www.wcpfc.int. 5 p.

WCPFC 2009. Tuna fishery yearbook. Western and Central Pacific Fisheries Convention. http://www.wcpfc.int. 115 p.

WCPFC 2010a. Conservation and management measure for North Pacific striped marlin. Western and Central Pacific Fisheries Convention. CMM 2010–01. http://www.wcpfc.int. 2 p.

WCPFC 2010b. Conservation and management measure for shark. Western and Central Pacific Fisheries Convention. CMM 2010–07. http://www.wcpfc.int. 3 p.

WPRFMC 2010. Pelagic fisheries of the Western Pacific Region 2009 annual report. Western Pacific Regional Fisheries Management Council. 302 p.

WCPFC-SC7 2011. (Draft) Strategic research plan 2012–2016. WCPFC-SC7–2011/GN-WP-05 http://www.wcpfc.int. 7 p.

Weilgart, L. S., 2007. The impact of anthropogenic ocean noise on cetaceans and implications for management. *Can. J. Zool.* **85,** 1091–1116.

Wessel, P., D. T. Sandwell, S.-S. Kim. 2010. The global seamount census. *Oceanography* **23,** 24–33.

White, M., I. Bashmachnikov, J. Aristegui, A. Martins. 2007. Physical processes and seamount productivity. In: T. J. Pitcher, T. Morato, P. J. B. Hart, M. R. Clark, N. Haggan, R. S. Santos (Eds.). *Seamounts: Ecology, Fisheries & Conservation.* Blackwell Publishing, Oxford: 65–84.

Williams, A., T. A. Schlacher, A. A. Rowden, F. Althaus, M. R. Clark, D. A. Bowden, R. Stewart, N. J. Bax, M. Consalvey, R. J. Kloser. 2010. Seamount megabenthic assemblages fail to recover from trawling impacts. *Marine Ecology* **31,** 183–199.

Wilson, R. R., R. S. Kaufmann. 1987. Seamount biota and biogeography. In: B. H. Keating, P. Fryer, R. Batiza, G. W. Boehlert (Eds.). *Seamounts, Islands and Atolls.* American Geophysical Union **43,** 355–377.

Woodworth-Jefcoats, P. A., J. L. Blanchard, J. P. Dunne, J. J Polovina. 2012. Projected Climate impacts on the pelagic ecosystem size structure and catches in the North Pacific over the 21st century. *Global Climate Change,* DOI: 10.1111/gcb12076.

Worm, B., Sandow, M., Oschlies, A., Lotze, H. K., Myers, R. A. 2005. Global patterns of predator diversity in the open oceans. *Science* **309,** 1365–1369.

Worm, B., E. B. Barbier, N. Beaumont, J. E. Duffy, C. Folke, B. S. Halpern, J. B. C. Jackson, H. K. Lotze, F. Micheli, S. R. Palumbi, E. Sala, K. A. Selkoe, J. J. Stachowicz, R. Watson. 2006. Impacts of Biodiversity Loss on Ocean Ecosystem Services. *Science* **314,** 787–790.

Worm, B., R. Hilborn, J. K. Baum, T. A. Branch, J. S. Collie, C. Costello, M. J. Fogarty, E. A. Fulton, J. A. Hutchings, S. Jennings, O. P. Jensen, H. K. Lotze, P. M. Mace, T. R. McClanahan, C. Minto, S. R. Palumbi, A. M. Parma, D. Ricard, A. A. Rosenberg, R. Watson, D. Zeller. 2009. Rebuilding global fisheries. *Science* **325,** 578–585.

Yasui, M. 1986. Albacore, *Thunnus alalunga,* pole-and-line fishery around the Emperor Seamounts. In: R. N. Uchida, S. Hayasi, G. W. Boehlert (Eds.). *Environment and resources of seamounts in the North Pacific.* NOAA Technical Report **43,** 37–40.

Yesson, C., M. R. Clark, M. L. Taylor, A. D. Rogers. 2011. The global distribution of seamounts based on 30 arc seconds bathymetry data. *Deep Sea Res. I* **58,** 442–453.

Young, J. W., M. J. Lansdell, R. A. Campbell, S. P. Cooper, F. Juanes, M. A. Guest. 2010. Feeding ecology and niche segregation in oceanic top predators off eastern Australia. *Marine Biology,* DOI: 10.1007/s00227–010–1500-y.

Chapter 14. EBM In Highly Impacted Coasts and Estuaries

EDWARD HOUDE

University of Maryland Center for Environmental Science

ANDREW KENNY

Centre for the Environment, Fisheries, and Aquaculture Science
United Kingdom

SHIJIE ZHOU

Commonwealth Scientific and Industrial Research Organization
Marine and Atmospheric Research

Contents

1. Introduction

Coastal ecosystems lie at the interface between terrestrial and aquatic habitats and are both naturally stressed and often highly impacted by human activities. Various definitions have been proposed for the coastal zone. The Millennium Ecosystem Assessment (MA, 2005) defined coastal ecosystems as areas extending seaward from the coast to the middle of the continental shelf, and inland to include all areas strongly influenced by proximity to the ocean. The MA further defined these complex systems as encompassing areas having water depths to 50 m below mean sea level and extending to terrestrial areas elevated to 50 m above the high tide level, or extending landward up to 100 km from shore. In the aquatic realm, coastal ecosystems include estuaries, coral reefs, intertidal zones, seagrass communities, shipping routes, and coastal aquaculture sites. In recent decades we have come to appreciate the connectedness of terrestrial and coastal ecosystems (Ruttenberg and Granek, 2011) and the need to account for it in Ecosystem-Based Management (EBM).

The Sea, Volume 16, edited by Michael J. Fogarty and James J. McCarthy
ISBN 978-0-674-07270-1

To conserve values and productivity of coastal ecosystems it is widely recognized that there is a need for collaborative, integrative science and management actions that support policies to conserve values and services of coastal ecosystems (UNEP and IOC-UNESCO, 2009). The many competing human interests that rely on sustainable ecosystem goods and services in coastal ecosystems must be balanced by providing equitable access to diverse stakeholders (Agardy et al., 2011). Within the aquatic portion of the coastal zone, those interests and services include fisheries, transportation, recreation, tourism, energy generation, and a reliable supply of high-quality water for human uses. In the terrestrial portion, additional uses and services also are common, e.g., agriculture, forestry, ports.

EBM as a goal has been broadly recommended but nowhere in marine ecosystems is it more needed than in estuaries, coastal embayments and coastal seas. These systems and services they provide are the most highly impacted and degraded by human activity and also are susceptible to looming effects of climate change and ongoing threats of severe weather, flooding, and tsunamis (Halpern et al., 2008). Agardy et al. (2011) provide a strong rationale and justification for EBM in the coastal ocean, emphasizing the need for management that sustains critical ecosystem structure, functions and processes for "a joined approach that marries the seemingly competing interests for ocean and coastal resources and space, such as environmental tourism, fisheries and energy generation, within a robust framework and a spatial planning perspective."

1.1 High impact regions

The global coastline is > 1.6 million km in length, and 123 nations claim a share of coastlines and coastal ecosystems (MA, 2005). The "coastal zone" includes a narrow band of land dominated by ocean influences (e.g., tides, marine aerosols, etc.). There are > 1200 major estuaries worldwide (UNEP, 2006) that comprise a 500,000 km^2 area and receive ~80% of the world's freshwater discharge (MA, 2005) before delivering it to the sea. In addition, there are countless small estuarine systems and subsytems. More than one-third of the world's people reside in coastal areas that constitute only 4% of the global terrestrial habitat (UNEP, 2006). Most of the world's largest cities (> 2.5 million inhabitants) lie within the coastal zone. It is not surprising that such high density of humans has led to stress. Coastal ecosystems, estuaries, and embayments provide key habitats and nurseries for biological communities: e.g., coral and other reefs, mangroves, seagrass beds, lagoons, mudflats, beaches, dunes, saltmarshes. Many, perhaps most, of these coastal and estuarine ecosystems are in decline and their potential to deliver critical ecosystem services is diminished (Lotze et al., 2006).

A broad goal of many programs with EBM objectives is integrated coastal management. Such programs must be comprehensive and management must be strategically adaptive and responsive to uncertainties in ecosystem behavior. In heavily impacted estuarine and coastal ecosystems, the complexities and uncertainties of management actions are compounded because, in addition to the primarily regulatory and mitigative approaches common offshore, nearshore EBM programs usually have rebuilding components that seek to restore structure, habitat, and productivity.

Worldwide, many coastal ecosystems are under stress. In this chapter, we review issues and problems faced by emerging EBM in highly impacted coastal and estuarine areas. Our emphasis is centered on European, North American and Australian coastal ecosystems, and on EBM approaches that are implemented or under development in these regions. While many impacts and stresses are common to all coastal ecosystems, there are contrasts. Some of the contrasts are a consequence of the different histories of dense human settlement and stresses, which have been common for millennia in Europe, for four centuries in North America, but for only two centuries in Australia. Accumulation of stress on coastal ecosystems occurred at a relatively slow pace in Europe and North America until the mid-19th century when impacts accelerated. European settlement of Australia did not begin until the 19th century, but its effects accumulated rapidly with the advent of industrial-scale agriculture, establishment of cities on major coastal waterways, and major engineering projects in watersheds (e.g., dams). Consequently, while the timelines differ, there has been a rapid convergence of similar stresses and their effects across the globe.

1.2 Characterizing human activities in estuarine and coastal ecosystems

Humans have always benefited from services provided by coastal ecosystems. In recent decades as stresses increased commensurate with increasing human impacts, the need for precautionary, coastal and marine spatial management, often characterized as EBM, has emerged. In the past century, traditional pressures on coastal and estuarine ecosystems, e.g., fishing and transportation, have been compounded by broader demand from emerging stakeholder interests that increasingly drive stress, e.g., pollution, shoreline development and modifications, energy development (e.g., oil, gas, windfarms), sedimentation, tourism, and recreational activities that include boating, swimming and beach uses. Importantly, human activities throughout watersheds shape the character and productivity of estuaries. For example, forestry, agriculture, and flood control practices alter water flow and delivery (dams and diversions), and excessive nutrient loading of nitrogen and phosphorus often leads to eutrophication. At the water's edge, shoreline development and habitat alteration, along with accompanying sedimentation, are problematic. Continued urbanization adds to loadings and stresses. Impacted coastal ecosystems generally have altered biodiversity and shifts in species represented in fisheries production. Degraded coastal ecosystems that are denuded of seagrasses, fringing marshes, forest buffers and other terrestrial vegetation are more vulnerable to storms and floods and have reduced appeal for tourism and recreation. Highly modified and stressed coastal estuaries may be particularly susceptible to invasive species. Eutrophication can fuel harmful algal blooms and lead to hypoxic "dead zones" that are increasingly common since the mid 20th century.

The high economic values of services provided by ecosystems at the interface of coast, land, and watersheds were catalogued by Barbier et al. (2011) who argued that these high values must be seriously considered in developing coastal management policies. There is a need for collective governance and jurisdictional responsibility to represent the interests of multiple stakeholders. (Paterson et al., 2011). EBM may demand tradeoffs because of multiple competing and conflicting objectives. EBM is place-based, but to be effective in moderate to large coastal ecosystems it will require that management extends across political boundaries and national interests. As Agardy et al. (2011) stated, the objective of EBM in coastal regions is to manage multiple human activities for a common outcome. Adaptive management is a natural and necessary framework for EBM and is particularly important in fast-changing, highly impacted coastal zones.

2. Background

Below, we briefly describe the geography and character of the three regions highlighted in this chapter. We focus on coastal areas in Europe, Australia and North America. Their heavily inhabited coasts (Figure 14.1) are marked by a diversity of ecosystems, including estuaries, bays, inlets, barrier islands, fjords, tidal flats, lagoons, salt marshes, mangrove swamps, coral reefs, deltas and dunes. The impressive economic values of ecosystem services provided by these features and systems (Barbier et al., 2011) must be figured into coastal management policies.

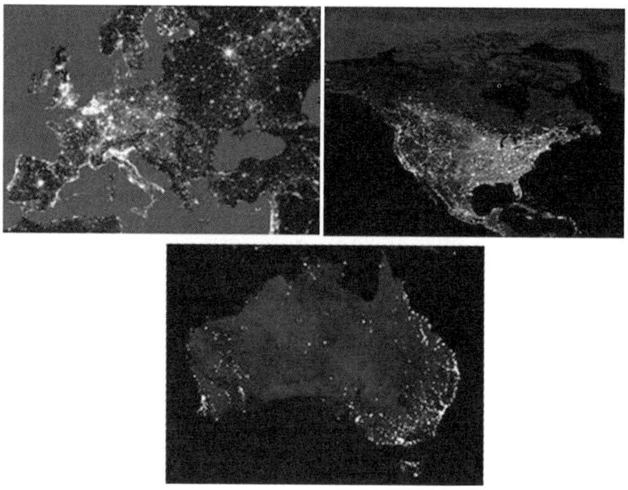

Fig. 14.1 – (See color insert) Satellite night images of Europe, North America, and Australia, revealing areas of urbanization along the coasts. (Google Earth)

2.1 Europe

The landmass of Europe is surrounded by a complex network of regional seas interconnected by narrow straits, each supporting diverse islands, embayments and estuaries which often receive significant freshwater inputs from major river basin catchments. The principal European sea areas are generally defined as the NE Atlantic (comprising the North Sea, Irish Sea, SW approaches, Bay of Biscay and Iberian Peninsula), the Baltic Sea (comprising, the Gulf of Bothnia, Gulf of Finland and of Riga), the Mediterranean Sea (comprising the Alboran Sea, Balearic Sea, Ligurian Sea, Tyrrhenian Sea, Adriatic Sea, Ionian Sea, Aegean Sea and the Sea of Crete) and the Black Sea (Figure 14.2). All support distinctive sub-regional ecosystems driven by local ecological conditions and the influence of diverse human activities. The EU has a coastline of 68,000 km, which increases to 185,000 km if the EEA member countries Turkey, Iceland and Norway are included. Almost half of the EU's population lives less than 50 km from the sea, the majority being concentrated in urban areas along the coast. In 2001, 70 million people, 14% of the EU population, lived within 500 m of the coast (EEA, 2010).

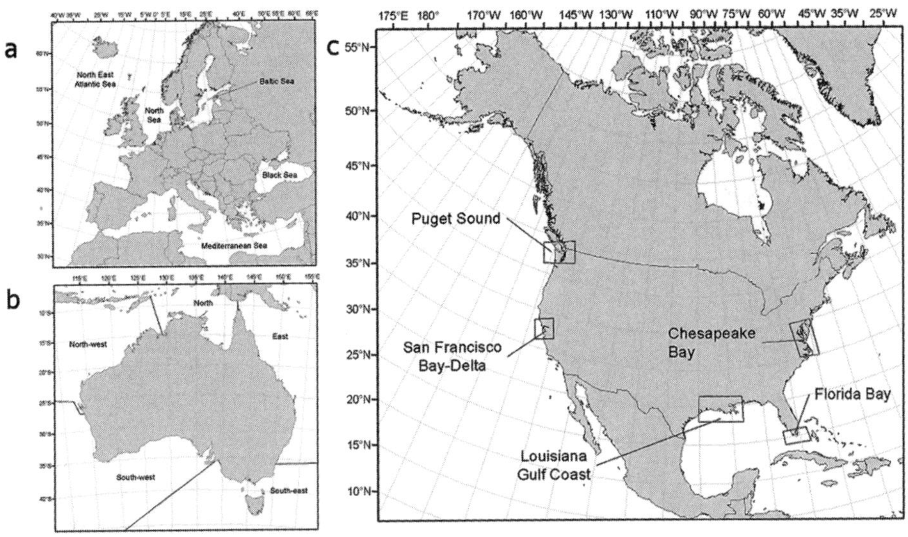

Fig. 14.2 – (See color insert) a. Europe. b. Australia. c. North America. The chapter focuses on EBM in highly impacted coastal areas of these three broad geographical regions. (Map courtesy of Sean Lucey, NEFSC)

2.2 *Australia*

Australia's ocean territory, third largest in the world, covers 14.7 million km^2 and exceeds the terrestrial area. The coastline is 59,700 km long, with more than 1000 estuaries, and borders the Pacific, Indian and Southern Oceans (Figure 14.2). The marine environment spans climatic zones from the tropical north to temperate south, and includes external territories in the polar zone. Coasts and estuaries constitute a dynamic, transition zone between the land and the sea, from which Australians derive many economic, social and environmental benefits. More than 50% of Australians reside within seven km of the coast. In addition, millions of domestic and international tourists visit coastal areas. In the decade from 2002 to 2011, Australia's population increased by 14.7% (ABS, 2011), with growth in coastal zones higher than the national average (ABS, 2010).

Australian State governments manage coastal waters within three nautical miles, while the Commonwealth Government manages waters from 3 to 200 nautical miles (the EEZ). Management of coastal resources is largely the responsibility of the States but, by the definition of the Millennium Ecosystem Assessment (MA, 2005), coastal zones extend beyond the state and territorial waters and the Commonwealth Government bears responsibilities for many coastal issues, including defining maritime boundaries, overseeing national conservation goals, and maritime safety.

2.3 North America

The North American continent has a long and diverse coastline. The Canadian coastline is the longest of any nation (approximately 200,000 km), while the combined coastlines of the United States (150,000 km) and Mexico (50,000 km) are nearly as long (Figure 14.2). Each of the North American nations has sovereignty over coastal ecosystems on the Atlantic and Pacific Coasts. The coastline of Alaska alone (approximately 50,000 km) and its valuable natural and mineral resources are indicative of the scale and importance of EBM needs in North American coastal waters.

More than half of the approximately 300 million U.S. residents live along the coast and that number will increase by 14 million in 2020 (JOCI, 2011). In the U.S., more than 300 seaports are located in coastal cities. The narrow 3-nautical mile band of the coastal U.S. supports a large fraction of the nation's marine fish landings and value. Not only numbers but also diversity of coastal stakeholders is increasing in North America. Historically, relatively few stakeholder groups, primarily fishing and farming communities, and shipping interests, were prominent and sectoral management was adequate to mostly insure relatively healthy ecosystems. Urbanization took hold in the 19th century and intensive agriculture replaced forested watersheds. The value of extractive and exploitative activities in the North American coastal zone, while difficult to estimate, is conservatively stated for the U.S. fishing industry alone to be valued at $45 billion in annual income that supports 1.5 million jobs (JOCI, 2011). In 2007, it is estimated that coastal counties in the U.S. supported 69 million jobs and contributed $8 trillion to the economy.

3. Provisioning Services

Provisioning services attributable to coastal ecosystems are valuable contributors to the economies of coastal nations. In this section we highlight valuable services identified for Europe, Australia and North America. These services clearly have high value and must be conserved and enhanced, but they also can represent threats to the highly impacted coastal zone.

3.1 Europe

3.1a Fisheries

Although membership in the European Union has expanded in recent years, EU fisheries production (catch and landings) has gradually declined (by 21% between 1992 and 2006) and is now at about 5.3 million tonnes.[1] In 1990, EU catches represented 7% of world fisheries, which made the EU the third largest world producer after China and Peru. However, in 2005, EU production accounted for only 6% of

[1] http://circa.europa.eu/irc/opoce/fact_sheets/info/data/policies/fisheries/article_7225_en.htm

world catches. In 2006, 75% of EU catches were caught in the North-East Atlantic (including the North Sea), 11% in the Mediterranean and Black Sea, 8% in the mid-eastern Atlantic and 3% in the western Indian Ocean.

3.1b Aquaculture

A rapid increase in European aquaculture occurred over the past 15 years. Today, European marine aquaculture produces 1.8 million tonnes of fish, shellfish and crustaceans, generating a turnover of around EUR 3.2 billion, while supporting 65,000 jobs. The largest aquaculture producer in Europe is Norway where Atlantic salmon (*Salmo salar*) accounts for about 40% of Europe's total aquaculture production. Spain is second largest with its production dominated by blue mussel *Mytilus edulis*, followed by France's production dominated by Pacific cupped oyster *Crassostrea gigas* (EEA, 2010).

3.1c Energy

In December 2008, the EU adopted an ambitious climate change and energy package which commits European countries to increase their share of renewable energy to 20% of Europe's total energy by 2020 (EEA, 2010). Reaching this target requires new facilities and energy platforms of which wind turbines will contribute significantly over the next decade. In 2009, wind energy accounted for 4.8% of the EU's total electricity consumption (Figure 14.3), but is expected to grow in most countries and triple by 2020 (EEA, 2010). At present, most offshore installations are located in relatively shallow parts of the North Sea and Baltic Sea where wind energy potential is greatest (Figure 14.3 inset). In addition, oil and gas production continues to provide a significant supply of energy within the European Economic Area (EEA) and is the world's fourth largest producer of oil and gas.[2] In 2008, it was estimated there were about 14 billion barrels of oil reserves and about 38 billion barrels of oil-equivalent gas remaining in the EEA.

3.1d Shipping

Shipping of goods by sea among European countries comprises about 60 % of all maritime transport and, with well developed coastal routes linked to inland waterways, is a major alternative to road transport. In total, 1,720 million tonnes of cargo were moved by sea in 2007, amounting to approximately 40% of all EU cargo transportation. Liquid bulk, which includes liquefied gas, crude oil and oil products, accounted for about half that tonnage. Intensive shipping traffic is confined to specific routes. Transporting goods by ship (especially chemicals) carries risk of accidents that could have serious environmental impacts (EEA, 2010).

3.2 Australia

3.2a Fisheries

While Australia has the third largest EEZ, its waters rank only 52nd globally in terms of volume of fish landed. However, Australia's selective fisheries production focuses on high-value export species such as lobsters, prawn, tuna, salmon and

[2] http://www.ogp.org.uk/pubs/438.pdf

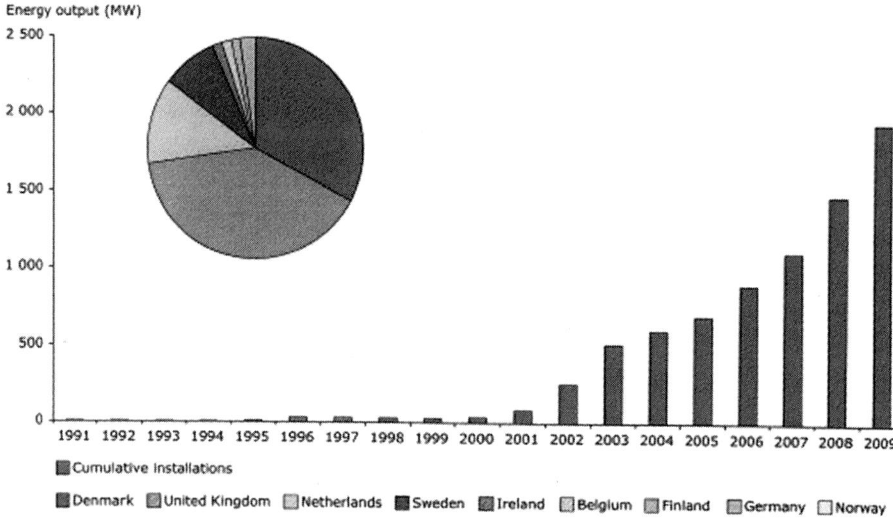

Fig. 14.3 – (See color insert) European wind energy production. (EEA, 2010)

abalone, largely from the nearshore and coastal zones. In 2009–10, the value of fisheries landings decreased by 3% from the previous year, to $1.34 billion (Australian), while volume decreased slightly by 1% to 171,512 tonnes (ABARES, 2011).

3.2b Aquaculture

The Australian aquaculture industry is small by global standards, contributing less than 0.1% to global production. The volume of aquaculture production in 2009–10 was 73,542 tonnes, which is 30% of total Australian fisheries production. Its value was $870.4 million, 40% of the gross value of Australian fisheries production (ABARES, 2011). Prawns, tuna, salmonids, edible oysters, and pearl oysters accounted for 87% of the value. The value of farmed salmonids rose by 13% to $369.1 million in 2009–10. Farmed salmonids constitute the largest aquaculture species group and also the most valuable fisheries product in Australia.

3.2c Energy

Australia has about 2% of the world's gas reserves and 0.2% of crude oil reserves. Large reserves of gas and significant reserves of oil lie under the seabed of coastal waters, for which there is both domestic and global demand. In 2008, Australia was the 18th largest producer of natural gas, the 7th largest exporter of natural gas in the world and the 3rd largest natural gas exporter in the Asia-Pacific Region. The oil and gas industries are now concentrated in the southeast (Bass Strait) and in the north and north-west regions.

3.2d Shipping

Import and export activities in Australia rely heavily on marine transport facilities, including ports that are primarily located in urbanized areas along the coast. About 97% of the volume of Australia's goods is transported through the network of ports. In 2008–09, approximately 800 million tonnes of cargo were moved through Australian wharves by 4200 vessels. Port facilities are under constant pressure to expand and increase capability to support and load ever larger vessels to transport materials overseas.

3.3 North America

3.3a Fisheries

Fisheries constitute a major provisioning service in coastal regions that provide food and recreational opportunities. Nearshore coastal (0–3 miles offshore) fisheries in the U.S. accounted for 33% of landed weight and 38% of value in 2010 (NOAA, 2010) despite only encompassing a small fraction of the 200-mile EEZ. Many of North America's most valuable fisheries for shellfish, e.g., shrimp, scallops, lobsters, crabs, and anadromous fishes, e.g., salmons, striped bass, are primarily exploited in the ecologically sensitive coastal zone.

3.3b Aquaculture

Marine aquaculture in coastal ecosystems is increasing in North America, although freshwater plus marine culture accounts for only 1.5% of global aquaculture production (FAO, 2010). The U.S. ranks 13th globally in aquaculture production. In 2009, 39,000 tonnes were marine species (NOAA, 2010), with 54% of that production from bivalve molluscs (oysters, clams, mussels). Marine aquaculture is growing in Canada where salmon farming is an important enterprise on both coasts, producing 109,000 tonnes in 2009 (http://www.dfo-mpo.gc.ca/aquaculture/ref/stats/aqua-ff-fc-2009-eng.htm).

3.3c Energy

Development of traditional (oil and natural gas) and renewable (e.g., wind) energy resources are becoming increasingly important in coastal waters of the U.S. About 25% of total U.S. crude oil production occurs in state and federal marine waters (http://stateofthecoast.noaa.gov/economy.html). Energy production must be balanced against the risks from activities and accidents in sensitive coastal ecosystems. Emerging wind energy could support a substantial fraction (possibly 20%) of U.S. energy demand by 2030 (http://www.nrel.gov/docs/fy11osti/52664.pdf), although no offshore wind farms were operating in 2012.

3.3d Shipping

Coastal waters of the U.S. support 25,300 miles of waterways used to transport > 2 billion tonnes of cargo annually. U.S. ports host > 8% of annual global vessel calls, second only to China. And, tourism on cruise vessels in North American coastal waters hosted 2.8 million passengers on > 4,200 cruises in 2011 (www.marad.dot.gov/data.statistics). Marine transport accounts for 44.5% of the value and 78% of the tonnage of U.S. international trade, while directly or indirectly supporting 13 million jobs (www.cmts.gov/downloads.CMTS_Ocean_Policy_Response_Paper.pdf). Ports and coastal marine waters on the west and east coasts of Canada also host valuable marine transport industries worth $404 billion CA in 2009 (http://www.statcan.gc.ca/pub/54–205-x/2009000/aftertoc-aprestdm1-eng.htm).

4. Threats and Challenges

Maintaining high economic values of provisioning services described above and conserving the supporting, regulating, and cultural services of coastal ecosystems are issues of great concern in the highly impacted coastal zone. It is here that effects of human activities and other stresses pose the greatest threat to marine ecosystems. Impacts, which can be cumulative, are driven by diverse human activities and pressures, such as fishing and aquaculture, land-based agricultural activities, nutrient and contaminant loading, chemical pollution from manufacturing and processing industries, shipping, coastline development, recreation, exploration and development of oil, gas and other mineral resources (Table 14.1). Further, significant coastal pressures include introductions of alien species, marine litter, noise, urbanization and tourism, destruction of habitats and construction of offshore structures, many of which are exacerbated by climate change.

4.1 Europe

While Europe's marine ecosystems are under considerable threat, there are substantial ecological, economic, cultural and 'lifestyle' differences across Europe that result in significant geographic and demographic variability. The variability in urbanization is clearly seen in a night-time satellite image (Figure 14.1) that shows clusters of densely populated, industrialized areas in central Europe bordering the southern North Sea and parts of the Adriatic Sea, whereas areas east of the Baltic Sea and Black Sea are more sparsely populated. The pressures from human activities therefore differ significantly among regions. Areas of low population in some instances correspond to intensive agricultural land use with associated high fertilizer application and nutrient loading to marine ecosystems.

4.1a Overfishing

Fishing is one of the most extensive coastal pressures in Europe. Despite a decline in effort in recent years and a corresponding general decline in marine fish catches

TABLE 14.1.
Relative threat levels (present and future) of stressors to coastal and estuarine ecosystems in Europe, Australia, and North America.

Stresssor or Threat	Europe	Australia	North America
Fishing and Overfishing	3 ↓	1.5 ↓	2 ↔
Impacts of Aquaculture	2 ↑	1 ↔	1 ↑
Hazardous Materials	2 ↓	1 ↔	2 ↔
Eutrophication and Harmful Algal Blooms	2.5 ↔	1 ↔	2 ↔
Biodiversity Loss	?? ↑	1 ↔	2 ↑
Invasive Species	2.5 ↑	1 ↔	1.5 ↑
Ports and Transportation	2.5 ↑	2 ↑	2.5 ↔
Impacts of Energy Development	1.5 ↑	1 ↔	1.5 ↔
Freshwater and Flood Risk	2 ↑	1 ↔	2 ↑
Coastal Development	3 ↑	1.5 ↑	2.5 ↑
Climate Change and Sea Level Rise	2 ↑	1 ↑	2 ↑

Key. Level of threat: 1 = low; 2 = moderate; 3 = high. Future trends in threat: ↔ little or no change; ↑ increasing threat; ↓ diminishing threat throughout Europe since 1985 (EEA, 2011), fishing pressure in many instances has exceeded sustainable levels and safe biological limits (SBL). A significant problem has been a lack of timely progress in reducing capacity of European fishing fleets (EEA, 2011). Consequently, 30% of Europe's commercial fish stocks are now fished beyond SBL and, in 2010, 70% of commercial stocks were fished above their Maximum Sustainable Yield (MSY) (EEA, 2011). Furthermore, by-catches, destruction of sea-floor habitats and illegal, unreported and unregulated fishing are pressures largely unaccounted for in current fisheries management. However, some of the most productive European fisheries are in heavily modified systems with relatively high nutrient inputs, for example the southern North Sea and Adriatic, which now experience excessive primary production. The overlap of fishing intensity in the southern North Sea with primary production is notable (Figure 14.4). The North Sea is one of the world's most productive regional seas, but it is also a region where fish stocks are in greatest need of rebuilding. Some areas of the North Sea (e.g., the southern part) are 'naturally' productive and resilient, and may recover rapidly if stocks are appropriately managed (effort and catch controls).

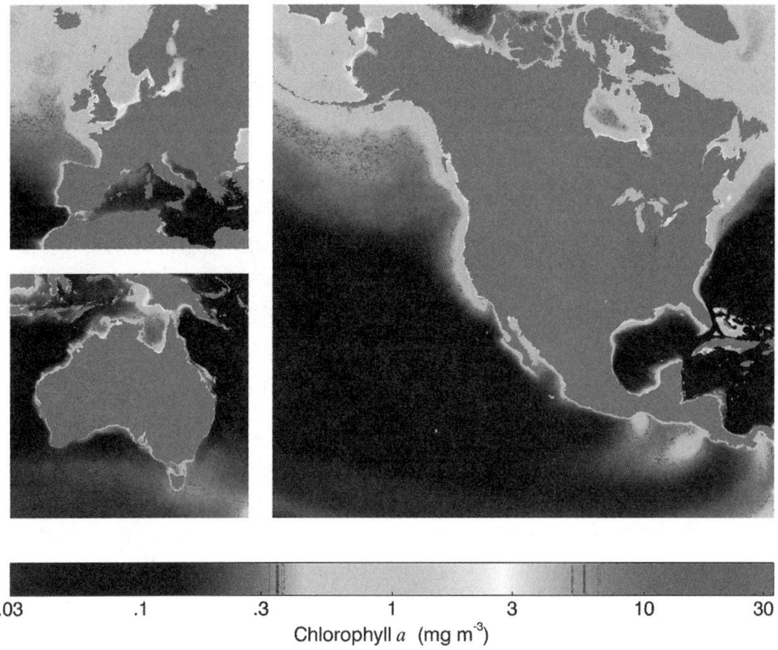

Fig. 14.4 – (See color insert) Chlorophyll distributions in waters of a) Europe, b) Australia, and c) North America, indicating the high primary productivity of near-shore coastal areas. (Map courtesy of Kim Hyde, NEFSC)

4.1b Ecological impacts of aquaculture

European aquaculture production has increased significantly over the past 15 years resulting in discharges of nutrients, antibiotics and fungicides used in the therapeutic treatments of fish diseases. In addition, there is potential for 'genetic pollution' of wild stocks by the escape of farmed fish. There also is the threat of increased fishing mortality on wild stocks of forage fishes used for aquaculture feed. The combined effects of inshore aquaculture and intensive agricultural practices in the catchments have given rise to widespread excessive nutrient loading that can promote eutrophication and depletion of oxygen, and mortality of organisms, especially in bottom waters.

4.1c Hazardous substances and eutrophication

Terrestrial sources of pollutants (e.g. chemical contaminants and nutrients) entering the marine environment are greatest where rivers discharge through heavily urbanized, industrialized and agriculturally developed parts of Europe. Land-based pollutants are a major threat to coastal marine ecosystems, accounting for about 80% of all such inputs to European marine ecosystems (EEA, 2010). For example, the German Bight in the southern North Sea receives freshwater from the rivers Rhine and Elbe. Seabed sediments and water near the river plumes show elevated levels of mercury and high levels of nutrients that contribute to elevated primary production (Figure 14.4). Indeed, despite measures to reduce nutrient loading into European seas, 85% and 80% of monitoring stations, respectively, show no change in nitrogen and phosphorus levels (EEA, 2010). Oxygen depletion is particularly serious in the Baltic and Black Seas.

While the direct discharge of toxic chemicals and substances is on a downward temporal trend in Europe, high concentrations sometimes occur in fish and shellfish. Pollution also includes illegal oil discharges and accidental oil spills from ships. Recent stringent controls limiting discharges of certain hazardous substances to the marine environment (e.g. the Water Framework Directive—see below) are having a positive effect on reducing total contaminant inputs. However, the legacy of past polluting activities remains a significant threat that is likely to be exacerbated by long-term trends in storm effects on demersal habitats caused by climate change and increasing trends in coastal development activities (OSPAR Commission, 2010).

4.1d Shipping, oil and chemical spills

The Göteborg Declaration concluded that the most significant risks to marine ecosystems were those from shipping and fishing activities (Ministry of Sustainable Development Sweden, 2006). Direct pressures from shipping come mainly from emissions to the air of engine exhaust gases (SO_x, NO_x), illegal discharges of oil and other hazardous substances, spills of oil and other hazardous substances from maritime accidents, alien species spread through ballast water discharges, inputs from anti-fouling paints, litter, and noise.

A significant indirect pressure from shipping is the need for dredging to provide deep-water access into ports and harbors, a trend that is increasing as size of ships increases. Underwater noise from ships' propellers and engines is an emerging issue (EEA, 2010). In addition, construction activities, airguns used in seismic surveys, and high-intensity military sonars all contribute to elevated underwater noise. Shipping noise now dominates the low frequency background noise in much of the world's oceans and is growing by 3–5 decibels per decade (EEA, 2010). Such noise may impact marine life, particularly marine mammals (Talpalar and Grossman, 2005).

4.1e Invasive species

The cumulative number of non-indigenous marine species recorded in European coastal seas has risen dramatically in the last decade (EEA, 2010), reaching 1376 by May 2009. Of these, 108 species were recorded in the Baltic Sea, 164 in the Black Sea, 376 in the North East Atlantic and 931 in the Mediterranean Sea. The function and structure of marine ecosystems are threatened by these non-indigenous (or alien) species originating from ships' ballast water discharges and hulls of ships. Between 3,000–4,000 million tonnes of untreated ballast water are discharged from ships each year in ports and coastal regions (EEA, 2010). Guidelines issued by the International Maritime Organization encourage the exchange of ballast water at sea, but in many instances it is only safe to do so within relatively sheltered confines of the coast, increasing the risk of non-native species being introduced into coastal waters.

4.1f Coastal development

Europe's coastal zones are under multiple pressures and face many economic, social and environmental problems (EEA, 2006a). The most significant include the destruction and degradation of coastal habitats to make way for housing, industry, agricultural land, and infrastructure for tourism and transport.

4.1g Climate change

Climate change is increasing the vulnerability of ecosystems to existing levels of human activity. Sea surface temperature in Europe's regional seas has increased up to six times faster than the global average in the past 25 years. The consequences of this increase include reduced Arctic sea-ice coverage, sea-level rise, and increasing ocean acidification due to rising atmospheric CO_2 levels. The temperature increase is also leading to changes in plankton communities and some fish species assemblages, with a tendency for colder-water species to be pushed further north and into deeper water (Drinkwater, 2005). In the future, less sea ice will allow new shipping routes in the Arctic which will result in new economic opportunities, but also new environmental pressures.

4.2 Australia

As in Europe and North America, Australia's coastline is under increasing stress from a multitude of pressures (Table 14.1). A deficiency in knowledge of Australia's marine environment, including biodiversity, poses a great challenge to achieving environmental, social, and economic objectives of policies and management that seek to halt and reverse impacts on marine ecosystems. Evaluating human impacts on estuaries and coastal waters is difficult because, with eight State and Territorial governments, Australia lacks a systematic, national-level approach for monitoring environmental trends.

The National Land and Water Resources Audit assessed the condition of about 1,000 estuaries around Australia and found that 50% were near-pristine, 22% were largely unmodified, 19% had been modified and 9% had been extensively modified (NLWRA, 2002). Most of the near-pristine estuaries were located away from population centers. In December 2011, the Australian Department of Sustainability, Environment, Water, Population and Communities (SEWPaC) released a report on the Australia State of the Environment (SoE) that presents a comprehensive review of the state and trends of the environment, pressures on it, and drivers of those pressures (SoEC, 2011).

4.2a Overfishing

Of the many human-induced pressures that have impacted marine ecosystems in Australia, fishing is perhaps the biggest concern. Regions subject to high levels of commercial and recreational fishing include coastal waters off New South Wales in the East and off Southwest Australia, isolated pockets in the Northwest, and waters that span the Victoria and South Australia borders (Hayes et al., 2011)

Australia views fishing impacts on ecosystems seriously. In 2010, the Commonwealth government reviewed the status of 96 fish stocks (Woodhams et al., 2011). Of these stocks, 19.8% were classified as overfished or subject to overfishing, a percentage that is slowly declining.

Although the impact of fishing extends beyond direct harvest, results of an evaluation of sustainability of by-catch species impacted by fisheries managed by the Australian Fishery Management Authority (AFMA) indicated that, except for particularly vulnerable species with long life spans and large body sizes, fishing has low impact on non-target species (e.g., Zhou and Griffiths, 2008; Zhou et al., 2009, 2011). By-catches are declining in some fisheries. For example, by-catch of large sharks, rays, and turtles has been greatly reduced by installing turtle excluder devices (TED) in trawl fisheries for prawns. The turtle by-catch was reduced from 5500 in 1999 to 19 in 2008.

When high exploitation of targeted species continues while bycatch declines, unintended ecological effects of selective fishing may emerge, leading to changes in ecosystem structure and biodiversity (Zhou et al., 2010). Overfished stocks may be further depressed by competition and interactions from increases in abundance of other species. The challenge for EBM is to promote profitable and sustainable fishing while ensuring the sustainability of Australia's marine ecosystems.

4.2b Ecological impacts of aquaculture

As in Europe and North America, environmental risks from aquaculture include nutrient loading of estuarine systems, exotic diseases, and possible threats to wild populations through genetic contamination. There is evidence that aquaculture in Australia has been the source of major disease outbreaks in wild populations and use of chemicals has impacted threatened species such as sharks and seals. However, with appropriate management and verification, impacts of Australian aquaculture can be constrained to an acceptable level with only small environmental degradation (SoEC, 2011).

4.2c Hazardous substances and eutrophication

Household, commercial and industrial discharges and runoff from urban and developed areas carry pollutants into coastal and estuarine areas. Historically, Australian coastal waters were low in nutrients. Since European settlement, nutrient discharges have significantly increased. On the Queensland coast, adjacent to the Great Barrier Reef, major river catchments now deliver 2–10 times more nutrients and sediments than they did before European settlement. These rivers also deliver pesticides to reef and lagoon waters. In the Brisbane River, nitrate and phosphate loads increased more than 10-fold since 1950, while sediment loads increased 4-fold since 1918. Poor water quality has resulted in coral and seagrass losses in several areas as well as increases in nuisance macroalgae and occasional algal blooms (NRMMC, 2006). Recently, ocean acidification has been recognized as a significant risk.

4.2d Shipping, oil and chemical spills

In Australia, ports are primarily located in urbanized areas along the coast. These facilities are under constant pressure to expand and support ever larger vessels. Ship operations can produce waste and noise, strike reefs and habitats, transport pests and non-indigenous species, and pollute coastal waters. Wastes include oils, chemicals, sewage, garbage, and toxic compounds released from anti-fouling paints and ballast water. A major risk from marine transport is oil or chemical spills. The rapidly expanding oil and gas industry must comply with national laws, including the Offshore Petroleum and Greenhouse Gas Storage Act 2006. Under this legislation, companies must prepare legally binding environmental plans, including oil spill contingency plans (SoEC, 2011). Since 1973, Australia has had a National Plan for responding to maritime oil spills, which have decreased from 353 in 1998–99 to 140 in 2008–09 (AMSA, 2010).

Direct impacts of oil and gas infrastructure include seabed structures such as wellheads, anchors and pipelines; indirect effects of shipping traffic include risks from accidents and spills (SoEC, 2011). There is also significant coastal impact from associated infrastructure, including industrial sites and ports, residential bases for the workforce, onsite maintenance, transport and related industries.

4.2e Invasive species, pests and disease

The coastal regions in Australia are generally not heavily impacted by pests, invasive non-indigenous species, and outbreaks of disease (SoEC, 2011). While numerous introduced species are recognized, their ecological significance is unknown. The south-east region is most impacted by pests and disease. Pests include starfish, sea urchins, plankton, algae, molluscs, crustaceans and worms. Port Phillip Bay is one of the most invaded marine ecosystems in the Southern Hemisphere. Outbreaks of harmful native species, mainly toxic algal blooms, also are pervasive in this region. The zooplankter *Noctiluca* (often responsible for 'red tides') has recently become widespread in the south-east region. A severe outbreak of abalone viral ganglioneuritis has caused mortality of abalone in the region, with serious ecological consequences.

4.2f Coastal development

Australia-wide, most of the urban coast has been developed since 1980 and threats are intensifying. Human population increases and industrial development have created a major impact on the marine environment. For example, in Moreton Bay, Queensland, 1200 ha of mangroves and 600 ha of salt marsh were lost from 1974 to 1987 (NRMMC, 2006). Additionally, there have been complete losses of habitat in newly urbanized areas and the loss of coastal indigenous heritage places.

The mining industries in Australia, including iron ore, nickel, uranium, zinc, coal, natural gas, have enjoyed a boom over the past two decades. Many mining locations are in the coastal zone, on both land and water sites. Some mining operations include refineries, smelters and power generation stations that can degrade coastal ecosystems.

In June 2012, the United Nations Educational, Scientific and Cultural Organization (UNESCO) released a damning report criticizing management of the Great Barrier Reef and warned that it could be downgraded to a world heritage site "in danger" unless Australia makes major changes to its supervision. The report expresses "extreme concern" at the rapid rate of coastal development and warns that no new developments should proceed that could affect the reef until a thorough assessment is carried out.

4.2g Climate change

Australia is vulnerable to climate change because of its extensive coastline and because most people live in the coastal zone. Thousands of kilometers of the coastline are at risk from threat of rising sea level and extreme weather events associated with climate change (DCC, 2009). More intense storms, larger wave and storm surges, altered precipitation patterns, temperature increase, and ocean acidification are threatening Australia's marine biodiversity. There is strong evidence that the world's iconic Great Barrier Reef is stressed by temperature increase and acidification. Calcification rates of many corals have fallen by 15% since 1990, which is unprecedented in the 400 years of coral core records (De'ath et al., 2009).

Of the 22 "big ecological questions" identified by scientists as impediments to effective environmental management in Australia (Morton et al., 2009), over half are directed at climate change and associated processes of ocean acidification and coastal inundation. A range of ecological models and approaches have been developed in Australia to predict impacts of climate change from physical, biological, ecological, social, and economic perspectives (Plagányi et al., 2011).

4.3 North America

In North America, it is in the coastal zone where humans place the biggest demand on marine ecosystem services and pose the greatest threats. Management problems come from multiple stressors, many of which originate in the watersheds or airsheds linked to coastal ecosystems (Table 14.1). For example, excessive nutrient loading that fuels eutrophication and hypoxic dead zones in Chesapeake Bay and the Gulf of Mexico originates hundreds of kilometers distant and is attributable to inland agriculture. Although threats are widespread, the overall condition of North American estuaries and coastal ecosystems, based on a narrow set of water quality and fish condition metrics, was judged to be deteriorating at a slower rate than in past decades (Bricker et al., 2007) and by some measures is better today than two or three decades ago (USEPA, 2012). Improvements largely derive from better water quality attained through reductions in sediment, pollutant and contaminant loading.

4.3a Overfishing

Coastal regions in N. America have suffered from direct effects of overfishing, but also from compounding effects of other human activities in coastal waters and in the watershed (e.g., dams and diversions that impact spawning by anadromous fishes). In heavily utilized coastal ecosystems, overfishing has led to reduced yields and values, high by-catches, losses of biodiversity, and associated habitat destruction. Moreover, human activities unrelated to fishing, for example deteriorating water quality and habitat loss, have lowered fisheries productivity (e.g., lost nursery functions for anadromous fishes) in estuaries on the east and west coast of N. America). In the U.S., recognition of threats to fisheries and impacts of fisheries on coastal ecosystems are driving many plans for precautionary and adaptive Ecosystem-Based Fisheries Management (EBFM).

4.3b Ecological impacts of aquaculture

Although aquaculture can bring many benefits, its services come with risks. The challenges and threats noted above for Europe and Australia are equally applicable in North America, i.e., diseases and parasites, escape of cultured organisms, damage to coastal environments and deteriorating water quality. Additionally, legal conflicts over spatial allocation of aquaculture sites in estuaries and embayments and resulting restrictions of use by the public are emerging issues.

4.3c Hazardous substances and eutrophication

Water quality has improved marginally in many estuaries and coastal waters of the U.S. in recent decades (USEPA, 2012). Still, urban sewage treatment and discharge of industrial and agricultural contaminants remain major threats to coastal water quality. Excessive nutrient (nitrogen and phosphorus) loading to waterways from agricultural fertilizers and sewage effluents are among the major drivers of eutrophication, harmful algal blooms, and associated hypoxia (i.e., dead zones) in N. American estuaries.

4.3d Shipping, oil and chemical spills

As in Europe and Australia, increases in traffic by bigger vessels that require frequent dredging to deepen waterways for navigation, in addition to expansion of shoreline port facilities, are stressful to coastal ecosystems. Discharge of wastes, accidental spills of oil, and transport of invasive species in ballast water are among the many problems associated with shipping in coastal areas.

Oil spills and leakage from vessels and blowouts of oil wells in coastal waters of N. America are continuing threats despite many precautions to avoid them. Spilled oil in coastal ecosystems is especially problematic when it washes ashore, oiling beaches and destroying coastal marsh vegetation and biota. Major threats from oil spills have been centered in the Gulf of Mexico in recent decades but plans to expand petroleum exploration in coastal waters of N. America will elevate the threat of damage to ecosystems.

4.3e Invasive non-indigenous species

Invasive species threaten most N. American coastal ecosystems. The San Francisco Bay and Delta (case study below) is especially notable in that many, or most, of the organisms now common in this estuary were introduced, either intentionally or accidentally. Ballast water discharges, as described above for European coastal regions, are a primary source of invasive species. The invasives problem is spreading rapidly and is a threat to biodiversity and productivity of estuarine ecosystems.

4.3f Threatened and endangered species

Many large and vulnerable taxa, including sturgeons, salmons, and river herrings (alosines) on both coasts of N. America, are now threatened or endangered from lost habitat, deteriorating water quality, and overfishing. These threats are greater in coastal areas than offshore. For example, sea turtles on the Pacific and Atlantic coasts and Gulf of Mexico are threatened by incidental mortality in trawl fisheries. Overexploitation (sturgeons), habitat destruction (salmonids), inaccessibility or loss of spawning habitats (salmonids, shads, eels), endangerment from boating activities (manatee *Trichechus manatus*) and poor water quality (delta smelt *Hypomesus transpacificus*) are threats to biodiversity in N. American estuaries. Some organisms, notably coastal fish-eating birds (bald eagle *Haliaeetus leucocephalus,* osprey *Pandion haliaetus,* brown pelican *Pelecanus occidentalis*) have recovered in N. America during the past half century after suspensison of use of pesticides such as DDT.

4.3g Freshwater deficits

Too little or highly variable supplies of freshwater threaten coastal ecosystems, especially on the west coast of N. America. Seasonal cycles of productivity in estuaries and coastal ecosystems are cued to amounts, quality, and timing of freshwater inputs, as are spawning seasons and recruitment potential of anadromous and estuarine fishes. Amounts of freshwater discharging into the ocean are regulated or declining because of dams, reservoirs, and flow diversions. Engineering that channels or diverts natural flow is justified to reduce flood risk or to secure water supplies for domestic and agricultural uses, but these actions often are harmful to coastal ecosystems.

4.3h Coastal development

In N. America, the inhabited coastline (Figure 14.1) developed rapidly since the 17th century. In addition to urbanization, other threats and challenges emerged, including deforestation, proliferating highways and vehicles, and manufacturing industries that are concentrated along the coast. As in developed areas throughout the world, the shorelines of most urbanized areas in N. America have been hardened with asphalt and concrete, neritic habitat has been lost, impermeable surfaces and contaminated runoff have increased, and both quality and levels of biological productivity are declining.

The increasing human presence and industries along the coast reduce the value of ecosystem services, including the ability to buffer shorelines from storm surges that increase flood risk and erosion losses. Loss of natural features reduces biological productivity and regulation of nutrients, and diminishes the role of such features in reducing turbidity and purifying contaminated water. In North American estuaries, for example Chesapeake Bay, there are many instances of cascading effects of coastal development on erosion, high sediment loads and associated turbidity, leading to losses of submerged aquatic vegetation, reduction in benthic productivity, and shifts from hard-bottom to soft-bottom benthic communities.

4.3i Climate change

In N. America and globally, sea level rise attributed to climate change is perhaps the most serious and certain threat to people inhabiting the coastal zone. Inundation from storm surges and floods from coastal storms are increasingly common. Extensive levee systems and flow diversions, for example in coastal Louisiana and the California Delta, can mitigate and partly offset the effects of rising sea level, but often at the expense of further losses of key habitats in the coastal zone. Although the projected rates and extent of temperature increases, precipitation changes, and salinity shifts are poorly known on a regional basis, it is probable that effects will add substantial stress to most coastal ecosystems in N. America.

5. EBM in Action

Globally, there is broad realization that marine ecosystems are under stress and that broad, holistic management approaches, such as those embraced in EBM are needed. Policies, plans and programs are evolving. In Europe and the EC, much of the policy appears to be driven by top-down enactments such as the Marine Strategy Framework Directive (MSFD) (EC, 2008) and reform of the European Common Fisheries Policy. In Australia, national policies and regional actions by States are evolving to achieve integrated management. In N. America, much of the EBM activity in the coastal zone is local and regional although a new National Ocean Policy will play a bigger role (Presidential Executive Order #13547, 2010; see Cochrane et al. this volume; Samhouri et al. this volume).

5.1 Europe

As an increasing number of European coastal ecosystems either approach or exceed their capacity to sustain present levels of exploitation a noticeable shift has occurred in policies, which now favor development of more eco-efficient and sustainable economies. For example, there has been recent expansion in development of renewable energy sources and sustainable tourism, thus promoting more environmentally friendly technologies, products and services. This is expected to shift the balance of principal environmental pressures acting on coastal ecosystems, e.g. an increase in offshore renewable energy but a decrease in land-based inputs of chemical contaminants. Ensuring the combined pressures do not cause a tipping point in the carrying capacity of coastal ecosystems will depend on the implementation of an effective EBM approach.

5.1a European policies

The European Council in Göteborg (2001) adopted the first EU Sustainable Development Strategy (SDS) (EU, 2006) to balance the growing need to protect the quality of the environment (including coastal marine ecosystems) with the need to encourage economic development throughout Europe. The SDS has two main parts. The first sets out objectives and policy measures to tackle key unsustainable trends in the EU and globally. The priorities are to:

- combat climate change;
- ensure sustainable transport;
- address threats to public health, such as chemical pollution, unsafe food and infectious diseases;
- manage natural resources more responsibly and stop biodiversity decline;

- combat poverty and social exclusion; and
- meet the challenge of an ageing population.

The second and arguably more ambitious part of the SDS revises the process for policy development. It calls for a new approach to ensure that the EU's economic, social and environmental policies reinforce one another and are in balance. However, tensions exist between the need to sustain economic and social well-being while sustaining the ecological and environmental services upon which society depends.

Marine environmental policy has evolved rapidly in the EU over the last two decades. The EU has implemented two new 'holistic' or strategic policies (Maritime Strategy—Blue Book) and environmental legislation (Marine Strategy Framework Directive). The new European Marine Strategy Framework Directive (MSFD) entered into force in July 2008 and provides the practical mechanism for implementing an ecosystem approach to manage Europe's seas. Its major objective requires Member States to develop indicators of Good Environmental Status for marine ecosystems by 2020. Much effort has been expended to create ecosystem state indicators, but there is concern that the limited research base on social and economic processes and indicators will preclude full implementation of the Directive. With respect to its Common Fisheries Policy, the EC has identified the need to link fishery policy to the MSFD and to emphasize that a healthy marine ecosystem is a prerequisite for sustainable fisheries. This is an important departure from previous 'sectoral' thinking. The new policy demands a more holistic and interdisciplinary approach to research, monitoring and assessment to achieve EBM objectives enshrined in the MSFD.

In 2008, the European Commissioner for DG Fisheries and Maritime Affairs[3] stated that "*the goal of ecological sustainability must be placed before economic and social sustainability, since it is the precondition which makes them possible*". Environmentally responsible strategies for use of space at sea and on the coast are being developed under the Habitats Directive (through Natura 2000 network protected sites) and more generally under Maritime Spatial Planning and Integrated Coastal Zone Management. At present only 8% and 11% of coastal habitats and species, respectively, and 10% and 2% of marine habitats and species, respectively, are in favorable conservation status. Most coastal habitats and species have unfavorable conservation status or are unassessed. This situation exists despite the obligation to promote conservation and sustainable use of living resources as central to delivery of the ecosystem approach. New spatial management approaches for demersal fishing are being considered which rely more on assessing risks to coastal ecosystems from fishing by defining areas of high fishing activity using Vessel Monitoring System (VMS) data (Hiddink et al., 2006; Jennings et al., 2012). Using this knowledge and an understanding of seabed habitat vulnerability, areas of seabed can be assigned to risk categories, which will help to locate MPAs and also help to define fishing areas that are of low risk to the ecosystem.

[3]http://europa.eu/rapid/pressReleasesAction.do?reference=IP/08/1339&format=HTML&aged=0&language=EN&guiLanguage=en

Progress towards delivering EBM has been slower in Europe than in some other developed countries. Impediments include the short-term incompatibility of objectives, the relatively poor state of managed fish stocks, the influences of differing national priorities and a decision-making process that remains centralized at all levels despite the formation of regional fisheries advisory committees (RACs).

5.1b European regional seas

The following paragraphs explore through specific cases the historic development of pressures and stress on biotic resources and measures taken to resolve conflicts between sectors by the Regional Sea Commissions.

Baltic Sea. The Helsinki Commission (HELCOM) is responsible for implementing the Convention on the Protection of the Marine Environment of the Baltic Sea Area (Helsinki Commission, 2004) and has worked for over 30 years to reduce pollution. The focus of the initial Helsinki Convention signed in 1974 was largely on prevention and elimination of pollution by hazardous chemicals. During the 1970s when serious environmental problems emerged, including the collapse of seal and sea eagle populations caused by PCBs and DDTs, HELCOM's mission broadened to include combating eutrophication, protecting biodiversity, and ensuring the environmental safety of maritime activities within an EBM framework.

The catchment of the Baltic Sea is more than 1.7 million km^2, with a human population of approximately 85 million. The high population density, together with a well-developed agricultural sector and industrial activities that include fossil fuel extraction for energy production and transport, result in large inputs of nutrients, mainly nitrogen and phosphorus. The Baltic Sea, unlike other European regional seas, suffers from blooms of cyanobacteria. The increased input of nutrients and organic matter, mostly from agricultural practices, has increased the prevalence and persistence of hypoxia that impacts benthic communities. The area affected by hypoxia has increased from 10,000 km^2 in the early 20th century to 40,000 km^2 today. Eutrophication is one of the most pressing pollution issues for the Baltic Sea (Langmead et al., 2007).

A significant ecosystem shift occurred in the Baltic Sea in the late 1980s (Möllmann et al., 2009) when cod (*Gadus morhua*) fisheries collapsed, apparently a consequence of climate change that caused shifts in salinity and temperature. In parallel, nutrient levels increased from land-based inputs which fueled eutrophication and low oxygen conditions. At high fishing levels the cod stock became depleted and the biological community changed from a cod-dominated ecosystem to one where herring *Clupea harengus* and sprat *Sprattus sprattus* dominated (MacKenzie and Köster, 2004). Furthermore, the introduction of the alien gelatinous ctenophore *Mnemiopsis leidyi* is of particular concern (Lehtiniemi et al., 2010).

North-east Atlantic. OSPAR is the Convention through which 15 Governments of the western coasts and catchments of Europe, together with the European Community, cooperate to protect the marine environment of the North-East Atlantic. The 15 Governments are Belgium, Denmark, Finland, France, Germany, Iceland, Ireland, Luxembourg, The Netherlands, Norway, Portugal, Spain, Sweden, Switzerland and United Kingdom. The protocol began in 1972 with the Oslo Convention against dumping of materials or substances at sea and expanded to cover diffuse land-based sources (mainly from rivers) and discharges from offshore industries by the Paris Convention of 1974. The two conventions were unified, updated and extended by the 1992 OSPAR Convention. A new annex on biodiversity and ecosystems was adopted in 1998 to cover non-polluting human activities that adversely affect the sea.

In the North Sea, the combined consequences of increased sea temperatures and fishing pressure are of particular concern (Langmead et al., 2007). Scientific advice and policy measures have not prevented overexploitation of several fish species. The decline in cod stocks in recent years is thought to be primarily from high fishing pressure combined with increasing temperature that is forcing a northward shift of the copepod *Calanus finmarchicus,* an important prey of larval cod (Beaugrand et al., 2002). A decrease or even collapse of cod in the North Sea, the southern margin of its distribution, might be triggered by climate change alone (Beaugrand and Kirby, 2010).

Eutrophication is a major concern in coastal waters of the southern North Sea. Elevated primary production persists despite reductions in nutrient loads. Chemical pollution in the marine environment is also a concern, with an enormous number of contaminants present for which there is little knowledge of their effects on biota. However, riverine concentrations of most metals and organic contaminants have started to decline under more stringent regulation and control of land-based emissions.

Mediterranean. The Mediterranean was the first region to adopt an Action Plan (MAP) in 1975 after creation of the Regional Seas Programme in 1974. The Convention for the Protection of the Mediterranean Sea Against Pollution (Barcelona Convention) was adopted in 1976 (amended 1995) and entered into force in 1978 (amended 2004). Six Regional Activity Centres (RACs) are responsible for implementation of components of MAP under supervision of the Coordinating Unit (MEDU). A protocol designating Specially Protected Areas and biodiversity was put in force in 1986.

With the opening of the Suez Canal the Mediterranean Sea became directly linked to the Red Sea and now has about 17% of its species originating from the Red Sea (UNEP, 2006). Overall, the Mediterranean Sea is regarded as a biodiversity hot spot. Two remarkable ecosystems, the *Posidonia* and coral beds, occupy shallow coastal zones in the western basin. The Mediterranean's status as a biodiversity hot spot is increasingly compromised by human activity resulting in habitat and species losses. Mediterranean habitats commonly suffer from invasive species, coastal development, eutrophication, fishing and aquaculture.

Black Sea. The Convention on Protection of the Black Sea against Pollution was adopted in 1992. Establishment of the Black Sea Environment Programme in 1993 was followed by the Strategic Action Plan (SAP) for the Rehabilitation and Protection of the Black Sea, adopted in 1996. Terrestrial sources of pollution, the introduction of alien species and inadequate management of resources are major issues highlighted in the SAP, which represents a step towards attaining sustainable development in the Black Sea region. In April 2009, the SAP was revised by parties to the Convention (Bulgaria, Georgia, Romania, the Russian Federation, Turkey, Ukraine), reconfirming commitment to protect the Black Sea ecosystem.

The collapse of anchovy *Engraulis encrasicolus* in the Black Sea in the late 1980s was originally explained by invasion of the gelatinous ctenophore *M. leidyi.* However, recent model-based research indicated that the story is more complicated and that overfishing played a major role in this ecosystem shift. The Black Sea in the mid 1980s supported a high anchovy biomass, but by the late 1980s the fishery had collapsed, caused by a combination of high fishing mortality, increases in nutrient loadings, and introduction of the alien ctenophore (Oguz, 2007; Langmead et al., 2007). Following economic collapse of the socialist republics in the early 1990s, nutrient loading was reduced, and the Black Sea ecosystem began to recover (Langmead et al., 2007). The reduced nutrients, among other causes, created less favorable environmental conditions for *M. leidyi.* Also, the accidental introduction of another ctenophore species, *Beroe ovata,* a predator on *M. leidyi* (Mutlu, 2009), further improved ecosystem conditions and led to a recovery of the anchovy fishery. Unfortunately, *M. leidyi* has recently spread to the Mediterranean, the North Sea and the Baltic Sea.

5.2 Australia

5.2a Australian policies

Australia, regarded as a world leader in environmental policy, began developing such policies in the early 1980s around the concept of Ecologically Sustainable Development (ESD), which is a framework for EBM in Australia. Since then, more than 200 statutes, regulations and policies that address management of Australia's coasts and oceans have been developed (Vince, 2008).

The Fisheries Management Act of 1991 is the most important legislation affecting management of both commercial and recreational fisheries. It includes a wide array of controls and measures aimed at habitat and species protection. Principles of ESD are a major component of the Act. Four principles include: 1) intergenerational equity; 2) conservation of biological diversity and ecological integrity; 3) the precautionary principle; and 4) improved valuation, pricing and incentive mechanisms. Further, the Commonwealth developed "Guidelines for the Ecologically Sustainable Management of Fisheries" which has provided direction to fisheries under provisions of the Environment Protection and Biodiversity Conservation Act (EPBC).

In 1995, the Commonwealth published the Commonwealth Coastal Policy to address growing concern over degradation of coastal lands and pollution of coastal waters. The Policy recognized and addressed a multitude of problems in coastal zone management in Australia.

In 1998, the Commonwealth released Australia's Oceans Policy, which was recognized internally and internationally as a milestone in marine resource management. The policy outlined a new, national approach for EBM. It sought to integrate sectoral and jurisdictional interests through the establishment of new institutions and adoption of Regional Marine Plans. After a decade of implementation, the original focus of the policy's implementation process has changed, however, and the experience has demonstrated the major challenge of achieving a truly national integrated approach (Vince, 2008; SoEC, 2011).

In 1999, the Commonwealth unveiled its Environment Protection and Biodiversity Conservation (EPBC) Act. The EPBC Act is the Australian Government's centerpiece of broad environmental legislation. It provides a legal framework to protect and manage key flora, fauna, ecological communities and heritage places. The primary objective of the Act is to protect terrestrial and aquatic environments and to conserve Australian biodiversity. The Act stresses the importance of promoting ESD through the conservation and ecologically sustainable use of natural resources.

In addition to these key policies and legislation, many specific policies have been created at State and Territorial level or directed to particular topics. For instance, at the Commonwealth level, there are several policies related to fisheries management (MBDWG, 2008), including harvest strategy policies for target species, a fisheries ESD program, a plan for IUU fishing, ecological risk assessment and management, and policies for recreational fishing, indigenous fisheries, and by-catch.

Although Australia has a population of 23 million people, less than 60% of California's, it has eight independent states and territories, each with authority to legislate over matters not controlled by the Commonwealth government. Marine and coastal zone management is shared by the Commonwealth, States, Northern Territory and Local Governments. The complex jurisdictional structure has created inefficiencies in human resources and policy conflicts and often results in jurisdictional and sectoral conflict (Foster et al., 2004; Vince, 2008). Recognizing the problems, Australia's governments agreed in 2004 to work collaboratively in a Framework for a National Approach to Integrated Oceans Management. Its principles guide Australian governments in the coordination of planning, management and policy development to deliver ESD of the ocean and its resources.

At the State level, the governments have direct control over their internal waters and most aspects of their nearshore waters. The States have varying levels of control over many resources in the EEZ. The Coastal Waters Acts of 1980 gave authority to the states and the Northern Territory seaward to three nautical miles. As a result, a fishery may be managed either by a state, by the Australian Government, or by both a state government and the Australian Government. Similarly, the Australian Government delegates the assessment of various environmental impacts to state agencies under a system of joint authority. The states and territories also directly control stressful land-based activities in their jurisdictions that impact coastal ecosystems. An overarching framework for nationally integrated management would be a major contribution to improving management of the marine environment (SoEC, 2011).

Another issue is that policies may be implemented that lack scientific and technical support. Australia has been a step ahead of many countries in implementing EBFM that relies on scientific understanding of ecosystem structure, processes, functions, and interactions that are poorly known in marine ecosystems. Fishery management policies traditionally have depended on single species stock assessments, which may be insufficient in the new EBM context. For example, the Commonwealth established a fishery by-catch policy in 2000 (DAFF, 2000). Its goal is to ensure that by-catch species and populations are maintained by reducing by-catch of all ecological components. The relatively new Commonwealth Fisheries Harvest Strategy Policy (DAFF, 2007) establishes the fishery management objective as maximum economic yield (for the fishing fleets) from fisheries that focus on a few high-value species. The two policies tend to contradict each other and fail to account for long-term effects on marine ecosystems and fishery production. Maximizing catches of a few high-valued species while minimizing the impact on many low-valued species will ultimately result in changes to community structure, biodiversity, and ecosystem services (Zhou et al., 2010; Garcia et al., 2011).

A more successful example of marine EBM to achieve Ecologically Sustainable Development (ESD) in Australia is the Great Barrier Reef (GBR). The integrated coastal management in the GBR Marine Park and World Heritage Area is largely due to the Great Barrier Reef Marine Park Act, in place since 1975, and the 25-Year Strategic Plan for the Great Barrier Reef World Heritage Area, established in 1994 (see Kritzer et al., this volume, for more details).

5.2b Ecosystem Based Fishery Management

Progress in developing EBFM and research to support it are outstanding achievements in Australia. EBFM deals with the aggregate management of all fisheries-related activities and impacts within an ecosystem or bioregion. In particular, focus has broadened from concerns limited to target species and resources to concerns about non-target species, including protected species, habitats, and ecological communities. Smith et al. (2007) discussed scientific tools developed in Australia to meet the needs of EBFM.

Considerable research has been directed to Ecological Risk Assessment for Effects of Fishing (ERAEF) methods since the 1990s. ERAEF takes all impacted ecological components into consideration, including target species, by-product and by-catch species, threatened, endangered and protected species (TEP species), habitats, and ecological communities. Methods range from qualitative (Fletcher, 2005; Astles et al., 2006; Walker, 2005), semi-quantitative (Milton, 2001; Stobutzki et al., 2001), to fully quantitative approaches (Zhou and Griffiths, 2008; Zhou et al., 2009, 2011). Hobday et al. (2011) described a hierarchical framework for ERAEF. Methods that support EBFM, particularly the Productivity-Susceptibility Analysis (PSA) and the Sustainability Assessment for Fishing Effect (SAFE) have been widely applied in Australia and internationally.

5.2c Integrated Coasts Management (ICM)

Australia will benefit from a comprehensive or consistent approach to assess changes in coastal ecosystems (ASEC, 2006). A broad spectrum of research to develop understanding of ecological, processes in the coastal zone is needed for ICM (Wang et al., 2011). Coral reef ecology and management is a major research emphasis in the coastal zone of Australia. Research on ecological and tourism values of protecting reef resources has attracted substantial funding. Some key activities involve the GBR Marine Park Authority's evaluation of reef vulnerability to climate change, including increasing temperature and acidification. The Australian Institute of Marine Science (AIMS) is investigating response of corals to higher sea temperatures and capacity of corals to adapt.

5.2d Marine Bioregional Plans (MBPS)

In 2006, the Australian Government initiated a four-year program to develop MBPs, (http://www.environment.gov.au/coasts/mbp). Five marine planning regions were identified: the South West, the North West, the North, the East and the

South East Marine Regions. The MBPs provide a regional framework for decision making within the context of the Environmental Protection and Biodiversity Protection Act. A MBP describes each Region's conservation values, important features, species, communities, and ecological processes. It identifies regional priorities for action, based on threats to conservation values and long-term policy goals. MBPs include strategic guidance for decision-makers.

Marine bioregional planning is also the process that Australia uses to identify areas for inclusion in the National Representative System of Marine Protected Areas (NRSMPA). At the state level, there have been attempts at MPA planning, but the interjurisdictional actions needed to protect biodiversity values and ecological processes have been lacking. Although several states have conservation programs, many MPA designations have resulted from piecemeal or ad hoc decision-making and do not reflect ecosystem-based or region-wide needs for conservation. By 2008, marine parks and reserves in high-protection categories (IUCN categories I and II) covered about 5.8% of States waters and 4.2% of Commonwealth waters.

Below, we briefly describe the coastal environment and management of five Australian marine regions that were designated for development of MBPs (SoEC, 2011).

South-west Region. This region stretches from the eastern end of Kangaroo Island in South Australia to Shark Bay in Western Australia. By global standards, it is an area of high biodiversity and home to many endemic species, characterized by a mix of tropical, subtropical and temperate taxa. The status of marine biodiversity (SoEC, 2011) is considered good. There are, however, localized coastal areas of historical human impact where effects persist, including pollution and dredging of seagrass beds in Cockburn Sound, Perth; pollution-induced losses of seagrasses in Gulf St. Vincent, Adelaide; and pollution of Albany harbors in Western Australia.

The major marine industries in the Region include commercial fishing, marine-based tourism, shipping, petroleum exploration and production, defence activities and aquaculture. Twenty ports in the Region provide vital services. Petroleum production commenced in 2006 and exploration continues in the Perth Basin, the Mentelle Basin and the Bight Basin.

Management of this region includes establishment of one of the largest marine reserves in Australia, the Great Australian Bight Marine Park. This park is comprised of adjoining Australian and South Australian Government protected areas, with both governments managing the Park cooperatively. New marine reserves are being planned to form a regional network of marine protected areas as part of the developing National Representative System of Marine Protected Areas (NRSMPA).

North-west Region. This region extends from Shark Bay in Western Australia to the border between Western Australian and Northern Territory. It includes tropical and sub-tropical environments. Overall, biodiversity is very good. Much of the region is remote and many habitats are in very good or nearly pristine condition. These include large gulfs and bays, fringing coral reefs, seagrass beds, offshore shoals and islands, canyons, and shelf-break ecosystems. Some of the world's most extensive undisturbed tropical and subtropical habitats occur in shallow waters of the region. Nonetheless, there are localized areas near towns where habitats are in poor condition, and where ports and shipping activities have heavily impacted coral and mangrove habitats. Deepwater corals and sponges offshore on the North West Shelf are heavily degraded and only slowly recovering from impacts of historical trawling.

Indigenous coastal communities continue to maintain special links to their "sea country," an Aboriginal expression that defines cultural and natural values of coastal land and sea environments for which an Aboriginal group has traditional rights and responsibilities. Commercial fisheries are conducted in both Commonwealth and State waters of the region. Resource-dependent industries, as well as human communities dependent on them, are undergoing rapid growth fuelled by demand for minerals and petroleum products in east Asia. Australia's two largest ports (in terms of tonnage), Dampier and Port Hedland, are in the region and are expanding. Petroleum production continues to expand on the North West Shelf, including development of petroleum reserves (predominantly natural gas) in the Region's north.

There are several MPAs in the region and new MPAs, including an MPA network, are being developed.

North Region. The North Region is a tropical environment stretching from the Western Australian–Northern Territory border to Cape York in Queensland. The region has mostly shallow waters low in nutrients. There are few marine species unique to the region, but it is home to globally threatened taxa including sea turtles, dugong and sawfish. The tropical habitats of the region are remote and pristine, and biodiversity is very good. However, pressures of coastal development are evident in some areas, such as Darwin Harbour and Melville Bay where a localized, biologically dead area has been created by bauxite mining wastes. Rivers are mostly unmodified. Exceptions are the Ord River, modified by the Ord River Dam, that results in substantial impacts on estuarine habitats in the Cambridge Gulf and in the MacArthur River, which is diverted because of open-cut mining.

The Region is an important corridor for trade and communication between the south Pacific and the north-eastern Indian Ocean. Today, the Region supports a wide range of human uses and activities. Indigenous communities continue to maintain special links with their "sea country" and own most land in the Region.

Marine industries include commercial and recreational fishing, shipping, petroleum exploration and production, defence activities and aquaculture. The northern prawn fishery was once Australia's most valuable fishery. Other important fisheries include the offshore fisheries for shark and mackerel, and fisheries for Spanish mackerel, snapper and groper. The Region also supports important recreational fishing. Explorations in the north-west part of the Region have located major oil and gas deposits.

There are no MPAs in the Region and the Northern Territory has no legislation dedicated to design, declare or manage MPAs (SoEC, 2011). However, research and planning are under way to establish MPAs.

East Region. The East Region covers the east coast of Queensland and New South Wales, including the Great Barrier Reef, Torres Strait, the Coral Sea, Sydney Harbour, and the many smaller islands, bays and estuaries. Overall, biodiversity of the Region is in good condition, despite considerable pressure from land-based pollution (SoEC, 2011). Many habitats of the central and southern part of the Region are degraded and in poor condition, mainly attributable to human population pressures in coastal areas, beach modifications, loss of seagrass beds and corals, historical effects of heavy trawling, and altered river flows for agriculture, which have modified freshwater flow to estuaries and bays.

The Region is the most heavily populated coastline in Australia. Most human activity is closest to major population centers in New South Wales and south-east Queensland. Major marine industries include shipping, commercial and recreational fishing, and tourism. Other pressures include offshore oil, gas and mineral exploration, sea dumping and the laying of submarine communication cables. The otter trawl and eastern tuna and billfish fisheries are the largest Commonwealth fisheries targeting demersal fish, prawns, scallops, tuna, billfish and sharks. The Region is also a popular destination for recreational and charter fishing. Major international sea routes and Australia's largest importing ports are in the Region and handle nearly 45% of Australia's sea freight trade.

In addition to the iconic Great Barrier Reef Marine Park, there are several marine reserves in the Region, and new MPAs will be established.

South-east Region. The South-east Region extends from southern New South Wales to eastern South Australia and includes waters in Victoria, Tasmania and Macquarie Island. The pressures of human settlements, shipping, fishing and development have degraded habitats of many inshore waters. Overall, biodiversity is poor. This is the only region where a habitat type has become functionally extinct as a consequence of human activities—the oyster reefs that formerly dominated many estuaries and small bays were exterminated by mining and fishing practices by the end of the 19th century.

The South-east Region has many MPAs, including State waters in New South Wales, Victoria, Tasmania, and South Australia. These reserves contribute to the National Representative System of Marine Protected Areas (NRSMPA).

5.3 North America

5.3a North American policies

Policies that prioritize issues and needs for more integrated management in the coastal zone have evolved in the past decade. In the U.S., earlier legislation, e.g., the Coastal Zone Management Act, first authorized in 1972 (CZMA, 1972), laid the framework for many principles of EBM. The Act was promulgated to preserve, protect, develop and, where possible, restore or enhance the resources of the Nation's coastal zone. During the same period, the National Environmental Policy Act (NEPA, 1970) was enacted to protect environments from adverse government actions and to promote efforts that prevent or eliminate damage to the environment and biosphere. This broad Act, while not directed specifically at marine or coastal zones, has been effective in regulating adverse activities in stressed coastal ecosystems. Many goals of NEPA are in a broad sense those of EBM today.

The need for coordinated and integrated management of the oceans was emphasized in the Pew Oceans Commission Report (POC, 2003) and the Report of the U.S. Commission on Ocean Policy (USCOP, 2004). Both reports called for a move toward EBM. Both were advocates for development of place-based management and new governance structures, e.g., regional management bodies to conduct integrated EBM. These reports and others issued by the Joint Oceanographic Commission Initiative (e.g., JOCI, 2011) have recommended regional management planning for marine ecosystems, with substantial emphasis on highly impacted coastal ecosystems. Additionally, the U.S. Joint Subcommittee on Ocean Science and Technology proposed a national research agenda for the oceans that includes science to support EBM as an area of emphasis (JSOST, 2007).

Recently, the U.S. took an important step towards EBM with issuance of the Oceans Policy Task Force (2010) report. Among its nine objectives, developing EBM is the number one priority. Coastal and marine spatial planning (CMSP) is at the heart of this policy that proposes regional councils to plan and conduct EBM. This report and proposed policies were followed by Presidential Executive Order #13547 (2010) that is intended to drive implementation. A draft implementation plan advocates both EBM and CMSP, highlighting actions needed in the coastal zone (NOC, 2012).

In Canada, existing policies address needs for management of marine ecosystems, including coastal and estuarine areas. These policies were promulgated in the Canadian Oceans Act (1997) and Canada's Oceans Strategy (2002) that include proposed Coastal Management Areas (CMAs) and management plans that constitute a vision for EBM. The Canadian Oceans Strategy recognizes that management planning within the CMAs must be more rigorous and will be more difficult to achieve than plans for Large Offshore Marine Areas (LOMAs) that already are developed. Rosenberg et al. (2009) noted the lack of inshore EBM programs in Canada. A community-based Atlantic Coastal Action Plan, developed in 1991, is a broad ecosystem initiative seeking to implement EBM goals in coastal communities of eastern Canada (McNeil et al., 2006). Canada's Oceans Action Plan (2005) is a commitment to integrated management and principles of EBM, although it is not a tactical plan to support immediate actions. Native and Aboriginal peoples, and their heritage, are prominently recognized in integrated planning for management of Canada's coastal ecosystems, including the Arctic and Great Lakes.

In Mexico, marine EBM programs have been instituted primarily at the regional or local scale and typically are collaborations among local governments, stakeholders and NGOs. Implementation of marine reserves and protected areas is a common EBM approach. Ezcura et al. (2009) provide an example in the Gulf of California (Sea of Cortez) where declines in freshwater flow to the Gulf, bad fishing practices, and losses of coastal wetlands are being addressed through research and management under broad government guidelines, but primarily through local and regional research and management programs.

In the Caribbean region, goals and objectives compatible with EBM are expressed in the Cartagena Convention, signed in 1981, and the Specially Protected Areas and Wildlife Agreement (SPAW, 1990). The Agreement covers terrestrial and aquatic ecosystems, mostly those in the coastal zone. Signatories are Caribbean island nations, Mexico, Central American nations, nations on the north coast of South America, the U.S., and European nations with island territories, e.g., France, Netherlands, U.K. The SPAW aims to protect and sustainably manage habitats of special ecological value, and to protect and preserve threatened or endangered species. The SPAW protocol was adopted as international law in 2000.

5.3b Regional examples

Regional programs to manage coastal and estuarine ecosystems emerged on both coasts of North America in the late 20th century. Among numerous programs explicitly embodying principles of EBM are those conducted on the Chesapeake Bay, Puget Sound, San Francisco Bay and Delta, the coastal marshes and wetlands of Louisiana and Gulf of Mexico, and the Everglades-Florida Bay (Figure 14.2). At first glance, these five coastal ecosystems might seem to share little in common but each is closely linked to the watersheds and terrestrial ecosystems in their catchments. They share some similar stresses from pressures of human activity in the waters and in the watersheds (Table 14.1). Unlike offshore areas, the supply of freshwater and its management (amounts, delivery, quality, allocation, conservation) often are of primary importance in these heavily utilized ecosystems.

Chesapeake Bay. The Chesapeake Bay is the largest estuary in North America. In 2010, nearly 20 million people lived in its 166,000 km^2 drainage basin. Among the many problems facing the Bay, the foremost is eutrophication and associated effects (Kemp et al., 2005). The Chesapeake Bay Program (CBP) dates from 1983 (http://www.chesapeakebay.net/historyofcbp.aspx?menuitem=14904). From its outset, it was billed as an ecosystem management program (http://www.chesapeakebay.net/) and has served as a model for many other programs. The CBP seeks to manage the Bay for sustainability, while enhancing and restoring its ecosystem services. The goals, successes and failures of the CBP, and its approach to develop EBM have been described many times (Boesch, 2006; Boesch and Goldman, 2009).

The Chesapeake Bay (Figure 14.2) is a large (330 km long, 11,000 km^2 surface area), shallow estuarine system in a watershed that includes jurisdictions of six U.S. states. The Bay historically was rich in resources and its commercial and recreational fisheries even today yield upwards of 200,000 tonnes annually (Houde, 2011). Still, threats to the Bay's productivity indicate a need for EBM to address its many problems (Kemp et al., 2005; CBFEAP, 2006; Boesch and Goldman, 2009). Dramatic declines in seagrasses, oysters, and some anadromous fishes, combined with eutrophication and attendant hypoxia, are problems that have stubbornly resisted alleviation via management actions.

Reducing nutrient loading (nitrogen and phosphorus) is the primary goal of the CBP. Excess nutrients drive eutrophication and attendant problems (e.g., loss of submerged aquatic vegetation, hypoxia, loss of benthic habitat and resources). A near-tripling in human population within the watershed in the past three decades, combined with intensive agriculture in the catchment are sources of excess nutrients. Urbanization, shoreline development, and poor land-use practices lead to heavy sediment and nutrient loading, exacerbating eutrophication despite the well organized CBP that embraces principles of EBM. Overexploitation and unwise fishing practices are problematic in the Bay, although more manageable than the excess nutrient problem. Fisheries landings are less diverse than historically and have experienced a multi-decadal decline (CBFEAP, 2006; Green, 2011; Houde, 2011). The specter of climate change, with temperature increases, sea-level rise, altered hydrology, and salinity shifts will further impact the Bay ecosystem.

The lead agency in the CBP is the U.S. Environmental Protection Agency (US EPA), with many federal and state partners (Boesch and Goldman, 2009; Green,

2011). The CBP is comprehensive but, until recently, was poorly integrated across the many sectors and jurisdictions that must cooperate to conduct effective EBM. Goals of a "Chesapeake 2000" agreement (http://www.chesapeakebay.net/content/publications/cbp_12081.pdf) set restoration targets for 2010 that mostly were not met. Presidential Executive Order #13508 (EO 2009) ushered in a new era for the CBP, with a more committed federal government presence, a new Action Plan (Federal Leadership Committee, 2010) (http://executiveorder.chesapeakebay.net/file.axd?file=2010%2f9%2fAction+Plan+Briefing.pdf) and rigorous goals and timelines for compliance. The Plan espouses spatially explicit, local actions embedded in a broader strategic management plan. A Federal Leadership Committee (2010), instituted to provide stronger leadership to the CBP, is led by the US EPA with numerous federal agency partners.

While not fully successful in reaching its goals, the CBP has progressed. In fact, it could be argued that without the CBP the Bay's ecosystem services would have collapsed under multiple stresses. The program achieved reductions in per capita loadings of nutrients, but has failed to reach its nutrient reduction goal, largely because the human population in the watershed has tripled. While commercial and recreational fisheries remain productive, the status of individual fisheries varies (Houde, 2011). Oysters *Crassostrea virginica,* in decline well before the CBP, continue to decline. Blue crab *Callinectes sapidus,* the Bay's most valuable fishery, was stabilized by coordinated and effective management. Striped bass *Morone saxatilis* was rescued in the 1980s from overexploitation and collapse. A strategic "Fisheries Ecosystem Plan" is guiding development of EBFM in Chesapeake Bay (CBFEAP, 2006).

Puget Sound. The Puget Sound Partnership (PSP) is an integrated ecosystem program with adaptive EBM principles at its core and a goal to achieve objectives by 2020. The PSP was created by the Washington State Governor and Legislature in 2007 (Washington State Senate Bill 5372 (http://apps.leg.wa.gov/billinfo/summary.aspx?bill=5372&year=2008). Among the issues being addressed are effects of stress on watersheds, shoreline development, human health, fisheries, contaminants, transportation, flooding, and recreational use.

Puget Sound is a large coastal ecosystem in western North America (Figure 14.2), shared by the United States and Canada. The Sound region includes lands and waters extending from the Cascade and Olympic Mountain ranges to marine waters of the Straits of Juan de Fuca and seaward. Its marine waters cover 2,330 km^2 with a shoreline of 4,000 km. The watershed encompasses 41,550 km^2 of freshwater, marine and upland habitats with a large and growing human population. The fjord-like Puget Sound is embraced by forested regions with diverse freshwater sources, but also supports large urban areas (e.g., Seattle, Vancouver). Historically, the Sound's complex habitats (Sound Science, 2007) supported abundant Pacific salmon, and its watershed served as spawning and nursery habitat. Salmon stocks are now greatly diminished and many animals, including the orca (killer whale *Orcinus orca*) and seabirds, are at low abundance. The orca and some salmonids are reduced to threatened or endangered status.

The many stresses impacting Puget Sound are primarily attributable to rapid increases in human settlement, urbanization, and associated modifications of habitats (McClure and Ruckelshaus, 2007; Sound Science, 2007; Ruckelshaus et al.,

2009). The human population exceeds 7 million in the Puget Sound-Georgian Basin. Toxic wastes and contaminant loads in organisms are substantial, leading to advisories to reduce consumption of fish. Large marine mammals (e.g., seals and orcas) carry significant PCB burdens. Tributary spawning habitats of salmonids have been engineered and diverted, and dams have prevented access to spawning sites. More than 80% of wetlands and marshes have been lost. Increased nutrient and contaminant loading are worsened by complex circulation patterns that extend residence time of water in the Sound. Other impacts include hypoxia in parts of the Sound ecosystem. More than 90 species of invasive plants and animals are established. Climate change, with threats of sea level rise and temperature increases, will place additional stress on humans and other living resources.

The PSP embraces four strategies: 1) freshwater and terrestrial protection and restoration; 2) marine and nearshore protection and restoration; 3) pollution prevention and cleanup; 4) strategic leadership and collaboration. The PSP embodies a mix of public and private entities, including state and federal government partners (McClure and Ruckelshaus, 2007; Ruckelshaus et al., 2009). A comprehensive plan, the Action Agenda (2009), was instituted in 2009 and revised in 2011. The PSP itself lacks regulatory authority, but agencies with authority are represented in the Partnership. Lead agency partners are the U.S. Environmental Protection Agency (US EPA), the State of Washington, and NOAA who provide leadership and funding to support restoration and research.

The PSP's broad ecosystem perspective has goals that are unusually clear among EBM programs. The recognition of connectedness and needs to 1) consider consequences of human activities on food webs and habitats, 2) consider impacts in a broad context of ecosystem health, and 3) conduct integrative ecosystem analysis to support management plans are strengths of the Program.

San Francisco Bay and Delta. The San Francisco Bay (Figure 14.2) and its deltaic system of freshwaters are among the most highly impacted of coastal-estuarine ecosystems. Most problems emanate from the scarce, variable, and uncertain freshwater supply, and debates over how to allocate it for human uses and other needs of the ecosystem. Conserving and restoring the Bay and Delta and its ecosystem services is central to goals of adaptive EBM plans (Healey et al., 2008; Lund et al., 2010).

The Delta lies in the Central Valley of California and is linked to the San Francisco Bay by a complex network of human-altered waterways, diversions, levees, and conveyances designed to deliver scarce freshwater throughout central and southern California to support agriculture and domestic water needs. The Bay-Delta ecosystem stretches from the mountains east of Sacramento and the San Joaquin Valley to the Golden Gate at the entrance to the Pacific Ocean. Major cities, including San Francisco and Sacramento, are located on the Bay or in the watershed. The ecosystem has been stressed by human pressures that began more than 150 years ago with mining for gold and the massive sediment loadings that resulted. An elaborate levee system protects against floods, and pumping operations and diversions support agriculture and domestic water demand. Water quality has declined. Climate change brings new threats, including rising sea level and higher temperatures to this already stressed system (Robinson, 2009; Cloern et al., 2011).

The present-day Bay occupies only 70% of its mid-19th century area. In the Delta, most natural tidal marsh has disappeared and has been replaced in large part by non-native vegetation. Natural salinity and tidal regimes no longer operate in the engineered waterways (Lund et al., 2010). The historical connectivity, including critical migratory pathways for anadromous fishes, has been lost (Moyle et al., 2010). While commercial landings are small, recreational fisheries are important. There are some valuable commercial fisheries, e.g., Pacific herring *Clupea harengus,* dungeness crab *Metacarcinus magister.*

The cumulative stresses have led to loss of connectivity in the Delta and loss of nursery habitats for fishes and other organisms, especially threatened salmonids (Moyle et al., 2010). Threatened chinook salmon *Oncorhynchus tshawytscha* use the Delta and tributaries of the Sacramento and San Joaquin Rivers as spawning and nursery areas. Within the Delta itself, the endangered delta smelt *Hypomesus transpacificus* is a victim of lost habitat and deteriorating water quality. A precipitous decline in abundance and reproductive success of pelagic taxa, mostly fishes, termed the "pelagic organisms decline (POD)", began more than a decade ago and is attributed to a climate regime shift and associated environmental stress (Sommer et al., 2007; Cloern et al., 2011). Threatened and endangered species are intensively managed to protect and restore them as mandated by federal and California Endangered Species Acts.

Additional sources of stress are flood and earthquake risk to levees. Toxic contaminants, including pesticides and other industrial or agricultural toxic wastes, e.g., selenium and mercury, are pervasive, and high nutrient loads that fuel noxious algal blooms are becoming problematic.

The Bay-Delta ecosystem is highly susceptible to invasive species. There are more than 200 introduced, non-native species in the Bay-Delta. Invasives include species purposefully introduced (e.g., striped bass and American shad *Alosa sapidissima*), but most are incidentally introduced, presumably from ballast water discharges (e.g., the overbite clam *Corbula amurensis,* many fishes and plankton organisms). These introductions have had generally undesirable effects on food-web processes and community structure (Kimmerer, 2004; Healey et al., 2008).

A multifaceted, multi-agency EBM plan to restore the Sacramento-San Joaquin Rivers and the Bay-Delta ecosystem is under development (Delta Plan, *in prep*). Within the Bay itself, the San Francisco Bay Plan (1969) constrains development, dredging, and filling and is administered/enforced by the Bay Conservation and Development Commission (BCDC). The Delta Reform Act of 2009 created the Delta Stewardship Council that is charged to develop a Delta Plan based on EBM principles to achieve two co-equal goals: 1) a secure water supply for human use and 2) protection and restoration of the ecosystem. A complex partnership includes federal, California, and local agencies responsible for water, mineral and gas resources, living resources, environment, human well-being, agriculture, transportation, and recreation.

Louisiana coast. The coast of the Gulf of Mexico, stretching from Texas to Florida, includes 28,500 km of coastline (Figure 14.2) that supports some of the most valuable fisheries and energy resources in the U.S. Coastal Louisiana itself has a 12,800 km coastline that is susceptible to the multiple stresses now imposed on it. Traditional human occupations, such as fishing and agriculture in rural coastal communities, have been supplemented with newer marine industries, including oil and gas industries served by a navigation and transportation infrastructure in the coastal wetlands and Mississippi River, and by an extensive system of dredged channels (Boesch, 2001, 2006). The Mississippi River and city of New Orleans serve as the entry port to the heartland of the U.S. The port of New Orleans is the single largest in the U.S. (224 million tons of cargo annually). Wetlands and marshes dominate the coast and are critical habitats for major U.S. fisheries, e.g., shrimp and oysters. Nearly 25% of U.S. commercial fisheries products are landed in Louisiana, the nation's top producer of shrimp and oysters (ERTF, 2011).

Louisiana's coast and extensive wetlands constitute a closely coupled ecosystem strongly tied to the Mississippi River and its vast catchment (3.2 million km^2) in the central U.S. In addition to fisheries, the coastal wetlands ecosystem supports abundant waterfowl and wildlife. The threatened wetlands continue to experience losses. Wetlands are criss-crossed with navigation channels. Extensive levees and diversion systems control flows to reduce flood risk. Widespread changes in hydrology have accelerated erosion, subsidence, and loss of wetlands. An estimated 4,850 km^2 were lost from 1930 to 2000, with a sudden loss of 520 km^2 in 2005, in the aftermath of hurricanes Katrina and Rita. High nutrient loading in the Mississippi River, much of it originating far upriver in the agricultural heartland, has contributed to eutrophication and associated hypoxia, precipitating a summer "dead zone" that can cover from 10,000 to 20,000 km^2 (Rabalais, 2011). Despite these stresses, the coastal ecosystem remains rich and productive.

Most problems of coastal Louisiana derive from marsh and wetlands losses related to engineered controls over Mississippi River flow. Boesch (2006) summarized the issues and both Rabalais (2011) and Barbier (2011) have provided up-to-date perspectives. Channelling, diversions, and the extensive levee systems no longer promote natural deposition of Mississippi River sediments that historically built and sustained marshes but, instead, exacerbate erosion processes. Flood risk and storm surges from high seasonal flows and hurricanes also are exacerbated. Effects of climate change, especially sea level rise, will constrain maintenance and restoration of marshes. Loss of coastal wetlands threatens traditional occupations and ways of life (CPRA, 2012).

The Deepwater Horizon oil well blowout and resultant oil spill in the offshore Gulf of Mexico in 2010 (http://response.restoration.noaa.gov/deepwaterhorizon) highlighted yet another threat to the Gulf coastal ecosystem that must be considered in plans for EBM. Contamination of coastal wetlands by petroleum spills can destroy wetland vegetation; kill shellfish, birds and fish; and have major economic impacts on fishing, recreation and tourism (Barbier, 2011). While disasters the magnitude of Deepwater Horizon are rare, they are costly, longlasting, and inflict major damage to ecosystems and their services.

The cumulative impacts of stressors on the Gulf and Louisiana coastal ecosystem require management that is broad-based, adaptive, and reliant on EBM principles. A Presidential Executive Order #13554 (2010) established the Gulf Coast Ecosystem Restoration Task Force with EBM as its goal to protect and conserve the vulnerable shorelines and resources of the Gulf of Mexico. The federal Coastal Wetland Planning, Protection and Restoration Act (CWPPRA, 1990) initiated programs in Louisiana aimed at restoring the coastal ecosystem. After hurricanes

Katrina and Rita, the Louisiana Coastal Protection and Restoration Authority (CPRA) and other State and federal agencies, e.g., US Army Corps of Engineers, developed a Master Plan to restore the Louisiana coast and to protect people from flood risk (CPRA, 2007). A major revision in 2012 (CPRA, 2012) is a strategic and adaptive EBM plan. It is broad and integrative, spatially explicit, and dependent on structural and non-structural approaches to reduce flood risk while enhancing the health and quality of natural resources, and promoting compatible industries on the Louisiana coast.

Everglades/Florida Bay ecosystems. More than 8 million people now live in south Florida's subtropical landscape and ecosystems that have been greatly altered by rapid human settlement, extensive development of agriculture, diversion of freshwater flow to reduce flood risks, and exploitation of natural resources. Attempts are under way to reverse human impacts, restore ecosystems and their services, and implement long-term plans that embrace EBM.

The Florida Everglades ecosystem is unique, often described as a river of grass draining freshwater in sheet flow over a distance of 170 km from south-central Florida to the marine waters of Florida Bay and the Florida Keys (Figure 14.2). The vast Everglades once covered 12,200 km^2 but has been altered by humans in many ways, primarily by levees, dikes and diversions that drain the system to reduce flood risk, but also to support agriculture, urban development, and highway and railroad construction. The current Everglades is half the size of its pristine state (CERP, 1999). The Everglades merges at its southern end with the 2,200 km^2 marine Florida Bay. Habitat and water quality of the Bay ecosystem began a notable decline in the last two decades of the 20th century. Although causes are not fully understood, the declines in the shallow seagrass- and sponge-dominated Florida Bay followed reductions in amounts and quality of freshwater delivered from the adjoining Everglades. Declines of seagrasses, the appearance of unwanted algal blooms, and declines in habitat and abundance of economically important fish, shrimp, and spiny lobsters occurred in Florida Bay (Hunt and Nuttle, 2007; Herbert et al., 2011).

Everglades: Many stresses and threats to the marine Florida Bay ecosystem are linked to activities in the freshwater Everglades. These include human encroachment, agriculture, sprawling urban-suburban development, flood control, canals and diversions of flow, and a nearly 70% reduction in annual freshwater flow to Florida Bay (CERP, 1999). Excessive nutrient loading and contaminants from agriculture are problematic. Water levels have dropped drastically in the Everglades. These stresses have led to declines in fauna, including the large colonies of wading birds that were emblematic of the Everglades. Additionally, the Everglades is now plagued by many non-native species that include introduced fishes, reptiles and birds. The introduced Burmese python *Python molurus bivittatus* is a serious predation threat to large vertebrates (Dorcas et al., 2012).

Florida Bay: The Florida Bay marine ecosystem depends on freshwater flow from the Everglades to maintain its health and productivity. From the mid-20th century, freshwater flow (amounts and seasonality) no longer followed its historical pattern, and amounts were reduced by many human interventions. Salinities in parts of the Bay increased, nutrient profiles were altered, and seagrasses declined.

Sponges and habitat they provided (e.g., to spiny lobster *Panulirus argus*) declined from poor water quality and associated diseases (Hunt and Nuttle, 2007). A growing human presence (commercial and recreational fishing, boating and recreational activity) and increases in human habitation in the Florida Keys added to stresses on the Bay ecosystem. Effects of climate change and sea level rise will further threaten its structure and productivity.

The Comprehensive Everglades Restoration Plan (CERP, 1999), billed as the world's largest ecosystem restoration effort, has been implemented. Although envisioned in the mid-20th century, the CERP was not authorized until 2000 (WRDA, 2000). Its goal is to capture and redirect freshwater through the Everglades that now is diverted into man-made canals leading to the Atlantic Ocean and Gulf of Mexico. CERP's many projects primarily focus on restoring the freshwater ecosystem of South Florida through implementation of adaptive EBM. The Florida Bay Science Program (Hunt and Nuttle, 2007) recognizes that actions implemented by CERP (CERP, 2011) will benefit the marine ecosystem of Florida Bay.

Primary agencies involved are the U.S. Army Corps of Engineers, U.S. Department of the Interior, and the South Florida Water Management District, with other federal and state agencies, tribal governments, and local governments serving as partners. In CERP, the requirement for adaptive management in an EBM context is prominent and a strength of the program. With renewal of the Water Resources Development Act (WRDA, 2007), many new hydrological restoration projects will be implemented to benefit the Everglades, adjoining Florida Bay, and other South Florida marine ecosystems.

6. Key Findings and Conclusions

The concentration of human activity, level of impact, and the many stakeholder interests in the coastal zone, embayments and estuaries increasingly requires an adaptive EBM approach to conserve and sustain these ecosystems and their valuable services. For these heavily impacted ecosystems, the threats driving the need for EBM extend to the closely connected watersheds, and even airsheds, of their catchments. Concerns related to fisheries stresses and ineffective fisheries management that often dominate the call for EBM in offshore regions are also important nearshore, but stresses from fisheries are not always the primary driver in the nearshore ecosystems. In the U.S., the call for EBM in the nearshore zone usually is spearheaded by concerns other than fisheries management. In Europe, the recent emphasis on renewable energy production in the marine environment, particularly wind energy, is now emerging as a dominant factor influencing management decisions, and energy-related stresses from oil and gas production, as well as wind energy, also are important in Australia and North America. The environmental consequences of large, wind energy developments are poorly known and may be expected to have local and regional effects on climate and ecosystems (Zhou et al. 2012). There is strong evidence that the most heavily modified coastal systems in Europe tend to coincide with regions and catchments that have the longest periods of industrial and economic development (e.g., parts of northern and central Europe). Ecologically Sustainable Development (ESD), initiated in Australia during the early 1980s, continues to be the framework for its EBM policies. A key global concern is the requirement for integrated and adaptive EBM in highly impacted coastal zones. This includes management of freshwater availability, quality and allocation to serve the competing demands of diverse stakeholders and the need to protect human lives and well-being, and concerns over sustaining biodiversity and productivity.

Our review has focused on the developed regions of Europe, Australia, and North America. In these regions, there are many similarities in stressors (Table 14.1), management policies, and research. Although other global regions were not included in our review, particularly Asian and African countries and regions with developing economies, we suspect that the ecosystem services and many threats to them in highly impacted coastal areas are similar to those in Europe, Australia and North America. However, the levels of threats and the priorities for EBM actions may be quite different.

We undertook a subjective assessment of risk, based on our knowledge of human activities and the probable magnitude of effects on coastal marine ecosystems at present and in the future (Table 14.1). The potential future risk summarizes our view of trends in each activity along with their known environmental impacts. In addition, we considered the effectiveness of EBM approaches, e.g., emission controls of hazardous substances have been effective, while controlling fishing effort has been less effective. Although a measure of past success (or failure) in implementing management measures does not necessarily provide a realistic estimate of future management success, we nevertheless incorporated past success or failure into our overall assessment of potential future risk (Table 14.1).

The overall threat to coastal ecosystems from multiple stressors is lowest in Australia but higher in Europe and North America where the coastline is more developed and more activity occurs (Table 14.1). A recent synthetic study reported on the multiple pressures impacting Key Ecological Features (KEFs) in the Australian EEZ (Hayes et al., 2011). A simple composite indicated that the highest potential for human-induced disturbance generally correlates with human population density and proximity to land (Figure 14.5). It is noteworthy that overall level of risk to coastal ecosystems in the three broad regions we reviewed is predicted to increase, but with notable shifts among the top-ranked activities that cause stress. For example, trends in shipping, aquaculture, coastal development, renewable energy supply and climate change are all expected to increase or remain at relatively high levels, so their footprints or their magnitude of impact will increase. By contrast, the prospects are reasonably positive for controlling or reducing ecosystem risks posed by fishing, eutrophication, and emission of hazardous substances.

Humans clearly have the means to induce massive changes in coastal ecosystems. Some of these changes benefit humans in the provisioning services we derive. Fishing is the prime example. Its provisioning benefits accrue to humans but must be regulated to avoid unacceptable disturbance, stress, and decline in the other beneficial services provided by the ecosystem. Most of the pressures and threats reviewed in this chapter are stressors on coastal ecosystems that impose a risk to ecosystem productivity and resilience. The level at which these activities can occur yet still allow productive fisheries often is unknown, but must be understood if EBM is to be effective.

This chapter would not be complete without acknowledging that some human impacts on marine ecosystems are beneficial to human economies and well-being.

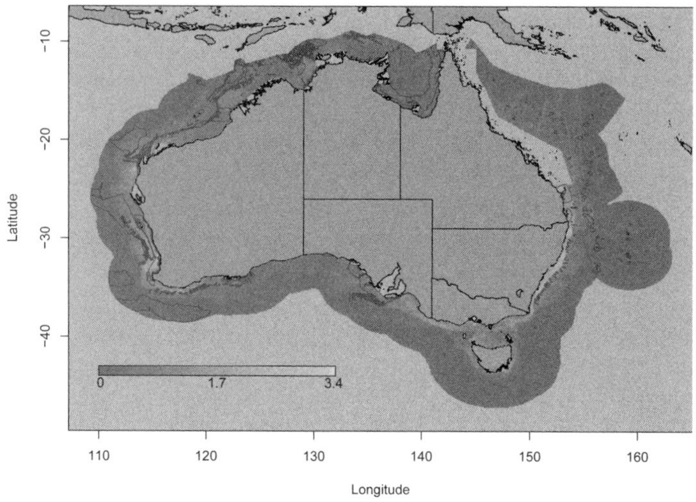

Fig. 14.5 – (See color insert) Standardized composite of human pressures in Australia's Exclusive Economic Zone. Features in purple are KEFs (see text). (Source: Hayes et al. 2011)

For example, the link between nutrient input from terrestrial systems and primary and secondary production of estuaries and coastal waters is established (Nixon et al., 1986; Lavelle et al., 2005). The high productivity of estuaries and coastal ecosystems often is dependent on nutrients in river discharge. While legitimate concerns over eutrophication and resulting hypoxia are well known, production and landings of fish and crustaceans often are elevated in these nutrient-enhanced systems (Breitburg et al., 2009). A case study of the Logan River in southeast Queensland demonstrated that high river discharge has a strong positive effect on production of coastal fisheries (Loneragan and Bunn, 1999). If nutrient loadings can be controlled to insure that coastal ecosystems do not shift to undesirable, highly eutrophic states, nutrient discharges from land potentially can increase production of coastal fisheries.

Perhaps the greatest challenge we face in implementing an EBM approach is in securing delivery of the provisioning services of coastal ecosystems, e.g., fisheries, to society while protecting other goods and services that insure long-term resilience and sustainability, e.g., biodiversity and habitat diversity. Integrated ecosystem assessments and more extensive use of ecosystem modeling are necessary to develop EBM plans that resource managers can implement with reasonable assurance of success. Dealing with tradeoffs will become increasingly necessary. We will be compelled to make decisions that balance the direct provisioning services to humans gained from estuaries and coastal ecosystems with the indirect services that support structure, function, and long-term sustainability. In essence, humanity requires tools which can better define and predict an optimum balance between nature and human needs.

Currently, EBM lacks integration and coordination at the global level. Even at national levels, management systems are weak with respect to integrating activities of the various federal, state and local government systems that plan, regulate and manage coastal and estuarine waters. Institutions and agencies with authority to manage activities in coastal systems may have conflicting missions or low incentive to manage collaboratively, although that approach ultimately is necessary for successful EBM. Successful EBM builds on integrated approaches that involve a wide spectrum of stakeholders and societal interests. New governance structures may be necessary to achieve the integration that is a priority for EBM in the coastal zone.

In summary, our key findings, conclusions, and recommendations are:

1) The dependency of, and demand for, ecosystem services in heavily impacted estuaries and coastal ecosystems is higher than in offshore ecosystems. The scope and breadth of stakeholder interests is high, as are the cumulative impacts of stresses generated by multiple stakeholder activities.
2) Suites of EBM assessment and management tools are evolving that may improve the structure and functioning of highly impacted coastal ecosystems. While these tools and approaches will serve to improve ecosystem structure and function, they will not restore them to their pristine states. Ongoing changes and new threats, especially climate change, add to uncertainties of management success, but also strengthen the argument for precautionary and integrative EBM.
3) In estuaries and nearshore ecosystems, programs espousing EBM usually include a substantial component aimed at ecosystem restoration or mitigation, in addition to regulatory actions that dominate in offshore regions. For example, dam removals, restoration of historical freshwater flow, planting of seagrasses, reduction of sewage or contaminant inputs, and habitat restoration are actions that can reverse damage to coastal ecosystems and restore services.
4) It is now widely recognized that an adaptive management approach, which takes into account the natural variability in productivity and carrying capacity of coastal ecosystems, is needed to successfully implement and conduct EBM. The need for adaptive EBM is greatest in heavily stressed nearshore ecosystems, but the challenge to effectively implement it also is greatest, given the expectations of numerous and diverse stakeholders and their competing interests.
5) In the U.S., national policies provide guidance and a framework for developing EBM plans for marine resources and ecosystems, but most programs to manage and restore coastal and estuarine ecosystems evolve from actions initiated by regional and local jurisdictions. Major programs in the U.S. and Australia generally are partnerships of federal, state and local agencies and institutions. This approach contrasts with that in Europe where many national programs to implement EBM issue from top-down directives that are multi-national in scope (e.g., MSFD).
6) In heavily impacted estuaries and coastal ecosystems, EBM must build a strong link to management in connected terrestrial ecosystems. Many of the biggest impacts and drivers of stress in coastal aquatic ecosystems originate in their watersheds. To be highly effective, EBM must rely on broadly integrative evaluation, analysis, monitoring and modeling that includes all key components of ecosystems.

7) New, institutional structures may be required to overcome impediments created by multiple agencies with sometimes conflicting jurisdictional authority to manage coastal ecosystems. Management institutions with broad authority will be required to implement adaptive EBM that fully and equitably addresses multi-stakeholder demands for coastal ecosystem goods and services.
8) Effective EBM will require tradeoffs in management for ecosystem services, particularly as human populations continue to grow. Demands from many and diverse stakeholders in the nearshore zone present a management challenge. For example, expected levels of provisioning services and goods from fisheries may have to go unrealized to conserve biodiversity or to insure human safety from flood risk (e.g., by building dams and flood-control structures). By its nature, spatial allocations and management (CMSM) in estuaries and coastal systems will require tradeoffs that include discretionary spatial allocations, zoning, and designations of marine reserves. These tradeoffs are more probable in highly impacted coastal ecosytems than offshore.
9) In the U.S., fishing is an important driver of stress, but not the major activity requiring an EBM approach in highly impacted estuaries and coastal areas. Variability in freshwater—its quantity, quality, timing of delivery, spatial distribution, and allocation are common stresses. Disruptions in natural freshwater flows and quality, including those attributable to climate change, are root causes of cumulative impacts and loss of ecosystem services in nearshore areas and will become increasingly stressful in future decades.
10) In Europe, the current major stressors and threats to coastal ecosystems are from fishing, shipping, eutrophication, and coastal developments, but in the future it is expected that aquaculture, renewable energy and climate change will become increasingly important. While in relatively good condition, Australia's coastal waters continue to be impacted by coastal development, fishing, aquaculture, oil and gas extraction, and climate change. Rising sea level, increased severity and incidence of extreme weather events, ocean acidification, and shifts in biodiversity and productivity, all are likely to have major impacts in coming decades in coastal seas around the globe.
11) Fishing is a unique human activity with direct consequences for marine ecosystems. Fisheries intentionally and selectively remove organisms and biomass from ecosystems and are most intensive in the coastal zone. It is impossible to fish without disturbing the ecosystem. Hence, adaptive EBM (or EBFM) is highly desirable to conserve productivity, sustainable yields, and biodiversity. For water and habitat, an EBM goal is to minimize unintended consequences and thus conserve the ecosystem's supporting and regulating services, while for fisheries the aim is to optimize the extractive provisioning service.
12) Cumulative impacts of multiple stressors are the source of many problems in estuaries and the coastal zone. Indicators and reference points that set targets and thresholds, and quantify impacts are needed to evaluate the cumulative effects on status and trends of coastal ecosystems and to develop plans to regulate, mitigate, or restore ecosystem services.

13) There is a widely recognized need to document effectiveness of EBM in nearshore and heavily impacted estuarine ecosystems and communicate it to stakeholders. In this regard, ecosystem report cards are becoming a common communication tool. If done well, such reports can effectively transmit information on status and trends of coastal ecosystems as well as identifying where management is succeeding or failing.
14) Most programs embracing EBM are in their infancy and have not yet achieved their goals. Accordingly, conditions in some coastal, estuarine ecosystems have not yet improved. But, what would be the present state and condition of ecosystems such as Chesapeake Bay lacking an approach that embraces EBM? It is probable that their condition would be further diminished. The EBM programs now evolving are not simple or quick solutions, but they do provide a framework for broadly integrative and spatially explicit approaches that are needed.

References

ABARES. 2011. Australian fisheries statistics 2010. Australian Bureau of Agricultural and Resource Economics and Sciences, Canberra, August.

ABS. 2010. Regional Population Growth, Australia, 2008–09. Australian Bureau of Statistics, cat. no. 3218.0, ABS, Canberra.

ABS. 2011. Australian Demographic Statistics. Australian Bureau of Statistics. http://www.abs.gov.au.

Action Agenda. 2009. Puget Sound Partnership. (http://www.psp.wa.gov/downloads/AA2009/Action_Agenda_FINAL_063009.pdf) (http://www.psp.wa.gov/action_agenda_2011_draft-content.php).

Agardy, T., J. Davis, K. Sherwood and O. Vestergaard. 2011. *Taking steps toward marine and coastal ecosystem-based management. An introductory guide.* United Nations Environmental Program. UNEP Regional Seas Reports and Studies No. 189.

AMSA. 2010. National Plan to Combat Pollution of the Sea by Oil and other Noxious and Hazardous Substances—annual report 2009–10. Australian Maritime Safety Authority, Canberra.

ASEC. 2006. *Australia State of the Environment 2006.* Australian State of the Environment Committee. Independent report to the Australian Government Minister for the Environment and Heritage, Department of the Environment and Heritage, Canberra.

Astles, K. L., M. G. Holloway, A. Steffe, M. Green, C. Ganassin and P. J. Gibbs. 2006. An ecological method for qualitative risk assessment and its use in the management of fisheries in New South Wales, Australia. *Fish. Res.* 82:290–303.

Barbier, E. B. 2011. Coastal wetland restoration and the Deepwater Horizon oil spill. *Vanderbilt Law Review* 64:1821–1849.

Barbier, E. B., S. D. Hacker, C. Kennedy, E. W. Koch, A. C. Stier and B. R. Silliman. 2011. The value of estuarine and coastal ecosystem services. *Ecological Monographs* 81:169–193.

BDCP. *In prep.* Bay Delta Conservation Plan. (http://baydeltaconservationplan.com/Home.aspx).

Beaugrand, G., F. Ibañez, J. A. Lindley and P. C. Reid. 2002. Diversity of calanoid copepods in the North Atlantic and adjacent seas: species associations and biogeography. *Marine Ecology Progress Series* 232:179–195.

Beaugrand, G. and R. R. Kirby. 2010. Climate, plankton and cod. *Global Change Biology* 16:1268–1280.

Boesch, D. F. 2001. Science and integrated drainage basin-coastal management: the Chesapeake Bay and the Mississippi Delta. pp. 37–50. In *Science and integrated coastal management.* von Bodungen, B. and K. Turner (eds.). Dahlem University Press, Berlin.

Boesch, D. F. 2006. Scientific requirements for ecosystem-based management in the restoration of Chesapeake Bay and coastal Louisiana. *Ecological Engineering* 26:6–26.

Boesch, D. F. and E. B. Goldman. 2009. Chesapeake Bay, USA. pp. 268–293. In *Ecosystem-based management for the oceans.* McLeod, K. and H. Leslie (eds.). Island Press.

Borja, A., M. Elliott, J. Carstensen, A.-S. Heiskanen and W. van de Bund. 2010. Marine management—Towards an integrated implementation of the European Marine Strategy Framework and the Water Framework Directives. *Marine Pollution Bulletin* 60:2175–2186.

Breitburg, D. L., D. W. Hondorp, L. A. Davias and R. J. Diaz. 2009. Hypoxia, nitrogen, and fisheries: integrating effects across local and global landscapes. *Annual Review of Marine Science* 1:329–349.

Bricker, S., B. Longstaff, W. Dennison, A. Jones, K. Boicourt, C. Wicks and J. Woerner. 2007. *Effects of nutrient enrichment on the nation's estuaries: a decade of change.* NOAA Coastal Ocean Program, Decision Analysis Series 26. National Center for Coastal Ocean Science, Silver Spring, MD, USA. 328 pp.

Canada's Oceans Action Plan. 2005. Canada's Ocean Action Plan. (http://www.dfo-mpo.gc.ca/oceans/publications/oap-pao/index-eng.asp).

Canadian Oceans Act. 1997. Oceans Act, S.C. 1996, c. 31. (http://laws-lois.justice.gc.ca/eng/acts/O-2.4/page-1.html#h-1).

Canada's Oceans Strategy. 2002. Canada's Oceans Strategy. (http://www.dfo-mpo.gc.ca/oceans/publications/cos-soc/index-eng.asp).

CBFEAP 2006. Fisheries ecosystem planning for Chesapeake Bay. American Fisheries Society, Trends in Fisheries Science and Management 3, Bethesda, MD.

Chesapeake Bay Action Plan, 2011, released in Sept 2010. (http://executiveorder.chesapeakebay.net/file.axd?file=2010%2f9%2fAction+Plan+Briefing.pdf).

CERP. 1999. Comprehensive Everglades Restoration Plan. Central and Southern Florida Project comprehensive review study. Final Integrated Feasibility Report and Programmatic Environmental Impact Statement. U.S. Army Corps of Engineers, Jacksonville District. (http://www.evergladesplan.org/docs/comp_plan_apr99/summary.pdf).

CERP. 2011. Comprehensive Everglades Restoration Project. 2010 Report to Congress. U.S. Army Corps of Engineers and associated/affiliated agencies. (http://www.evergladesplan.org/pm/pm_docs/rtc_2010/rtc_2010_final.pdf).

Cloern, J. E., N. Knowles, L. B. Brown, D. Cayan, M. D. Dettinger, T. L. Morgan, D. H. Schoellhamer, M. T. Stacey, M. van der Wegen, R. W. Wagner and A. D. Jassby. 2011. Projected evolution of California's San Francisco Bay-Delta-River system in a century of climate change. *PLoS ONE* 6:e24465. 13 pp.

CPRA. 2007. Integrated ecosystem restoration and hurricane protection: Louisiana's comprehensive master plan for a sustainable coast. Coastal Protection and Restoration Authority, State of Louisiana, Baton Rouge.

CPRA. 2012. Louisiana's comprehensive master plan for a sustainable coast. Coastal Protection and Restoration Authority, State of Louisiana, Baton Rouge. (http://www.coastalmasterplan.louisiana.gov/).

CZMA. 1972. Coastal Zone Management Act. U.S. Congress, Public Law 92–583, U.S.C. 1451–1456.

CWPPRA. 1990. Coastal Wetlands Planning, Protection and Restoration Act. U.S. Congress. Public Law 101–646.

DAFF. 2000. Commonwealth Policy on Fisheries Bycatch. Australian Department of Agriculture, Fisheries and Forestry. June 2000.

DAFF. 2007. Commonwealth Fisheries Harvest Strategy—Policy and Guidelines. Australian Department of Agriculture, Fisheries and Forestry. September 2007.

DCC. 2009. Australian Climate Change Science: A National Framework. Department of Climate Change, Canberra.

De'ath, G., J. M. Lough, and K. E. Fabricius. 2009. Declining coral calcification on the Great Barrier Reef. *Science* 323:116–119.

Delta Plan. *In prep.* Fifth staff draft Delta Plan, August 2011. (http://www.deltacouncil.ca.gov/sites/default/files/documents/files/Fifth_Staff_Draft_Delta_Plan_080211.pdf).

Delta Reform Act of 2009. 2009. The Delta Reform Act. California State Legislature, Bill SB 1, Public Resources. (http://www.leginfo.ca.gov/pub/09–10/bill/sen/sb_0001–0050/sbx7_1_bill_20091023_introduced.pdf).

DEWHA 2008. The North-West Marine Bioregional Plan: Bioregional profile—a description of the ecosystems, conservation values and uses of the North-west Marine Region. Department of the Environment, Water, Heritage, and the Arts, Kingston, Tasmania.

Dorcas, M. E., J. D. Willson, R. N. Reed, R. W. Snow, M. R. Rochford, M. A. Miller, W. E. Meshaka, P. T. Andreadis, F. J. Mazzotti, C. M. Romagosa and K. M. Hart. 2012. Severe mammal declines coincide with proliferation of invasive Burmese pythons in Everglades National Park. *Proceedings of the National Academy of Science, USA* 109:2418–2422.

Drinkwater, K. (2005). The response of Atlantic cod (*Gadus morhua*) to future climate change. *ICES Journal of Marine Science* 62:1327–1337.

EC. 2008. Directive 2008/56/EC of the European Parliament and of the Council Establishing a Framework for Community Action in the Field of Marine Environmental Policy (Marine Strategy Framework Directive). EU, Brussels.

EC. 2008. Establishing a framework for community action in the field of marine environmental policy (Marine Strategy Framework Directive). Directive 2008/56/EC of the European Parliament and of the Council of 17 June 2008.

EEA. 2006a. The changing faces of Europe's coastal areas. EEA Report No 6/2006. European Environment Agency, Copenhagen. Available at: www.eea.europa. eu/publications/eea_report_2006_6.

EEA. 2010. The European environment state and outlook 2010. EEA, Copenhagen, ISBN 978–92–9213–158–6, 58 pp.

EEA. 2011. Status of European marine fish stocks (CSI 032)—Assessment published (online) Sep 2011. See http://www.eea.europa.eu/data-and-maps/indicators/status-of-marine-fish-stocks/status-of-marine-fish-stocks-8.

Ehler, C. and F. Douvere. 2009. Marine spatial planning: a step-by-step approach toward ecosystem-based management. Intergovernmental Oceanographic Commission and Man and the Biosphere Programme. IOC Manual and Guides No. 53, ICAM Dossier No. 6. Paris: UNESCO. 2009 (English).

EO. 2009. Chesapeake Bay protection and restoration. Presidential Executive Order 13508. The White House, Washington, D.C.

ERTF. 2011. Gulf of Mexico regional ecosystem restoration strategy (preliminary). Gulf Coast Ecosystem Restoration Task Force. Report. (http://www.epa.gov/gcertf/pdfs/gcertfenlishver.pdf)

EU. 2006. Review of the EU sustainable development strategy (EU SDS)—renewed strategy. Council of the European Union. 10117/06. 29 pp.

Ezcura, E., O. Aburto-Oropeza, M. de los Angeles Carvajal, R. Cudney-Bueno and J. Torre. 2009. Gulf of California, Mexico. pp. 227–252. In *Ecosystem-based management for the oceans.* McLeod, K. and H. Leslie (eds.). Island Press.

FAO. 2010. The state of world fisheries and aquaculture, 2010. Food and Agriculture Organization of the United Nations, Rome, Italy. (http://www.fao.org/docrep/013/i1820e/i1820e00.htm).

Federal Leadership Committee. 2010. Strategy for protecting and restoring the Chesapeake Bay watershed. U.S. Environmental Protection Agency, EPA-903-R-10–003.

Fletcher, W. J. 2005. The application of qualitative risk assessment methodology to prioritize issues for fisheries management. *ICES J. Mar. Sci.* 62, 1576–1587.

Foster, E., M. Haward and S. Coffen-Smout. 2004. Implementing integrated oceans management: Australia's south east regional marine plan (SERMP) and Canada's eastern Scotian shelf integrated management (ESSIM) initiative. *Marine Policy* 29: 391–405.

Garcia, S. M. (Ed.), J. Kolding, J. Rice, M.-J. Rochet, S. Zhou, T. Arimoto, J. Beyer, L. Borges, A. Bundy, D. Dunn, N. Graham, M. Hall, M. Heino, R. Law, M. Makino, A. D. Rijnsdorp, F. Simard, A. D. M. Smith and D. Symons. 2011. Selective fishing and balanced harvest in relation to fisheries and ecosystem sustainability. IUCN and EBCD Report. Gland, Switzerland and Brussels, Belgium.

Granek, E. F., S. Polasky, C. V. Kappel, D. J. Reed, plus 13 co-authors. 2009. Ecosystem services as a common language for coastal ecosystem-based management. *Conservation Biology* 24:207–216.

Green, S. L. 2011. *Organizing professionals and stakeholders to conduct ecosystem based fisheries management: a case study in Chesapeake Bay.* PhD Dissertation, University of Delaware, Newark, DE.

Halpern, B. S., S. Walbridge, K. A. Selkoe, C. V. Kappel, F. Micheli, C. D'Agrosa, J. F. Bruno, K. S. Casey, C. Ebert, H. E. Fox, R. Fujita, D. Heinemann, H. S. Lenihan, E. M. P. Madin, M. T. Perry, E. R. Selig, M. Spalding, R. Steneck and R. Watson. 2008. A global map of human impact on marine ecosystems. *Science* 319:948–952.

Halpern, B. S., J. Diamond, S. Gaines, S. Gelcich, plus 15 co-authors. 2012. Near-term priorities for the science, policy, and practice of coastal and marine spatial planning (CMSP). *Marine Policy* 36:198–205.

Harvey, N. and B. Caton. 2011. *Coastal management in Australia.* University of Adelaide Press.

Hayes, K. R., D. Clifford, C. Moeseneder, M. Palmer and T. Taranto. 2011. National indicators of marine ecosystem health: Mapping project. Australian Government Department of Environment, Water, Heritage and Arts. CSIRO Wealth from Ocean project reference R00658–03–002.

Healey, M. C., M. D. Dettinger, and R. B. Norgaard (eds.). 2008. The state of Bay-Delta science. CALFED Science Program, Sacramento, CA. (http://science.calwater.ca.gov/pdf/publications/sbds/sbds_final_update_122408.pdf).

Helsinki Commission. 2004. Convention on the protection of the marine environment of the Baltic Sea area 1992 (Helsinki Convention). Baltic Marine Environment Protection Commission, HELCOM Secretariate, Finland. 45 pp.

Herbert, D. A., W. B. Perry, B. J. Cosby and J. W. Fourqurean. 2011. Projected reorganization of Florida Bay seagrass communities in response to the increased freshwater flow of Everglades restoration. *Estuaries and Coasts* 34:973–992.

Hiddink, J. G., T. Hutton, S. Jennings and M. J. Kaiser. 2006. Predicting the effects of area closures and fishing effort restrictions on the production, biomass, and species richness of benthic invertebrate communities. *Journal of Marine Science,* 64:453–463.

Hobday, A. J., A. D. M. Smith, I. C. Stobutzki, C. Bulman, R. Daley, J. M. Dambacher, R. A. Deng, A. Dowdney, M. Fuller, D. Furlani, S. P. Griffiths, D. Johnson, R. Kenyon, I. A. Knuckey, S. D. Ling, R. Pitcher, K. J. Sainsbury, M. Sporcic, T. Smith, C. Turnbull, T. I. Walker, S. E. Wayte, H. Webb, A. Williams, B. S. Wise and S. Zhou. 2011. Ecological risk assessment for the effects of fishing. *Fisheries Research* 108: 372–384.

Houde, E. D. 2011. *Managing the Chesapeake's fisheries: a work in progress. Chesapeake Perspectives.* Maryland Sea Grant, College Park, MD.

HSC. 2009. Inquiry into climate change and environmental impacts on coastal communities. House Standing Committee on Climate Change, Water, Environment and the Arts Committee, Canberra.

Hunt, J. H. and W. Nuttle (eds.). 2007. Florida Bay science program: a synthesis of research on Florida Bay. Florida Fish and Wildlife Research Institute, Technical Report TR-11. 148 pp.

Jennings, S., J. Lee and J. G. Hiddink. 2012. Assessing fishery footprints and the trade-offs between landings value, habitat sensitivity, and fishing impacts to inform marine spatial planning and an ecosystem approach. *ICES Journal of Marine Science;* doi:10.1093/icesjms/fss050.

JOCI. 2011. America's ocean future: Insuring healthy oceans to support a vibrant economy. Joint Ocean Commission Initiative, Washington, D.C. (http://www.jointoceancommission.org/resource-center/1-Reports/2011–06–07_JOCI_Americas_Ocean_Future.pdf).

JSOST. 2007. Charting the course for ocean science in the United States for the next decade. U.S. Joint Subcommittee on Ocean Science and Technology, Washington, D.C.

Kemp, W. M., W. R. Boynton, J. E. Adolf, D. F. Boesch, plus fourteen co-authors. 2005. Eutrophication of Chesapeake Bay: historical trends and ecological interactions. *Marine Ecology Progress Series* 303:1–29.

Kimmerer, W. J. 2004. Open-water processes of the San Francisco Estuary: from physical forcing to biological responses. San Francisco Estuary and Watershed Science 2(1):1–142 1–142. (online journal; http://escholarship.ucop.edu/uc/item/9bp499mv#page-1).

Langmead, O., A. McQuatters-Gollop and L. D. Mee. (eds.). 2007. *European lifestyles and marine ecosystems: Exploring challenges for managing Europe's seas.* University of Plymouth Marine Institute, Plymouth, UK. 43 pp.

Lavelle, P., R. Dugdale, R. Scholes, A. A. Berhe, E.Carpenter, L.Codispoti, A. M. Izac, J. Lemoalle, F. Luizao, M. Scholes, P. Tréguer, and B. Ward. 2005. Chapter 12. Nutrient cycling. In *Ecosystems and human well-being, Vol. 1. Current state and trends: findings of the Condition and Trends Working Group.* Edited by Hassan, R., R. Scholes and N. Ash. Island Press, USA.

Ledoux, L., J. E. Vermaat, L. M. Bouwer, W. Salomons and R. K. Turner. 2005. ELOISE research and the implementation of EU policy in the coastal zone. pp. 1–19. In *Managing European coasts: past, present, and future.* Vermaat, J. E. et al. (eds.) Springer-Verlag Berlin Heidelberg.

Lehtiniemi M., J. Javidpour, A. Lehmann and K. Myrberg. 2010. Spreading and reproduction of the American comb jelly *Mnemiopsis leidyi* in the Baltic Sea. Presentation at ICES Annual Science Conference, Nantes.

Leslie, H. M. and K. L. McLeod. 2007. Confronting the challenges of implementing marine ecosystem-based management. *Frontiers in Ecology and the Environment* 5:540–548.

Levin P. S., M. J. Fogarty, S. A. Murawski and D. Fluharty. 2009. Integrated ecosystem assessments: Developing the scientific basis for ecosystem-based management of the ocean. *PLoS Biol* 7:e1000014.

Loneragan, N. R. and S. E. Bunn. 1999. River flows and estuarine ecosystems: Implications for coastal fisheries from a review and a case study of the Logan River, southeast Queensland. *Australian Journal of Ecology* 24:431–440.

Lotze, H. K., H. S. Lenihan, B. J. Bourque, R. H. Bradbury, R. G. Cooke, M. C. Kay, S. M. Kidwell, M. X. Kirby, C. H. Peterson and J. B. C. Jackson. 2006. Depletion, degradation, and recovery potential of estuaries and coastal seas. *Science* 312:1809.

Lund, J. R., E. Hanak, W. E. Fleenor, W. A. Bennett, R. E. Howitt, J. F. Mount and P. B. Moyle. 2010. *Comparing futures for the Sacramento-San Joaquin Delta.* University of California Press.

MA. 2005. *The Millenium Ecosystem Assessment. Volume 1. Current State and Trends.* Island Press, Washington, D.C.

MacKenzie, B. R. and F. W. Köster. 2004. Fish production and climate: sprat in the Baltic Sea. *Ecology* 85:784–794.

MBDWG. 2008. A National Approach to Addressing Marine Biodiversity Decline. Marine Biodiversity Decline Working Group. Report to the Natural Resource Management Ministerial Council.

McClure, M. and M. Ruckelshaus. 2007. Collaborative science: moving ecosystem-based management forward in Puget Sound. *Fisheries* 32:458–461.

McNeil, T. C., F. R. Rousseau and L. P. Hildebrand. 2006. Community-based environmental management in Atlantic Canada: the impacts and spheres of influence of the Atlantic Coastal Actions Program. *Environmental Monitoring and Assessment* 113:367–383.

MEFEPO 2011. Key operational challenges to the introduction of an ecosystem approach to fisheries management: Workshop report on the project website: http://www.liv.ac.uk/mefepo/Reports_and_outputs.htm.

Milton D. A. 2001. Assessing the susceptibility to fishing of populations of rare trawl bycatch: sea snakes caught by Australia's northern prawn fishery. *Biological Conservation* 101:281–290.

Ministry of Sustainable Development Sweden. 2006. Göteborg declaration: North Sea ministerial meeting on the environmental impact of shipping and fisheries : Göteborg, Sweden 4 and 5 May 2006. ISBN 9163301121, 9789163301124. 75 pp.

Möllmann, C., R. Diekmann, B. Muller-Karulis, G. Kornilovs, M. Plikshs and P. Axe. 2009. Reorganization of a large marine ecosystem due to atmospheric and anthropogenic pressure: a discontinuous regime shift in the Central Baltic Sea. *Global Change Biology* 15:1377–1393.

Morton, S. R., O. Hoegh-Guldberg, D. B. Lindenmeyer, M. Harriss Olson, L. Hughes, M. T. McCulloch, S. McIntyre, H. A. Nix, H, S. M. Prober, D. A. Saunders, A. N. Andersen, M. A. Burgman, E. C. Lefroy, W. M. Lonsdale, I. Lowe, A. J. McMichael, J. S. Parslow, W. Steffen, J. E. Williams and J. C. Z. Woinarski. 2009. The big ecological questions inhibiting effective environmental management in Australia. *Austral. Ecology* 34:1–9.

Moyle, P. B., J. R. Lund, W. A. Bennett and W. E. Fleenor. 2010. Habitat variability and complexity in the upper San Francisco Estuary. San Francisco Estuary & Watershed *Science* 8(3):1–24 (online journal; http://escholarship.org/uc/item/0kf0d32x).

Mutlu, E. 2009. Recent distribution and size structure of gelatinous organisms in the southern Black Sea and their interactions with fish catches. *Mar. Biol.* 135:935–957.

NOC. 2012. National Ocean Policy implementation plan (draft). National Ocean Council. The Whitehouse, Washington, D.C., USA. (http://www.whitehouse.gov/sites/default/files/microsites/ceq/national_ocean_policy_draft_implementation_plan_01–12–12.pdf).

NEPA. 1970. The National Environmental Protection Act. U.S. Congress, Public Law 91–190, 42, U.S.C 4321–4347.

Nixon, S. W., C. A. Oviatt, J. Frithsen and B. Sullivan. 1986. Nutrients and the productivity of estuarine and coastal marine ecosystems. *Journal of the Limnological Society of South Africa* 12:43–71.

NLWRA. 2002. Catchment, river and estuary condition in Australia. National Land and Water Resources Audit, Canberra.

NOAA. 2010. Fisheries of the United States- -2010. NOAA Fisheries, Office of Science and Technology, Silver Spring, MD, USA. (http://www.st.nmfs.noaa.gov/st1/fus/fus10/index.html).

NOAA. 2011. Ecosystems Sciences and Management Working Group. (http://www.sab.noaa.gov/Meetings/2011/march/SAB_Mtg_Pres_Mar11_Sanchirico_FINAL_3042011.pdfhttp://www.sab.noaa.gov/Meetings/2011/march/SAB_Mtg_Pres_Mar11_Sanchirico_FINAL_3042011.pdf).

NRMMC. 2006. Australia's National Programme of Action for the Protection of the Marine Environment from Land-Based Activities. Natural Resource Management Ministerial Council. (http://www.deh.gov.au/coasts/pollution/npa/index.html). Australian Department of the Environment and Heritage, Canberra, Australia.

Ocean Policy Task Force. 2010. Final recommendations of the Interagency Ocean Policy Task force. The White House Council on Environmental Quality, Washington, D.C. (http://www.whitehouse.gov/files/documents/OPTF_FinalRecs.pdf).

Oguz, T. 2007. Nonlinear response of Black Sea pelagic fish stocks to over-exploitation. *Marine Ecology Progress Series* 345:211–228.

Olsson, P., C. Folke and T. P. Hughes. 2008. Navigating the transition to ecosystem-based management of the Great Barrier Reef, Australia. *Proceedings of the National Academy of Science, USA* 105:9489–9494.

Orth, R. J., T. J. B. Carruthers, W. C. Dennison, C. J. Duarte, J. W. Fourqurean, K. L. Heck, Jr., A. R. Hughes, G. A. Kendrick, W. J. Kenworthy, S. Olyarnick, F. T. Short, M. Waycott and S. L. Williams. 2006. A global crisis for seagrass ecosystems. *Bioscience* 56:987–996.

OSPAR. 2010. Quality Status Report 2010. OSPAR Commission, London, 176 pp.

Paterson, D. M., N. D. Hanley, K. Black, E. C. Defew and M. Solan (eds.). 2011. Biodiversity, ecosystems and coastal zone management: linking science and policy. Theme Section. *Marine Ecology Progress Series* 434:201–301.

Plagányi, E. E., J. D. Bell, R. H. Bustamante, J. M. Dambacher, D. M. Dennis, C. M. Dichmont, L. X. C. Dutra, E. A. Fulton, A. J. Hobday, E. I. Putten, F. Smith, A. D. M. Smith and S. Zhou. 2011. Modelling climate change effects on Australian and Pacific aquatic ecosystems: a review of analytical tools and management implications. *Marine and Freshwater Research* 62:1132–1147.

POC. 2003. America's living ocean: charting a course for sea change. A report to the nation. Pew Oceans Commission, Pew Trusts, Washington, D.C.

Presidential Executive Order (EO #13547). 2010. Stewardship of the ocean, our coasts, and the Great Lakes. The White House, Washington, D.C. (http://www.whitehouse.gov/the-press-office/executive-order-stewardship-ocean-our-coasts-and-great-lakes).

Presidential Executive Order (EO #13508). 2009. Chesapeake Bay protection and restoration. The White House, Washington, D.C. (http://edocket.access.gpo.gov/2009/pdf/E9–11547.pdf).

Presidential Executive Order (EO #13554). 2010. Establishing the Gulf Coast ecosystem restoration task force. The White House, Washington, D.C. (http://edocket.access.gpo.gov/2010/pdf/2010–25578.pdf).

Rabalais, N. N. 2011. Troubled waters of the Gulf of Mexico. *Oceanography* 24:200–211.

Robinson, D. 2009. Acclimating to a new Bay area: ecosystem-based approaches to management for the San Francisco Bay. San Francisco Bay Conservation and Development Commission. (http://www.bcdc.ca.gov/publications/EBM_Report.pdf).

Rosenberg, A. A., M. L. Mooney-Seus, I. Kiessling, C. B. Mogensen, R. O'Boyle and J. Peacey. 2009. Lessons from national-level implementation across the world. pp. 294–313. In *Ecosystem-based management for the oceans*. McLeod, K. and H. Leslie (eds.). Island Press, Washington, D.C.

Ruckelshaus, M., T. Essington and P. Levin. 2009. Puget Sound, Washington, USA. pp. 201–226. In *Ecosystem-based management for the oceans*. McLeod, K. and H. Leslie (eds.). Island Press, Washington, D.C.

Ruttenberg, B. I. and E. F. Granek. 2011. Bridging the marine-terrestrial disconnect to improve marine coastal zone science and management. *Marine Ecology Progress Series* 434:203–212.

San Francisco Bay Plan. 1969. San Francisco Bay Conservation and Development Commission. (http://www.bcdc.ca.gov/laws_plans/sfbay_plan.shmtl).

Smith, A. D. M., E. J. Fulton, A. J. Hobday, D.C. Smith and P. Shoulder. 2007. Scientific tools to support the practical implementation of ecosystem-based fisheries management. *ICES Journal of Marine Science* 64:633–639.

SoEC. 2011. Australia state of the environment 2011. State of the Environment Committee. Independent report to the Australian Government Minister for Sustainability, Environment, Water, Population and Communities. Canberra: DSEWPaC, 2011.

Sommer, T., C. Armor, R. Baxter, R. Breuer, L. Brown, et al. 2007. The collapse of pelagic fishes in the upper San Francisco Estuary. *Fisheries* 32:270–277.

Sound Science. 2007. Sound science: synthesizing ecological and socioeconomic information about the Puget Sound ecosystem. Sound Science Collaborative Team. NOAA, Northwest Fisheries Science Center, Seattle, WA. (http://www.nwfsc.noaa.gov/research/shared/sound_science/documents/SoundScience07.pdf).

SPAW, 1990. Specially Protected Areas and Wildlife Protocol. Caribbean Environmental Program, United Nations Environmental Program (http://www.cep.unep.org/cartagena-convention/spaw-protocol).

Stobutzki I. C., M. J. Miller and D. T. Brewer. 2001. Sustainability of fishery bycatch: a process for assessing highly diverse and numerous bycatch. *Environmental Conservation* 28:167–181.

Tallis, H., S. E. Lester, M. Ruckelshaus, M. Plummer, plus 19 co-authors. 2012. New metrics for managing and sustaining the ocean's bounty. *Marine Policy* 36:303–306.

Talpalar, A. E., and Y. Grossman. 2005. Sonar versus whales: Noise may disrupt neural activity in deep-diving cetaceans. *UHM* 32:135–139.

UNEP. 2006. Marine and coastal ecosystems and human well-being. Synthesis: a synthesis report based on the findings of the Millenium Ecosystem Assessment. United Nations Environmental Program.

UNEP and IOC-UNESCO. 2009. An assessment of assessments. Start-up phase of a Regular Process for global reporting and assessment of the state of the marine environment including socio-economic aspects. United Nations Environmental Program, International Oceanographic Commission and United Nations Educational, Scientific and Cultural Organizaton. (http://www.unep.org/DEWA/pdf/AoA/AoA+SDM+layout_ENGLISH+LR.pdf).

USCOP. 2004. An ocean blueprint for the twenty-first century. United States Commission on Ocean Policy, Washington, D.C.

USEPA. 2012. National coastal condition report IV. U.S. Environmental Protection Agency, EPA-842-R-10–003. Washington, D.C. (http://water.epa.gov/type/oceb/assessmonitor/nccr/upload/NCCR4-Report.pdf).

Vince, J. 2008. Ten years of implementing Australia's Oceans Policy: From an integrated approach to an environmental policy focus. *Maritime Studies* 159:1–10.

Walker T. I. 2005. Chapter 13. Management measures. In: Musick, J. A., R. Bonfil, (eds.). Management techniques for elasmobranch fisheries. Vol. 474. FAO Fisheries Technical Paper 474:216–242, 251.

Wang, X. H., X. Xu, S. G. Pearson, G. Xue, R. J. Morrison, D. Liu and P. Shi. 2011. Integrated coastal zone management research in Australia and China. *Labour and Management in Development Journal* 11:1–17.

Washington State Senate Bill 5372. Creating the Puget Sound Partnership. (http://apps.leg.wa.gov/billinfo/summary.aspx?bill=5372&year=2008).

Wilson, M. A., R. Costanza, R. Boumans and S. Liu. 2005. Integrated assessment and valuation of ecosystem goods and services provided by coastal systems. pp. 1–24. In *The intertidal ecosystem: the value of Ireland's shores.* Wilson, J. G. (ed.). Royal Irish Acacemy, Dublin.

Woodhams, J, I. Stobutzki, S. Vieira, R. Curtotti and G. A. Begg (eds.). 2011. Fishery status reports 2010: status of fish stocks and fisheries managed by the Australian Government. Australian Bureau of Agricultural and Resource Economics and Sciences, Canberra.

WRDA. 2000. Water Resources Development Act of 2000. U.S. Congress, Public Law 106–541. (http://www.fws.gov/habitatconservation/Omnibus/WRDA2000.pdf).

WRDA. 2007. Water Resources Development Act of 2007. U.S. Congress, H. R. 1495. (http://www.crcl.org/images/LCA_WRDA_Authorization.pdf).

Zhou, L., Y. Tian, S. B. Roy, C. Thorncroft, L. F. Bosart and Y. Hu. 2012. Impacts of wind farms on land surface temperature. *Nature Climate Change* 2:539–543.

Zhou, S. 2008. Fishery by-catch and discards: a positive perspective from ecosystem-based fishery management. *Fish and Fisheries* 9:308–315.

Zhou, S. and S. P. Griffiths. 2008. Sustainability assessment for fishing effects (SAFE): a new quantitative ecological risk assessment method and its application to elasmobranch bycatch in an Australian trawl fishery. *Fisheries Research* 91:56–68.

Zhou, S., S. P. Griffiths and M. Miller. 2009. Sustainability assessment for fishing effects (SAFE) on highly diverse and data-limited fish bycatch in a tropical prawn trawl fishery. *Marine and Freshwater Research* 60:563–570.

Zhou, S., A. D. M. Smith, A. E. Punt, A. J. Richardson, M. Gibbs, E. A. Fulton, S. Pascoe, C. Bulman, P. Bayliss and K. Sainsbury. 2010. Ecosystem-based fisheries management requires a change to the selective fishing philosophy. *Proceedings of the National Aacademy of Science, USA* 107:9485–9489.

Zhou, S., A. D. M. Smith and M. Fuller. 2011. Quantitative ecological risk assessment for fishing effects on diverse data-poor non-target species in a multi-sector and multi-gear fishery. *Fisheries Research* 112:168–178.

Chapter 15. Marine Ecosystem-Based Management: Past, Present, and the Future

MICHAEL J. FOGARTY

National Oceanic and Atmospheric Administration

JAMES J. MCCARTHY

Harvard University

Contents

1. Introduction

It is abundantly clear that development and implementation of marine Ecosystem-Based Management (mEBM) is rapidly gaining traction on the world stage. It is no less evident that much remains to be done. Tangible signs of progress can now be seen, as recognition of the necessity of adopting holistic approaches to management is acknowledged, appropriate regulatory and governance structures are established, and the scientific framework for mEBM crystallizes. The contributions to this volume provide ample evidence of advances on each of these fronts. Among the many challenges still facing us are the integration of the social and natural science components of the coupled Social-Ecological System (SES) approach needed to support mEBM (Charles, this volume); the ongoing process of formal adoption of mEBM as the centerpiece for ocean management by governments and regulatory bodies throughout the world (Cochrane et al. this volume); designation of spatial management units throughout the world ocean (Rice et al., this volume); and developing strategies to cope with the unquestionable complexity of the entire enterprise.

Any evaluation of the need for holistic ocean management in different regions, and our capability to implement this approach, must deal with the intertwined importance of scientific understanding, culture, and societal practices, including management structures past and present. The differences in approach to mEBM in the case studies in this volume in no small measure are a reflection of this. Selected ecotypes span the range from systems dominated by very low human population

The Sea, Volume 16, edited by Michael J. Fogarty and James J. McCarthy
ISBN 978-0-674-07270-1

levels, industrialization, and extreme environmental conditions at the poles to densely populated areas with relatively low levels of industrialized activity in much of the tropics. The combined threat of both high human population densities and heavy industrialization in much of the temperate zones, particularly in coastal and estuarine settings, brings with it multifaceted challenges for integrated management. The nature of governance and regulatory frameworks for the array of ecotypes examined here range from international control at the poles and on the open ocean to a complex mosaic of nested local, regional, and multinational structures in the tropics and the temperate zone. These extremes involve very different challenges to implementing mEBM. In the former, competing national interests must be reconciled in forging a management structure through high diplomacy. In the latter, management actions on a range of spatial scales must be coordinated by different management entities involving quite different challenges. Management actions taken at one level will necessarily affect the others, and an overarching management framework with nested spatial considerations is essential. Scientific capacity unquestionably plays a deciding role in how mEBM will develop. Currently, it is only in the most affluent parts of the world where the scientific infrastructure is in place to support advanced assessment and management activities. The World Bank's contribution to the development of scientific capacity in the developing world signals an important change to existing structures and limitations in support of science.

In the remainder of this closing section, we examine a selected set of issues of critical importance to the implementation of mEBM. We focus on a few key elements that we believe will strongly shape the form of mEBM as the process continues to unfold. These include emerging issues such as climate change, the need for fuller integration of the social and natural sciences, and the development of strategies for coping with complexity.

2. Ecosystem Services in Time and Space

Because of the very long history of human dependence on food from the sea, it is perhaps not surprising that fisheries and the ecosystem impacts attributable to fishing are prominent throughout this volume. Our reliance on fishery resources is perhaps the most extensive of the ecosystem services considered here and it also has the richest historical record. Fish and shellfish provide over three billion people with at least 15% of their protein needs; global consumption recently exceeded 18 kg per person per year (FAO 2012), a record high and a critical dimension of food security in many parts of the world. However, adverse impacts of fishing activity on coastal and marine ecosystems is a serious and longstanding problem. Evidence of localized depletion of coastal marine resources can be traced back at least 13,000–23,000 years (Lotze et al. this volume). Global marine fish landings from capture fisheries have now essentially leveled off. In contrast, marine aquaculture production is rapidly increasing, although it still does not compare to global levels of freshwater aquaculture. As marine fishery resources were depleted in Europe by the 15th century, the search for new fisheries served as a critical catalyst for exploration and trade, ultimately leading to European settlement in North America. This points to the broader import of the global expansion of fishing over the last several centuries.

As early as the 14th century, it was recognized that a broader ecosystem perspective was required to effectively manage fishery resources. For example, in 1376, a petition was placed before the Commons in Britain to ban the Wondyrchoun, a type of trawl/dredge, because: "*. . . the great and long iron of the Wondrychoun presses so hard on the ground when fishing that it destroys the . . . plants growing on the bottom under the water, and also the spat of oysters, mussels, and other fish, by which the large fish are accustomed to live and be nourished*" (Anon. 1921). These observations reflect a deeply intuitive understanding of ecological factors potentially affecting delivery of a key ecosystem service.

This extensive history ultimately sets the stage for very active developments in Ecosystem-Based Fishery Management (EBFM), which are represented in the contributions to this volume. However, unless the development of EBFM is ultimately framed within the broader domain of mEBM, the prospect of continued conflict with other human activities in the coastal zone is inevitable (Samhourie et al., this volume). Management plans developed for any ocean use sector such as fisheries will work at cross-purposes with plans for other sectors. As shown in Chapter 1 of this volume, the development of integrated ocean management plans occupies the nexus of the roadmap to mEBM, integrating elements from sectoral management plans (see Fogarty and McCarthy Figure 1.2) and providing a vehicle for identifying and resolving conflict and competing interests

The rapid expansion of human population size in coastal areas by the 18th century and continuing to the present has entailed reliance on a broadening array of services, and associated threats and impacts to these services (Houde et al., this volume). Extraction of natural products (e.g. mangrove products; Kritzer et al., this volume) has increased greatly over this period with attendant effects on other underlying services including habitat, nutrient supplies, filtration of contaminants, water supplies, and shoreline and coastal protection. These problems are now ubiquitous throughout the world, again with the exception of the polar regions. The effects of these stressors and the magnitude of the resulting threats and impacts fueled the development of sophisticated approaches to integrated coastal zone management. Various forms of spatial management strategies, including protected areas and zoning emerged naturally from these considerations (Rice et al., this volume) along with other management tools for abatement and remediation of the effects for different stressors. The recognized importance of spatial management as part of the arsenal supporting mEBM is clear in the now widespread distribution of spatial management units throughout the world ocean. Over 5800 Marine Protected Areas (MPAs) are now in place in coastal and marine waters (http://www.iucn.org/knowledge/publications_doc/publications/?6500/Global-ocean-protection—present-status-and-future-possibilities/). However, the level of enforcement and the types of protection afforded by these MPAs spans a very broad spectrum. In addition, because many MPAs have been implemented under different, and often non-coordinated management authorities, the full benefits of implementing this management measure may not be fully realized, or may in fact lead to unwarranted complexity.

The emergence of interest in renewable energy from ocean-based sources and locations, including winds, waves and tides, is the newest entrée to the marine provisioning services. It has the potential to meet a critical societal need to reduce carbon emissions and generate clean energy. Almost alone among the other provisioning services, it is now taking shape in concert with the overall development of mEBM, affording an opportunity to avoid past mistakes in devising management strategies, and providing an important opportunity to ensure that its development entails consideration of historical and current patterns of ocean use that involve different sectors. Because these energy installations will entail the pre-emptive use of space, the potential for conflict with other sectors must be squarely faced and tradeoffs evaluated (Guerry and Tallis; Altman et al., this volume).

3. Future Threats to Ecosystem Services

Among the most pressing threats we face is the specter of climate change. Strategies for coping with climate change must be embedded in our overall approach to mEBM. The effects of past climate events can be seen in the paleo-ecological record and in the history of global fisheries. As noted above, historical documentation of catch in some of the major fisheries of Europe and Asia extend over many centuries and in these records rest clues to the waxing and waning of fish stocks in response to climate and the effects of human activities such as fishing. They also offer some clues to future dynamics in these systems. The potential consequences of present and future climate change for earth systems, the effects of other anthropogenic impacts on marine ecosystems, and the possible interaction between climate change and human activities affecting the oceans are now broadly recognized and are of vital concern.

Climate change will fundamentally alter oceanographic structures and features with direct implications for marine ecosystems and human societies. Climate change predictions consistently indicate increases in atmospheric temperatures and regional changes in precipitation patterns. Temperature changes are projected to be greatest at higher latitudes. This differential warming is expected to result in a reduction in latitudinal temperature gradients that may result in changes in the intensity of global and local wind fields. Precipitation is also projected to increase at higher latitudes. Sea level rise due to thermal expansion of seawater and glacial melt, and the Arctic habitat lost as sea ice thins and shrinks in areas present real threats to many of the systems considered in this volume (Gaiches et al., this volume).

Oceanographic features including currents, fronts, upwelling and downwelling zones and ice cover in polar regions will be strongly influenced by these changes in temperature, salinity, and winds. The potential implications of ocean acidification for ecosystem structure and function are now under active investigation. Collectively, these changes hold the potential to strongly impact the delivery of critical ecosystem services. Shifts in distribution of marine populations can be expected with increasing water temperatures, strongly affecting community composition of marine ecosystems. Changes in distribution with respect to latitude or depth will affect the availability of fish and invertebrate species to economically important regional fisheries.

These impacts will play out in very different ways in the systems we consider in the volume. Below we list just a few examples with the recognition that the changes induced will take place on many fronts and affect many ecosystem services. The loss of sea ice in polar systems will fundamentally alter a key structural ecosystem element on which many organisms depend (Gaichas et al., this volume). Loss of sea ice will also fundamentally change access to key services not currently used including shipping and transportation in future ice-free channels, possible extraction of mineral and hydrocarbon resources now under ice, and extraction of fishery resources in now unaccessible locations. In open ocean systems, inundation of low lying island nations is a distinct prospect and measures for relocation are now being planned (Polovina et al. this volume). Temperate shelf systems will likely experience radical alteration in the composition of ecological communities (Samhouri et al., this volume). Widespread coral bleaching with increasing water temperatures may irrevocably change the structure of tropical ecosystems and the diverse services they supply including food, tourism and recreation, and other cultural services (Kritzer et al., this volume)

4. Coupled Social-Ecological Systems

The need to strengthen ties between the natural and social sciences is critical for the further development and implementation of mEBM (Charles, this volume). Coupled human-resource systems are increasingly recognized as a globally prevalent form of complex adaptive systems with deep significance to ecosystem-based management. Complex adaptive systems hold the potential for rapid shifts in the state of coupled systems, requiring a strong commitment to precautionary management and the need for management institutions to move rapidly and adaptively in response to changing conditions. This general framework appears to hold considerable promise as a vehicle for the fuller integration of the social and natural sciences in support of mEBM. There is now ample evidence that effective engagement of human communities directly tied to the utilization of ecosystem services is critical to the ultimate success of any management strategy. Understanding the perspectives of stakeholders and effective communication with those most affected by potential management actions can help increase prospects for obtaining buy-in for management and reducing the prospects for unintended consequences if user groups deliberately circumvent selected management strategies (see Plummer and Levin, this volume).

5. Managing Complexity

Whether we can cope with the daunting complexity of ecosystems is indeed a legitimate concern. Here, we argue that advances in tactical management can be made now using existing models. The development of models in support of marine ecosystem-based management can be arrayed along a continuum of complexity involving tradeoffs in realism, mechanistic detail, and parameter and/or model uncertainty (Fulton and Link, this volume). Models reflecting this overall spectrum of complexity are now being used to support mEBM as part of an Integrated Ecosystem Management (IEA) framework for management strategy evaluation and the formulation of tactical management advice (see Plummer and Levin; Samhouri et al., this volume). The emphasis on identifying social and ecological indicators to portray the status of coupled human-ecological systems in the IEA structure also allows the development and application of simpler approaches to assessment. Recall that the economy, essentially a complex human ecology, is often evaluated using a select and fairly limited number of indicators and metrics (e.g. employment, spending, and savings rates; leading economic indicators from the financial sector, etc.) and corrective actions typically use a relatively simple set of tools (controlling money supply, interest rates, and so on). Although more complex macro- and micro-economic models are actively employed to guide regulatory actions, often as in ecological models, forecast skill is low (see Silver 2012). The indicator approach used in the context of IEAs in very much in this spirit (see Samhouri et al, this volume). By classifying indicators as pressure and state variables and connecting the two, we can assess the impact of modifying pressure variables subject to human control and evaluating the resulting response observed in social and ecological state variables. Determining the functional relationships between pressure and state variable opens an opportunity to establish targets and limits for the selection of appropriate management actions.

It must be further recognized that while the complexity of the coupled human-resource system is unquestionably high, adoption of mEBM offers opportunities for ultimately decreasing administrative and regulatory complexity by consolidating a very large number of individual sector plans into integrated management plans for defined ecological regions. This can help avoid duplication of effort and increase the likelihood that separate management efforts for different sectors or activities will work synergistically rather than antagonistically.

6. Conclusions

Perhaps one of the most important and difficult issues involved in establishing mEBM is the need to directly confront tradeoffs. In many instances we currently lack information on the weights that society would place on alternative courses of action and outcomes. When tradeoffs have been ignored or weights that reflect societal preferences have not been specified, the decision-making process is stymied. Here, the critical necessity of building on insights and research tools of the social sciences is most evident. As shown by Guerry and Tallis and Altman et al. (this volume), decision support tools for evaluating tradeoffs are now well developed. The challenge is to resolve the tradeoffs equitably. This requires deft application of the arts of communication, negotiation, and compromise. In the end, it is essentially a political issue and therefore requires perspectives and approaches from political science.

Although the tradeoff issue can be framed as a purely decision-theoretic construct (e.g Keeney and Raiffa 1993), the complexity of the decision problem, the likely uncertainty in key underlying issues, and the intercession of human values may require an alternative approach. Herbert Simon, Nobel laureate in economics, developed an alternative approach, 'satisficing' (Simon 1996) in which simpler and more qualitative approaches are employed. Simon questioned whether a complex optimization and decision rule framework was in fact preferable to simpler heuristic methods when the full costs involved with the former are considered. Simon further argued that we often cannot make a full evaluation of all alternatives and that frequently all the necessary information to make 'optimal' decisions is not available. A satisficing solution is one that yields a defensible outcome that is satisfactory to the end users. It is one that is good enough to meet objectives without pretense to optimality. Conceived in this way, perhaps the ultimate approach to decision making can be simplified and made more transparent to stakeholders as part of the overall mEBM. It is likely to resonate more fully with those affected by decisions because it more fully follows approaches adopted by most of us in everyday life. This does not in any sense reduce the difficulty of negotiation or of finding compromise but it does provide a flexible and adaptable structure for decision-making. Our exploration of the demand on ecosystem services in different areas, the threats to continued supply of these services, and the likely impacts of these threats is intended to set the stage for understanding where these tradeoffs are necessary and how to accommodate them.

Acknowledgments

We extend our deepest gratitude to the contributors to this volume. They have individually assembled pieces to the puzzle that is mEBM and collectively have shown how the parts fit together.

References

Anon. 1921. The history of trawling. *Fish Trades Gazette,* March 19, 1921, pp. 22–69.

FAO (Food and Agriculture organization of the United Nations). 2012. State of World Fisheries and Aquaculture. http://www.fao.org/docrep/016/i2727e/i2727e01.pdf

Keeeny, R. L. and H. and Raiffa. 1993. Decisions with multiple objectives: Preferences and value tradeoffs. Cambridge University Press, Cambridge. 569 pp.

Silver , N. 2012. *The Signal and the Noise: Why So Many Predictions Fail—but Some Don't.* The Penguin Press. New York.

Simon, H. A. 1996. The Sciences of the Artificial. 3d Ed. MIT Press, Cambridge, MA

ACRONYMS AND ABBREVIATIONS

ACAP	Arctic Contaminants Action Program
ACIA	Arctic Climate Impact Assessment
AEPS	Arctic Environmental Protection Strategy
AFMA	Australian Fishery Management Authority
AMAP	Arctic Monitoring and Assessment Program
AMO	Atlantic Multidecadal Oscillation
ARIES	Artificial Intelligence for Ecosystem Services
ATCM	Antarctic Treaty Consultative Meeting
ATS	Antarctic Treaty System
BCC	Benguela Current Commission
BCLME	Benguela Current Large Marine Ecosystem
BENEFIT	Benguela Current Environment Fisheries Interaction and Training [Programme]
Bmsy	Biomass at maximum sustainable yield
BRP	Biological Reference Point
CAFF	Conservation of Arctic Flora and Fauna
CBD	Convention on Biological Diversity
CBP	Chesapeake Bay Program
CCAMLR	Commission for the Conservation of Antarctic Marine Living Resources
CCAS	Convention for the Conservation of Antarctic Seals
CCLME	California Current Large Marine Ecosystem
CDS	Catch Documentation Scheme
CECM	Executive Committee of the Council of Ministers [Cuba]
CEP	Committee for Environmental Protection [Antarctic Treaty]
CERP	Comprehensive Everglades Restoration Plan
CITES	Convention on International Trade in Endangered Species of Wild Fauna and Flora
CITMA	Ministry of Science, Technology and the Environment [Cuba]
CLME	Caribbean Large Marine Ecosystem [and Adjacent Areas]
CMA	Coastal Management Area
CMSP	Coastal and Marine Spatial Planning
CNAP	National Center for Protected Areas [Cuba]
COTS	Crown of Thorns Starfish
CP	Contracting Party
CPUE	Catch Per Unit Effort
CRAMRA	Convention on the Regulation of Antarctic Mineral Resource Activities
CRCP	Coral Reef Conservation Project
CSIRO	Commonwealth Scientific and Industrial Research Organization [Australia]
CZMA	Coastal Zone Management Act [USA]

CZMAI	Coastal Zone Management Authority and Institute [Belize]
DFO	Department of Fisheries and Oceans [Canada]
DOC	Dissolved Organic Carbon
DPSIR	Driver-Pressure-State-Impact Response
DSS	Decision Support System
E2E	End-to-End [model]
EA	Ecosystem (or Environmental) Accounting
EAC	East Australian Current
EAF	Ecosystem Approach to Fisheries
EAFM	Ecosystem Approach to Fisheries Management
EBFA	Ecosystem-Based Fisheries Assessment
EBFM	Ecosystem-Based Fisheries Management
EBIFM	Ecosystem-Based Integrated Fishery Management [Korea]
EBSA	Ecologically or Biologically Significant Areas
EcoQO	Ecosystem Quality Objective
EEA	European Environmental Agency
EEZ	Exclusive Economic Zone
EIA	Environmental Impact Assessment [Norway]
ENGO	Environmental Non-Governmental Organization
ENSO	El Niño-Southern Oscillation
EPA	Environmental Protection Agency [USA]
EPBC	Environment Protection and Biodiversity Conservation [Act] [Australia]
EPPR	Emergency Prevention, Preparedness and Response
ERAEF	Ecological Risk Assessment for the Effects of Fishing
ERI	Ecosystem Risk Index
ESA	Endangered Species Act [USA]
ESAM	Extended Single-Species Assessment Model
ESD	Ecologically Sustainable Development
EU	Equatorial Upwelling
EU	European Union
EwE	Ecopath with Ecosim
FAO	Food and Agriculture Organization
FC	Fishing Commission [NAFO]
FEP	Fisheries Ecosystem Plan [USA]
FLR	Fisheries Library in R
FMP	Fishery Management Plan [USA]
FRI	Fishery Risk Index
GADGET	Globally Applicable Area-Disaggregated General Ecosystem Toolbox
GARM	Groundfish Assessment Review Meeting [USA]
GBR	Great Barrier Reef
GBRMP	Great Barrier Reef Marine Park
GCM	General Circulation Model
GES	Good Environmental Status
GIS	Geographic Information System
HAB	Harmful Algal Bloom

HELCOM	Helsinki Commission
HRA	Habitat Risk Assessment
HWB	Human Well-Being
IATTC	Inter-American Tropical Tuna Commission
ICCAT	International Commission for the Conservation of Atlantic Tunas
ICES	International Council for the Exploration of the Sea
ICM	Integrated Coastal Management
ICZM	Integrated Coastal Zone Management
IEA OR IA	Integrated Ecosystem Assessment
IFMP	Integrated Fisheries Management Plan
IFRAME	Integrated Fisheries Risk Analysis Method for Ecosystems
IMO	International Maritime Organization
InVEST	Integrated Valuation of Environmental Services and Tradeoffs
IOD	Indian Ocean Dipole
IPBES	[United Nations] Intergovernmental [Science-Policy] Platform on Biodiversity and Ecosystem Services
IPOA	International Plan of Action
ISA	International Seabed Authority
ITQ	Individual Transferable Quota
IUCN	International Union for the Conservation of Nature
IUU	Illegal, Unreported and Unregulated
IWC	International Whaling Commission
JOCI	Joint Oceanographic Commission Iniative
KEF	Key Ecological Feature
KEMFRI	Kenya Marine and Fisheries Research Institute
LME	Large Marine Ecosystem
LMMA	Locally Managed Marine Areas
LOMA	Large Ocean Management Area
LOS OR UNCLOS	United Nations Convention on the Law of the Sea
MA	Millennium Ecosystem Assessment
MACC	Marine and Coastal Committee [Australia]
MAFMC	Mid-Atlantic Fishery Management Council [USA]
MAP	Mediterranean Action Plan
MARCO	Mid-Atlantic Regional Council on the Ocean [USA]
MAREANO	Marine Area Database for Norwegian Coastal and Sea Areas
MARPOL	International Convention for the Prevention of Pollution from Ships
MBP	Marine Bioregional Plan
MCMC	Markov Chain Monte Carlo [Sampling]
ME	Ministry of Environment [Korea]
MEA	Millennium Ecosystem Assessment
mEBM	marine Ecosystem-Based Management
MEY	Maximum Economic Yield
MICE	Model of Intermediate Complexity for Ecosystem Assessments

MIDAS	Marine Integrated Decision Analysis System
MIFAFF	Ministry of Food, Agriculture, Forestry and Fisheries [Korea]
MIMES	Multiscale Integrated Models of Ecosystem Services
MLTM	Ministry of Land, Transport and Maritime Affairs [Korea]
MMSY	Multi-species Maximum Sustainable Yield
MOY	Maximum Optimum Yield
MPA	Marine Protected Area
MRM	Minimum Realistic Model
MSC	Marine Stewardship Council
MSE	Management Strategy Evaluation
MSFD	Marine Strategy Framework Directive [European Union]
MSP	Marine Spatial Planning
MSY	Maximum Sustainable Yield
MTE	Management Tactic Evaluation
NAFO	Northwest Atlantic Fisheries Organization
NAO	North Atlantic Oscillation
NatCap	Natural Capital Project
NEAFC	North East Atlantic Fisheries Commission
NEFMC	New England Fishery Management Council
NEPA	National Environmental Policy Act [USA]
NESLME	North East Shelf Large Marine Ecosystem
NFRDI	National Fisheries Research and Development Institute [Korea]
NGO	Non-Governmental Organization
NL	Newfoundland and Labrador
NMFS	National Marine Fisheries Service [USA]
NOAA	National Oceanic and Atmospheric Administration [USA]
NOP	National Ocean Policy [USA]
NPSG	North Pacific Subtropical Gyre
NPZD	Nutrient-Phytoplankton-Zooplankton-Detritus
NRC	National Research Council [USA]
NRMMC	Natural Resource Management Ministerial Council [Australia]
NROC	Northeast Regional Ocean Council [USA]
OECD	Organization for Economic Cooperation and Development
OMZ	Oxygen Minimum Zone
ORI	Objective Risk Index
OSMOSE	Object-oriented Simulator of Marine Ecosystems [Model]
OSPAR	Oslo/Paris Convention for the Protection of the Marine Environment of the Northeast Atlantic
PAH	Polycyclic Aromatic Hydrocarbons
PAME	Protection of the Arctic Marine Environment
PBGB	Placentia Bay-Grand Banks
PFMC	Pacific Fishery Management Council [USA]
POC	Pew Oceans Commission
PSA	Productivity-Susceptibility Analysis

PSP	Puget Sound Partnership
PSSA	Particularly Sensitive Sea Area
QSR	Quality Status Report
RA	Risk Assessment
RAC	Regional Activity Center [Mediterranean]
REDD+	Reducing Emissions from Deforestation and Forest Degradation
RFMO	Regional Fishery Management Organization
ROMS	Regional Ocean Modeling System
RPB	Regional Planning Body [USA]
SAFE	Sustainability Assessment for Fishing Effects
SAM	Southern Annular Mode
SAP	Strategic Action Plan [Black Sea]
SAV	Submerged Aquatic Vegetation
SBL	Safe Biological Limits
SC	Scientific Council [NAFO]
SCAR	Scientific Committee on Antarctic Research
SDS	Sustainable Development Strategy [European Union]
SDWG	Sustainable Development Working Group [Arctic Council]
SEA	Strategic Environmental Assessment
SEAFO	South East Atlantic Fisheries Organization [USA]
SEAPODYM	Spatial Ecosystem and Population Dynamics Model
SEC	South Equatorial Current
SEEA	System of Environment-Economic Accounting [United Nations]
SES	Social-Ecological Systems
SESSF	South East Scalefish and Shark Fishery [Australia]
SIA	Social Impact Assessment
SPAW	Specially Protected Areas and Wildlife
SPC	Secretariat of the Pacific Community
SPRFMO	South Pacific Regional Fisheries Management Organization
SPSG	South Pacific Subtropical Gyre
SRI	Species Risk Index
SRSD	Special Region of Sustainable Development
SFF	Sustainable Fisheries Framework
SSMU	Small Scale Management Unit
SST	Sea Surface Temperature
TAC	Total Allowable Catch
TED	Turtle Excluder Device
TEK	Traditional Ecological Knowledge
TEP	Threatened, Endangered and Protected
TSS	Traffic Separation Scheme
UNCED	United Nations Conference on Environment and Development
UNCLOS OR LOS	United Nations Convention on the Law of the Sea
UNCSD	United Nations Conference on Sustainable Development

UNEP	United Nations Environment Program
UNESCO	United Nations Educational, Scientific and Cultural Organization
USCOP	United States Commission on Ocean Policy
VME	Vulnerable Marine Ecosystem
VMS	Vessel Monitoring System
WCA	West Coast Aquatic [Vancouver Island, Canada]
WCED	World Commission on Environment and Development
WCGA	West Coast Governors' Alliance [USA]
WCPFC	Western and Central Pacific Fisheries Commission
WCR	Wider Caribbean Region
WECAFC	Western Central Atlantic Fisheries Commission
WG	Working Group
WGEAFM	Working Group on Ecosystem Approaches to Fisheries Management [NAFO]
WGFMS	Working Group of Fishery Managers and Scientists [NAFO]
WHA	World Heritage Area
WPRFMC	Western Pacific Regional Fishery Management Council
WSSD	World Summit on Sustainable Development
WWF	World Wildlife Fund
ZBREUP	Zone Under Special Regime of Use and Protection [Cuba]
ZBRMIC	Zone Under Special Regime of Integrated Coastal Management [Cuba]

INDEX

The letter *f* following a page number denotes a figure; *t* denotes a table.